Universality of
Nonclassical Nonlinearity

Pier Paolo Delsanto

Editor

Universality of Nonclassical Nonlinearity

Applications to Non-Destructive Evaluations and Ultrasonics

 Springer

Pier Paolo Delsanto
Dip. Fisica
24 Corso Duca degli Abruzzi
Torino 10129
Italy

Library of Congress Control Number: 2006925905

ISBN-10: 0387-33860-8

ISBN-13: 9780387338606

9 8 7 6 5 4 3 2 1

springer.com

Alles verändert sich, aber dahinter ruht ein Ewiges.

J.W. von Goethe

Preface

Variety is perhaps the most amazing attribute of Nature, with an almost endless array of different molecules and aggregates and tens of millions of distinct forms of life. Yet, in spite of this bewildering diversity, there are some common patterns, henceforth called "phenomenological universalities," that are found over and over again in completely different contexts. A quest for universalities is not only interesting per se, but can also yield practical applications. If several fields share a common mathematical or conceptual background, a cross-fertilization among them may lead to quick progress, even if ultimately the specific details of any individual application must be considered independently.

We all know that we live in a nonlinear world, although scientists have often tried to linearize it, sometimes as a first step towards understanding its complexity, often because, before the advent of ubiquitous high computational power, a linear approximation was the only viable alternative. In this book we use the term "nonclassical nonlinearity" with reference to a particularly intriguing kind of phenomenology, which has been extensively investigated in the last decade in the fields of elasticity and geomaterials and whose universality has been conjectured. Its signatures are hystereticity, discrete memory, and other effects which, in the case of continuum mechanics, have been called "fast" and "slow" dynamics.

Because what is currently "classical" in a field may still be considered nonclassical in another, we purposely omit here a general definition of nonclassical nonlinearity, leaving it to be defined throughout the book in the context in which it is used. For example, hysteretic phenomena have long been studied in ferromagnetism and cannot certainly be called nonclassical there.

The first part of the book is devoted to a review of (nonclassical) nonlinearity in several disciplines, ranging from physics and mechanics to biology, medicine, and social sciences. It also includes a discussion of the general mathematical background. The second and third parts refer to applications to nondestructive evaluation and ultrasonics. Part 2 is concerned with models and methods for performing numerical simulations (e.g., of the ultrasonic wave propagation in heterogeneous and nonlinear media). Part 3 describes experimental techniques and reviews specific applications of industrial interest, for example, in the fields of aeronautics, geomaterials, civil engineering, and NDE in general.

Acknowledgments

The book and most of the work behind it was made possible due to the generous support of the European Science Foundation, under the ESF/PESC Programme NATEMIS 2000-5. In fact 19 out of 32 chapters were written by authors who participated in the Programme. It is my pleasure to thank J. Engelbrecht, who induced me *sponte coacta* to edit the book and the members of the NATEMIS Steering Committee, J. Gallego Juarez, C. Hedberg, S. Hirsekorn, A. Ravasoo, Z. Prevorovsky, and K. Van den Abeele, who endorsed the idea. My gratitude also goes to my colleagues and local collaborators, G.P. Pescarmona, M. Scalerandi, F. Bosia, A. Gliozzi, M. Griffa, M. Hirsekorn, M. Nobili, S. Delsanto, and L. Morra, for their constant help. Last, but not least, I wish to thank Ms. S. Balestracci for her efficient and cheerful assistance.

Contents

Part I
The Universality of Nonclassical Nonlinearity

1

Towards a Top-Down Approach in Materials Science

P.P. Delsanto[1,3] and S. Hirsekorn[2]

[1]Politecnico di Torino, Dipartimento di Fisica, C.so Duca degli Abruzzi 24,10129 Torino, Italy.
[2]Fraunhofer Institut für zerstörungsfreie Prüfverfahren (IZFP), Universität, Geb. 37, 66123 Saarbrücken, Germany.
[3]To whom correspondence should be addressed: P.P. Delsanto, INFM – Politecnico di Torino, Dipartimento di Fisica, C.so Duca degli Abruzzi 24,10129 Torino, Italy, pier.delsanto@polito.it.

Abstract
Materials science or, more specifically, materials characterization represents an extremely vast interdisciplinary research arena involving scientists with very different backgrounds working at many different applications spanning more than ten orders of magnitude in the size of the specimens. From their collective work, some general patterns have emerged, such as scaling, nonclassical nonlinearity, and other "universalities." A quest for universal laws is not only interesting per se, but can also yield practical applications. If several fields share a common mathematical or conceptual background, cross-fertilization among them may lead to quick progress, even if ultimately the specific details of any individual application must be considered independently. The idea behind the proposed top-down approach is of course, not to replace, but to complement current investigations by searching for solutions that often mutatis mutandis already exist, but are confined to a different network of researchers.

In the present contribution we start from the conjecture, based on many experimental observations, of the existence of a nonlinear mesoscopic elasticity universality class. We search for the basic mathematical roots of nonclassical nonlinearity, in order to explain its universality, classify it, and correlate it with the underlying meso- or microscopic interaction mechanisms. In our discussions we explicitly consider two quite different kinds of specimens: a two-bonded-elements structure and a thin multigrained bar. It is remarkable that although the former includes only one interface and the latter very many interstices, the same "interaction box" formalism can be applied to both. The generality of the proposed formalism suggests that a similar approach may be adopted in completely different contexts, for example, in biological, biomedical, and social sciences.

Keywords: Cross-fertilization, fast dynamics, hysteresis, materials science, nonclassical nonlinearity, numerical simulations, scaling, slow dynamics, ultrasonic NDE, universality

1. Introduction

Materials science is undoubtedly one of the hottest research fields nowadays, inasmuch as progress in mechanical or biomedical engineering often depends on the availability or development of more advanced (or cheaper) material components with optimized properties and geometry. This is particularly true for nanostructured materials, which represent a revolution of traditional material design, via atomic-level structural control to tailor the engineering properties. At the same time materials science is an extremely vast interdisciplinary research arena, involving scientists with very different backgrounds working at an almost endless array of different applications spanning more than ten orders of magnitude in the size of the specimens.

Therefore, the field of materials science has progressed mostly as a consequence of the pursuit of specific applications or of the background of the researcher or of the available facilities and/or other external factors. This bottom-up approach has been extremely successful, not only for the wealth of applications which it has fostered, but also for the many advances in the understanding of the materials' properties and underlying physical laws. Among the most interesting findings we recall here the relevance of scaling laws in the mechanics and physics of solids,[1] the unexpected nonlinearity of many phenomena,[2] and the complexity, whether in the materials' structure or in their mechanical behavior.[3]

In this context many patterns have been discovered that are remarkably similar, although they concern completely different phenomenologies. This is hardly surprising, because often the background mathematics is the same. We call them "universalities"[4] in the sense that they refer to a "transversal" generality (not to a uniformly general behavior within a given class of phenomena). The inverse may also be true; that is, we conjecture that for each mathematical niche some related effect may be found in nature (whether in physics, chemistry, biology, or even sociological sciences) if only one looks hard enough. Such a niche theory is commonplace in biology (where it is explained as a consequence of evolution). However, many examples of its applicability may be found in other fields as well. For example, in elementary particles physics, any reaction that is not explicitly forbidden by some quantum mechanical rule does in fact occur (even if at a minimal rate). Likewise in crystallography virtually any symmetry class that is mathematically predictable is found to be represented in nature.

A quest for universal laws or, at least, more general patterns is not only interesting per se, but can also yield practical applications. If several fields share a common mathematical or conceptual background, a cross-fertilization among them may lead to quick progress, even if ultimately the specific details of any individual application must be considered independently.

The idea behind such a top-down approach is, of course, not to replace, but to complement bottom-up investigations by searching for solutions that often mutatis mutandis already exist, but are confined to another circle of researchers. By identifying the common conceptual background among different problems, it becomes easier to increase the communication and transfer of information. A specific example of the proposed top-down approach is provided by the "Interaction Box Formalism," which is discussed in the next section. In the third section some assumptions are considered that, if verified for the particular application being studied, may greatly increase the predictive power of the formalism. Finally, in Section 4, the application of the proposed approach to two problems of interest for nonlinear ultrasonic NDE is discussed.

2. The Interaction Box Formalism

Let us consider any kind of nonlinear cause–effect relationship, in which both the cause $C(t)$ and the effect $E(C(t))$ are time dependent. The input $C(t)$ may be, according to the field of application, any kind of induced field, pulse, signal, excitation, or perturbation. For simplicity we assume a harmonic time dependence of the cause

$$C(t) = A \, \sin(\omega t) \tag{1.1}$$

with an initial phase $\varphi_0 = 0$. In order to keep the treatment as general as possible, we represent the correlation between E and C by means of an unspecified interaction box. $E(C(t))$ represents the response of the system.[4,5]

Because, due to some sort of Heisenberg principle, too much generality yields too little information, we assume a certain periodicity in the system, for example, represented by the boundaries of the interaction box, which allows us to write E as a Fourier expansion in the variable ωt,

$$E\left(C(\omega t)\right) = E_0 + \sum_{j=1}^{\infty} E_j \sin(j\omega t + \varphi_j). \tag{1.2}$$

In Eq. (1.2) the amplitudes and phases, E_j and φ_j, of the various harmonics depend, of course, on the input amplitude A. The Fourier expansion in the variable ω_t implies only higher, but not subharmonic generation. The latter can be easily included by an extended expansion on the base of rational fractions of the variable ωt.[6] It is clear from Eq. (1.2) that E is a one-value function of t, but not of C. In fact by developing E as a function of

$$x = \frac{C}{A} = \sin(\omega t) \tag{1.3}$$

one finds (see Appendix A)

$$E(x) = E_{\text{av}} + G(x) \pm H(x), \tag{1.4}$$

where

$$E_{\text{av}} = E_0 + \sum_{j=1}^{\infty} E_{2j} \sin(\varphi_{2j}), \tag{1.5}$$

$$G(x) = \sum_{n=1}^{\infty} \sum_{j=n}^{\infty} g_{nj} P_j x^n, \tag{1.6}$$

$$H(x) = \sqrt{1 - x^2} \sum_{n=1}^{\infty} \sum_{j=n}^{\infty} h_{nj} Q_j x^{n-1}, \tag{1.7}$$

$$P_j = E_j \cos(\varphi_j), \quad Q_j = E_j \sin(\varphi_j) \text{ if } j \text{ is odd; and}$$
$$P_j = E_j \sin(\varphi_j), \quad Q_j = E_j \cos(\varphi_j) \text{ if } j \text{ is even.} \tag{1.8}$$

The explicit expressions of g_{nj} and h_{nj} are reported in Appendix A.

If the input amplitude A is not too large, only a limited number N of harmonics are nonnegligible in Eq. (1.2). For example, for $N = 3$ we have from Eqs. (A7) and (A8),

$$G(x) \approx (P_1 - 3P_3)x - 2P_2 x^2 - 4P_3 x^3, \tag{1.9}$$

$$H(x) \approx \sqrt{1 - x^2}(Q_1 + Q_3 + 2Q_2 x - 4Q_3 x^2). \tag{1.10}$$

It is clear from Eq. (1.4) that E is, in general, a two-values function of x, which results in a hysteretic loop in the plot E versus x of a complete cycle (i.e., that there is a branching of the response at $x = \pm 1$ in two functions, thus generating a hysteretic cycle). From a mathematical point of view this is simply the consequence of $\cos(\omega t) = \pm\sqrt{1 - x^2}$. A physical interpretation of the hystereticity may be, according to the harmonics involved, a trivial time delay between the input and the response, or some kind of inertia in the system, which is dynamically affected by the input excitation, or some combination of both. From Eq. (1.7) it is easy to see that the no-hystereticity condition, $H(x) = 0$ for any value of x, requires that all Q_j vanish. This implies that for all nonnegligible harmonics $\sin(\varphi_j) = 0$ if j is odd and $\cos(\varphi_j) = 0$ if j is even. The term E_{av} in Eq. (1.4) represents the equilibrium point, and thus $\Delta E = E - E_{av}$ is the deviation from it.

It may be interesting to see how the hysteretic loops change according to the values of the amplitudes and phases of the nonnegligible harmonics. Eight different cases are reported here. In Figures 1.1a and 1.1b the response is linear because no higher harmonics are present; in Figure 1.1a there is no delay between input and output, whereas a time delay ($\varphi_1 \neq 0$) induces a (trivial) loop in Figure 1.1b. In Figures 1.1c and 1.1d, a first-order nonlinearity, the second harmonic, is added to the linear nonhysteretic response ($E_2 \neq 0$), in Figure 1.1c without a hysteretic loop because the analyticity conditions are satisfied (see above) and in Figure 1.1d with a hysteresis in the second harmonic because $\varphi_2 = \pi/4$. Figures 1.1e and 1.1f are similar to Figures 1.1c and 1.1d except that the nonlinearity is of the second order (i.e., in the third harmonic). We note that if only the nth harmonic causes hysteresis, $(n - 1)$ nodes are present in the loop (see Figures 1.1b, 1.1d, 1.1f, 1.1g). If more harmonics contribute to the hysteretic behavior, however, some of the nodes may disappear and/or the characteristic cuspidal shape may emerge at both ends of the loop (at $x = \pm 1$).

3. Input–Output Correlation

In order to obtain other explicit predictions from the interaction box formalism (besides the hystereticity) one needs to introduce additional assumptions. We assume here that the value of the function G for the same value of the cause C does not depend on the input amplitude A, at least up to a certain value $\overline{A}$.[4,5] The range of validity (and physical interpretation) of this assumption depends of course on the special problem considered. Thus any discussion of it is postponed to the next section, where a particular problem of interest in the field of materials science is discussed.

Considering only a finite number N of harmonics, the proposed assumption may be written as [see Eq. (1.6)]

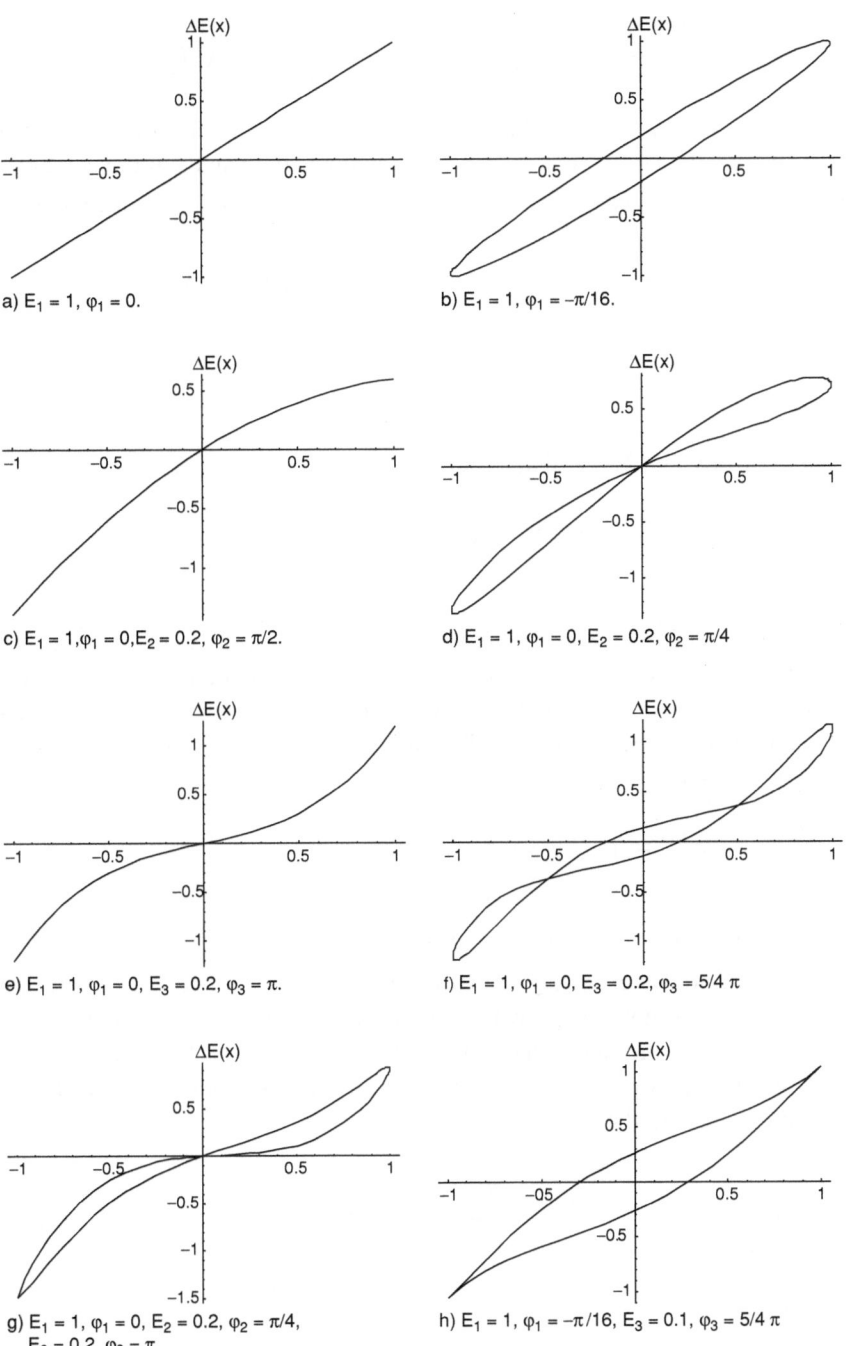

Fig. 1.1. Plots of the response relative to the equilibrium point ΔE versus the excitation x for eight different cases; the values of the amplitudes and phases of the contributing harmonics are: (a) $E_1 = 1, \varphi_1 = 0$, (b) $E_1 = 1, \varphi_1 = -\pi/16$, (c) $E_1 = 1, \varphi_1 = 0, E_2 = 0.2, \varphi_2 = \pi/2$, (d) $E_1 = 1, \varphi_1 = 0, E_2 = 0.2$, $\varphi_2 = \pi/4$, (e) $E_1 = 1, \varphi_1 = 0, E_3 = 0.2, \varphi_3 = \pi$, (f) $E_1 = 1, \varphi_1 = 0, E_3 = 0.2, \varphi_3 = 5/4\pi$, (g) $E_1 = 1$, $\varphi_1 = 0, E_2 = 0.2, \varphi_2 = \pi/4, E_3 = 0.2, \varphi_3 = \pi$, (h) $E_1 = 1, \varphi_1 = -\pi/16, E_3 = 0.1, \varphi = 5/4\pi$.

$$\sum_{n=1}^{N}\sum_{j=n}^{N} g_{nj} P_j \left(\frac{C}{A}\right)^n = \sum_{n=1}^{N}\sum_{j=n}^{N} g_{nj} \overline{P}_j \left(\frac{C}{\overline{A}}\right)^n \qquad (1.11)$$

for any $|C| < A$, where A may be any input amplitude smaller than $\overline{A}$. Here and in the following we define $\overline{P}_j, \overline{Q}_j, \overline{E}_j, \overline{\varphi}_j$, and so on as the values of the corresponding variables for $A = \overline{A}$. Because Eq. (1.11) must hold for any $|C| \le A$, it follows

$$\sum_{j=n}^{N} g_{nj} P_j = y^n \sum_{j=n}^{N} g_{nj} \overline{P}_j, n = 1, \ldots, N, \qquad (1.12)$$

where

$$y = \frac{A}{\overline{A}}. \qquad (1.13)$$

The system of Eqs. (1.12) is satisfied only for the set of solutions

$$P_n = \sum_{i=n}^{N}\sum_{j=i}^{N} \pi_{ij}^{(n)} \overline{P}_j y^i, \quad n = 1, \ldots, N, \qquad (1.14)$$

where $\pi_{ij}^{(n)}$ are numerical coefficients that vanish unless both i and j have the same parity as n. They can be easily calculated starting from the equation for $n = N$, then the equation for $n = N - 2$, which after substituting the computed value of P_N has the only unknown P_{N-2}, and so on. Likewise for P_{N-1}, P_{N-3}, and so on. For $N = 3$ the nonzero $\pi_{ij}^{(n)}$ are

$$\pi_{nn}^{(n)} = 1 \quad for \quad n = 1, 2, 3; \quad \pi_{13}^{(1)} = -\pi_{33}^{(1)} = 3, \qquad (1.15)$$

so that

$$P_1 = (\overline{P}_1 + 3\overline{P}_3)y - 3\overline{P}_3 y^3, \quad P_2 = \overline{P}_2 y^2, \quad P_3 = \overline{P}_3 y^3. \qquad (1.16)$$

In the limit of no hysteresis ($Q_n = 0$) we have $E_n = |P_n|$ [Eqs. (1.8)]. Because we do not expect any abrupt change when the phases φ_n approach the analyticity condition, we may assume the same functional dependence on $y = A/\overline{A}$ for E_n and consequently also for Q_n. A power series expansion in y and the interdependence between P_n, Q_n, and E_n lead to the results

$$Q_n = \sum_{i=n}^{N}\sum_{j=i}^{N} \pi_{ij}^{(n)} \overline{Q}_j y^i, \qquad (1.17)$$

$$\frac{Q_n}{\overline{Q}_n} = \frac{P_n}{\overline{P}_n}, \quad n = 1, \ldots, N. \qquad (1.18)$$

4. Applications to Ultrasonic NDE

A case of special interest for applications of the interaction box formalism to ultrasonic NDE is the propagation of ultrasound in heterogeneous media.[5] Recent experiments on rocks and other materials, such as soil, cement, concrete, and damaged elastic materials, have led to the discovery of nonlinear hysteretic effects in their elastic behavior.[7] These observations have suggested the existence of a nonlinear mesoscopic elasticity universality class, to which all the aforementioned materials belong.[2]

We show in this section that such a universality of nonclassical nonlinear (NCNL) effects may be explained in the framework of the interaction box formalism. For the sake of clarity we define here as nonclassical the hysteretic behavior of the nonlinear response of a specimen, when it is due to a variation of its material properties between the phases of increasing and decreasing applied stress, and not to a trivial time delay as discussed in Section 2. The same term (NC) may also be used for other kinds of hysteretic cycles, such as strain versus temperature or humidity.[8] The relevance of NCNL elasticity for NDE purposes stems from the very large enhancement of NCNL effects in the presence of small or even micro damages in the specimen.

In order to illustrate the applicability of the interaction box formalism to the analysis of ultrasonic wave propagation, we consider two problems that are very different from an experimental point of view, but may be analyzed, at least up to a certain extent, in a unified fashion. The first one is the development of techniques to characterize the bond quality (or strength) of adherent joints. Binding forces are nonlinear and cause a nonlinear modulation of transmitted and reflected ultrasonic waves. As a consequence, the generated higher harmonics of an insonified monochromatic wave yield information about the adhesive bonds and possible presence of flaws in the interface region.[9]

The second problem concerns the propagation of an ultrasonic wave through a multi-grained material specimen.[10] If the specimen consists of a thin bar, the problem can be simplified by means of a one-dimensional schematization, in which longer segments representing the grains alternate with shorter ones associated with the interstices. The grains are usually assumed for simplicity to be linearly elastic, whereas all the nonlinearity (classical and nonclassical) is confined to the interstices, hence called Hysteretic Mesoscopic Units (HMU). The HMUs behave either rigidly or elastically, depending on the local pressure P. Thus the NCNL behavior of the specimen is solely attributed to transitions (or jumps) between different interstitial states.

The two problems can be treated synoptically in the framework of the interaction box formalism, assuming, for example, a monochromatic ultrasonic wave $C(t)$ as input. As a result of its interaction with the specimen (the two bonded elements in the first case and the multigrained bar in the second), a certain output $E(C(t))$ is generated. If the interactions are nonlinear, E includes, in addition to the fundamental harmonic, also a number (in principle infinite) of higher harmonics. From the point of view of our formalism, it is irrelevant whether E is obtained for transmission or reflection (pulse–echo mode).

A completely different treatment of the two problems is possible in the framework of the Local Interaction Simulation Approach (LISA)[11, 12] or other finite difference or finite element techniques.[13] We do not report here the details of the LISA implementation nor the results, because they are discussed in later chapters.[14] It is interesting, however, to compare conceptually these two very different approaches, as we do in the following.

LISA or other virtual experiments solve a direct problem, in which the physical properties of the interaction box are assumed to be known. Because they are not (in fact a very large amount of effort is currently being devoted to the study of the underlying molecular dynamics), a phenomenological approach based on the Preisach–Mayergoyz (PM) space formalism[15] is commonly adopted. This approach consists of a completely arbitrary model, in which bistate transitions are assumed to occur under preassigned conditions between the two adherent joints (in the first problem) or at a very large number of arbitrarily distributed interstices (the HMUs) in the second one. The arbitrariness of the model is, however, compensated in both cases by a large statistics and parameter fitting. Taking the PM space parameters for a dynamic calculation from an inversion of results of quasistatic experiments does not seem to us to be very plausible, because the range of variation of stress is generally orders of magnitude larger in quasistatic than in dynamic resonant experiments.

By contrast the proposed interaction box formalism aims at the solution of an inverse problem,[16] that is, predicting the basic properties of the interaction box from the knowledge of the effect, (i.e., the amplitudes and phases of the higher harmonics). More realistically, the interaction box formalism may be thought of as another phenomenological approach with a very small number of parameters, if only a few higher harmonics need to be considered. Also, the amplitudes of the higher harmonics may be easily measured experimentally. At the present stage, however, more assumptions need to be introduced (e.g., about the loading and unloading history) before the interaction box formalism may be considered as a tool for practical applications. Another open question is the range of validity of the assumptions made in Section 3. Comparisons with the results of real and/or virtual experiments are necessary before any realistic conclusion may be reached. The assumptions may also greatly depend on the special problem to be considered.

An application of the interaction box formalism which seems to be almost within reach is the spectral analysis of the PM spaces, adopted by different competing models, for the simulation of NCNL effects in nonlinear mesoscopic elastic materials. The corresponding formalism is described in Chapter 14 for the case of two bonded elements with a single interface.

5. Conclusions

Out of a vast collection of experiments on the elastic behavior of rocks and other materials such as soil, cement, concrete, and damaged elastic materials, the conjecture has emerged of a nonlinear mesoscopic elasticity universality class, to which all the aforementioned materials belong. Common characteristics of all these materials are two sets of nonclassical nonlinear effects called "fast" and "slow dynamics," respectively, to the study and exploitation of which most of Parts 2 and 3 of this book are devoted.

We wish here to go two steps further. We conjecture first that all materials are, in general, nonclassical nonlinear, except under special conditions (which are derived in Section 2), although in many applications NCNL effects may be totally negligible for practical purposes. Second, we conjecture that similar NCNL effects are to be expected in the phenomenology of all sciences, for example, biology and social sciences. Most of the other chapters of Part 1 are, in fact, concerned with the search and discussion of NCNL effects in other fields.

In order to illustrate the conjectured universality, an interaction box formalism is introduced, which, under a minimal amount of assumptions, demonstrates the hystereticity of any cyclical excitation of a completely general (and unspecified) system (the "box"). In Section 3 additional assumptions are introduced which, if verified for a specific phenomenology, may provide an interesting new tool for the characterization and/or classification of the system. If in a specific situation these assumptions are not correct, others might be introduced that require suitable modifications of the formalism.

Finally, the application of the interaction box formalism is discussed for two problems of NDE interest: the propagation of ultrasonic waves (a) across an interface between two adherent joints, and (b) along a thin bar of a multigrained material. Both have been widely studied as direct problems in which the physical properties of the propagation medium are assumed to be known. The models that have been proposed to perform the numerical simulations based on different PM-space protocols are phenomenological and the results depend strongly on the choice of protocol and parameters (most of which need to be fitted or fixed arbitrarily). However, the results are generally good and reproduce well (at least qualitatively) most of the observed phenomenology. Much of Part 2 of this book is devoted to the description of the various models and corresponding results.

By contrast, the interaction box formalism is based on a priori considerations and quite general assumptions. If few higher harmonics are nonnegligible (which is usually the case in practical applications), very few parameters are needed (their amplitudes and phases), and the amplitudes are easy to measure experimentally. Also, calculations are far easier and cheaper to perform. However, a great amount of top down work (i.e., by introducing additional assumptions according to the specific problem to be considered and investigating their consequences and model predictions) is still necessary before the method may become useful for applications. The extension of the interaction box formalism to predict the downwards shift of the resonance frequency[17] and to bridge the gap among models at different levels (nano, meso, and macro)[18] are currently in progress. Comparisons between the results of LISA simulations and interaction box formalism calculations are also planned.

Acknowledgements

This work has been supported by the ESF-PESC program NATEMIS. We are indebted to many colleagues and collaborators for fruitful discussions. Among them are F. Pastrone, J. Engelbrecht, G.P. Pescarmona, A. Berezovsky, N. Pugno, M. Scalerandi, F. Bosia, M. Griffa, A. Gliozzi, S. Delsanto, M. Hirsekorn, and M. Nobili.

References

[1] A. Carpinteri, Scaling laws and renormalization groups for strength and toughness of disordered materials, *Int. J. Solids Structures*, 31, 291–302 (1994); A. Carpinteri and N. Pugno, Are scaling laws on strength of solids related to mechanics or to geometry?, *Nature Mat.*, vol. 4, 421–3 (2005).

[2] R.A. Guyer and P.A. Johnson, Nonlinear mesoscopic elasticity: Evidence for a new class of materials, *Phys. Today* **30**(4), 30–36 (1999).

[3] See in several issues of *Mat. Res. Soc. Bull.*, e.g., **24**(2), (1999); **25**(5), (2000); **26**(3), (2001); **27**(1), (2002).

[4] S. Hirsekorn and P.P. Delsanto, On the universality of nonclassical nonlinear phenomena and their classification, *Appl. Phys. Lett.* **84**, 1413–1415 (2004).

[5] P.P. Delsanto and S. Hirsekorn, A unified treatment of nonclassical nonlinear effects in the propagation of ultrasound in heterogeneous media, *Ultrasonics* **42**, 1005–1010 (2004).

[6] F. Bosia, S. Hirsekorn, and N. Pugno, work in progress; see also Chapter 30.

[7] See Chapters 4, 5, 25, and 26.

[8] T.J. Ulrich and T.W. Darling, Observations of anomalous elastic behavior in rock at low temperatures, *Geophys. Res. Lett.* **28**, 2293–2296 (2001); see also Chapters 17 and 21.

[9] S. Hirsekorn, Nonlinear transfer of ultrasound by adhesive joints—A theoretical description, *Ultrasonics* **39**, 57–68 (2001); see also Chapter 15.

[10] See Chapters 11 and 16.

[11] P.P. Delsanto, R.B. Mignogna, R.S. Schechter, and M. Scalerandi. In: *New Perspective on Problems in Classical and Quantum Physics*, edited by P.P. Delsanto and A.W. Saenz (Gordon Breach, New York, 1998) Vol. 2, pp. 51–74.

[12] P.P. Delsanto and M. Scalerandi, Modeling nonclassical nonlinearity, conditioning, and slow dynamics effects in mesoscopic elastic materials, *Phys. Rev. B* **68**, 064107–064116, (2003).

[13] See Chapters 11–13.

[14] See Chapters 16–18.

[15] See Chapters 11, 14–16, and 18.

[16] See Chapters 22 and 32.

[17] P.P. Delsanto, M. Nobili, and N. Pugno, work in progress.

[18] P.P. Delsanto, M. Griffa, C.A. Condat, S. Delsanto, and L. Morra, Bridging the gap between mesoscopic and macroscopic models: The case of multicellular tumor spheroids, *Phys. Rev. Lett.* **94**, 148105, 1–4 (2005).

[19] L.S. Gradshteyn and L.M. Ryhbik, *Table of Integrals, Series and Products* (Academic Press, London, 2000).

A. Derivation of the Explicit Formulas for $G(x)$ and $H(x)$

After suitable manipulations of well-known trigonometric formulas[19] one can express $\sin(j\tau)$ and $\cos(j\tau)$, both for odd and even values of $j \geq 1$, as linear combinations of powers of $x = \sin \tau$ with or without a single factor $c = \cos(\tau) = \pm\sqrt{1 - x^2}$:

$$\sin\left((2j - m)\tau\right) = c^{1-m} \sum_{p=0}^{j-1} a_{2j-m,p} x^{2p+1}, \quad m = 0, 1; \qquad \text{(A1a)}$$

$$\cos\left((2j - m)\tau\right) = c^{m} \sum_{p=0}^{j-1} b_{2j-m,p} x^{2p}, \quad m = 0, 1; \qquad \text{(A1b)}$$

where

$$a_{2j-m,p} = (-1)^p 2^{2p-m+1} \left(\frac{2j-1}{j+p}\right)^m \binom{j+p}{2p+1},$$ (A2a)

$$b_{2j-m,p} = (-1)^p 2^{2p} \left(\frac{j}{j+p}\right)^{1-m} \binom{j+p-m}{2p}.$$ (A2b)

These results substituted into Eqs. (1.2) and (1.8) yield ($\tau = \omega t$):

$$E(x) = E_0 + \sum_{j=1}^{\infty} \left[(E_j \cos(\varphi_j)) \sin(j\tau) + (E_j \sin(\varphi_j)) \cos(j\tau) \right]$$

$$= E_0 + \sum_{j=1}^{\infty} \left[Q_{2j} \sin(2j\tau) + P_{2j} \cos(2j\tau) + P_{2j-1} \sin((2j-1)\tau) \right.$$

$$\left. + Q_{2j-1} \cos((2j-1)\tau) \right] = E_{\mathrm{av}} + G(x) \pm H(x),$$ (A3)

where

$$E_{\mathrm{av}} = E_0 + \sum_{j=1}^{\infty} P_{2j} b_{2j,0},$$ (A4)

$$G(x) = \sum_{j=1}^{\infty} \left[P_{2j} \sum_{p=1}^{j} b_{2j,p} x^{2p} + P_{2j-1} \sum_{p=0}^{j-1} a_{2j-1,p} x^{2p+1} \right] = \sum_{i=1}^{\infty} P_i \sum_{n=1}^{i} g_{ni} x^n,$$ (A5)

$$H(x) = \sqrt{1-x^2} \sum_{j=1}^{\infty} \left[Q_{2j} \sum_{p=0}^{j-1} a_{2j,p} x^{2p+1} + Q_{2j-1} \sum_{p=0}^{j-1} b_{2j-1,p} x^{2p} \right]$$

$$= \sqrt{1-x^2} \sum_{i=1}^{\infty} Q_i \sum_{n=1}^{i} h_{ni} x^{n-1},$$ (A6)

and, from Eqs. (A2)

$$g_{nj} = (-1)^{IP(n/2)} 2^n \frac{j}{j+n} \left(\frac{(j+n)/2}{n}\right),$$ (A7)

$$h_{nj} = (-1)^{IP((n-1)/2)} 2^{n-1} \frac{j}{j+n} \left(\frac{(j+n)/2-1}{n-1}\right),$$ (A8)

if n and j have the same parity, otherwise $g_{nj} = h_{nj} = 0$. Here, $IP(a)$ means the integer part of a.

In Eq. (A4) the second term on the right-hand side corresponds to the contribution of the missing term with P_{2j} in Eq. (A5) when $p = 0$. Because it does not depend on x it is included in the equilibrium point E_{av}. With $b_{2j,0} = 1$ from Eq. (A2b), Eqs. (A4) and (1.5) are the same. Likewise Eqs. (A5) and (A6) yield Eqs. (1.6) and (1.7), and Eqs. (1.9) and (1.10) easily follow from Eqs. (A5) to (A8) for the finite number $N = 3$ of contributing harmonics.

2

Nonlinearity and Complexity in Elastic Wave Motion

Franco Pastrone

Dipartimento di Matematica, Università di Torino, Italy-franco.pastrone@unito.it.

Abstract

The goal of this chapter is to provide some elementary basic ideas about nonlinear wave motion in complex elastic structures. The path consists of three steps: the basic concepts of the nonclassical generalized continua, which encompass the theory of elasticity; the meaning of complexity in the theory of elastic structures; and elastic nonlinear wave propagation. Microstructures and waves in microstructured solids are introduced to provide a meaningful example of the general scheme, whose results are of some interest in the theory and in the applications.

Keywords: Complex structures, generalized continua, microstructures, nonlinear elasticity, nonlinear waves

1. Introduction

A mathematical theory of elasticity, which takes into account nonlinearity, complexity, and nonlinear wave propagation, can be developed only if it would be encompassed in a more general model of continua, that we can call a nonclassical theory of continua or theory of generalized continua.

In fact, the nonlinear theory of elasticity ought more correctly to be called the exact theory of elastic bodies and it is a part of the more general continuum mechanics as developed in the 1950s and 1960s by Truesdell, Toupin, and Noll, in their monumental books (1960, 1965). The mathematical theory of elasticity has been the explicit title of many fundamental books, among whom we quote Love (1944, 4th ed.), and, almost one century later, Marsden (1983), both of them giving the basis of the study of deformation and strain in elastic continua, and Truesdell and Wang (1973), where a modern general treatment of the theory of elasticity is presented, including the theory of wave propagation.

The fact that it is usually called nonlinear elasticity is a heritage of the historical development of physics, which is essentially linear because of the second law of Newton (proportionality of the acceleration exploited to the force impressed) and the rule of the parallelogram for the composition of forces. Moreover, the theory of linear elasticity has been greatly successful for almost two centuries, both from a mathematical point of view and the experiments and applications performed in engineering. Even now, many researchers think of elasticity as a linear theory that sometimes can be extended to a nonlinear theory, ad hoc constructed. The problem of buckling of structures and of propagation of waves in fluids pointed out that linear models were not sufficient to explain simple phenomena. The pioneering work of Murnaghan in The United States and Signorini in Italy in the 1930s opened the way to the foundation of classical continuum mechanics in the 1950s due mainly to Truesdell and his group.

In this framework, the exact theory of elasticity is based upon two well-known assumptions.

(i) The deformation tensor (or strain tensor) is defined by

$$\varepsilon = \frac{1}{2}(\nabla \mathbf{u} + \nabla \mathbf{u}^T + \nabla \mathbf{u} \nabla \mathbf{u}^T)$$

or other similar tensors, where $\nabla \mathbf{u}$ is the displacement gradient and the superpose T means transpose.

(ii) The stress–strain relations, namely, the constitutive equations, are given by

$$T = T(F, \mathbf{X}),$$

where T is the Cauchy's stress tensor, F the deformation gradient ($F = \nabla \mathbf{u} + I$, I being the identity), and $\mathbf{X}$ a point of the body. Other similar expressions can be obtained from this one, involving different stresses, such as Piola–Kirchhoff's stress. If there is no dependence on $\mathbf{X}$, the material is said to be homogeneous. It must be remarked that in general T is a nonlinear function of its argument F. In some theories, such as materials with memory, T is not even a function, but a functional. In some cases it can encompass also hysteretic phenomena, such as in the theory of hypoelasticity.

Sometimes, the assumption (i) is referred to as geometrical nonlinearity and (ii) as physical nonlinearity.

Obviously, the linear theory can be obtained via a linearization of the strain, by means of the linear strain tensor and linearized constitutive equations:

$$e = \frac{1}{2}(\nabla \mathbf{u} + \nabla \mathbf{u}^T); \qquad T = Ae.$$

The linear constitutive equations are called the generalized Hooke's law, where A is the elasticity tensor, namely, a fourth-order tensor whose elements are constant if the material is homogeneous and in the isotropic case it depends only on two elastic constants, for instance, the Lamè's constants or other constants well known in the literature such as Young's modulus, Poisson's ratio, and so on.

The study of free vibrations of solid bodies has been developed since Poisson, Clebsch, Lamè, Cauchy, Green, Christoffel, and Lord Raleigh, mainly during the nineteenth century, and many succeeding scientists, but until more recent years, the model was the linear theory of elasticity and waves were seen merely as vibrations, that is, with small variations in deformations and stresses. Relevant developments were obtained in the study of waves in fluids, electromagnetic waves, and optical waves, but my task is restricted to the problem of waves in elastic bodies.

Nonlinear wave propagation was introduced by the pioneering works of Hadamard, Hugoniot, Rankine, and a long list of authors, but one modern approach, the singular surface method, is due mainly to Truesdell, as quoted both in CFT and in NFTM, Thomas (1961), Chen (1973), Hayes (1963), Hayes and Rivlin (1972), Jeffrey (1976, 1980) and many others. According to this theory, a wave is a perturbation traveling through a body and at each time the body is divided in two parts: the unperturbed part and the perturbed part, the separation being a surface (the wave front) that moves with a certain speed in the body. Different families of waves are classified according to the fact that there are discontinuities across this surface of derivatives of the fields involved: so we have acceleration waves, shock waves, or higher-order waves, as briefly discussed in Section 4. It is noticeable that Christoffel probably was the first to discuss wave propagation as the advance through the medium of a surface of discontinuity (the wave front; see Love (1994), p. 18), which is the starting point to approach the problem in nonlinear theories as briefly discussed above.

Another approach was introduced by Whitham (1974) and extensively used by Engelbrecht et al., as discussed in Chapter 3 of this volume, to which we refer for more details. It consists mainly in the concept of a hierarchy of waves; namely, we choose one or more scale parameters, such that when it (or they) vanishes or goes to infinity different wave operators prevail on the other ones and from asymptotical analysis we can point out the role of the different scale depending on the features. For instance, the dispersive effects due to the microstructure can be examined, or how the waves, if we choose the parameter as the ratio of the characteristic scale of the microstructure over the wavelength of the excitation, are sensible to the influence of the micro- or of the macrostructure when this parameter is small or large.

2. Nonclassical Continua

The work of Truesdell, Toupin, and Noll not only gave a firm framework to classical continuum mechanics (even from the title of the monumental treatise by Truesdell and Toupin (1960): *The Classical Field Theories*), but also opened new perspectives toward the creation of new theories of generalized, or nonclassical, continua.

The origin of possible generalizations can be found in the works of Cauchy, Voigt, Boltzman, and the Cosserat brothers, but surely, as said by Forest and Sievert (2003): "The sixties have definitively been the Golden Age of the Mechanics of generalized continua with the milestones [1]–[3] {Toupin, Mindlin, Eringen, ...}" and "They provided us with a rigorous and almost exhaustive corpus of balance and constitutive equations for generalized continua" [Forest and Sievert (2003), p. 71].

The problems that arise are: (i) what do we mean by generalized continua?; (ii) which relation can be established with the theory of complexity?; what is the relevance of wave propagation in this contest? In the following sections an attempt is made to give a brief answer to such questions.

The first problem is not trivial and we refer to Maugin ($2004_{1,2}$) for a deeper analysis. According to him, and we agree, "the notion itself of generalization depends also (subjectivity) on the education and knowledge of the author" [Maugin (2004_1)].

In the literature very often we read the statement: "the generalized Hooke's law," that means the constitutive equations of an anisotropic homogeneous linear elastic body, where the generalization is due to the anisotropy: a very weak generalization indeed. A stronger generalization, even in the case of linear elasticity, is represented by the assumption of inhomogeneity: in fact, the dependence of the elasticity coefficients on the material point $\mathbf{X}$ of the body changes the nature of the problem substantially.

In classic continuum mechanics the Cauchy axiom on the tractions inside the body implies the symmetry of the Cauchy stress tensor. This axiom has the physical meaning that there are no internal couples and no internal microstructures that can be described by additional internal degrees of freedom. There is another underlying axiom requiring that the body be a flat manifold, or, equivalently a Euclidean space, as well as the physical space. The flatness of the body manifold can be also seen as the consequence of the requirement that the so-called compatibility conditions are satisfied such as to ensure that the strain tensor is related to smooth displacement fields. These conditions, in finite elasticity, imply that the Riemann tensor vanishes at each point; namely, the curvature is zero everywhere. In linear elasticity the compatibility conditions are called the St. Venant conditions and do not have a clear geometrical interpretation.

Again referring to Maugin (2004$_2$), generalization arises when we relax one or more such assumptions. In this sense, not only the generalized Hooke's law is weak, but also linear piezoelectricity and linear thermodynamics are such, whereas inhomogeneity and dissipation represent a step toward a stronger generalization.

As an example of continua, which do not satisfy the assumption of a flat material body manifold, we can recall the theory of defects and specifically the theory of solids with a continuous distribution of dislocations. This problem is strictly connected with the theory of material inhomogeneities and we can deal with different mathematical possibilities, each one related to different physical properties: we can assume that the connection of the body manifold has a symmetric part that makes the curvature vanish, and a skew-symmetric part that is responsible for the torsion of the manifold, as typical in the theory of continuous distribution of dislocations, or that the body is an Einstein–Cartan space equipped with both curvature and torsion, as in general relativity, which can be seen as a theory of the second-order gradient of the space–time metric and the body is the physical universe.

The symmetry of the Cauchy stress fails to be true if we assume the existence of couple stresses, or if we introduce the concept of internal structures, adding internal degrees of freedom, both of mechanical and nonmechanical nature. One talks of microstructures, granular media, Cosserat continua, micromorphic continua, ferroelectric crystals, ferromagnetic media with intrinsic spin, and so on. Other kinds of generalization have been introduced relaxing the assumption of locality, which is strictly related to the Cauchy postulate, leading to the so-called nonlocal theories. In such models, one can take into account the effects of edges, apex points through the hyperstress, the capillarity effects, and resonance between length scales.

Two aspects must be put in evidence: from the mathematical point of view, in nonclassical elasticity the constitutive functions depend not only on the gradient of deformation F, but also on higher-order gradients or on gradients of the physical fields coupled to the deformations. From the physical point of view, the existence of characteristic lengths in wave propagation is related to the rise of dispersive terms and one can notice the "competition" between nonlinearity and dispersion.

Maugin, in several papers, proposed an intrinsic material formulation of continuum mechanics, based on the notion of Eshelby's stress tensor "since this formulation automatically captures the relevant new ingredients" [Maugin (2004$_2$)].

I do not want to go further in this direction, in as much as, for our purposes, it is enough to keep in mind that as the generalized theory of elasticity, or nonlinear nonclassical elasticity, we mean the particular case of a generalized continuum where the above-mentioned assumptions of classical theory are relaxed and in particular we deal with bodies whose constitutive equations depend not only on F, but also on other variables apt to describe the internal structure of the material; for instance, internal scales can be taken into account, due to the presence of microstructures, the Cauchy stress tensor is not symmetric, and couple stresses are present. These ingredients are not necessarily present all together; the presence of one or more of them generates different generalizations, as discussed in Maugin (2004$_{1,2}$).

3. Complex Materials

At this point the connection with the theory of complex structures is quite evident. Even according to the Chambers dictionary, as reminded by Engelbrecht (1993)$_2$, "*simple* is that which consists of one thing or element, *complex* what is composed of more than one or of many parts." Hence we can say that a simple material is a classical continuum, being constituted by a single structure. More exactly, according to Noll's theory, a simple material is a material such that the constitutive functionals reduce to functions depending on the actual values of the deformation gradient, the temperature, and the gradient of temperature. It is a material without memory and with a "simple" internal structure. Elastic materials are particular cases of this one, when there is no dependence on temperature and its gradient. When we deal with nonclassical generalized continua, the material is no simpler not in the sense of the dictionary, nor in the sense of complexity, not even in Noll's sense. Hence a body with some kind of internal structure is complex and if it is made of elastic material it is a nonclassical elastic body. Let me remark that nonlinearity is always assumed.

More recently, the problem of complexity revealed its importance also in continuum mechanics and, in particular, in nonlinearly elastic structures. The problem of scale-depending phenomena became more and more relevant and different names have been given according to the different microscales used: mesomechanics, nanomechanics, and microstructure theory are terms widely used, sometimes with a kind of overlapping of models. Let us call a complex system a material continuum that exhibits such a structure and in this sense we can give the name of NCNL elasticity to the theories that model complex elastic structures. The influence of micro- (or meso- or nano-) structures on the behavior of the macrostructures is of great importance in applications in very different fields: fluids with bubbles, microcrack distribution in solids, crystal fluids, dislocations and disclinations, granular solids, porous media, and so on. In many cases the mathematical theory does not provide suggestions as to how to perform experiments in order to exhibit the existence, consistency, and influence of the microstructure over the macrobody. In this sense we can talk of nonclassical nonlinear elasticity, in the sense that some of the pillars of the classical exact theory of elasticity are relaxed. For instance, as in our case, the Cauchy stress tensor is no longer symmetric because of the presence of microstructures described by additional internal degrees of freedom, which implies the presence of applied couples in bulk and at the surface.

4. Nonlinear Waves

There are two main approaches to the study of nonlinear wave propagation. The first, much more general, is based on the theory of singular surfaces. This theory, though, as already said, developed since the twentieth century by Christoffel and after by Duhem, Hadamard, Hugoniot, Rankine, and many others, has been clearly established in an exact contest for classical continua by Truesdell and Noll, hence by Thomas, Jeffrey, Chen, and Hayes, just to recall a few names already quoted in the previous sections.

The idea is that when a wave travels through a body, two subregions can be identified: the unperturbed part, where the wave has not yet arrived, and the perturbed part, where the perturbation is acting, separated by a surface, whose physical meaning is that one of the wave fronts moves through the body with a certain speed. Mathematically, it is assumed that through this surface, at each time, some field derivatives, both in space and time, have a discontinuity. If the second derivatives are discontinuous and the first derivatives are continuous, we call the wave an "acceleration wave;" if the first derivative is discontinuous we have a "shock wave." We can even assume that the field itself is discontinuous across the surface and is called a "dislocation wave," but this case is difficult to treat and beyond our goal.

Because the discontinuities must be somehow connected, because the body is not going to break, the so-called geometrical and kinematical compatibility conditions can be derived, introducing the speed of propagation and the amplitude of the wave, which are the main features of the wave and they are unknowns and must be determined. In the case of acceleration waves, applying such conditions to the field equations one obtains the Hadamard–Hugoniot–Rankine–$\cdots$ conditions, namely, equations involving the jumps, hence the wave speeds and amplitudes, such that the problem reduces to an algebraic eigenvalue problem; the eigenvalues and the eigenvectors are the squares of the wave speeds and the amplitudes, respectively. Under suitable conditions on the acoustic tensor, which appears naturally in nonlinear hyperelasticity as the set of second-order derivatives of the strain energy function with respect to the coordinates, it is in principle possible to evaluate those unknowns. Moreover, using each set of eigenvalues and the corresponding eigenvectors, one can derive the equations of evolution of the amplitudes and try to evaluate it along each wave.

This method is somehow similar to the method of characteristics of hyperbolic systems, but it is more general and we do not need to deal with hyperbolic systems; indeed we say that the field equations system is hyperbolic if there exist exactly as many real speeds as the number of equations, even if it is not hyperbolic in the sense of PDE classification.

Similarly, a general analysis for shock waves can be developed, as well as for wave propagation in rods, shells, and plates. Actually, it seems feasible to apply this method also to microstructured solids and to complex materials, as it shown in the next section.

Another approach, due mainly to Whitham (1974) and Engelbrecht (1983), and we refer to Engelbrecht (1993$_{1,2}$) and to Chapter 3 of this volume for a more detailed discussion, consists in assuming nonlinear but explicit forms of the constitutive equations, including dispersive and dissipative terms, such that the governing equations assume a typical form [see Engelbrecht (1993$_2$) formula (2.30)], namely a system quasilinear in the leading terms, but strongly nonlinear in the remaining parts, such that it should be possible to find an associated linear, strictly hyperbolic system that would provide solutions of the linearized problem useful for discussions of the corresponding solutions of the nonlinear system. Let us remark that the famous KDV equation is encompassed in this approach.

The nonlinear system in general cannot be solved by direct methods. One possibility is to carry out an asymptotic analysis, as done by Taniuti and Nishihara (1983), or to resort to other methods like the approximate equation method, as by Whitham (1974), Engelbrecht (1993$_1$), the method of hierarchy of waves, again Whitham (1974), Engelbrecht (1993$_1$,1997), and in this volume).

These methods have been successfully applied to many different problems in solid mechanics, in fluids, and in materials with different kinds of internal structures and different dimensions, and, among other results, the possibility of existence of solitons can be proved.

5. Nonlinear Waves in Microstructured Solids

In this section an example is provided of possible applications of the theory of nonlinear wave propagation in complex materials, within the framework of nonclassical nonlinear elasticity.

The linear theory of elastic microstructured solids, including vibrations, has been established by Mindlin (1964), thereafter by Eringen (1966) and Kunin (1983), whereas general treatises on the theory of microstructures are due to Maugin (1980), Capriz (1989), and Eringen (2000, 2002).

I restrict my example to the so-called vectorial microstructures, which can encompass many relevant models, such as Cosserat media, granular solids, micromorphic bodies, and solids with microcracks. Some interesting results have been obtained in the one-dimensional case, when a scalar function describes the microstructure and both dispersion and dissipation can be taken into account, within, obviously, the nonlinearity. A wide class of phenomena can be described by means of microstructural models of solids and fluids, where the microstructure can be described by vector fields over the body. In principle, there are no restrictions on the number of vector fields, which are unknown variables of the problem, but there are obvious restrictions due to the possible physical meaning of each vector field.

Let B be the body, as a manifold embedded in a three-dimensional affine space, $\mathbf{X}$ a point of this body in its reference configuration C^*, and $\mathbf{x}$ the corresponding point in the actual configuration C. As usual, the displacement is given, in terms of the position vector $\mathbf{r} = \mathbf{r}(X^h, t) = \mathbf{x} - \mathbf{o}$, $\mathbf{o}$ a fixed point in the physical space and $\mathbf{R} = \mathbf{X} - \mathbf{o}$, by $\mathbf{u} = \mathbf{r} - \mathbf{R}$. Commas denote partial derivatives with respect to X^h and superposed dots denote partial derivatives with respect to time; for example,

$$\mathbf{r},_h = \frac{\partial \mathbf{u}}{\partial X^h}; \qquad \dot{\mathbf{r}} = \frac{\partial \mathbf{u}}{\partial t}.$$

By microstructure we mean that at each point $x \in C$ it is possible to apply a microscope and discover a "small world." We assume that this "small world" is a manifold of dimension n, and we label this micromanifold M_x. By means of a mathematical procedure, elsewhere called "magnification process" (see Pastrone (2005), one can associate this microscopic world with a set of vectors,

$$\mathbf{d}_K = \mathbf{d}_K(X^h, t) \qquad K = 1, 2, \ldots, N,$$

which are a particular case of more sophisticated possible models, but general enough for our purposes. Such vectors can be called "directors," according to the usual language of polar continua, and they are apt to provide a description of some properties of the microstructure as they act at the macroscopic level. In fact, each director can belong to a vector space of dimension 1, 2, ..., according to which physical property it is related and how we performed the magnification process.

The kinetic energy density of the body is defined as a quadratic form in the velocities $\dot{\mathbf{r}}, \dot{\mathbf{d}}_K$:

$$T = \frac{1}{2}[\rho(X^h)\dot{\mathbf{r}} \cdot \dot{\mathbf{r}} + \rho(X^h)I^{HK}\dot{\mathbf{d}}_H \cdot \dot{\mathbf{d}}_K]. \tag{2.1}$$

In Eq. (2.1), ρ is the usual three-dimensional mass density in the reference configuration; I^{HK} are the coefficients of the inertia terms. We assign a strain energy density function

$$W = W(\mathbf{r},_I; \mathbf{d},_J; \mathbf{d}_{J,h}; X^h) \tag{2.2}$$

whose existence follows from the assumption that the total power P_T is given by $P_T = dW/dt$ and the total energy by:

$$E = \int_B (T - W)\rho \, dX^1 dX^2 dX^3 - \int_B \rho W_b \, dX^1 dX^2 dX^3, \tag{2.3}$$

where W_b is the potential of the external body forces, which depends on $\mathbf{r}$ and X^h only.

We avoid internal constraints and separate the problem of the boundary conditions. The equations of motion can be derived as the Euler–Lagrange equations of the energy functional

$$\mathcal{E} = \int_{t_0}^{t_1} E \, dt : \qquad \delta\mathcal{E} = 0 \text{ for any admissible motion} \qquad \Rightarrow$$

$$\left(\frac{\partial W}{\partial \mathbf{r}_{,i}}\right)_{,i} - \frac{\partial W}{\partial \mathbf{r}} = \frac{d}{dt}\frac{\partial T}{\partial \dot{\mathbf{r}}} \tag{2.4}$$

$$\left(\frac{\partial W}{\partial \mathbf{d}_{K_{i,}}}\right)_{,i} - \frac{\partial W}{\partial \mathbf{d}_K} = \frac{d}{dt}\frac{\partial T}{\partial \dot{\mathbf{d}}_K}.$$

We can also take account of the dissipation, but we refer to Pastrone (2005) for a more detailed discussion.

In the following are a few general results that can be derived avoiding heavy calculations. We simplify the notation, introducing a Euclidean vector space E, of dimension M, whose elements are given as ordered sets: $\mathbf{p} \in E, \mathbf{p} \equiv \{\mathbf{r}, \mathbf{d}_1, \ldots, \mathbf{d}_N\}$; hence E is generated by the vectors $\mathbf{r}, \mathbf{d}_K$ and its dimension depends on the dimensions of the spaces to which the directors belong. A useful inner product can be defined by:

$$\forall \, \mathbf{p}^1, \mathbf{p}^2 \in E, \langle \mathbf{p}^1, \mathbf{p}^2 \rangle \equiv \mathbf{r}^1 \cdot \mathbf{r}^2 + \delta^{HK}\mathbf{d}_H^1 \cdot \mathbf{d}_K^2. \tag{2.5}$$

Now we deal briefly with wave propagation, according to the finite discontinuity surface model recalled in Section 4. A surface Σ moving through the body B, of equation $\phi i(X^i, t) = 0$, is called an acceleration wave if the field $\mathbf{p}$ and its first derivatives $\mathbf{p}_{,i}$ and $\dot{\mathbf{p}}$ are continuous on Σ, but some second derivative has finite discontinuities there.

We assume some familiarity with the theory of singular surfaces and the usual kinematical conditions of compatibility yield:

$$[|\mathbf{p}_{,ij}|] = \mathbf{A}n_i n_j, \qquad [|\dot{\mathbf{p}}|] = \mathbf{A}v^2, \tag{2.6}$$

where the double brackets mean jump across the surface Σ, the vector field $\mathbf{A}$ represents the amplitude vector of the wave, n_i are the components of the unit vector normal to Σ, and v the wave speed. If we apply the jump condition to the equation of motion written in terms of the field variables $\mathbf{p}$:

$$\left(\frac{\partial W}{\partial \mathbf{p}_{,i}}\right)_{,i} - \frac{\partial W}{\partial \mathbf{p}} = K\ddot{\mathbf{p}}, \tag{2.7}$$

where K is the linear transformation naturally induced by the kinetic energy we obtain the Hugoniot–Hadamard conditions in the compact form:

$$\mathcal{Q}\mathbf{A} = KA v^2, \tag{2.8}$$

where $\mathcal{Q}$ is the acoustic tensor, given by:

$$\mathcal{Q} \equiv \frac{\partial^2 W}{\partial \mathbf{p}_{,i}\partial \mathbf{p}_{,j}}n^i n^j. \tag{2.9}$$

Because the acoustic tensor is symmetric, (2.8) represents an eigenvalue problem such that, if $\mathcal{Q}$ is positive semidefinite, that is,

$$\langle \mathbf{A}, \mathcal{Q}\mathbf{A} \rangle \geq 0, \forall \mathbf{A} \in E, \mathbf{A} \neq 0, \tag{2.10}$$

the eigenvalues of (2.8) are M and all positive, hence there exist M real velocities $v_{(S)}$, $S = 1, 2, \ldots, M$. If the condition (2.10) is valid for all directions, we recover the so-called strong ellipticity condition in the static case. Hence we can claim that the stability of equilibrium implies the acoustic tensor is positive semidefinite and consequently we have exactly 12 acceleration wave speeds. Conversely, if the body loses the ability to propagate some acceleration waves the corresponding equilibrium configuration is unstable.

Exploring in more detail the general equations and conditions of this section, one could obtain explicit results valid in many particular cases, some of which are encompassed in this model. Shock waves can also be investigated recovering some expected results. For instance, using this formalism it is easy to find that in linear theory the velocities of acceleration and shock waves are the same. Other results about infinitesimal vibrations and normal modes can be reached but we choose to stop here.

6. The One-Dimensional Case

According to different assumptions on the geometry and kinematics of the microstructures one can recover many particular cases often discussed in the literature, usually each one being introduced independently.

One-dimensional microstructured bodies have been extensively studied in a series of papers [see Pastrone et al. (2004) and references cited therein] dealing with different particular models. Their common general features can be easily derived in the present contest. The body is a one-dimensional manifold, with a material coordinate x and a unit vector $\mathbf{e}$, such that the vector fields $\mathbf{r}$ and $\mathbf{d}$ can be written as $\mathbf{r} = r(x, t)\,\mathbf{e}$ and $\mathbf{d} = \psi(x, t)\backslash \mathbf{e}$. Hence we deal with the scalar functions $r = r(x, t)$, $\psi = \psi(x, t)$ only.

The strain energy function $W = W(u, u_x, \psi, \psi_x, x)$ is an assigned smooth function and the kinetic energy is a quadratic form in $\dot{u}$, $\dot{\psi}$:

$$T = \frac{1}{2}(\rho \dot{u}^2 + I \dot{\psi}^2), \tag{2.11}$$

where $\rho = \rho(x, t)$ is a one-dimensional mass density and I an inertia term connected with the microstructure, which can have different explicit forms according to the kind of microstructure one can represent with this model (i.e., microcrack density, dislocation density, voids, etc.). If we assume dissipation, we introduce dissipative stresses such that the field equations take the form [see Pastrone et al. (2004)]:

$$\rho u_{tt} = \left(\frac{\partial W}{\partial u_x}\right)_x - \frac{\partial W}{\partial u} + D_1 \tag{2.12}_1$$

$$\rho \psi_{tt} = \left(\frac{\partial W}{\partial \psi_x}\right)_x - \frac{\partial W}{\partial \psi} + D_2, \tag{2.12}_2$$

where u is a displacement field, the subscripts mean derivatives with respect to time t or to the spatial coordinate x, and the terms D_1, D_2 summarize the dissipation. In some cases it is assumed $I = 0$ (namely, the microstructure has no inertia). The case with no dissipation ($D = 0$) has been studied in Engelbrecht and Pastrone (2003).

The particular choice of the strain energy function W gives rise to different nonlinear models. The effects of nonlinearity, dispersion, and dissipation can be quite evident in the evolution of traveling waves in complex structures and they have been analyzed in some simpler cases [see Engelbrecht et al., (1999), Porubov and Pastrone (2004), and Pastrone (2005)].

For instance, if we assume that W is given by [see Pastrone (2005)]:

$$W = \frac{1}{2}\alpha u_x^2 + \frac{1}{3}\beta u_x^3 - A\psi u_x + \frac{1}{2}B\psi^2 + \frac{1}{2}C\psi_x^2, \tag{2.13}$$

where α, β, A, B, C, are material constants related to the usual elastic moduli in a known way [see Porubov and Pastrone (2004)].

The field equations may be further simplified by the so-called slaving procedure, namely, by considering some parameters that can dominate the behavior of the wave, and can be in competition among themselves, such that nonlinearity, dissipation, and dispersion can be taken into account with different weights. In other words, sometimes they can be considered "negligibly small," sometimes not, according to the different effects we want to point out. Moreover we can eliminate the variable ψ in Eq. (2.12), when the energy function (2.13) is used, such that we obtain the evolution equation.

$$v_{tt} - v_{xx} - \varepsilon\alpha_1(v^2)_{xx} - \gamma\,\alpha_2 v_{xxt} + \delta(\alpha_3 v_{xxxx} - \alpha_4 v_{xxtt}) + \gamma\,\delta(\alpha_5 v_{xxxxt} + \alpha_6 v_{xxttt}) = 0, \tag{2.14}$$

where $\alpha_1, \dots, \alpha_6$ are material constants uniquely related to the previous ones; $\varepsilon, \delta, \gamma$ are the leading scale parameters related to nonlinearity, dispersion, and dissipation, respectively.

In Porubov and Pastrone (2004) and Pastrone et al. (2004) it has been proved that if there is no dissipation and nonlinearity and dispersion are balanced, that is, $\gamma = 0, \varepsilon = O(\delta)$, there exists a solution of the kind of a bell-shape solitary traveling wave; if dispersion is weak and nonlinearity is balanced by the dissipation, that is, $\delta < \varepsilon$ and $\gamma = O(\varepsilon)$, there exists a kink-shaped traveling solution, and other similar results can be proved.

The wave hierarchy approach can be also used, as proved in Engelbrecht et al. (2003) and Pastrone et al. (2004); the typical form of the wave equations obtained in many one-dimensional problems is

$$u_{tt} + \lambda u_{xx} = [N(u) + M(u)]_{xx} + D, \tag{2.15}$$

where u is a displacement field, the subscripts mean derivatives with respect to time t or to the spatial coordinate x, N is the nonlinear part, M the dispersive part, and D is the dissipation. Usually M and N are expressed in terms of second-order derivatives, at least. For more details and a broader analysis we refer to Engelbrecht et al., (1999), Engelbrecht and Pastrone (2003), and elsewhere in this volume).

The model equations demonstrate the fundamental influence of a microstructure on the wave motion, the dispersive character of motion, and the influence of the scale parameters. This permits us to relate the experimental measurements directly to theoretically introduced parameters, as clearly shown by Berezovski in Engelbrecht (2004) for the changes of the wave speeds.

Conclusions

Nonlinear elasticity as a mathematical well-established theory finds its origin in the 1930s and was completely developed during the 1960s, as briefly said in the first two sections. Also nonclassical theory was known long ago to the community of scientists working in this field, even if with slightly different names: generalized elasticity, second-gradient theory, dislocation theory, microstructures, and so on. In this short chapter I wanted to add a new ingredient: complexity. As I tried to explain, complexity is a term that covers many different subjects; even the meaning can change according to the field where it is used: informatics, biology, dynamical systems, economy, or engineering. The example introduced in Sections 5 and 6 clearly shows the idea of complexity as it can be introduced in materials science, in addition to the usual approach, and it is hoped that it has been shown that interesting results can be obtained mixing all such ingredients such as nonlinear elasticity, nonclassical models, and complexity, which, with the help of powerful tools such as wave propagation theory (but one could also bring into this cake the bifurcation theory for further results), can give rise to a framework of great generality and within the reach of actual and possible results.

Acknowledgments

This research has been supported by the Italian MIUR-COFIN Project 2000: "Mathematical Models in Material Sciences" and by a grant of the ESF programme NATEMIS. The author is grateful to J. Engelbrecht and G. Maugin for helpful discussions.

References

Capriz, G. (1989), *Continua with Microstructure,* Springer Tracts Nat. Phil., v. 34, Springer, Berlin, New York.

Chen, P.J. (1973), Growth and decay of waves in solids, in *Flugge's Handbuch der Physic,* vol. Via/3, Springer, Berlin.

Cosserat E. and F. (1909), *Théorie des corpes déformables,* Hermann, Paris.

Coulson, C.A. and Jeffrey, A. (1977), *Waves,* Longman Math. Texts, Essex, UK.

Engelbrecht, J. (1983), *Nonlinear Wave Processes of Deformation in Solids*, Pitman, London.

Engelbrecht, J. (1993)$_1$, Qualitative Aspects of Nonlinear Wave Motion: Complexity and Simplicity, *Appl. Mech. Rev.*, **46**, no 12, part 1, 509–518.

Engelbrecht, J. (1993)$_2$, Complexity and simplicity, *Proc. Estonian Acad. Sci. Phys. Math*, **42/1**, 107–118.

Engelbrecht, J. (1997), *Nonlinear Wave Dynamics, Complexity and Simplicity*, Kluwer, Dordrecht, The Netherlands.

Engelbrecht, J. and Braun, M., (1998), Nonlinear waves in nonlocal media, *Appl. Mech. Rev.*, **51**, No. 8, 475–488.

Engelbrecht, J. and Pastrone, F. (2003), Waves in microstructured solids with nonlinearities in microscale, *Proc. Estonian Acad. Sci. Phys. Math*, **52/1**, 12–20.

Engelbrecht, J., Berezovski, A., Pastrone, F. and Braun, M. (2004), Waves in microstructured materials and dispersion, *Phil. Mag*, **85**, Nos 33–35, 4127–4141.

Engelbrecht, J., Cermelli, P. and Pastrone, F. (1999), Wave hierarchy in microstructured solids, in *Geometry, Continua and Microstructure* (G. Maugin ed.) Hermann, Paris.

Ericksen, J.L. (1972), Wave propagation in thin elastic shells, *Arch Rat. Mech. Anal.*, **43**, 167–178.

Eringen, A.C. (1966), Linear theory of micropolar elasticity, *J. Math. Mech.*, **15**, 909–923.

Eringen, A.C. (2000), *Microcontinuum Field Theories. I – Foundations*, Springer, Berlin, New York.

Eringen, A.C. (2002), *Nonlocal Continuum Field Theories,* Springer, Berlin, New York.

Forest, S. and Sievert, R. (2003), Elastoviscoplastic constitutive framework for generalized continua, *Acta Mech.*, **160**, 71–111.

Hayes, M.A., (1963), Wave propagation and uniqueness in prestressed elastic solids. *Proc. Roy. Soc. Lond.*, **A274**, 500–506.

Hayes, M.A. and Rivlin, R.S. (1972), Propagation of sinusoidal small-amplitude waves in a deformed viscoelastic solid – II, *J. Acoust. Soc. Amer.*, **51**, 1652–1663.

Jeffrey, A. (1976), *Quasilinear Hyperbolic Systems and Waves*, Pitman, London.

Jeffrey, A. (1980), Lectures on nonlinear wave propagation, in *Wave Propagation* (Corso CIME, Bressanone), Liguori, Bologna, 7–97.

Kunin, I.A. (1983), *Elastic Media with Microstructure*, 2 Vol., Springer, Berlin.

Love, A.E.H. (1944), *A Treatise on the Mathematical Theory of Elasticity* (4th ed.), Dover, New York.

Marsden, J.E. and Hugues, T. (1983), *Mathematical Foundations of Elasticity*, Prentice Hall, Englewood Cliffs, NJ.

Maugin, G.A. (1980), *Acta Mech.*, **35**, pp.1–70.

Maugin, G.A. (1993), *Material Inhomogeneities in Elasticity*, Chapman & Hall, London.

Maugin, G.A. (1999), *Nonlinear Waves in Elastic Crystals*, Oxford Univ. Press, Oxford, UK.

Maugin, G.A. (2004a), Introduction a la mecanique des milieux continus generalises et ses applications, *Proc. Colloque National MECAMAT 2004* (Aussois, Janvier 2004), eds. P. Babin, R. Dendievel, S. Forest, J.F. Ganghoffer, A. Zeghadi, and M.H. Zoberman, pp. 47–54, Association Mécamat, Paris [also on CD-Rom].

Maugin, G.A. (2004b), Generalized continuum mechanics: Three paths, *ICTAM'04, Warsaw* (15–21 Aug. 2004), Proc. CD-Rom ISBN 83-89697-01-1, Eds. W. Gutkowski and T.A. Kowaleski, FSM3L_11347, 2 pages.

Mindlin, R.D. (1964). Microstructure in linear elasticity, *Arch. Rat. Mech. Anal.*, **1**, 51–78.

Pastrone, F. (2003), Waves in solids with vectorial microstructure, *Proc. Estonian Acad. Sci. Phys. Math*, **52/1**, 21–29.

Pastrone, F. (2005), Wave propagation in microstructured solids, *Math. Mech. Solids*, **10**, 349–357.

Pastrone, F., Cermelli, P. and Porubov, A.V. (2004), Nonlinear waves in 1-D solids with microstructure, *Mater. Phys. Mech.*, 9–16.

Porubov, A.V. (2003), *Amplification of Nonlinear Strain Waves in Solids*, World Scientific, Singapore.

Porubov, A.V., Pastrone, F. (2004), Nonlinear bell-shaped and kink-shaped strain waves in microstructured solids, *Intl. J. Nonlinear Mech.*, **39**, 1289–1299.

Taniuti, T. and Nishihara, K. (1983), *Nonlinear Waves,* Pitman, London (in Japanese, 1977).

Thomas, T.J. (1961), *Plastic Flow and Fracture in Solids*, Academic, New York.

Toupin, A. (1962), Elastic materials with coupled-stresses, *Arch. Rat. Mech. Anal.*, **11**, 385–414.

Toupin, A. (1964), Theories of elasticity with coupled-stress, *Arch. Rat. Mech. Anal.*, **17**, 85–112.

Truesdell, C.A. and Noll, W. (1965), The nonlinear field theories of mechanics, in *Flugge's Handbuch der Physik*, vol. III/3, Springer, Berlin, 1–602.

Truesdell, C.A. and Toupin, R. (1960), The classical field theories, in *Flugge's Handbuch der Physik*, vol. III/1, Springer, Berlin, 226–793.

Truesdell, C.A. and Wang, C.C. (1973), *Introduction to Rational Elasticity*, Noordhoff, Leyden.

Witham, G.B. (1974), *Linear and Nonlinear Waves*, Wiley, New York.

3

Hierarchies of Waves in Nonclassical Materials

Jüri Engelbrecht,[1] Franco Pastrone,[2] Manfred Braun,[3] and Arkadi Berezovski[4]

[1] Centre for Nonlinear Studies, Institute of Cybernetics at Tallinn University of Technology, Akadeemia tee 21, Tallinn, Estonia je@ioc.ee.
[2] Department of Mathematics, University of Turin, Via Carlo Alberto 10, 10123 Torino, Italy franco.pastrone@unito.it.
[3] Chair of Mechanics, University Duisburg-Essen, 47048 Duisburg, Germany m.braun@uni-duisburg.de.
[4] Centre for Nonlinear Studies, Institute of Cybernetics at Tallinn University of Technology, Akadeemia tee 21, Tallinn, Estonia Arkadi.Berezovski@cs.ioc.ee.

Abstract

Wave propagation in microstructured materials is directly affected by the existence of internal space scales in the compound matter. To allow for microstructures the classical continuum theory has to be generalized. In this chapter, the coupled balance laws for macro- and microstructure based on the Mindlin model are formulated. Using the slaving principles relating macro- and microdisplacements, the governing equations are derived for single- and two-scale (scale within scale) cases. These equations exhibit hierarchical properties assigning the wave operators to internal scales. In terms of macrodisplacements, higher-order dispersive terms appear that are related to the scale of the microstructure and reflect the effects of microinertia. The dispersion relations of propagating waves are established and compared with approximations resulting from hierarchical models and also with some simplified models. Linear theory is based on a quadratic free energy function, whereas in nonlinear theory cubic terms also are taken into account. The corresponding governing equation includes nonlinearities in both macro- and microscale. This consistent modeling opens up new possibilities to NonDestructive Testing (NDT) of material properties.

Keywords: Dispersion, microstructure, wave hierarchy, wave propagation

1. Introduction

Materials used in contemporary advanced technologies are often characterized by their complex structure satisfying many requirements in practice. This concerns polycrystalline solids, ceramic composites, alloys, functionally graded materials, granular materials, and the like. Often one should also account for damage effects; that is, materials are still usable when they have microcracks. In all these materials there exists an intrinsic space-scale, such as the lattice period, the size of a crystallite or a grain, or the distance between microcracks. This scale-dependence should be taken into account in establishing the governing equations. The classical theory of continuous media is built up using the assumption of smoothness of continua. The continua (materials) we are interested in contain irregularities with one or more internal scales and therefore the notion "microstructured materials" is used. The complex dynamic behavior of such microstructured materials cannot be explained by the classical theory of continua.

A more detailed description and distinction of classical and nonclassical theories of continua is given by Pastrone in Chapter 2 of this volume. Here we restrict ourselves to basic principles needed for modeling dynamical processes.

The cornerstones for describing dynamical processes of microstructured materials at intensive and high-speed deformations are the following.

(i)   Nonclassical theory of continua able to account for internal scales

(ii) Hierarchical structure of waves due to the scales in materials

(iii) Nonlinearities caused by large deformation, depending on the character of the stress–strain relations

Within the theories of continua the problems of irregularities of media were predicted a long time ago by the Cosserats and Voigt, and more recently by Mindlin (1964), Eringen (1966), and others. The elegant mathematical theories of continua with voids or with vector microstructure, of continua with spins, of Cosserat or micromorphic continua, and so on have been elaborated since; see the overviews by Capriz (1989) and Eringen (1999). Every irregularity (or inclusion) creates an additional stress field around itself. Consequently the most general approach in modeling should be the presentation of all the conservation laws and constitutive equations taking this stress field into account.

The straightforward modeling of microstructured solids leads to assigning all the physical properties to every volume element dV in a solid thus introducing the dependence on material coordinates X^k. Then the governing equations implicitly include space-dependent parameters but, due to the complexity of the system, can be solved only numerically. Another probably much more effective way is to separate macro- and microstructure in continua. Then the conservation laws for both structures should be separately formulated (Mindlin 1964; Eringen 1966; Eringen 1999), or the microstructural quantities are separately taken into account in one set of conservation laws (Maugin 1993). The last case uses the concept of pseudomomentum and material inhomogeneity force. Separating the macro- and microstructure gives two possibilities: either to consider both structures inertial or to suppose the microstructural quantities to behave noninertially. The first case is exactly what has been done by Mindlin (1964) and Eringen (1966, 1999); the second case leads to the formalism of internal variables (Maugin 1990; Maugin and Muschik 1994).

The second pillar mentioned above is the hierarchy of waves. The concept of hierarchy of waves is introduced by Whitham (1974).

High intensities of external forces and high deformation rates dictate the need to consider nonlinearities in governing equations. One should distinguish between geometrical (large deformation) and physical (stress–strain relation) nonlinearities; see Engelbrecht (1997). Physical nonlinearities are also called material nonlinearities and may be described by the approximation of the strain energy including not only second-order but also higher-order terms. These problems for microstructured solids have been analyzed, for example, by Erofeyev (2003); see also references therein. The nonlinear theory also needs a clear distinction between material and spatial coordinates.

In terms of wave characteristics, there are many physical effects due to microstructure and its possible structural changes in the wave field. In addition, the influence of nonlinearities causes nonadditivity of other physical effects. Leaving aside more complicated effects such as phase transition, kinetic localization of damage, shear bands, and so on, even the basic dissipative and dispersive effects are strongly influenced by nonlinearities. Stressing the importance of both dissipation and dispersion, in this chapter we clarify the role of dispersion only.

Dispersive and nonlinear effects combined may lead to the celebrated solitary waves. The Korteweg–de Vries equation includes quadratic nonlinearity and cubic dispersion and has served for more than 100 years as a model case for the balance of dispersion and nonlinearity. The soliton concept has formed a new paradigm in mathematical physics. When we come to microstructured materials the situation is not so simple. There still seem to be discrepancies between various mathematical models concerning the dispersion relation. In this context the discrete modeling of crystal lattices is also used (Brillouin 1953; Askar 1985; Maugin 1999). The continuum models (Erofeyev 2003; Porubov 2003) have been elaborated with various levels of accuracy.

We have previously analyzed dissipative effects in microstructured materials (Engelbrecht et al. 1999), nonlinearities in microscale (Engelbrecht and Pastrone 2003), and general dispersive effects (Engelbrecht et al. 2004).

Here in this chapter we concentrate our attention on the description of dispersive effects in microstructured solids following the consistent theory of nonclassical continua. This allows us to unite two important concepts, namely the influence of microstructure on dispersion from one side and the concept of hierarchies from another side. The third pillar—nonlinearities—takes more space. Its consistent description will be published elsewhere. Here we touch this problem only briefly.

The chapter is organized as follows. Section 2 involves the derivation of the basic single-scaled model. Mindlin (1964) assumption on strain in microstructure is used and the governing equations derived using the Euler–Lagrange formalism. It is shown that the model is consistent within the framework of pseudomomentum (Maugin 1993). Section 3 describes the modeling for the case of two-scale (scale within the scale) microstructure. In Section 4 the hierarchy of waves is explained following Whitham (1974) idea. Section 5 is devoted to the dispersion analysis of waves and Section 6 to nonlinear models. A discussion and further prospects are given in Section 7.

2. The Basic Single-Scaled Model

2.1 Governing Equations

Here we follow Mindlin (1964) who has interpreted the microstructure "as a molecule of a polymer, a crystallite of a polycrystal or a grain of a granular material." This microelement is taken as a deformable cell. Note that if this cell is rigid, then the Cosserat model follows. The displacement u of a material particle in terms of macrostructure is defined by its components $u_i \equiv x_i - X_i$, where $x_i, X_i (i = 1, 2, 3)$ are the components of the spatial and material position vectors, respectively. Within each material volume (particle) there is a microvolume, and the microdisplacement u' is defined by its components $u'_i \equiv x'_i - X'_i$, where the origin of the coordinates x'_i moves with the displacement u. The displacement gradient is assumed to be small. This leads to the basic assumption of Mindlin (1964) that "the microdisplacement can be expressed as a sum of products of specified functions of x'_i and arbitrary functions of x_i and t." The first approximation is then

$$u'_j = x'_k \, \varphi_{kj}(x_i, t). \tag{3.1}$$

The *microdeformation* is

$$\frac{\partial u'_j}{\partial x'_i} = \partial'_i u'_j = \varphi_{ij}. \tag{3.2}$$

Furthermore we consider the simplest 1-D case and drop the indices i, j dealing with u and φ only. The indices t and x used in the sequel denote differentiation.

The fundamental balance laws for microstructured materials can be formulated separately for macroscopic and microscopic scales (Eringen 1999). We show here how the balance laws can be derived from the Lagrangian (Mindlin 1964; Pastrone 2003)

$$\mathcal{L} = K - W, \tag{3.3}$$

formed from the kinetic and potential energies

$$K = \frac{1}{2}\rho u_t^2 + \frac{1}{2}I\varphi_t^2, \qquad W = W(u_x, \varphi, \varphi_x), \tag{3.4}$$

where ρ is the density and I the microinertia.

The corresponding Euler–Lagrange equations have the general form

$$\left(\frac{\partial L}{\partial u_t}\right)_t + \left(\frac{\partial L}{\partial u_x}\right)_x - \frac{\partial L}{\partial u} = 0, \tag{3.5}$$

$$\left(\frac{\partial L}{\partial \varphi_t}\right)_t + \left(\frac{\partial L}{\partial \varphi_x}\right)_x - \frac{\partial L}{\partial \varphi} = 0. \tag{3.6}$$

Inserting the partial derivatives

$$\frac{\partial L}{\partial u_t} = \rho u_t, \qquad \frac{\partial L}{\partial u_x} = -\frac{\partial W}{\partial u_x}, \qquad \frac{\partial L}{\partial u} = 0,$$

$$\frac{\partial L}{\partial \varphi_t} = I\varphi_t, \qquad \frac{\partial L}{\partial \varphi_x} = -\frac{\partial W}{\partial \varphi_x}, \qquad \frac{\partial L}{\partial \varphi} = -\frac{\partial W}{\partial \varphi}, \tag{3.7}$$

into Eqs. (3.5) and (3.6) we obtain the equations of motion

$$\rho u_{tt} - \left(\frac{\partial W}{\partial u_x}\right)_x = 0, \qquad I\varphi_{tt} - \left(\frac{\partial W}{\partial \varphi_x}\right)_x + \frac{\partial W}{\partial \varphi} = 0. \tag{3.8}$$

Denoting

$$\sigma = \frac{\partial W}{\partial u_x}, \qquad \eta = \frac{\partial W}{\partial \varphi_x}, \qquad \tau = \frac{\partial W}{\partial \varphi}, \tag{3.9}$$

we recognize σ as the macrostress (Piola stress), η as the microstress, and τ as the interactive force.

The equations of motion (3.8) take now the form

$$\rho u_{tt} = \sigma_x, \tag{3.10}$$

$$I\varphi_{tt} = \eta_x - \tau. \tag{3.11}$$

These equations can be compared with the analogous equations deduced in a different way by Capriz (1989).

The simplest potential energy function describing the influence of a microstructure is a quadratic function

$$W = \frac{1}{2}\alpha u_x^2 + A\varphi u_x + \frac{1}{2}B\varphi^2 + \frac{1}{2}C\varphi_x^2 \tag{3.12}$$

with α, A, B, C denoting material constants. Inserting it into Eq. (3.9) and the result into Eqs. (3.10) and (3.11) the governing equations take the form

$$\rho u_{tt} = \alpha u_{xx} + A\varphi_x, \tag{3.13}$$

$$I\varphi_{tt} = C\varphi_{xx} - Au_x - B\varphi. \tag{3.14}$$

Equations (3.13) and (3.14) with proper initial and boundary conditions form the basis for further analysis.

2.2 Balance of Pseudomomentum

The analysis above is based on two balance laws of momentum, expressed by Eqs. (3.10) and (3.11). Following Maugin (1993) we show that the balance of pseudomomentum is a direct consequence of these equations.

We multiply Eq. (3.10) by u_x and Eq. (3.11) by φ_x, add the equations, and obtain

$$\rho u_{tt} u_x + I\varphi_{tt}\varphi_x = \sigma_x u_x + \eta_x \varphi_x - \tau \varphi_x. \tag{3.15}$$

Now the identical expressions

$$\rho u_t u_{xt} + I\varphi_t \varphi_{xt} = \frac{1}{2}\left(\rho u_t^2 + I\varphi_t^2\right)_x \tag{3.16}$$

are added on either side, leading to the equation

$$(\rho u_t u_x + I\varphi_t \varphi_x)_t = \frac{1}{2}\left(\rho u_t^2 + I\varphi_t^2\right)_x + \sigma_x u_x + \eta_x \varphi_x - \tau \varphi_x. \tag{3.17}$$

On the left-hand side, the pseudomomentum

$$\mathcal{P} = -\left(\rho u_t u_x + I\varphi_t \varphi_x\right) \tag{3.18}$$

is recognized, whereas the right-hand side of Eq. (3.17) can be expressed in terms of the Lagrangian density

$$\mathcal{L} = \frac{1}{2} \left(\rho u_t^2 + I \varphi_t^2 \right) - W(u_x, \varphi, \varphi_x). \tag{3.19}$$

Its derivative with respect to x is, on account of Eqs. (3.9),

$$\mathcal{L}_x = \frac{1}{2} \left(\rho u_t^2 + I \varphi_t^2 \right)_x - \sigma u_{xx} - \eta \varphi_{xx} - \tau \varphi_x. \tag{3.20}$$

Using these formulas Eq. (3.17) can be rewritten in the form

$$-\mathcal{P}_t = \mathcal{L}_x + (\sigma u_x + \eta \varphi_x)_x. \tag{3.21}$$

In the dynamic setting the Eshelby stress is defined by

$$b = -(\mathcal{L} + \sigma u_x + \eta \varphi_x). \tag{3.22}$$

Thus Eq. (3.21) can be represented in the form of balance of pseudomomentum

$$\mathcal{P}_t - b_x = 0. \tag{3.23}$$

In the special case of the quadratic function (3.12), the Eshelby stress assumes the form

$$b = -\frac{1}{2} \left(\rho u_t^2 + I \varphi_t^2 + \alpha u_x^2 - B \varphi^2 + C \varphi_x^2 \right). \tag{3.24}$$

The balance equation (3.23), however, holds independently of the constitutive equation for the strain energy density. The essential assumption is that there is no direct dependence of the strain energy density on the coordinate x, that is, that the material is homogeneous. Otherwise an inhomogeneous term would show up on the right-hand side of Eq. (3.23), thus turning it into an imbalance of pseudomomentum.

3. The Two-Scale Model

We follow the same idea as in Section 2 but generalize it for a two-scale situation. In physical terms it means that every deformable cell of the microstructure includes new deformable cells at a smaller scale. So instead of the system macrostructure–microstructure, the material is supposed to be composed by the macrostructure including microstructure 1 at a certain scale that includes microstructure 2 at some smaller scale. A qualitative sketch of such a material is shown in Figure 3.1. The displacements at the different scales are

$$u_j = u_j(x_i, t), \tag{3.25}$$
$$u_j' = x_k' \varphi_{kj}(x_i, t), \tag{3.26}$$
$$u_j'' = x_k'' \bar{\psi}_{kj}(x_i', t), \tag{3.27}$$

macrostructure   microstructure 1   microstructure2

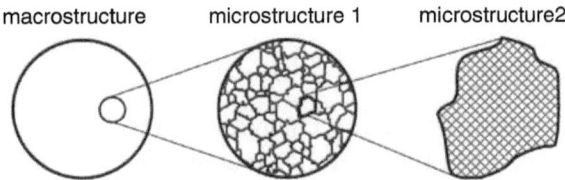

Fig. 3.1. Multiple scales of material structure.

respectively, where x'_k, x''_k correspond to the local coordinates within respective cells. As we are interested in motion on the macrolevel, we assume the relation (3.27) to be replaced by

$$u''_j = x''_k \, \psi_{kj}(x_i, t). \tag{3.28}$$

The gradients of these displacements,

$$\frac{\partial u'_j}{\partial x'_i} = \varphi_{ij}, \qquad \frac{\partial u''_j}{\partial x''_i} = \psi_{ij} \tag{3.29}$$

represent the deformations on the microlevels.

In the 1-D case the indices i, j, k can be dropped. In generalization of Eqs. (3.10) and (3.11) one has now the balance laws

$$\rho u_{tt} = \sigma_x, \tag{3.30}$$

$$I_1 \varphi_{tt} = \eta_{1x} - \tau_1, \tag{3.31}$$

$$I_2 \psi_{tt} = \eta_{2x} - \tau_2, \tag{3.32}$$

in which two microstresses η_1, η_2 and two interactive forces τ_1, τ_2 occur. The microinertias at the two scales are I_1, I_2. All the stress components and forces are determined from the free energy W by the relations

$$\sigma = \frac{\partial W}{\partial u_x}, \quad \eta_1 = \frac{\partial W}{\partial \varphi_x}, \quad \eta_2 = \frac{\partial W}{\partial \psi_x}, \quad \tau_1 = \frac{\partial W}{\partial \varphi}, \quad \tau_2 = \frac{\partial W}{\partial \psi}, \tag{3.33}$$

in generalization of Eq. (3.9).

In order to begin explaining the dispersive effects, we assume the quadratic free-energy function

$$W = \frac{1}{2} \left(\alpha u_x^2 + 2A_1 u_x \varphi + B_1 \varphi^2 + C_1 \varphi_x^2 + 2A_2 \varphi_x \psi + B_2 \psi^2 + C_2 \psi_x^2 \right), \tag{3.34}$$

where α and A_i, B_i, C_i $(i = 1, 2)$ denote material constants. Introducing (3.33) and (3.34) into (3.30)–(3.32), the governing equations assume the form

$$\rho u_{tt} = \alpha u_{xx} + A_1 \varphi_x, \tag{3.35}$$

$$I_1 \varphi_{tt} = C_1 \varphi_{xx} - A_1 u_x - B_1 \varphi + A_2 \psi_x, \tag{3.36}$$

$$I_2 \psi_{tt} = C_2 \psi_{xx} - A_2 \varphi_x - B_2 \psi, \tag{3.37}$$

generalizing (3.13) and (3.14).

4. Hierarchies of Waves

4.1 Preliminaries

Whitham (1974) has described certain complicated wave systems where a scale parameter δ plays an important role. Depending on its limit values, $\delta \to \infty$ or $\delta \to 0$, one or another wave operator governs the process asymptotically. Thus, the full system includes a hierarchy of waves with certain stability conditions; see Whitham (1974). Here we show that waves in microstructured materials exhibit the hierarchical behavior governed by a parameter that is the ratio of the characteristic scale of a microstructure and the wavelength of the excitation.

4.2 Single Scale

Let the scale of the microstructure be l and the excitation characterized by its amplitude U_0 and wavelength L. Then we can introduce the following dimensionless variables and parameters,

$$U = \frac{u}{U_0}, \quad X = \frac{x}{L}, \quad T = \frac{c_0 t}{L}, \quad \delta = \frac{l^2}{L^2}, \quad \epsilon = \frac{U_0}{L}, \tag{3.38}$$

where $c_0^2 = \alpha/\rho$. We also suppose that $I = \rho l^2 I^*$, $C = l^2 C^*$, where I^* is dimensionless and C^* has the dimension of stress. Note that I is scaled against ρ, so that any difference of densities is embedded in I^*.

Next, the system (3.13), (3.14) is rewritten in its dimensionless form and then the slaving principle (Christiansen et al. 1992; Porubov 2003) is used. It means, in principle, that we determine φ in terms of U_x using a series representation. Indeed, Eq. (3.14) yields

$$\varphi = -\frac{\epsilon A}{B} U_X - \frac{\delta}{B} \left(\alpha I^* \varphi_{TT} - C^* \varphi_{XX} \right). \tag{3.39}$$

If we consider $\varphi = \varphi_0 + \delta \varphi_1 + \cdots$, we get

$$\varphi_0 = -\frac{\epsilon A}{B} U_X, \tag{3.40}$$

$$\varphi_1 = \frac{\epsilon \alpha A I^*}{B^*} U_{XTT} - \frac{\epsilon A C^*}{B^2} U_{XXX}. \tag{3.41}$$

Inserting Eqs. (3.40) and (3.41) into the governing system in its dimensionless form we get finally, in terms of U, the partial differential equation (Engelbrecht and Pastrone 2003)

$$U_{TT} = \left(1 - \frac{c_A^2}{c_0^2} \right) U_{XX} + \frac{c_A^2}{c_B^2} \left(U_{TT} - \frac{c_1^2}{c_0^2} U_{XX} \right)_{XX}, \tag{3.42}$$

where $c_1^2 = C/I, c_A^2 = A^2/\rho B, c_B^2 = BL^2/I$. Note that c_B^2 involves the scales L and l, and c_A^2 includes the interaction effects between the macro- and microstructure (through the parameter A). This means that

$$\frac{c_A^2}{c_B^2} = \delta I^* \frac{A^2}{B^2}. \tag{3.43}$$

Equation (3.42) reflects the hierarchical nature of wave propagation in microstructured solids: if c_A^2/c_B^2 is small, then waves are governed by the properties of macrostructure. If, however, c_A^2/c_B^2 is large, then waves "feel" more microstructure. Note that in the absence of the interaction between macro- and microstructure (i.e., when $A = 0$), the wave operator in terms of U is simply $U_{TT} - U_{XX}$.

It is of interest to restore the dimensions in order to compare the various approximations. First, the system (3.13), (3.14) of two second-order equations can be represented also in the form of one fourth-order equation

$$u_{tt} = \left(c_0^2 - c_A^2\right) u_{xx} - p^2 \left(u_{tt} - c_0^2 u_{xx}\right)_{tt} + p^2 c_1^2 \left(u_{tt} - c_0^2 u_{xx}\right)_{xx}, \qquad (3.44)$$

where $p^2 = I/B$ is an inherent time constant. Equation (3.42), however, can be rewritten as

$$u_{tt} = \left(c_0^2 - c_A^2\right) u_{xx} - p^2 c_A^2 \left(u_{tt} - c_1^2 u_{xx}\right)_{xx}. \qquad (3.45)$$

It is obvious that the approximated model (3.45), which displays the hierarchical structure, neglects u_{tttt} completely and the influence of u_{ttxx} is different in (3.44) and (3.45). What is important is that in this approximation the effects of inertia of microstructure and wave velocity in pure microstructure are taken into account. There are certainly other approximations possible. From lattice theory (see, e.g., Maugin 1999), the governing equation in its simplest form is

$$u_{tt} = c_0^2 u_{xx} + \frac{1}{12} c_0^2 a^2 u_{xxxx}, \qquad (3.46)$$

where a is the distance between the particles. A similar equation for periodic structures is obtained by Santosa and Symes (1991). The equation must be compared with

$$u_{tt} = \left(c_0^2 - c_A^2\right) u_{xx} + p^2 c_A^2 c_1^2 u_{xxxx} \qquad (3.47)$$

resulting from Eq. (3.45) if the mixed derivative u_{ttxx} is discarded. If only the effect of microinertia is retained (Wang and Sun 2002) then in our notations the governing equation reads

$$u_{tt} = \left(c_0^2 - c_A^2\right) u_{xx} - p^2 c_A^2 u_{ttxx}. \qquad (3.48)$$

The dispersion analysis below (Section 5) shows the difference between the various models.

4.3 Multiple Scales

We apply now the same reasoning as above to the system (3.35)–(3.37), that is, the balance laws in terms of the macrodisplacement u and the microstrains φ and ψ of the two levels of microstructure. In order to do so, the following dimensionless variables are introduced,

$$U = \frac{U}{U_0}, \quad X = \frac{x}{L}, \quad T = \frac{c_0 t}{L}, \quad \epsilon = \frac{U_0}{L}, \quad \delta_1 = \frac{l_1^2}{L^2}, \quad \delta_2 = \frac{l_2^2}{L^2}, \qquad (3.49)$$

where U_0 and L are the amplitude and the characteristic length of the excitation, and l_1 and l_2 are the scales of the microstructures. The lengths l_1, l_2 control the order of magnitude of the corresponding material parameters which, therefore, are chosen as

$$I_1 = \rho l_1^2 I_1^*, \quad I_2 = \rho l_2^2 I_2^*, \quad C_1 = l_1^2 C_1^*, \quad C_2 = l_2^2 C_2^*, \quad A_2 = l_2 A_2^*. \quad (3.50)$$

Here I_1^*, I_2^* are dimensionless, and A_2^*, C_1^*, C_2^* have the dimensions of stress. Note that I_1 and I_2 are scaled against ρ and possible differences on densities of microstructures are embedded in I_1^* and I_2^*, respectively.

Substituting the parameters (3.49) and (3.50) into the governing equations (3.35)–(3.37) we obtain

$$U_{TT} = U_{XX} + \frac{A_1}{\epsilon \alpha} \varphi_X, \quad (3.51)$$

$$\varphi_{TT} = \frac{C_1^*}{\alpha I_1^*} \varphi_{XX} - \frac{1}{\delta_1} \frac{\epsilon A_1}{\alpha I_1^*} U_X - \frac{1}{\delta_1} \frac{B_1}{\alpha I_1^*} \varphi + \frac{\sqrt{\delta_2}}{\delta_1} \frac{A_2^*}{\alpha I_1^*} \psi_X, \quad (3.52)$$

$$\psi_{TT} = \frac{C_2^*}{\alpha I_2^*} \psi_{XX} - \frac{1}{\sqrt{\delta_2}} \frac{A_2^*}{\alpha I_2^*} \varphi_X - \frac{1}{\delta_2} \frac{B_2}{\alpha I_2^*} \psi. \quad (3.53)$$

The slaving principle (Christiansen et al. 1992; Porubov 2003) can now be used, taking into account two independent small parameters δ_1 and δ_2. To this end, first the variable ψ is determined from Eq. (3.53) and expressed in terms of ϕ and its derivatives. Then Eq. (3.53) is used to express ϕ in terms of derivatives of U. This expression is finally inserted into Eq. (3.51) resulting in a single differential equation for the displacement U.

In the first step, from Eq. (3.53) the expansion

$$\psi = -\sqrt{\delta_2} \frac{A_2^*}{B_2} \left[\varphi_x + \delta_2 \frac{\alpha I_2^*}{B_2} \left(\varphi_{TT} - \frac{C_2^*}{\alpha I_2^*} \varphi_{XX} \right)_X + \cdots \right] \quad (3.54)$$

is obtained. It is inserted into Eq. (3.52), which also is expanded and yields

$$\varphi = - \frac{\epsilon A_1}{B_1} U_X + \delta_1 \frac{\epsilon \alpha A_1 I_1^*}{B_1^2} \left(U_{TT} - \frac{C_1^*}{\alpha I_1^*} U_{XX} \right)_X$$
$$+ \delta_2 \frac{\epsilon A_1 (A_2^*)^2}{B_1^2 B_2} \left[U_{XXX} - \delta_2 \frac{\alpha I_2^*}{B_2} \left(U_{TT} - \frac{C_2^*}{\alpha I_2^*} U_{XX} \right)_{XXX} + \cdots \right]. \quad (3.55)$$

Finally this expression is inserted into Eq. (3.51) resulting in the partial differential equation

$$U_{TT} = \left(1 - \frac{A_1^2}{\alpha B_1} \right) U_{XX} + \delta_1 \frac{A_1^2 I_1^*}{B_1^2} \left(U_{TT} - \frac{C_1^*}{\alpha I_1^*} U_{XX} \right)_{XX}$$
$$+ \delta_2 \frac{A_1^2 (A_2^*)^2 I_2^*}{B_1^2 B_2^2} \left[\frac{B_2}{\alpha I_2^*} U_{XXXX} - \delta_2 \left(U_{TT} - \frac{C_2^*}{\alpha I_2^*} U_{XX} \right)_{XXXX} \right]. \quad (3.56)$$

Equation (3.56) is the demanded hierarchical equation in terms of the macrodisplacement U where the microstructures are reflected in special wave operators. In order to compare the result with the basic system (3.35)–(3.37), and also with the results of Section 4.2, we return to the original dimensional parameters. By resubstituting the variables and parameters from Eqs. (3.49) and (3.50) the partial differential equation is obtained in the form

$$u_{tt} = \left(c_0^2 - c_{A1}^2\right)u_{xx} + p_1^2 c_{A1}^2 \left[u_{tt} - \left(c_1^2 - c_{A2}^2\right)u_{xx}\right]_{xx}$$
$$- p_1^2 c_{A1}^2 p_2^2 c_{A2}^2 \left(u_{tt} - c_2^2 u_{xx}\right)_{xxxx}, \tag{3.57}$$

where the parameters

$$c_1^2 = \frac{C_1}{I_1}, c_{A1}^2 = \frac{A_1^2}{\rho B_1}, p_1^2 = \frac{I_1}{B_1}, c_2^2 = \frac{C_2}{I_2}, c_{A2}^2 = \frac{A_2^2}{I_1 B_2}, p_2^2 = \frac{I_2}{B_2} \tag{3.58}$$

have been introduced. The parameters c_i and c_{Ai} are velocities whereas the p_i denote time constants.

5. Dispersion

5.1 General

Internal scales of microstructured solids lead to dispersive effects. This is also quite clear from the governing equation derived in Section 4. The presence of higher-order derivatives in the governing equations indicates dispersion. Below we demonstrate how the various combinations of material parameters and wave characteristics are reflected in dispersion relations. We start from the models with dimensions and then introduce dimensionless wave number and frequency. The solution is assumed in the form of a wave

$$u(x,t) = \hat{u}\exp[i(kx - \omega t)], \tag{3.59}$$

with wave number k, frequency ω, and amplitude $\hat{u}$.

5.2 Single Scale

The corresponding mathematical models are presented in Section 4.2. Introducing now (3.59) into Eq. (3.44), the dispersion relation

$$\omega^2 = \left(c_0^2 - c_A^2\right)k^2 + p^2\left(\omega^2 - c_0^2 k^2\right)\left(\omega^2 - c_1^2 k^2\right) = 0 \tag{3.60}$$

is obtained. The parameters involved are a time constant p and three characteristic velocities c_0, c_1, and c_A. Instead of c_A the velocity $c_R = \sqrt{c_0^2 - c_A^2}$ could be introduced as a parameter because it has an obvious meaning for the given wave process. Waves of very low frequencies ($\omega \ll p^{-1}$) are propagated at the velocity c_R. The velocity c_A does not occur explicitly as a limit velocity.

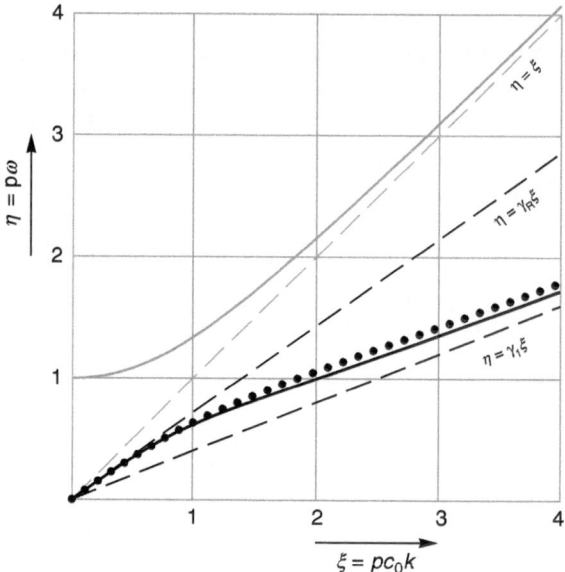

Fig. 3.2. Dispersion curves for $\gamma_1 = 0.4$ and $\gamma_A = 0.7$: ——— exact dispersion relation (3.62); •••••• approximate dispersion relation (3.63).

In order to reduce the number of independent variables we normalize the wave number, the frequency, and the propagation speeds by defining

$$\xi = pc_0 k, \qquad \eta = p\omega, \qquad \gamma_1 = \frac{c_1}{c_0}, \qquad \gamma_A = \frac{c_A}{c_0}. \tag{3.61}$$

Using these new quantities the full dispersion relation (3.60) assumes the form

$$\eta^2 = \left(1 - \gamma_A^2\right)\xi^2 + \left(\eta^2 - \xi^2\right)\left(\eta^2 - \gamma_1^2\xi^2\right). \tag{3.62}$$

In the same way, the approximate differential equation (3.45) yields the dimensionless dispersion relation

$$\eta^2 = \left(1 - \gamma_A^2\right)\xi^2 - \gamma_A^2\left(\eta^2 - \gamma_1^2\xi^2\right)\xi^2. \tag{3.63}$$

Eventually the simplified differential equations (3.47) and (3.48) yield

$$\eta^2 = \left(1 - \gamma_A^2\right)\xi^2 + \gamma_A^2\gamma_1^2\xi^4, \tag{3.64}$$

$$\eta^2 = \left(1 - \gamma_A^2\right)\xi^2 - \gamma_A^2\eta^2\xi^2, \tag{3.65}$$

respectively.

The full dispersion relation (3.62) represents two branches which, in general, are distinct (see Figures 3.2 and 3.3). The upper, or "optical" branch starts in the (ξ, η)-plane at $\eta = 1$ with zero slope, and the lower, or "acoustical" branch starts at the origin with slope $\gamma_R = c_R/c_0$. In the short wave limit $\xi \gg 1$ the branches asymptotically approach the lines $\eta = \xi$ and $\eta = \gamma_1\xi$. In the exceptional case $\gamma_A = 0$, $\gamma_1 < 1$ the

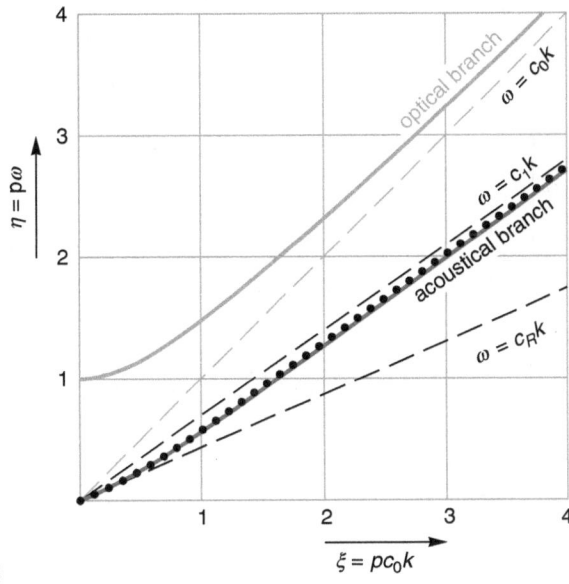

3

Fig. 3.3. Dispersion curves for $\gamma_1 = 0.7$ and $\gamma_A = 0.9$: ———— exact dispersion relation (3.62); • • • • • • approximate dispersion relation (3.63).

branches meet in one point. Because the free energy (3.12) should be positive definite, we always have $\gamma_A < 1$. There is, however, no physical restriction on the magnitude of γ_1. Figures 3.2 and 3.3 show examples where $\gamma_1 < \gamma_R < 1$ and $\gamma_R < \gamma_1 < 1$, respectively.

The important question is how the hierarchical model describes the situation. The corresponding dispersion relation (3.63) provides an approximation for the acoustical branch only. The curve starts at $\xi = 0$ with the slope γ_R and, for $\xi \to \infty$, tends asymptotically to the line $\eta = \gamma_1 \xi$ provided $\gamma_1 < 1$ and $\gamma_A > 0$ (see Figure 3.3). The special feature of this approximation is that it can be used over the whole range of wave numbers because it does not represent a short-wave or long-wave approximation. The underlying assumption is that the influence of the microstructure is small. In the case of Figure 3.3, the full and approximate dispersion relations agree pretty well. The approximation gets worse if the parameter γ_A tends to zero and, for $\gamma_A = 0$, degenerates to the nondispersive wave represented by $\eta = \xi$.

The simplified cases (3.64) and (3.65) give rather distorted results. The dispersion curves deviate strongly from the correct course (see Figure 3.4).

5.3 Multiple Scales

The underlying mathematical model and its hierarchical approximation are presented in Sections 3 and 4.3, respectively. Introducing now (3.59) into the full set of Eqs. (3.35)–(3.37), we obtain the dispersion relation

$$\left(c_0^2 k^2 - \omega^2\right)\left(c_1^2 k^2 - \omega^2 + \omega_1^2\right)\left(c_2^2 k^2 - \omega^2 + \omega_2^2\right)$$
$$- c_{A2}^2 \omega_2^2 k^2 \left(c_0^2 k^2 - \omega^2\right) - c_{A1}^2 \omega_1^2 k^2 \left(c_2^2 k^2 - \omega^2 + \omega_2^2\right) = 0, \qquad (3.66)$$

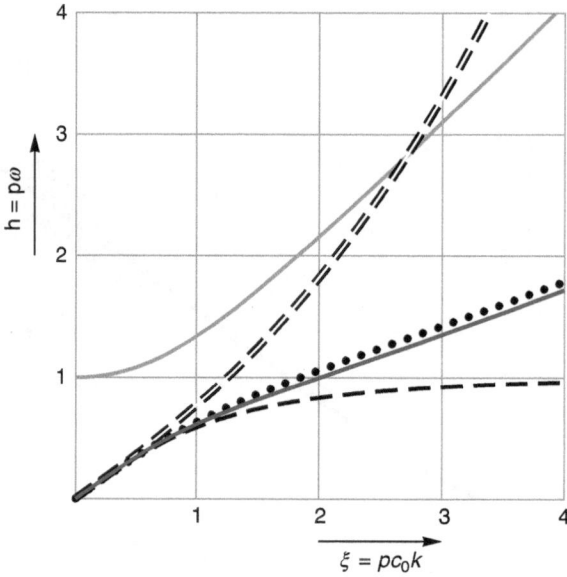

Fig. 3.4. Dispersion curves for $\gamma_1 = 0.4$ and $\gamma_A = 0.7$: ———— exact dispersion relation (3.62); •••••• approximate dispersion relation (3.63); — — — simplified dispersion relation (3.64); $===$ simplified dispersion relation (3.65).

where, for convenience, we have introduced the reciprocal time constants $\omega_1 = 1/p_1$ and $\omega_2 = 1/p_2$. In addition, the hierarchical governing equation (3.57) yields the dispersion relation

$$\left(c_0^2 - c_{A1}^2\right) k^2 - \omega^2 + \frac{c_{A1}^2}{\omega_1^2} k^2 \left[\left(c_1^2 - c_{A2}^2\right) k^2 - \omega^2\right]$$

$$+ \frac{c_{A1}^2 c_{A2}^2}{\omega_1^2 \omega_2^2} k^4 \left(c_2^2 k^2 - \omega^2\right) = 0. \tag{3.67}$$

In the further analysis, the dimensionless quantities

$$\xi = p_1 c_0 k, \qquad \eta = p_1 \omega \tag{3.68}$$

are used. Introducing them into the exact dispersion relation (3.66) yields the dimensionless form

$$\left(\xi^2 - \eta^2\right) \left(\gamma_1^2 \xi^2 - \eta^2 - \eta_1^2\right) \left(\gamma_2^2 \xi^2 - \eta^2 + \eta_2^2\right)$$

$$- \gamma_{A2}^2 \left(\eta_2^2/\eta_1^2\right) \xi^2 \left(\xi^2 - \eta^2\right) - \gamma_{A1}^2 \xi^2 \left(\gamma_2^2 \xi^2 - \eta^2 + \eta_2^2\right) = 0, \tag{3.69}$$

and the hierarchical approximation (3.67) is converted to

$$\left(1 - \gamma_A^2\right) \xi^2 - \eta^2 + \gamma_{A1}^2 \xi^2 \left[\left(\gamma_1^2 - \gamma_{A2}^2\right) \xi^2 - \eta^2\right]$$

$$+ \gamma_{A1}^2 \gamma_{A2}^2 \frac{\eta_1^2}{\eta_2^2} \xi^4 \left(\gamma_2^2 \xi^2 - \eta^2\right) = 0. \tag{3.70}$$

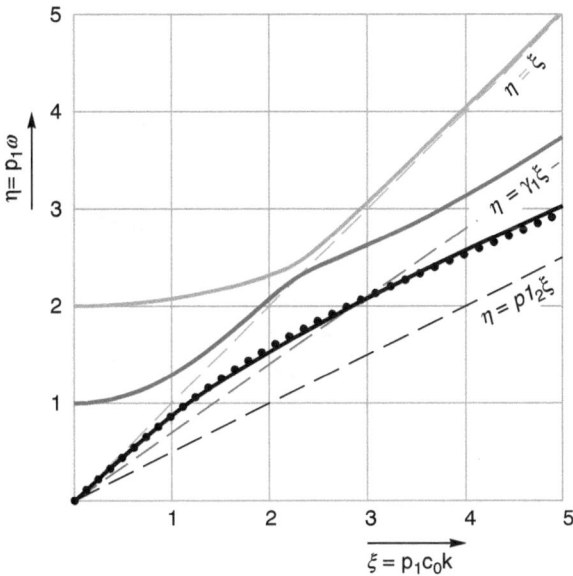

Fig. 3.5. Dispersion curves for $\gamma_1 = 0.7$, $\gamma_2 = 0.5$, $\gamma_{A1} = \gamma_{A2} = 0.4$, $\eta_2 = 2$; ———— exact dispersion relation (3.69); • • • • • • approximation (3.70).

The parameters in (3.69) and (3.70) denote velocity ratios and dimensionless fixed frequencies; viz.,

$$\gamma_1 = \frac{c_1}{c_0}, \quad \gamma_2 = \frac{c_2}{c_0}, \quad \gamma_{A1} = \frac{c_{A1}}{c_0}, \quad \gamma_{A2} = \frac{c_{A2}}{c_0}, \quad \eta_1 = 1, \quad \eta_2 = \frac{\omega_2}{\omega_1}. \quad (3.71)$$

These parameters are chosen in accordance with the corresponding dimensionless quantities of the single-scale case.

The full dispersion relation (3.69) now represents three branches, which are distinct if the "coupling coefficients" γ_{A1} and γ_{A2} are nonzero. The lowest, or "acoustical" branch starts at the origin whereas the two other branches in the long-wave limit represent standing waves of dimensionless frequencies $\eta_1 = 1$ and η_2. The approximate dispersion relation (3.70) obtained by the slaving principle approximates the acoustic branch. Figures 3.5 and 3.6 show two examples for different sets of the parameters. In both cases the approximation is acceptable. This depends, however, severely on the chosen parameters.

6. Nonlinearities

As said in the introduction, one should often account for large deformation or complicated stress–strain relations that lead to nonlinear mathematical models. This means that the full deformation tensor involves nonlinear terms and the free-energy function W depends on higher-order terms. In this case dispersion effects as described in Section 5 are combined with nonlinear effects. Here we present a brief description of the nonlinear theory based on our earlier results (Engelbrecht and Pastrone 2003; Berezovski et al. 2003; Janno and Engelbrecht 2004).

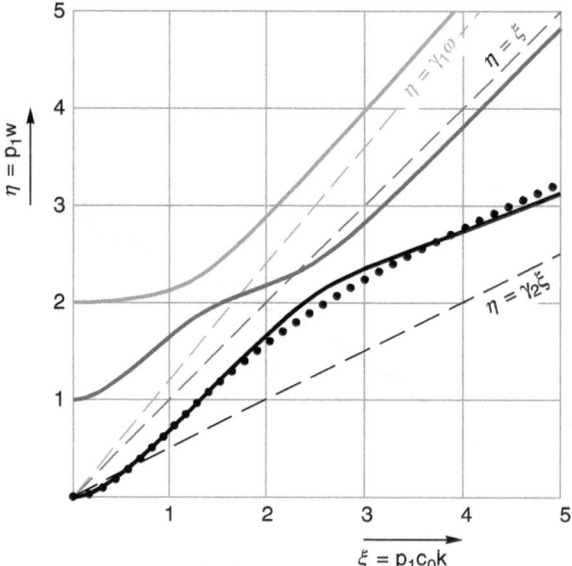

Fig. 3.6. Dispersion curves for $\gamma_1 = 1.2$, $\gamma_2 = 0.5$, $\gamma_{A1} = 1$, $\gamma_{A2} = 0.7$, $\eta_2 = 2$; ———— exact dispersion relation (3.69); • • • • • • approximation (3.70).

The analysis is restricted to the single-scale model (3.10), (3.11). Based on estimations (see Engelbrecht 1983) that physical nonlinearity is stronger than geometrical, we limit ourselves to extending the free energy function. So instead of (3.12) we assume

$$W = \frac{1}{2}\alpha u_x^2 + A\varphi u_x + \frac{1}{2}B\varphi^2 + \frac{1}{2}C\psi^2 + \frac{1}{6}Nu_x^3 + \frac{1}{6}M\varphi_x^3, \qquad (3.72)$$

including nonlinearities on both the macro- and the microlevel. Using the relations (3.9) for determining the macrostress, the microstress, and the interactive force, we obtain the equations

$$\rho u_{tt} = \alpha u_{xx} + N u_x u_{xx} + A\varphi_x \qquad (3.73)$$

$$I\varphi_{tt} = C\varphi_{xx} + M\varphi_x\varphi_{xx} - Au_x - B\varphi \qquad (3.74)$$

which now replace (3.13) and (3.14).

We introduce the same dimensionless variables and scaling as in Section 4 and, in addition, $M = M^*l^3$. Following the same scheme as before, we obtain the hierarchical equation

$$U_{TT} = \left(1 - \frac{c_A^2}{c_0^2}\right)U_{XX} + \frac{1}{2}k_1\left(U_X^2\right)_X$$

$$+ \frac{c_A^2}{c_B^2}\left(U_{TT} - \frac{c_1^2}{c_0^2}U_{XX}\right)_{XX} + \frac{1}{2}k_2\left(U_{XX}^2\right)_{XX} \qquad (3.75)$$

which generalizes (3.42). Here $k_1 = N\epsilon/\alpha$, $k_2 = \delta^{3/2}(A^3 M^*\epsilon)/(\alpha B^3)$ are the parameters expressing the strengths of physical nonlinearities on the macro- and microscale, respectively. It has been shown (Janno and Engelbrecht 2004) that this model may exhibit the balance between nonlinear and dispersive effects, and therefore solitary waves may exist. A similar situation arises for nonlinear waves in rods (Samsonov 2001; Porubov 2003) when the governing equation is of the type (3.75) with $k_1 \neq 0, k_2 = 0$.

7. Discussion

Nonclassical theory of continua takes the internal scales into account and is therefore able to describe microstructural effects. In the limit case we could intuitively understand that the microstructure is composed of (different) particles and so we are actually dealing with crystal lattices (Brillouin 1953; Maugin 1999). In crystal lattices the simplest case with identical particles leads to a dispersion relation (Brillouin 1953)

$$\omega(k) = \frac{2c_0}{a} \left| \sin \frac{a}{2} k \right| \tag{3.76}$$

in our notation where, as in Eq. (3.46), the length a denotes the distance between the particles. Comparing (3.76) with dispersion relations of Section 5, it is obvious that our model grasps the essential convexity of the dispersion curve.

The model we have used for describing the microstructure is rather general: it is based on Euler–Lagrange equations and it could be represented also in terms of the balance of pseudomomentum (Maugin 1993).

The main value of the model is the explicit description of the hierarchy in Whitham's sense: the model is composed of two (or more) wave operators and, depending on the characteristic length of the initial excitation, a certain wave operator predominates. The dispersion curves (Section 5) demonstrate the transformation from one operator with its wave speed to another. The presence of the microstructure reduces the dimensionless wave speed to $1 - c_A^2/c_0^2$, as can be seen from Eq. (3.42). Contrary to simplified models, the double dispersion with different terms U_{TTXX} and U_{XXXX} is of importance as in the case of waves in rods (Samsonov 2001; Porubov 2003). In physical terms, the influence of the microstructure on the wave motion is twofold: both the inertia and the elastic properties of the microstructure affect the dispersion of waves, in general. In the lattice theories when turned to continuum models the inertia effects are missing.

The multiscale model (3.56) actually prolongs the hierarchical properties of the single-scale model (3.42). Indeed, microstructure 1 is affected in the same way by microstructure 2 as it affects the macrostructure itself.

Indeed, the wave operators macro versus micro 1 and micro 1 versus micro 2 are related by similar sign convention, and the wave velocity in microstructure 1 is affected by properties of microstructure 2 in a similar way as the wave velocity in macro is affected by properties of microstructure 1. It is seen that higher-order dispersive terms $U_{XXXX}, U_{XXXXXX}, \ldots$ coincide with those derived from the lattice theory (Maugin 1999), but again mixed derivatives $U_{TTXX}, U_{TTXXXX}, \ldots$ reflecting the role of microinertias also enter the equations.

The proper modeling is certainly important for solving direct problems, given the initial excitation and calculating the wave field. No less important are the inverse problems when from given excitation and measured wave field the material properties are to be determined. The methods of NonDestructive Testing (NDT) of materials are all based on solving the inverse problems. Based on the essentially more accurate mathematical model described in this chapter, it should be possible to determine the properties of microstructured solids. The preliminary results in this direction are obtained by Janno and Engelbrecht (2005) with regard to Eq. (3.42). It is possible to determine three material parameters from three phase velocities measured at various wavelengths.

The results above are also supported by numerical calculations (Berezovski et al. 2003; Engelbrecht et al. 2004) but there are many studies in progress.

Acknowledgment

The authors appreciate the support obtained from the ESF-PESC programme NATEMIS for their studies. One of the authors (J. E.) acknowledges also the support by the A. v. Humboldt Foundation and the Estonian Science Foundation.

References

Askar, A., (1985), *Lattice Dynamical Foundations of Continuum Theories*. (Singapore: World Scientific).

Berezovski, A., Engelbrecht, J., and Maugin, G. A., (2003), Numerical simulation of two-dimensional wave propagation in functionally graded materials. *Eur. J. Mech. Solids*, **22**, 2, 257–265.

Brillouin, L., (1953), *Wave Propagation in Periodic Structures*. (Toronto and London: Dover).

Capriz, G., (1989), *Continua with Microstructure*. (New York: Springer).

Christiansen, P. L., Muto, V., and Rionero, S., (1992), Solitary wave solution to a system of Boussinesq-like equations. *Chaos Solitons Fractals*, **2**, 45–50.

Engelbrecht, J., (1983), *Nonlinear Wave Processes of Deformation in Solids*, (Boston: Pitman).

Engelbrecht, J., (1997), *Nonlinear Wave Dynamics. Complexity and Simplicity*, (Dordrecht: Kluwer).

Engelbrecht, J. and Pastrone, F., (2003), Waves in microstructured solids with strong nonlinearities in microscale. *Proc. Estonian Acad. Sci. Phys. Math.*, **52**, 12–20.

Engelbrecht, J., Berezovski, A., Pastrone, F., and Braun, M., (2004), Waves in microstructured materials and dispersion. *Phil. Mag.*, **85**, Nos 33–35, 4127–4141.

Engelbrecht, J., Cermelli, P., and Pastrone, F., (1999), Wave hierarchy in microstructured solids. *Geometry, Continua and Microstructure*, edited by G. A. Maugin (Paris: Hermann Publ.), pp. 99–111.

Eringen, A. C., (1966), Linear theory of micropolar elasticity. *J. Math. Mech.*, **15**, 909–923.

Eringen, A. C., (1999), *Microcontinuum Field Theories. I Foundations and Solids*. (New York: Springer).

Erofeyev, V. I., (2003), *Wave Processes in Solids with Microstructure*. (Singapore: World Scientific).

Janno, J. and Engelbrecht, J., (2005), Solitary waves in nonlinear microstructured materials, *J. Phys. A: Math. Gen.* **38**, 5159–5172.

Maugin, G. A., (1990), Internal variables and dissipative structures. *J. Nonequilib. Thermodyn.*, **15**, 173–192.

Maugin, G. A., (1993), *Material Inhomogeneities in Elasticity*. (London: Chapman & Hall).

Maugin, G. A., (1999), *Nonlinear Waves in Elastic Crystals*. (Oxford: Oxford University Press).

Maugin, G. A. and Muschik W., (1994), Thermodynamics with internal variables. *J. Nonequilib. Thermodyn.* **19**, 217–249 (Part I), 250–289 (Part II).

Mindlin, R. D., (1964), Micro-structure in linear elasticity. *Arch. Rat. Mech. Anal.*, **16**, 51–78.

Pastrone, F., (2003), Waves in solids with vectorial microstructure. *Proc. Estonian Acad. Sci.*, **52**, 1, 21–29.

Porubov, A. V., (2003), *Amplification of Nonlinear Strain Waves in Solids*. (Singapore: World Scientific).

Samsonov, A., (2001), *Strain Solutons in Solids and How to Construct Them.* (London: Chapman & Hall/CRC).

Santosa, F. and Symes W. W., (1991), A dispersive effective medium for wave propagation in periodic composites. *SIAM J. Appl. Math.*, **51**, 984–1005.

Wang, Z.-P. and Sun, C. T., (2002), Modeling micro-inertia in heterogeneous materials under dynamic loading. *Wave Motion*, **36**, 473–485.

Whitham, G. B., (1974), *Linear and Nonlinear Waves.* (New York: Wiley).

4

Nonequilibrium Nonlinear-Dynamics in Solids: State of the Art

State of the Art in Nonequilibrium Dynamics

P. A. Johnson

Geophysics Group, MS D443, Los Alamos National Laboratory, Los Alamos, New Mexico 87545 USA, 505-667-8936 office, 505-667-8487 fax, paj@lanl.gov

Abstract

Dynamic nonlinear elastic behavior, *nonequilibrium dynamics*, first observed as a curiosity in earth materials has now been observed in a great variety of solids. The primary manifestations of the behavior are characteristic wave distortion, and *slow dynamics*, a recovery process to equilibrium that takes place linearly with the logarithm of time, over hours to days after a wave disturbance. The link between the diverse materials that exhibit nonequilibrium dynamics appears to be the presence of soft regions, thought to be primarily "damage" at many scales, ranging from order 10^{-9} m to 10^{-1} m at least. The regions of soft matter may be distributed as in a rock sample, or isolated, as in a sample with a single crack. The precise physical origin of the behavior is clear in some cases such as granular media where the source of the nonequilibrium dynamics, grain-to-grain interaction, is understood. In other materials, it appears that the origin must be due fundamentally to shear sliding, related to crack and possibly dislocation dynamics, as well as less clear origins. Because the physical origins of the behavior are related to damage, damage diagnostics in solids, Nonlinear NonDestructive Evaluation, follows naturally. Nonequilibrium dynamics also plays a significant role in other areas such as earthquake strong ground motion and potentially to earthquake dynamics.

Keywords: Nonlinear Elastic Wave Spectroscopy (NEWS), NRUS, NWMS, SDD, Time Reversal, Earthquake triggering, nonlinear NDE, nonequilibrium dynamics, strong ground motion

1. Introduction

Over the last two decades, studies of nonlinear dynamics in materials, known as *non-classical* or *anomalous* that include rock, damaged materials some ceramics, sintered metals, granular media etc., have increased markedly (Ostrovsky and Johnson, 2001, Guyer and Johnson, 1999). These materials exhibit what we term *nonequilibrium dynamics* at elevated strain amplitudes ($> \sim 10^{-6}$). Specifically, when the material is disturbed by a wave, the modulus decreases. We call this *nonlinear fast dynamics*. Following this, it takes tens of minutes to hours to return to its equilibrium state. This is called *slow dynamics* (Johnson et al., 1996; TenCate and Shankland, 1996). Further, the apparent mixture of fast and slow dynamics known as *conditioning* that takes place during nonlinear fast dynamics provides additional complexity not observed in materials whose nonlinearity is due to anharmonicity. The nonequilibrium dynamics is due to mechanically "soft" inclusions (soft matter) in a "hard" matrix (e.g., Ostrovsky and Johnson, 2001). For instance, a crack in a solid will induce nonequilibrium dynamics, but a void will not; a sandstone exhibits nonequilibrium dynamics due to distributed soft inclusions, also known as the bond system, but a bar of aluminium does not. Experimental methods and theory have been developed to interrogate nonequilibrium dynamics in solids.

In this paper we briefly address underlying theory, then provide an overview of the primary methods to interrogate nonequilibrium dynamical behavior, termed Nonlinear Elastic Wave Spectroscopy (NEWS) (Johnson, 1999). We outline Nonlinear Resonant Ultrasound Spectroscopy (NRUS), Slow Dynamics Diagnostics (SDD), Nonlinear Wave Modulation Spectroscopy (NWMS) and Time Reversal Nonlinear Elastic Wave Spectroscopy (TR NEWS). Following the description of these methods, we will briefly describe nonlinear imaging methods currently in development and then provide an overview of new areas of research where nonequilibrium dynamics may be important. Next we address unsolved problems related to the origin of elastic nonlinear behavior, and briefly look into the future.

2. Theory

Fundamentally, elastic nonlinearity implies that the stress-strain relation (also known as the equation of state, EOS) is nonlinear. For such a relation, the one-dimensional stress (σ)-strain (ε) can be described by

$$\sigma = K_o \varepsilon (1 + \beta \varepsilon + \partial \varepsilon^2 + ...), \qquad (4.1)$$

where K_o is the linear modulus, and β and δ are the first and second order classical nonlinear parameters, normally of order 1–10 in value. At low dynamic wave amplitudes (strains of less than order 10^{-6} under ambient pressure), there is evidence that all (or at least most) solids behave in a manner according to the above equation (TenCate et al, 2004; Pasqualini, this volume). At ambient pressure and temperature conditions, for wave amplitudes above approximately 10^{-6} strain, the material EOS is thought to be hysteretic. A hysteretic EOS relation is,

$$\sigma = K_o \varepsilon \left(1 + \beta \varepsilon + \alpha \varepsilon, f \left(\frac{\partial \varepsilon}{\partial t} \right) \right), \qquad (4.2)$$

where α is the hysteretic nonlinear parameter and is dependent on the strain derivative $\frac{\partial \varepsilon}{\partial t}$ due to the hysteresis (e.g., Guyer et al., 1997). Eq. (4.2) is a practical estimate of the dynamics, especially for NDE applications, but does not capture the entirety of nonequilibrium dynamics: the slow dynamics and material conditioning (TenCate and Shankland, 1996; Johnson and Sutin, 2005) as outlined in Fig. 4.1. As previously noted, slow dynamics means the material takes time to return to its rest state modulus K_o relaxing as the logarithm of time. An example of conditioning is as follows: if a

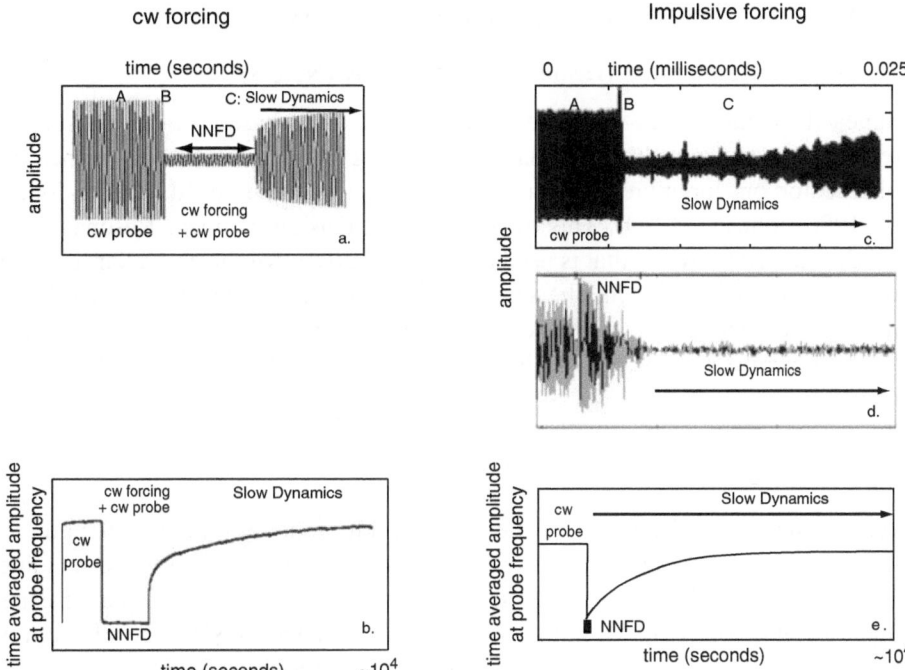

Fig. 4.1. *Nonequilibrium dynamics for two types of forcing.* The figure illustrates the full nonequilibrium dynamics that includes *nonlinear fast dynamics* (also known as *nonclassical* or *anomalous* nonlinear dynamics), *conditioning*, and *slow dynamics*. Figures (a) and (b) show how nonequilibrium dynamics are manifest when a low-amplitude, continuous-wave (*cw*) probe-wave is input into a sample in the presence of a large amplitude vibration. One sees in (a) the undisturbed probe wave (time A) and the corresponding time-average amplitude of the signal in (b) ["*cw* probe"]. At time B, a high-amplitude vibration begins and the probe-wave amplitude changes due to material nonlinearity (see Slow Dynamics Diagnostics section of this paper and Figure 4.5 for more). From the time the vibration is turned on until it is turned off, nonlinear fast dynamics, including conditioning, take place ("Nonclassical Nonlinear Fast Dynamics NNFD" in (a) and (b)). As soon as the large amplitude wave is terminated, one sees in (a) and (b) an instantaneous, partial recovery of the amplitude, and then a longer term recovery that is linear with the logarithm of time where slow dynamics is the sole process acting in the system. Figures (c-e) show the situation where the sample is disturbed by an impact, such as a tap, in the presence of the probe (time B in (c)). Figure (d) shows a zoom of (c) where one can observe the onset of the tap-induced vibration and its ring down ("NNFD" in (d-e)). After the vibration energy has dissipated, slow dynamics is the sole process operating in the system, the onset of which is shown in (c-d), and the long term behaviour is seen in (e).

rock sample is driven at fixed amplitude for a period of time, the modulus will decrease immediately with the onset of the wave, but then continue to decrease slightly to a new equilibrium value as long as the drive is maintained (TenCate and Shankland, 1996). Conditioning is a small effect in most materials as can be seen in Fig. 4.1b. It may or may not be correct to think of conditioning as a mix of fast and slow dynamics. In any case, Eq. (4.2) has been applied broadly to describe the material elastic nonlinearity. The rate-dependent effect of conditioning appears to have only a minor influence on estimates of α (e.g., Johnson and Sutin, 2005).

3. Nonlinear Elastic Wave Spectroscopy (NEWS)

3.1 Nonlinear Resonant Ultrasound Spectroscopy (NRUS)

Nonlinear Resonant Ultrasound Spectroscopy (NRUS) is based on the measurement of resonance frequency shift and material damping as a function of resonance peak amplitude for one or more resonance modes (e.g., Winkler et al., 1979; Johnson et al., 1996; Johnson, 1999). This method is an extension of linear Resonant Ultrasound Spectroscopy (RUS) that is used in industrial NDE (Migliori and Sarrao, 1999). In this type of measurement, the change in resonance frequency of a mode with drive amplitude is a measure of the wavespeed and modulus change. For instance, in a simple geometry such as a cylindrical bar driven at the fundamental mode, the wavespeed c is,

$$c = f\lambda = 2fL = \sqrt{\frac{K}{\rho}} \tag{4.3}$$

where f is resonance frequency of the fundamental mode, λ is the wavelength, L is the bar length, K is modulus and ρ is density. The equation becomes correspondingly more elaborate for more complicated sample geometries. A typical resonance experimental configuration is shown in Fig. 4.2. Figure 4.3 shows an NRUS result from two concrete samples, one virtually "intact" measured in the undamaged state, and one damaged. The hysteretic nonlinear parameter α can be extracted from the change in frequency with strain amplitude,

$$\frac{\Delta f}{f_0} = \alpha\varepsilon \tag{4.4}$$

where f_0 is the equilibrium frequency, Δf is the change in resonance frequency, α is the nonlinear parameter in Eq. (4.2), and ε is strain. α ranges from approximately

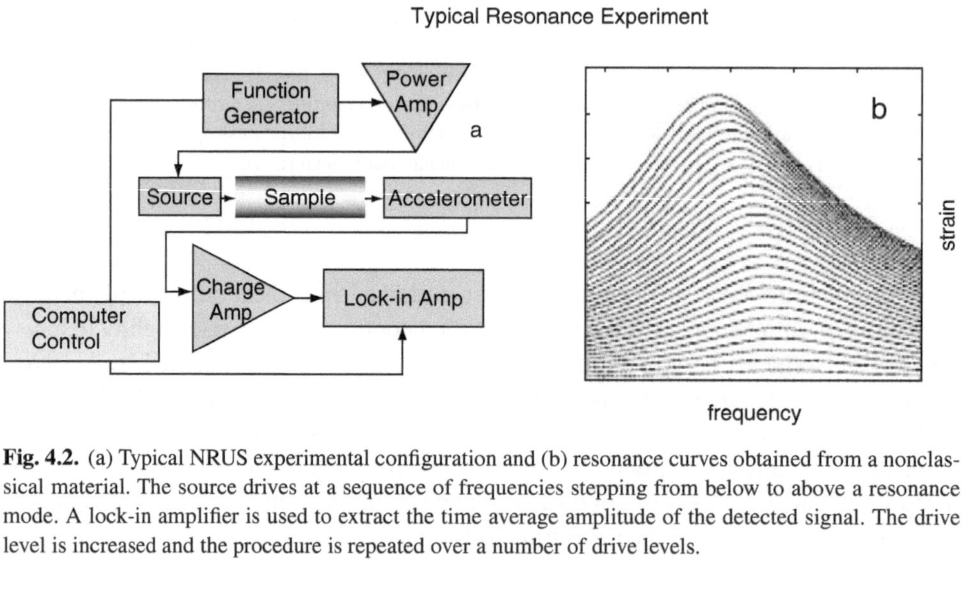

Typical Resonance Experiment

Fig. 4.2. (a) Typical NRUS experimental configuration and (b) resonance curves obtained from a nonclassical material. The source drives at a sequence of frequencies stepping from below to above a resonance mode. A lock-in amplifier is used to extract the time average amplitude of the detected signal. The drive level is increased and the procedure is repeated over a number of drive levels.

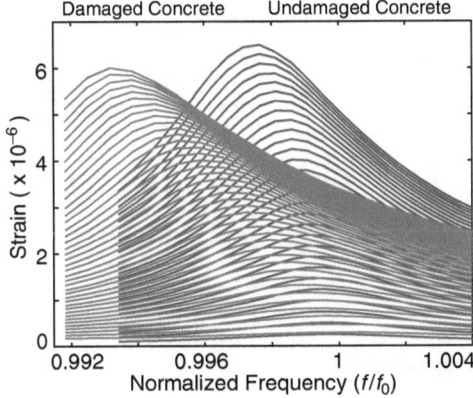

Fig. 4.3. NRUS measurements from intact and damaged concrete samples. Note that the undamaged sample exhibits a small amount of peak shift, meaning it is somewhat elastically nonlinear in its original state. The damaged sample is significantly more nonlinear in that the peak shift is pronounced and the material dissipation characteristics are increased, seen by an increase in the resonance peak width. The frequency axis is normalized to the low amplitude, equilibrium (linear elastic) value (figure courtesy of L. Byers and J. TenCate).

$10\text{-}10^4$. Simultaneous to modal frequency shift, nonlinear damping increases (e.g., Johnson and Sutin, 2005).

3.2 Slow Dynamics Diagnostics (SDD)

The phenomenon of slow dynamics (SD) were first observed in relatively homogeneously elastically nonlinear materials, such as rock and concrete, that have distributed nonlinear sources e.g. (Johnson et al., 1996; TenCate and Shankland, 1996), and has more recently been show to exist in a broad range of solids with both distributed and localized nonlinear sources [cracks, delaminations] (Johnson and Sutin, 2005). The process of SD recovery can be observed by applying RUS measurements at successive times after large-amplitude wave excitation, as well as by observation of the pure tone signal variation. Both methods are described below.

In the RUS variation of the SDD method we take advantage of both the amplitude and frequency of the recovery of a sample mode. In SDD, the equilibrium, low amplitude (linear) amplitude frequency response of the sample is first measured. The sample is then driven at large strain amplitude (order 5 microstrains) to induce material softening. Immediately upon termination of the drive, the RUS measurement recommences at very low strain amplitude ($\sim 10^{-7}$) for probing the recovery. An example of SDD in steel is shown in Fig. 4.4. We see in the left hand figure results from an undamaged sample. The results for the damaged sample are shown in the right-hand side of Fig. 4.4, evident by the initial change in frequency and the successive recovery. The sample recovery time shown is 141 seconds. Full recovery took approximately one hour.

For quick application, a variation of the SDD method known as the *slope amplifier* is useful. Figure 4.5 describes how the SDD slope amplifier works, and Fig. 4.6 shows SDD, slope-amplifier results obtained from an automotive bearing cap.

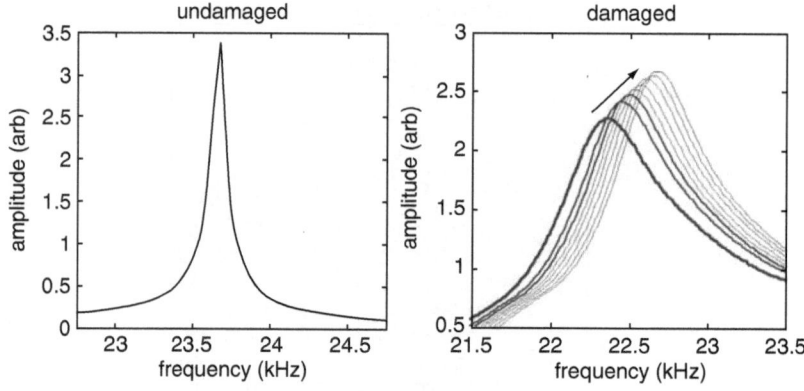

Fig. 4.4. Resonance response of a mode in an undamaged sample and the recovery process of SD in a damaged sample.

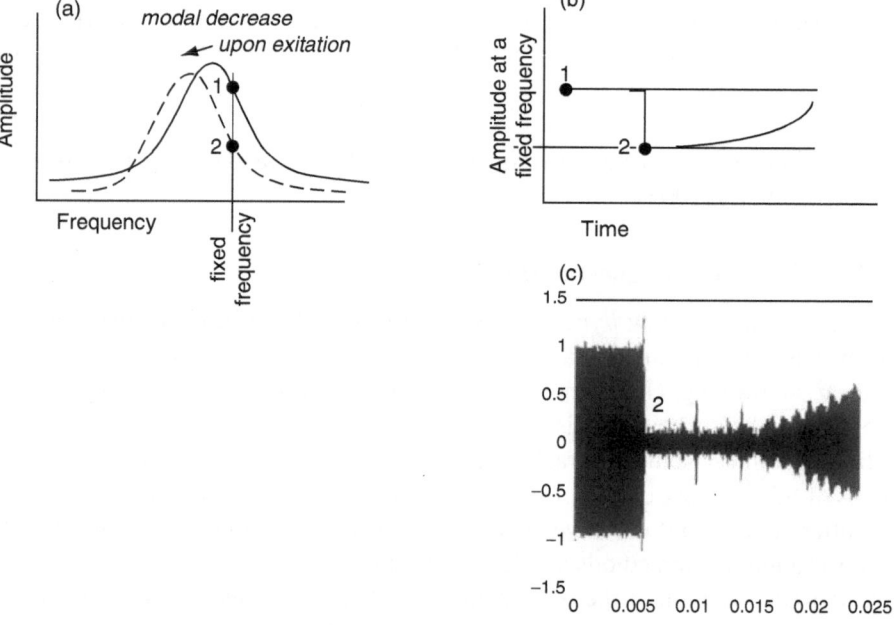

Fig. 4.5. Variation on the SDD technique: the slope amplifier. (a) A low-amplitude signal probes the sample near a modal peak. The signal amplitude [1] is controlled by the modal structure. The time-average amplitude behavior of the time signal is shown in (b). The sample is disturbed by a large amplitude signal induced by a tap for instance, and the modal peak shifts downward causing the probe wave amplitude to change in amplitude to position [2] and (a) and (b). Slow dynamics keeps the modal peak diminished in frequency and thus in amplitude. An actual example of SDD applied to a damaged solid is shown in (c).

One can follow the onset and partial or full recovery of a sample by applying the slope amplifier, capturing more detail than RUS which requires a minute or more for each resonance sweep. For example, Fig. 4.7 shows the onset and several hundred seconds of recovery in four materials.

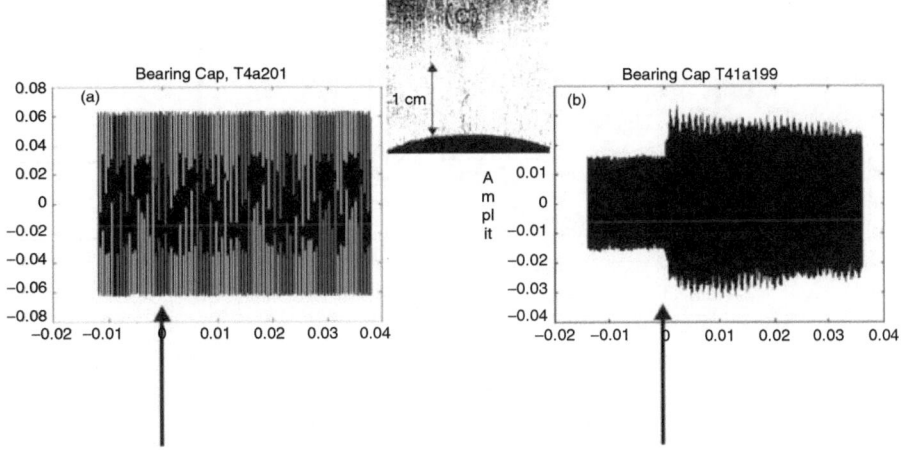

Fig. 4.6. Slope amplifier results in an undamaged (a) and damaged (b) automotive bearing cap. The sample crack is shown in (c). Note that in this experiment, the probe frequency was lower than the resonance peak frequency and therefore the amplitude increases when impacted the source. Arrows point to the time of impact. The data were high-pass filtered.

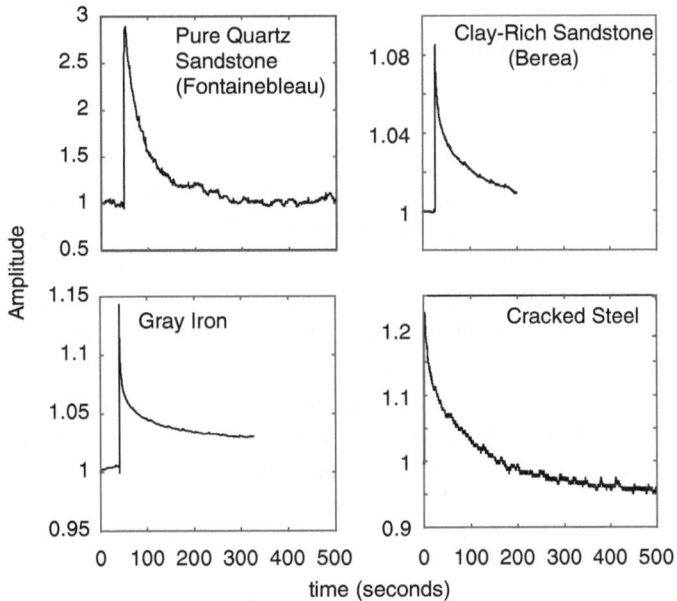

Fig. 4.7. Slow dynamics recovery in four materials applying the *slope amplifier*. Each curve represents the time-averaged amplitude of the low-amplitude, continues-wave probe, at a frequency just below a resonance mode. This method provides the means to capture all of the physical characteristics of the recovery.

3.3 Nonlinear Wave Modulation Spectroscopy (NWMS)

One of the simplest ways to evaluate nonlinear elastic properties of a material is to measure the modulation of an ultrasonic wave by low-frequency vibration. This method is termed here nonlinear wave modulation spectroscopy (NWMS) pioneered

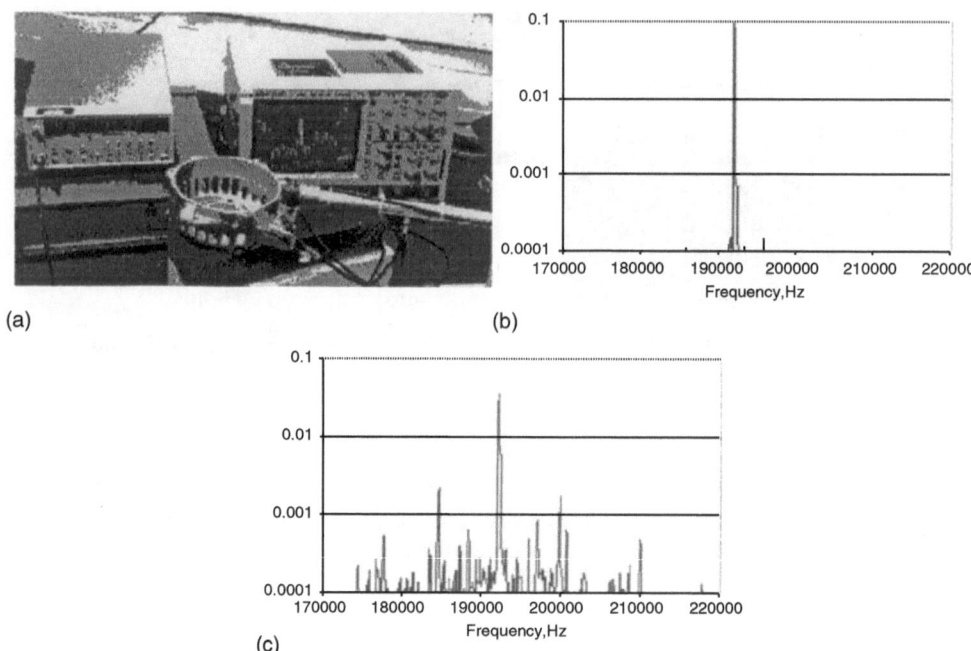

Fig. 4.8. NWMS test in an alternator housing. (a) Experimental configuration. (b) Result in undamaged sample. (c) Result from the damaged sample.

by the group at the Institute of Applied Physics in Nizhny Novgorod, Russia. The following experiment demonstrates a variation of NWMS in testing an alternator housing. The experimental setup is shown in Fig. (4.8a). Ultrasonic transducers were glued to the sides of each sample, and low-frequency, broadband vibration was generated with an instrumental hammer. The spectrum of the signal for an intact part is compared with that from a part with a tiny crack in Figs. 4.8b and 4.8c. Clearly the modulation (sideband components) identifies the damaged sample. Many groups have developed a multitude of variations of the technique, for instance Solodov's group at Moscow State University who have focused on interfaces and disbonding.

Imaging Nonlinear Scatterers

3.4 Imaging Applying NWMS

Nonlinear imaging is in its infancy in solids. One method, the NWMS nonlinear imaging method, described in detail in (Kazakov et al., 2002) is presented here followed by a method based on Time Reversal. In the NWMS method, a low frequency, continuous wave (*cw*) excitation is applied to the specimen simultaneous to a group of high frequency tonebursts (rather than a *cw* probe as in NWMS). In the experiment, the wavefield scattering from a hole, created by drilling, and a crack, created by cyclic loading, were measured in a small steel plate (Fig. 4.9a). Ultrasonic pulses with frequency 3MHz were used for imaging, and a low-frequency vibration of 10 Hz was produced by a shaker. Figure (4.9b) shows how the method works. Figure (4.10)

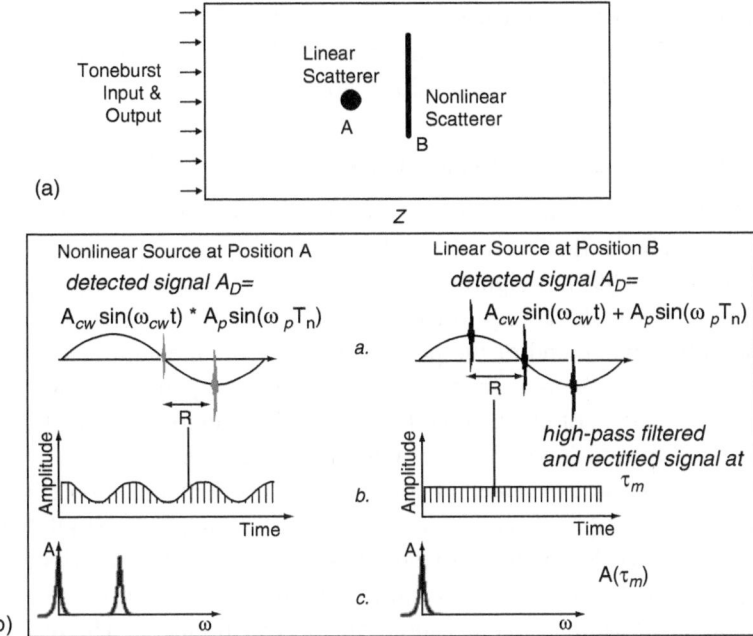

Fig. 4.9. Conceptual view of the data collection and the processing in the nonlinear imaging technique. (a) shows a sample with a hole [A] and a crack [B]. In (b) the left panel shows signals reflected from the crack at different times. The first pulse is reflected at the moment when the applied vibration stress reaches the maximum, minimizing the acoustic impedance of the crack. At this time the crack is compressed and the signal reflected from the crack has minimal amplitude. The second pulses is shown at time when the vibration stress is small and there is no change in the reflected impulse. The third pulse shows the extension phase of the vibration at the moment the crack is maximally opened, minimizing the acoustic impedance of the crack, causing the reflection signal to be higher than in previous phases. Thus, the pulse train is modulated. An FFT of the modulated pulse train gives the time delay of the reflection from the crack. Going through the same process for the hole we observe no modulation.

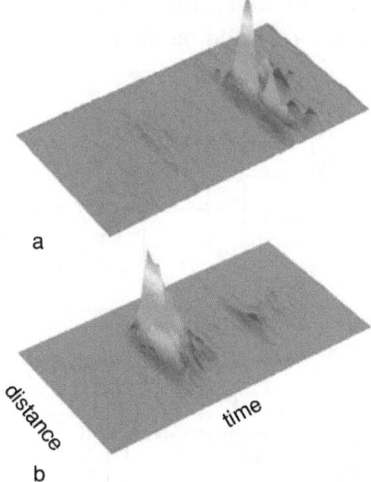

Fig. 4.10. Results of the pulse NWMS method in a sample of steel. The source and receiver are located on the left side (the axis marked "distance". (a) Image of crack applying the nonlinear method. Note hole is not imaged. (b) Standard pulse-echo result showing hole (large amplitude response) and crack, which is much smaller in amplitude do to the shadowing effect of the hole. Distance is obtained from the time axis by using the wavespeed.

shows the measurements, comparing a standard, pulse-echo measurement to the non-linear method. The method provides the means to isolate a nonlinear scatterer. It has not been demonstrated in three-dimensions, however.

3.5 Time Reversal Nonlinear Elastic Wave Spectroscopy (TR NEWS)

Much of the seminal research in Time Reverse Acoustics (TRA) has been carried out by the group located at the University of Paris VII (Laboratoire Ondes et Acoustique, ESPCI) (e.g., Fink, 1997). A significant aspect of TR in regards to elastic nonlinearity is that it provides one the ability to focus an ultrasonic wave, regardless of the position of the initial source and of the heterogeneity of the medium in which the wave propagates. Currently, we are exploiting the focusing properties of TR and the elastic nonlinear properties of cracks together to develop methods for crack and damage location (e.g., Sutin and Johnson, 2005; Ulrich et al., 2006).

Figure 4.11 shows an experimental configuration for demonstration experiments of TR conducted in a sample of sandstone, and TR NEWS in a glass parallelepiped sample (Sutin et al, 2004; Sutin and Johnson, 2004). The method is described as follows. A pulse was applied to the first transmitter. The detected signal measured from the opposite side of the sample was measured by a laser vibrometer. The recorded signal was time reversed as shown in Fig. 4.12a. The TR signal was then re-radiated. The TR focused signal was recorded by the laser vibrometer and analyzed. A typical TR focused signal is shown in Fig. 4.12b. The spatial distribution of the focused signal is shown in Fig. (4.12c).

The significant elastic nonlinearity due to the presence of a crack can be used for crack location. By scanning the surface using the laser vibrometer in tandem with TR focusing at each scan point, then analyzing for nonlinear response at that point, one can determine if damage exists in the scanned area. The feasibility of this technique was evaluated in an experiment where the TRA focusing was conducted along a single line scan in the glass sample with and without damage present. Figure 4.13 presents the results. A small, 3mm crack oriented parallel to the glass surface is located at the

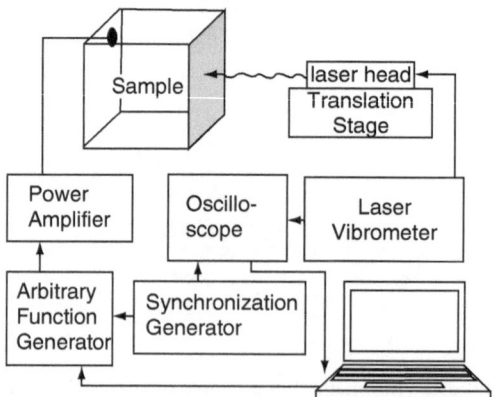

Fig. 4.11. TR NEWS experimental configuration.

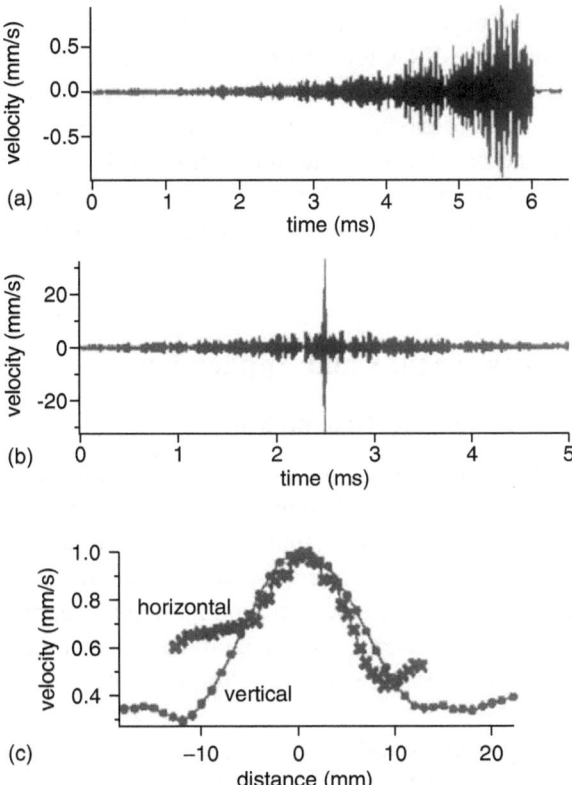

Fig. 4.12. Linear Time Reversal (TR) results in a sandstone sample. (a) Detected, time-reversed signal. (b) re-emitted and detected TR signal. (c) Spatial distribution of TR detected signal in the sample measured along perpendicular lines crossing at the TR focal point. Amplitudes are normalized to the maximum (from Sutin et al., 2004).

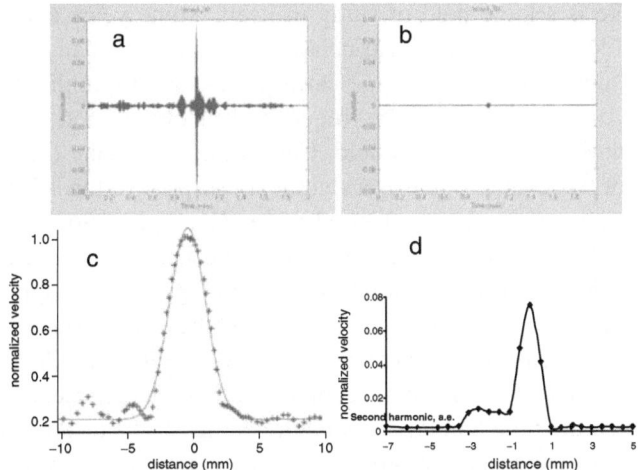

Fig. 4.13. TR NEWS results in a cracked solid. (a) The TR signal bandpass-filtered around the second harmonic, detected above the crack with a laser vibrometer and, (b), 30 mm away from the crack. The fact that a signal is observed means that strong nonlinear response exists at the crack as we expect. By scanning along the crack, it can be imaged. (c) Spatial distribution of the fundamental and (d), the second harmonic TR signal above the crack, normed to the maximum amplitude. The crack was about 2 mm in diameter (see Sutin and Johnson, 2005).

surface. For observation of nonlinear effects, narrow band filtering was used to detect the second harmonic of the TRA focused signal (Fig. 4.13). It can be seen that that the amplitude of the second harmonic of the signal detected above the crack is much higher than the amplitude of the harmonic from the intact surface signal. This experiment demonstrates the feasibility of the application of TR NEWS for crack imaging (see Sutin et al., 2004, for details). Other methods of nonlinear imaging are in development as well, in particular one using modal analysis (Van Den Abeele, this volume).

4.   Some New Areas of Study and Application

4.1 Granular Media and Strong Ground Motion

Much recent effort has gone into nonlinear studies of granular media by the group at Université du Mans (France) and a collaborative effort between Los Alamos and Université de Marne-la-Vallée (France). Granular media is another member of the large class exhibiting nonequilibrium dynamics. It is interesting material from the perspective of earth processes, particularly strong ground motion and earthquake physics, and because it is a much simpler system to understand than many others of the class: One can apply Hertz-Mindlin theory (or some variation thereof) in order to understand its elastic behavior.

In earthquake strong-ground-motion, broad-frequency band waves propagate from hypocentral depths of order 10 km to the earth's surface. Sediments at the surface, composed of granular media, can respond by ringing at their resonance modes. The shear modes are particularly dangerous because, if they couple into building (or other structure) modes, damage or failure of the structure can take place.

Predicting the elastic linear and nonlinear behavior of near surface sediments during an earthquake is a large field of study in itself. Some years ago, it was demonstrated that significant elastic nonlinear behavior manifest by changes in surface-layer resonances may take place during large earthquakes, due to nonlinear response. For instance, a 75% decrease in resonance frequency was observed at one site in the Los Angeles Basin during the 1994 Parkfield, California earthquake (Field et al., 1997). The magnitude of this change came as a significant surprise to the seismic community. Currently, in collaboration with the United States Geological Society, the University of Memphis, the University of Texas at Austin, the University of Massachusetts at Amherst and the Massachusetts Institute of Technology, we are applying an active, large-vibrator source to *in situ* characterization of the near surface layers in an attempt to induce and measure nonlinear response. We observe significant nonlinear response of a near-surface layer from a recent, preliminary experiment at Garner Valley California, located near the San Andreas Fault southeast of Los Angeles. Figure 4.14 shows results of one experiment where the vibrator source was driven in compression in an NRUS-type experiment. A decrease in resonance frequency of order 25% over a strain interval of approximately 10^{-6}-10^{-4} was observed. Slow dynamics appeared to be present as well; however due to experimental difficulties, the observation was unconvincing. Future experiments are planned, and at least one follow-on experiment

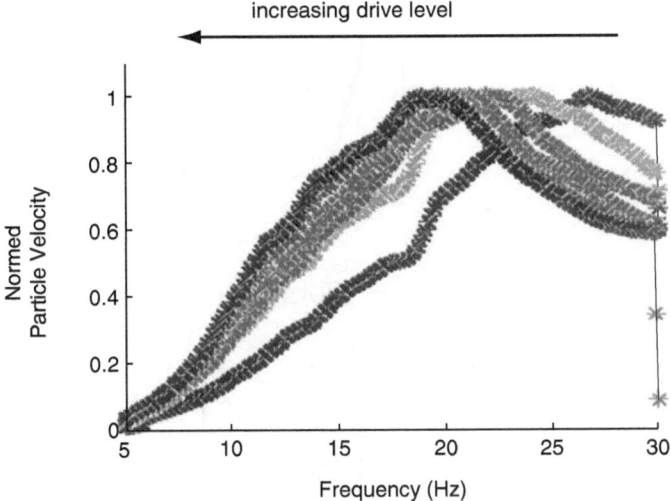

Fig. 4.14. NRUS field experiment at Garner Valley, California. The curves are more complex than in lab studies; however, the dominant frequency peak shift is significant, from approximately 27 to 19.5 Hz over a strain interval of about 10^{-6} to 10^{-4}. The resonating layer is 3-m thick. The shear wave resonance shift is even larger (from Pearce et al., 2004).

will have been completed by the time this paper goes to press. Note that the results indicate that nonequilibrium dynamics takes place over frequencies from order 1 Hz to hundreds of kHz if we compare them to laboratory experiments described above.

4.2 Granular Media and Earthquake Triggering

Recently, we speculated that a phenomenon known as *dynamic earthquake triggering* (Gomberg et al. 2001; 2004) could be due, at least in part, by nonequilibrium dynamics (Johnson and Jia, 2005). Normally, an earthquake exhibits precursors known as foreshocks, followed by a main shock (the magnitude of which is reported for an earthquake—the associated smaller earthquakes are not normally reported to the public), followed by aftershocks. Under certain, and apparently rare conditions (Gomberg et al., 2001), some of the aftershocks can take place at hundreds of kilometers from a mainshock at the time or soon after the seismic wave from the mainshock impinges on a distant fault. This is the phenomenon of *dynamic earthquake triggering*. It has been a puzzle for a number of years as to why dynamic triggering takes place because wave strains at these distances tend to be order 10^{-6} and is it difficult to understand how such small strains could be responsible for this phenomenon.

We speculate that if the conditions are right, triggering may be due to nonlinear softening and weakening of the fault core (the gouge material, granular in nature, that is created by a fault as it progressively slips over the history of the fault). Our conceptual model is that the dynamic wave temporarily reduces the core modulus. The modulus reduction is accompanied by a material strength reduction sufficient to induce fault slip, thereby triggering events. Figure 4.15 shows how this may happen. Taking Eq. 1, (or any nonlinear material softening theory for that matter) we relate the material

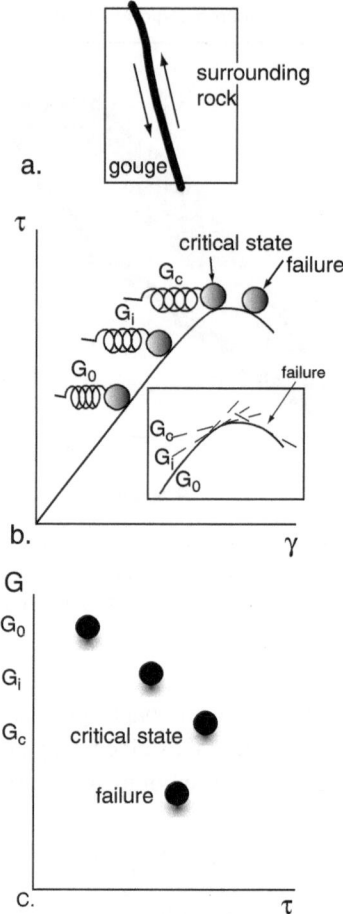

Fig. 4.15. How nonlinear dynamics takes a fault in a critical state to failure. (a) the physical system. (b-c) shear-stress τ, shear-strain curve γ and shear modulus G of the fault gouge. A wave with nonlinear amplitudes causes the modulus to decrease. If the gouge is in a critical state, near failure, the seismic wave can cause modulus softening and failure.

softening to weakening. We consider the competent fault blocks containing the softer fault core in Fig. 4.15a, and the effect of a seismic wave impinging on the system that is in a *critical state*, near failure (Figs. 4.15b, 4.15c). In order for triggering to take place, we speculate that the *triggering conditions* must be met: (a) experiments indicate that strain amplitude must be above 10^{-6}; (b) the fault must be in a *critical state*, near failure; and (c) the confining pressure in the fault core must be small. Small confining pressures are shown to be necessary based on numerous tests that indicate elastic nonlinearity decreases with confining pressure (e.g., Zinszner et al., 1997). Confining pressures could be low at earthquake nucleation depths (order 10 km), if fluid pressures are high. There is observational evidence that suggests this is the case in some faults (e.g., Nur and Booker, 1992; Miller, 1996). We suggest that this is why most seismic waves, even from large events, do not cause triggering (except near the earthquake source)—the triggering conditions are not met: their strain amplitudes tend to be 10^{-7}-10^{-6} at regional distances, and the other conditions may be quite rare. Only large events that focus sufficiently large strains cause triggering beyond what is traditionally deemed the aftershock zone. Field observations support this (J. Gomberg, personal communication, 2005). In these cases, the fault core must meet the triggering conditions, where the fault core can be instantaneously taken through instability to failure by the impingement of the seismic wave.

5. On the origin on Nonequilibrium Dynamics

5.1 Regimes of nonlinear dynamics

Many aspects of nonequilibrium dynamics remain to be understood. One aspect is the process of slow dynamics and how it relates to nonlinear fast dynamics and conditioning. That issue is progressively being addressed by various groups (e.g., Tencate et al., 2004; Pasqualini, this volume). We now understand that there are clear regimes of elastic behavior, contrary to what was thought by some by some of us in the past. These claims were based on the fact that no linear elastic regime seemed to be observed (e.g., Zinszner et al., 1997, Guyer and Johnson, 1999). This erroneous interpretation was due to thermal contamination at very low strain levels that masked the dynamic elastic behavior. In the lowest amplitude regime, the materials behave linearly—there is no modulus dependence on strain amplitude (Pasqualini, this volume). In the next regime, the materials act as a classical nonlinear oscillator that can be described by Landau theory (up to strains of roughly $1\text{-}3\text{x}10^{-6}$ at ambient conditions), and above this, nonequilibrium dynamics emerges. Figure 4.16 shows observations for many rocks over under

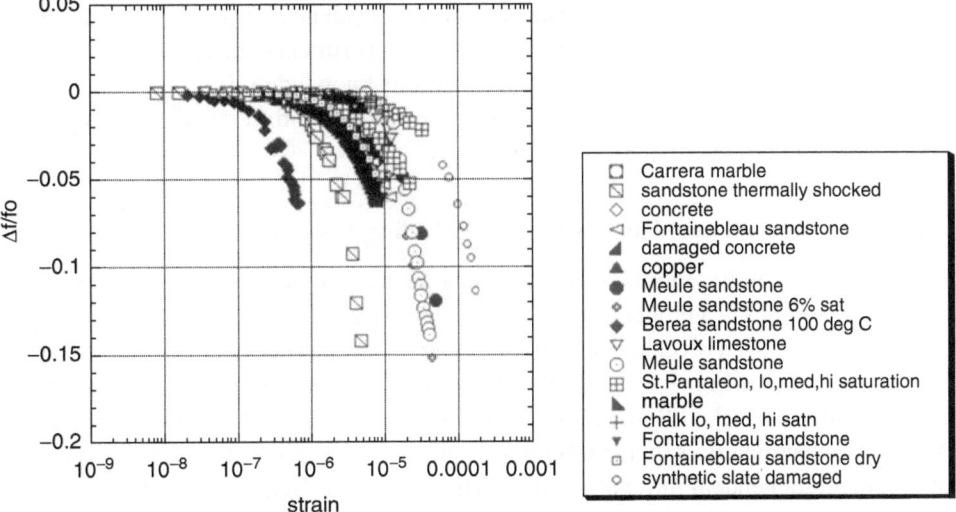

Fig. 4.16. NRUS experimental data taken over a range of saturation, temperature, and "damage" conditions. Based on recent, very careful experiments in Berea and Fontainebleau sandstones, it is currently believed that there are three regimes on elastic behavior (TenCat et al., 2004). One, at low amplitude, that is elastically linear. Following this is elastic behavior that is Landau in nature, and, above this strain, nonequilibrium behavior exists (data taken by B. Zinszner M. Masson, P. Rasolofasoan and P. Johnson at the Institut Francais du Petrole; Slate data from K. Van Den Abeele, Catholic University Leuven Campus Kortrijk).

many saturation conditions extracted from data presented in Johnson et al., (1996). The regimes are material dependent as one might guess, and this is clear from Fig. 4.16 as well. This last and most interesting regime (nonequilibrium) remains to be carefully understood in terms of what is physically taking place (e.g., is it simply a mix of nonlinear fast dynamics an slow dynamics?), and to develop a verifiable physics-based theory that describes all aspects of it. The original P-M Space theory does not account for conditioning or slow dynamics (e.g., Guyer et al., 1997). The variations of the P-M space theory that include conditioning and slow dynamics, although effective at modeling observed behaviors, are *ad hoc*, based on thermal fluctuations (e.g., Scalerandi et al., 2003). Physical based models such as a recently-proposed ratchet-model (Vakhnenko et al., 2004) are physically-based but the physics are as yet verified by experiment and will be hard to do so. Models indicating that thermal heating and diffusion are the source of nonequilibrium dynamics are questionable if one invokes three dimensional thermal diffusion, which is the case (Zaitsev et al., 2002), as shown by Pasqualini (in preparation, 2006).

Some have suggested that fluids are responsible for nonequilibrium dynamics. Van Den Abeele et al. (2000) have shown that fluids act to modify the internal forces in porous media and thereby influence the nonlinear behavior, but are not responsible for the underlying behavior. In any case, some materials in the class are dry with no means for fluid penetration (e.g., gray iron, alumina ceramic are two examples described in Johnson and Sutin, 2005).

A very important issue that has not been explored experimentally is whether shear sliding is the fundamental cause, at least in some cases, of nonequilibrium dynamics, as some of us currently believe. Many controlled experiments have been conducted with longitudinal or bulk modes. For instance, there is evidence suggesting the nonlinear response in shear is larger than in bulk mode but experiments aimed at isolating shear from other effects have not been conducted to our knowledge. Because I believe the physical origin is shear sliding, such experiments would aid tremendously in helping verify such a hypothesis and developing theory.

5.2 Slow and Nonlinear Fast Dynamics Including Dissipation

Results were recently reported of the first systematic study of nonlinear fast dynamics and slow dynamics in a number of solids (Johnson and Sutin, 2005). Observations were presented from seven diverse materials showing results of nonlinear fast dynamics and slow dynamics (see Fig. 4.17). The materials include samples of gray iron, alumina ceramic, quartzite, cracked Pyrex glas, marble, sintered metal, and perovskite ceramic. It was shown that materials that exhibit nonequilibrium behavior have very similar ratios of amplitude-dependent internal-friction to the resonance-frequency shift as a function of strain amplitude. The ratios range between 0.28 and 0.63 (except for cracked Pyrex glass, which exhibits a ratio of 1.1), and the ratio appears to be a material characteristic. The ratio of internal friction to resonance frequency shift as a function of time during slow dynamics is time independent, ranging from 0.23 - 0.43 for the materials studied (within the error bars they are approximately the same). The above relations relating nonlinear attenuation and frequency shift in slow and fast dynamics demand

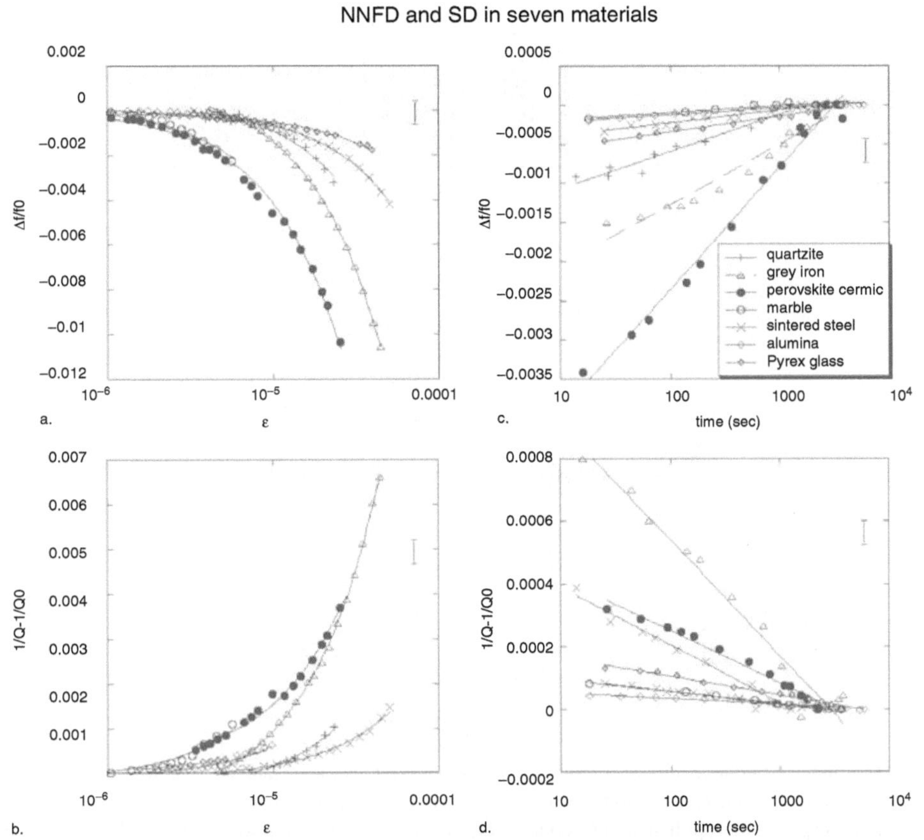

NNFD and SD in seven materials

Fig. 4.17. Nonequilibrium behavior in seven materials obtained from NRUS measurements (from JS). (a) and (c) show the change is resonance frequency normalized to the linear resonance frequency for nonlinear fast and slow dynamics respectively. (b) and (d) show the behavior in wave dissipation, $1/Q$, for nonlinear fast and slow dynamics respectively, where Q_0 is the linear value and Q is the nonlinear value. The error bars shown are based are extremely conservative: they are obtained from the frequency sampling rate of the experiment, carried through all of the calculations. The lines in the figures are present to guide the eye (they are linear fits in the case of fast dynamics and logarithmic fits for slow dynamics).

more study to see if the relation between nonlinear dissipation and frequency shift (modulus change) are always material dependent, and what that implies for a theoretical description.

We note that some characteristics of slow dynamics have yet to be studied. The slope amplifier offers a means to carefully study the recovery process. For instance, we may see physical processes that could aid in theory development that have been overlooked in the past. Figure 4.18 hints at one interesting behavior where we may be observing something like isolated cascade slip events during recovery.

5.3 On developing a physics-based model of nonequilibrium dynamics

One suggested approach to addressing a generalized, physics-based model is to start with specific systems and look for similarities in the underlying physics. For instance,

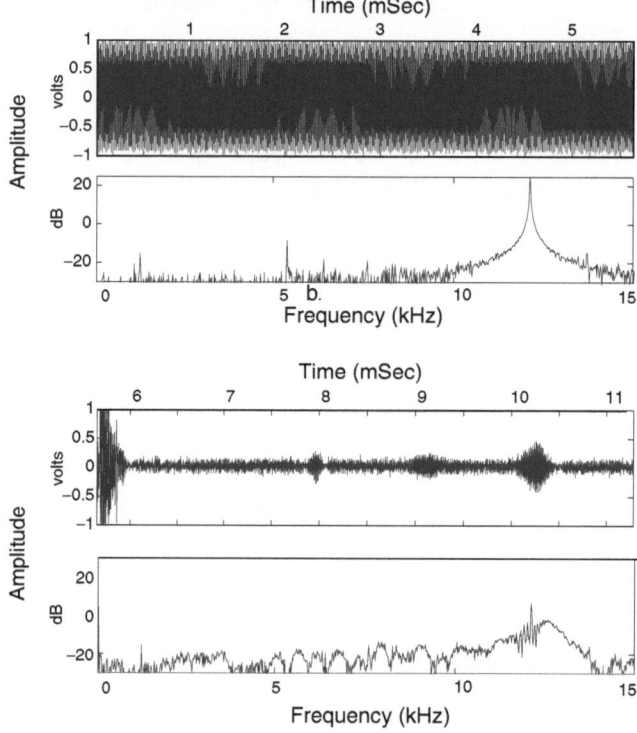

Fig. 4.18. Interesting characteristics of slow dynamics in Fontainebleau sandstone applying a low amplitude probe and an impulse. The sample was a bar of sandstone approximately 2x20 cm in dimension. A low amplitude probe was applied at one end and detected at the other. A tap was introduced to induce slow dynamics. The top two panels show the probe time-series and corresponding Fourier transform before the tap. The bottom panels show the time and frequency response during, and for a number of milliseconds after, the tap. Note the "humps" in amplitude. What are these? Could they be large slip events or some other as yet, unidentified process that takes place during slow dynamics? This general topic requires more study and may aid in a new theoretical description of nonequilibrium dynamics.

in granular media, Hertz-Mindlin theory can be applied as noted above; for cracks in metals, crack dynamics and dislocation-point defect interaction may be applied. The underlying shear-mechanisms may ultimately be related in a broader model. Incorporating in slow dynamics may be the most difficult aspect (see below). Such an approach could presumably evolve to a generalized nonequilibrium theory for all, or most of the materials in the class.

5.4 Universal behavior?

A discussion has taken place in the literature as to whether or not the observed behaviours of nonequilibtium dynamics are universal (see, e.g., Guyer and Johnson, 1999; Hirsekorn and Delsanto, 2004; Hirsekorn and Delsanto this volume). I and others would argue that they are not in the sense of critical phenomena such as a phase change, but they are in the sense that the nonlinear signatures are identical across a large number of very different materials (Johnson and Sutin, 2005).

6. Summary and Conclusions

In summary, we know much from experiments regarding the behaviours and the breath of the class of materials that exhibit nonequilibrium dynamics. Fundamentally, we believe that the nonlinear response is attributable to the damage features in the material at many scales. A proper theory containing the underlying physics is still to be addressed, however. Nonetheless, many applications based on nonlinear methods, including applications to earthquake strong ground motion and elsewhere, but especially in regard to NonDestructive testing have been developed. One can determine whether or not a material exhibits nonequilibrium behaviour and simultaneously if it is damaged, if there is disbonding, etc. However, without knowledge of the physical basis, the relation between the nonlinear response (often in the form of the hysteretic nonlinear parameter α) and damage quantity or other features responsible for the nonlinearity as in granular media, must be obtained empirically. With a physics-based model we will have the means to relate the nonlinear response to damage quantity, with the caveat that, in solids with macroscale damage, only some portion actively contributes to the nonlinear response (we know this from the NWMS imaging experiment mentioned above where crack tips seemed to make the primary contribution to the nonlinear response). Imaging of nonlinear scatterers is in its infancy and should progress significantly in the near future. Much work remains in regards to understanding the relation of slow and fast dynamics, the details of slow dynamics. There are enormous and fascinating opportunities for new research that can provide insight into nonequilibrium dynamics.

7. Acknowledgements

This work is supported by Institutional Support (LDRD) at Los Alamos and by the US Department of Energy Office of Basic Energy Science. Thanks to many colleagues, including Jim TenCate, Alexander Sutin, Salman Habib, Robert Guyer, Fred Pearce, Joan Gomberg, Donatella Pasqualini, Xiaoping Jia, T.J. Ulrich, Marco Scalerandi, P.P. Delsanto, and Katrin Heitmann.

References

Field, E. H., Johnson, P. A., Beresnev, I. and Zeng, Y., 1997, Nonlinear ground-motion amplification be sediments during the 1994 Northridge earthquake, *Nature* **390**: 599–602.

Fink, M., 1997, Time reversed acoustics *Physics Today* **50**: 34–40.

Gomberg, J., Reasenberg, P. A., Bodin P. and Harris R.A., 2001, Earthquake triggering by seismic waves following the Landers and Hector Mine earthquakes, *Nature* **411**: 462–466.

Gomberg, J., Bodin, P., Larson K., and Dragert, H., 2004, Earthquake nucleation by transient deformations caused by the M = 7.9 Denali, Alaska earthquake, *Nature* **427**: 621–624.

Guyer, R. A., and Johnson, P. A., 1999, The astonishing case of mesoscopic elastic nonlinearity, *Physics Today* **52**: 30–35.

Guyer, R. A., McCall, K. R., Boitnott, G. N., Hilbert, L. B., and Plona, T. J., 1997, Quantitative implementation of Preisach-Mayergoyz space to find static and dynamic elastic moduli in rock, *J. Geophys. Res.* **102**: 5281–93.

Hirsekorn, S., and Delsanto, PP, 2004, On the universality of nonclassical nonlinear phenomena and their classification, *Applied Phys. Lett.* **84**: 1413–1415.

Johnson, P., 1999, The new wave in acoustic testing, *Materials World, the J. Inst. Materials* **7**: 544–546.

Johnson, P.A., and Jia, X, 2005, Nonlinear dynamics, granular media and dynamic earthquake triggering, *Nature, in review.*

Johnson, P. and Sutin, A., 2005, Slow dynamics in diverse solids, *J. Acoust. Soc Am.* **117**: 124–130.

Johnson, P. A. and A. Sutin, 2005, QNDE 2004, Nonlinear elastic wave NDE I. Nonlinear resonant ultrasound spectroscopy (NRUS) and slow dynamics diagnostics (SDD), Review of Progress in Quantitative Nondestructive Evaluation, Volume 24B, Thompson, D and Chimenti D., Eds., 377–384.

Johnson, P. A., Zinszner, B., and P. N. J. Rasolofosaon, 1996, Resonance and nonlinear elastic phenomena in rock, *J. Geophys. Res.* **101**: 11553–11564.

Kazakov, V. V., Sutin, A. and Johnson, P. A., 2002, Sensitive imaging of an elastic nonlinear wave source in a solid, *Applied Physics Letters* **81**, 646–648.

Korotkov, A.S., Slavinsky, M.M., Sutin, A.M. *Acoustical Physics.* **40**, 84 (1994).

Migliori, A. M.. and Sarrao, J., 1997, "Resonant Ultrasound Spectroscopy", John Wiley.

Miller, S. A., 1996, Fluid-mediated influence of adjacent thrusting on the seismic cycle at Parkfield, *Nature* **382**: 799–802.

Maev, RG; Solodov, I.Yu., 2000, Nonlinear acoustic spectroscopy of cracked flaws and disbonds: fundamentals, techniques and applications, Review of Progress in Quantitative Nondestructive Evaluation, 25–30 July 1999, Montreal, Que., Canada, AIP Conference Proceedings 2000, **509-B**, 1409–1416.

Nur, A. and Booker, J., 1977, Aftershocks caused by pore fluid flow?, *Science* **175**: 885–887.

Ostrovsky, L. and Johnson, P. A., 2001, Dynamic nonlinear elasticity in geomaterials, *Rivista del Nuovo Cimenta* **24**, 1–46.

Pasqualini, D., this volume.

Pasqualini, D., 2006, The role of thermal diffusion in nonequilibrium dynamics, *in preparation.*

Pearce, F., Bodin, P. Brackman, T., Lawrence, Z., Gomberg, J., Steidl, J., Menq, F., Guyer, R., Stokoe, K., and Johnson, P. A., 2004, Nonlinear soil response induced in situ by an active source at Garner Valley, *Eos Trans. American Geophysical Union* **85**, Fall Meeting Supplement, Abstract S42A-04.

Sutin, A., TenCate, J. and Johnson, P. A., 2004, Single-channel time reversal in elastic solids, *J. Acoust. Soc. Am.,* **116**, 2779–2784.

Sutin, A. and Johnson, P. A., 2005, Nonlinear elastic wave NDE II. Nonlinear wave modulation spectroscopy and nonlinear time reversed acoustics, Review of Progress in Quantitative Nondestructive Evaluation, Volume 24B, Thompson, D and Chimenti D., Eds., 385–393.

TenCate J. and Shankland T., 1996, Slow dynamics in the nonlinear response of Berea sandstone, *J. Geophys. Res.* **23**, 3019–3022.

Ulrich, T. J., Johnson, P. A., Guyer, R. A., 2006, Investigating interaction dynamics of elastic waves with a complex nonlinear scatterer applying the Time Reverse mirror, *Appl. Phys. Lett.*, in press.

Ulrich, T. J., Sutin, A. and Johnson, P. A., 2006, Imaging Nonlinear Scatterers Applying the Time Reversal Mirror, *J. Acoust. Soc. Am.* **119:** 1514–1518.

Van Den Abeele, K. E.-A., Johnson, P. A. and Sutin, A., 2000, Nonlinear Elastic Wave Spectroscopy (NEWS) techniques to discern material damage. Part I: Nonlinear Wave Modulation Spectroscopy (NWMS), *Research on NonDestructive Evaluation* **12:** 17–30.

Van Den Abeele, K. E.-A., Carmeliet, J., TenCate, J. and Johnson P. A., 2000, Nonlinear Elastic Wave Spectroscopy (NEWS) techniques to discern material damage. Part II.

Zaitsev, V., Gusev, V. and Castagnede, B., 2002, Thermoelastic mechanism for logarithmic slow dynamics and memory in elastic wave interactions with individual cracks, *Phys. Rev. Lett.* **84:** 159–162.

5

Nonlinear Rock Mechanics

G.E. Exadaktylos

Mining Design Laboratory, Department of Mineral Resources Engineering Department, Technical University of Crete, University Campus, Akrotiri GR-73100, Chania, Greece, Tel: +30 28210 37690, Fax: +30 28210 37891, e-mail: exadakty@mred.tuc.gr, http://minelab.mred.tuc.gr.

Abstract

Rock mechanics is a rapidly evolving scientific discipline that is concerned with the development of experimental and theoretical tools to study and predict the behavior of intact (or damaged) and discontinuous (fractured) rocks under the influence of chemo-thermo-poro-mechanical effects under static or dynamic conditions. Nonlinearity is inherent in many rock mechanical problems. Some indicative examples are briefly listed herein. In physical nonlinearity, few, if any, rocks are truly "elastic" and even fewer are "linear" or "Hookean." Natural or stress-induced nonlinear directional response (anisotropy) is possible. In addition, coupled thermal, fluid flow, and mechanical effects or processes may give considerable nonlinearities in the response of porous rocks. In geometric nonlinearity, many structures undergo very large deformations in normal or in damaged conditions (e.g., buildings and other manmade structures after major earthquakes (See chapter 4)). In constraints, nonlinearity, contact between deformable rocks (e.g., contact of lips of faults), or rock structure may occur such that the common surface is unknown. A central point of any rock mechanical problem is the constitutive description of the rock. In this chapter the basic ingredients of a nonlinear constitutive mechanical theory for rocks based on experimental evidence is outlined and tested by exploiting triaxial compression experiments of a sandstone.

Keywords: Damage, fracture mechanics, hypoelasticity, Mohr–Coulomb, nonlinearity, plasticity, rocks, sandstone, triaxial compression

1. Introduction

Rocks are granular, porous, heterogeneous, anisotropic natural materials formed under certain geological processes (i.e., sedimentary, magmatic, or metamorphic) during a rather extended (geological) time scale (Figures 5.1a,b), hence their behavior is more complex as compared to concretes, ceramics, metals, and other manmade materials. Rocks are composed of a vast variety of minerals and occur in an almost infinite range of conditions, from crystalline solids to aggregations of independent particles. In general, rocks exhibit elasticity, plasticity, damage (Van den Abeele and Windels, this volume), cracking, elastic hysteresis, and memory (Pascualini, this volume;

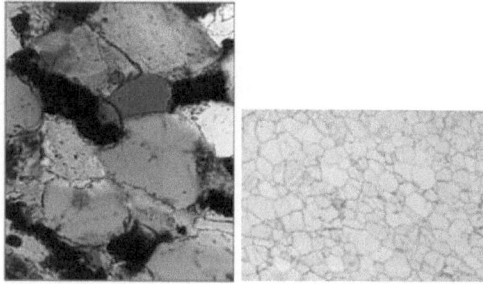

Fig. 5.1. Typical granular microstructure (fabric) of rocks observed with an optical microscope: (a) heterogeneous Berea sandstone with pores and microcracks (Guyer and Johnson, 1999) and (b) low porosity homogeneous Gioia marble rock microstructure with twins in calcite crystals (courtesy of P. Tiano).

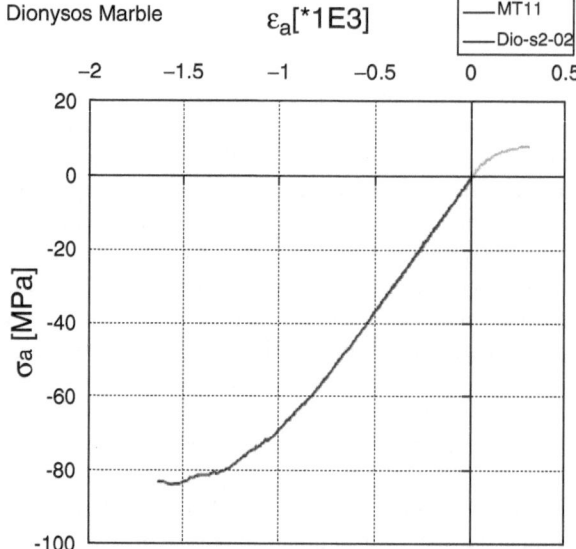

Fig. 5.2. Typical axial stress–axial strain curve of a rock in uniaxial tension–compression (Dionysos marble) (courtesy of I. Vardoulakis).

TenCate et al., this volume), dilatancy, creep, pressure, and rate dependency, nonequilibrium nonlinear dynamics (Johnson, this volume; TenCate et al., this volume), size effects, and anisotropy, among others. Another important property of geologic materials that is attributed to the presence of healed or open microcracks, pore topology, and other defects such as soft inclusions, is that their uniaxial tensile strength is much smaller (one order of magnitude) than their uniaxial compressive strength (Figure 5.2). This is a clear manifestation of "brittleness." The convention of positive tension and elongation is assumed unless stated otherwise.

Due to all these phenomena that accompany rock mechanical behavior, modern rock mechanics should rely upon all the up-to-date developments of the fundamental theories of continuum and discontinuum mechanics, elasticity, strength of materials, damage mechanics, plasticity, and fracture mechanics in order to present robust models for rocks subjected to static or to dynamic loads. The linking of all these theories under the umbrella of a unique theory capable of describing the rock mechanical behavior at all scales (i.e., in the range of 1e 6 m to 1e 4 m) and boundary conditions encountered in practice, is one of the fascinating challenges of the future.

In the following paragraphs a brief account of nonlinearities accompanying rock behavior under static mechanical loads is given, and an example of calibration of a new nonlinear mechanical model on a series of uniaxial and triaxial compression experimental data of a heterogeneous sandstone is illustrated.

2. Elasticity and Plasticity of Rocks

The study of the elasticity of rocks is the first step towards the construction of robust models. For easy reference a few definitions and basic properties of elastic, hyper-, and hypoelastic constitutive equations are mentioned in this section [for an extensive review see (Truesdell and Noll, 1965; Chen nad Han, 1988; Vardoulakis and Sulem, 1995)]. A material is called *elastic* if: (a) it possesses only one ground state, that is, a state that is undeformed and is also stress-free, and if (b) the stress σ_{ij} is a function of the deformation gradient or strain $\varepsilon_{kl}^{(el)}$, that is,

$$\sigma_{ij} = T_{ij}\left(\varepsilon_{kl}^{(el)}\right), \qquad i, j, k, l = 1, 2, 3, \tag{5.1}$$

wherein superscript (el) indicates elastic strains and the usual notation and rules for tensors are followed (e.g., Frederick and Chang, 1972). The elastic material defined by (5.1) is called a "Cauchy elastic material." From this equation we observe that in *closed stress paths* in stress space elastic materials are characterized by zero residual strain. In the small-strain linear Cauchy elasticity in isothermal or adiabatic conditions, the stress–strain relationship may be stated in the following way,

$$\sigma_{ij} = C_{ijkl}\varepsilon_{kl}^{(el)}, \qquad C_{ijkl} = \text{constants} \tag{5.2}$$

More restrictive is the definition of the *hyperelastic* or Green elastic materials. In hyperelasticity we postulate a strain energy density function

$$w^{(el)} = w^{(el)}(\varepsilon_{ij}^{(el)}) \tag{5.3}$$

such that,

$$\sigma_{ij} = \frac{\partial w^{(el)}}{\partial \varepsilon_{ij}^{(el)}}. \tag{5.4}$$

The above relationship means that the stress tensor is derived from the gradient of the strain potential function, or alternatively that the stress is normal to the surface $w^{(el)} = const$. Thus we conclude that equation Eq. (5.2) for isotropic elastic materials follows from the form (5.4) for hyperelastic materials. The converse is not generally true. If the material is hyperelastic along a closed strain path the total specific work done by the stresses is null. This is not generally true for (Cauchy) elastic materials. However, in closed stress paths in stress space both elastic materials and hyperelastic materials are characterized by zero residual strain. We observe that both the constitutive equations of isotropic elastic materials 5.3 and for isotropic hyperelastic materials (5.6) lead through formal material time-differentiation to equations of the rate form

$$\dot{\sigma}_{ij} = C_{ijkl}^{(el)} \dot{\varepsilon}_{kl}^{(el)}, \tag{5.5}$$

where the dot indicates differentiation w.r.t. time. Truesdell and Noll (1965) defined a class of materials, which they call "hypoelastic materials", that obey rate constitutive equations like the one above, which are linear in $\dot{\varepsilon}_{ij}^{(el)}$, with the additional restriction that the corresponding fourth-order constitutive tensor is an isotropic tensor function of the stress. Hypoelastic constitutive models are used to describe the mechanical behavior of a class of materials in which the state of stress depends on the current state of strain as well as on the stress path followed to reach that state. Hypoelasticity equations are derived from hyperelasticity. In general, however, hypoelastic constitutive equations are neither integrable to a finite form (5.1) nor connected to a strain energy function through a constitutive equation of the form of (5.4). Thus, hypoelastic equations will lead in general to residual strain, if integrated along closed stress paths, and to violations of the second law of thermodynamics if integrated along closed strain paths.

In elastoplastic constitutive equations it is often assumed that the background elasticity is a Hooke-hypoelasticity; that is,

$$C_{ijkl}^{(el)} = G \left\{ \delta_{ik}\delta_{jl} + \delta_{il}\delta_{jk} + \frac{2v}{1 - 2v} \delta_{ij}\delta_{kl} \right\} \tag{5.6}$$

with constant secant shear modulus G and constant secant Poisson's ratio v, where δ_{ik} is the Kronecker delta. The next modeling step is based on the study of rock plasticity and strength. Incremental plasticity theory is based on a few fundamental postulates. Plasticity models are written as rate-independent models or as rate-dependent models. A rate-independent model is one in which the constitutive response does not depend on the rate of deformation: the response of many rocks at low temperatures relative to their melting temperature and at low strain rates is effectively rate independent. In a rate-dependent model the response does depend on the rate at which the material is strained. Examples of such models are the simple "creep" models and the rate-dependent plasticity model that is used to describe the behavior of rocks at higher strain rates. Because these models have similar forms, their numerical treatment is based on the same technique. A basic assumption of elastic–plastic models is that the deformation can be divided into an elastic part and an inelastic (plastic) part. This decomposition can be used directly to formulate the plasticity model. Historically, an additive strain rate decomposition is employed (Hill, 1950),

$$\dot{\varepsilon}_{ij} = \dot{\varepsilon}_{ij}^{(el)} + \dot{\varepsilon}_{ij}^{(pl)}, \tag{5.7}$$

where the superscript pl indicates plastic strains. For rate-independent materials we may use instead of the rate of deformation the incremental deformation; that is,

$$\Delta\varepsilon_{ij} = \dot{\varepsilon}_{ij}. \tag{5.8}$$

The elastic part of the response is assumed to be derivable from an elastic model presented previously. The cohesional, frictional, and dilatational properties of rocks up to failure may be modeled within the frame of elastoplasticity theory with strain-hardening yield surface and nonassociative flow rule. The yield surface F that defines the limit to this region of purely elastic response and plastic potential Q that defines the plastic part of strain may be expressed as follows,

$$F = F\left(\sigma_{ij}, \psi\right), \quad Q = Q\left(\sigma_{ij}, \psi\right), \tag{5.9}$$

in which ψ denotes a hardening parameter, that is, a measure of plastic deformation. The hardening parameter or parameters are state variables that are introduced to allow the models to describe some of the complexity of the inelastic response of real materials. In the simplest plasticity model (*perfect plasticity*) the yield surface acts as a limit surface and there are no hardening parameters at all: no part of the model evolves during the deformation. Complex plasticity models usually include a large number of hardening parameters.

Plastic strain rates are generated when the state of stress lies on the yield surface and if loading of that yield surface is taking place, that is to say, the following "consistency criterion" is satisfied,

$$F = 0, \quad \dot{F} > 0, \quad \text{and} \quad \dot{\psi} > 0. \tag{5.10}$$

When the material is flowing inelastically the inelastic part of the deformation is defined by the flow rule, which we can write in incremental form as follows,[1]

$$\Delta \varepsilon_{ij}^{(pl)} = \Delta \psi \frac{\partial Q}{\partial \sigma_{ij}}, \quad \psi \geq 0. \tag{5.11}$$

The rate form of the flow rule is essential to incremental plasticity theory, because it allows the history dependence of the response to be modeled. The plastic potential and the yield function may be identical, that is, $Q = F$, only if the measured dilatation and strength responses of rock are identical. Such models are called *associated flow* plasticity models. Associated flow models are useful for materials in which dislocation motion provides the fundamental mechanisms of plastic flow when there are no sudden changes in the direction of the plastic strain rate at a point. They are generally not accurate for materials in which the inelastic deformation is primarily caused by frictional mechanisms as in the case for geomaterials. For a plastic potential that is an isotropic function of the stress tensor, Eq. (5.11) describes a *co-axial flow rule*; that is, the principal axes of plastic strain rate coincide with the principal axes of stress.

The feature that distinguishes the inelastic behavior of nonmetallic porous materials, such as concrete, rocks, and soils, from that of metals is the occurrence of plastic volume changes. Metals undergo no change in volume as a result of plastic deformation. Rocks, on the other hand, may either dilate (increase in volume) or compact (decrease in volume) as a result of plastic deformation. From the macroscopic point of view and

---

[1] The inequality $\psi \geq 0$ is essential in plasticity and defines the irreversible character of plastic deformations.

for granular media under shear, irreversible shear strains $g^{(pl)}$ and irreversible volume changes $v^{(pl)}$ are linked together. This is usually expressed by the well-known phenomenological dilatancy constraint (Vardoulakis and Sulem, 1995)

$$\dot{v}^{(pl)} = d\left(g^{(pl)}\right)\dot{g}^{(pl)}, \qquad \dot{v}^{(pl)} = \dot{\varepsilon}_{kk}^{(pl)}; \tag{5.12}$$

$d\left(g^{(pl)}\right)$ is called the "mobilized dilatancy coefficient." That is to say, there is great class of deformations where there is no need to treat irreversible volume changes separately from irreversible shear deformations. Within the frame of nonassociative flow theory of plasticity one may chose the deviatoric plastic strain $\dot{g}^{(pl)}$ as the hardening parameter as proposed by Kachanov (1974), which can be interpreted as the average interparticle slip. This strain invariant may be expressed as follows,

$$\dot{g}^{(pl)} = \sqrt{2\dot{e}_{ij}^{(pl)}\,\dot{e}_{ji}^{(pl)}}, \qquad \dot{e}_{ji}^{(pl)} = \dot{\varepsilon}_{ji}^{(pl)} - \dot{\varepsilon}_{kk}^{(pl)}\delta_{ij}/3. \tag{5.13}$$

Thus, one may set

$$\Delta\psi \equiv \Delta g^{(pl)}. \tag{5.14}$$

3. Experimental Evidence

In this work we employ a database of uniaxial and triaxial compression experiments on Serena (or Firenzuola) sandstone intact cylindrical specimens with diameter $D = 5\,\text{cm}$ and height $H = 10\,\text{cm}$ that were performed at SINTEF (Norway). The mineralogical setup and basic physical properties of this type of sandstone are displayed in Table 5.1. In all the tests carried out in the frame of this work a certain number of unloading–reloading cycles were performed in order to study the elasticity of the test specimens. During its test the axial force (F), the engineering axial strain (ε_a), and the engineering radial (or lateral) strain (ε_r) were recorded by LVDTs and stored on a computer. The axial stress (σ_a) was computed from the formula

$$\sigma_a = \frac{F}{\pi D^2/4}. \tag{5.15}$$

For the cylindrical samples subjected to axial loading and under small strains the volumetric strain (ε_v) was computed from the formula

$$\varepsilon_v = (2\varepsilon_r + \varepsilon_a), \tag{5.16}$$

where ε_r and ε_a are the axial and radial strains, respectively.

As illustrated in Figure 5.3 the tested rock exhibits strong stress dependence of the elastic (unloading–reloading) curves, which are characterized by appreciable nonlinearity. Thus, the simple secant-modulus calibration procedure by virtue of relation (5.6) with rather linear unloading–reloading curves cannot be readily applied. Vardoulakis et al. (1998) developed a hypoelastic model for marble that accounts for stress dependency of the Young's modulus but assumes a constant Poisson ratio of the marble. In a next section a new hypoelastic model based on damage mechanics is developed that considers as variables a secant elastic modulus and Poisson ratio of intact rocks.

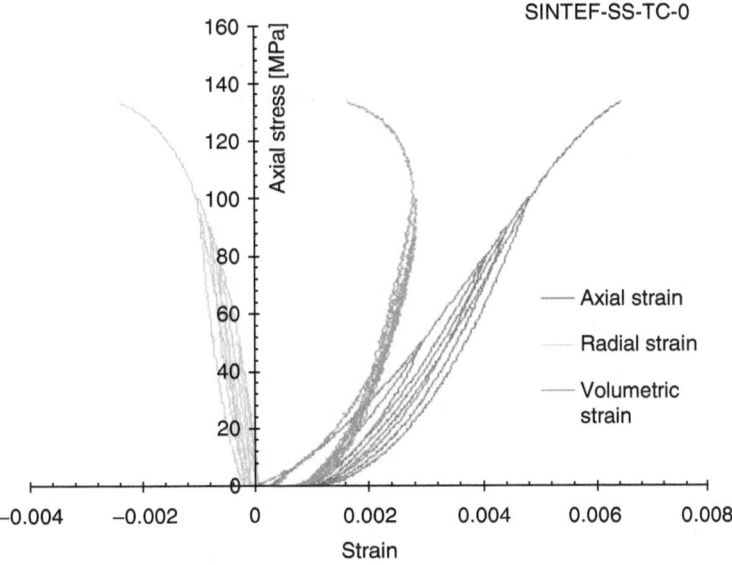

Fig. 5.3. Axial stress versus axial, radial, and volumetric strains for Serena sandstone specimen SS-TC-0 at zero confining stress (courtesy of E. Papamihos)

Table 5.1. Mineralogical and petrophysical properties of the tested Serena sandstone

Mineral or physical property	Serena (Firenzuola) sandstone
Calcite [%]	21 – 8
Dolomite [%]	7 – 0
Quartz [%]	32 – 36
Potassium Feldspar [%]	7
Plagioclase [%]	13 – 15
Phylosilicates [%]	20 – 34
Total porosity [%]	9.76
Bulk density [Mg/m^3]	2.57

In the sequel, the observed mechanical behavior of sandstone in Uniaxial Compression (UC) and Triaxial Compression (TC) is described with simple mathematical relations. Note that in this section we deviate momentarily from the assumed stress sign convention and we assume compressive stresses as positive. First, by considering only the loading branch of the UC data, the path of a rock sample to failure can be followed by plotting the measured axial and radial strains versus the applied axial stress. For example, the graphs of axial stress versus axial strain and radial strain versus axial strain for the uniaxial compression test SS-TC-0 are displayed in the Figures 5.4a and b, respectively. We remark here that the high unconfined compressive strength exhibited by this sandstone is due to the high content of quartz (see Table 1). The data taken from primary loading loops are fitted by polynomials of the form

$$\sigma_a = a_1 x + a_2 x^2 + a_3 x^3 + \cdots,$$
$$1000 \cdot \varepsilon_r = b_1 x + b_2 x^2 + b_3 x^3 + \cdots, \qquad (5.17)$$
$$x = 1000 \cdot \varepsilon_a.$$

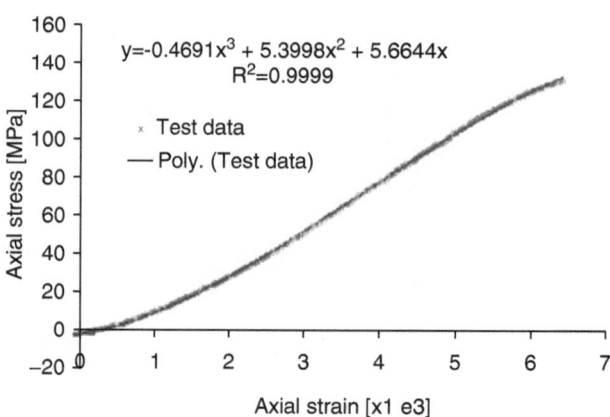

Fig. 5.4. Loading branches of (a) axial stress–axial strain and (b) radial strain–axial strain curves of Serena sandstone specimen SS-TC-0 in UC and fitted polynomial curves.

The nonlinearity of sandstone is manifested by the dependence of the tangent modulus of deformability and lateral strain factor on the applied stress. In fact, differentiating formulae (5.17) with respect to x or ε_a we obtain the following expression for the tangent moduli,

$$E_t = \frac{\partial \sigma_a}{\partial \varepsilon_a} = a_1 + 2a_2 x + 3a_3 x^2 + \cdots ,$$

$$\nu_t = -\frac{\partial \varepsilon_r}{\partial \varepsilon_a} = -b_1 - 2b_2 x - 3b_3 x^2 + \cdots . \qquad (5.18)$$

In the case of test SS-TC-0 six unloading–reloading cycles were performed before the peak stress at failure in order to infer its elastic properties. From the graphs displayed in Figures 5.5a,b it may be observed that the unloading–reloading curves corresponding to $\sigma_a - \varepsilon_a^{(el)}$ and to $\varepsilon_r^{(el)} - \varepsilon_a^{(el)}$ display nonlinearity and hysteresis. Neglecting hysteresis for the sake of simplicity, each of these loops is best-fitted by second-degree polynomials.

The recorded peak stresses during the four uniaxial and triaxial compression tests are plotted in Figure 5.6a in the form of Mohr circles; that is,

$$\sigma = \frac{\sigma_1 + \sigma_3}{2} + \frac{\sigma_1 - \sigma_3}{2} \cos 2\theta, \quad \tau = \frac{\sigma_1 - \sigma_3}{2} \sin 2\theta, \qquad (5.19)$$

wherein σ_1, σ_3 denote the principal stresses at failure (i.e., axial and confining, respectively) and θ is the angle subtended between the horizontal line and the outward normal to the plane in which the normal and shear stresses (σ, τ), respectively, act. According to the celebrated Mohr–Coulomb (MC) linear failure criterion (Jaeger and Cook, 1976),

$$|\tau| = c + \tan \varphi \sigma, \qquad (5.20)$$

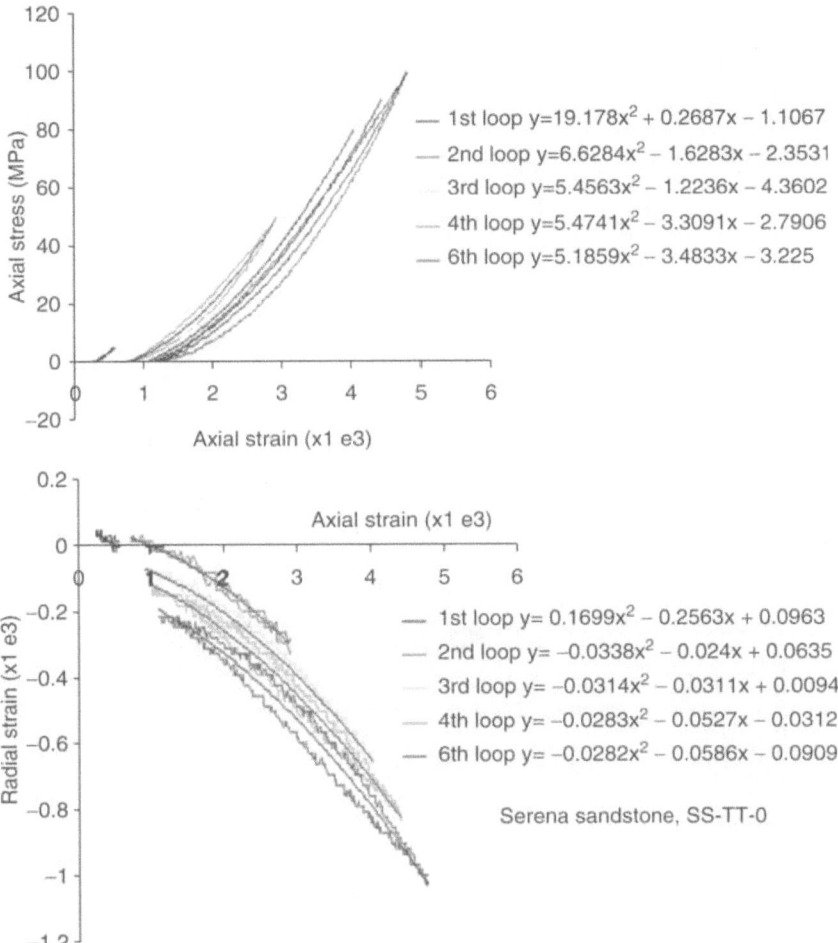

Serena sandstone, SS-TC-0

1st loop y=19.178x^2 + 0.2687x − 1.1067
2nd loop y=6.6284x^2 − 1.6283x − 2.3531
3rd loop y=5.4563x^2 − 1.2236x − 4.3602
4th loop y=5.4741x^2 − 3.3091x − 2.7906
6th loop y=5.1859x^2 − 3.4833x − 3.225

1st loop y= 0.1699x^2 − 0.2563x + 0.0963
2nd loop y= −0.0338x^2 − 0.024x + 0.0635
3rd loop y= −0.0314x^2 − 0.0311x + 0.0094
4th loop y= −0.0283x^2 − 0.0527x − 0.0312
6th loop y= −0.0282x^2 − 0.0586x − 0.0909

Serena sandstone, SS-TT-0

Fig. 5.5. Unloading–reloading loops for Serena sandstone in UC: (a) $\sigma_a - \varepsilon_a^{(el)}$, (b) $\varepsilon_r^{(el)} - \varepsilon_a^{(el)}$.

where the cohesion c and the internal friction angle φ of Serena sandstone are derived by passing a straight line that is tangent to all Mohr circles (e.g., Figure 5.6a). The values of these properties have been found to be c = 23.5 MPa and $\varphi = 53°$. The photos in Figure 5.6b illustrate the failure modes exhibited by three sandstone specimens subjected to different confining pressures. The high friction angle of the sandstone is manifested with the low angle subtended between the vertical axis and the shear crack exhibited by the specimens at the moment of failure[2] (Figure 5.6b).

---

[2] As is well known this angle denoted by the symbol β is given by $\beta = \pi/4 - \varphi/2$.

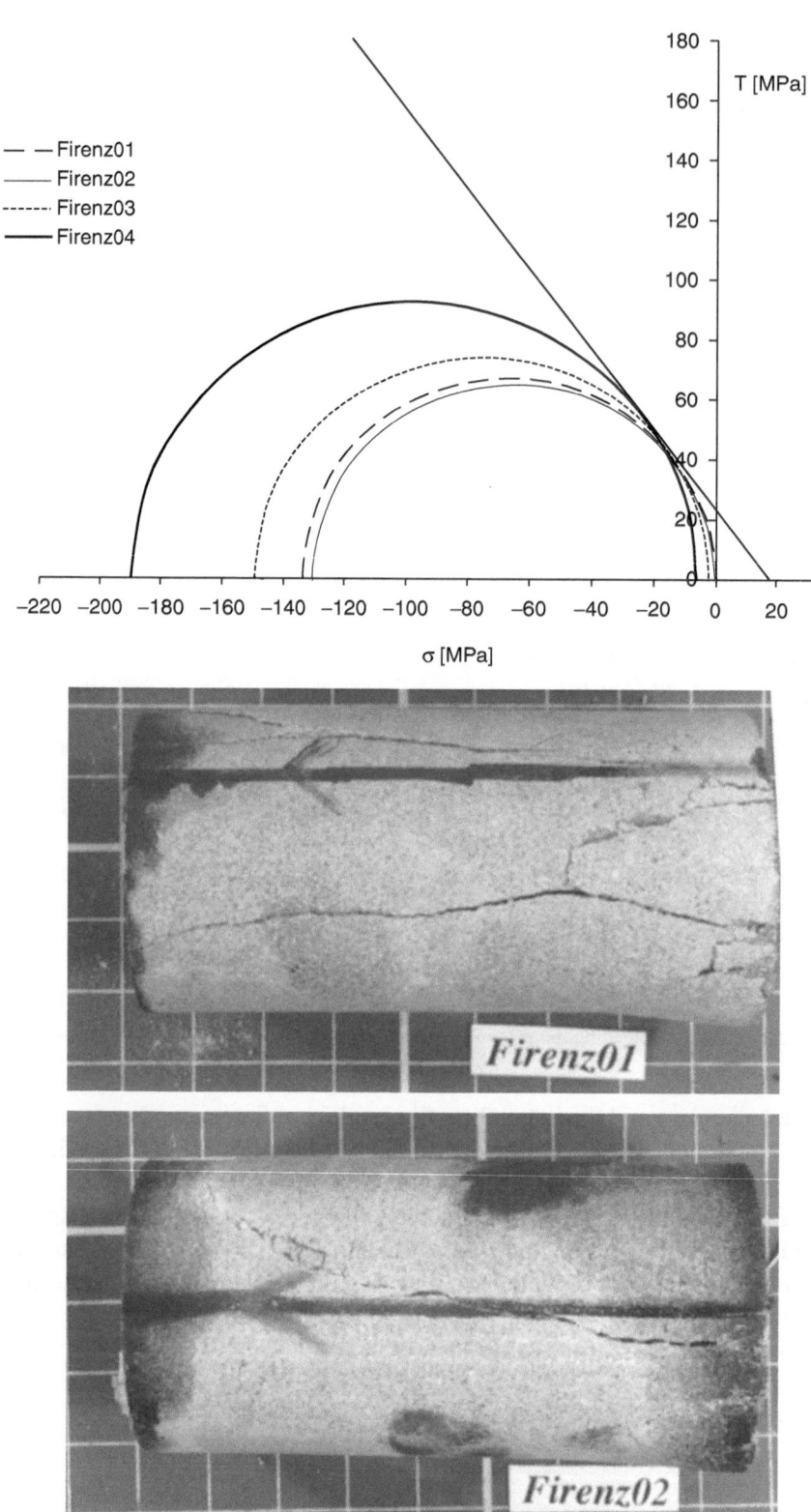

Fig. 5.6. Continued.

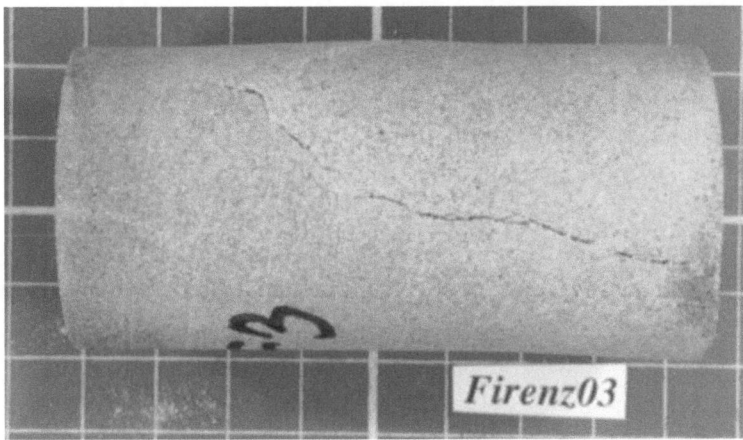

Fig. 5.6. (a) Mohr circles and fitted linear Mohr–Coulomb failure envelope; (b) photos of sandstone cylindrical specimens broken in uniaxial and triaxial compression tests (courtesy of E. Papamihos).

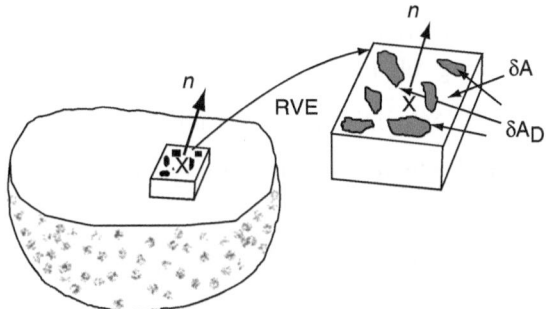

Fig. 5.7. Representative Elementary Volume (REV) of damaged rock.

4. Calibration of a Nonlinear Model on Experimental Data

4.1 A Hypoelastic-Damage Model for Sandstone

In this section a new hypoelastic theory for intact rocks is developed to account for this stress dependency of both elastic moduli of Serena sandstone found in the experiments.

Elasticity of intact rocks is determined by the elastic stiffness of the uncracked rock and the geometry (density and orientation) of microcracks. It may be assumed that the geometry of microcracks—which may be approximated in any plane by the area of intersections of cracks with that plane—can be modeled through a continuum variable at the mesoscale (i.e., grain scale). In order to manipulate a dimensionless quantity the crack area δA_D is scaled with the size of the area of the Representative Elementary Volume (RVE). This size is of primary importance in the definition of a continuous variable in the sense of continuum mechanics. This continuum damage variable is similar to the plastic strain of classical plasticity that at a given point represents the average of many grain slips.

If the area δA with outward unit normal n_j of the RVE with position vector x_i of Figure 5.7 is loaded by a force δF_i the usual apparent traction vector $\sigma_i = \sigma_{ij} n_j$ is

$$\sigma_i = \lim_{\delta A \to S} \frac{\delta F_i}{\delta A}, \quad i = 1, 2, 3, \tag{5.21}$$

where S is the representative area of the intact rock. The value of the dimensionless damage quantity $D(n_i, x_i)$ may be defined as follows,

$$D = \frac{\delta A_D}{\delta A}. \tag{5.22}$$

At this point we may introduce an effective traction vector $\sigma_i^{(e)}$ that is related to the surface that effectively resists the load, namely,

$$\sigma_i^{(e)} = \lim_{\delta A \to S} \frac{\delta F_i}{\delta A - \delta A_D}, \quad i = 1, 2, 3. \tag{5.23}$$

From relations (5.21)–(5.23) it follows that

$$\sigma_i^{(e)} = \frac{\sigma_i}{1 - D}, \quad i = 1, 2, 3. \tag{5.24}$$

According to the above definitions the elastic deformation of the intact rock can be described with the following relations.

- The relation $\sigma_{ij}^{(e)} - \varepsilon_{ij}^{(el)}$ which is obtained from elasticity
- The relation $\sigma_i - \sigma_i^{(e)}$ which is obtained by employing the concept of damage (Lemaitre, 1992).

It is convenient to decompose the stress tensor σ_{ij} into deviatoric and hydrostatic parts as follows,

$$\sigma_{ij} = s_{ij} + p\delta_{ij}, \tag{5.25}$$

wherein s_{ij} denotes the stress deviator and $p = \sigma_{kk}/3$ is the mean pressure. Furthermore, we introduce the stress invariants

$$I_{1\sigma} = \sigma_{kk}, \quad J_{2s} = \frac{1}{2}s_{ij}s_{ji}, \tag{5.26}$$

wherein $I_{1\sigma}$ is the first invariant of the spherical stress tensor and J_{2s} is the second invariant of the deviatoric stress tensor. The generalization of damage theory in three dimensions may be performed by assuming that microcracks and pores reduce[3] the apparent distortional and hydrostatic intensities of the stress tensor according to the relations

$$I_{1\sigma} = (1 - D_s) \cdot I_{1\sigma}^{(e)}, \quad T = (1 - D_c) \cdot T^{(e)}, \tag{5.27}$$

---

[3] It may be noted here that a general theory must allow for both enhancement and degradation of material properties due to mechanical loads. The former case corresponds to negative damage measures and describes pore and microcrack closure (healing) due to hydrostatic pressure and the latter corresponds to positive damage measures and describes microcrack opening and propagation. Both degradation and enhancement of the properties of a solid may be embraced under the term "material divagation" that is used to describe processes where the mechanical properties of a material change in time or wander from the values that characterize the material in a reference configuration. In general, divagation can result from any thermal, mechanical, chemical, or electrical process experienced by the material.

where $T = \sqrt{J_{2s}}$ denotes the deviatoric shearing stress intensity. In the above relations it is assumed that the scalar damage variables D_s, D_c are the spherical (hydrostatic) and distortional intensities of the damage tensor D_{ij}, respectively. From relations (5.25) and (5.27) the apparent versus the effective stress tensor relationship may be obtained:

$$\sigma_{ij} = (1 - D_c) \cdot \sigma_{ij}^{(e)} + \frac{1}{3} \cdot (D_c - D_s) \cdot \delta_{ij} \cdot \sigma_{kk}^{(e)}. \tag{5.28}$$

Next, we recall the finite-elasticity equations for the volumetric and deviatoric strains

$$\varepsilon_{kk}^{(el)} = \frac{p}{K_s} \; ; \; e_{ij}^{(el)} = \frac{s_{ij}}{2G_s} \tag{5.29}$$

and

$$e_{ij}^{(el)} = \varepsilon_{ij}^{(el)} - \frac{1}{3}\delta_{ij}\varepsilon_{kk}^{(el)}, \tag{5.30}$$

where K_s is the secant bulk modulus and G_s is the secant shear modulus of the rock material that are related to the secant Young modulus E_s and Poisson ratio ν_s through the formulae

$$K_s = \frac{E_s}{3(1 - 2\nu_s)}, \quad G_s = \frac{E_s}{2(1 + \nu_s)}. \tag{5.31}$$

Alternatively, we may also find the relations,

$$\nu_s = \frac{3K_s - 2G_s}{2(3K_s + G_s)}, \quad E_s = \frac{9\,K_s\,G_s}{3K_s + G_s}. \tag{5.32}$$

Equations (5.27) and (5.29) lead to the following relations,

$$K_s = K_0\,(1 - D_s), \quad G_s = G_0\,(1 - D_c). \tag{5.33}$$

From (5.29) and (5.30) one may derive rate-type elasticity (hypoelasticity) equations that are obtained through formal material time differentiation

$$\dot{p} = K_s\dot{\varepsilon}_{kk}^{(el)} + p\frac{\dot{K}_s}{K_s}; \quad \dot{s}_{ij} = 2G_s\dot{e}_{kk}^{(el)} + s_{ij}\frac{\dot{G}_s}{G_s}, \tag{5.34}$$

where $e_{ij}^{(el)}$ denotes the elastic strain deviator, and $\varepsilon_{kk}^{(el)}$ is the elastic volumetric strain. By recourse to formal differentiation of formulae (5.38)

$$\frac{\dot{K}_s}{K_s} = -\frac{\dot{D}_s}{1 - D_s}; \quad \frac{\dot{G}_s}{G_s} = -\frac{\dot{D}_c}{1 - D_c} \tag{5.35}$$

and relations (5.28) we extract the relation between the apparent stress and the elastic strain increments

$$\dot{\sigma}_{ij} = 2G_s\dot{\varepsilon}_{ij}^{(el)} + \left(K_s - \frac{2}{3}G_s\right)\delta_{ij}\,\dot{\varepsilon}_{kk}^{(el)} - s_{ij}\frac{\dot{D}_c}{1 - D_c} - p\,\delta_{ij}\frac{\dot{D}_s}{1 - D_s}. \tag{5.36}$$

The above incremental expression may be set into the equivalent compact form

$$\dot{\sigma}_{ij} = C_{ijkl}^{(el)}\dot{\varepsilon}_{kl}^{(el)}, \tag{5.37}$$

where $C_{ijkl}^{(el)}$ is the "tangent elastic stiffness matrix" that according to the above damage model is now an anisotropic fourth rank tensor.

4.2 A Plasticity Model for Sandstone

Herein it is assumed that the "isotropic hardening rule" of rocks in compression is the same as that in tension. That is, past the initial yield state, friction is mobilized and increases as a function of plastic shear strain until it reaches saturation at some peak value. This friction-hardening phase is consequently described as an isotropic hardening phase as shown in Figure 5.8 [state i (initial yield) to state f (failure)] in which q is the strength of the uncracked rock matter (it is called *tensile limit* of the material).

The yield curves for this model are linear with their slopes to be steeper than the initial i yield curve. This is expected because the mean orientation of the active cracks $f(g^{(pl)})$ is changing as the rock proceeds from initial yield to failure according to the rough model of Figure 5.9. Thus $f\left(g^{(pl)}\right)$ represents a stress orientation coefficient in terms of fracture mechanics or a friction coefficient in terms of MC yield criterion.

For the calibration of the MC yield surface based on the UC test results:

$$
\begin{aligned}
F = \sqrt{J_{2s}} &\left[\sin\left(\alpha_{so} + \frac{\pi}{3}\right) + \frac{1}{\sqrt{3}} \cos\left(\alpha_{so} + \frac{\pi}{3}\right) \sin\varphi_m \right] \\
&- \left(q - \frac{1}{3}I_{1\sigma}\right) \sin\varphi_m = 0,
\end{aligned}
\tag{5.38}
$$

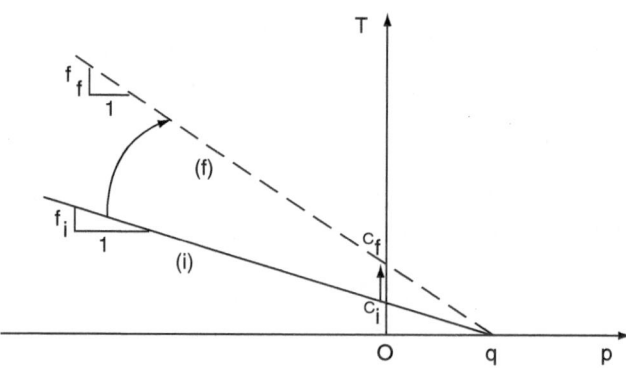

Fig. 5.8. Motion of the yield surface in $(T-p)$ stress space. $(i-f)$ isotropic friction hardening phase with constant q.

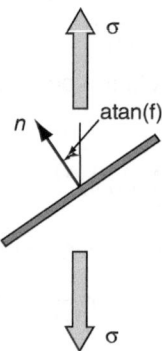

Fig. 5.9. Physical meaning of the stress inclination parameter f, that is, the angle subtended between the outward unit normal vector n on a straight microcrack with the tensile (or compressive) stress (σ) axis.

where α_{so} is the stress invariant angle of similarity (i.e., the third stress invariant), we set $\alpha_{so} = \pi/3$ (Chen and Han, 1988) which means

$$F = 0 \quad \Leftrightarrow \quad T\left(\frac{\sqrt{3}}{2} - \frac{1}{2\sqrt{3}}\sin\varphi_m\right) - (q - p)\sin\varphi_m = 0, \qquad (5.39)$$

with the mean normal stress defined as follows (compressive stresses assumed here negative)

$$p = \frac{1}{3}(2\sigma_r + \sigma_z) \quad \Rightarrow \quad p = \frac{1}{3}(-\sigma_a). \qquad (5.40)$$

The friction coefficient is denoted by the symbol f_c and is defined by the ratio

$$F = 0 \quad \Rightarrow \quad f_c = \frac{T}{q - p}. \qquad (5.41)$$

The mobilized internal friction angle ϕ_m and the mobilized cohesion c_m for the constant q MC-model read as follows

$$\sin\phi_m = \frac{3 f_c}{2\sqrt{3} + f_c}, \quad c_m = q\tan\phi_m. \qquad (5.42)$$

4.3 Constitutive Elastoplastic Model of Serena Sandstone

Returning to the sandstone UC test, in the uniaxial compression case we have (assuming compressive stresses as positive)

$$p = \frac{1}{3}\sigma_a, \quad s_a = \frac{2}{3}\sigma_a, \quad s_r = -\frac{1}{3}\sigma_a \qquad (5.43)$$

and the incremental stress–strain relations (5.36) take the form

$$\begin{aligned}
d\sigma_a &= 2G_s d\varepsilon_a^{(el)} + \left(K_s - \tfrac{2}{3}G_s\right)\left(d\varepsilon_a^{(el)} + 2d\varepsilon_r^{(el)}\right) \\
&\quad - s_a \frac{\dot{D}_c}{1-D_c} - p\frac{\dot{D}_s}{1-D_s}, \\
0 &= 2G_s d\varepsilon_r^{(el)} + \left(K_s - \tfrac{2}{3}G_s\right)\left(d\varepsilon_a^{(el)} + 2d\varepsilon_r^{(el)}\right) \\
&\quad - s_r \frac{\dot{D}_c}{1-D_c} - p\frac{\dot{D}_s}{1-D_s}.
\end{aligned} \qquad (5.44)$$

The empirical relations of bulk and shear secant moduli of Serena sandstone are derived from the unloading–reloading test data and the above relations (5.44) (Figure 5.10a). The two "enhancing" functions at hand may be approximated in first order for every loop by the linear relations

$$\begin{aligned}
D_s\left(\varepsilon_a^{(pl)}\right) &= -41.5\left(\varepsilon_a^{(el)} - \varepsilon_a^{(pl)}\right), \\
D_c\left(\varepsilon_a^{(pl)}\right) &= -30\left(\varepsilon_a^{(el)} - \varepsilon_a^{(pl)}\right).
\end{aligned} \qquad (5.45)$$

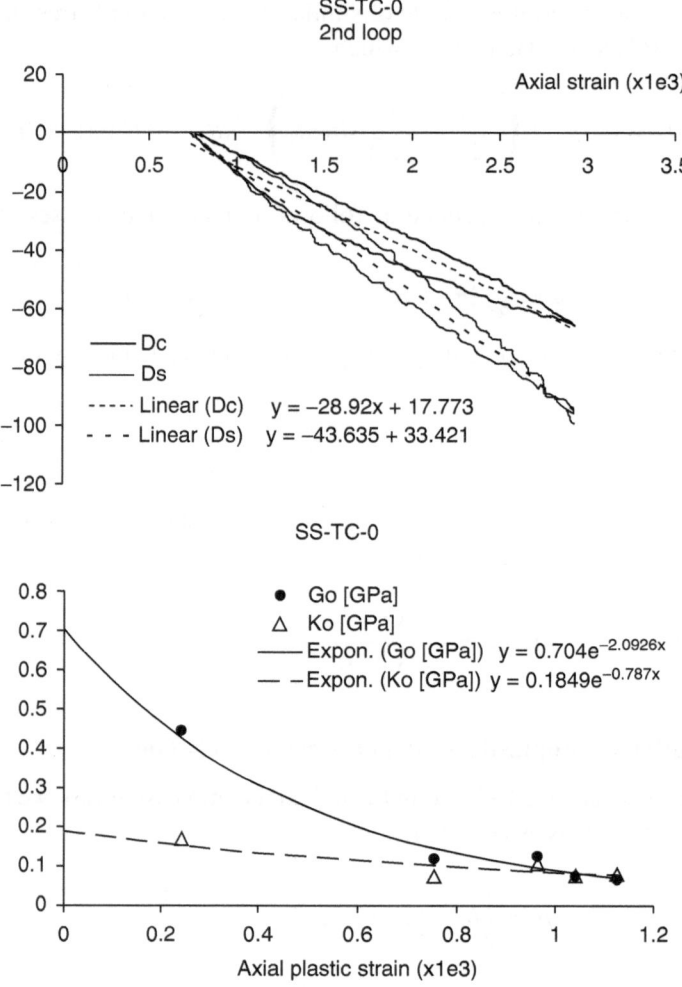

Fig. 5.10. (a) Dependence of damage scalar variables of sandstone on axial strain in the second loop, and (b) best exponential fit dependence of initial elastic moduli of each unloading–reloading loop on the axial plastic strain.

Then based on the above damage theory the moduli of sandstone were expressed as a function of the damage or enhancing functions as follows

$$K_s = K_0 \left(\varepsilon_a^{(pl)} \right) \left[1 - D_s \left(\varepsilon_a^{(pl)} \right) \right],$$
$$G_s = G_0 \left(\varepsilon_a^{(pl)} \right) \left[1 - D_c \left(\varepsilon_a^{(pl)} \right) \right],$$

(5.46)

where the initial elastic moduli were found to be negative exponential functions of the axial plastic strain (Figure 5.10b),

$$K_0 = 0.185 \, e^{-0.787 \, \varepsilon_a^{(pl)}},$$
$$G_0 = 0.704 \, e^{-2.093 \, \varepsilon_a^{(pl)}}.$$

(5.47)

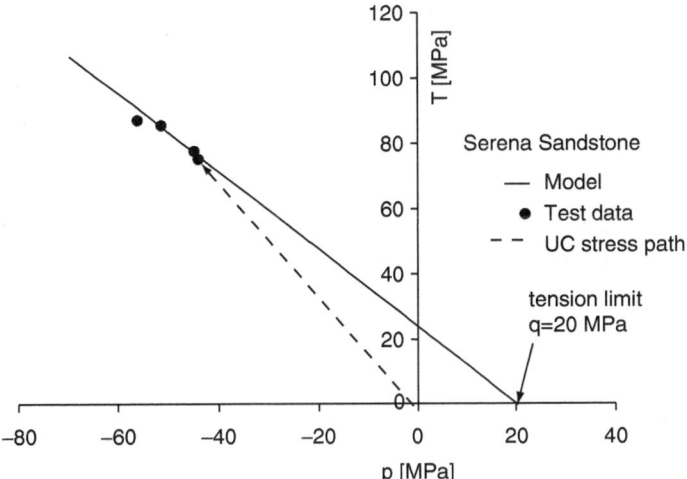

Fig. 5.11. Determination of the tension limit by best fitting a straight line on the uniaxial and triaxial compression test data at failure in the (p, T) stress space and UC stress path.

It is worth noticing that the above relations indicate that enhancement is linked or coupled with the plasticity exhibited by the sandstone.

The tension limit q of this type of the sandstone is found by fitting a straight line on the uniaxial and triaxial compression test data in the p-P space, as it is displayed in Figure 5.11. We can evaluate at each point of the stress–strain curves the plastic strains and plot the friction coefficient, the mobilized friction angle, and the mobilized cohesion as a function of the plastic shear strain intensity. The latter quantity is calculated as follows,

$$\dot{g} = \dot{g}^{(el)} + \dot{g}^{(pl)} = \frac{\dot{T}}{G} + \dot{g}^{(pl)} \quad \Rightarrow \quad \dot{g}^{(pl)} = \dot{g} - \frac{\dot{T}}{G}. \qquad (5.48)$$

For the flow rule, by assuming coaxiality of stresses and strains, we employ a MC expression of the form

$$Q = T \left(\frac{\sqrt{3}}{2} - \frac{1}{2\sqrt{3}} \sin \psi_m \right) + p \, \sin \psi_m = 0. \qquad (5.49)$$

Hence, the mobilized dilatancy angle ψ_m is calculated from the following relationship,

$$\sin \psi_m = \frac{3d}{2\sqrt{3} + d}. \qquad (5.50)$$

An algorithm has been constructed based on the set of equations (5.44)–(5.47) describing the elasticity of the rock, as well as the set of equations (5.7)–(5.14), (5.17), (5.18), (5.38)–(5.42) and (5.48)–(5.50) describing its plastic behavior, in order to calculate the dependence of basic mechanical parameters on the amount of plastic shear strain intensity that is used as a load parameter. Figure 5.12 displays the dilation response, whereas Figure 5.13 shows the typical variation of the mobilized friction and

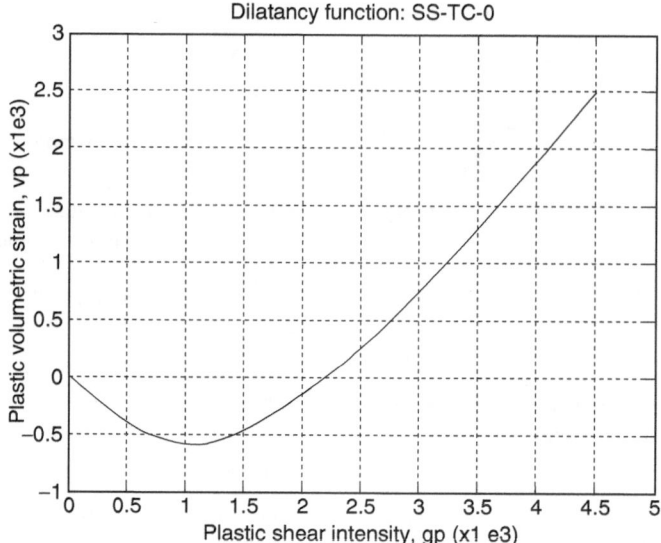

Fig. 5.12. Dilation response of Serena sandstone specimen SS-TC-0 in UC.

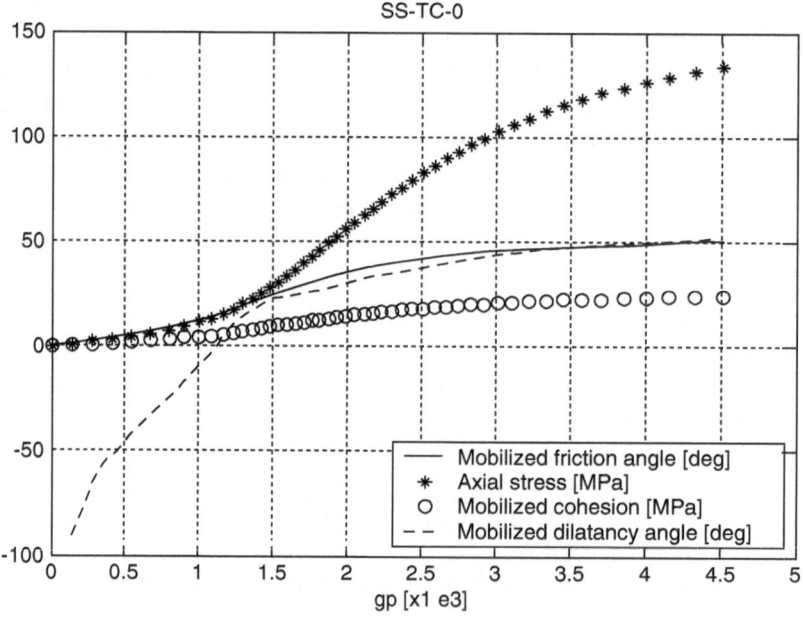

Fig. 5.13. Plots of mobilized friction and dilatancy angles, axial stress, and mobilized cohesion of Serena sandstone specimen SS-TC-0 in UC. The convention of compression positive is assumed.

dilatancy angles, cohesion, and axial stress as functions of the plastic shear strain intensity for the Serena sandstone, that are predicted by the assumed elastoplastic model. From the latter plot it may be seen that after some plasticity is developed in the specimen, the mobilized friction and dilatancy angles coincide, indicating an associated flow rule. Also, as was expected, the peak cohesion and peak internal friction angle predicted by this model are in good agreement with the respective values of these strength properties derived from the linear MC failure envelope (e.g., Figure 5.6a.)

The predictability of the general model proposed above may be validated in a further step against additional test data (e.g., from the triaxial tests and tension tests or others). This validation procedure will reveal the weaknesses of the model that must be further elaborated in an iterative manner, until we obtain its most general applicability.

5. Summary and Conclusions

Carefully designed simple rock mechanics experiments performed on sandstone revealed their nonlinearity both in the elastic and plastic regimes. This necessitated the formulation of a nonlinear theory based on elasticity, damage mechanics, and plasticity theories. In a subsequent stage this theory was calibrated applying triaxial compression tests on Serena sandstone. Future work will include validation of the proposed model against more element or structural tests in the tensile and compressive regimes on the same type of stone. The hysteresis displayed by the sandstone was not considered in this first attempt. This is a topic of major interest that may be attacked by virtue of the theory of fracture mechanics in the near future. Fracture mechanics may also be used as a powerful tool to describe (a) the brittleness displayed by the rocks (i.e., their approximate tenfold decrease in compressive strength properties when they are subjected to tensile loads), and (b) the size effect that is manifested by the considerable reduction of their strength with the increase of the size of the structure. That is to say, modern rock mechanics should rely upon all the up-to-date developments of the fundamental theories of elasticity, strength of materials, damage mechanics, plasticity, and fracture mechanics in order to present robust models for rocks subjected to static or to dynamic loads. The linking of all these theories under the umbrella of a unique theory capable of describing the rock mechanical behavior at all scales (i.e., in the range 1e-6 m—1e 4 m) and boundary conditions encountered in praxis, is one of the fascinating challenges of the future.

Acknowledgments

This work is a result of research supported by funds of the MCDUR Project (G6RD-CT-2000-00266) and DIAS Project (DIAS-EVK4-CT-2002-00080) of the European Union.

References

Chen, W.F. and Han, D.J., 1988, *Plasticity for Structural Engineers*. Springer-Verlag, New York.
Frederick, D. and Chang, T.S., 1972, *Continuum Mechanics*, Scientific, Cambridge.
Guyer, R. and Johnson, P., 1999, Nonlinear mesoscopic elasticity: Evidence for a new class of materials, *Phys. Today*, 30–36.
Hill, R., 1950, *The Mathematical Theory of Plasticity*, Clarendon Press, Oxford.
Jaeger, J.C. and Cook, N.G.W., 1976, *Fundamentals of Rock Mechanics*, Chapman and Hall, London.

Johnson, P.A., 2005, Nonequilibrium non-linear dynamics in solids: State of the art in methods and applications, this volume.

Kachanov, L.M., 1974, *Fundamentals of the Theory of Plasticity*, MIR, Moscow.

Lemaitre, J., 1992, *A Course on Damage Mechanics* Springer-Verlag, Berlin.

Truesdell, C. and Noll, W., 1965, *The Non-Linear Field Theories of Mechanics*, Springer-Verlag, Berlin.

Vardoulakis, I. and Sulem, J., 1995, *Bifurcation Analysis in Geomechanics*, Blackie Academic & Professional, Berlin.

Vardoulakis, I., Kourkoulis, S.K. and Exadaktylos, G.E., 1998, Elasticity of marble, In *Recent Advances in Mechanics*, Kounadis A.N. & Gdoutos E.E. eds., Xanthi, Greece, July 10-12.

6

Universality in Nonlinear Structural Dynamics

Alberto Carpinteri and Nicola Pugno[1]

Department of Structural Engineering and Geotechnics, Politecnico di Torino, Corso Duca degli Abruzzi 24, 10129, Torino, Italy.
[1]To whom correspondence should be addressed: nicola.pugno@polito.it

Abstract

The aim of this chapter is the investigation of the universal behavior of nonlinear vibrating structures. The approach is formulated in terms of the black box interaction formalism, only recently introduced (see Chapter 1), but including subharmonics and multidegrees of freedom. As a prototype of the nonlinear box, we focus attention on a multicracked cantilever beam. The cause of the vibration (i.e., the input in the box) is represented by a harmonic force excitation; the effect (i.e., the output), by the tip displacement. Universality corresponds to zero-, high-, and subharmonic generations, describing complex phenomena such as period doublings and transition towards deterministic chaos. Applications to damage detection and structural monitoring seem to be promising.

Keywords: Chaos, complexity, cracks, dynamics, nonlinear, universality

1. Introduction

The study of the nonlinear dynamics of structures represents a powerful tool for damage detection. Vibration-based inspection of structural behavior offers an effective tool of nondestructive monitoring. The analysis of the dynamic response of a structure to excitation forces and the monitoring of alterations, which may occur during its lifetime, can be employed as a global integrity-assessment technique to detect, for example, the presence of a crack or play in joints. The damage assessment problem in cracked structures has been extensively studied in the last decade [1–5], highlighting that the vibration-based inspection is a valid method to detect, localize and quantify cracks especially in one-dimensional structures. Dealing with the presence of a crack in a beam, previous studies have demonstrated that a transverse crack can change its state (from open to closed and vice versa) when the structure, subjected to an external load, vibrates [3, 4]. As a consequence, a nonlinear dynamic behavior is introduced.

The aim of this chapter is the investigation of the universal behavior in the complex oscillatory behavior of damaged nonlinear structures. In particular, we have focused our attention on a cantilever beam with several breathing transverse cracks and subjected to harmonic excitation perpendicularly to its axis. The method, that is an extension of the high-harmonic analysis presented in [4] to subharmonic and zero-frequency components, allows us to capture the complex behavior of the nonlinear structure, for example, the occurrence of period doubling, as experimentally observed [5]. The approach is written in terms of the black box interaction formalism, recently developed for high-harmonics (see [6] and Chapter 1), but including subharmonics and multidegrees of freedom, as described in References [7,8].

2. Nonlinear Dynamics of Structures

2.1 The Interaction Box Formalism

Consider a nonlinear structure, having several degrees of freedom and subjected to the multicomponent cause $\{C\}$ of the vibration, that is, a set of harmonic forces/couples with multiple angular frequencies rather than a fundamental angular frequency ω. The effect $\{E\}$, dual to the cause $\{C\}$, that is, the structural displacements (translations and rotations), must satisfy:

$$[M]\{\ddot{E}\} + [D]\{\dot{E}\} + [K]\{E\} + \{B\,(\{E\})\} = \{C\}, \qquad (6.1)$$

where $[M]$, $[D]$, and $[K]$ represent the mass, damping, and stiffness matrices respectively, and $\{B\}$ is the nonlinear component of the box (or structure) [6]. See the Appendix. For free vibrations $\{C\} = \{0\}$.

In general the cause $\{C\}$ can be put in the following form,

$$\{C\} = \sum_{j=0}^{N} \left(\{C_S\}_j \sin j\,\omega t + \{C_C\}_j \cos j\,\omega t\right). \qquad (6.2)$$

Assuming as the period of the effect a multiple s of the period of the cause (usually $s = 1$), and according to Fourier analysis, we can write:

$$\{E\,(t)\} = \sum_{j=0}^{N} \left(\{E_S\}_j \sin j\frac{\omega}{s}t + \{E_C\}_j \cos j\frac{\omega}{s}t\right), \qquad (6.3)$$

in which an s different from the unity parameter describes subharmonic generation ([7, 8]; $s = 1$ in [4] and [6]) and N should be large enough (theoretically infinite) to reach a good approximation. Introducing the time dependence for $\{E\,(t)\}$ of Eq. (6.3) into the nonlinear box part $\{B\,(E)\}$ yields:

$$\{B\,(E\,(t))\} = \sum_{j=0}^{N} \left(\{B_S\}_j \sin j\frac{\omega}{s}t + \{B_C\}_j \cos j\frac{\omega}{s}t\right), \qquad (6.4)$$

where $\{B_{S,C}\}$ are constants related to $\{E_{S,C}\}$.

Introducing Eqs. (6.2–6.4) into Eq. (6.1) and balancing the harmonics with the same angular frequency, would formally solve the problem, correlating cause and effect. A algebraic system of nonlinear equations is derived in the form of

$$\begin{bmatrix} [K] - \dfrac{j^2\omega^2}{s^2}[M] & -\dfrac{j\omega}{s}[D] \\ \dfrac{j\omega}{s}[D] & [K] - \dfrac{j^2\omega^2}{s^2}[M] \end{bmatrix} \begin{Bmatrix} \{E_S\}_j \\ \{E_C\}_j \end{Bmatrix} = \begin{Bmatrix} \{C_S\}_j \\ \{C_C\}_j \end{Bmatrix} - \begin{Bmatrix} \{B_S(\{E_S\},\{E_C\})\}_j \\ \{B_C(\{E_S\},\{E_C\})\}_j \end{Bmatrix},$$

(6.5a)

or, in compact form:

$$[A(j)]\{E(j)\} = \{C(j)\} - \{B(j)\},$$

(6.5b)

where $j = 0, 1, \ldots, N$.

For a monochromatic single cause:

$$C_{ij} = C\delta_{js}\delta_{ip},$$

(6.6)

p being the node position corresponding to the point where the sinusoidal cause of intensity C is applied.

Each of the N systems in Eqs. (6.5) can be easily solved numerically using an iterative procedure, starting assuming $\{B(j)\} = \{0\}$ and then evaluating $\{B(j)\}$ according to the solutions for $\{E(j)\}$ derived at the previous step, until a satisfactory convergence is reached.

2.2 Application to Cracked Structures

To quantify universal behaviors in nonlinear dynamics of structures, we refer to a multicracked beam. The cracks "breathe" during the vibration and thus cause a variation of the stiffness of the structure, that is, a nonlinearity. Mathematical details are reported in [4, 7] and briefly summarized in the appendix. We consider here just two different and simple cases: a weakly or a strongly damaged structure. An extensive parametrical investigation can be found in [8]. Only in the latter case, the so-called period doubling phenomenon, experimentally observed by Brandon and Sudraud [5], clearly appears. The beam considered here is the same as that described in the mentioned experimental analysis. It is 270 mm long and has a transversal rectangular cross-section of base and height, respectively, of 13 and 5 mm. The material is UHMW-ethylene, with a Young's modulus of 8.61×10^8 N/m^2 and a density of 935 kg/m^3. We have assumed a modal damping of 0.002. It is discretized with 20 finite elements. We have found that values of $s = 4$ and $N = 16$ give a good approximation; that is, for larger values of s and N, substantially coincident solutions are obtained. The first natural frequency of the undamaged structure is $f_u = 10.6$ Hz. A monochromatic cause, a force at the tip, is considered. The effect is assumed to be the tip displacement.

For each of the two considered structures (Figures 6.1a and 6.2a) the time history of the applied force and of the free-end displacement (Figures 6.1b and 6.2b), are shown as well as the zero-, high-, and subharmonic components of the free-end displacement (Figures 6.1c and 6.2c). Obviously, in a hypothetical linear (i.e., here undamaged) structure, the response is linear by definition with only one harmonic component at the same frequency of the monochromatic excitation (Case 0).

Case 1. In the weakly nonlinear structure of Figure 6.1a, the response converges and it appears only weakly nonlinear, as depicted in Figure 6.1b. The harmonic components in the structural response are the zero–one (presence of a negative offset in

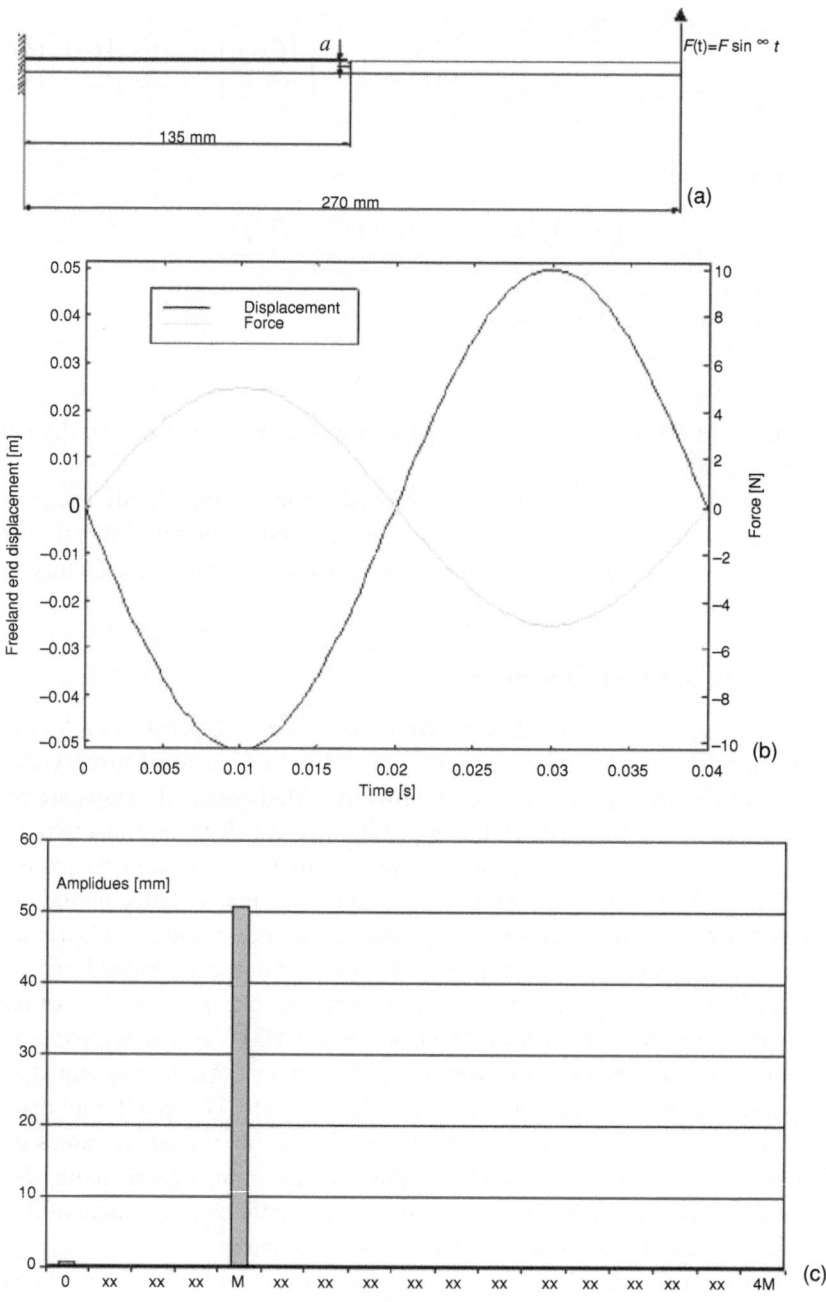

Fig. 6.1. (a) Structure I – Damaged structure and characteristics of the excitation ($a = 2.4\,\mathrm{mm}$, $C = F = 5N$, $f = \omega/2\pi = 25\,\mathrm{Hz}$); (b) time history of the free end displacement and of the applied force; (c) zero-(offset), sub-, and super-harmonic components for the free end displacement (i.e., $\sqrt{E^2_{S\,20\,j} + E^2_{C\,20\,j}}$ for $j = 0, 1, \ldots, 16$).

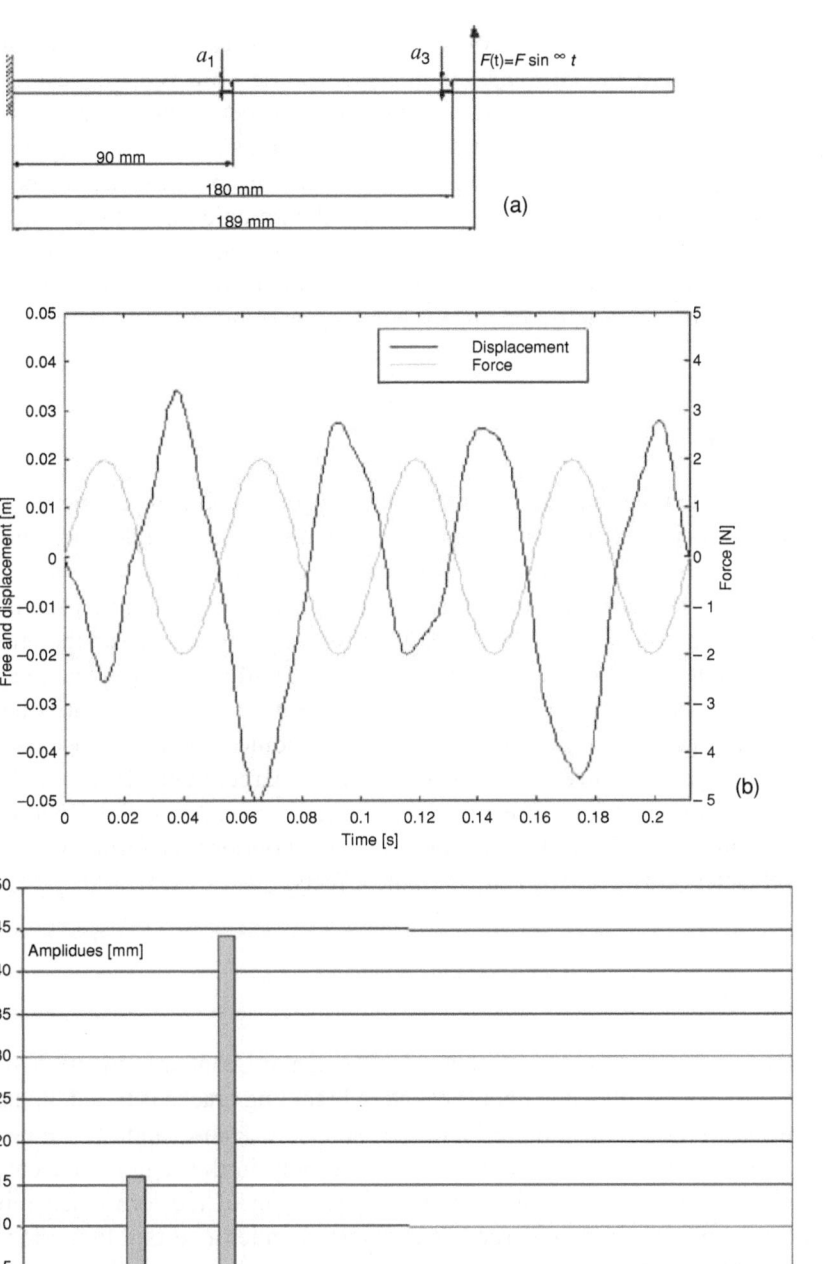

Fig. 6.2. (a) Structure II – Damaged structure and characteristics of the excitation ($a_1 = 4.25$ mm, $a_2 = 4.25$ mm, $C = F = 2N$, $f = \omega/2\pi = 18.9$ Hz); (b) time history of the free end displacement and of the applied force; (c) zero- (offset), sub-, and super-harmonic components for the free end displacement (i.e., $\sqrt{E^2_{S\ 20\ j} + E^2_{C\ 20\ j}}$ for $j = 0, 1, \ldots, 16$).

the displacement of the free-end, downwards in Figure 6.1a) and the superharmonic ones (Figure 6.1c). No subharmonic components can be observed, suggesting that a threshold value of the nonlinearity has to be reached for subharmonic generation.

Case 2. In the strongly nonlinear structure of Figure 6.2a the nonlinearity increases, as depicted in Figure 6.2b. The harmonic components in the structural response are the zero–one, the superharmonic as well as the subharmonic ones (Figure 6.2c). The threshold nonlinearity has been reached. It should be emphasized that a strong presence of the $\omega/2$ component causes the period doubling of the response. The free-end vibrates practically with a period doubled with respect to that of the excitation.

3. Conclusions

From the reported numerical examples (for an extensive numerical parametrical investigation see [8]) a universal behavior emerges. Obviously, for a single monochromatic cause, if the nonlinearity is zero, the effect can be caught with $N = s = 1$. On the other hand, if a weak nonlinearity is considered, only offset and high-harmonic components can be observed in the effect. As a consequence, for this case, the effect can be easily caught using classical Fourier series ($s = 1$) with $N > 1$ [and large enough, in the sense that effect $(N) \cong$ effect$(N' > N)$]. Moreover, if the nonlinearity becomes stronger, (i.e., larger than a given threshold), subharmonic generation appears. In this case, the effect can be deduced by using a parameter s larger than 1 [$s > 1$ and $N > 1$ large enough, in the sense that effect $(s, N) \cong$ effect$(s' > s, N' > N)$]. Theoretically, values of s tending to infinite (Fourier series become Fourier transforms) would allow one to catch deterministic chaos and aperiodic effects.

A. Appendix

The mathematical model used for the considered beam of Leonhard Euler (1707–1783) with several transverse one-side nonpropagating breathing cracks is based on the finite element model. According to the principle of Ademare Jean-Claude Barré de Saint-Vénant (1797–1886) the stress field is influenced only in the region adjacent to the crack. The element stiffness matrix, with the exception of the terms that represent the cracked element, may be regarded as unchanged under a certain limitation of the element size. The additional stress energy of a crack leads to a flexibility coefficient expressed by a stress intensity factor derived by means of the theorem of Carlo Alberto Castigliano (1847–1884) in the linear elastic regime.

Neglecting shear action (Euler beam), the strain energy of an element without a crack can be obtained as

$$W^{(0)} = \frac{1}{2EI} \int_0^l (M + Pz)^2 \mathrm{d}z = \frac{1}{2EI}\left(M^2 + P^2l^3/3 + MPl^2\right), \qquad \text{(A1)}$$

where E is the Young modulus of the material constituting the finite element; $I = bh^3/12$ is the moment of inertia of its cross-section, having base b and height h; and M and P are the bending moment and shear load acting at the ends of the finite element of length l. The additional energy due to the crack is:

$$W^{(1)} = b \int_0^a \left[\left(K_I^2(x) + K_{II}^2(x) \right) / E' + (1 + v) K_{III}^2(x)/E \right] dx, \quad \text{(A2)}$$

where $E' = E$ for plane stress, $E' = E/(1+v)$ for plane strain, and v is the Poisson ratio. $K_{I,II,III}$ are the stress intensity factors for opening, sliding, and tearing-type cracks, of depth a, respectively. Taking into account only bending (i.e., the predominant load):

$$W^{(1)} = b \int_0^a \left[(K_{IM}(x) + K_{IP}(x))^2 + K_{IIP}^2(x) \right] / E' dx, \quad \text{(A3)}$$

with:

$$K_{IM} = \left(6M \big/ bh^2 \right) \sqrt{\pi a} F_I(w)$$

$$K_{IP} = \left(3Pl \big/ bh^2 \right) \sqrt{\pi a} F_I(w) \quad \text{(A4)}$$

$$K_{IIP} = (P/bh) \sqrt{\pi a} F_{II}(w),$$

where $w = a/h$ and:

$$F_I(w) = \sqrt{2/(\pi w)} \tan(\pi w/2)(0.923 + 0.199(1 - \sin(\pi w/2)^4))/\cos(\pi w/2)$$
$$F_{II}(w) = (3w - 2w^2)(1.122 - 0.561w + 0.085w^2 + 0.18w^3)/\sqrt{1 - w}. \quad \text{(A5)}$$

The term $c_{ik}^{(0)}$ of the flexibility matrix $[C_e^{(0)}]$ for an element without a crack can be written as

$$c_{ik}^{(0)} = \frac{\partial^2 W^{(0)}}{\partial P_i \partial P_k} \quad i, k = 1, 2 \quad P_1 = P, \quad P_2 = M. \quad \text{(A6)}$$

The term $c_{ik}^{(1)}$ of the additional flexibility matrix $[C_e^{(1)}]$ due to the crack can be obtained as

$$c_{ik}^{(1)} = \frac{\partial^2 W^{(1)}}{\partial P_i \partial P_k} \quad i, k = 1, 2 \quad P_1 = P, \quad P_2 = M. \quad \text{(A7)}$$

The term c_{ik} of the total flexibility matrix $[C_e]$ for the damaged element is:

$$c_{ik} = c_{ik}^{(0)} + c_{ik}^{(1)}, \quad \text{(A8)}$$

From the equilibrium condition

$$\left(P_i \ M_i \ P_{i+1} \ M_{i+1} \right)^T = [T] \left(P_{i+1} \ M_{i+1} \right)^T, \quad \text{(A9)}$$

where

$$[T] = \begin{bmatrix} -1 & -l & 1 & 0 \\ 0 & -1 & 0 & 1 \end{bmatrix}^T .$$ (A10)

Applying the theorem of Enrico Betti (1823–1892), the stiffness matrix of the undamaged element can be written as

$$[K_e] = [T] \left[C_e^{(0)} \right]^{-1} [T]^T ,$$ (A11)

or

$$[K_e] = \frac{EI}{l^3} \begin{bmatrix} 12 & 6l & -12 & 6l \\ 6l & 4l^2 & -6l & 2l^2 \\ -12 & -6l & 12 & -6l \\ 6l & 2l^2 & -6l & 4l^2 \end{bmatrix} ,$$ (A12)

and the stiffness matrix of the cracked element may be derived as

$$[K_{de}] = [T][C_e]^{-1} [T]^T .$$ (A13)

In order to evaluate the dynamic response of the cracked beam when acted upon by an applied force, it is supposed that the crack does not affect the mass matrix. Therefore, for a single element, the mass matrix can be formulated directly:

$$[M_e] = [M_{de}] = \frac{ml}{420} \begin{bmatrix} 156 & 22l & 54 & -13l \\ 22l & 4l^2 & 13l & -3l^2 \\ 54 & 13l & 156 & -22l \\ -13l & -3l^2 & -22l & 4l^2 \end{bmatrix} ,$$ (A14)

where m is the mass for unity length of the beam.

Assuming that the damping matrix $[D]$ is not affected by the crack, it can be calculated through the inversion of the modeshape matrix $[\phi]$ relative to the undamaged structure:

$$[D] = \left([\phi]^T \right)^{-1} [d] [\phi]^{-1} ,$$ (A15)

where $[d]$ is the following matrix,

$$[d] = 2 \begin{bmatrix} \zeta_1 \omega_1 M_1 & 0 & 0 & \dots & 0 \\ 0 & \zeta_2 \omega_2 M_2 & 0 & \dots & 0 \\ \dots & \dots & \dots \dots & \dots \\ \dots & \dots & \dots \dots & \dots \\ 0 & \dots & \dots & 0 & \zeta_n \omega_n M_n \end{bmatrix} ,$$ (A16)

in which ζ_i is the modal damping ratio, ω_i is the ith natural frequency and M_i is the ith modal mass relative to the undamaged beam.

Accordingly, the mass, damping, and stiffness matrices of the structure are deduced by expansion and summation of the element matrices.

Regarding the nonlinearity imposed by the presence of the cracks:

$$\{B(E)\} = \sum_m \lfloor \Delta K^{(m)} \rfloor f^{(m)}(\{E\})\{E\},$$ (A17)

where $[K] + \sum_m [\Delta K^{(m)}]$ is the stiffness matrix of the undamaged beam and $[\Delta K^{(m)}]$ is half of the variation in stiffness introduced when the mth crack is fully open (see [7]). According to this notation, $f^{(m)}(\{E\})$ ranges between -1 and $+1$ and models the transition between the conditions of mth crack fully open and fully closed, depending on the curvature of the corresponding cracked element [4]. Considering the function $f^{(m)}(\{E\})$ as linear versus the curvature of the corresponding cracked element [4, 7] implies

$$f^{(m)}(\{E\}) = \frac{E_{m_k} - E_{m_h}}{\left| E_{m_k} - E_{m_h} \right|_{\max}} = \Lambda_m \left(E_{m_k} - E_{m_h} \right), \quad (A18)$$

where the numerator reports the difference of the rotations for the mth element. Correspondingly:

$$\{B(j)\} = \sum_m \begin{bmatrix} [\Delta K^{(m)}] & [0] \\ [0] & [\Delta K^{(m)}] \end{bmatrix} \left\{ \begin{Bmatrix} H_j^{(m)} \\ K_j^{(m)} \end{Bmatrix} \right\}, \quad (A19)$$

where

$$
\begin{aligned}
H_{ij}^{(m)} &= \frac{\Lambda_m}{2} \sum_{j_1, j_2 : j_1 + j_2 = j} \left\{ \left(E_{Sm_k j_1} - E_{Sm_h j_1} \right) E_{Ci j_2} \right. \\
&\quad \left. + \left(E_{Cm_k j_1} - E_{Cm_h j_1} \right) E_{Si j_2} \right\} + \frac{\Lambda_m}{2} \sum_{j_1, j_2 : j_1 - j_2 = \pm j} \\
&\quad \pm \left\{ \left(E_{Sm_k j_1} - E_{Sm_h j_1} \right) E_{Ci j_2} - \left(E_{Cm_k j_1} - E_{Cm_h j_1} \right) E_{Si j_2} \right\}, \quad \text{(A20a)}
\end{aligned}
$$

$$
\begin{aligned}
K_{ij}^{(m)} &= \frac{\Lambda_m}{2} \sum_{j_1, j_2 : j_1 + j_2 = j} \left\{ -\left(E_{Sm_k j_1} - E_{Sm_h j_1} \right) E_{Si j_2} \right. \\
&\quad \left. + \left(E_{Cm_k j_1} - E_{Cm_h j_1} \right) E_{Ci j_2} \right\} + \frac{\Lambda_m}{2} \sum_{j_1, j_2 : j_1 - j_2 = \pm j} \\
&\quad \pm \left\{ \left(E_{Sm_k j_1} - E_{Sm_h j_1} \right) E_{Si j_2} - \left(E_{Cm_k j_1} - E_{Cm_h j_1} \right) E_{Ci j_2} \right\}. \quad \text{(A20b)}
\end{aligned}
$$

References

[1] P. Gudmundson, The dynamic behaviour of slender structures with cross-sectional cracks, *J. Mech. Phys. Solids* **31**, 329–345 (1983).

[2] W. Ostachowicz and M. Krawczuk, Vibration analysis of a cracked beam, *Comput. Structures* **36**, 245–250 (1990).

[3] R. Ruotolo, C. Surace, P. Crespo, and D. Storer, Harmonic analysis of the vibrations of a cantilevered beam with a closing crack, *Comput. Structures* **61**, 1057–1074 (1996).

[4] N. Pugno, C. Surace, C. and R. Ruotolo, Evaluation of the non-linear dynamic response to harmonic excitation of a beam with several breathing cracks, *J. Sound Vibration* **235**, 749–762 (2000).

[5] J.A. Brandon and C. Sudraud, An experimental investigation into the topological stability of a cracked cantilever beam, *J. Sound Vibration* **211**, 555–569 (1998).

[6] S. Hirsekorn and P.P. Delsanto, On the universality of nonclassical nonlinear phenomena and their classification, *Appl. Phys. Lett.* **84**, 1413–1415 (2004).

[7] A. Carpinteri and Pugno N., Towards chaos in vibrating damaged structures—Part I: Theory and period doubling, *J. Appl. Mech.* **72**, 511–518 (2005).

[8] A Carpinteri and N. Pugno, Towards chaos in vibrating damaged structures—Part II: Parametrical investigation, *J. Appl. Mech.* **72**, 519–526 (2005).

7

The Evolutionary Advantage of Being Conservative: The Role of Hysteresis

Gian Piero Pescarmona

Department of Genetics, Biology and Biochemistry, University of Torino Via Santena 5 bis 10126 Torino, Italy.

Abstract

In the quest for universal laws underlying the behavior of living organisms it is necessary to remember that all forms of life are the result of an endless evolutionary process. Evolution itself (in an ever-changing environment) requires continuous cycles of random mutations and selection of the fittest to the new environment. Selection means that for each winner there will be many losers that will disappear. Competition for essential nutrients is the battlefield: the winner will be able to take up enough nutrients to duplicate itself. Selection is an all or none process. The single individual can prevail over the competitors and duplicate or not, but it cannot duplicate partially. The whole picture is full of uncertainty: the amount of energy required to duplicate depends on local factors such as nutrient concentration or the number and strength of the competitors. Due to the limited energy and nutrient availability, any living organism had to develop strategies to save energy and to afford duplication only in favorable conditions. The knowledge of the environment as a tool for successful competition is a scale-insensitive feature of living organisms. The evolutionary winner, the fittest, is the one that knew most about the environment. Hysteretic behavior, as a memory of the more recent past, may represent a significant advantage in the struggle for survival in our overcrowded environments. A selection of examples of hysteretic behaviors at the cellular and molecular levels and their chemical mechanisms are presented to support the hypothesis of universality of nonclassical nonlinear phenomena.

Keywords: Competition, enzymes, evolution, hysteresis, living systems, universality

1. An Evolutionary Approach to the Role of Hysteresis in Biological Systems

1.1 Introduction

Recently it has been claimed that a broad category of materials (rocks, soil, cement, concrete, damaged materials, etc.) share nonclassical nonlinear elastic behavior including stress–strain hysteresis and discrete memory in quasistatic experiments. These materials have in common a heterogeneous structure with soft "bond" elements, where the elastic nonlinearity originates, contained in hard matter (e.g., a rock sample). The bond system normally comprises a small fraction of the total material volume, and can be localized (e.g., a crack in a solid) or distributed (as in a rock). A model has been presented in which the soft elements are treated as hysteretic or reversible elastic units connected in a one-dimensional lattice to elastic elements (grains), which make up the hard matrix.[1] From the experimental point of view the hysteretic behavior of these materials has been evidenced by the anomalous elastic effects due to mechanical stresses and their dependence on temperature.[2] By modeling the structure as a mixture of grains and interstices it is possible to simulate to a good extent the behavior of these materials.[3]

Pre-eminent features of a universality class are its applicability to a whole class of objects with similar structure and its scale independence. In this chapter we argue in favor of the applicability of this class to living systems on the basis of their heterogeneous chemical structure (comparable to a mixture of grains and interstices) and some evolutionary considerations pointing to the great selective advantage of different forms of hysteretic (memory-carrying) behaviors.

1.2 The General Features of Living Objects

The main difference between living organisms and inanimate objects resides in the fact that living organisms are intrinsically fragile and have a finite lifespan. This apparent disadvantage is fully compensated by a corresponding advantage: fragility also means the opportunity to mutate their patterns of molecules (DNA or proteins); finite lifespan yields the ability of living organisms to reproduce themselves. The younger will replace the older, in some cases with a higher fitness to the environment, and therefore death will not necessarily decrease the size of the population, which will depend strictly on the ability of the local environment to supply enough nutrients for its growth and survival. As our environment is populated by numberless species of living organisms (from viruses to humans) all sharing the same physical space and competing for the same essential nutrients, the whole picture is highly dynamic and knowledge of the basic features of life has to be introduced to explain the role of hysteresis in biological phenomena.

1.2.1 The Time Arrow/Dissipative Systems

All forms of life are dissipative, in the sense that life takes place only in environments with an excess of energy: in our case, sunlight. This energy surplus is required to power all of our functions, including the continuous replacement of degraded structural molecules. Sunlight is restlessly flowing over us day after day and makes our life possible. Sunlight follows a one-way flow and cannot go back; its irreversibility is the physical basis for our time irreversibility. Dissipative systems, as they are continuously getting energy from the environment and using it to do something (it doesn't matter what), are always asymmetric.[4] We have to identify and quantify these intrinsic asymmetries and then evaluate which are the environmental asymmetries (nutrients and oxygen) that can supply enough energy to drive the intrinsic ones.

Ionic asymmetry is shared by all living organisms; it involves both cations such as Na^+, K^+, Ca^{++}, H^+ and anions such as Cl^-, $Pi(HPO_4^{--})$, and protein carboxyl groups.[5] As biological membranes are not fully impermeable to ions, the ion gradients between the inner and outer compartments would spontaneously disappear. Their persistence (with a very small shift from the average value) means that gradients are sustained by an active process driven by different types of ion-sensitive-ATPases.[6]

ATP synthesis requires a supply of nutrients, both reducing (carbon skeletons: monosaccharides, fatty acids (FA), amino acids (AA), or the light-dependent reduced cofactors in photosynthetic organisms) and oxidizing (O_2, nitrate, sulfate). As electrons spontaneously flow from less (C, H) to more electronegative atoms such as oxygen, life is possible wherever we have asymmetry in the electron distribution: more electrons than expected from electronegativity on C or H (e.g., CH_4 instead of CO_2 and H_2O), less on oxygen (e.g., O_2 instead of H_2O and CO_2). For any given organism it is always possible to identify the local conditions (reducing and oxidizing nutrient availability) necessary for its life. After billions of years organisms have adapted themselves to almost any environment, and have developed every type of machinery (enzymes, organelles) to take advantage of the locally available nutrients; indeed no form of life is possible in the absence of nutrients in whatever form. Identification of nutrient type, concentration, and diffusion rate in the environment is therefore the first step in the definition of any form of life.[4]

1.2.2 Competition

As biological systems tend to expand exponentially in a finite environment, they—sooner or later—become limited in their growth by the scarcity of some essential factor (nutrient) and the competition for the limiting nutrients will locally drive the selection. Since life started to evolve billions years ago, all possible nutrient sources have been exploited and life is now possible only at the expense of other forms of life. In a closed environment, such as the earth's crust, nutrients have to be considered always limiting and available for a limited time Figure 7.1. Any molecular trick, able to improve nutrient uptake, will be of great evolutionary advantage.

1.2.3 Selection

In a competitive world, at any time, somebody survives and somebody dies on the basis of its fitness to the local environment. The survivor is what we see and can describe.

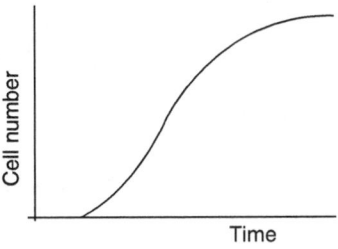

Fig. 7.1. In a closed environment (e.g., bacteria in a flask) the cells grow until the nutrient supply becomes limiting and growth stops. If more than one population is present, the efficiency in taking up nutrients becomes a powerful evolutionary advantage.

As the structure of living organisms is the structure of the fittest for any given environment, a proper description of the niche where they live should be included when dealing with their structures and metabolic pathways. As a matter of fact any other environment will lead to the selection of a different set of organisms. The way dissipative systems can afford evolution is simple but expensive from the energetic point of view. To cope with external changes, organisms have to change themselves, and they can only do that randomly, using the limited local supply of nutrients to build up new molecules instead of the older ones; afterwards, if the changes increase the fitness, the organism will survive and spread, otherwise it will disappear and nobody will be able to record its changes. If we assume that life is the final result of previous numberless cycles of random changes/selection in a competitive world, where essential molecules (nutrients) are limiting and energy is in excess, then it is quite clear that it is possible to identify many "universal" behaviors shared by all organisms, the most efficient taking up nutrients in a competitive environment.[4]

1.3 The Chemical Texture of Life

Although in materials such as rocks nonclassical nonlinear elastic behavior can be demonstrated only when powerful stresses are applied, in biological systems the strength of the bonds involved in structure stabilization is much lower, and therefore nonclassical nonlinear elastic behavior should be the rule rather than the exception. We describe briefly here the properties of molecules involved in cell and tissue formation.

Figure 7.2 shows a representative example of a biological structure and the type of weak bonds involved in its stabilization. The membrane phospholipids double leaflet is stabilized by van der Waals forces ($<1\,\mathrm{kcal\,mol^{-1}}$) in the hydrophobic core and hydrogen bonds (1–$7\,\mathrm{kcal\,mol^{-1}}$) with water at the inner and outer sides. The strength of these bonds depends also on the ionic environment, which can change upon cell

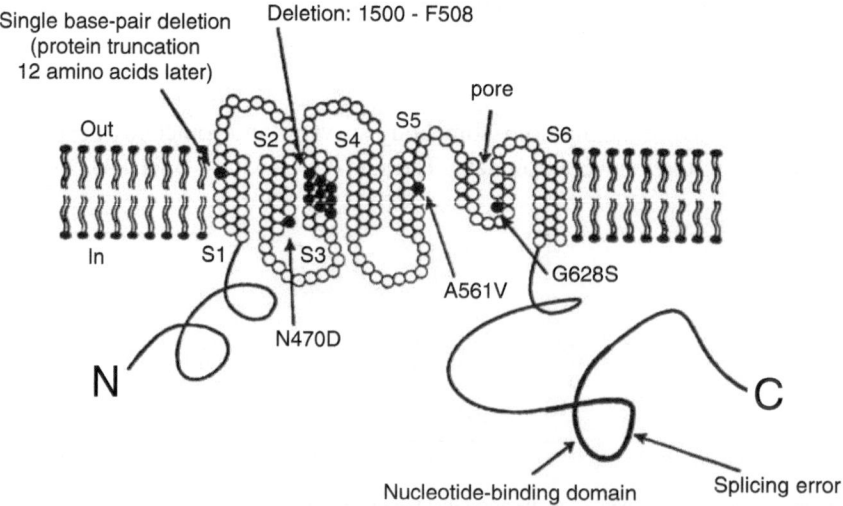

Fig. 7.2. A schematic picture of a transmembrane protein showing the sites of genetic defects impairing its function.

activation. The conformation of transmembrane proteins depends on hydrogen bonds, hydrophobic bonds (2–$3\,\text{kcal mol}^{-1}$), and ionic bonds ($1^{-6}\,\text{kcal mol}^{-1}$). Charges responsible for ionic bonds depend on local pH and on the charge of surrounding groups: charge changes can induce protein shape changes and vice versa (Figure 7.4). Usually the protein configuration is stabilized by a complex pattern of weak bonds, and the deletion or substitution of even one single amino acid, if involved in specific bonds, may lead to significant configuration change and loss of function.

1.3.1 Asymmetry

Asymmetry in the cell is mostly linked to the existence of a cell membrane separating the self from the not-self. Membrane asymmetries include a different content of phospholipids in the inner and outer leaflets, a different electric charge (membrane potential), and an ionic asymmetry that involves both cations such as Na^+, K^+, Ca^{++}, H^+ and anions such as Cl^-, $Pi(HPO_4^{--})$, and protein carboxyl groups.[5]

Biological membranes are not fully impermeable to ions, and the ion gradient between the inner and outer compartments drives a passive flux through specific channels (proteins) whose properties depend strongly on local conditions (pH, membrane potential, cholesterol). See Figure 7.3. Mechanical and thermal stresses can also affect permeability through these channels. As a matter of fact, ionic asymmetry is very stable (with a very small shift from the average value), independently from ion influx, due to a very active pumping process driven by different types of ion-sensitive-ATPases.[6] Both passive ion channels and active pumps show a hysteretic behavior.[7,8]

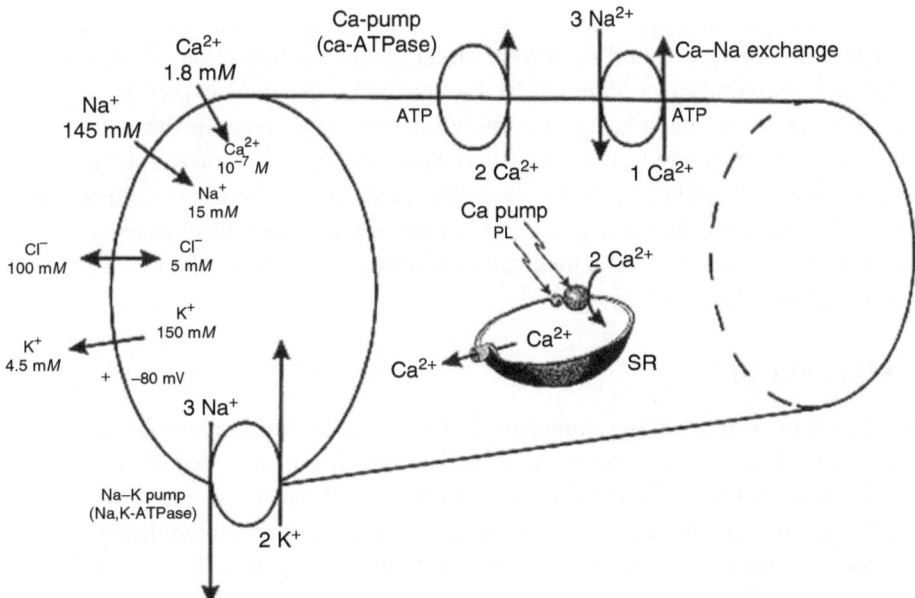

Fig. 7.3. Intracellular and extracellular ion distribution. ATP can drive active ions pumping against the gradient, whereas passive flux according to the gradient is gated by channels modulated by voltage or different ligands

1.3.2 Context Sensitive

If heterogeneity and fragility can be regarded as defects in industrial materials, being the source of unexpected and unwanted behaviors, in the case of biological structures the main property of a living organism (from microbes to man) is its sensitivity to the environment. Without the extreme sensitivity at our fingertips, our hand would be continuously hurt, burnt, and unable to perform its tasks. This extreme sensitivity is acquired through the use of very soft structures such as our organic molecules: nucleic acids, proteins, lipids, and the complex structures derived from their combination. Membranes, cell organelles, and tissues' extracellular matrix all are made of different mixtures of a limited number of molecules (glucose, amino acids, fatty acids), embedded in 75 to 80% water. The shape stability of these molecules depends, as expected, on the temperature. Homeothermic organisms like mammals guarantee a proper functioning of the body by keeping its temperature constantly around 37°C. Fever over 39°C induces the unfolding of a group of proteins (HSP, Heat Shock Protein) with a dramatic change of their function and mounting of the "inflammatory response."[9] A number of local factors can modify protein folding as their tertiary structure (that is responsible for their surface shape and of soft bonds between "grains") relies on many weak bonds. pH changes, ion binding (Ca^{++}, Na^{+++}, K^{+}), and interactions with other proteins or membranes can heavily modify the chemical properties of these molecules affecting their binding to physiological substrates and ligands. In most cases protein conformational changes are reversible and depend on local conditions and ligand concentration.

Hemoglobin binds O_2 in the lung (high pO_2) and releases it in the tissues (low pO_2), undergoing every few seconds a shift from low to high ($\rightarrow$) and from high to low pO_2 ($\leftarrow$) (Figure 7.4). Hemoglobin binding to O_2 doesn't show any hysteretic behavior because any lag in the O_2 release ($\leftarrow$) would strongly impair the tissue's O_2 uptake and function, and therefore would negatively affect the natural selection. In hemoglobin S (HbS), on the contrary (see Figure 7.5), the hysteretic behavior is present, prompting us to wonder about the selective advantage in this case. HbS in the deoxygenated form can polymerize into long filaments; red cells assume a sickle (S) shape and stop the blood flow in capillaries. In this case efficiency in O_2 release means efficiency in blocking blood flow. Therefore evolution has chosen a compromise: to reduce tissues' oxygen release (via a hysteretic mechanism), but to avoid total ischemia secondary to red cell sickling.[11]

1.3.3 Short Lived

The limited lifespan of living organisms is the result of the limited lifespan of all its components: tissues, cells, and molecules. The half-life of a molecule (i.e., the time needed to replace half of the molecules originally present) usually ranges from hours to days for structural molecules such as proteins and from minutes to hours for metabolic fuels such as glucose. Protein turnover rate is limited by specific constraints linked to the complexity of protein synthesis, that requires DNA transcription to mRNA, mRNA translation to a protein, and posttranslational modifications. As a whole, this process takes about six hours, whereas protein degradation may be very fast (minutes). Whenever a fast protein-mediated response is needed, as in the case of response to hypoxia

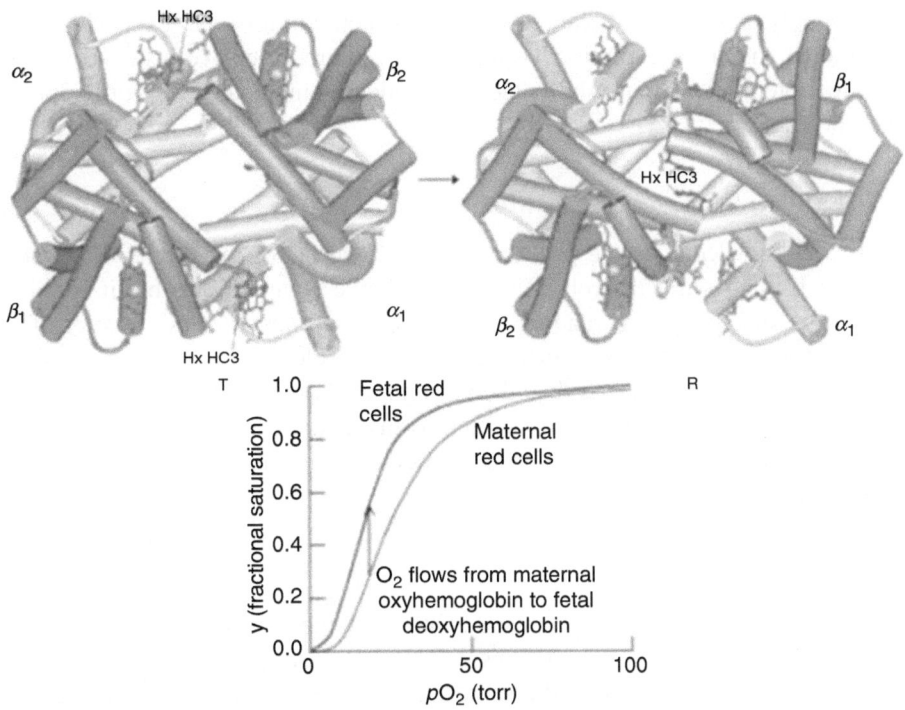

Fig. 7.4. Hemoglobin is a classical example of a reversible ligand (O_2) binding to a tetrameric protein with allosteric behavior mediated by conformational changes (upper). The dissociation curve (lower) can be shifted to the right or to the left according to local conditions (pH, 2,3-BPG, CO_2) and to small changes in amino acid patterns (fetal Hb versus adult Hb).[10]

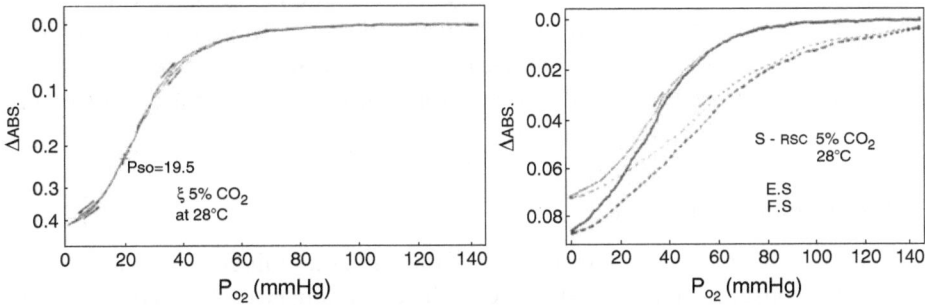

Fig. 7.5. Normal Hb does not show any hysteretic behavior that would impair O_2 transport by slowing uptake in the lung and release in the peripheral tissues (left panel). In the case of Hb S the dissociation curve shows a hysteretic behavior that dramatically impairs the O_2 release to the tissues, but avoids ischemia (right panel). Δ_{abs} is the percentage of deoxygenated Hb.

mediated by HIF (hypoxia-induced factor), the protein is continuously and constitutively synthesized, but the active protein level is kept very low by a continuous degradation mediated by ubiquitination.[12] The process is very active and the HIF half-life is about five minutes. The process guarantees a timely correspondence between the local oxygen supply and the metabolic response of the cell, but it is very expensive. Whenever possible, therefore, evolution has replaced the more expensive systems, where the signaling protein is continuously replaced, introducing molecules with different types of memory.

1.4 The Molecular Bases Of Cell Memory

Memory, as knowledge of the past that allows forecasts of the future, offers strong selective advantages in relatively stable or repetitive environments but it could be a disadvantage in rapidly and randomly changing environments. As a matter of fact, recording past experience has an energetic cost; to be selected by evolution any type of memory must have a cost lower than the advantages it offers in terms of better fitness. For man, language, in its spoken or written forms, is the conscious form of memory, but many additional unconscious forms of memory, spanning through different size and time scales, exist.

1.4.1 DNA

DNA has existed for billions of years and contains all the information required to drive the birth and the growth of a new organism. DNA is transcribed to mRNA and mRNA is translated into proteins. With the same DNA it is possible to get many different proteins through alternative splicing of mRNA, and proteins can undergo heavy post-translational modifications. These two processes allow, for example, the development of different organs in the same organism in response to different local environments. DNA can change very slowly and anyway the changes will be incorporated into the next generations according to the generation time (years for man, minutes for bacteria). The genetic information for some higher organisms will therefore evolve too slowly to cope with environmental changes, and survival will require migration or modification of the environment (houses, heating, agriculture). The selection process at the gene level requires short times (months, years) for viruses and bacteria but for complex organisms may require a thousand years.

1.4.2 Internal Milieu

Multicellular organisms and the stability of their internal milieu (blood and extracellular fluids, constant temperature) is one of the evolutionary tricks used to reduce the need of DNA evolution when its rate cannot cope with the rate of external changes. Stability of complex organisms includes the creation of stores of nutrients such as intracellular glycogen (glucose store) or bone (phosphate store). In our experience modeling cell life without such internal nutrient stores, because nutrient uptake is usually a strong limiting factor, increases dramatically the amplitude of metabolic oscillations and therefore the cell death rate.[13]

Stability of the internal milieu avoids the need for the genetic evolution of a large number of genes as a function of environmental changes. Keeping the internal milieu stable means keeping the memory of the fittest environment and avoiding the cost of multiple cycles of random mutation and selection of each of the proteins involved in basic cellular metabolism. Most of the physiological functions involved in the stability of the metabolism of complex organisms are cyclic (because of feedback mechanisms) over different time scales and include breathing, circulation, daily fat accumulation, fat storage for hibernation, and menstrual cycle for females (nutrient accumulation for a new life). The time scale is days or months.

1.4.3 Hormones, Cytokines, Receptors (Turnover)

After every meal blood glucose increases, peaks at two hours, and slowly decreases. Insulin release from the pancreas follows the same kinetics. Insulin level is under the strict control of blood glucose. Apparently glucose and insulin have the same information content, but in vitro treatment of fibroblasts with glucose induces differentiation to cartilage, with glucose + insulin to adipocytes.[14] Insulin incorporates the information that glucose will be supplied for at least some hours, and the cell, on the basis of this information, will break down part of the excess glucose to transform part of the glucose into fatty acids. Insulin belongs to the class of signaling molecules with polypeptide structure. Many other hormones and cytokines (messengers for intercellular communication) and their cellular receptors have a similar structure. As the synthesis of this sort of short protein requires at least four hours, they always carry the information that the local conditions for their synthesis must be present for at least four hours. Because the environment is the whole organism, the time scale is hours. These processes are responsible for the circadian rhythms of many physiological functions.

1.4.4 Proteins' Hysteretic Behavior

At the cellular level the time scale of the events is in the range from seconds (metabolic oscillations) to milliseconds (membrane potentials, ion channels opening). Moreover, at the microscopic level, cells are extremely heterogeneous and out of phase as far as their electric and metabolic behavior is concerned.[15] At this level the only possible form of memory is a protein configuration change. As already described, the protein tertiary structure and configuration is based on multiple very weak bonds sensitive to pH, ion and ligand concentration, other proteins, posttranslational modifications, membrane translocation, and so on (Figures 7.3 and 7.4). In Section 1.5 we describe the current approach to the role of hysteresis in protein function and its future perspectives, and in Section 1.6 we briefly describe its historical background.

1.5 The Advantage of Being Hysteretic

The memory of the past at any level has allowed evolution of life as we know it now, preventing death due to unexpected local environmental changes. At the cellular level a slow response to rapid changes in the microenvironment confers robustness to the whole system preventing the interpretation of random local changes (noise) as signals.

1.5.1 From Dose–Response to a Switch

Most of the work done by biochemists on the kinetics of chemical reactions has been performed on enzymes: soluble enzymes at the beginning and then polymeric

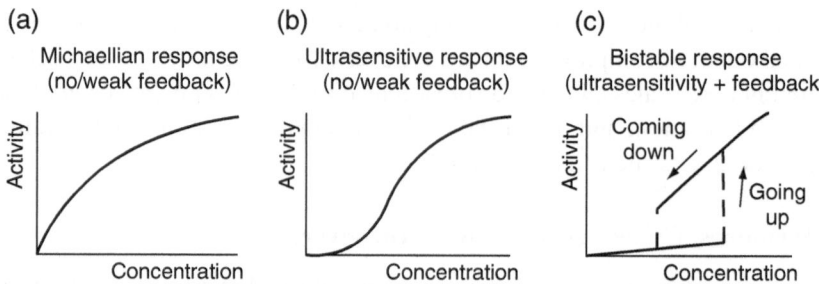

Fig. 7.6. Evolution of protein–ligand interaction from the simplest to the most complex (a)→(c).[16]

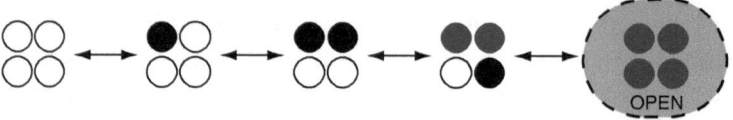

Fig. 7.7. A tetrameric protein showing the stepwise change mediated by the passage of subunits from a tense (empty) to a relaxed (filled) configuration.

enzymes, multienzymatic complexes, membrane-bound enzymes, and so on. All this information applies to any protein interacting with some ligand (membrane receptor, ion channels, ion pumps, etc.) through weak bonds. A preliminary assumption is that any chemical reaction can be split in many steps and that with increasing the ligand concentration, the activity increases until it saturates all the protein binding sites (maximal enzyme activity) and protein concentration becomes limiting (Figure 7.6a). Enzymes are not able to change the reaction direction, which depends on the free energy difference, ΔG, but can accelerate it simply by a factor of thousands. The need for an efficient catalyst arises from the limited availability of nutrients in any environment, and at the beginning the species with the most efficient enzymes can overcome all others.

According to Figure 7.6a the cell responds to every stimulus, even to one single molecule: in the case where the local noise is oscillating from 1 to 5 molecules, any cell that responds to any weak stimulus in this range would waste a lot of energy following decoy stimuli. Let us assume that by genetic mutation some cells develop a type b response with minimal sensitivity to 6, 9, and 12 molecules. The cell with the highest sensitivity (6) will be able to sense 100% of the signals and the others (9,12) will sense only part of them. This is a very strong selective advantage and after a while the cell (6) will be the only survivor.

The molecular origins of sigmoid curves (Figure 7.6b) are usually complex and depend on multiple interactions between similar or different proteins. In hemoglobin (a tetrameric protein) the shift from T to R forms (Figure 7.7) depends on oxygen binding to each monomer, leading to a conformational and affinity change of the adjacent subunits; in proteins such as membrane calcium channels (one pore with multiple transmembrane segments; Figure 7.2) the configuration change leading to channel opening follows a type b kinetic with a very steep rise, configuring a switch response.[17] This type of response depends on calcium concentration and shows a threshold. It can be

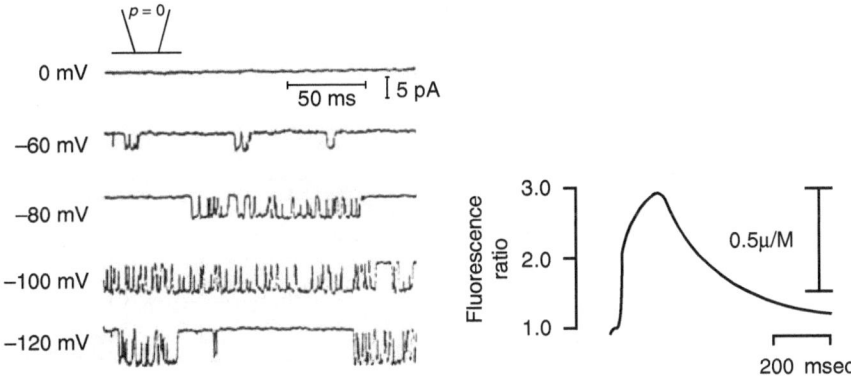

Fig. 7.8. (a) Measurement of single-voltage gated channel activity with patch-clamp technique. The single channel displays different opening frequencies according to the applied voltage. (b) The average calcium influx with time depends on the total number of open channels at any time.

easily modeled by imposing a multistep configurational change and assuming an elastic behavior of the protein: any hit with the ligand will shift the configuration towards the open form; the R configuration will last for a fixed time, and then will return to the previous state.[18] Changing the ligand concentration and the relaxation time it is possible to simulate any experimental result with this type of all-or-none channels (Figure 7.8a).

The average calcium influx on thousands of cells (each with a thousand channels) has lost the all-or-none response pattern and smoothly changes with time, showing dose-dependence of the total calcium influx (Figure 7.8b).

A similar all-or-none behavior has been described in one step, the phosphorylation mediated by Janus Kinase (JNK), of the complex cascade downstream of the TNF alpha receptor, regulating the cell fate in the sense of apoptosis or proliferation according to local conditions (Figure 7.9).

Also in this case the average behavior over a large cell population displays a typical sigmoid response to progesterone as a stimulus (completely superimposable to those shown in Figure 7.4 for hemoglobin), but if we go down to the single-cell level we are able to identify an all-or-none behavior[19] of the single JNK molecule (Figure 7.10).

In the case of tissues such as muscle, where the function is cooperative and depends on the average behavior of all cells together, the state of the single channels does not affect the function too much, however, in the case of oocytes, the all-or-none switch will force the cell into the replication cycle. It has recently been demonstrated that the JNK system in these cells is hysteretic and bi-stable (Figure 7.6c) and that its behavior doesn't correspond to a very steep sigmoid curve (Figure 7.6b).[19]

1.5.2 Cell Cycle Oscillator/No Way Back

Forty years ago, Monod and Jacob hypothesized that bi-stable signaling systems might provide cells with the sort of long-term memory required to maintain differentiation long after a differentiation-inducing stimulus is removed.[20] The work of Bagowski and Ferrell (2001) indicates that, in oocytes, JNK activation is one such memory system.

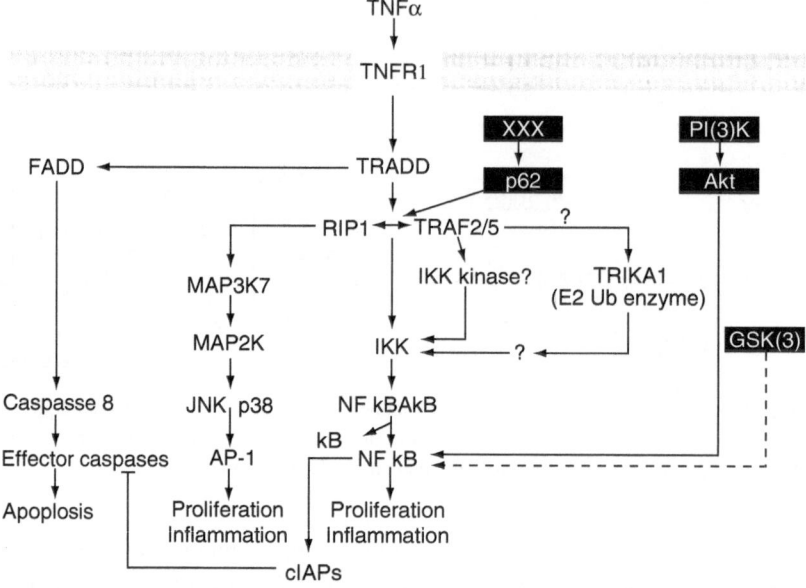

Fig. 7.9. The intricate network of intracellular signaling downstream of the TNF receptor, where the enzyme JNK plays a fundamental role.

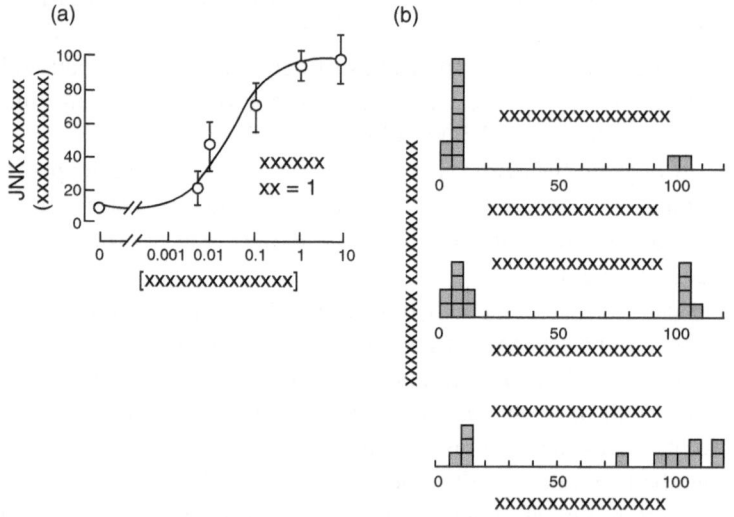

Fig. 7.10. JNK activation in progesterone-treated oocytes. (a) The data of the whole cell population are well fitted by a Michaelian stimulus–response curve with a Hill coefficient of 1. (b) Measured JNK activities of individual oocytes. Each box represents one individual oocyte (redrawn from Reference 19).

JNK activation in oocytes is physiologically triggered by progesterone, but JNK remains active throughout early embryogenesis, long after the initiating progesterone stimulus has ceased.[19] The evidence presented in this paper provides a mechanistic explanation for how this long-term activation of JNK is achieved. Other bi-stable signaling systems are present in the oocyte. The Mos/MEK1/p42 MAPK cascade appears to be bi-stable,[21] and, by virtue of its autocatalytic character, Cdc2-cyclin B may also be part of a bi-stable system.[22,23] Thus, bi-stability and all-or-none signal transduction systems may be common, at least in processes such as oocyte maturation and early embryogenesis that are driven by combinations of essentially digital, all-or-none responses that arise out of interconnected bi-stable signaling pathways.

What it is of interest from the evolutionary point of view is that, although all the systems share the same all-or-none behavior, the specific mechanisms through which the feedback is produced and the bi-stability is achieved are all different and occur on different time scales. In the case of p42 MAPK, the positive feedback is provided by translational regulation;[21] in the case of JNK, JNK activation is due to posttranslational positive feedback; in the case of Cdc2, both translational and posttranslational feedback may be important. Thus, it appears that nature has converged upon the same basic scheme for producing decisive digital responses, but has done so through a variety of different specific mechanisms. This convergent evolution underscores the likely importance of hysteresis and bi-stability as "universal" behavior in a very basic process such as progression of a cell cycle, in addition to the many others cited in this chapter.

1.6 Hysteresis in Biochemistry: A Historical Survey

1.6.1 The Ancestor

The first paper where the word hysteresis was used in a biochemical context dates back to 1970.[24] "**Hysteretic enzymes** are defined as those enzymes which **respond slowly** (in terms of some kinetic characteristic) **to a rapid change in ligand**, either substrate or modifier, concentration. Such slow changes, defined in terms of their rate relative to the over-all catalytic reaction, result in a lag in the response of the enzyme to changes in the ligand level." Molecular mechanisms responsible for this behavior include ligand-induced isomerization of the enzyme, displacement of tightly bound ligands by other ligands, or polymerization and depolymerization, membrane translocation. Such enzymes frequently play a key role in metabolic regulation and the knowledge of their behavior is of great value in discussing the regulation of complex processes in vivo. This method of approach, however, at that time turned out to be of limited usefulness "since the equations for complex systems can easily become much too cumbersome to deal with."[24]

At that time computers were not as widespread as now and were much less powerful and therefore complex simulations were not possible. The main conclusion was that any shift from the expected dependence on substrate concentration (hyperbolic) of an enzymatic reaction should depend on some more complex interaction of the protein with the environment (other small molecules, other proteins, or complex structures such as membranes).

1.6.2 Examples on the Molecular Scale

With time, more and more proteins have been included in the "hysteretic" category, but their inclusion relied mostly on the existence of a lag phase in the kinetic response to substrate (sigmoid curve). They include:

- Many key enzymes in main metabolic pathways: G6P phosphatase,[25] liver fructose 1,6-bisphosphatase,[26] dihydrofolate reductases,[27] carnitine palmitoyltransferase,[28] and so on.
- Thermal hysteresis proteins (THPs) have been found in vertebrates, invertebrates, plants, bacteria, and fungi. They can depress the freezing point of water (in the presence of ice crystals) in a noncolligative manner by binding to the surface of nascent ice crystals. The THPs comprise a disparate group of proteins with a variety of tertiary structures and often no common sequence similarities or structural motifs. THPs represent one more example of parallel and convergent evolution with different proteins being adapted for an hysteretic behavior.[29]

1.6.3 Examples on the Cellular Scale

In human erythrocytes it has recently been demonstrated that voltage-dependent nonselective cation (NSVDC) channels exist in two states of activation depending on the initial conditions for the activation. The hysteretic behavior, which in patch-clamp experiments has been found for the individual channel unit, is thus retained at the cellular level and can be demonstrated with red cells in suspension.[8]

1.6.4 Examples on the Tissue Scale

As molecular interactions between the cells share the same properties of those between molecules inside the cells (cooperativity based on weak interactions) it is not surprising that hysteretic behaviors have been described also at the level of tissue functions in:

Heart: Adaptation of QT interval to heart rate changes[30]
Eye: Accomodative hysteresis as a function of target-dark focus separation[31]
Lung: Lung and alveolar wall elastic and hysteretic behavior[32]
Brain: Robust persistent neural activity in a model integrator with multiple hysteretic dendrites per neuron[33]
Ligament: Stress–strain responses under tensile and compressive loading conditions[34]

1.6.5 Examples on the Whole Organism Scale

Most of the functions of living organisms, no matter how complex, display a circadian rhythmicity that depends on networks of individual feedback between the environmental (night/day) signals and internal clocks. Training of internal clocks to a new environment requires an adaptation time that varies from species to species and from tissue to tissue.[35] The hysteresis of internal clocks is at a higher level of complexity, but is based on similar behavior at lower levels. Cell division, for example, in many mammalian tissues is associated with specific times of the day, but how the circadian clock controls this timing has not yet been fully understood. It has been shown[36] that in the regenerating liver (of mice) the circadian clock controls the expression of cell cycle-related genes that in turn modulate the expression of active Cyclin B1-Cdc2 kinase, a key regulator of mitosis, whose behavior has already been described in Section 1.5.2. Among the genes involved in this pathway, expression of *wee*1 is directly regulated by the molecular components of the circadian clockwork. In contrast, the circadian clockwork oscillates independently of the cell cycle in single cells. Thus, the intracellular circadian clockwork can control the cell-division cycle directly and unidirectionally in proliferating cells.

The complex network of interactions (vessels, nutrition, hormones, circadian rhythms) supporting the feeding of the single parts of an organism, or, for example, a tumor, justifies the assumption of an oscillating nutrient supply to a tumor proposed by Condat and coworkers in this book as one of the major determinants of hysteretic behavior in cancer growth.[37]

2. Conclusions

Living systems, as a whole, show hysteretic behavior in many cases, but it is impossible to find evidence for a strict structure–behavior relationship. In fact a tetrameric protein can display any type of response (linear, sigmoid, bi-stable) according to small molecular changes induced by selective pressure, and, when required, hysteretic behaviors can be developed at any level (molecular, intracellular, intercellular) and at different time scales. In summary, although hysteretic behavior in materials depends on their mesoscopic structure, in living organisms the structure has been selected by evolution to fit the requirements for a hysteretic behavior. As most of the biological structures (macromolecules, organelles, cells) have a mesoscopic structure stabilized by weak bonds and are very sensitive to environmental (pH, ions, ligands) changes, a hysteretic behavior may be easily induced. In conclusion: hysteretic behaviors exist in living systems and are easily inducible, but the real question is in which cases they offer an evolutionary advantage. In a world where on any scale all the events are interconnected and regulated by both positive and negative interactions, most of the local conditions are not stable but oscillating. Whenever these oscillations are not harmful for life they can be regarded as "noise" and a hysteretic behavior would be the best choice, whereas when they reach a specific threshold they become a "signal," requiring a response. The hysteretic behavior at the molecular level seems to be the tool chosen by evolution to set the threshold between noise and signal, and in a relatively stable context as it is now on the earth's crust to be "conservative" and relatively insensitive to local changes can be an advantage.

Acknowledgments

This work has been supported in part by Fondazione Internazionale Ricerche Medicina Sperimentale (FIRMS), Compagnia di SanPaolo, and Ministero dell'Istruzione, dell'Universita'e della Ricerca.

References

[1] See Chapters 11, 16, and 18.
[2] See Chapter 17.

[3] P.P Delsanto and M. Scalerandi, Modeling nonclassical nonlinearity, conditioning, and slow dynamics effects in mesoscopic elastic materials, *Phys. Rev. B* **68**, 641071–641079 (2003).

[4] G.P. Pescarmona, The life context: cells, nutrients and signals, in: *Recent Research Development in Biophysical Chemistry*, edited by C.A. Condat and A. Baruzzi (Research Signpost, Kerela, India, 2002), pp. 69–90.

[5] T.M. Devlin, *Biochemistry* (Wiley-Liss, New York, 1997).

[6] N. Sperelakis, *Cell physiology Source Book* (Academic Press, 1998) pp. 171.

[7] O. Scharff, B. Foder and U. Skibsted, *Biochim Biophys Acta.* **730**, 295–305 (1983).

[8] P. Bennekou, T.L. Barksmann, L.R. Jensen, B.I. Kristensen and P. Christophersen, Voltage activation and hysteresis of the non-selective voltage-dependent channel in the intact human red cell, *Bioelectrochemistry* **62**, 181–5 (2004).

[9] W. van Eden, R. van der Zee and B. Prakken, Heat-shock proteins induce T-cell regulation of chronic inflammation, *Nat Rev Immunol.* **5**, 318–30 (2005).

[10] J.M. Berg, J.L. Tymoczko and L. Stryer, *Biochemistry* (New York, W. H. Freeman and Co.; 2002). Chapter 10.2

[11] H. Mizukami, A.G. Beaudoin, D.E. Bartnicki DE and B. Adams, Hysteresis-like behavior of oxygen association-dissociation equilibrium curves of sickle cells determined by a new new method, *Proc Soc Exp Biol Med.* **154**, 304–9 (1977).

[12] I. Groulx and S. Lee, Oxygen-dependent ubiquitination and degradation of hypoxia-inducible factor requires nuclear-cytoplasmic trafficking of the von Hippel-Lindau tumor suppressor protein, *Mol Cell Biol.* **22**, 5319–36 (2002).

[13] M. Scalerandi, G.P. Pescarmona, P.P. Delsanto and B. Capogrosso Sansone, Local interaction simulation approach for the response of the vascular system to metabolic changes of cell behavior, *Phys Rev E* **63**, 11901–11910 (2001).

[14] M.F. Pittenger, A.M. Mackay, S.C. Beck, R.K. Jaiswal, R. Douglas, J.D. Mosca, M.A. Moorman, D.W. Simonetti, S. Craig and D.R. Marshak, Multilineage potential of adult human mesenchymal stem cells, *Science* **284**, 143–147 (1999).

[15] J.M. Levsky and R.H. Singer, Gene expression and the myth of the average cell, *Trends Cell Boil.* **13**, 4–6 (2003).

[16] J.R. Pomerening, E.D. Sontag and J.E. Jr. Ferrell, Building a cell cycle oscillator: hysteresis and bistability in the activation of Cdc2, *Nat Cell Boil.* **5**, 346–351 (2003).

[17] N. Sperelakis, *Cell Physiology Source Book* (Academic Press, 1998) chapters 26, 28.

[18] G.P. Pescarmona, data not published.

[19] C.P. Bagowski and J.E. Jr Ferrell, Bistability in the JNK cascade, *Current Biology* **11**, 1176–1182 (2001).

[20] J. Monod and F. Jacob, General conclusions: teleonomic mechanisms in cellular Metabolism, growth, and differentiation, *Cold Spring Symp Quant Biol*, **26**, 389–401 (1961).

[21] Ferrell JE Jr and E.M. Machleder, The biochemical basis of an all-or-none cell fate switch in Xenopus oocytes, *Science* **280**, 895–898 (1998).

[22] M.J. Solomon, M. Glotzer, T.H. Lee, M. Philipee and M.W. Kirschner, Cyclin activation of p34cdc2, *Cell* **63**, 1013–1024 (1990).

[23] I. Hoffmann, P.R. Clarke, M.J. Marcote, E. Karsenti and G. Draetta, Phosphorylation and activation of human cdc25-C by cdc2–cyclin B and its involvement in the self-amplification of MPF at mitosis, *EMBO J* **12**, 53–63 (1993).

[24] C. Frieden, Kinetic aspects of regulation of metabolic processes. The hysteretic enzyme concept, *J Biol Chem.* **245**, 5788–99 (1970).

[25] K.L. Nelson-Rossow, K.A. Sukalski and R.C. Nordlie, Hysteresis at near-physiologic substrate concentrations underlies apparent sigmoid kinetics of the glucose-6-phosphatase system, *Biochim Biophys Acta.* **1163**, 297–302 (1993).

[26] G.B. Van den Berg, H. Vaandrager-Verduin, T.J. Van Berkel and J.F. Koster, Hysteretic behaviour of rat liver fructose 1,6-biophosphatase induced by zinc ions, *Arch Boichen Biophys.* **219**, 277–85 (1982).

[27] J.R. Appleman, W.A. Beard, T.J. Delcamp, N.J. Prendergast, J.H. Freisheim and R.L. Blakley, Atypical transient state kinetics of recombinant human dihydrofolate reductase produced by hysteretic behavior. Comparsion with dihydrofolate reductases from other sources, *J Biol. Chem.* **264**, 2625–33 (1989).

[28] A.C. Lloyd, C.A. Carrpenter and E.D. Saggerson, Intertissue differences in the hysteretic behavior of camitine palmitoyltransferase in the presence of malonyl-CoA, *Biochem J.* **237**, 289–91 (1986).

[29] J. Barrett, Thermal hysteresis proteins, *Int J Biochem Cell Biol.* **33**, 105–17 (2001).

[30] E. Pueyo, P. Smetana, P. Laguna and M. Malik, estimation of the QT/RR hysteresis lag, *J Electrocardiol.* **36** Suppl, 187–90 (2003)

[31] S.M. Ebenholtz, Estimation of the QT/RR hysteresis lag, *Vision Res.* **32**, 925–9 (1992).

[32] K.K. Brewer, H. Sakai, A.M. Alencar, A. Majumdar, S.P. Hold, K.R. Lutchen, E.P. Ingentino and B. Suki, Lung and alveolar wall elastic and hysteretic behavior in rats: effects of in vivo elastase treatment, *J Appl Physiol* **95**, 1926-36 (2003).

[33] M.S. Goldman, J.H. Levine, G. Major, D.W. Tank and H.S. Seung, Robust persistent neural activity in a model integrator with multiple hysteretic dendrites per neuron, *Cereb Cortex.* **13**, 1185–95 (2003).

[34] M. Pini, H.W. Wiskott, S.S. Scherrer, J. Botsis and U.C. Belser, Mechanical characterization of bovine periodontal ligament, *J Periodontal Res.* **37**, 237–44 (2002).

[35] T. Roenneberg and M. Merrow, The network of time: understanding the molecular circadian system, *Curr Biol.* **13**, R198–207 (2003).

[36] T. Matsuo, S. Yamaguchi, S. Mitsui, A. Emi, F. Shimodz and H. Okamura, Control Mechasnism of the Circadian Clock for Timing of Cell Division in vivo, *Science*, **302**, 225–259 (2003).

[37] See Chapter 8.

8

Cancer Growth: A Nonclassical Nonlinear Phenomenon?

C.A. Condat,[1,2,4] B.F. Gregor,[3] Y. Mansury,[3] and T.S. Deisboeck[3]

[1]Department of Physics, University of Puerto Rico, Mayagüez, PR 00681, USA.
[2]CONICET and FaMAF, Universidad Nacional de Córdoba, 5000-Córdoba, Argentina.
[3]Complex Biosystems Modeling Laboratory, Harvard-MIT (HST) Athinoula A. Martinos Center for Biomedical Imaging, Massachusetts General Hospital, Charlestown, MA 02129.
[4]To whom correspondence should be addressed; e-mail: cacondat@yahoo.com.

Abstract
Using a general growth model, based on scaling and energy conservation, we show results which indicate that cancer growth belongs to the universality class of nonclassical nonlinear phenomena. We introduce a generalized remanence, which permits us to evaluate the intensity of nonclassical effects, with implications ranging from experimental to clinical cancer research.

Keywords: Cancer, cycles, hysteresis, mesoscopic, multicellular spheroid, nonclassical, nonlinear, remanence, scaling

1. Introduction

The concept of nonclassical nonlinear (NCNL) phenomena was first introduced in the field of rock elasticity, where such unusual elastic properties as extreme nonlinearity, hysteresis, and discrete memory were observed. In particular, modeling the elastic properties of rocks requires the addition of nonanalytical terms to the corresponding nonlinear wave equation [1]. The universal features of NCNL phenomena have been very recently identified by Hirsekorn and Delsanto [2]. Here, we show that the methods described in Reference [2] may also help us to better understand biomedical grand challenges such as cancer growth and eventually may even allow us to design a more successful treatment regimen. Certain mesoscopic models of cancer growth are explicitly nonanalytic due to the use of thresholds for cellular proliferation and cell death [3–5]. In this chapter we show that NCNL behavior is already present in a very simple "macroscopic" model of cancer growth. Cautiously extrapolated to the biological situation, this finding may support the notion that nonclassical nonlinearity is an intrinsic feature of tumoral evolution.

And in fact, empirical evidence indicates that cancer growth is a strongly nonlinear process [15, 16]. These nonlinearities have been incorporated into current models [3–8]. To test the possible nonclassical aspects of cancer growth, we follow the prescription in Reference [2], investigating the reaction of a neoplasm to a time-dependent (e.g., diffusive) chemical "input" signal. In principle, this could be done by analyzing its evolution under the application of an antitumoral therapeutic course but, in practice, accurate data could be difficult to obtain and interpret, due to the inherent complexities of in vivo cancer growth. For this reason, we consider here a simplified problem, investigating the evolution of a multicellular tumor spheroid (MTS) when the availability of nutrients and/or oxygen oscillates with time. An MTS is a spherical aggregation of tumor cells grown in vitro under strictly controlled environmental conditions [9–13].

It is important to remark that, even though we focus our analysis on an experimental situation where the NCNL effects should be directly observable, oscillations in the nutrient supply, due to local fluctuations in the blood flow, are natural features of the environment into which the tumor grows. Moreover, a living organism is the result of an ongoing selection process; therefore, we expect the tumor behavior to have already incorporated the fittest response to environmental oscillations. Other instances of NCNL phenomena in living systems are discussed elsewhere in this book [14, 15].

We model the growing tumor using the general model for ontogenetic growth developed by West, Brown, and Enquist (WBE) [16], which has recently been applied both to the growth of real cancers in vivo and MTSs [17, 18]. This model yields the time dependence of the total mass of live MTS cells, which contain both transitory and oscillating components. We identify the NCNL elements in the oscillatory component, evaluating their meaning and relevance.

2. The Model

We start from the generalized version of the WBE model used in Reference [16] to investigate MTS growth. In this model, the incoming rate B of energy flow, which is assumed to depend only on the organism mass, is used both for cell maintenance and proliferation. Here, we introduce an externally controlled time-dependent factor $F(t)$ in the incoming energy flow, and write $B = (E_c/\mu_o)F(t)m^p$, where E_c is the metabolic energy needed to create a cell (= replication or proliferation) and μ_o is the average tumor cell mass. Because dead cells neither consume nutrients nor reproduce, we take into account only live cancer cells. Under these conditions, the equation describing the change in the live tumor mass m is,

$$\frac{dm}{dt} = F(t)m^p - b(t)m. \tag{8.1}$$

Here the exponent p depends on the energy distribution mechanism: for instance, $p = 3/4$ corresponds to a fractal-like distribution network [16], whereas $p = 2/3$ corresponds to pure diffusion. The power law nature of this term has been recognized for a long time [19], although the value of the exponent is still the subject of controversy [20, 21]. The last term represents the energy allocated to cellular metabolism or maintenance; there the factor $b(t)$ stands for the ratio between the metabolic rate of a single cell and E_c. The difference between the first and second terms corresponds to the energy being allocated to cell proliferation.

Equation (8.1) can be easily solved by making the transformation,

$$y(t) = 1 - [m(t)/M]^{1-p}, \tag{8.2}$$

where M is a constant mass characteristic of the system under study. Transformation (8.2) linearizes Eq. (8.1), which now reads,

$$\frac{dy}{dt} + (1-p)by = (1-p)b - \frac{1-p}{M^{1-p}}F(t). \tag{8.3}$$

By writing $\alpha = (1-p)b$, and introducing the new function $X(t)$,

$$y(t) = e^{-\alpha t}X(t), \tag{8.4}$$

we can integrate Eq. (8.3),

$$X(t) = X(0) + (1-p)\int_0^t e^{\alpha t'}\left[b - M^{1-p}F(t')\right]dt'. \tag{8.5}$$

Here $X(0) = y(0) = 1 - [m(0)/M]^{1-p}$. We can now investigate different forms for the tumor's nourishment "protocol" $F(t)$. If $F(t)$ is a constant, we obtain the universal solution of Reference [16]. More useful to our purposes is to consider a sinusoidally varying function, which allows us to discuss possible NLNC effects. The rationale for choosing a sinusoidal form is to reflect periodic variation in the nutrient supply as conferred through, for example, cardiac oscillations in vivo. Specifically, we choose

$$F(t) = a + c\,\sin(\omega t), \tag{8.6}$$

with $c \leq a$. Because in an experiment the nourishment period $T = 2\pi/\omega$ will be usually long compared with the typical diffusive times in the system, we can consider that all cells are being nourished in phase. By substituting Eq. (8.6) into Eq. (8.5), we obtain,

$$X(t) = X(0) + \frac{ac}{a(\alpha^2 + \omega^2)}\left[e^{\alpha t}(\omega\cos\omega t + \alpha\sin\omega t) - \omega\right]. \tag{8.7}$$

Therefore, the total mass evolves according to

$$m(t) = M\left[\Theta(t) + 1 - \frac{ac}{a(\alpha^2 + \omega^2)}(\omega\cos\omega t + \alpha\sin\omega t)\right]^{1/{1-p}}, \tag{8.8}$$

where $\Theta(t)$ is a transitory component that decays over a time $\tau_T = \alpha^{-1}$,

$$\Theta(t) = \left\{\frac{ac\omega}{a\left(\alpha^2 + \omega^2\right)} - \left[1 - \left(\frac{m(0)}{M}\right)^{1-p}\right]\right\}\exp\left(-\frac{t}{\tau_T}\right). \tag{8.9}$$

The parameter τ_T is a characteristic time for the reaction of the tumor system to perturbations in the nourishment protocol. The response to "slow" sinusoidal perturbations whose frequencies ω satisfy $\omega\tau_T \ll 1$ will quasistatically follow the perturbation and the growth pattern will closely match the modifications in the nourishment protocol. On the other hand, "fast" perturbations, such that $\omega\tau_T \gg 1$, cannot be followed by the cell system and will generate a delay and an overall decrease in the magnitude of the response. Below we discuss the impact of modifications in the factor $\omega\tau_T$ on the hysteretic cycle.

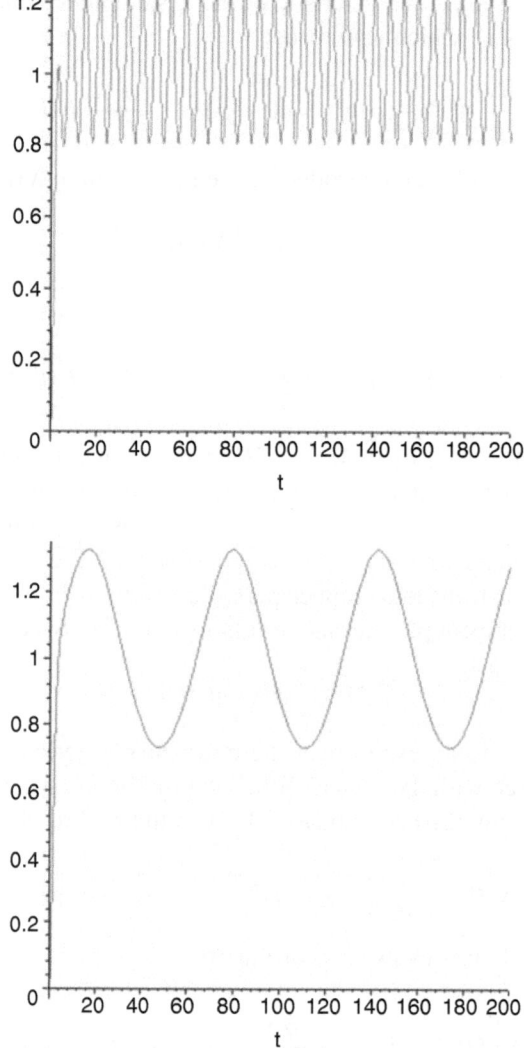

Fig. 8.1. Total tumor mass as a function of time, for diffusive feeding ($p = 2/3$). Here we chose $M = 1$, $c/a = 0.1$, and (a) $\omega/\alpha = 1$ and (b) $\omega/\alpha = 0.1$.

The mass evolution for a system with $p = 2/3$ is shown in Figure 8.1 for $m(0) = 0$. We observe that for $\omega/d = 1$ sharply defined oscillations appear in the total mass. After the transitory has died down, the mass fluctuates as

$$m_D(t) = M \left[1 - \frac{\alpha c}{a(\alpha^2 + \omega^2)^{1/2}} \sin(\omega t + \phi) \right]^{1/1-p}, \qquad (8.10)$$

with the phase $\phi = \text{arctg}(\omega/\alpha)$.

2.1 Special Case: $p = 2/3$

The preceding description holds for arbitrary values of the parameter p in the interval $(0,1)$. The particular value we choose depends on the nutrient transport mechanism. In the case of in vivo tumors, especially after the onset of angiogenic development, the nutrients arrive through the "neo" vascular net, which, according to West and coworkers, suggests $p = 3/4$ [16]. MTSs, on the other hand, directly absorb nutrients that diffuse through the culture medium; for a nurturing process that occurs through nutrient diffusion towards the tumor surface, $p = 2/3$ and $\alpha = b/3$. In this case, the output function (the mass) can be written as

$$m_D(t) = m_0 + \sum_{j=1}^{3} m_j \sin(j\omega t + \varphi_j). \tag{8.11}$$

Here, $\varphi_j = j\phi + (\pi/2)\delta_{j2}$ (δ_{ij} is Kronecker's symbol) and the m_js are given by

$$m_0 = M\left(1 + \frac{3B^2}{2}\right) \qquad m_1 = -3MB\left(1 + \frac{B^2}{4}\right) \tag{8.12a}$$

and

$$m_2 = \frac{-3MB^2}{2} \qquad m_3 = \frac{MB^3}{4}, \tag{8.12b}$$

with

$$B = (c/a)(1 + v^2)^{-1/2}. \tag{8.13}$$

Here, we have introduced the dimensionless input frequency $v = \omega/\alpha$ (Note that $1/\alpha = \tau_T$, the characteristic response time). Following Reference [2] we define $x = \sin \omega t$ and write the output relative to the equilibrium position as

$$\Delta m(x) = G(x) \pm H(x), \tag{8.14}$$

where

$$G(x) = x\left[P_1 - 2xP_2 + (3 - 4x^2)P_3\right] \tag{8.15}$$

and

$$H(x) = \sqrt{1 - x^2}\left[Q_1 + 2xQ_2 + \left(1 - 4x^2\right)Q_3\right]. \tag{8.16}$$

The coefficients P_i and Q_i can be readily evaluated yielding

$$P_1 = -\frac{3Ma}{c}B^2\left(1 + \frac{B^2}{4}\right) \qquad P_2 = \frac{3M}{2}B^2\left[1 - 2\left(\frac{aB}{c}\right)^2\right] \tag{8.17a}$$

$$P_3 = \frac{Ma}{4c}B^4\left[-3 + \left(\frac{2aB}{c}\right)^2\right] \qquad Q_1 = -\frac{3Ma}{c}vB^2\left(1 + \frac{B^2}{4}\right) \tag{8.17b}$$

$$Q_2 = \frac{3Ma^2}{c^2}vB^4 \qquad Q_3 = \frac{Ma}{4c}vB^4\left[3 - v^2\left(\frac{2aB}{c}\right)^2\right]. \tag{8.17c}$$

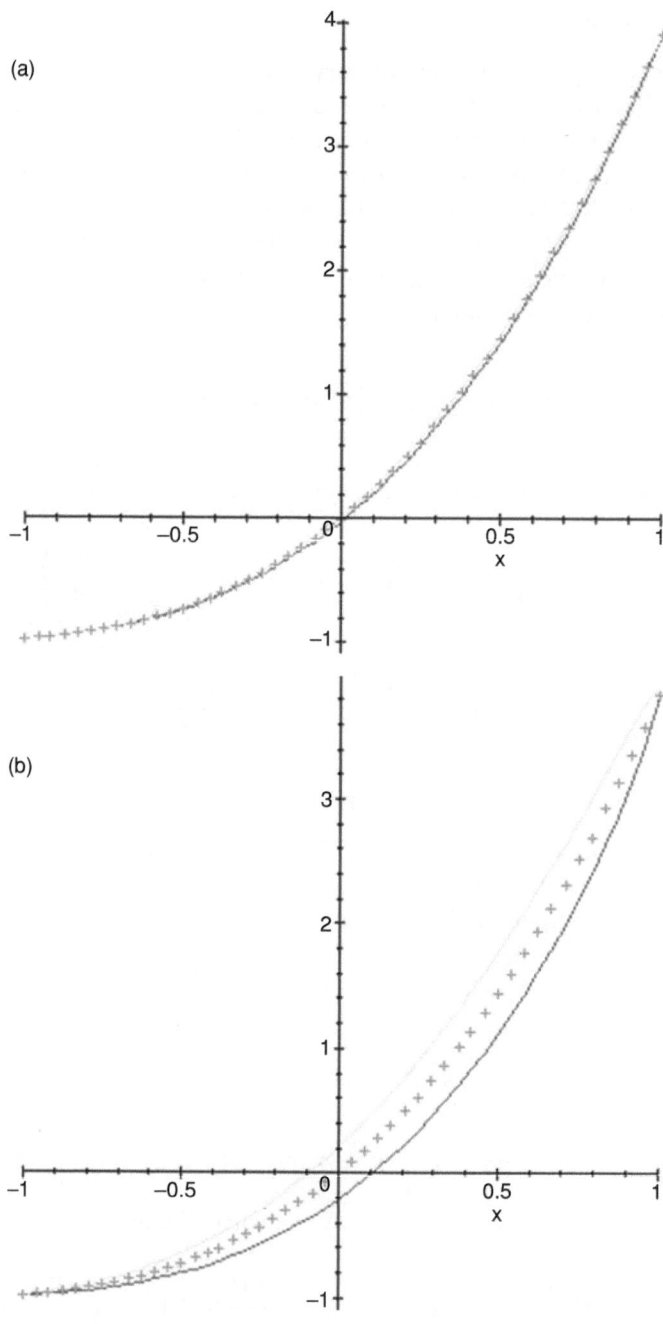

Fig. 8.2. Continued.

Large values of Q_j are associated with a strong hysteretic response, but nonlinearity becomes relevant only when there are one or more values of P_j or Q_j ($j > 1$) that are nonnegligible when compared to the largest between P_1 and Q_1. Some typical cycles are shown in Figure 8.2. In this figure we note the following.

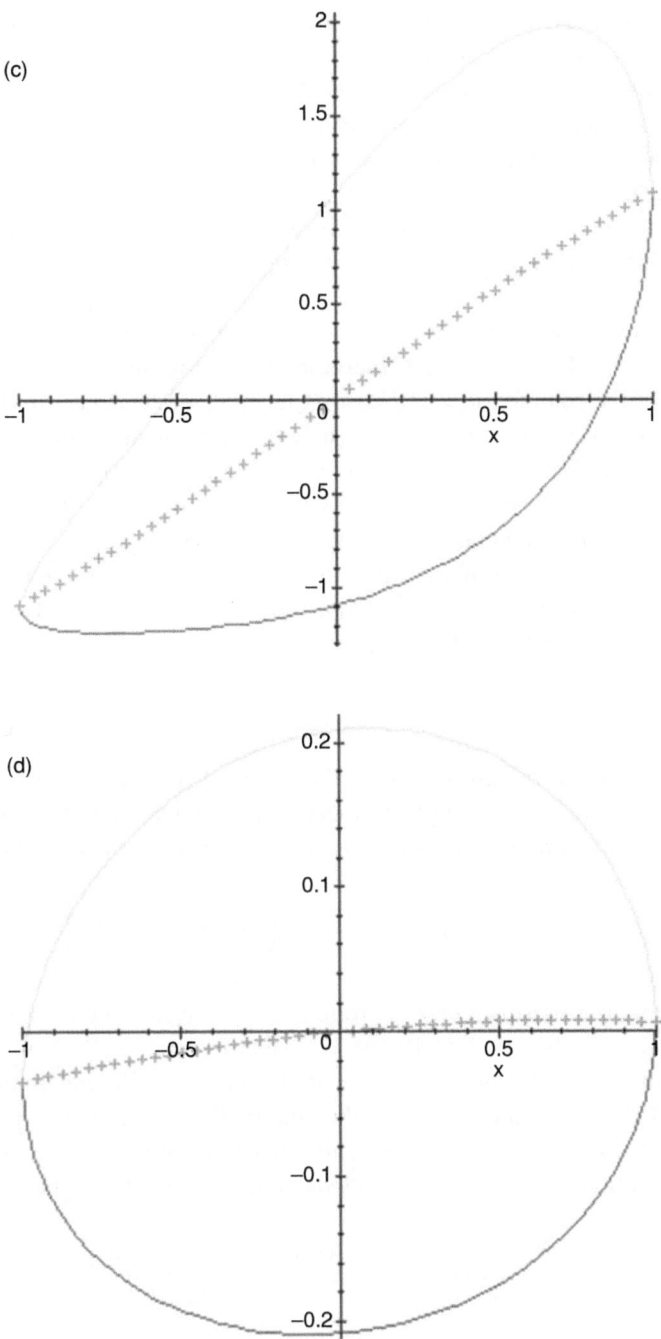

Fig. 8.2. Hysteretic response of a tumor to a sinusoidal feeding signal. Here $p = 2/3$, $M = 1$, $c/a = 0.7$, and $\omega/\alpha = 0.01$ (a), 0.1 (b), 1 (c), and 10 (d). The dotted line corresponds to $G(x)$; the upper and lower solid lines correspond, respectively, to $G(x) - H(x)$ (green upper line) and $G(x) + H(x)$ (blue lower line).

There is a strong asymmetry between the positive and negative x regions. In the quasistatic growth limit, $\nu \ll 1$, the time-averaged live cell mass is much larger than the stationary live cell mass in the absence of the oscillatory input. Note that a positive value of x means an above-average nutrient/oxygen influx.

The width of the hysteretic cycle goes through a maximum at $\nu = 1$ (please note the different scales).

For small values of ν the cycle becomes very thin, which should have been expected, in as much as the phase shift $\phi = \text{arctg}(\nu)$ is very small. Nonlinearity is strong, though: $P_2 \approx 0.3 P_1$.

For large values of ν the phase shift $\phi \to \pi/2$, which translates into the rounded cycle shown in (d). Mass growth lags increased nutrient influx by $\pi/2$.

Hysteresis in the rounded $\nu = 10$ case is mostly due to delays in the system response: it is easy to see that Q_1 is at least two orders of magnitude larger than the remaining P_j, Q_j. This agrees with the observation in Reference [2] that symmetric hysteretic cycles can be usually associated with delays, not with NCNL effects.

One measure of the "nonclassical strength" of a system, that is, the magnitude of its NCNL reaction to an external perturbation is the *generalized remanence R*, which is the displacement of the output from equilibrium when the magnitude of the input signal vanishes; that is, at $x = 0$,

$$R = |\Delta m(0)| = \left| \sum_{j=0} Q_{2j+1} \right|. \tag{8.18}$$

In the context of cancer growth, R is the excess live cancer cell mass when the nutrient/oxygen input goes through its mean value. If we take $p = 2/3$ and write $M = (a/b)^3$ (the maximum attainable mass for the $c = 0$ case), we obtain

$$R(\nu) = M \left(\frac{c}{a} \right) \frac{\nu}{1 + \nu^2} \left[3 + \left(\frac{c}{a} \right)^2 \left(\frac{\nu}{1 + \nu^2} \right)^2 \right]. \tag{8.19}$$

This is an important result. The hysteretic effects vanish in the $\nu \to 0$ and $\nu \to \infty$ limits, reaching a well-defined peak at $\nu = 1$, that is, $\omega = \alpha$. See Figure 8.3, where we plot the dimensionless ratio $(b/a)^3 R = R/M$ against ν. The maximum remanence occurs when the input frequency equals the "internal" frequency $\alpha = (\tau_T)^{-1}$, that is, when $\omega \tau_T = 1$. Alternatively, we can state that the peak location is proportional to the cell consumption rate $b = 3\alpha$. It depends on this parameter alone. The maximum remanence is

$$R_{\max} = R(\nu = 1) = \frac{3Mc}{2a} \left[1 + \frac{1}{12} \left(\frac{c}{a} \right)^2 \right]. \tag{8.20}$$

3. Discussion

By considering periodic oscillations in the nourishment protocol of a tumor, and analyzing its growth with a generalization of the WBE model, we have been able to show that cancer growth has well-defined NCNL features; these are more obvious when

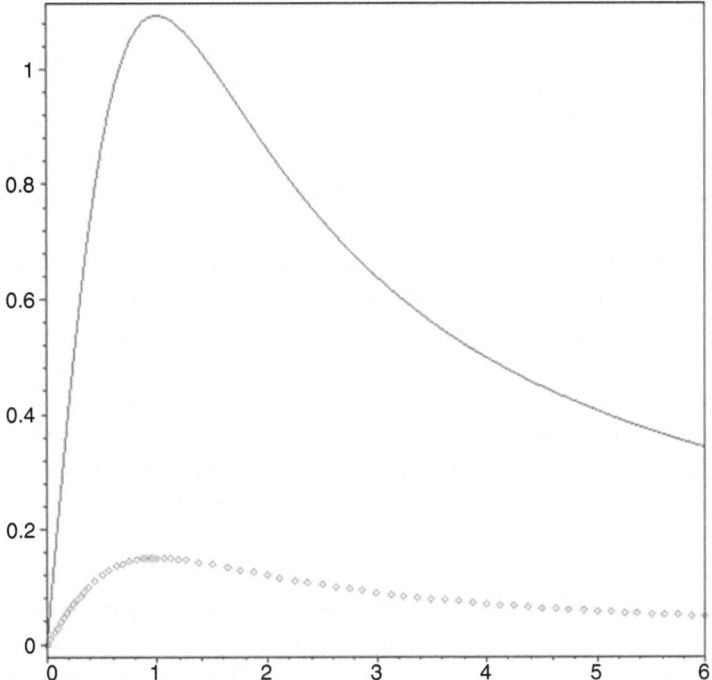

Fig. 8.3. Generalized remanence as a function of dimensionless input frequency $v = \omega/\alpha$ for $p = 2/3, a/c = 0.1$ (solid line), and 0.7 (dotted line).

the input frequency is comparable with the inverse of the intrinsic response time of the system (Figure 8.2c). For smaller frequencies, nonlinearity plays an important role, but hysteresis is weak (Figures 8.2a,b), whereas for high frequencies, the wide hysteretic cycles are associated with response delays, not with nonlinearity (Figure 8.2d).

In this chapter we propose that the NCNL features of MTS growth can be experimentally identified by the use of a time-varying nourishment protocol as may be attained through, for example, peristaltic pump systems. However, NCNL features should be inherent to cancer growth and do not depend on whether we perturb the tumor. There is an interesting byproduct of our model: by noting that the location of the remanence peak is at $\omega = b/3$, we observe that the proposed dynamical experiment generates an independent value of the WBE "steady-state" parameter b. This is not entirely unexpected, because the response to an oscillatory perturbation of a physical system often yields information about a steady-state property.

The effects of a changing periodic input could be spontaneously present in cancer in vivo: some studies indicate that tumor growth is affected by circadian rhythms [21, 23]. On the other hand, cancer evolution is likely to be insensitive to fast oscillatory processes, such as those associated with the pulsations of blood vessels. Such processes will satisfy the inequality $\omega\tau_T \gg 1$: the external input changes too fast for the system to respond and the hysteretic cycle becomes a circle of negligible radius.

The particular application we considered here corresponds to diffusion-controlled feeding, but we can consider other energy input mechanisms. This may lead to different values for the exponent p and, consequently, to some quantitative changes in our predictions, but the existence of nonclassical nonlinear features should be a robust feature of cancer growth, completely independent of the value we select for p.

Cancer growth is a very complex phenomenon. Our model is extremely simplified, but it clearly exhibits phenomena that are intrinsic to tumor growth. In a more realistic model cell death and other cellular processes must be taken into account in addition; nonetheless, we expect the central predictions of this chapter to survive the introduction of further complexity. An approximation in our model is that we have not introduced any explicit delay between changes in the input and cellular behavior (the observed response delays are emergent properties of the model). This is a very good approximation when $\omega\tau_T < 1$ (the cell system accommodates to the changing environment), but it is probably a poor approximation if $\omega\tau_T \Im 1$. In this case changes in the nourishment pattern will average out before the cell system can respond. For instance, we expect the sharp oscillations apparent in Figure 8.1b to be partially smoothed out. Some preliminary calculations show, however, that the NCNL response is strengthened by the introduction of explicit delays.

Finally, if cautiously extrapolated to a more clinically relevant situation, our model may prove useful as well. For instance, chemotherapy treatment cycles are supposedly administered in intervals sufficiently long to allow the tumor cell system to quasistatically follow the external changes. The possibility of using convenient variations of the model presented here to predict optimal cycle lengths with regard to the regimen's antitumor efficacy is currently under study.

Acknowledgments

This work has been supported in part by NIH grants CA085139 and CA113004 and by the Harvard–MIT (HST) Athinoula A. Martinos Center for Biomedical Imaging and the Department of Radiology at Massachusetts General Hospital. Y.M. is the recipient of a NCI-Training Grant Fellowship from the National Institutes of Health (CA09502). This research was also supported by CONICET and SECyT-UNC, Argentina.

References

[1] R.A. Guyer and P. Johnson, Nonlinear mesoscopic elasticity: Evidence for a new class of materials, *Phys. Today* **52**, 30–36 (1999).

[2] S. Hirsekorn and P.P. Delsanto, On the universality of nonclassical nonlinear phenomena and their classification, *Appl. Phys. Lett.* **84**, 1413–1415 (2004).

[3] M. Scalerandi, A. Romano, G.P. Pescarmona, P.P. Delsanto, and C.A. Condat, Nutrient competition as a determinant for cancer growth, *Phys. Rev. E* **59**, 2206–2217 (1999).

[4] B. Capogrosso Sansone, M. Scalerandi, and C.A. Condat, Diffusion with evolving sources and competing sinks: Development of angiogenesis, *Phys. Rev. E* **65**, 011902, 1–9 (2001).

[5] M. Scalerandi, B. Capogrosso Sansone, C. Benati, and C. A. Condat, Competition effects in the dynamics of tumor cords, *Phys. Rev. E* **65**, 051918, 1–10 (2002).

[6] *A Survey of Models for Immune-Immune System Dynamics*, edited by J.A. Adam and N. Bellomo (Birkhäuser, Boston, 1997).

[7] R. Kansal, S. Torquato, G.R. Harsh, IV, E.A. Chiocca, and T.S. Deisboeck, Simulated brain tumor growth dynamics using three-dimensional cellular automata, *J. Theor. Biol.* **203**, 367–382 (2000).

[8] S.C. Ferreira, Jr., M.L. Martins, and M.J. Vilela, Reaction-diffusion model for the growth of avascular tumor, *Phys. Rev. E* **65**, 021907, 1–8 (2002).

[9] G. Hamilton, Multicellular spheroids as an in vitro tumor model, *Cancer Lett.* **131**, 29–34 (1998).

[10] K.E. Thomson and H.M. Byrne, Modelling the internalization of labelled cells in tumour spheroids, *Bull. Math. Biol.* **61**, 601–623 (1999).

[11] W. Mueller-Klieser, Tumor biology and experimental therapeutics, *Crit. Rev. Oncol/Hematol.* **36**, 123–139 (2000).

[12] R. Chignola, A. Schenetti, E. Chiesa, R. Foroni, S. Sartoris, A. Brendolan, G. Tridente, G. Andrighetto, and L. Liberati, Forecasting the growth of multicell tumour spheroids: implications for the dynamic growth of solid tumours, *Cell Prolif.* **33**, 219–229 (2000).

[13] J.M. Kelm, N.E. Timmins, C.J. Brown, M. Fussenegger, and L.K. Nielsen, A method for the generation of homogeneous multicellular tumor spheroids applicable to a wide variety of cell types, *Biotech. Bioeng.* **83**, 173–180 (2003).

[14] G.P. Pescarmona, The evolutionary advantage of being conservative: The role of hysteresis, this volume, Chapter 7.

[15] A.J. Banchio and C.A. Condat, Seasonality and harvesting, revisited, this volume, Chapter 9.

[16] G.B. West, J.H. Brown, and B.J. Enquist, A general model for ontogenetic growth, *Nature* **413**, 628–631 (2001).

[17] C. Guiot, P.G. Degiorgis, P.P. Delsanto, P. Gabriele, and T.S. Deisboeck, Does tumor growth follow a "universal law"?, *J. Theor. Biol.* **225**, 147–151 (2003).

[18] P.P. Delsanto, C. Guiot, P.G. Degiorgis, C.A. Condat, Y. Mansury, and T. Deisboeck, Growth model for multicellular tumor spheroids, *Appl. Phys. Lett.* **85**, 4225–4227 (2004).

[19] M. Kleiber, Body size and metabolism, *Hilgardia* **6**, 315–353 (1932).

[20] P.S. Dodds, D.H. Rothman, and J.S. Weitz, Re-examination of the "3/4–law" of metabolism, *J. Theor. Biol.* **209**, 9–27 (2001).

[21] A.M. Makarieva, V.G. Gorshkov, and B.-L. Li, A note on metabolic rate dependence on body size in plants and animals, *J. Theor. Biol.* **221**, 301–307 (2003).

[22] *Circadian Cancer Therapy*, edited by W.J.M. Hrushesky (CRC, Boca Raton, FL, 1994).

[23] T. Roenneberg and M. Merrow, The network of time: Understanding the molecular circadian system, *Curr Biol.* **13**, R198–207 (2003).

9

Seasonality and Harvesting, Revisited

A.J. Banchio[1] and C.A. Condat[1,2,3]

[1]CONICET and FaMAF, Universidad Nacional de Córdoba, 5000-Córdoba, Argentina.
[2]Department of Physics, University of Puerto Rico, Mayagüez, PR 00681, USA.
[3]To whom correspondence should be addressed; e-mail: cacondat@yahoo.com.

Abstract
Ecological processes are often influenced by seasonal variations. These variations may generate complex responses, which can be advantageously analyzed using the concept of nonlinear nonclassical universality. This procedure is illustrated by considering periodic fluctuations in single species harvesting and seasonal changes in the environment carrying capacity. It is shown that the ratio between the perturbation period and the reaction time of the associated autonomous system plays a crucial role in the response, whose spectral components are investigated in detail.

Keywords: Carrying capacity, cycles, ecology, harvesting, hysteresis, nonclassical, nonlinearity, periodicity, recovery time, remanence, seasons

1. Introduction

The evolution of ecological phenomena is strongly influenced by seasonal variations, which introduce periodically varying inputs. A challenge for the ecologist is to understand the response of a given system to these inputs. A host of single species and competing species population models have been developed to predict how periodic variations affect different biosystems [1, 2]. The literature on the dynamical behavior of ecosystems subject to periodic perturbations is indeed quite extensive, ranging from general environmental perturbations [3] to predator–prey models with a seasonal functional response [4] to multistrain epidemic models [5]. Some biological species, such as the flour beetle, have been subjected to very detailed experimental studies and modeling in order to probe the influence of periodic oscillations in relevant environmental parameters [2, 6].

Strongly nonlinear models for the evolution of interacting species lead to rich dynamical structures, but their complexity makes it convenient to enlarge as much as possible the arsenal of analytical methods. In this connection, we note that the recently discovered universality of nonclassical nonlinear (NCNL) phenomena offers a potent alternative tool to investigate and predict phenomena related to periodic perturbations [7]. Starting from the observation of nonlinear hysteretic effects in the elastic behavior of rocks and other materials [8,9], Delsanto and Hirsekorn were able to define a "nonlinear mesoscopic elastic universality class," to which such materials as rock, soil, and concrete belong [7]. Using a "response box" formalism, they were able to classify the response of NCNL systems according to their spectral contents.

The main objective of this contribution is to show that relevant ecological phenomena belong in the NCNL universality class, even if their nature is completely different from that of rock elasticity. We take advantage of universality to borrow hysteresis diagrams and the concept of remanence from the theory of magnetism, and use these elements to study two simple paradigms of ecological processes. First, we consider the periodic harvesting of a single species, and then we investigate the evolution of a population subject to a periodically varying carrying capacity.

Due to its obvious economic importance related to forest management, insect control, fisheries, and so on, the problem of harvesting has been the subject of many studies in the past [1, 10]. With some modifications, these analyses can be adapted to the associated problems of seeding and pest control. Here we obtain the spectral response of the harvested population in the case of periodic harvesting, showing that the strongest NCNL effects occur when the harvesting frequency is approximately the inverse of the system recovery time. Similar results are obtained when it is the carrying capacity that oscillates periodically, although the NCNL effects are stronger.

2. The Model

The evolution of a population subject to harvesting can be described using a modified logistic equation,

$$\frac{dP}{dt} = kP\left[1 - \frac{P}{N(t)}\right] - J(P,t),$$

(9.1)

where $P(t)$ is the total population of the species under consideration and k is a positive constant. $N(t)$ is the carrying capacity of the environment, which may be subject to seasonal variations, and $k(1 - P/N)$ is the per capita birth rate [1]. The harvesting rate $J(P,t)$ may depend explicitly on time and on the instantaneous population, although in this work we assume that it depends only on t.

If N and J are both time- and population-independent, Eq. (9.1) is autonomous and it is easy to find that there are two equilibrium solutions, $P_s^{\pm}$,

$$P_s^{\pm} = \frac{1}{2}\left(N \pm \sqrt{N^2 - \frac{4N}{k}J}\right).$$

(9.2)

Because only real solutions are of our interest, there is a maximum limit $J_M = kN/4$ for the harvesting rate. If $J = J_M$, both solutions coalesce to $P_s^{\pm} = N/2$. If $J < J_M$, and $P(0) > P_s^-$, the population evolves towards the effective carrying capacity P_s^+ ($< N$). However, if $J < J_M$, but $P(0) < P_s^-$, the population dies out.

After a small perturbation, the population returns to its equilibrium value P_s^+ over the *recovery time*

$$\tau_k = \frac{\sqrt{N/k}}{\sqrt{kN - 4J}}.$$

(9.3)

This recovery time grows monotonically as the harvesting rate is increased, starting from $\tau_k = 1/k$ for $J = 0$ (a higher harvesting rate leads to a slower population recovery).

In this work we consider the effect of a periodicity either in the harvesting term or in the carrying capacity. In this case, following Reference [7], we write the output function (the instantaneous population) as a Fourier expansion,

$$P^+(t) = P_0 + \sum_{j=1}^{\infty} P_j \sin(j\omega t + \phi_j). \tag{9.4}$$

By defining $x = \sin \omega t$, we can write the output relative to the equilibrium position as [7],

$$\Delta P(x) = P^+(t) - P_0 \equiv G(x) \pm H(x), \tag{9.5}$$

where

$$G(x) = x[S_1 - 2x S_2 + (3 - 4x^2)S_3 - \cdots] \tag{9.6}$$

and

$$H(x) = \sqrt{1 - x^2}[Q_1 + 2x Q_2 + (1 - 4x^2)Q_3 + \cdots]. \tag{9.7}$$

The values of the coefficients S_i and Q_i depend on the details of the input signal: $S_n = P_n \cos \phi_n$, $Q_n = P_n \sin \phi_n$ if n is odd, and $S_n = P_n \sin \phi_n$, $Q_n = P_n \cos \phi_n$ if n is even [7].

3. Periodicity in Harvesting: Weak NCNL Effects

Suppose that the environment carrying capacity is constant, but the harvesting rate $J(t)$ contains a periodic component with period T [11],

$$J(t) = a + c \sin(\omega t). \tag{9.8}$$

Here a and c are, respectively, the amplitudes of the constant and periodic harvesting components, and $\omega = 2\pi/T$, where T is the harvesting period. In this formulation the harvesting yield is externally fixed; that is, it does not depend on the population. We note that an alternative harvesting strategy would be to consider a population-dependent yield, for example, writing $J(t) = P(t)(a + c \sin \omega t)$.

If the harvesting period is much larger than the recovery time τ_k of the autonomous (i.e., $c = 0$) problem, $\omega \tau_k \ll 1$, the system evolves quasistatically, and the time-dependent solution will be well approximated by the stationary solution corresponding to the instantaneous value of the harvesting term

$$P_s^{\pm}(t) \approx \frac{1}{2}\left(N \pm \sqrt{N^2 - \frac{4N}{k}[a + c \sin(\omega t)]}\right). \tag{9.9}$$

The square root can be expanded in a power series and then the powers of the sine expressed as a combination of harmonics. Comparing with Eq. (9.4), we readily obtain analytical expressions for the parameters P_j and ϕ_j, which are valid in the $\omega \tau_k \ll 1$ regime. Defining

$$M = \frac{N}{2}\sqrt{1 - \frac{4a}{kN}} \tag{9.10}$$

and

$$\xi = \frac{4c}{kN - 4a}, \tag{9.11}$$

the first few $P_j s$ are, up to order ξ^4,

$$P_0 \cong M\left(1 - \frac{1}{16}\xi^2 - \frac{15}{1024}\xi^4\right) \tag{9.12a}$$

$$P_1 \cong -\frac{M}{2}\left(\xi + \frac{3}{32}\xi^3\right) \tag{9.12b}$$

$$P_2 \cong \frac{M}{16}\left(\xi^2 + \frac{5}{16}\xi^4\right) \tag{9.12c}$$

$$P_3 \cong \frac{M}{64}\xi^3 \ . \tag{9.12d}$$

We also obtain $\phi_1 = \phi_3 = 0$, and $\phi_2 = \phi_4 = \pi/2$. Note that as the amplitude c of the periodic harvesting is increased, more intensity is transferred from P_0 to the higher P_i components.

If $\omega\tau_k \gg 1$, the internal dynamics of the system are too slow to respond to the perturbation, and the solution is approximately given by Eq. (9.2). For arbitrary values of $\omega\tau_k$, Eq. (9.1) must be solved numerically. Once the solution is obtained, it can be Fourier-expanded and the coefficients P_j and ϕ_j evaluated. We solve Eq. (9.1) by using a simple Euler forward integration scheme, verifying that the results do not depend on the time step selection. To obtain the parameters for the Fourier expansion of the population $P(t)$, we consider one period of the stationary solution, using the numerical data to compute the sine and cosine fast Fourier transforms. It is then straightforward to compute the amplitudes and phases P_j and ϕ_j.

For all evaluations we have chosen $N = 50$ and $k = 0.2$. The results for the amplitudes and phases are shown in Figure 9.1. In the limit $\omega\tau_k \ll 1$, the results are given by Eqs. (9.12), although more terms in the expansion are needed to obtain a good approximation for P_3. The amplitudes $P_j(j > 0)$ decrease as $\omega\tau$ increases, tending to zero in the $\omega\tau_k \to \infty$ limit: the system has no time to react to input changes and the response amplitude averages to zero. The phase shift, on the other hand, increases monotonically with the frequency of the applied signal. It is also clear that the delay increases with j. The magnitude of NCNL effects results from a combination of high values of $P_j(j > 1)$ and large phase shifts. Therefore, we expect NCNL effects to be strongest at intermediate values of $\omega\tau_k$. To investigate this, we follow Reference [7] and plot (see Figure 9.2) the hysteresis diagrams corresponding to various values of the harvesting frequency.

The central lines in Figures 9.2a to 9.2c correspond to $G(x)$, and the upper and lower lines correspond, respectively, to $G + H$ and $G - H$ [see Eqs. (9.6) and (9.7)]. The diagram shapes can be interpreted using Figure 9.1. For instance, the coefficients $P_j(j > 0)$ are comparatively large, whereas ϕ_1 is very small, and $\phi_2 \approx 1$ when

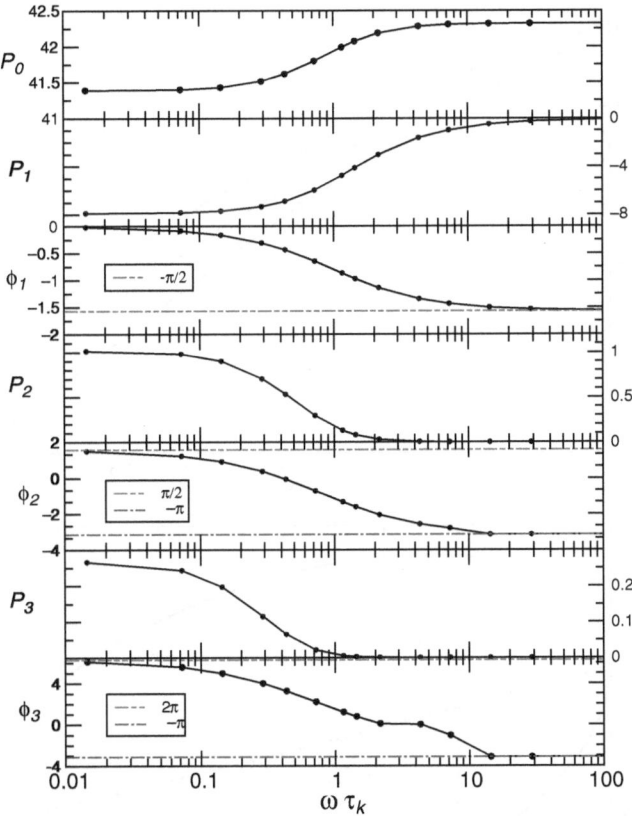

Fig. 9.1. Amplitudes and phases of the response to a sinusoidal harvesting variation. Here $N = 50$, $k = 0.2$, $a = 1.3$, and $c = 1$. Eq. (9.12a) yields $Lim_{(\omega \to 0)} P_0 = 41.446$. Note that the vertical scales for the different panels are alternatively on the left and right hand sides of the figure.

$\omega \tau_k = 0.14$ (Figure 9.2a). The large value of $x S_1 = x P_1 \cos \phi_1$ is responsible for the high $G(x)$ tilt, and the $2x^2 S_2 = 2x^2 P_2 \sin \phi_2$ term is responsible for its bending. The pearlike cycle shape is due to the high value of $x Q_2 = x P_2 \cos \phi_2$ combined with the small value of Q_1. The decreasing value of the output $\Delta P(x)$ with increasing $\omega \tau_k$ is due to the monotonic decrease in $P_j (j > 0)$.

Figure 9.2b can be understood by noting that, because $\omega \tau_k = 2.17$, $\phi_1 \approx -2\pi/3$. Therefore, $S_1 \approx -3 \cos 2\pi/3$ has less than half the value corresponding to the case of Figure 9.2a, and the tilt of G is consequently smaller. On the other hand, $Q_1 \approx 3 \sin 2\pi/3$ is rather large, generating a very wide cycle, whose symmetry is guaranteed by the smallness of $Q_j (j > 1)$.

For Figure 9.2c, $\omega \tau_k = 7.22$, $\phi_1 \approx -\pi/2$, and all $P_j (j > 1)$ are negligible. Therefore G is very small and $H \approx (1 - x^2)^{1/2}$. The three cases are shown together in Figure 9.2d, which helps us to compare their features.

It is interesting to note that the hysteresis cycles have a negative tilt. This is at variance with the results presented in References [7] and [12] and simply indicates that the oscillating harvesting term initially introduces a negative input in the system. The

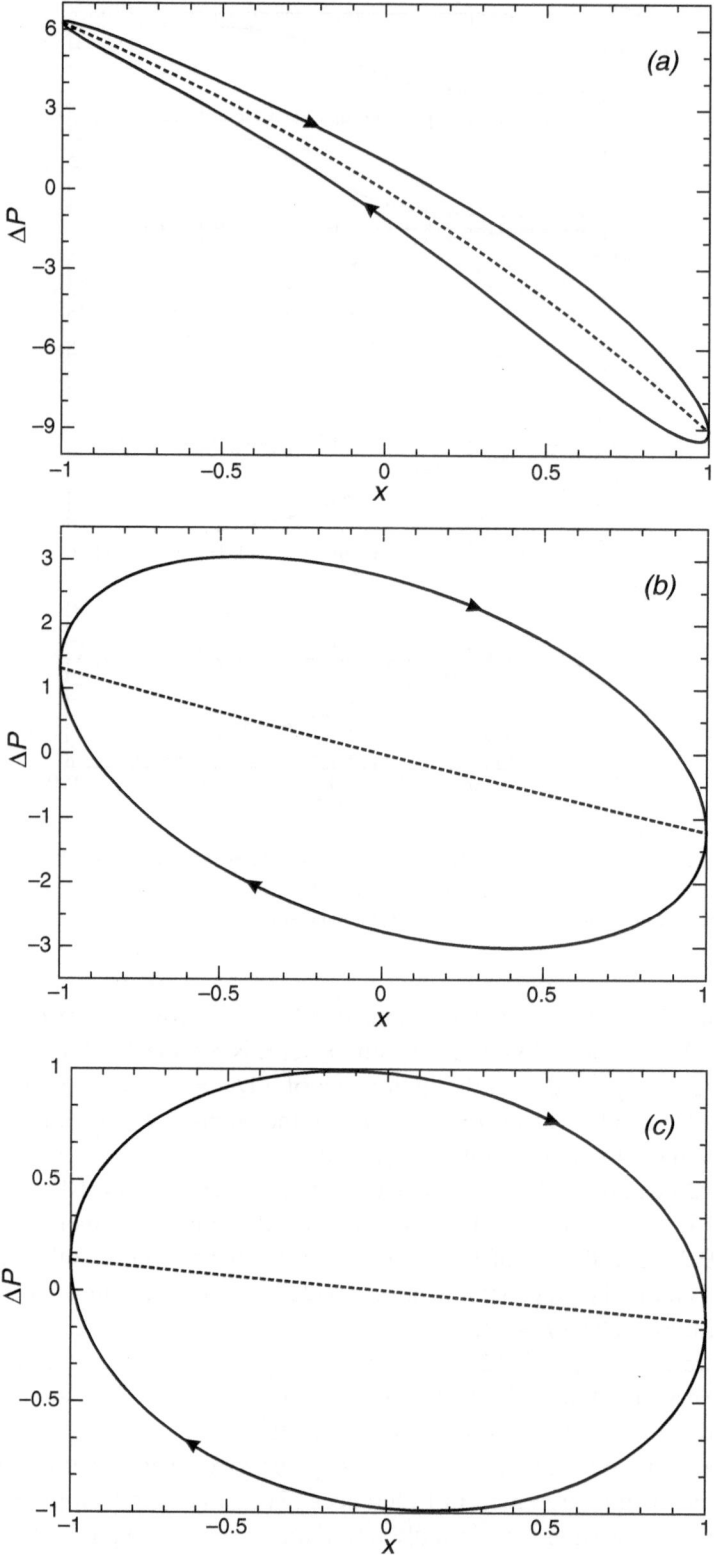

Fig. 9.2. Continued.

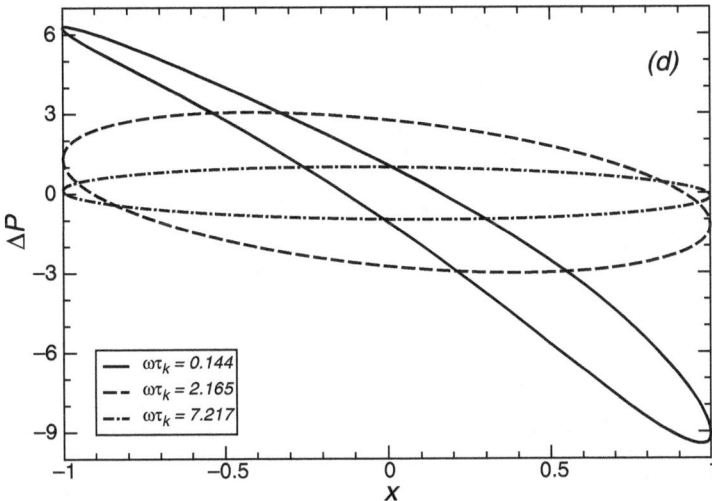

Fig. 9.2. Hysteresis diagrams corresponding to a sinusoidal harvesting variation. We used the parameters of Fig. 9.1 and $\omega\tau_k = 0.14$ (a), 2.17 (b), and 7.22 (c). The three results are presented together in (d).

problem of seasonal seeding (e.g., periodically seeding trout in a stream), can be modeled by considering the coefficients a and c in Eq. (9.8) to be negative. In such a case there is an initially positive input into the system and the hysteresis graphs are again positively tilted.

As in the discussion of cancer growth [12], a suitable measure of the NCNL reaction to an external perturbation is the *generalized remanence R*, which is defined as the displacement of the output from equilibrium when the driving field vanishes; that is, at $x = 0$,

$$R = |H(0)| = \left|\sum_{j=0} Q_{2j+1}\right|. \tag{9.13}$$

The remanence is plotted in Figure 9.3 as a function of the dimensionless parameter $\omega\tau_k$. We see a well-defined maximum when $\omega\tau_k \approx 1$. It vanishes in the small and large frequency limits, which correspond, respectively, to regions where the phase shift and the amplitude of the higher components become negligibly small. For annual periodic harvesting, the system goes through a cycle once a year, so that the remanence indicates the population depletion three months after the collection reaches its maximum. For a given value of ω, the magnitude of the remanence will depend on the recovery time: intense harvesting results in a larger τ_k, and thus in a displacement towards the right in Figure 9.3.

4. Periodicity in Carrying Capacity: Strong NCNL Effects

Next, we tackle the problem of an oscillating carrying capacity in the absence of harvesting,

$$\frac{dP}{dt} = kP\left[1 - \frac{P}{N_0 + b\sin(\omega t)}\right], \tag{9.14}$$

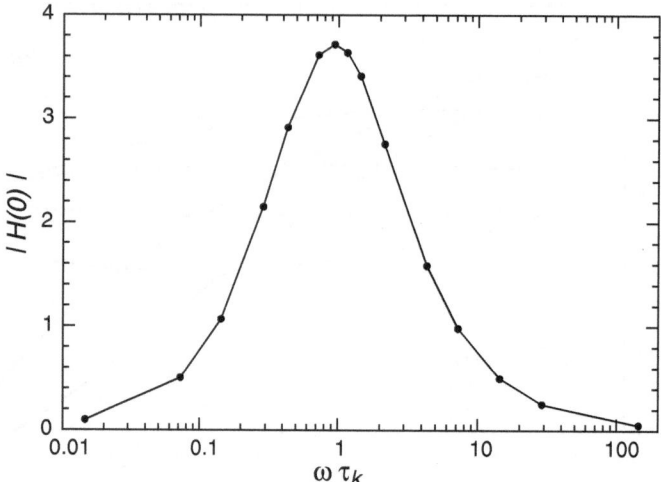

Fig. 9.3. Remanence corresponding to a sinusoidal harvesting variation for the parameters of Fig. 9.1.

with $b \leq N_0$. It is easy to introduce harvesting, but we prefer to neglect it to keep the interpretation as simple as possible. If $b = 0$, the equilibrium solution is $P_{eq} = N_0$. The response time of the linearized system is simply $\tau_k = 1/k$. Equation (9.14) is of the Bernoulli type, and can be integrated using standard methods [13]. If the process starts at $t = 0$, we obtain

$$P(t) = e^{kt}\left[\frac{1}{N_0} + k\int_0^t \frac{e^{ks}}{N_0 + b\sin(\omega s)}ds\right]^{-1}. \tag{9.15}$$

Although we can evaluate the integral numerically, we use instead the NCNL methods, which permit a deeper analysis. Under quasistatic conditions, $\omega\tau_k \ll 1$, the solution after the initial transitory is given by

$$P(t) \approx N_0 + b\sin(\omega t). \tag{9.16}$$

Comparing with,

$$P(t) = P_0 + \sum P_n \sin(n\omega t + \phi_n), \tag{9.17}$$

we find $P_0 = N_0$, $P_1 = b$, $P_n = 0$ ($n \geq 1$), and $\phi_1 = 0$. The population adapts continuously to the instantaneous carrying capacity.

In the opposite limit, $\omega\tau_k \gg 1$, the system records an effective carrying capacity N_{eff},

$$N_{eff} = \left[\frac{1}{T}\int_0^T \frac{dt}{N_0 + b\sin(\omega t)}\right]^{-1} = \sqrt{N_0^2 - b^2}, \tag{9.18}$$

which vanishes when $b \to N_0$: the rapid oscillations do not permit the system recovery, even if periodically the instantaneous carrying capacity reaches a large value, $N = 2N_0$.

The parameters P_n and ϕ_n for arbitrary values of $\omega\tau_k$ are obtained numerically. They are represented, for $n < 4$, in Figure 9.4, where we chose $N = 50$, $k = 0.2$, and $b = 20$. We note immediately several important differences with Figure 9.1:

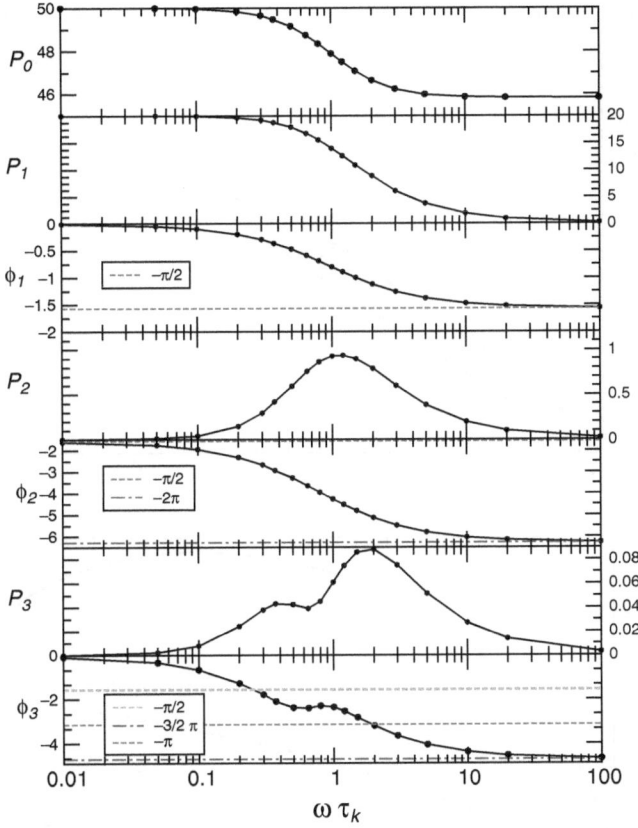

Fig. 9.4. Amplitudes and phases of the response to a sinusoidal variation in the carrying capacity. Here $N = 50, k = 0.2$, and $b = 20$. Eqs. (9.15) and (9.16) yield $Lim\ (\omega \to 0)\ P_0 = 50$.

(a) P_0 decreases with increasing $\omega\tau_k$. For large values of $\omega\tau_k$, $P_0 \to N_{\text{eff}}(\approx 45.83)$.

(b) P_1 is positive and quite large for small values of $\omega\tau_k$. In particular, $\text{Lim}_{\omega \to 0} P_1 = b$, because the periodic input directly translates into the output first harmonic.

(c) P_2 and P_3 are not monotonic, but exhibit well-defined maxima at intermediate values of $\omega\tau_k$. This affects the shape of the hysteresis cycles.

(d) The phases ϕ_1 and ϕ_2 become more negative with increasing $\omega\tau_k$, but ϕ_3 has a hump about $\omega\tau_k \approx 1$.

The hysteresis cycles are shown in Figure 9.5 for representative values of $\omega\tau_k$ and the parameter values used in Figure 9.4. As a consequence of the large value of P_1 when $\omega\tau_k = 0.1$ (Figure 9.5a), we obtain a very steep growth of $G(x)$: note that $xS_1 = xP_1 \cos\phi_1 \approx 20x$. In Figure 9.5b we see that $G(x)$ is markedly concave, the concavity being due to the $-2x^2S_2 \approx -1.3x^2$ term. The large value of $H(x)$—and the big remanence—follow from $Q_1 \approx 8$. For $\omega\tau_k = 10$ (Figure 9.5c), $\phi_1 \approx -\pi/2$ and $\phi_2 \approx -2\pi$. Consequently, $G(x)$ is very small. The asymmetry in $H(x)$ is due to the relatively large value of $Q_2(\approx 0.2)$. Note that, as in Figure 9.2, the central lines correspond to $G(x)$, but, because $H < 0$, the upper lines correspond to $G{-}H$.

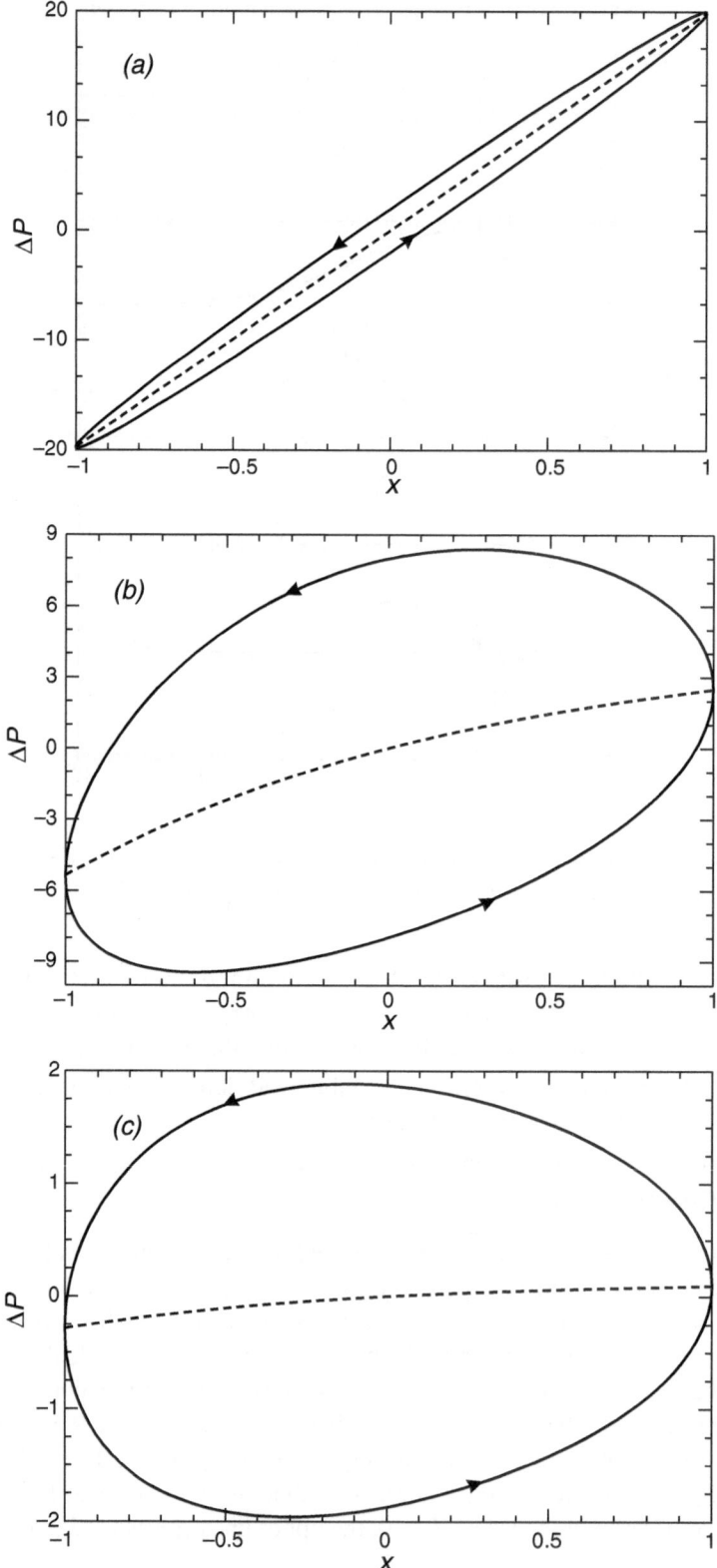

Fig. 9.5. Continued.

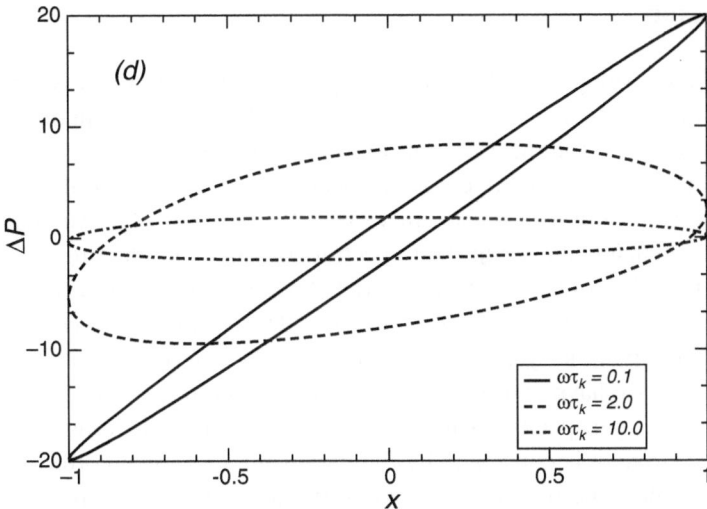

Fig. 9.5. Hysteresis diagrams corresponding to a sinusoidal variation in the carrying capacity. We used the parameters of Fig. 9.4 and $\omega\tau_k = 0.1$ (a), 2 (b), and 10 (c). The three results are presented together in (d).

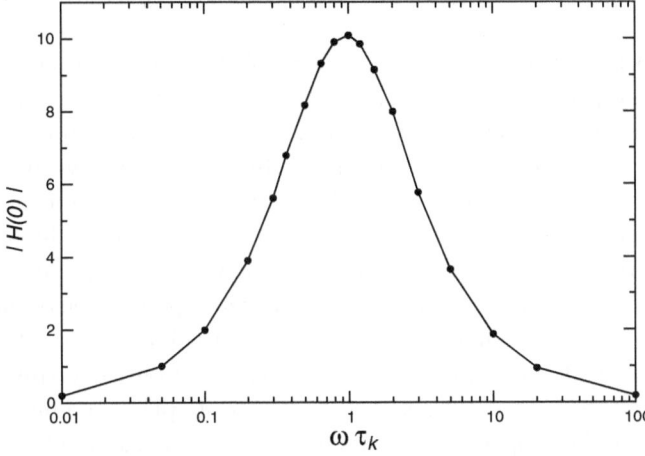

Fig. 9.6. Remanence corresponding to a sinusoidal variation in the carrying capacity for the parameters of Fig. 9.4. Note the well-defined peak at $\omega\tau_k \approx 1$.

Figure 9.5c clearly exhibits the strong dependence of the evolution of a population subject to seasonal carrying capacity variations on its intrinsic response time τ_k. If the system has a short recovery time, $\omega\tau_k \ll 1$, it follows the perturbation closely: population variations are very strong and in phase with the carrying capacity oscillations. As the species' reaction time is increased, population variations weaken and become more phase-shifted with respect to the input oscillations; the retardation increases monotonically with the recovery time, reaching a maximum equal to $T/4$ when $\tau_k \gg T$. The maximum remanence (Figure 9.6) occurs at intermediate values of τ_k, for which NCNL effects are strongest. Indeed, the maximum remanence occurs precisely at $\omega\tau_k = 1$. It is qualitatively similar to the remanence in the harvesting problem, but substantially larger.

5. Discussion

The formalism developed in Reference [7] provides us with a general framework to discuss the magnitude and relative importance of the various components of the response of a complex nonlinear system to a periodic perturbation. In this chapter we have shown that two common ecological phenomena can be described as belonging to the nonclassical nonlinear universality class. We have therefore studied them using the NCNL formalism, analyzing their spectral content and obtaining the corresponding hysteresis cycles. In particular, we were led naturally to the concept of remanence, which is well known in the study of magnetic materials, and which here quantifies the changes remaining in the system population when the external signal vanishes.

The examples we have considered, which correspond to periodic variations in harvesting and carrying capacity, exhibit, respectively, weak and strong NCNL features. The response strength is controlled by the parameter P_1, which corresponds to oscillations having the same frequency, although usually not the same phase, as the input signal. The bending, deformation, and rotation of the hysteresis diagrams are related to the coefficients P_2, P_3, and so on, of the higher harmonics and to the phase shifts. Strong NCNL effects are characterized not only by a high remanence, but also by a nonmonotonic dependence of P_2, P_3, ... and ϕ_3, ... with the all-important parameter $\omega\tau_k$.

We should bear in mind that the emergence of hysteretic cycles does not by itself mean that the system behaves nonclassically. In some cases, the cycles result from a simple, "classical," response delay. A signature of nonclassical effect is the influence of higher harmonics on the cycle shape. In the cases analyzed in this chapter, hysteresis vanishes in the limits $\omega\tau_k \to 0$ and $\omega\tau_k \to \infty$, but for moderately large values of $\omega\tau_k$, although the cycle subsists, the influence of higher harmonics is weak and, consequently, the system response is approximately classical. As we have shown, nonclassical effects are maximized when the natural response time of the unperturbed system is commensurate with the period of the perturbation, that is, when $\omega\tau_k \approx 1$.

Of course, the models considered in this chapter offer only oversimplified descriptions of real-life systems, but our goal was to choose simple paradigms to demonstrate some of the possibilities opened by the inclusion of ecological problems in the NCNL universality class. We conclude with two remarks.

The applicability of the concept of NCNL to living systems is not restricted to ecological problems. Whenever nutrition or another key element for the survival of a given species undergoes temporal oscillations, it is likely that the evolution of the resulting population can be analyzed using the NCNL formalism. Work in this direction has already started in the case of cancer cells [12]. The formalism could also be advantageously applied to investigate cell processes occurring in the presence of natural periodic variations, for example, intracellular calcium oscillations [14].

NCNL properties are explicitly brought out by the application of periodic perturbations, but they should also be present in the case of other perturbations. More theoretical work is required to determine precisely the class of nonlinear problems that should be expected to exhibit NCNL properties, which appears to be vast, and to develop the tools to analyze the case of nonperiodic perturbations.

Acknowledgments

This research has been supported by Fundación Antorchas, CONICET, and SECyT-UNC, Argentina.

References

[1] See, for example, J.D. Murray, *Mathematical Biology*, 3rd Ed., Vol. 1 (Springer Verlag, Berlin, 2002), and references therein.

[2] P. Turchin, *Complex Population Dynamics: A Theoretical/Empirical Synthesis* (Princeton University Press, Princeton, NJ, 2003).

[3] R.K. Upadhyay, V. Rai, and S.R.K. Iyengar, How do ecosystems respond to external perturbations?, *Chaos, Solitons Fractals* **11**, 1963–1982 (2000).

[4] S.M. Moghadas and M.E. Alexander, Dynamics of a generalized Gauss-type predator-prey model with a seasonal functional response, *Chaos, Solitons Fractals* **23**, 55–65 (2005).

[5] M. Kamo and A. Sasaki, The effect of cross-immunity and seasonal forcing in a multi-strain epidemic model, *Physica D* **165**, 228–241 (2002).

[6] S.M. Henson and J.M. Cushing, The effect of periodic habitat fluctuations on a nonlinear insect population model, *J. Math. Biol.* **36**, 201–226 (1997).

[7] S. Hirsekorn and P.P. Delsanto, On the universality of nonclassical nonlinear phenomena and their classification, *Appl. Phys. Lett.* **84**, 1413–1415 (2004).

[8] R.A. Guyer and P. Johnson, Nonlinear mesoscopic elasticity: Evidence for a new class of materials, *Phys. Today* **52**, 30–36 (1999).

[9] P.P. Delsanto and M. Scalerandi, Modeling nonclassical nonlinearity, conditioning, and slow dynamics effects in mesoscopic elastic materials, *Phys. Rev. B* **68**, 064107, 1–9, (2003).

[10] D. Ludwig, B. Walker, and C.S. Holling, *Sustainability, Stability and Resilience*, *Conservation Ecology* [online] **1**(1),7, 1997, and references therein, http://www.consecol.org/vol1/iss1/art7.

[11] A pedagogical introduction can be found in the article by A. Cross, G. McGuire, and C. Tashian, http://www.tashian.com/car/docs/harvesting.

[12] C.A. Condat, B.F. Gregor, Y. Mansury, and T.S. Deisboeck, Cancer growth is a nonclassical nonlinear phenomenon, this book, Chapter 8.

[13] M. Fan and K. Wang, Optimal harvesting policy for single population with periodic coefficients, *Math. Biosciences* **152**, 165–177 (1998).

[14] A. Goldbeter, *Biochemical Oscillations and Cellular Rhythms* (Cambridge University Press, Cambridge, UK, 1996).

10

Nonlinear Dynamical Systems in Economics

Paolo Patelli

T-13 and CNLS, University of California, Los Alamos National Laboratory, Los Alamos, New Mexico 87545, USA.
Santa Fe Institute, 1399 Hyde Park Rd, Santa Fe, New Mexico, 87501, USA. Correspondence: Paolo Patelli, Theoretical Division: T-13/CNLS, University of California, Los Alamos National Laboratory, Los Alamos, New Mexico 87545, USA; e-mail: paolo@lanl.gov.

Abstract
This chapter describes how economic theories approach the concept of nonlinearity and gives several examples. Although economics is a complex nonlinear system, nonlinearity is a recent concept in economics. Yet there are several phenomena that can be described only by nonlinear models. Power law distributions, economic scaling, out of equilibrium systems, and self-organizing criticality are evidences of nonlinearity. Different economic theories that deal with nonlinearity are reported. These models are the first nonlinear models introduced in economics and they are purely theoretical models without any connection to empirical evidence. Finally some empirical evidence of nonlinearity and the recent nonlinear models based on them and on bounded rational economic agents are described.

Keywords: Economics, linear and nonlinear models, neoclassical economics theory, nonlinear dynamical systems, nonlinearity, power law, scale free distributions, scaling, self-organizing criticality

1. Nonlinearity in Economics

Economics is a quite recent science. During the second half of the eighteenth century, for the first time Adam Smith and Ricardo tried to explain the economical behavior making economics a distinguished science. But it was only during the middle of the nineteenth century that a mathematical formalism was used to explain economic problems. The mathematical tools developed and used in physics were applied directly to economics. Physics concepts were also applied to economics, in particular the mechanistic worldview. According to this view if one knows the initial condition of a phenomenon and the rules that govern it, then one can also determine the future dynamics of the observed phenomenon. Unlike in physics, where the mechanistic approach was overcome by quantum mechanics, in economics the mechanistic approach still exists in present economics theory, known as mainstream or neoclassical economic theory. Neoclassical economic theory is based on scientific determinism, where the economic agents are rational, assessing all information, and they all behave in the same way maximizing their utility function. Based on these unrealistic hypotheses, this theory is not able to describe real economic behaviors. Most of the models developed by neoclassical economic theory are linear models and describe only time-independent properties at the equilibrium. The general approach used by neoclassical economics in model development is called the top/down approach: the models are pure mathematical and theoretical without any connection to empirical evidence. It was not until the 1980s and 1990s that neoclassical economic theory considered the dynamical properties of economic systems (Benhabib 1992 and Lorenz 1997), and started to develop nonlinear models. Examples of these nonlinear models are presented in the next section. The nonlinear models developed by neoclassical economics are just extensions of existing linear models.

Unlike neoclassical theory, the starting point of all models developed by econometrists is always empirical evidence. In econometrics the models are based on empirical data and their goal is to reproduce real data. Although econometrists started using simple linear models, they also built measurements and tools to determine if the empirical data series have chaotic behavior. This interest in chaotic behavior led them to use also nonlinear models to obtain short-term forecasts.

In the 1950s Herbert Simon introduced a big innovation. Unlike the neoclassical economic theory, Simon introduced the concept of bounded rationality. He argued that economic agents often fail to maximize profits, making decisions without assessing all information and long-term effects, contrary to the neoclassical theory hypotheses. These ideas are the foundations for the economy as a complex evolving system.

The next sections give some examples of how economics has been dealing with nonlinearity. In particular I start with some models from neoclassical economic theory such as Pohjola, Chiarella, and Day. I end up describing some empirical evidence of nonlinearity and more recent nonlinear concepts, such as power law, scaling, and self-organizing criticality.

2. From Linearity to Nonlinearity

Although most of the economics phenomena are complex and nonlinear, many of the theoretical models used to describe these phenomena are linear. Only at the end of the 1980s did economists start to develop nonlinear models. These models are extensions of linear dynamics models to which some nonlinear features are applied. Many of the assumptions used in expanding linear models are very artificial and ad hoc.

An example of how economics moved from linear to nonlinear dynamical systems is the extension of the Goodwin model (Goodwin 1967). In 1967 Goodwin introduced a very simple, conservative, dynamical model based on the linear Lotka–Volterra predator–prey model. Goodwin described a two-sector model: the first sector consists of capitalists and the second one of workers. Workers spend all their income and capitalists save all their income. The employment rate represents the prey and the labor income share is the predator. When there is no employment the labor income share tends to zero, and vice versa: when the labor income share tends to zero, the employment rate increases. The resulting dynamics implies that the "capitalistic economy" is oscillating permanently. In 1981 Pohjola developed a discrete-time version of the original model (Pohjola 1981). This extension of the linear Goodwin model is a nonlinear model showing chaotic behavior. Chiarella (1990) introduced instead a lag structure in the growth rate of real wages, transforming the model into a nonlinear dissipative system.

Other examples are the standard neoclassical growth model (Day 1982), and the IS-LM framework, in which the investment function possess some strange properties (Day and Shafer 1991).

3. Hysteresis in Economics

Hysteresis is a phenomenon used not only in physics (e.g., see the first chapter in this book), but also in economics. In economics hysteresis can be introduced at the micro and macro levels. A classical example of hysteretic behavior in microeconomics is due to the presence of sunk costs (Baldwin 1989). In order to enter a new market, a firm has to pay an entry cost that cannot be recouped. This entry cost is treated as a sunk cost by the firms that are already in the market. The firm will enter the market only if the revenues cover the sunk cost. A price level above the sum of variable and sunk cost will trigger the entry of new firms in the market. If a firm is active in the market and the price decreases below the entry price level, the firm will not leave the market. The firm will stay in the market until the price is greater than the variable production cost. Thus the number of firms active in the market is a variable that depends not only on current prices but also on the firm's history.

A firm i decides at time t if it will produce one unit of output or none: $x_{i,t} \in 0, 1$. There are two different cost components for the production of an output unit: the variable cost c_i and the fixed cost f_i. If the company has not produced in the previous period, the company is subject to the fixed cost in the next period. So the unit cost C_i at time t is $C_{i,t} = c_i$ if $x_{i,t} = x_{i,t-1} = 1$, $C_{i,t} = c_i + f_i$ if $x_{i,t-1} = 0$ and $x_{i,t} = 1$, and $C_{i,t} = 0$ if $x_{i,t} = 0$. The firm i will decide to produce an output unit $x_{i,t}$ if the price at time t exceeds the production cost: $p_t \geq C_{i,t}$. The supply function will follow the hysteretic loop in Figure 10.1.

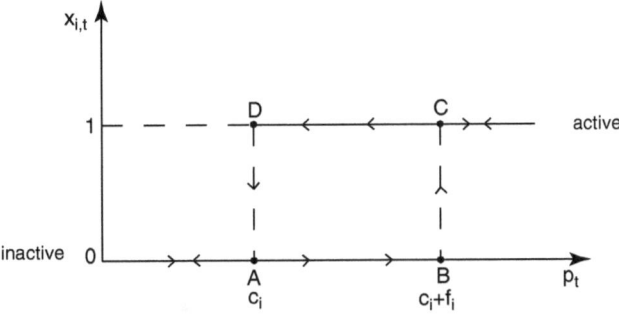

Fig. 10.1. Supply with sunk-costs. The firm produces 0 or 1 unit of output, c_i is the variable cost, and f_i is the fixed cost of market entry.

At the macro level, the main economic problem is to aggregate the microbehavior of heterogeneous firms $i = 1, \ldots, n$. This can be done using the aggregation procedure of Mayergoyz (1986) introduced to economics by Amable (Amable et al. 1991). This procedure, developed in physics to model magnetic properties, has been used to describe the hysteretic nonlinear behavior in rocks, as one can see in different chapters in this book (Chapters 1, 4, 11). We call C_i the cost level that triggers the entry of the firm in the market and c_i the level that triggers the firm's exit. Every firm is defined by the pair (C_i, c_i); all possible combinations of $C - i$ and c_i cover the area above the 45 line in the first quadrant because $C_i \geq i$. The combinations of C_i and c_j on the 45 line are the nonhysteretic firms with $f_i = 0$.

Figure 10.2, panel 1, shows the PM-space when there is a price increase. The initial condition is with price $p = 0$, therefore no firms are active in the market. Let F_a be the set of active firms and F_i the set of inactive firms. A price increase triggers the entry of new firms in the market expanding the grey area F_a of active firms. If there is a subsequent price decrease firms start to leave the market. A firm leaves the market if the price p is lower than its variable cost c_i. Depicted in Figure 10.1 panel 2 is the effect of the price decrease on the number of active firms F_a, the grey area. Panels 3 and 4 of Figure 10.2 show the effect of a successive price increase and price decrease

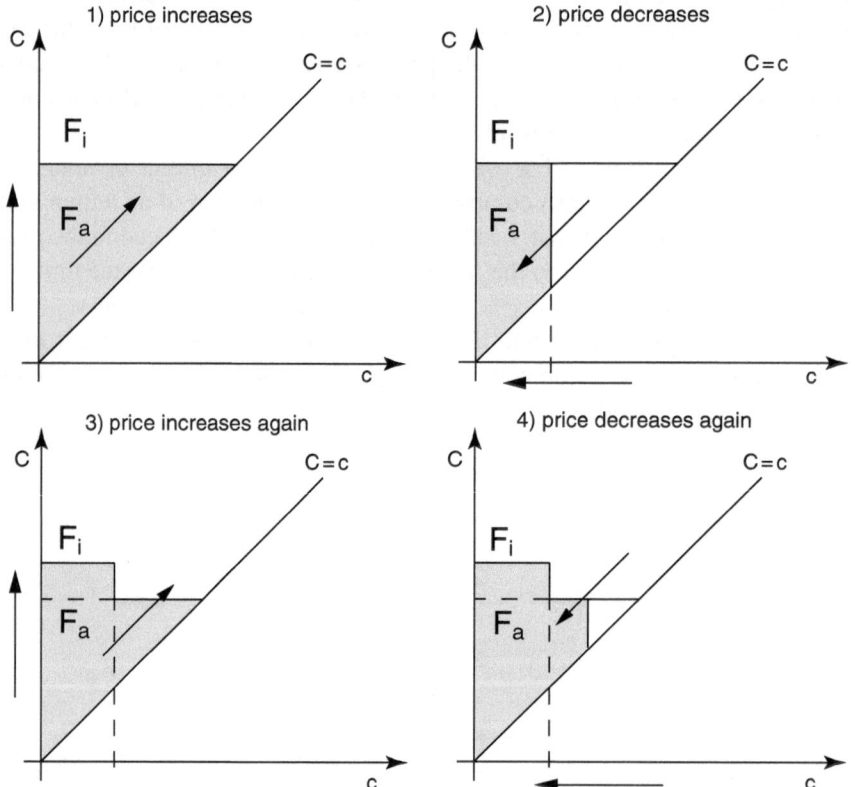

Fig. 10.2. PM space. Each point represent a firm identified by the variable cost c_i and the market entry cost C_i.

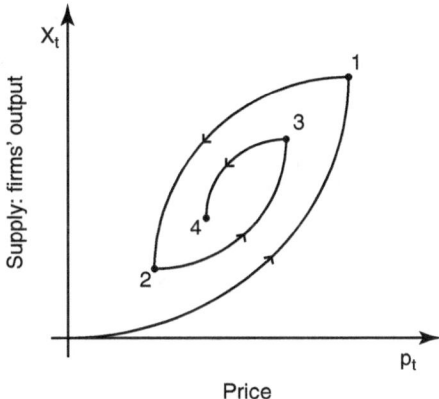

Fig. 10.3. Hystertic loop of firms' aggregate output.

on the number of firms active in the market. The aggregation of microhysteretic behavior, such as the ones described for the single firm, leads to a hysteretic behavior at the macro level. The firm distribution in the cost space is very important: a uniform distribution implies a continuous macroeconomic loop such as shown in Figure 10.3. Information on the density distribution determines the curvature of the branches in Figure 10.3. Hysteresis is a source of path-dependent multiple equilibria and multibranch nonlinearity. In economics there are other examples of hysteretic systems such as the work of Baldwin and Krugman (1989) on exchange rates, Cristophe (1997) on the value of U.S. multinational corporations, and Ljungqvest (1994) on hysteresis in international trade. In empirical economic research an econometric model that attempts to describe hysteretic macrodynamics was proposed by Piscitelli et al. (2000).

4. Scaling in Economics

In general, systems out of equilibrium and nonlinearity are associated with power law distributions. This is valid also in economics where there is much empirical evidence of power law distributions. A power law is a relation of the form $f(x) = kx^{-\alpha}$, where $x > 0$ and k and α are constants. Power laws are a necessary and sufficient condition for scale free behavior. A scale free behavior suggests that the same underlying mechanism is responsible for similar behavior at different scales.

Pareto introduced the first example of the power law in economics. He observed that in Italy 80% of the land was owned by 20% of the population. This observation led Pareto to investigate the phenomenon more in general and conclude that, in all countries, wealth distribution follows a power law, $P(x) = kx^{-\alpha}$ (Pareto 1896). Figure 10.4 shows a rank plot of the annual income of the 500 richest people in the world for the year 2002. The parameter α is close to -1. One model that can describe the Pareto distribution is a simple nonlinear stochastic process: a multiplicative process with a reflecting barrier, $x(t + 1) = x(t)a(t)$, with $x(0)$ and $a(t)$ positive random numbers and a barrier that reflects $x(t)$ when it becomes negative.

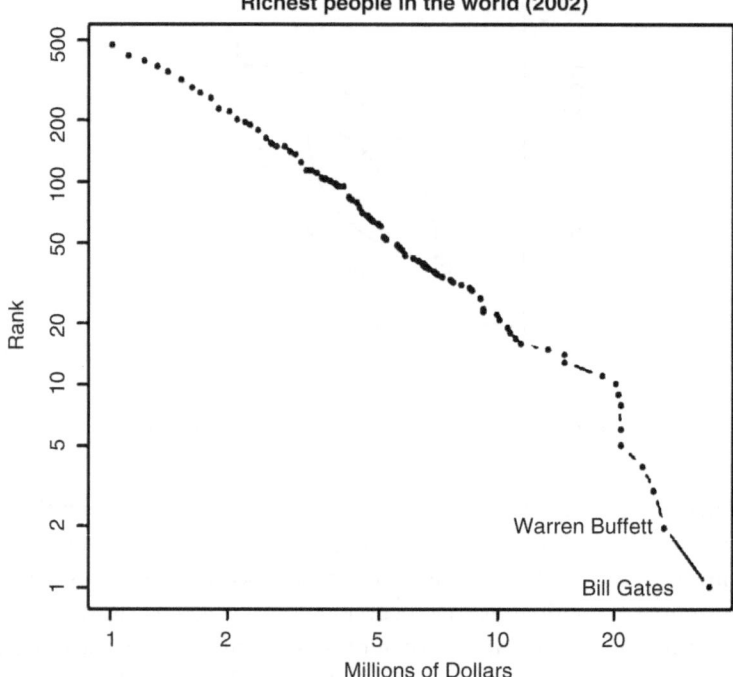

Fig. 10.4. World richest people wealth rank plot based on 2002 Forbes data. The power law exponent alpha is close to −1.

Another famous power law is the Zipf law, named after the linguist George Zipf (Zipf 1949, 1932 and Axtell 2001). He observed statistical regularity in the distribution of words used in any large text. If we rank the words according to the frequency f of their occurrence, as a function of the rank r we find a power law distribution (Zipf's law) $f(r) \sim r^{-\alpha}$ with $\alpha \sim 1$. Zipf also observed that a similar power law can describe the distribution of cities' sizes (Gabaix 1999). If we rank the cities by population size and plot the rank versus their frequency in a log–log scale, we obtain a power law with slope $\alpha = -1$.

There are many examples of power laws in finance. The two most important ones are the power laws in stock returns and cluster volatility. The autocorrelation function of absolute price changes (volatility) is a long memory process that decays as $\tau^{-\alpha}$ with $0.2 < \alpha < 0.5$ (Ding et al. 1993, Poon and Granger 2003, and Mantegna and Stanley 1999). Signed prices are uncorrelated but absolute price changes are strongly positively autocorrelated. Figure 10.5 shows the autocorrelation function of the absolute difference of log prices for the "Vodafone" stock listed on the London Stock Exchange. The autocorrelation coefficients follow a power law with exponent α close to −0.4. This phenomenon is called *cluster volatility*. Cluster volatility is not consistent with the mainstream economic equilibrium models and efficient market hypothesis because price changes are supposed to occur in the presence of new information. But Cutler et al. (1989) showed that there is no significant relation between news and price changes.

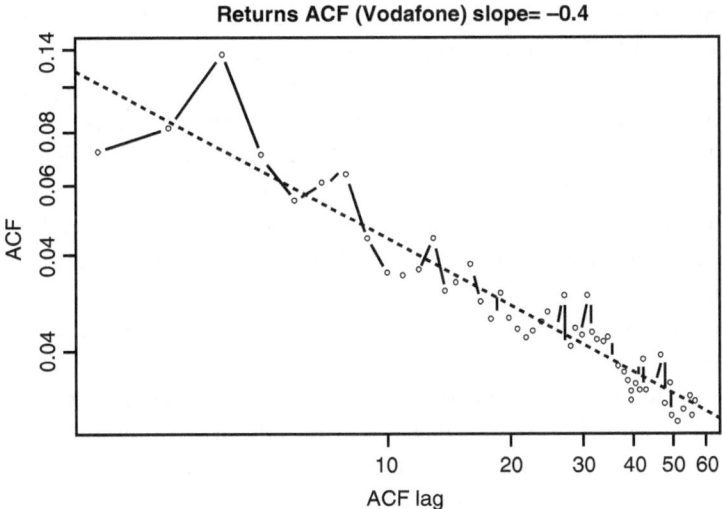

Fig. 10.5. Autocorrelation function of Vodafone log returns in double logarithmic scale. The power law exponent α is close to -0.4.

Another example of the power law in finance is the returns distribution. We define returns as the log price changes, $r(t) = \log p(t - \tau)$. Several empirical studies (Ghashghaie et al. 1996, Fama 1965, Lux 1996, Mantegna and Stanley 1995 and 1999, Mandelbrot 1963, and Plerou et al. 1999) show that returns on short time scales are distributed as a power law with exponent $1.5 < \alpha < 6$.

The volatility long memory autocorrelation has an important consequence on risk measures and option pricing. Risk can be estimated using a volatility measure. But if the absolute returns are strongly correlated because of the cluster volatility, then the estimated variance from the sample is lower than the real variance causing an underestimation of risk. Option pricing is also affected by the power law distribution of returns.

4.1 Process That Generate Power Law Distribution

There are many ways to generate power laws. Nonlinear stochastic processes can produce power laws. As we mentioned previously, the Pareto power law distribution can be described by a multiplicative random process with reflecting barrier.

Another example is the preferential attachment process introduced by Yule, which has been used to explain the power law of the node degree distribution in a graph. The best known example of a graph with such properties is the World Wide Web (WWW), where the distribution of the number of links that a page receives is distributed as a power law (Barabasi et al. 1999). For this example a preferential attachment process can be described in the following steps: we add a new Web page and we link it to the Web pages present in the network randomly. Each Web page in the WWW has a probability to be chosen that is proportional to the number of Web pages that are pointing to it. Therefore if many pages link a Web page, this page has a higher probability to be chosen. In economics the firm's growth follows a power law distribution (Gibrat's law), and this process can be described using preferential attachment exactly in the same way as the WWW example (Ijiri and Simon 1977).

Scale free distributions are also associated with out of equilibrium phenomena that are close to particular phase-transition regimes or "critical points." The parameters of such systems have to be carefully tuned in order to reach the critical point, where we observe that the system relaxes following a scale free distribution. An example in physics is the power law distribution of water bubbles' size at the boiling point. The critical boiling point can be reached only if the value of the system temperature is properly chosen.

There are, however, some systems that tune themselves to the critical point. At the critical point there is no natural scale for the fluctuation's size. The scale free fluctuations arise while tuning in a self-organized manner: this phenomenon is called self-organizing criticality (Bak et al. 1987). An example of a large nonlinear dissipative system that shows self-organizing criticality is a sand pile. Imagine dropping sand over a clean table. The sand is going to pile up until it starts creating avalanches and falling off the table (dissipation). The avalanches' sizes are distributed as a power law.

Economists have applied the self-organizing criticality concept to different fields in economics. A well-known example is the Scheinkman and Woodford model (Scheinkman and Woodford 1994). They used self-organizing criticality to explain the fat tails in the inventory distributions of production chains. Nirei incorporated the concept in a real business cycle model. There are also applications of self-organizing criticality to financial markets, currency markets, and industrial organizations (Focardi et al. 2002).

5. Conclusions

Economics is a complex nonlinear discipline in continuous evolution. Heterogeneity among agents, simple nonoptimal behaviors, and complex interactions are perhaps the real sources of nonlinearity of the observed data.

As described in this chapter, there are nonlinear models but they are not yet very successful. The "failure" of these models is in part due to two facts. First, these models are just extensions of already existing simple linear models. Second, the models are developed without taking into account empirical observations. From the economic data point of view it is very hard to establish the existence of deterministic chaos in the real time series. If chaotic behavior is observed then long-term predictions are infeasible. Vice versa, if we consider a short time scale, nonlinear models can provide better predictions than linear ones.

Even very simple nonlinear dynamical systems generate very complex dynamics. If we choose the right nonlinearity, then we can model almost any particular dynamical phenomena. But data (over)fitting is not the goal of economics. The real goal is instead the abstraction of complex phenomena in a parsimonious model.

References

B. Amable et al. Strong hysteresis: An application to foreign trade. OFCE Working Paper 9103, 1991.

R. Axtell. Zipf distribution of U.S. firm size. *Science*, 293:1818–1820, 2001.

P. Bak, C. Tang, and K. Wiesenfeld. Self-organized critically—An explanation of 1/f noise. *Phys. Rev. Lett.*, 59(4):381–384, 1987.

R.E. Baldwin. Sunk-cost hysteresis. NBER Working Paper, 2911, 1989.

R.E. Baldwin and P. Krugman. Persistent trade effect of large exchange rate shocks. *Quart. J. Econ.*, 104, 635–654, 1989.

A.L. Barabasi, R. Albert, and H. Jeong. Mean-field theory for scale random networks. *Physica A*, 272:173–189, 1999.

J. Benhabib, editor, *Cycles and Chaos in Economic Equilibrium*, Princeton University Press, Princeton, NJ, 1992.

C. Chiarella. *The Elements of a Nonlinear Theory of Economic Dynamics*, Springer–Verlag, Berlin–Heidelberg–New York, 1990.

S.E. Cristophe. Hysteresis and the value of the U.S. multinational corporation. *J. Business* 70, 3:435–462, 1997.

D.M. Cutler, I.M. Poterba, and L.H. Summers. What moves stock prices?. *J. Portfolio Manag.* 15:4–12, 1989.

R.H. Day. Irregular growth cycles. *Amer. Econ. Rev.* 72:406–414, 1982.

R.H. Day and W. Shafer. Keynsian chaos. *J. Macroecon. Behav. Organization*, 16:37–83, 1991.

Z. Ding, C.W.J. Granger, and R.F. Engle. A long memory property of stock returns and a new model. *J. Empirical Finance*, 1:83, 1993.

E.F. Fama. The behavior of stock-market prices. *J. Business*, 38(1):34–105, 1965.

S. Focardi, S. Cincotti, and M. Marchesi. Self-organization and market crashes. *J. Econ. Behav. Organization*, 49(2):241–267, 2002.

X. Gabaix. Zipf's law for cities: An explanation. *Quart. J. Econ.*, 114(3):739–768, 1999.

S. Ghashghaie, W. Breymann, J. Peinke, P. Talkner, and Y. Dodge. Turbulent cascades in foreign exchange markets. *Nature*, 381(6585):767–770, 1996.

R.M. Goodwin, A growth cycle. In Feinstein, C.H. editor: *Socialism, Capitalism and Economic Growth*. Cambridge University Press, Cambridge, UK, 1967.

Y. Ijiri and H.A. Simon. *Skew Distributions and the Sizes of Business Firms*.

L. Ljungqvist. Hysteresis in international trade: A general equilibrium analysis. *J. Intl. Money Finance*, 13, 387–99, 1994.

H.W. Lorenz, *Nonlinear Dynamical Economics and Chaotic Motion*. Springer–Verlag, New York, 1997.

T. Lux. The stable Paretian hypothesis and the frequency of large returns: An examination of major German stocks. *Appl. Financial Econ.*, 6(6):463–475, 1996.

B. Mandelbrot. The variation of certain speculative prices. *J. Business*, 36(4):394–419, 1963.

R.N. Mantegna and H.E. Stanley. Scaling behavior in the dynamics of an economic index. *Nature*, 376(6535):46–49, 1995.

R.N. Mantegna and H.E. Stanley. *Introduction to Econophysics: Correlations and Complexity in Finance*. Cambridge University Press, Cambridge, UK, 1999.

I.D. Mayergoyz. Mathematical models in hysteresis. *IEEE Trans. Magnetics*, 22:603–8, 1986.

M. Nirei. Inventory Dynamics and Self-Organized Criticality. Research in Economics, 54: 375–383, 2000.

V. Pareto. *Cours d'Economie Politique*. Droz, Geneva, 1896.

L. Piscitelli et al. A test for strong hysteresis. *Comput. Econ.*, 15:59–78, 2000.

V. Plerou, P. Gopikrishnan, L.A.N. Amaral, M. Meyer, and H.E. Stanley. Scaling of the distribution of price fluctuations of individual companies. *Phys. Rev. E*, 60(6):6519–6529, 1999. Part A.

M.J. Pohjola. Stable and chaotic growth: The dynamics of a discrete version of Goodwin's growth cycle model. *Zeitshrift fur Nationalokonomie* 41:27–38, 1981.

S.H. Poon and C.W.J. Granger. Forecasting volatility in financial markets: A review. *J. Econ. Lit.*, 41(2):478–539, 2003.

J. Scheinkman and M. Woodford. Self-organized criticality and economic fluctuations. *Amer. Econ. Rev.*, 84(5): 417–421, 1994.

G. Yule. A mathematical theory of evolution based on the conclusions of Dr. J.C. Willis FRS. Philosophical Transactions of the Royal Society of London Series B-Biological Sciences 213: 21, 1924.

G. Zipf. *Selective Studies and the Principle of Relative Frequency in Language*. Harvard University Press, Cambridge, 1932.

G. Zipf. *Human Behavior and the Principle of Least Effort*. Addison-Wesley, Cambridge, 1949.

Part II
Applications to NDE and Ultrasonics: Models and Numerical Simulations

11

Micropotential Model for Stress–Strain Hysteresis and PM Space

V. Aleshin[1] and K. Van Den Abeele

K.U.Leuven Campus Kortrijk, Interdisciplinary Research Center, E. Sabbelaan 53, B-8500
Kortrijk, Belgium.
[1]To whom correspondence should be addressed: aleshinv@mail.ru

Abstract

In this chapter we present and discuss a physical mechanism of stress–strain hysteresis in
microcracked solids. Four steps are considered. Step 1: Starting from a random collection of
noninteracting microcracks with rough surfaces we built up an expression for the potential energy of such a particular system. This potential consists of three terms: adhesion energy of crack
surfaces, strain energy of the surrounding bulk material, and strain energy of local deformations
arising due to alterations of the crack apertures. Step 2: We examine and establish the conditions
under which this potential contains two separated minima, representing a bi-stable system. Step
3: Stress–strain hysteresis is found to result from transitions in the double-well potential due to
external impact. Finally, in Step 4, we show that the Preisach–Mayergoyz (PM) formalism can
be applied to the considered system, yielding predictions of static and dynamic nonlinearity.
The principal difference between the proposed approach and phenomenological models is that
the parameters of hysteretic elements and their distributions are not arbitrary but are deduced
from an analysis of the micro-geometry of the crack network.

Keywords: Double-well potentials, microcracked solids, stress–strain hysteresis

1. Introduction

It is reasonably safe to state that pure, uniform, and regular materials are exceptional
in common life. Ninety-nine percent of the time, mankind produces, treats, and uses
materials that have impurities, irregularities, inclusions, and defects that are inherent properties of their microstructure. Most well-known examples are geomaterials,
such as volcanic (granite, basalt, etc.) and sedimentary rocks (sandstones and lime-
stones, slates, etc.), as well as artificial cementitious materials such as concrete, mortar,
and roofing slate used in construction. Polycrystalline metals and alloys constitute a
third exemplary class of materials with an outspoken grainy microstructure. In general, all solids that are not single crystals can be regarded as materials with inherent
random structure at a mesoscopic scale, that is, at a scale which significantly exceeds
the atomic size but is still small compared to macroscopic dimensions.

Internal defects can be roughly categorized in three types: 1-D dislocations, 2-D internal contacts/cracks, and 3-D pores/voids. Among these, the second type of defects is by far the most essential in terms of material performance. Indeed, influence of dislocations is negligible if we speak about seismology or building construction, whereas pores and voids usually contribute to the most interesting material properties much less than cracks and contacts do (e.g., failure loads, acoustic and static nonlinearities, sound attenuation, etc.). This makes microcracked solids an extremely important class of materials, and justifies the fact that an accurate description of their mechanical properties is critical.

2. State of the Art

Researchers working in the field of nonlinear elasticity and acoustics of microstructured solids can be conditionally classified into two groups. One group proposes basic physical microscopic mechanisms that qualitatively explain experimentally discovered macroscopic features. The other group tries to build phenomenological models without asking what physics actually accounts for the declared properties, and seeks quantitative signatures in the model–data comparison. The latter approach is close to a pure engineering application. Both approaches have experienced a considerable breakthrough in the last decades. An up-to-date review is presented by Ostrovsky and Johnson,[1] Nazarov et al.,[2] and also Chapter 4 of this book.

2.1 Physical Mechanisms of Nonlinearity

As early as 1960 to 1970, there was considerable interest in the mechanism behind the nonlinear attenuation of seismic waves in rocks. The oldest explanation proposed in classical papers by Walsh[3] and others[4,5] consisted in hysteretic frictional sliding between grain contacts and grain boundaries. For nearly 30 years hysteretic friction was believed to be the dominant mechanism. In 1994, Tutuncu and coworkers pointed out certain discrepancies of the traditional approach under high ambient pressure.[6]

The alternative explanation of Tutuncu et al. consists in adhesion hysteresis, and accounts for hysteresis and losses even if the contacts are strongly compressed and no intergranular friction is possible. The explanation was, however, only qualitative, and the quantitative theory that predicts the shape of the stress–strain hysteresis from statistical data on microgeometry was not built. Along the same line, Pecorari[7] recently proposed an advanced model of adhesion hysteresis for an interface between two rough surfaces (see also Chapter 19). These results offered a meaningful example of elementary bonds with a nonunique constitutive behavior resembling that of a two-state relay used in phenomenological models, and with dynamics similar to the acoustic nonlinear properties of rocks known from macroscopic experiments.

Even though appealing as such, adhesion hysteresis, as a microscopic mechanism, can be dominant only for relatively small strains (acoustics), that is, when cracks activated in this process typically have apertures comparable to the atomic size. However, as soon as the deformation achieves millistrains, much wider cracks will be opened and closed, and therefore hysteretic jumps of a few nanometers induced by adhesion will be negligible at these deformation levels.

Apart from adhesion hysteresis and friction, several alternative physical mechanisms may also contribute to the specific hysteretic constitutive behavior of materials with internal cracks. Examples are the stick–slip motion of asperities[8] and the movement of dislocations.[9] The latter concept was used by Nazarov et al. in the late 1980s (recently summarized[2] in 2003) as an explanation of hysteresis of polycrystalline metals and geomaterials. In addition, Kim et al.[10] demonstrated that hysteresis in the stress–strain relation may follow directly from the elastoplasic loading and elastic unloading of a single rough contact.

In this chapter we present a supplementary approach for the description of nonlinearity both at small (acoustics) and high (static tests) strains (for details see Aleshin and Van Den Abeele[11]). We combine a micromodel for adhesion with statistical data on the microgeometry of a material. Doing so, we build up the potential energy of a microcrack system as a function of a distribution of the crack apertures and the total strain in the material. A detailed analysis of the potential shows that there are certain conditions for which some of the microcracks have two minima, depending on the external impact. The existence of hysteresis is attributed to transitions of the system between these equilibrium positions. The double-well potential is the result of the sum of the adhesion energy and the strain energy of the intact material, and causes the macroscopic stress–strain hysteresis and nonlinear effects in microcracked materials.

2.2 Phenomenological Models

The more phenomenological approach to dealing with hysteresis and its effects on macroscopic static and dynamic processes was brought into scope by Ortin[12] and, independently, by Guyer et al.[13] who applied the well-known formalism of the Preisach space (proposed for ferromagnetics in 1935 by Preisach[14] and generalized for a rate-independent hysteresis of arbitrary nature by Krasnosel'skii and Pokrovskii[15]) to the problem of elasticity of geomaterials. This phenomenological approach, referred to as the PM-space formalism, uses a two-state relay for "open" and "closed" cracks. It inspired a new wave of interest in the subject through a large number of papers on the description and links between static and dynamic hysteresis,[16] hysteretic acoustical nonlinearity in damaged materials,[17,18] complicated triaxial static stress tests,[19] and tensorial representation of stress–strain hysteresis.[20] Without concentrating on the physical nature of hysteresis and micromechanics, these studies solved certain practical problems of material characterization and produced plausible constitutive relations, similar to those required by engineers to model static and dynamic processes in complicated 3-D structures. However, the negligence of the underlying physical nature of these materials and the disregard of the frictional and adhesion properties of the contacts, inevitably led to a deficiency in the description of particular experimentally observed features: temperature[21] and humidity[22] effects on the nonlinearity, slow dynamics,[1,21] and so on. This is obvious because phenomenological models are intended to solve particular problems only and will never capture the whole picture.

The methodology we present in this chapter leads to the construction of a PM-space and its density, just as in the phenomenological approach. However, the parameters of the elementary hysteretic units are not arbitrary. They can be calculated from statistics of a realistic crack network. On the other hand, we use only one of the possible mechanisms for hysteresis and make some severe assumptions in the analysis, so that the obtained description has a qualitative character only.

3.   Typical Geometry of a Crack Network and Definition of Its Statistics

Modern visualization techniques enable us to view crack networks in microcracked materials via electron or optical microscopy. One of these techniques[23] consists in filling the cracks with Wood's metal and in subsequently polishing a section of a sample. The quality of visualization is adequate for statistical analysis of the microgeometry of the section, and corresponding methods of stereology yield 3-D parameters of the crack network.[24] Two parameters are of particular interest: the total area of crack S_V per unit of volume and the total length L_V of all crack ends (tips) per unit of volume. Typical values of S_V were measured by Nemati and Stroeven[24] for concrete. Values of L_V have never been measured, however, in principle, this parameter can be deduced from a similar analysis as well. Figure 11.1 illustrates the definition S_V and L_V graphically.

Crack networks typically have a fractal nature, implying that their parameters depend on the magnification. However, being a physical fractal, the fractal scaling law breaks down for scales approaching the atomic size, and the definition of S_V and L_V becomes unambiguous in that case.

At the highest magnification[23] crack networks resemble a collection of plane cracks. Of course, this may be attributed to the limitations of the experimental technique (i.e., Wood's metal can not penetrate in too-narrow cracks). In spite of this, and even though

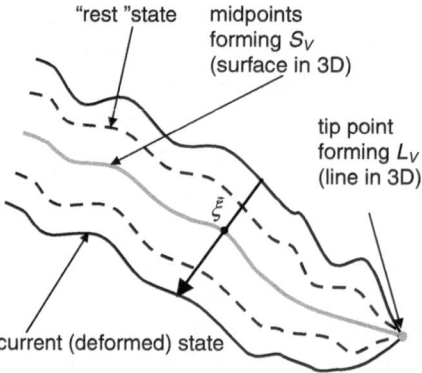

Fig. 11.1. Illustration of the definition of the local rest aperture and the local deformed aperture for an elementary quasiplane crack. The midpoints define the total crack area (per unit of volume) and the tip points the total length L_V of the tips (per unit of volume). Both are shown in gray.

images of microcracks in other materials may look different, we assume that the smallest cracks are almost plane and clearly separated one from another.

Having accepted this (rather severe) assumption of the existence of the elementary quasiplain crack, we define its aperture as the separation vector $\vec{\xi}$ (see Figure 11.1). averaged over the elementary crack, and its roughness D as the standard deviation of its modulus $\xi = \left|\vec{\xi}\right|$. Considering the case of pure uniaxial deformations, and neglecting all shearing motions of the crack surfaces (another strong assumption), it is sufficient to consider one projection of this vector only.

In order to construct the micropotential of an elementary crack we need information about the following parameters: the current/actual crack aperture a, and the initial crack roughness D and aperture A in the absence of adhesion and external forces ("rest aperture"). The parameters A and D of an individual elementary crack are considered to be its inherent characteristics. We assume that they may be different for different elementary cracks and we therefore introduce a corresponding distribution $f(A, D)$. In the usual sense, $f(A, D)dAdD$ is the probability of finding the element in $(A, A + dA) \times (D, D+dD)$ and $\int f(A, D)dAdD = 1$. The distribution is regarded as a given and unchanging characteristic of a material (which can be determined by visualization techniques of crack networks).

4. Micropotential of an Elementary Crack

Now let us consider an individual elementary crack with inherent parameters A and D. We build up an expression for its potential energy as a function of its actual aperture a. Because some of the points belonging to opposite crack surfaces can be in atomic contact, the adhesion energy of the crack will be changing with varying apertures. Furthermore, aperture variations impose local deformation of the surrounding intact material which yields a corresponding energy storage. We treat these two contributions (adhesion potential and local strain energy) consecutively in the following subsections.

4.1 Adhesion Potential for Cracks with Rough Correlated Surfaces

The semiphenomenological consideration that we set out in this section is different from the traditional contact mechanics of rough surfaces with adhesion, dating back to the 1966 paper by Greenwood and Williamson.[25] The reason for the difference is that contact mechanics generally treats uncorrelated surfaces, whereas uncorrelated roughness completely eliminates the effect of adhesion.[26] In the case of internal cracks considered in this chapter, the surfaces are essentially correlated (conforming), because the cracking originated, for instance, from processes of solidification, differential stresses, or thermal loading where the faces or "coasts" of the crack up to that time formed an intact entirety. For such contacts the adhesion effects and the pull-off forces are substantial.

Instead of a proper development of the contact mechanics for correlated cracks, we present an estimated calculation based on empirical facts. As a start, we recollect that researchers[27,28] have found that rock joints (another example of a contact with correlated surfaces) possess a lognormal distribution of apertures.

For the adhesion energy per unit area of a quasiplain contact $W(a)$ we can therefore write that

$$W(a) = \int_0^\infty w(\xi)\varphi(\xi)d\xi, \tag{11.1}$$

where $w(\xi)$ is the adhesion potential for a purely planar contact and $\varphi(\xi)$ is the postulated lognormal distribution:

$$\varphi(\xi) = \frac{e^{h/2}}{a\sqrt{\pi h}} \exp\left(-\frac{1}{h}\ln^2\left(\frac{\xi}{a}e^{3h/4}\right)\right). \tag{11.2}$$

Here h is a parameter that implicitly relates the standard deviation D to the current aperture:

$$D = a\, e^{-3h/4}\sqrt{e^{2h} - e^{3h/2}}. \tag{11.3}$$

For the adhesion potential $w(\xi)$ of parallel surfaces we use the expression given by Israelachvili[29] which is the result of integration of the Lennard–Jones potential (also referred to as the 6–12 potential or the potential of the Van der Waals force):

$$w(\xi) = \frac{H}{12\pi}\left(-\frac{1}{\xi^2} + \frac{a_0^6}{60}\frac{1}{\xi^8}\right), \tag{11.4}$$

with H the Hamaker constant of the material and a_0 the atomic size. Bringing together Eqs. (11.1)–(11.4) one simply obtains:

$$W(a) = \frac{H}{12\pi}\gamma^3\left(-\frac{1}{a^2} + \frac{a_0^6}{60}\frac{\gamma^{33}}{a^8}\right), \tag{11.5}$$

where $\gamma = 1 + (D/a)^2$.

This primitive contact mechanics theory, despite its simplified character, reflects some principal features of correlated contacts.

- The distribution $\varphi(\xi)$ of the gap ξ is highly asymmetric at comparable a and D (partial contact), implying that small separations are preferable to larger ones. This means that the surfaces are statistically conforming.
- The implicit use of the potential for interacting planes also reflects the fact that we deal with correlated contacting shapes, whereas the traditional contact mechanics, following Greenwood and Williamson,[25] treats interactions of spherical asperities (Hertz problem).
- The resulting potential [Eq. (11.5)] always has one minimum; that is, there is always a nonzero pull-off force. Its value is largest for perfectly conforming contacts ($D = 0$) and smaller for less and less correlated surfaces. The presence of the pull-off force also means that the material is consolidated.

- The minimum of the potential $w(\xi)$ [Eq. (11.4)] or $W(a)$ for $D = 0$ yields a surface energy $\Gamma = 15^{1/3} H/(16\pi a_0^2)$. Moreover, when we approximate the potential $w(\xi)$ at its minimum by means of a parabola, we can estimate the Young modulus: $E = 15^{1/2} H/(\pi a_0^3)$ for a material with perfect contacts (for instance, a strongly consolidated geomaterial with weak stress-strain hysteresis). If we assume a typical value for the Hamaker constant and the atomic size ($H = 10^{-19}$ J and $a_0 = 3 \ 10^{-10}$ m), we obtain that $\Gamma \approx 5 \cdot 10^{-2}$ J/m^2 and $E \approx 5 \cdot 10^9$ Pa, which are indeed reasonable values too.

4.2 Estimation of the Stabilizing Potential of Crack Tips

So far, we have simply modeled the crack network as a system of adhesive planes. Such a representation inevitably leads to the following consequence: if all cracks are widely open and the adhesive force is negligible, the material becomes unconsolidated and is destroyed. However, in reality, the condition in which all cracks are open is not sufficient for failure because the material is essentially held together as a whole by the crack tips. More precisely, every applied tension that widens the crack aperture leads to an increase of the potential energy due to the high stress concentration in the vicinity of the crack tips and extra deformation parallel to crack surface. Figure 11.2 illustrates this situation in a qualitative manner: tension applied to a sample tends to tear up the material in near-tips zones and stretches the layers parallel to the crack surface (as well as deforming them in the vertical direction).

The stress analysis of a nontrivial crack network is extremely complicated, and here we limit ourselves to a simple and rough estimation for the additional potential energy, based on the theory of Hertz contact (see, for instance, Landau and Lifshitz, 1986[30]).

We consider the contact of two balls of radii R and R', having elastic moduli E and E' and Poisson ratios ν and ν' respectively. Assuming a compression force F, the radius of contact b between the two balls then is given by:

$$b = F^{1/3} \left(K R R' / \left(R + R' \right) \right)^{1/3}, \tag{11.6}$$

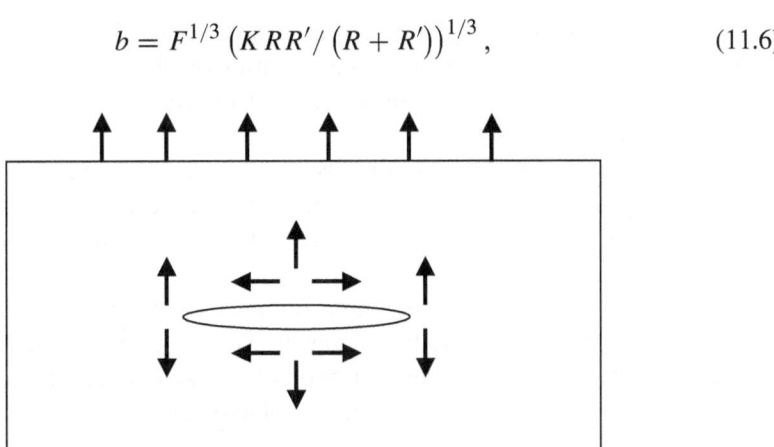

Fig. 11.2. Qualitative scheme of tensile deformations in the neighborhood of an isolated crack.

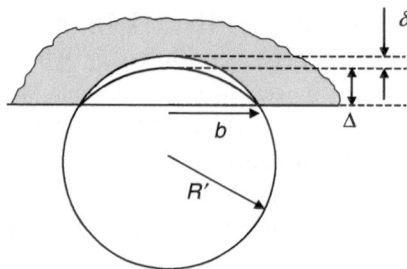

Fig. 11.3. Modification of the Hertzian problem for interaction of an elastic half-space with a spherical dent of depth Δ and a rigid sphere of a radius R'.

and the displacement of the balls' centers by:

$$\delta = F^{2/3} K^{2/3} \left(1/R + 1/R'\right)^{1/3}, \tag{11.7}$$

where

$$K = \tfrac{3}{4} \left(\left(1 - \nu^2\right)/E + \left(1 - \nu'^2\right)/E\right). \tag{11.8}$$

In the usual Hertz theory, the radii R and R' are considered to be constant and the radius b of the contact is the variable parameter. For our purposes, the approach is somewhat different. Here, we assume that the contact radius b remains constant during the indentation. Furthermore, we assume $E' = \infty$ (rigid material), and consider the radius R' to be a function of δ such that b remains constant (see Figure 11.3). This situation actually describes the interaction of a half-space containing a spherically dented area of constant radius b, with a rigid sphere with variable radius R'. Combining Eqs. (11.6)–(11.8), we obtain the following expressions for the force and the potential energy:

$$F(\delta) = \frac{4}{3} \frac{E}{1 - \nu^2} b\, \delta, \qquad P(\delta) = \frac{2}{3} \frac{E}{1 - \nu^2} b\, \delta^2. \tag{11.9}$$

Note that all values related to the rigid sphere have disappeared. Consequently, the potential energy characterizes the elastic deformation of a half-space with a dent of depth Δ and surface area πb^2, whose dent is slightly displaced in the direction perpendicular to the surface over a distance δ.

Following elementary calculus, it is easy to verify that the average depth for a small spherical dent equals 2/3 of its maximum depth. Because a crack is formed by two half-spaces, we associate $2\Delta/3$ with $A/2, 2(\Delta + \delta)/3$ with $a/2$ and multiply $P(\delta)$ by 2.

If a representative volume V contains n round cracks with radius b, the total surface and perimeter length (tips length) amount to $S_V = n\pi b^2/V$ and $L_V = 2n\pi b/V$. Hence, $b = 2S_V/L_V$ and the concentration of cracks n/V equals $L_V^2/(4\pi S_V)$. As a result, the total potential energy per unit of volume reads:

$$T(a) = 2P(\delta)\frac{n}{V} = \frac{3}{8\pi} \frac{E}{1 - \nu^2} L_V \left(a - A\right)^2. \tag{11.10}$$

This form can be used as a rough estimation of the extra potential energy associated with the presence of tips in the crack network. Despite the ad hoc assumptions and its qualitative character, it describes the physical situation in a rather plausible way. Note that the extra potential and the corresponding strengthening of the material associated with the presence of tips are essentially proportional to L_V. In fact, the above consideration and derivation only provides evidence for the fact that an extra term in the potential due to the tips should have the dependence $T(a) \sim L_V(a - A)^2$.

In deriving Eq. (11.10) we assumed that the straining does not produce additional damage (because crack radius b was a constant). This way, all considerations proposed in this chapter are only valid at moderate stresses that do not change the degree of damage of the material.

5. Lagrangian Formalism for a Material Containing Identical Elementary Cracks

We first formulate the micropotential approach for the simple case when all elementary cracks are identical (have identical A and D). After examining some principal features in the evolution of such a system we proceed to a real case of a nontrivial distribution $f(A, D)$.

We introduce the displacement $u(x, t)$ and the strain $\partial u / \partial x \equiv \varepsilon(x, t)$ as functions of the spatial coordinate x and time t. The total deformation of the material can be expressed as the result of two mechanisms: deformation (strain ε^*) of the intact material zones and changes of the crack apertures. Therefore we write:

$$\varepsilon = \varepsilon^* + S_V(a - A). \tag{11.11}$$

Here $S(a - A)$ is the change of the crack volume caused by the aperture variation, summed over all cracks in the volume V ($S_V = S/V$). In the same equation, ε^* denotes the mean (uniform) strain of the intact material in the zones away from the cracks. In the vicinity of the cracks, the strain is largely distorted by the deformations of the crack surfaces as discussed in the previous section. However, because we believe the cracks to be rare events, we may assume that the zones of extra distortion are small too. Assuming Hooke's law for the intact material, the total potential energy Π for a representative unit of volume is then given by

$$\Pi = \tfrac{1}{2}E\varepsilon^{*2} + S_V W(a) + T(a). \tag{11.12}$$

We repeat that this form is reasonable only in those cases where the strain energy $\tfrac{1}{2}E\varepsilon^{*2}$ is distributed in the intact zones and strain energy $T(a)$ is localized in the vicinity of cracks. Only then is it permissible to sum up the two strain energies without cross-terms.

Combining Eq. (11.11) and Eq. (11.12), we build up the Lagrangian of the system:

$$L(u, u_x, u_t, a, a_x, a_t) = \tfrac{1}{2}\rho\, u_t^2 - \tfrac{1}{2}E(u_x - S_V(a - A))^2 - S_V W(a) - T(a), \tag{11.13}$$

where ρ is the equilibrium density. The Lagrange equations:

$$\begin{cases} \frac{\partial}{\partial t}\left(\frac{\partial L}{\partial u_t}\right) + \frac{\partial}{\partial x}\left(\frac{\partial L}{\partial u_x}\right) - \frac{\partial L}{\partial u} = 0 \\[2mm] \frac{\partial}{\partial t}\left(\frac{\partial L}{\partial a_t}\right) + \frac{\partial}{\partial x}\left(\frac{\partial L}{\partial a_x}\right) - \frac{\partial L}{\partial a} = 0 \end{cases}, \tag{11.14}$$

hence yield:

$$\begin{cases} \rho \, u_{tt} = E u_{xx} - E S_V a_x \\ \partial L / \partial a = 0 \end{cases}. \tag{11.15}$$

The above evolution equations were obtained by means of a purely mechanical treatment, that is, without taking into account temperature effects. The first equation is the common wave equation modified for the presence of microcracks though S_V. Because A is considered to be an x-independent constant (identical crack opening at rest), and the stress represents the generalized force corresponding to the generalized coordinate u, we obtain the following expression for the stress–strain relation from Eq. (11.12),

$$\sigma = -\partial L / \partial u_x = E u_x - E S_V (a - A). \tag{11.16}$$

The second equation puts a constraint on the possible aperture values a of the cracks. More specifically, we find that the aperture a must correspond to an extremum of the function:

$$U(a) = W(a) + \tfrac{1}{2} E S_V (1 + \alpha) \left(a^2 - 2Aa \right) - E u_x a, \tag{11.17}$$

in which we have collected all a-dependent terms of the Lagrangian and introduced the parameter α which incorporates the effects of the crack tips through L_V:

$$\alpha = \frac{3}{4\pi} \frac{1}{1 - \nu^2} \frac{L_V}{S_V^2}. \tag{11.18}$$

However, because the proportionality constant in $T(a)$ is only approximately known (see previous Section 4.2), we regard α as a phenomenological constant.

Obviously, the extremum of Eq. (11.17) should be a minimum, otherwise the solution will be unstable. Furthermore, the solution of the above system of Lagrangian equations [Eqs. (11.15)] also depends directly on the number of minima of $U(a)$(and their values), where $u_x = \varepsilon$ can be considered as a parameter. A thorough investigation of the minima of this expression is thus essential for the further interpretation of the results.

Before we actually examine the occurrence of minima of $U(a)$, we eliminate A (which is constant for all elementary cracks) by introducing the new variable s:

$$s = \varepsilon + S_V (1 + \alpha) A. \tag{11.19}$$

With this,

$$U(a) = W(a) + E S_V (1 + \alpha) a^2 \big/ 2 - E s a. \tag{11.20}$$

The potential now only depends on the argument a and the parameters D [through $W(a)$, see Eq. 11.5] and s. All other values (E, S_V, α, H, a_0) are considered to be material constants.

6.   The Bi-Stable State of an Elementary Crack

The complicated form of the adhesion potential $W(a)$ [Eq. (11.5)] forces us to analyze $U(a)$ numerically. Doing so, we come to the following observations.

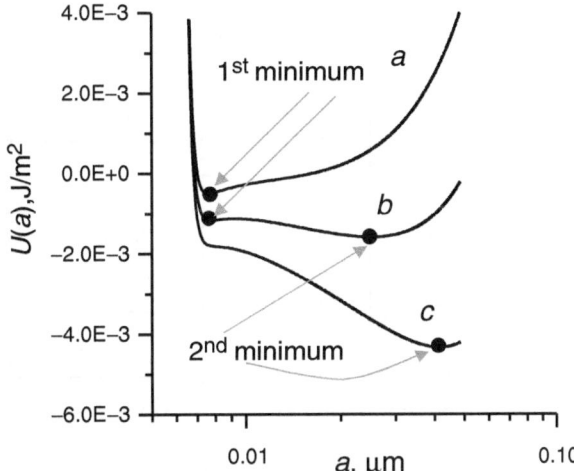

Fig. 11.4. Three possible configurations of the potential $U(a)$ depending on the value of the strain: (a) the occurrence of only the first minimum (in the adhesive zone) when $s < s_c$; (b) the appearance of a double-well potential with both first and second minima present when $s_c < s < s_o$; (c) the existence of only the second minimum when $s_o < s$. The system is said to be "closed" in case (a), always, and case (b), if $\varepsilon_t > 0$. It is "open" in state (c), always, and (b), if $\varepsilon_t < 0$.

If the roughness parameter D exceeds a critical value D_{cr}, the potential $U(a)$ has only one minimum for any value of the "strain" s. Large values of D indeed correspond to cracks with weak adhesion properties, and their contribution to the potential is too weak to form an additional extremum. On the other hand, if the adhesion is strong enough ($D < D_{cr}$), one minimum or two minima may exist depending on the magnitude of s (see Figure 11.4).

We consider the latter case in more detail. Let us first of all suppose that s is high and negative (strongly compressed material, curve a in Figure 11.4). The contribution of $-Esa$ to the potential is large and results in the fact that there is only one minimum located at small aperture values a ($a \sim a_{min}$ of the adhesive potential; see Figure 11.3). The system is said to be closed. If s increases and exceeds a critical value s_c, the contribution of the term $-Esa$ becomes smaller and an additional minimum appears (curve b in Figure 11.4). However, in our mechanical model described by Eq. (11.15), the system stays in the first minimum (closed state), despite the fact that the second state (open) is also possible. Actually this is true only if the thermal movement is negligible compared to the typical potential barrier, otherwise random jumps to the open state and back are possible (allowing transitions between two minima). When increasing the strain further, above another critical value s_o, a situation is achieved in which only the second minimum exists (Figure 11.4, curve c). The system now switches to the open state and stays there for all higher strain values.

If we decrease the strain from highly positive to highly negative magnitudes, the evolution of a behaves completely analogously to the case of a strain increase with the only difference that in a two-well configuration the elementary crack remains open (again, in reality only if the temperature is negligible) until the second minimum disappears, and only the minimum at small aperture values exists.

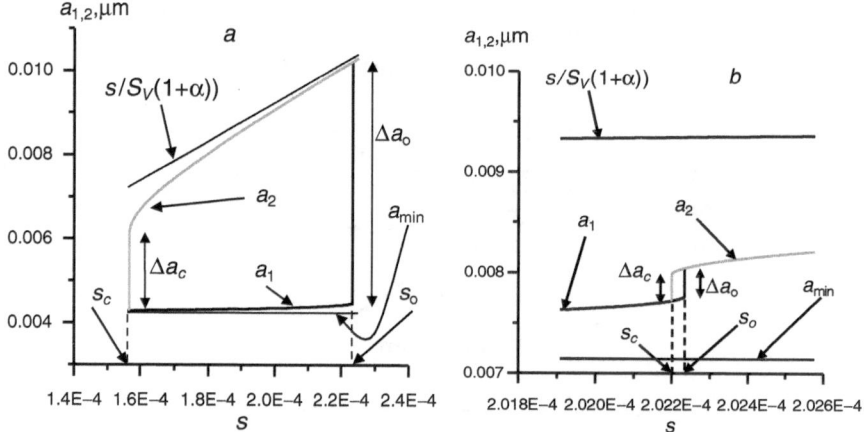

Fig. 11.5. Illustration of the s-dependence of the apertures $a_1 = a_1(s, D)$ (thick black curve representing the closed state) and $a_2 = a_2(s, D)$ (thick gray curve representing the open state) of the $U(a)$ potential's minima (assuming $S_V = 10^3\,\mathrm{m}^{-1}$, $L_V = 10^8\,\mathrm{m}^{-2}$) for two values of D: (a) $D = 3.91 \cdot 10^{-3}\,\mu\mathrm{m}$, (b) $D = 7.25 \cdot 10^{-3}\,\mu\mathrm{m}$. Here $s = \varepsilon + S_V(1 + \alpha)A$ [Eq. (11.19)].

Because the aperture a directly influences the stress–strain relation Eq. (11.16), it is interesting to carefully examine its strain dependence. As mentioned above, the aperture a is always equal to either the value of the first minimum a_1 or of the second minimum a_2 of the potential $U(a)$. Figure 11.5 illustrates the dependencies $a_1 = a_1(s, D)$ and $a_2 = a_2(s, D)$ that were obtained numerically for two values of the roughness parameter D. On the plot we identified the critical "strain" values s_c and s_o, and the hysteretic jumps Δa_o and Δa_c in the aperture of the crack (thin double arrows). In the case where we considered a moderate roughness (case a, with $D = 3.91 \cdot 10^{-3}\,\mu\mathrm{m}$) the difference in strains $\Delta s = s_o - s_c$ equals $6.67 \cdot 10^{-5}$, indicating that this is a rather hard (or strong) element. In the case of high roughness (case b, with $D = 7.25 \cdot 10^{-3}\,\mu\mathrm{m}$) the difference $\Delta s = s_o - s_c$ reduces to $3.64 \cdot 10^{-8}$, representing a soft (or weak) element. Here, the terms "hardness" and "softness" of an element actually refer to its sensitivity to strain variations: a soft element can be switched by a small variation and a hard one needs larger strain alterations to switch.

Figure 11.5 also shows the asymptotic dependencies $a_{\min}$ (i.e., the coordinate of the adhesion potential's $W(a)$ minimum) and $s/(S_V(1 + \alpha))$ for highly negative and positive strains, respectively (thin black lines for both asymptotes). Clearly, when $a_2 \gg a_1$ (and correspondingly Δs is large as in Figure 11.5a), the adhesion and the elastic components of $U(a)$ [Eq. (11.20)] can be minimized independently, and the resulting minima are reasonable approximations for the true values; that is,

$$a_1(s, D) \approx a_{\min}(D), \qquad a_2(s, D) \approx s/\left(S_V(1 + \alpha)\right). \qquad (11.21)$$

These asymptotic dependencies are only valid for cracks with large enough Δs. In Figure 11.5b we indeed notice a considerable deviation of the numerical results from the asymptotes for low Δs.

In addition, we can remark that the elementary hysteresis shown in Figure 11.5b is quite similar to the one used by Scalerandi et al.[31] discussed in Chapters 16 and

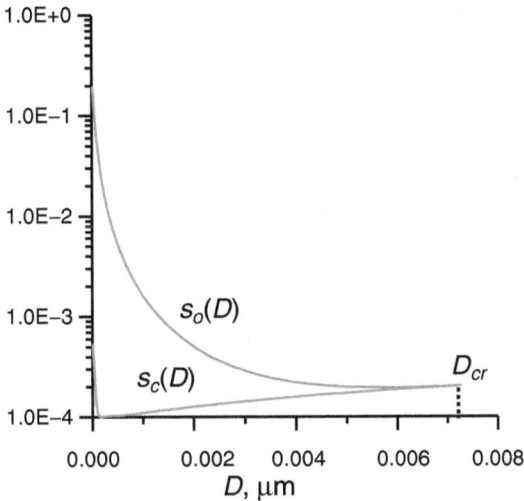

Fig. 11.6. The dependencies of the closing and the opening strains (s_c and s_o) on the roughness parameter D. The difference $s_o - s_c$ (which is characteristic for the "strength" or "hardness" of a crack) diminishes for increasing D. Elementary cracks with $D > D_{cr} = 7.30 \cdot 10^{-3}$ µm are not hysteretic at all because $\Delta s = 0$ and $\Delta a_o = \Delta a_c = 0$ (Figure 11.7).

17 of this book, where it appeared from rather intuitive considerations: elements in a closed state are assumed to be rigid (horizontal line in the stress–strain curve) and open elements have additional elasticity due to the soft interstices between the elastic grains (resulting in an inclined line in the corresponding stress–strain relation). In the present model, which is based on physical laws, a discontinuity is found to occur both at opening and at closing, whereas Scalerandi et al. only treated the case of a stress–strain jump at closure.

The dependencies of the critical strains s_c and s_o on the roughness parameter D are shown in Figure 11.6, whereas Figure 11.7 illustrates the hysteretic jumps $\Delta a_o(D)$ and $\Delta a_c(D)$. Again, we remark that the difference $\Delta s = s_o - s_c = \varepsilon_o - \varepsilon_c$ can be interpreted as a characteristic of the "strength" of a crack: "weak" cracks switch at small strain excursions and "strong" ones change states at large strain amplitudes. The highest difference Δs is reached for zero roughness (maximum $\Delta s \approx 0.2$ in our example for the chosen parameters S_V, L_V, H, E). The more roughened the coasts of a crack are, the weaker it is. When the roughness D exceeds a critical value D_{cr} (which equals $7.30 \cdot 10^{-3}$ µm in our example), $\Delta s = 0$ and $\Delta a_o = \Delta a_c = 0$, and consequently the crack becomes nonhysteretic. In this case only the second minimum of $U(a)$ is present.

Returning again to the expression for the physical strain $u_x = \varepsilon = s - (1 + \alpha)S_V A$ [Eq. (11.19)], we find the true opening and closing strains ε_o and ε_c for any given parameters A and D by means of the following mapping.

$$\varepsilon_o(A, D) = s_o(D) - (1 + \alpha) S_V A, \qquad \varepsilon_c(A, D) = s_c(D) - (1 + \alpha) S_V A.$$
$$(11.22)$$

The above discussion enables us to easily track the evolution of the crack system [solution of Eqs. (11.15)]. Suppose we know the strain protocol $\varepsilon(x, t)$, at any given

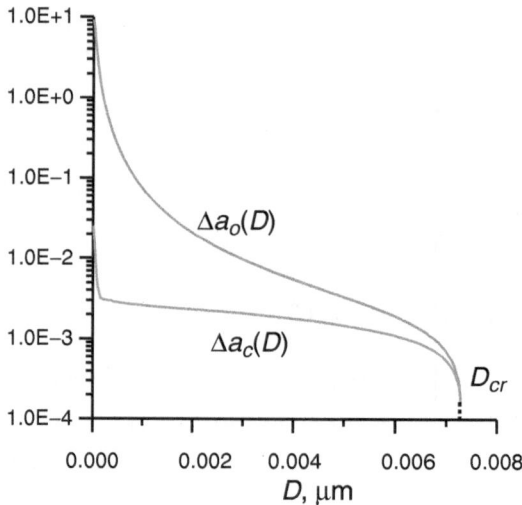

Fig. 11.7. Roughness (D) dependence of hysteretic jumps Δa_o and Δa_c.

moment t for the current position x. If ε_t is positive (opening), then $a = a_1(s, D)$ if $\varepsilon < \varepsilon_o$ and $a = a_2(s, D)$ if $\varepsilon > \varepsilon_o$. On the other hand, if the strain decreases (i.e., $\varepsilon_t < 0$), then $a = a_2(s, D)$ if $\varepsilon > \varepsilon_c$ and $a = a_1(s, D)$ if $\varepsilon < \varepsilon_c$. The resultant value of a subsequently determines the stress by means of Eq. (11.16), and by solving the wave equation [the first equation of Eqs. (11.15)] we can obtain the solution for the next time layer $t + \Delta t$.

7. Evolution of a System with a Statistical Ensemble of Different Cracks

Returning now to the original problem of the dynamics of a system with all different elementary cracks, we accept another strong assumption consisting in the complete independence of all cracks. This means that we can write the full Lagrangian as the sum of partial Lagrangians that are related to the corresponding subsystems. It is possible to show[11] that the evolution equations in this case take on the form:

$$\rho\, u_{tt} = E u_{xx} - E S_V \int f(A, D) a_x^{AD} dA\, dD, \qquad (11.23)$$

$$a^{AD} \text{ is a minimum of } U^{AD}\left(a^{AD}\right) = W(a^{AD}) + \tfrac{1}{2} E S_V\,(1 + \alpha)\left(a^{AD^2} - 2A a^{AD}\right)$$
$$- E S_V u_x a^{AD}, \qquad (11.24)$$

where a^{AD} is the aperture of an elementary crack with parameters A and D. The minimization problem [Eq. (11.24)] is obviously the same as the minimization of the previously considered potential $U(a)$ [Eq. (11.17)], and therefore all results of Section 6 hold.

Equation (11.23) can be rewritten in the form $\rho u_{tt} = \sigma_x$, where

$$\sigma = E u_x - E S_V \int f(A, D) \left(a^{AD} - A \right) dA \, dD. \tag{11.25}$$

For every crack with parameters A and D, and at every actual value of the strain $\varepsilon = u_x$ we use the above-formulated evolution rules to find out whether its state is open or closed: a crack opens if ε becomes larger than ε_o when the strain is increasing ($\varepsilon_t > 0$) and closes for decreasing strains if $\varepsilon < \varepsilon_c$. The numerical analysis of the potential $U(a)$ outlined in the previous section enables us to compute the values $\varepsilon_o(A, D)$, $\varepsilon_c(A, D)$, $a_1(A, D)$, and $a_2(A, D)$ with any desired precision. Taking a_1 for the closed state and a_2 for the open one, we compute the integral in Eq. (11.25) and obtain the stress–strain relation necessary to solve the evolution equation. If for a given crack $D > D_{cr}$, only the solu0tion a_2 is applicable.

At this point it becomes obvious that our microcracked system consisting of a statistical ensemble of differing cracks amounts to a slight variation of the well-known Preisach–Mayergoyz (PM) or, keeping the original terminology of Mayergoyz,[33] Preisach–Krasnoselskii (PK) formalism. The principal difference is that all parameters of the problem in our description are derived from the crack's statistics, albeit in a greatly simplified manner. There are several other differences of secondary importance.

- In the present framework, we arrived at a $(\varepsilon_o, \varepsilon_c)$ PM-space, whereas traditionally (σ_o, σ_c) coordinates are used. Phenomenologically there is no preference between these two approaches, even though our present study shows that $(\varepsilon_o, \varepsilon_c)$-spaces correspond to completely independent cracks. The dynamics resulting from a $(\varepsilon_o, \varepsilon_c)$-space were analyzed in Reference [17] in more detail.

- The shape of the hysteresis for an elementary unit (Figure 11.5) is different from the traditional rectangular one[32] (see also Chapters 12 and 21). Correspondingly, an elementary unit always contributes to the force term σ_x in the evolution equation $\rho u_{tt} = \sigma_x$, whereas in the traditional case it has a nonzero contribution only if switching. The hysteresis curves found here are similar to the ones assumed by Scalerandi et al. (Chapters 16 and 17).

The latter consideration deserves a separate discussion. In the usual phenomenological models, the hysteretic stress–strain nonlinearity is supplemented with a nonhysteretic (sometimes called classical) term, which is typically represented by a polynomial expansion of the stress as function of the strain. This term is introduced, in particular, to reflect the experimental fact that a microcracked material is stiffer at higher compression. In the present approach this is satisfied automatically, because elementary hysteretic curves (Figure 11.5) are more inclined in the open state than in the closed one. As the nonzero strain-derivative contributes to $d\sigma/d\varepsilon$ with a minus sign [Eq. (11.25)], the material becomes softer when more and more cracks are open. The difference with the traditional models is, however, that we do not attribute this feature to classical nonlinearity (anharmonicity of the interatomic potential in a single crystal) but also to the contact nonlinearity in the same way as for the hysteresis itself.

Figure 11.8a illustrates the density distribution $\rho(\varepsilon_o, \varepsilon_c)$ of the hysteretic elements obtained using the above framework for a uniform distribution $f(A, D)$, limited to

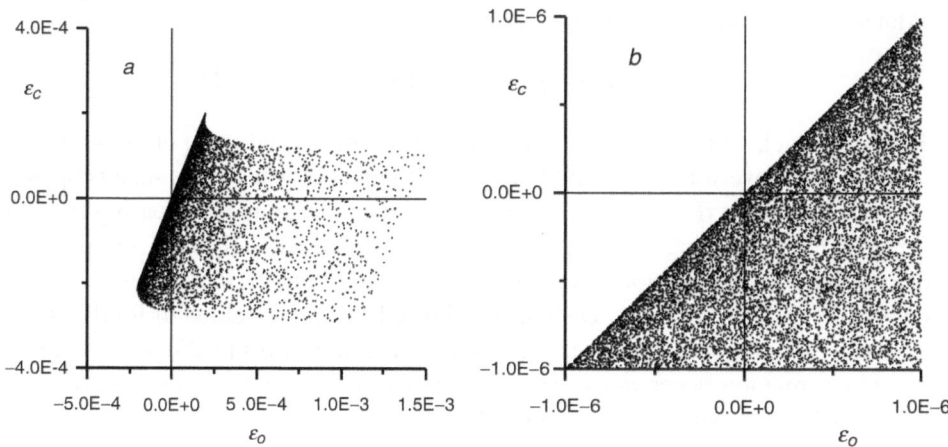

Fig. 11.8. The density $\rho(\varepsilon_o, \varepsilon_c)$ of hysteretic elements (PKM distributions): (a) general distribution obtained in the framework of the micropotential model (only points with $\varepsilon_\perp < 10^{-3}$ are plotted): (b) the fragment of the PKM space in (a) with $-10^{-6} < \varepsilon_o, \varepsilon_c < 10^{-6}$.

elements with $\varepsilon_\perp < 10^{-3}$ (elements with $\varepsilon_\perp < 10^{-3}$ are rare and most probably not physical, because such high strains would definitely produce extra damage to the material; see last comment of Section 4). Figure 11.8b represents a fragment of the same PM-space with $10^{-6} < \varepsilon_o, \varepsilon_c < 10^{-6}$. This fragment is of particular interest for dynamics, because elementary units in this sub-space will have a dominant contribution to the stress–strain relation for typical acoustical strains.

The PM distribution (with different underlying elementary hysteretic units) has been used in many studies to describe a multitude of static and dynamic properties of non-linear materials. However, it is clear that some recently observed universal features of microcracked media cannot be attributed to hysteresis only. In particular, we mention the observation of slow dynamics (see review by Ostrovsky and Johnson[1] and Chapter 4 of this book). Slow dynamics is manifested by a temporary increase of the nonlinearity strength immediately after an intense dynamic excitation and by a subsequent logarithmic relaxation of the nonlinearity to its equilibrium value. The issue of slow dynamics can, in principle, be explained in the framework of the present concept in the following manner: the application of a high loading will generally lead to an active destruction of the material in the vicinity of the crack tips, and therefore increases S_V. If we assume that the average enlargement of the linear dimensions of the cracks is β ($\beta > 1$), the area increases β^2 times, whereas the increase of L_V is not larger than β (it can even be smaller because the crack tip lines may straighten during this process). Consequently, the combination $S_V(1 + \alpha) \approx S_V \alpha \sim L_V / S_V$ in Eq. (11.24) decreases, making the crack network weaker (which can be confirmed through an analysis of the potential). As a consequence D_{cr} increases, and the sample therefore will have more hysteretic cracks. In addition, we recall that the increased area S_V is a proportionality coefficient for the nonlinearity itself [Eq. (11.25)]. Thus, given the increase in S_V and a consequent weakening of the crack network, it is conceivable that the material becomes more nonlinear after high loading. Furthermore, if we conjecture that thermal jumps and creep due to the adhesion force can heal and reunite the damaged crack tips again, it is also plausible that the material will relax again in the course of time to its equilibrium state. Certainly, in the present form, our theory does not take into account the effects of extra cracking and slow healing, but they can be thought of as a possible future improvement of the model.

Another interesting and more or less universal feature concerns the typical excitation amplitude dependence of the resonant frequency in sandstones, reported by Ten-Cate et al.,[33] in which the existence of two ranges of amplitudes is revealed (see also Chapter 26): for small amplitudes (lower than about 10^{-6}) the dependence is observed to be quadratic and for higher strains (amplitude $\varepsilon_A > 10^{-6}$) it becomes linear. In the framework of the PM-space, a quadratic amplitude dependence of the resonant frequency implies a linear increase of the density of the hysteretic contributions proportional to $\varepsilon_\perp$, whereas a linear decrease of the resonance frequency is obtained when the density of the hysteretic elements is approximately constant. Such a situation either requires a sparse concentration of elements just below the diagonal $\varepsilon_c = \varepsilon_o$, or the occurrence of infinitesimally small hysteretic jumps for the elements located close to the diagonal. However, both conditions are generally not found in the present model. An alternative explanation for the existence of the quadratic regime may consist in the presence of spontaneous thermal jumps, which constantly erase the memory of the states of the elements near the diagonal. Such a model was proposed by Tournat and Gusev[34] (Chapter 21) as an extension of the PM approach for nonzero temperature. In fact, at small strains the potential barriers distinguishing one metastable state from another are small and can be comparable to kT.

8.   Conclusions

The theory presented in this chapter deals with the development of a physics-based model form for the potential energy of crack networks to explain the hysteretic elastic behavior of microcracked materials. The most important geometrical parameters necessary for this description are the average area of cracks per unit of volume, and the distribution of asperities and rest apertures over the crack network. The resultant potential energy consists of three terms: the strain energy of the surrounding intact material, the adhesive potential of the crack surfaces, and an additional stabilizing contribution due to the crack tips.

Starting from the expression of the potential energy, the influence of microcracks on material behavior was obtained using a purely mechanical approach (by means of the Lagrange equations). Hysteresis in the stress–strain relation, usually introduced in a phenomenological and artificial way, is attributed in the present description to the occurrence of metastable states of the crack network, which follows from the double-well potential of the system. As a result, a physical basis has been provided for the general phenomenological theory of velocity-independent hysteresis (PM-formalism) together with its typical consequences for statics and dynamics.

An explanation of related phenomena, such as slow dynamics and the existence of several different ranges in the strain dependence of the nonlinearity (which are strikingly universal for a wide range of materials) can be obtained by introducing a nonzero temperature into the model and by considering damage evolution.

As a final conclusion, we itemize the major assumptions made in the analysis.

Key assumptions (Rejection of either one of these assumptions is not possible in the framework of the present theory).

- Existence of elementary cracks, for which averaging of the asperities and apertures can be performed
- Independence of elementary cracks, additivity of partial potential energies
- Neglect of branching lines in the crack network

Important assumptions (These assumptions can be changed or rejected, but this will require a major reconsideration of the theory's components).

- The model of adhesion following the Lennard–Jones potential and the lognormal distribution of apertures
- Account of crack tips using the modified theory of Hertzian contacts
- Localization of strain energy of tips in the vicinity of cracks
- Zero temperature
- Neglect of shear effects
- No additional damage during straining
- One-dimensional wave equation and disregard of crack orientation

Acknowledgments

The authors gratefully acknowledge the support of the Flemish Fund for Scientific Research. (G.0206.02 and G.0257.02), the provisions of the European Science Foundation Programme NATEMIS, and the European FP5 and FP6 Grants DIAS (EVK4-CT-2002–00080) and AERONEWS (AST3-CT-2003–502927).

References

[1] L.A. Ostrovsky and P.A. Johnson, Dynamic nonlinear elasticity in geomaterials, *Riv. Nuovo Cimento* **24**(7) (2001).

[2] V.E. Nazarov, A.V. Radostin, L.A. Ostrovsky, and I.A. Soustova, Wave processes in media with hysteretic nonlinearity. Part I. *Acoust. Phys* **49**(3), 344–353 (2003).

[3] J.B. Walsh, Seismic wave attenuation in rock due to friction, *J. Geophys. Res.* **71**, 2591–2599 (1966).

[4] G. Mavko, Frictional attenuation: An inherent amplitude dependence, *J. Geophys. Res.* **84**, 4769–4777 (1979).

[5] W.P. Mason, Internal friction mechanism that produces an attenuation in the earth's crust proportional to the frequency, *J. Geophys. Res.* **74**, 4963–4966 (1979).

[6] M.M. Sharma and A.N. Tutuncu, Grain adhesion hysteresis: A mechanism for attenuation of seismic waves, *Geoph. Res. Lett.* **21**(21), 2323–2326 (1994).

[7] C. Pecorari, Adhesion and nonlinear scattering by rough surfaces in contact: Beyond the phenomenology of the Preisach–Mayergoyz framework, *J. Acoust. Soc. Am.* **116**(4), 1938–1947 (2004).

[8] R. Mindlin and H. Deresiewicz, Elastic spheres in contact under varying oblique forces, *J. Appl. Mech.* **20**, 327 (1953).

[9] A. Granato and K. Lucke, Theory of mechanical damping due to dislocations, *J. Appl. Phys.* **27**(6), 583–593 (1956).

[10] J.-Y. Kim, A. Baltazar, and S.I. Rokhlin, Ultrasonic assessment of rough surface contact between solids from elastoplastic loading-unloading hysteretic cycle, *J. Mech. Phys. Solids* **52**(8), 1911–1934 (2004).

[11] V. Aleshin and K. Van Den Abeele, Micro-potential model for stress–strain hysteresis of microcracked materials, *J. Mech. Phys. Solids* (2005), **53**(4), 795–824 (2005).

[12] J. Ortin, Preisach modeling of hysteresis for a pseudoelastic Cu-Zn-Al single crystal. *J. Appl. Phys.* **71**(3), 1454 (1992).

[13] R.A. Guyer, K.R. McCall, and G.N. Boitnott, Hysteresis, discrete memory and nonlinear wave propagation in rock: A new paradigm, *Phys. Rev. Let.*, **74**, 3491–3494 (1995).

[14] F. Preisach, Über die magnetische Nachwirkung. *Z. Phys.* **94**, 277 (1935).

[15] M.A. Krasnosel'skii and A.V. Pokrovskii, *Systems with Hysteresis* (Nauka, Moscow, 1983 or Springer, Berlin, 1989).

[16] R.A. Guyer, K.R. McCall, G.N. Boitnott, L.B. Hilbert Jr., and T.J. Plona, Quantitative implementation of Preisach–Mayergoyz space to find static and dynamic elastic moduli in rock. *J. Geophys. Res.* **102**(B3), 5281–5293 (1997).

[17] V. Aleshin, V. Gusev, and V. Zaitsev, Propagation of acoustic waves of nonsimplex form in a material with hysteretic quadratic nonlinearity: Analysis and numerical simulations, *J. Comput. Acoust.* **12**(3), 319–354 (2004).

[18] K. Van Den Abeele, F. Schubert, V. Aleshin, F. Windels, and J. Carmeliet, Resonant bar simulations in media with localized damage, *Ultrasonics* **42**, 1017 (2004).

[19] M. Boudjema, I.B. Santos, K.R. McCall, R.A. Guyer, and G.N. Boitnott, *Nonlinear Proc. Geophys.* **10**, 589–597 (2003).

[20] K. Helbig, A formalism for the consistent description of non-linear elasticity of anisotropic media, *Revue de l'Institut Français du Pétrole*, **53**(5), 693–708 (1998).

[21] J. TenCate, E. Smith, and R. Guyer, Universal slow dynamics in granular solids, *Phys. Rev. Lett.* **85**(5), 1020–1023 (2000).

[22] K. Van Den Abeele, J. Carmeliet, P.A. Johnson, and B. Zinster. Influence of water saturation on the nonlinear elastic mesoscopic response in Earth materials and the implications to the mechanisms of nonlinearity, *J. Geophys. Res.* **107**(B6) 2121 (2002).

[23] A. Carpinteri, B. Chiaia, and K. Nemati, Complex fracture energy dissipation in concrete under different loading conditions, *Mech. Mater.* **26**, 93–108 (1997).

[24] K.M. Nemati and P. Stroeven, Stereological analysis of micromechanical behavior of concrete, *Materials and Structures/Matériaux et Constructions* **34**(242), 486–494 (2001).

[25] J.A. Greenwood and J.B.P. Williamson, Contact of nominally flat surfaces, *Proc. Roy. Soc. London A*, **295**(1442), 300–319 (1966).

[26] K. Kendall, *Molecular Adhesion and it Applications* (Kluwer, New York, 2001).

[27] A.A. Keller, High resolution CAT imaging of fractures in consolidated materials, *Int. J. Rock Mech. Min. Sci.* **34**(3/4), 358–375 (1997).

[28] R.A. Johns, J.D. Steude, L.M. Castanier, and P.V. Roberts, Nondestructive measurements of fracture aperture in crystalline rock cores using X-ray computed tomography, *J. Geophys. Res.* **98**, 1889–1900 (1993).

[29] J.N. Israelachvili, *Intermolecular and Surface Forces* (Academic Press, London, 1995).

[30] L.D. Landau, and E.M. Lifshitz, *Theory of Elasticity* (Pergamon, New York, 1986).

[31] M. Scalerandi, P.P. Delsanto, V. Agostini, K. Van Den Abeele, and P.A. Johnson, Local interaction simulation approach to modelling nonclassical, nonlinear elastic behavior in solids, *J. Acoust. Soc. Am.* **113**, 3049 (2003).

[32] I.D. Mayergoyz, Hysteretic models form the mathematical and control theory points of view, *J. Appl. Phys.* **51**(1), 3803–3805 (1985).

[33] J.A. TenCate, D. Pasqualini, S. Habib, K. Heitmann, D. Higdon, and P.A. Johnson, Nonlinear and nonequilibrium dynamics in geomaterials, *Phys. Rev. Lett.* **93**(6), 065501 (2004).

[34] V. Tournat and V.E. Gusev, Frequency-dependent nonlinearity in the Preisach–Arrhenius model for acoustic hysteresis, *Proc. of 5th World Congress on Ultrasonics*, Paris, France (2003).

12

Multiscale Approach and Simulations of Wave Propagation and Resonance in Media with Localized Microdamage: 1-D and 2-D Cases

Koen Van Den Abeele[1] and Sigfried Vanaverbeke

Katholieke Universiteit Leuven, Interdisciplinair Research Center (IRC), KULAK, E. Sabbelaan 53, B-8500 Kortrijk, Belgium.
[1]To whom correspondence should be addressed: Koen Van Den Abeele, Katholieke Universiteit Leuven, Interdisciplinair Research Center (IRC), KULAK, E. Sabbelaan 53, B-8500 Kortrijk, Belgium; e-mail: koen.vandenabeele@kuleuven-kortrijk.be.

Abstract

The development of a numerical framework in support of observations of nonlinear elastic wave spectroscopy is of high importance and may guide the analysis of nonlinear features towards new imaging techniques. In this contribution, we outline and review a multiscale model for one- and two-dimensional problems of wave propagation and resonance in microcracked materials. At the microlevel, we simulate microcracks by triggerlike elements with a two-state nonlinear stress–strain relation. We scale up the microscopic state relation on the mesoscopic level by means of a scalar Preisach approach. For multidimensional hysteretic elasticity, this approach is generalized by simply decomposing the stress-tensor into its eigenstress components. Finally we use a staggered grid formulation to predict the macroscopic response to an arbitrary excitation signal. Using this multiscale model, we investigate the influence of a microdamaged zone on the resonance signatures (resonance frequency and damping), and on the propagation characteristics (wave speed, generation of harmonics). Numerical results are presented for one- and two-dimensional problems as a function of the excitation amplitudes of the input signal, and for different geometries and distributions of the hysteretic properties within the microdamaged zone. In 1-D, we specifically focus our attention on the influence of different Preisach-space distributions on the nonlinear characteristics, thereby clearly illustrating the fact that the expected power law dependencies in the dynamic modulus–strain relation and in the generation of higher harmonics are drastically altered by nonuniform PM-space distributions. In 2-D, we illustrate numerical simulations for nonlinear Rayleigh wave propagation along the surface of a microcracked solid and for the in-plane propagation of pulses in plates.

Keywords: Hysteresis, microdamage, nonlinear resonant ultrasound spectroscopy, nonlinear wave propagation, nonlinearity

1. Introduction

Nonlinearity and hysteresis in the quasistatic stress–strain equation are instantly recognizable manifestations observed in all damaged solids. It is reasonable to assume that this manifestation has major consequences on the dynamics of microcracked solids as well. In fact, several experiments conducted in the last decade have provided abundant evidence for this. Apart from a selection of references in this chapter [1–7], we also refer to Chapters 4, 5, 24–31 and the many references therein. In terms of modeling the dynamics, the consideration of nonlinearity, and in particular hysteresis, presents some significant challenges. In general, modeling the propagation of transient elastic waves in linear solid media can already be very convoluted, and closed-form analytical solutions can only be found for simple geometries and easy assumptions. If, on top of mode conversions, anisotropy, inhomogeneities, inclusions, and boundaries, one also takes into account the existence of nonlinear and nonunique state relations, it becomes clear that only numerical modeling techniques can be used.

Various numerical models have been developed over the years to account for the local inhomogeneous nature of solids. Finite element methods, boundary element methods, finite difference methods, and finite integration techniques are well-established techniques and are documented with impressive illustrations for the case of linear wave propagation and resonance simulations. Only a few researchers have actually reworked the existing codes to include the presence of nonunique stress–strain relations. However, many efforts in this field are currently limited to one-dimensional simulations. For reference we mention the work of Nazarov et al. [1,8], Delsanto and Scalerandi [9,10], Gusev and Aleshin [11], Schubert et al. [12,13], and Van Den Abeele et al. [14,15]. The numerical simulations of the group of Delsanto and Scalerandi are based on a nonlinear version of the Local Interaction Simulation Approach coupled to a spring model. The method is extensively illustrated for one-dimensional problems of wave propagation and wave resonances (see Chapter 17 of this book for an extensive overview), and the first results of particular two-dimensional problems are discussed in Chapter 18 of this book. Schubert and Van Den Abeele use the Elastodynamic Finite Integration Technique (EFIT) as the basis for their models. EFIT was originally developed by Fellinger et al. [16] and is now well established for 2-D and 3-D simulations in linear solids [17–19]. The EFIT procedure uses the integral form of the basic equations, and the discretization in terms of a staggered grid formulation provides a very stable and efficient numerical code. Schubert has altered the 1-D EFIT version for the inclusion of classical quadratic nonlinearity [12] and for the treatment of a single crack closure and opening using a dynamic grid management [13]. Van Den Abeele et al. on the other hand coupled the 1-D EFIT model to the general description of stress–strain hysteresis using a Preisach model [15] with a multiscale approach. Just recently, Bou Matar et al. suggested a modification of the numerical treatment of the multiscale approach by introducing a pseudospectral solver [20].

In this chapter, we first review the general description of stress–strain hysteresis using the PM approach and illustrate the impact of hysteretic nonlinearity on the wave propagation and wave resonance behavior in the one-dimensional case. To do this, we develop a multiscale approach: at the microlevel, microcracks and inhomogeneities are simulated by triggerlike elements with a two-state nonlinear stress–strain relation. The microscopic state relation is then scaled up onto the mesoscopic level by means of the Preisach-space model for hysteretic elasticity, and, finally, a staggered grid EFIT formulation is used to predict the macroscopic response to an arbitrary excitation signal.

After discussing some particular examples of the one-dimensional nonlinear wave propagation and resonance features, we extend the multiscale model to higher dimensions by means of a generalization of the Preisach-space model for multidimensional hysteretic elasticity. We illustrate the model by results of two-dimensional wave propagation for nonlinear Rayleigh waves and in plane pulse propagation in media with localized damage.

As a justification for this research, we mention that many historical monuments throughout Europe are suffering from severe microdamage spreading inwards from the stone surface and reaching several millimeters in depth. The investigation of the linear and nonlinear behavior of Rayleigh waves may help in defining new techniques to quantify and characterize in situ the extent and depth of the damage in comparison to virgin material. Similarly, safety of aircraft is a high-priority issue and therefore the development of appropriate quality control techniques for aeronautical components is extremely important. In our opinion nonlinear ultrasonic methods will increase the sensitivity for detecting incipient cracks at early stages of deterioration and fatigue processes. Numerical models can help to interpret the results, define new strategies, optimize detection positions, and the like. For more information on the above research topics, sponsored by the European Union, see http://minelab.mred.tuc.gr/dias/ and www.kuleuven-kortrijk.be/AERONEWS.

2. PKM Model for Hysteretic Stress–Strain Behavior

The quasistatic stress–strain relation of a wide class of microinhomogeneous materials generally exhibits the fact that the strain response to stress increase and stress decrease is nonunique, even in the purely elastic (nonplastic) regime. As a consequence of this type of mechanical hysteresis, interior loops can be observed with evidence for end point memory [21]. It has also been observed that such hysteretic loops widen when microdamage is progressively increasing [22], and that the shape of the loops may depend on water saturation [23], temperature, external pressure, and so on. Because the description of the hysteretic phenomena during a quasistatic experiment forms the basis of our dynamic multiscale model, we review briefly the key elements of the underlying theoretical model. The model is based on the work by Guyer et al. [24], who translated the ideas of Preisach [25] and Mayergoyz [26] to the field of rock elasticity and mechanics, ignoring rate dependence. An extension of this theoretical approach to rate-dependent hysteresis can be found in Chapter 21 of this book.

Let us consider a representative cell (interval in one dimension, area or volume in two and three dimensions) containing a large number of grains and bonds. We assume that the bonds have a two-state behavior, which we can associate with being "open" or "closed" depending on the pressure. The grains are the hard material portions and behave reversibly elasticly, with potential presence of "atomic" nonlinearity [acoustoelasticity, i.e., pressure dependence of the modulus $K_g(P)$]. Whereas the grains contribute to the overall strain (ε) in a traditional manner upon an arbitrary pressure change (pressures P are positive in compression, stresses σ are negative):

$$\Delta\varepsilon_g = -\int\limits_{P}^{P+\Delta P} \frac{dP'}{K_g(P')} = \int\limits_{-\sigma-\Delta\sigma}^{-\sigma} \frac{d\sigma'}{K_g(\sigma')}, \tag{12.1}$$

the bonds only contribute to a change in strain if they alter states from open to closed or vice versa due to the imposed stress change. To quantify the contribution from the bond system, we presume a simple phenomenological state relation in which the strain contribution of each bond (also called unit) changes from zero to a finite (constant) value γ depending on the actual pressure (P) [or stress (σ)] value and its history. While increasing pressure, the strain in the unit is zero for $P < P_c$ ($\sigma > \sigma_c$) (open state), and $-\gamma$ for $P > P_c$ ($\sigma < \sigma_c$) (closed state). When decreasing the stress, the strain equals $-\gamma$ for $P > P_o$ ($\sigma < \sigma_o$) (closed state) and zero for $P < P_o$ ($\sigma > \sigma_o$) (open state). Naturally, we have that $P_c > P_o$. For simplicity we assume that only the parameters P_o and P_c (or σ_o and σ_c) can vary from unit to unit. As a consequence, each unit within a representative cell can be represented in a stress–stress space according to their values P_o and P_c. This representation is commonly termed $P(K)M$-space, and can be dealt with mathematically by its density distribution $\rho(P_c, P_o)$ [15,24,27,28]. If the representative cell has N bonds we can express the contribution of the full bond system as

$$\Delta\varepsilon_b = -\hat{\gamma}\left(f_c(P + \Delta P) - f_c(P)\right) = \hat{\gamma}\left(f_c(-\sigma) - f_c(-\sigma - \Delta\sigma)\right) \tag{12.2}$$

with $\hat{\gamma} = \gamma N$ and $f_c(P)$ the fraction of units that are closed at pressure P, which can be calculated as an integral of ρ over the area of closed units in the stress–stress space. Using this notation, $\hat{\gamma}$ clearly denotes the strain contribution of the bond system when all units are in the closed state.

Using the infinitesimal ($\Delta P, \Delta\varepsilon$) relation,

$$\Delta\varepsilon = \Delta\varepsilon_g + \Delta\varepsilon_b = -\int\limits_{P}^{P+\Delta P} \frac{dP'}{K_g(P')} - \hat{\gamma}\left(f_c(P + \Delta P) - f_c(P)\right), \tag{12.3}$$

it is straightforward to predict the strain response for a given pressure protocol and known representations of the PKM density and acoustoelasticity. All is reduced to a careful administration of the open and closed units in the PKM-space and a proper numerical integration. Some examples are given in Figure 12.1 showing the influence of different types of PKM-spaces. Hysteresis and end point memory clearly imply that there can be an infinite number of stress–strain relations. One stress level corresponds to an infinite number of possibilities for the corresponding strain values.

The (inverse) modulus follows from Eq. (12.3) as follows.

$$K^{-1} = \lim_{\Delta\sigma \to 0}\left(\frac{\Delta\varepsilon}{\Delta\sigma}\right) = \frac{1}{K_g(-\sigma)} + \hat{\gamma}\frac{df_c}{d\sigma}(-\sigma). \tag{12.4}$$

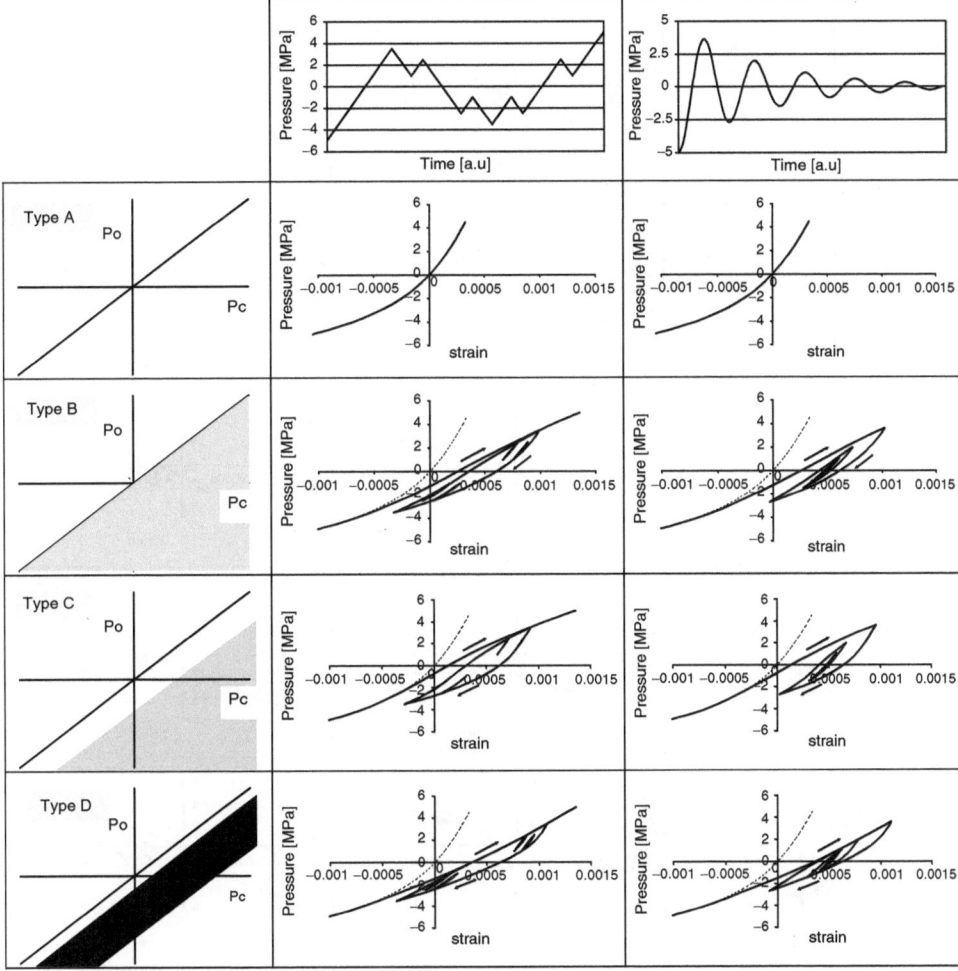

Fig. 12.1. PKM space densities and corresponding stress–strain relations for two types of protocols, shown on top. The considered PKM-spaces are confined to $-5\,\text{MPa} \leq P_o \leq P_c \leq 5\,\text{MPa}$. Gray scale corresponds to the density: (A) empty PKM-space (no hysteresis); (B) uniform PKM-space; (C) shifted PKM-space (offset = 1.5 MPa); (D) banded PKM-space (offset 1 = 0.5 MPa; offset 2 = 2 MPa); (E) split PKM-space; (F) upper PKM-space; (G) nonuniform PKM-space (a = 0 MPa, b = 3 MPa; c = 0.5 MPa^{-1}, m = 0); (H) nonuniform PKM-space (a = 0 MPa, b = 3 MPa; c = 0.5 MPa^{-1}, m = 1); (I) nonuniform PKM-space (a = 0 MPa, b = 3 MPa; c = 0.5 MPa^{-1}, m = 2). For Types G, H, I, we used the density given by Eq. (12.9). In all cases (except obviously for type A) $\hat{\gamma} = 0.001$. The classical nonlinearity satisfies $K_g(P) = 10^{10}(1 + \tanh(0.5P))$.

With reference to the PKM-space, the calculation of the modulus basically requires the integration of the density at $P(-\sigma)$ of those units that change state. For increasing pressure this integration is performed at $P_c = P$ over a vertical line interval; for decreasing pressures the integration occurs at $P_o = P$ over a horizontal line interval. The hysteresis in the modulus–stress relation obviously comes in through the second term in Eq. (12.4). It is exactly the above relation that we use in the model for dynamic wave propagation and wave resonance.

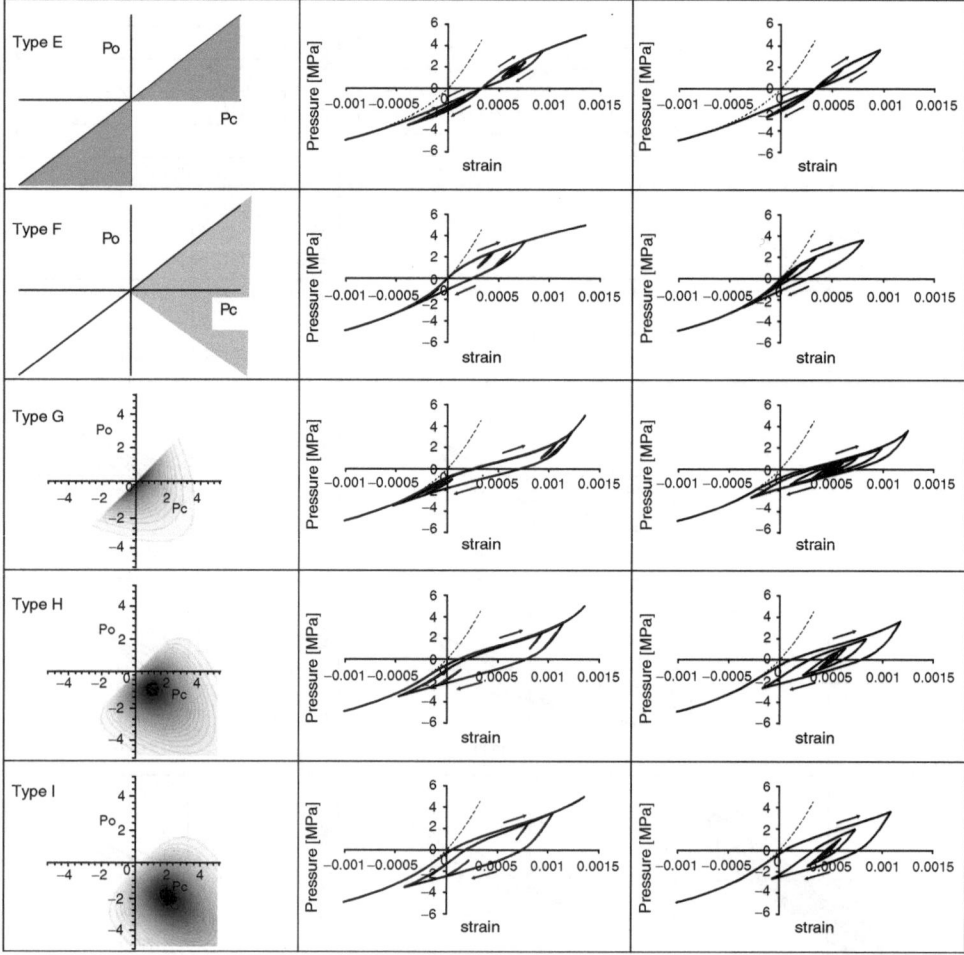

Fig. 12.1. *Continued.*

We remark that the PKM space described above is a static space, meaning that elements do not switch by themselves, for instance, due to thermally activated random transitions. Only "external" variations in stress impel the unit to change its state. A second remark is that this current representation is rate independent implying that there is no influence on the frequency of oscillation of the stress waves. However, some researchers are trying to overcome this caveat by proposing a generalized Preisach–Arrhenius approach (e.g., [29], and Chapter 21). And as a final remark, we mention the work of Scalerandi (e.g., [9] and Chapter 17) who treated hysteretic nonlinearity with a slightly different elementary unit, in which the two states of the unit can be either elastic or rigid (see also Chapter 11). Yet, in the absence of underlying physics, which should be provided by theories of molecular dynamics based on pseudopotential energy approaches and atomistic models, we assume that the hysteretic part can be modeled by the two-state (rigid–rigid) behavior, exactly as described above.

3. The Multiscale Concept

The multiscale concept used in the model is schematically visualized in Figure 12.2. We consider a macroscopic sample (for instance, a bar of 250 mm length) to be built up by a finite number of representative material elements or cells. Each cell (i.e., an interval in a one-dimensional representation) is composed of a large number of microscale units that can be thought of as individual grains, separated by bonds, grain boundaries, interphase zones, microcracks, and the like. The microscopic units have a typical length of 0.1 to 100 microns. A material cell thus represents a mesoscopic entity in between the macroscopic and the microscopic levels. Typical mesoscopic cell lengths are of the order of a few mm.

At the microscale level, we assume that the strain response of individual units can be grouped in two classes: the largest part of a unit behaves reversibly elastic, or "classical," according to a classical nonlinear pressure–strain relation (including the acoustoelastic characteristics), and a small section (about 1% in length of the elastic part) follows a "nonclassical" behavior which includes hysteresis effects (see Figure 12.2). Assuming the two-state rigid–rigid behavior for an elementary hysteretic unit, we can express the classical and nonclassical contributions of a single unit as follows.

$$\varepsilon_c = -\int^P \frac{dP'}{K_c(P')} \qquad \varepsilon_{\mathrm{nc}} = \begin{cases} 0 & \text{if } P < P_c \text{ and } \dot{P} > 0 \\ -\gamma & \text{if } P > P_c \text{ and } \dot{P} > 0 \\ -\gamma & \text{if } P > P_o \text{ and } \dot{P} < 0 \\ 0 & \text{if } P < P_o \text{ and } \dot{P} < 0 \end{cases}. \qquad (12.5)$$

Here, K_c $(= K_g$ defined earlier) is usually expressed by means of a traditional nonlinear powerlaw relation between modulus and pressure:

$$K_c(P) = K_o(1 + \beta P + \delta P^2 + \cdots) \qquad (12.6)$$

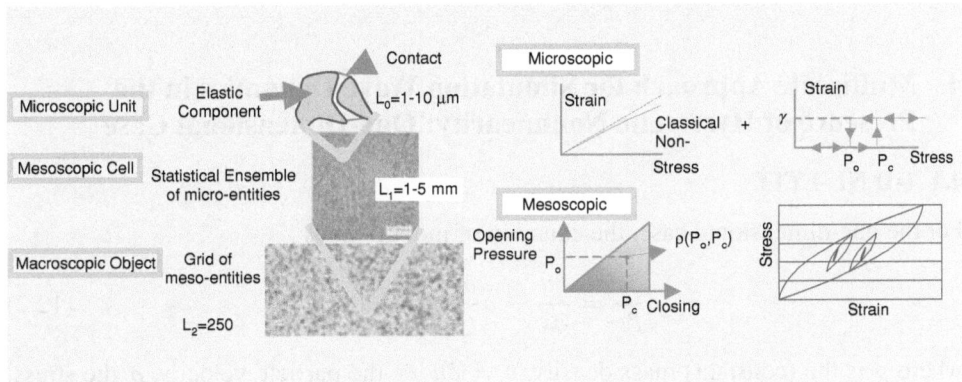

Fig. 12.2. Multiscale concept used in the simulations, illustrating the macroscopic, mesoscopic, and microscopic levels with the corresponding constitutive equations.

with K_o the linear modulus value, and β and δ combinations of third- and fourth-order elastic constants defining the classical nonlinearity. The finite (constant) strain value γ in the hysteretic component defines the degree of the hysteresis of a single unit.

As mentioned above, each material cell consists of a large number of elastic and hysteretic components, with generally differing characteristics. However, using a mean field approach, we can assume that all elastic parameters are equal in a material cell and that only the parameters P_o and P_c can vary from unit to unit. Thus, if a material cell consists of N hysteretic units, K_o, β, δ, and $\hat{\gamma} = N\gamma$ are considered to be "effective" material constants that are defined at the intermediate level. All hysteretic units within a material cell can be represented in a pressure–pressure space by means of a density distribution $\rho(P_c, P_o)$. The four effective constants, K_0, β, δ, and γ, together with the particular PKM-space-density form a unique signature of the material cell. However, they may of course differ from cell to cell.

To calculate the strain (respectively, modulus) response of each material cell following a stress increase/decrease from P to $P + \Delta P$, we can apply the incremental relationship (12.3) [respectively (12.4)] that has been illustrated above. Note that the difference in the microscopic stress–strain relation and the mesoscopic state relation is clearly visible in Figure 12.2. Due to the statistical distribution, the discontinuous state relation of the individual hysteretic units leads to a continuous state relation (with discontinuous derivatives) on the intermediate level.

The scaling up from the mesoscopic (or element) level to the macroscopic (or sample) level can be performed through any type of finite difference or finite element method. Here, we use the formalism of EFIT, which was originally developed by Fellinger et al. [16]. Basically, the technique implements a discretization of the incremental equation of state (Hooke's law in linear elastic systems) and of the equation of motion. In general 3-D problems, the EFIT procedure uses the integral form of these basic equations, not the differential form, and performs the integration over certain control volumes or integration cells. The diagonal stress component integration cells coincide with the material cells defined in the multiscale concept. Velocities are calculated at the surfaces of the material cell. In this chapter, we explicitly deal with the 1-D and 2-D problems.

4. Multiscale Approach for Simulating Wave Dynamics in the Presence of Hysteretic Nonlinearity: One-Dimensional Case

4.1 1-D NL-EFIT

For the one-dimensional case, the equation of motion reads

$$\rho \frac{\partial v}{\partial t} = \frac{\partial \sigma}{\partial x} - \rho \frac{\omega}{Q} v + F(x, t), \tag{12.7}$$

where ρ is the (constant) mass density, $v = \partial u / \partial t$ the particle velocity, σ the stress, and x the space coordinate in 1-D. Frequency-dependent ($\omega = 2\pi f$) attenuation is taken into account and is assumed to be proportional to the velocity; Q is the quality factor (inverse attenuation). F is an arbitrary external force as a function of x and

t, which generally is located at one of the two edges of the sample (at $x = 0$ or $x = L$).

The stress rate equation can be derived by calculating the temporal derivative of the equation of state that describes the stress–strain relation at the intermediate level. As mentioned above, the mesoscopic stress–strain relation contains contributions from classical (linear and nonlinear) elastic elements and from nonclassical hysteretic units within a material cell. The strain response due to a variation of pressure is given by Eq. (12.3). In this framework, the strain response is clearly a function of the previous stress history, σ_{his}. Using $\partial \varepsilon / \partial t = \partial v / \partial x$, we then rewrite the stress rate equation as follows.

$$\frac{\partial \sigma}{\partial t} = \frac{\partial \sigma}{\partial \varepsilon} \frac{\partial \varepsilon}{\partial t} = K(\sigma, \sigma_{his}) \frac{\partial \varepsilon}{\partial t} = K(\sigma, \sigma_{his}) \frac{\partial v}{\partial x}, \qquad (12.8)$$

where $\partial \sigma / \partial \varepsilon = K(\sigma, \sigma_{his})$ is the nonlinear elastic modulus in the presence of hysteresis [cfr. Eq. (12.4)]. As mentioned before, the function $f_c(\sigma)$ used in Eq. (12.4) represents the fraction of the PKM-space area occupied by closed units. This function is highly dependent on the previous history of the stress, and its derivative (essentially a line integral) needs to be updated at each time step taking into account previous maxima and minima of the local stress excursions.

Let us return to Eqs. (12.7) and (12.8) which are the basic velocity–stress equations needed for the EFIT discretization procedure. The EFIT discretization uses a staggered grid formalism, in which velocity and stress are determined at different positions inside a certain grid cell. If Δx denotes the representative length of the material cells, we calculate the stress at $(m + 1/2)\Delta x$, that is, in the middle of the material cells, whereas the velocity is determined at $m\Delta x$, that is, at the edges of each cell, with $m = 0, 1, 2, 3, \ldots, L/\Delta x$. Likewise, we also implement a staggered temporal grid. The velocity is calculated at full-time steps $k\Delta t$, whereas stress is calculated at half-time steps $(k + 1/2)\Delta t$ with $k = 0, 1, 2, 3, \ldots$.

Assuming the external force is a sinusoidal force acting at $x = 0$, the time discretization procedure of the 1-D EFIT code then yields the following sequence of steps.

- Step 0: Suppose $v^{[k-1]}$ and $\sigma^{[k-1/2]}$ are known at $t = (k-1)\Delta t$ and $t = (k-1/2)\Delta t$.
- Step 1: Express the temporal change of v at time $k - 1/2$ using Eq. (12.7):

$$\dot{v}^{[k-1/2]} = \frac{1}{\rho} \left(\frac{\partial \sigma}{\partial x} \right)^{[k-1/2]} - \frac{\omega}{Q} \left(\frac{v^{[k]} + v^{[k-1]}}{2} \right) + \frac{F_o}{\rho} \sin(\omega t) \cdot \delta(x = 0).$$

- Step 2: Update v at time k : $v^{[k]} = v^{[k-1]} + \dot{v}^{[k-1/2]} \Delta t$.
- Step 3: Express the temporal change of σ at time k using Eq. (12.8):

$$\dot{\sigma}^{[k]} = K^{[k]} \left(\frac{\partial v}{\partial x} \right)^{[k]}.$$

- Step 4: Update σ at time $k + 1/2$: $\sigma^{[k+1/2]} = \sigma^{[k-1/2]} + \dot{\sigma}^{[k]} \Delta t$.

This scheme looks extremely simple. Nonetheless, to calculate the modulus K as a function of stress (and stress history), $K^{[k]} = K(\sigma^{[k]}, \sigma_{his}^{[k]})$, the use of $\sigma^{[k]}$ as an argument is required, that is, the stress value at time $k \Delta t$. Because stress is only determined at half-time steps of Δt, it is necessary to use an appropriate interpolation procedure; for example, $\sigma^{[k]} = 0.5(\sigma^{[k-1/2]} + \sigma^{[k+1/2]})$. The problem is, however, that $\sigma^{[k+1/2]}$ is not yet available because it has to be calculated in Step 4.

To solve this problem, there are two possible solutions: we can use the approximation $\sigma^{[k]} \approx \sigma^{[k-1/2]}$ in Step 3 thereby reducing the accuracy of the numerical scheme to first order. Or, as a second solution, which avoids the loss of accuracy, we calculate K as a function of strain (and strain history) instead of stress (and stress history). Still, because Eq. (12.4) is expressed as function of stress increase (due to the PKM-space which is expressed in terms of pressures), the calculation of K for a certain strain increase/decrease $\Delta\varepsilon$ needs to be performed in an iterative manner starting from an initial guess of $\Delta\sigma$. Unless the PKM-space is expressed in terms of "opening strains" and "closing strains," this requires more calculation time which is a major disadvantage of using the second solution over the first.

After updating the stress components in Step 4, we can jump back to Step 1 and continue the leapfrog scheme with another iteration in time.

In the following section we discuss simulation results of the multiscale EFIT method for longitudinal wave propagation in a one-dimensional rod of infinite length, and for longitudinal resonances in a 1-D system of finite length L. We consider cases with either spatially uniform or localized nonlinearity. The nonlinearity is introduced through mesoscopic level PKM-spaces for which the density distribution may depend on the location (a zero density distribution represents a piece of linear or classically nonlinear material). Results are shown for various types of PKM density distributions of hysteretic units (types B to I, as in Figure 12.1) and are compared with linear calculations (empty PKM-space, type A).

As an initial condition, we assume that the material is at a certain ambient pressure (usually taken to be zero, i.e., rest-state), and that there is no initial velocity distribution in the rod at $t = 0$.

The boundary condition at one end of the rod ($x = 0$) is prescribed by a sinusoidal forcing with amplitude F and frequency f. For wave propagation simulations (high frequencies) we use a continuous wave excitation at 100 kHz, and implement an absorbing boundary layer at the other end of the rod to avoid reflections. The calculated signals are analyzed as a function of the forcing amplitude at a fixed distance from the source, and as a function of distance for variable forcing amplitudes.

For wave resonance simulations, at much lower forcing frequencies than for wave propagation, we consider a stress-free boundary at $x = L$, and use a continuous wave excitation. We study the steady-state response of the sample as a function of the frequency and the forcing amplitude.

4.2 Simulations for One-Dimensional Wave Propagation

In all simulations discussed in this section, we model an infinite bar as a bar of 500 mm length consisting of 5000 mesoscopic cells of 0.1 mm length each. To avoid reflections in the study of wave propagation, we extended the bar with 250 cells that form a perfectly matched absorbing boundary layer. The spatial distribution of density and (linear) modulus of the material cells is uniform: $\rho = 2600\,\mathrm{kg/m^3}$, and $K_0 = 10\,\mathrm{GPa}$. The quality factor is set to a high value in order not to unnecessarily complicate the analysis with the influence of attenuation ($Q = 8000$). Furthermore, we restrict the analysis to the influence of hysteresis, and ignore the contribution of classical nonlinearity; that is, $K_c(P; x) = K_0$ for all pressure excursions and at all x. The properties of hysteresis are defined by the parameter $\gamma(x)$ and the PKM density distribution $\rho_{\mathrm{PKM}}(P_c, P_o; x)$. If these properties are independent of x we speak about spatially uniform hysteresis. In that case, we take $\gamma(x) = \gamma = 2 \cdot 10^{-18}$. Otherwise, we consider spatially localized hysteresis, and we assume a gradual distribution of the hysteretic strength $\gamma(x) = 2 \cdot 10^{-18} \exp(-((x - 0.045)/0.015)^2)$, simulating a localized zone of nonlinearity centered at 45 mm in the bar. In any case, the PKM-space is constrained to finite pressure values ranging from $-5\,\mathrm{MPa}$ to $5\,\mathrm{Mpa}$, implying that there are no hysteretic units with characteristic values of P_c or P_o outside these stress limits. To show the influence of the PKM distribution on the dynamics, we consider various cases of nonuniform distributions of PKM densities:

$$\rho_{\mathrm{PKM}}(P_c, P_o; x) = \frac{R(P_c, P_o; x)}{N_{\mathrm{PKM}}(x)} \quad \text{with } N_{\mathrm{PKM}}(x) = \iint R(P_c, P_o; x)\,dP_o\,dP_c$$

(12.9)

and

$$R(P_c, P_o; x) = \exp\left[-\left(\frac{(P_c + P_o) - a(x)}{b(x)}\right) - c(x)(P_c - P_o)\right] \frac{(P_c - P_o)^{m(x)}}{\left(\frac{m(x)}{c(x)e}\right)^{m(x)}}.$$

The parameters a, b, c, and m may in general depend on x. However, here, we use the same values for all cells. Expression (12.9) enables us to vary the PKM-distribution ($P_c > P_o$) from a uniform density distribution (type B in Figure 12.1 for $a = 0, b = \infty$, and $m = 0$) to a fairly localized density distribution (see type G ($m = 0$), H ($m = 1$), I ($m = 2$) in Figure 12.1. The constants a and b determine the center and the width of the density distribution parallel to the diagonal. The constant c and the power m define the distribution away from the diagonal. The normalization constant N_{PKM} can be determined by requiring that the density over the entire PKM-space ($-5\,\mathrm{MPa} < \sigma_c < 5\,\mathrm{MPa}$ and $-5\,\mathrm{MPa} < \sigma_o < \sigma_c$) be equal to unity.

Apart from the material parameters, we also need to specify the forcing frequency and amplitude. We assume a source located at $x = 0$ which produces a 100 kHz sinusoidal signal of which the amplitude may vary over three orders of magnitude in the simulations. The numerical time step is 1/200 of the wave period. Finally, the virtual receiver is located at 100 mm from the source.

The numerical model calculates the spatial distribution of stress and velocity over the bar at each step in time according to the staggered scheme discussed above. Subsequent to a transitional process, the steady-state response of the sample is usually reached after a finite number of cycles of excitation. For the analysis, we perform a Fourier transformation on the steady-state response of the velocity signal (using exactly eight periods) at the receiver position, and we store the values of the fundamental velocity amplitude, and its harmonics (of order two up to five) as a function of the forcing amplitude. Strains can be deduced from the particle velocity by dividing by the linear wave speed ($c_o = 1961.16\,\text{m/s}$). With this, we can quantify the generation of harmonics as a function of the fundamental component, and the loss of energy due to hysteresis.

At first, we have verified that a spatially uniform hysteretic bar with uniform PKM distribution (type B) assures results consistent with previously reported analytical, quasianalytical, and numerical results [9,28,30]. There is no sign of even harmonics; all odd harmonics vary linearly with distance, quadraticles in strain, and linearly in frequency. Attenuation is increasing linearly with strain and distance.

Figure 12.3 illustrates the influence of the PKM distribution on the generation of the third harmonic for a spatially uniform hysteretic bar (all cells have the same nonlinearity). We considered Type A, B, C, D, G, H, and I distributions. For Type C, the shift was 550 Pa. For Type D, the band extends between 300 Pa and 9300 Pa, and for G, H, I, $a = 0\,\text{Pa}$, $b = 10^4\,\text{Pa}$, and $c = 10^{-4}\,\text{Pa}^{-1}$.

The results for the linear case (no hysteresis, dashed line) indicate the noise floor for the discrete Fast Fourier transform (the harmonics are down by at least 160 dB). The open squares (Type B) show the quadratic behavior of the third harmonic for the uniform PKM-spaces. In Figure 12.3a, we clearly observe the onset of nonlinearity in the shifted (Type C) and banded (Type D) PKM-space distributions, and the presence of a saturation level for the banded PKM space. Figure 12.3b also illustrates a saturation level for the nonuniform PKM-spaces at high strains (Types G, H, and I), and a clear nonquadratic dependence (rather order three or four) of the third harmonic for $m = 1$ and $m = 2$, corresponding to distributions with lesser units near the diagonal than away from the diagonal.

In addition, we considered the case of a localized zone of hysteresis situated between the source and the receiver. The nature of the PM-space distributions is identical for all elements, but the strength of the microscopic units varies according to a Gaussian function between 20 mm and 70 mm, centered at 45 mm. Figure 12.4 illustrates the growth of the third harmonic with distance for four PKM types and for three values of

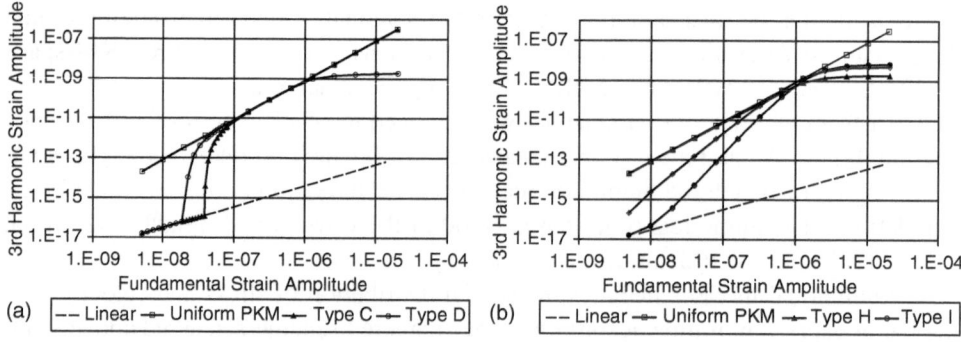

Fig. 12.3. Third harmonic generated in a spatially uniform hysteretic bar at 0.1 m from the source for various PKM-space distributions. (a) Linear, Type B, Type C, and Type D; (b) Linear, Type B, Type G, Type H, and Type I.

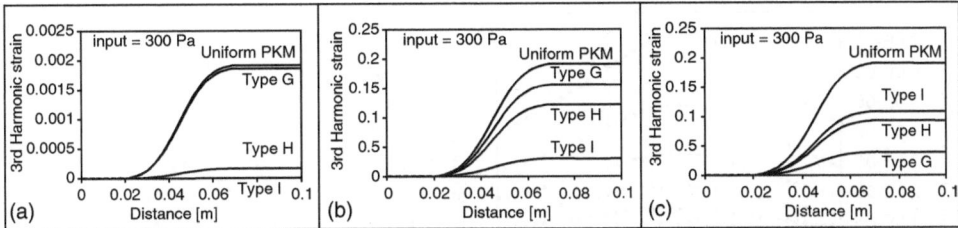

Fig. 12.4. Third harmonic generated in a bar with localized hysteretic nonlinearity (centered at 0.045 m from the source) versus distance for three different source amplitudes and for various PKM space distributions: Type B, Type G, Type H, and Type I.

the excitation. We observe that the third harmonic growth is limited to the nonlinear zone, and that the nonuniform PKM-spaces with $m = 1$ and $m = 2$ show evidence of a threshold phenomenon. The nonquadratic amplitude dependence of the third harmonic is again visible from the comparison of these graphs, as well as the saturation of the growth at large excitation levels.

4.3 Simulations for One-Dimensional Wave Resonance

Most of the above-used simulation parameters apply also for the simulations of wave resonances. However, we now assume a bar of finite length ($L = 250$ mm) with a free boundary at $x = L$. To achieve steady-state responses, we introduced a uniform quality factor $Q = 80$ (inverse attenuation). Because the frequencies for wave resonances are much lower, we can significantly increase the spatial grid size. The EFIT discretization parameters for simulations of the fundamental longitudinal resonance are taken as follows: $\Delta x = L/120$, and $\Delta t = 1/(384\,f)$.

Starting from the initial rest state, and implementing the forcing at $x = 0$ as the boundary condition, we use the numerical model to calculate the spatial distribution of stress and velocity over the bar at each step in time. Subsequent to a transitional process, the steady-state response of the sample is usually reached after a finite number of cycles of excitation. Here, we have taken $5Q$ cycles as a rule of thumb, with Q the quality factor. The following eight cycles of the steady-state velocity signal at the position $x = L$ are then used for the analysis. Using a Fast Fourier transform, we calculate and store the values of the fundamental velocity amplitude and its harmonics (of order two up to five) as a function of forcing frequency and forcing amplitude. With this information, we plot (fundamental and harmonic) resonance response curves at various driving forces, and perform subsequent analysis of the data. On one hand, we determine the peak coordinates and the width of the resonance line for each fundamental resonance curve (i.e., each sweep in frequency at a fixed forcing amplitude). The peak amplitude gives the maximal response amplitude $v_{1r}(F)$ for the given forcing F; the peak frequency is the resonance frequency $f_r(F)$ of the system at that forcing. The width of the peak at half power $w(F)$ can be translated into a value of $Q(= f_r(F)/w(F))$ which may also vary as a function of F. By plotting the relative changes in the resonance frequency (i.e., $(f_r(F_0) - f_r(F))/f_r(F_0) = \Delta f_r(F)/f_r(F_0)$ with F_0 an extremely small value that yields the linear resonance response) and the

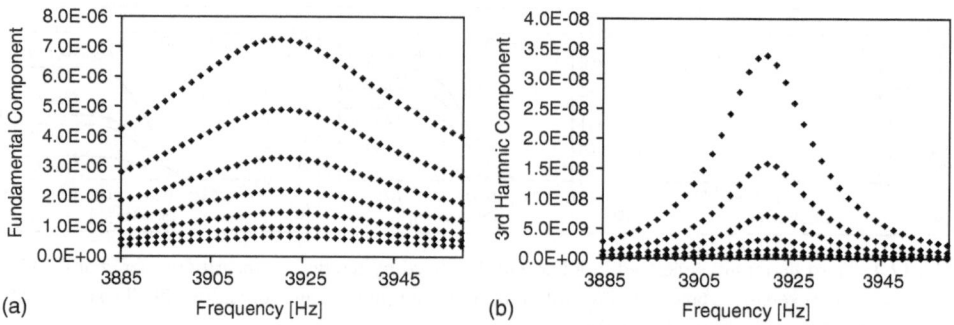

Fig. 12.5. Typical resonance curves of the fundamental and of the third harmonic strain response for a spatially uniform distribution of the hysteresis parameters along the bar (Type B PKM-space).

changes in the quality factor (i.e., $(Q(F_0) - Q(F))/Q(F_0) = \Delta Q(F)/Q(F_0)$) as a function of the peak amplitude $v_{1r}(F)$ measured at the same forcing, we can quantify the effect of hysteresis and nonlinearity on the modulus and the attenuation. The second part of the analysis is to determine the harmonic amplitudes $v_{ir}(F)$ ($i = 2, 3, 4, 5$) at the resonance frequency $f_r(F)$ for the different forcing values. Plotting these values against the peak amplitude $v_{1r}(F)$ of the fundamental component in the response quantifies the effect of hysteresis and nonlinearity on the generation of harmonics.

Figure 12.5 illustrates the typical resonance curves of the fundamental and of the third harmonic strain response, obtained by the multiscale approach for a spatially uniform distribution of the hysteresis parameters along the bar. Note that the even harmonics are not generated in purely hysteretic systems, because we ignored the classical nonlinearity contribution. Several sweeps were calculated at progressively increasing forcing amplitudes. We assumed a uniform PKM-space density distribution (Type B) for each material cell, that is, $a = 0, b = +\infty, c = 0$, and $m = 0$ independent of x. The parameter γ equals $2 \cdot 10^{-18}$.

Figure 12.6 illustrates the outcome of the usual analysis of the resonance curves for the relative change in frequency and in quality factor as well as for the third harmonic. We considered various types of PKM distributions (B, G, H, and I).

We remark that the Type B (uniform) PKM-space is giving us the expected behavior of a relative frequency shift and attenuation increase which are both linearly proportional to the strain. The third harmonic is quadratic in the strain [15,27,28]. The strain proportionality of the relative frequency shift is related to the fact that the ratio of the number of units that is activated, over the strain amplitude, is linear in the strain. For Type G distributions, with nonzero diagonal density, but decaying away from the diagonal, we observe a saturation in the observations at large excitation amplitude. It is even possible that the nonlinearity signature decreases with excitation. This is due to the fact that, relatively speaking, we activate more units at low amplitude than at high amplitude. Again, type H and I (with a zero diagonal density) clearly show a deviating power law dependence, which starts out as quadratic or even cubic, then changes to quasilinear before it finally turns over to saturation and decrease.

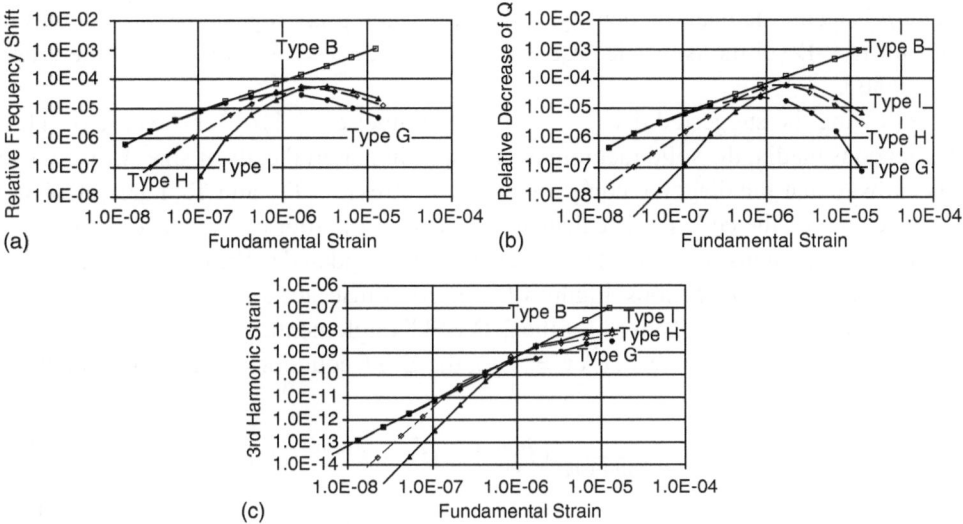

Fig. 12.6. Analysis of the amplitude-dependent resonance behavior for a spatially uniform distribution of hysteresis parameters along the bar in terms of relative frequency shift (A), relative increase in attenuation (Q^{-1}) (B) and third harmonic (C). Three Types of PKM space are considered: Type B, G, H, and I.

This proves that even a purely hysteretic system can produce a quadratic strain dependence of the relative frequency shift at low strains which turns into a linear shift at high strains (see Chapter 26 of this book), provided the PM-space distribution is nontrivial and accurately specified.

For an investigation and discussion of the influence of the localization and the extent of nonlinear zones, we refer to Chapter 23 of this book where we have compared the 1-D resonance results from the present multiscale model to analytical expressions obtained using the normal mode solution.

5. Multiscale Approach for Simulating Wave Dynamics in the Presence of Hysteretic Nonlinearity: Two-Dimensional Case

5.1 2-D NL-EFIT

The starting point of the two-dimensional EFIT method is again the Cauchy equation of motion together with the stress-rate equation. In 2-D problems, they can be expressed in terms of two-particle velocity components (v_x and v_y) and three stress components (T_{xx}, T_{yy}, and T_{xy}) as follows.

$$\rho\dot{v}_x = \frac{\partial T_{xx}}{\partial x} + \frac{\partial T_{xy}}{\partial y}; \quad \rho\dot{v}_y = \frac{\partial T_{xy}}{\partial x} + \frac{\partial T_{yy}}{\partial y} \tag{12.10a}$$

$$\dot{T}_{xx} = K_1\frac{\partial v_x}{\partial x} + K_2\frac{\partial v_y}{\partial y}; \quad \dot{T}_{yy} = K_1\frac{\partial v_y}{\partial y} + K_2\frac{\partial v_x}{\partial x}; \quad \dot{T}_{xy} = \mu\left(\frac{\partial v_y}{\partial x} + \frac{\partial v_x}{\partial y}\right). \tag{12.10b}$$

Here ρ is again the material density, and K_1, K_2, and μ are representations for elastic moduli in 2-D (in the isotropic case $K_1 = \lambda + 2\mu$, $K_2 = \lambda$ with λ and μ the Lamé constants).

Following the suggested discretization of Fellinger et al. [16] for the case of inhomogeneous media, the solid medium is then split in material cells of size Δx by Δy in such a way that the diagonal components of the stresses (T_{xx} and T_{yy}) are calculated in the center of the cell, T_{xy} is calculated at each of the four corners, and the particle velocity components (v_x and v_y) are determined and updated at the parallel sides of the cell in the x- and y-directions. Figure 12.7 exactly illustrates a set of four neighboring fundamental material cells used for the 2-D EFIT implementation.

If we consider a square grid ($\Delta x = \Delta y$), the EFIT equations take on the following form.

$$\dot{v}_x^{(n,m)}(t) = \frac{1}{\Delta x} \frac{2}{\rho^{(n,m)} + \rho^{(n+1,m)}} \left(T_{xx}^{(n+1,m)}(t) - T_{xx}^{(n,m)}(t) + T_{xy}^{(n,m)} - T_{xy}^{(n,m-1)} \right)$$

(12.11a)

$$\dot{v}_y^{(n,m)}(t) = \frac{1}{\Delta x} \frac{2}{\rho^{(n,m)} + \rho^{(n,m+1)}} \left(T_{yy}^{(n,m+1)}(t) - T_{yy}^{(n,m)}(t) + T_{xy}^{(n,m)} - T_{xy}^{(n-1,m)} \right).$$

(12.11b)

$$\dot{T}_{xx}^{(n,m)}(t) = \frac{1}{\Delta x} K_1^{(n,m)} \left[v_x^{(n,m)}(t) - v_x^{(n-1,m)}(t) \right]$$
$$+ \frac{1}{\Delta x} K_2^{(n,m)} \left[v_y^{(n,m)}(t) - v_y^{(n,m-1)}(t) \right]$$

(12.11c)

$$\dot{T}_{yy}^{(n,m)}(t) = \frac{1}{\Delta x} K_1^{(n,m)} \left[v_y^{(n,m)}(t) - v_y^{(n,m-1)}(t) \right]$$
$$+ \frac{1}{\Delta x} K_2^{(n,m)} \left[v_x^{(n,m)}(t) - v_x^{(n-1,m)}(t) \right]$$

(12.11d)

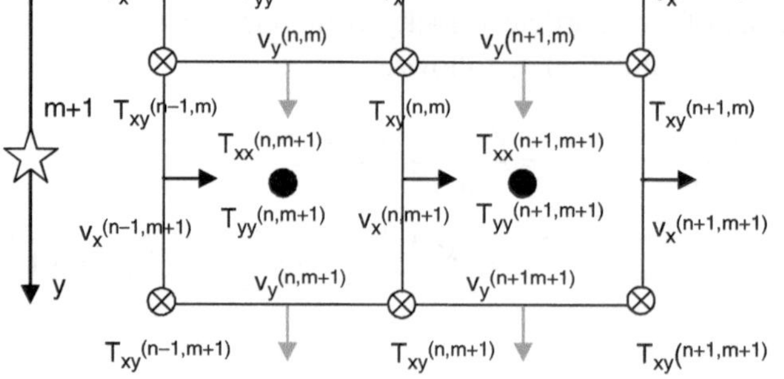

Fig. 12.7. Set of four elementary mesoscopic cells used in the 2-D EFIT model.

$$\dot{T}_{xy}^{(n,m)}(t) = \frac{\mu_{Har}}{\Delta x} \left[v_x^{(n,m+1)}(t) - v_x^{(n,m)}(t) + v_y^{(n+1,m)}(t) - v_y^{(n,m)}(t) \right]$$

$$(12.11e)$$

$$\text{with} \quad \frac{1}{\mu_{Har}} = \frac{1}{4} \left(\frac{1}{\mu^{(n,m)}} + \frac{1}{\mu^{(n+1,m)}} + \frac{1}{\mu^{(n,m+1)}} + \frac{1}{\mu^{(n+1,m+1)}} \right).$$

The update of the velocities and stresses is performed using central differences for the time derivatives yielding a common leapfrog scheme:

$$v_i^{(n,m)}(t) = v_i^{(n,m)}(t - \Delta t) + \Delta t \, \dot{v}_i^{(n,m)} \left(t - \frac{\Delta t}{2} \right) \qquad \text{for } i = x, y \quad (12.12a)$$

$$T_{ij}^{(n,m)} \left(t + \frac{\Delta t}{2} \right) = T_{ij}^{(n,m)} \left(t - \frac{\Delta t}{2} \right) + \Delta t \, \dot{T}_{ij}^{(n,m)}(t) \qquad \text{for } i, j = x, y$$

$$(12.12b)$$

As in the one-dimensional case, this scheme is straightforward for linear elastic solids. Yet, when microdamage is considered within the material cells, the stress–strain rate relations become more complicated and the moduli K_1, K_2, and μ may depend on the actual values of the three stress components acting on a material cell and on their history. The obvious extension of the scalar Preisach model to a vectorial Preisach model was turned down by Helbig because it is unable to account for the Poisson effect [31,32]. Helbig therefore suggested to reformulate the elastic tensor in terms of its eigensystem (eigenstiffnesses and eigenstrains). The eigenvectors correspond to the particular configuration in which the corresponding eigenstrain vector is perfectly collinear with the corresponding eigenstress vector. Indeed, by defining the volumetric, deviatoric, and shear stresses and strains:

$$T_V = \frac{T_{xx} + T_{yy}}{2}; \quad T_D = \frac{T_{xx} - T_{yy}}{2}; \quad T_S = T_{xy},$$

$$\varepsilon_V = \frac{\varepsilon_{xx} + \varepsilon_{yy}}{2}; \quad \varepsilon_D = \frac{\varepsilon_{xx} - \varepsilon_{yy}}{2}; \quad \varepsilon_S = \varepsilon_{xy}$$

$$(12.13)$$

the rate equations become

$$\dot{T}_V = K_V \dot{\varepsilon}_V; \quad \dot{T}_D = K_D \dot{\varepsilon}_D; \quad \dot{T}_S = K_S \dot{\varepsilon}_S$$

$$\text{with} \quad K_V = \frac{K_1 + K_2}{2}; \quad K_D = \frac{K_1 - K_2}{2}; \quad \text{and} \quad K_S = \mu. \quad (12.14)$$

In general, one could imagine that the new moduli K_V, K_D, and K_S may depend on any of the three components of the stress and their history. As a first approximation, however, we can assume that they only depend on the actual value and the history of the eigenstress component which appears in the scalar equations, that is, $K_V(T_V, T_{V,his})$, $K_D(T_D, T_{D,his})$, and $K_S(T_S, T_{S,his})$. Doing so, the three rate equations can be treated as scalar equations and the updates can be performed individually using the scalar Preisach model (as in the one-dimensional case) by attributing a statistical distribution of microscopic triggerlike units to each pair of eigenstress/eigenstrain. Helbig showed that, up to a certain strain level, the eigen-stress/eigenstrain system resulting from available measurement data do not substantially change [31]. Therefore one may assume that the output strain response of each stress projection is still collinear with the corresponding eigenstress/eigenstrain vector. A similar assumption was made by McCall et al. for the analysis of measurements of axial and radial strain responses to complicated mean stress and shear stress protocols [33].

Once the eigenstress components (T_V, T_D, T_S) are updated, it is easy to recalculate the Cartesian stress components (T_{xx}, T_{yy}, T_{xy}), and one can proceed with the common leapfrog scheme for the calculation of the dynamic nonlinear and hysteretic response.

In order not to complicate the analysis too much, we have limited ourselves in this chapter to the case in which the hysteretic nonlinearity is only affecting the volumetric stress–strain relation, and not the deviatoric or shear relations. We consider two examples. In the first example, we simulate the propagation of a Rayleigh wave through a medium with a localized surface deterioration. The second example deals with an in-plane investigation of localized damage in a plate using pulsed ultrasound.

5.2 Simulations for Two-Dimensional Nonlinear Rayleigh Wave Propagation

In this first example, we consider a solid medium measuring 120×50 mm^2, with a density of 1000 kg/m^3, an overall longitudinal velocity of 2000 m/s, and a shear velocity of 1155 m/s. No linear attenuation mechanism is assumed within the body. The top side of the medium is considered to be free (air). The solid body (SB) is surrounded by three absorbing boundary layers (left, right, and bottom) to avoid artificial reflections from the numerical boundaries. The source S is located on the free surface, at the left-hand upper corner of the solid body. Two receivers R_1 and R_2 are located at 20 mm and 100 mm from the source (see Figure 12.8). A surfacial zone (Zone), centered at 60 mm from the source, is considered to be microdamaged. Everywhere else the solid body is linear elastic ($\lambda_{SB} = 1.332$ GPa, $\mu_{SB} = 1.334$ GPa, and $\gamma_{SB} = 0$).

The excitation signal is modeled as an apodized sinusoidal force F_y with frequency ranging from 50 to 200 kHz, and is treated as a local source term for the v_y component in Eq. (12.11b) at position S on the free surface.

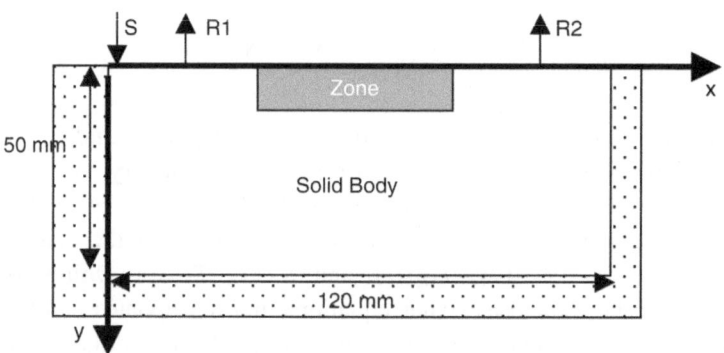

Fig. 12.8. Geometry of the 2-D Rayleigh wave propagation simulation.

Experimental work on progressively damaged materials has shown that microdamage can result in a reduction of the linear moduli, and/or an increase of the nonlinearity on the other hand [2,5–7]. We have investigated these effects in the model simulations by considering various cases with gradually varying linear and nonlinear hysteretic parameters in the "microdamaged" zone near the surface. In order to avoid reflections from discontinuous transitions in parameters between layers, we assume generalized Gaussian distributions for the shear modulus μ and of the hysteresis parameter γ in the following form.

$$\mu(x, y) = \mu_{SB}(1 - \mu_R \, e^{-\frac{1}{p}\left[\frac{x-x_s}{W_x}\right]^p - \left[\frac{y-y_s}{W_y}\right]^2}), \quad \gamma(x, y) = \gamma_{MAX} e^{-\frac{1}{p}\left[\frac{x-x_s}{W_x}\right]^p - \left[\frac{y-y_s}{W_y}\right]^2}.$$

$$(12.15)$$

For the calculations, we considered distributions of μ and γ that are centered in the x-direction at 60 mm from the source($x_s = 60$ mm) with a width equal to $W_x = 12$ mm, and centered in the y-direction at the surface ($y_s = 0$) with variable width W_y. The parameter p was set to 8. The maximum reduction of the modulus in the damaged zone (at the surface) is 20% ($\mu_R = 0.2$). The maximum strength of the hysteretic nonlinearity at the surface corresponds to $\gamma_{MAX} = 2 \cdot 10^{-18}$. The stress–strain relation of the material cells in the damaged zone is updated in the assumption of a uniform PM-distribution and only volumetric hysteresis is considered. Other possible driving scenarios for the hysteresis in 2-D are suggested in Chapter 18. No classical nonlinearity is assumed.

We investigated the influence of hysteresis and/or modulus reduction in two ways. First, we examined the effect on the phase velocity of the Rayleigh wave component along the surface. We determined the phase velocity at different excitation frequencies (f) from the difference of the phase spectra at R_1 and R_2 following the method described in Aki and Richards [34]. This procedure was repeated for various cases with and without modulus reduction, and with and without hysteresis.

The results of this investigation led to the following twofold conclusion (figures are omitted for sake of brevity): (a) the wave speed of a Rayleigh wave is highly dependent on the stiffness of the near-surface layers, and a modulus gradient introduces dispersion or frequency-dependent wave speeds. This is well known in seismology [34] and NDT as linear depth profiling, and is attributed to the difference in penetration depth of the Rayleigh wave at different frequencies. (b) The (microscopic) nonlinearity has little or no "measurable" extra effect on the wave speed of a propagating Rayleigh wave. This latter observation can also be readily explained by recalling that the nonlinearity is a second-order effect on the modulus reduction which is typically below 0.1% (e.g., Figure 12.6) [1–7]. The effect on the Rayleigh wave velocity is thus minimal, and extremely hard to measure in situ.

In the second set of simulations, we concentrated on the nonlinear properties of Rayleigh wave propagation by analyzing the level of the third harmonic spectral component of the Rayleigh wave's out-of-plane particle velocity v_y along the surface. It is well known that the third harmonic in a one-dimensional hysteretic nonlinear bar (in the absence of attenuation) accumulates linearly with distance, and is quadratic in the amplitude of the fundamental component [28]. We have found the same dependencies for the out-of-plane particle velocity of the Rayleigh wave (not shown for brevity) propagating along the surface in 2-D. With regard to the frequency dependence, the displacement amplitude u_{3f} is generally quadratic in the driving frequency [28]. Taking into account the extra frequency dependence for velocity components ($v = \omega u$), we may expect a cubic frequency dependence for the third harmonic particle velocity $v_{y,3f}$ at a fixed position and for a fixed excitation level.

In Figure 12.9a, we compare the results of layered hysteresis without modulus reduction for different depths of the microdamage region (by assuming different parameters for the width W_y). The results first of all show the predicted cubic relationship with frequency. As can be expected, we observe that the third harmonic component diminishes with decreasing depth of the microdamaged zone. However, Figure 12.9b (a normalized version of Figure 12.9a) clearly illustrates that the reduction is highly frequency dependent. Compared to the value at $W_y = 2.5$ mm, the third harmonic for $W_y = 0.2$ mm reduced by more than 80% at 50 kHz and only by 45% at 200 kHz. This is again an expected result related to the frequency-dependent penetration depth of Rayleigh waves.

This yields a second conclusion: the amplitude of the third harmonic of a Rayleigh wave is highly sensitive to the microscopic nonlinearity of the near-surface layers and a nonlinearity gradient introduces a new type of dispersion on the harmonics.

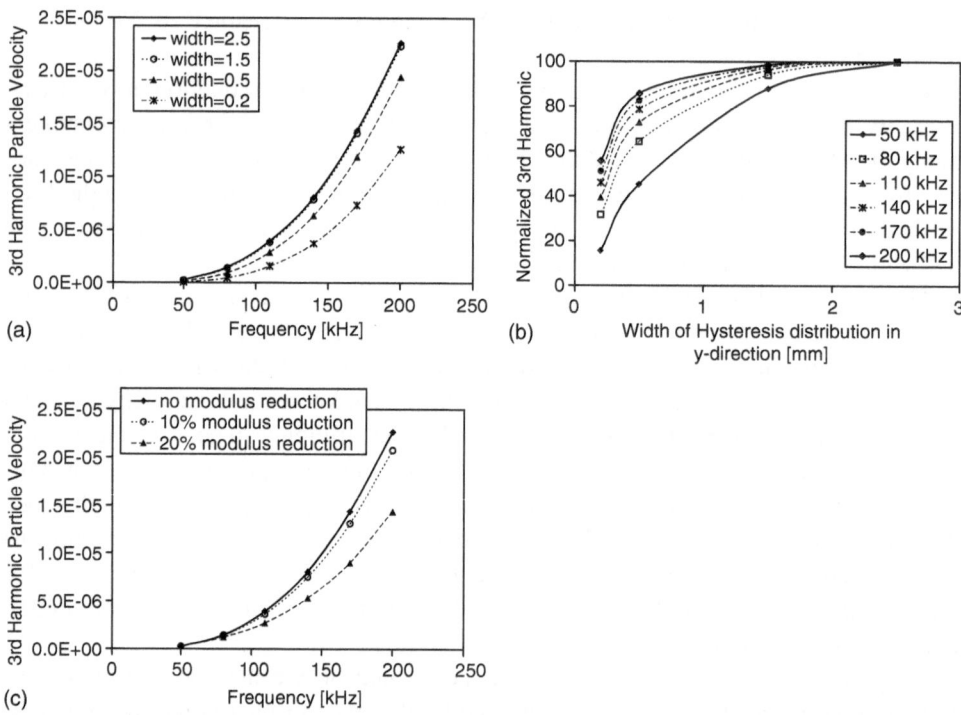

Fig. 12.9. (a) Frequency dependence of the third harmonic particle velocity component v_y in the Rayleigh wave propagation for different depths of the microdamaged zone without modulus reduction; (b) same results as in Figure 12.9a, but plotted in normalized form as function of the depth; (c) frequency dependence of the third harmonic particle velocity component v_y in the Rayleigh wave propagation for different levels of modulus reduction.

Finally we consider a modulus reduction on top of a hysteretic nonlinearity. Figure 12.9c illustrates the frequency dependence of $v_{y,3f}$ at R_2 for a zone with hysteretic nonlinearity and different levels of modulus reduction. A reduction of the modulus in the microdamaged zone leads to a reduction in the observation of the third harmonic of the particle velocity v_y too.

5.3 Simulations for Two-Dimensional In-Plane Wave Propagation

In this second 2-D example, we consider a solid plate of aluminum with a density of $2700 \, \text{kg/m}^3$, an overall longitudinal velocity of 6198 m/s, and a shear velocity of 3122 m/s. Again, no linear attenuation mechanism is assumed within the medium. The dimensions of the plate are $30 \times 50 \, \text{cm}$ (Figure 12.10). The damaged zone is modeled by imposing an 80% reduction of the shear modulus and a hysteretic modulus–stress relation with uniform (P_o, P_c)-distribution and hysteretic parameter $\gamma = 2 \cdot 10^{-18}$, uniformly distributed over a square region of $3 \times 3 \, \text{cm}$, centered in the middle of the plate. Everywhere else, the medium is assumed to be linearly elastic. Again, we limit ourselves to the case in which the hysteretic nonlinearity is only affecting the volumetric stress–strain relation, and not the deviatoric or shear relations. This time we apply rigid boundary conditions at the boundaries of the plate. The excitation signal is introduced in Eq. (12.11a) as an external time-apodized stress, with a 2-D spatial Gaussian distribution applied over a region of $10 \times 50 \, \text{mm}$ centered at $(x_s = 125 \, \text{mm}, \; y_s = 150 \, \text{mm})$.

Figure 12.11, for instance, shows the T_{xx} stress component at the location of the receiver over a timescale of $2.25 \cdot 10^{-4}$ s for a source amplitude $A = 2 \cdot 10^5$ Pa. The response for $A = 5 \cdot 10^4$ Pa looks very much the same, and from the normalized signals it is not easy to discern the effects of the nonlinearity. The nonlinear signatures can only be disclosed by analyzing the signals in the frequency domain, either in their entire form, partially, or by using moving windows.

For a quick preview of the harmonic generation, we prefer to analyze the harmonic content of the time signals by windowing them with a Gaussian apodization function with maximum at $t = 5.48 \cdot 10^{-5}$ s. This isolates the part of the signals that

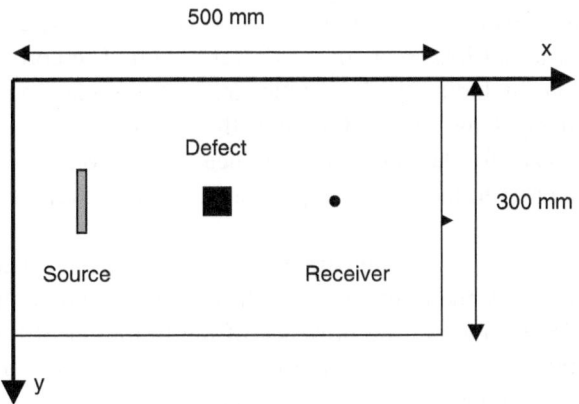

Fig. 12.10. Geometry of the 2-D simulations.

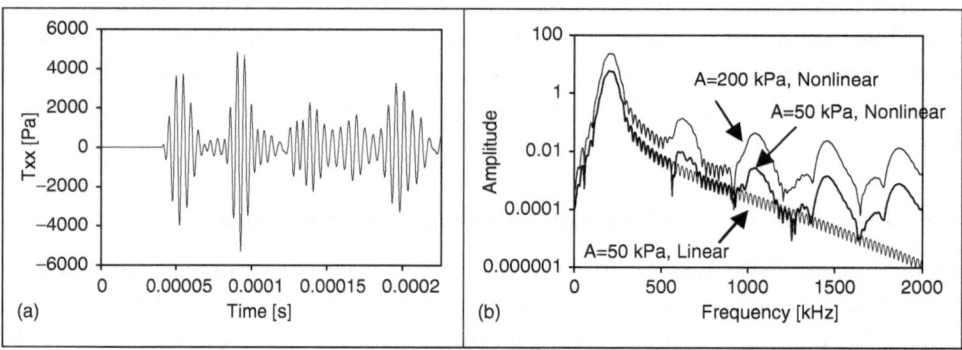

Fig. 12.11. (a) Stress signal at the receiver for $A = 2 \cdot 10^5$ Pa; (b) FFT of the stress signals from simulations including hysteresis at $A = 5 \cdot 10^4$ Pa and $A = 2 \cdot 10^5$ Pa, and from a linear simulation with $A = 5 \cdot 10^4$ Pa.

corresponds to the first arrival of the longitudinal wave at the receiver. The Fourier transforms for Gaussian windowed signals at $A = 5 \cdot 10^4$ and $A = 2 \cdot 10^5$ Pa are shown in figure 12.12b. In Figure 12.12b we have also added the FFT of the signal for a linear simulation at $5 \cdot 10^4$ Pa. The generation of harmonics is clearly evident. Note that hysteretic nonlinearity creates only odd harmonics in the time signals and that the well-known quadratic dependence of the harmonics on the amplitude of the fundamental holds [28].

A more detailed, time-resolved picture of the development of nonlinearity can be obtained through wavelet analysis [35]. Wavelet analysis decomposes a function into a sum of shifted and scaled copies of a suitably chosen wavelet kernel. The wavelet theory covers the limitations in Fourier analysis by windowing the signal into variable-sized regions, using long time intervals for information on low frequencies and shorter intervals for high frequencies. The result of applying the continuous wavelet transform on a given time signal is a wavelet map in which the amplitude of the wavelet coefficients is plotted as a function of time and the pseudofrequency of the wavelets [35, 36]. We calculated the continuous wavelet transform of the stress signals at the receiver position for $A = 5 \cdot 10^4$ Pa and $A = 2 \cdot 10^5$ Pa using the complex Morlet wavelet [36]. The results of this calculation are illustrated in Figure 12.12. The two upper panels (a) and (b) display the wavelet maps for $A = 5 \cdot 10^4$ Pa and $A = 2 \cdot 10^5$ Pa resulting from simulations without hysteresis in the defect region. The corresponding wavelet maps for simulations including the hysteretic stress–strain relation are shown in plots (c) and (d) respectively. Panels (e) and (f) show the absolute value of the difference between the wavelet maps for the linear and nonlinear simulations at $A = 5 \cdot 10^4$ Pa and $A = 2 \cdot 10^5$ Pa. In order to facilitate comparison, all plots are on the same logarithmic scale from 0 to 8.

The goal of Figure 12.12 is to illustrate the changes in the received time signals that are brought about by the introduction of hysteresis. When comparing figures (e) and (f), we can clearly observe the presence of the third and fifth harmonics of the fundamental at 600 and 1000 KHz in the first two to three pulse arrivals and in the last one. The increase in the strength of the harmonics is about one order of magnitude (factor 16) according to the quadratic dependence on fundamental amplitude.

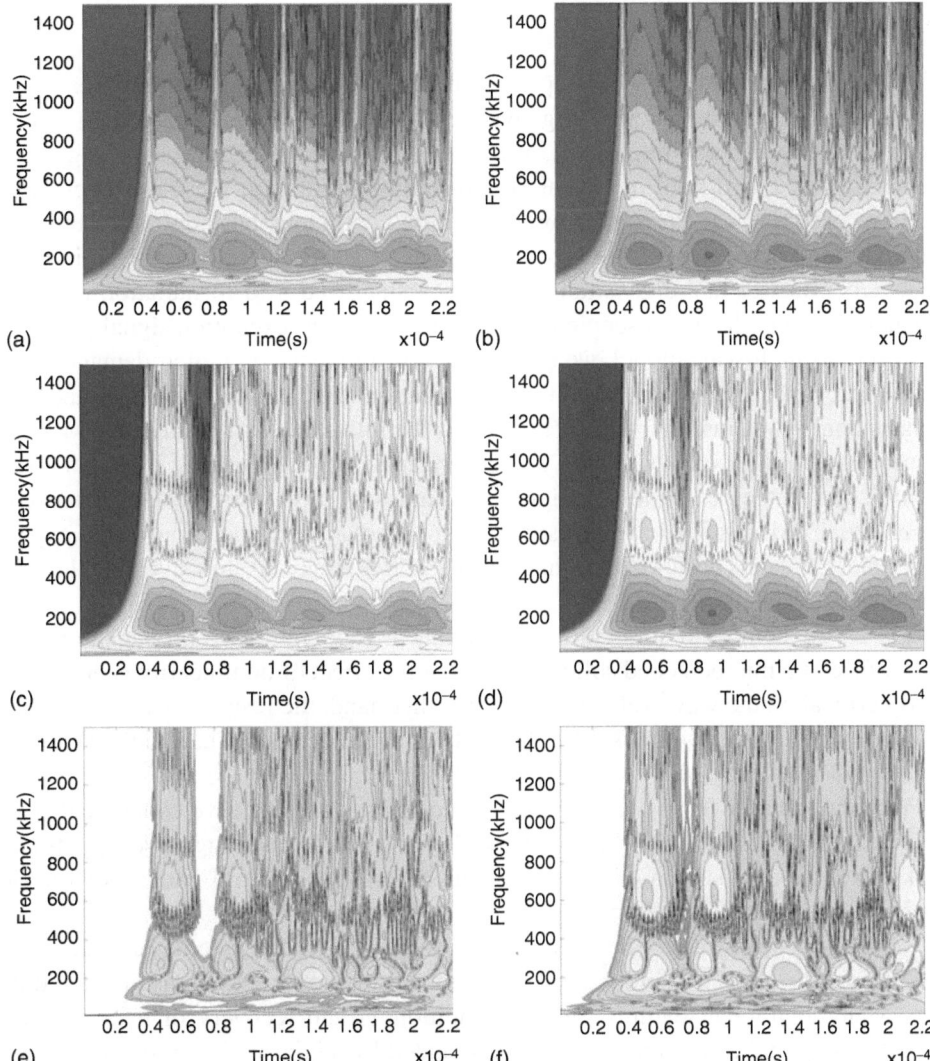

Fig. 12.12. Continuous wavelet transform of the stress signals at $A = 5 \cdot 10^4$ Pa (left) and $A = 2 \cdot 10^5$ Pa (right) for simulations without hysteresis (a) and (b) and including hysteresis (c) and (d). Figures (e) and (f) show the absolute value of the difference between the wavelet at $A = 5 \cdot 10^4$ Pa (c–a) and $A = 2 \cdot 10^5$ Pa (d)–(b). All plots are on the same logarithmic scale from 0 to 8.

In the second half of the time window, one observes a temporary decrease in the amplitude of the harmonics. This reduction is a consequence of the arrival of shear waves at the receiver which are generated through mode conversion in the defect region and because of wave reflection at the boundaries of the computational domain. Shear waves are less affected by the nonlinearity because we did not include shear hysteresis in our simulation. We are currently further investigating the influence of the various types of hysteresis on 2-D wave propagation.

6. Conclusions

We have illustrated a multiscale approach to model wave propagation and resonance phenomena in materials containing hysteretic (multibranch, or nonunique) stress–strain relations. At the microlevel, we have simulated microcracks by triggerlike elements with a two-state nonlinear stress–strain relation. On the mesoscopic level we have used a scalar Preisach approach with a continuous but nontrivial statistical distribution to achieve the stress–strain relations of a representative volume in terms of its eigenstresses and eigenstrains. Finally, a staggered grid formulation is implemented to predict the macroscopic response to an arbitrary excitation signal.

Using this multiscale model, we investigated the influence of a microdamaged zone on the resonance signatures, and on the propagation characteristics for one- and two-dimensional problems. In 1-D, we illustrated the influence of a nontrivial PM-space distribution on the expected power law dependencies in the dynamic modulus–strain relation and in the generation of higher harmonics. It turns out that the existence of a nontrivial PM distribution may serve as an alternative explanation of the typical experimental results discussed in Chapter 26, in cases where no slow dynamical effects are to be considered.

In 2-D, the numerical simulations for nonlinear Rayleigh wave propagation along the surface of a microcracked solid showed that the effect of nonlinearity is not particularly critical on the wave velocities. On the other hand, the results encourage the use of depth profiling methods based on the analysis of the generated harmonic content as function of frequency.

From the example of in-plane propagation of pulses in plates, we conclude that the nonlinear effects are definitely of second order, and that, even though two signals may look alike, the comparison of the wavetrains in terms of spectral content or wavelet analysis can be extremely clarifying.

We are perfectly aware of the fact that the model implemented here is a basic model which uses the simplest double rigid states model, without fancy transitions, and the like. Many additional features such as conditioning and recovery of material parameters during and after high strains can be added by slight modifications of the elementary assumptions (see, for instance, Chapters 16–18). However, in view of their practical use in inverse modeling, we believe that the current models already put very high limitations on achievability because of the many parameters that can be varied in direct modeling. For real NDT applications one should primarily focus on effect of the main parameters on the nonlinear signatures without overloading the models with third-order effects.

Acknowledgments

The authors gratefully acknowledge the support of the Flemish Fund for Scientific Research. (G.0206.02 and G.0257.02), the provisions of the European Science Foundation Programme NATEMIS, and the European FP5 and FP6 Grants DIAS (EVK4-CT-2002-00080) and AERONEWS (AST3-CT-2003-502927).

References

[1] Nazarov VE, Ostrovsky LA, Soustova IA, and Sutin AM (1988) *Sov Phys Acoust* 34(3): 284–289.

[2] Nagy PB and Adler L (1992) in DO Thompson, DE Chimenti (Eds), *Review of Progress in Quantitative Nondestructive Evaluation*, Vol 11, Plenum, New York, pp. 2025.

[3] Johnson PA, Zinszner B, and Rasolofosaon PNJ (1996) *J Geophys Res* 101: 11553–11564.

[4] Van Den Abeele K, Johnson PA, and Sutin AM (2000) *Res Nondest Eval* 12/1: 17–30.

[5] Van Den Abeele K, Carmeliet J, TenCate JA, and Johnson PA (2000) *Res Nondestruct Eval* 12(1): 31–42.

[6] Van Den Abeele K and De Visscher J (2000) *Cement Concrete Res*. 30(9): 1453–1464.

[7] Van Den Abeele K, Van De Velde K, and Carmeliet J (2001) *Polymer Composites* 22(4): 555–567.

[8] Nazarov, VE, Radostin AV, Ostrovskii, LA, and Soustova, IA (2003) *Acoust Phys* 49(4): 344–353, 444–448.

[9] Scalerandi M, Agostini V, Delsanto PP, Van Den Abeele K, and Johnson PA (2003) *J Acoust Soc Am* 113: 3049–3059.

[10] Delsanto PP and Scalerandi M (2003) *Phys Rev B* 68: art. No. 064107.

[11] Gusev V and Aleshin V (2002) *J Acoust Soc Am* 112: 2666–2679.

[12] Schubert F (2001) Elastic wave modelling in nonlinear mesoscopic materials by using finite integration techniques with dynamic grid management, *Proc WCU*, Rome.

[13] Schubert F and Koehler B (2000) Modeling of linear and nonlinear elastic wave propagation using finite integration techniques in Cartesian and curvilinear coordinates, *Euromech* 419, Prague.

[14] Van Den Abeele K, Johnson PA, Guyer RA, and McCall KR (1997) *J Acoust Soc Am* 101(4): 1885–1898.

[15] Van Den Abeele K, Schubert F, Aleshin V, Windels F, and Carmeliet J (2004) *Ultrasonics* 42: 1017–1024.

[16] Fellinger P, Marklein R, Langenberg KJ, and Klaholz S (1995) *Wave Motion* 21(1): 47–66.

[17] Schubert F and Koehler B (1998) in: *Nondestructive Characterization of Materials VIII*, Plenum, New York, pp. 567.

[18] Schubert F, Koehler B, and Peiffer A (2001) *J Comp Acoust* 9, 1127–1146.

[19] Schubert F and Koehler B (2001) *J Comp Acoust* 9, 1543–1560.

[20] Bou Matar O, Dos Santos S, Fortineau F, Haumesser L, and Vander Meulen F, private communication.

[21] Johnson PA, Rasolofosaon PNJ (1996) *Nonlinear Proc. Geophys.* 3: 77–88.

[22] Yang B and Mall S. (2003) *Intl J Damage Mech*. 2003; 12: 45–64.

[23] Van Den Abeele K, Carmeliet J, Johnson PA, and Zinszner B (2003) *J Geophys Res* 107(B6): Art. No. 2121.

[24] Guyer RA, McCall KR, and Boitnott GN (1995) *Phys Rev Lett* 74: 3491–3494.

[25] Preisach F (1935) *Z Phys* 94: 277–302.

[26] Mayergoyz ID (1985) *J Appl Phys* 57(8): 3803–3805.

[27] McCall KR and Guyer RA (1994) *J Geophys Res* 99(B12): 23887–23897.

[28] Van Den Abeele K, Johnson PA, Guyer RA, and McCall KR (1997) *J Acoust Soc Am* 101: 1885–1898

[29] Gusev V and Tournat V (2005) Amplitude- and frequency-dependent nonlinearities in the presence of thermally induced transitions in the Preisach model of acoustic hysteresis, *Phys Rev. B* 72–5, 054104.

[30] Aleshin V, Gusev V, and Zaitsev VY (2004) *J Comp Acoust* 12: 319–354.

[31] Helbig K (1998) *Rev de l'IFP* 53, pp. 693–708.

[32] Helbig K and Rasolofosaon PNJ (2000) Anisotropy 2000 Fractures, converted waves and case studies, 9IWSA transactions, SEG.

[33] Boudjema M, Santos IB, McCall KR, Guyer RA, and Boitnott GN (2002) *Nonlinear Proc. Geophys.* 10(6): 589–597.

[34] Aki K and Richards PG (2002) *Quantitative Seismology*, University Science, Sausalito, CA.

[35] Siqueira L, Gatts C, da Silva RR, and Rebello J (2004) *Ultrasonics* 41: 785–797.

[36] Misiti M, Misiti Y, Oppenheim G, and Poggi JM (1996) *Wavelet Toolbox for Use with MATLAB*, Version 1, The Mathworks Inc. Natuk, MA.

13

Modeling and Numerical Simulation of Nonclassical Effects of Waves, Including Phase Transition Fronts

Arkadi Berezovski,[1,3] Jüri Engelbrecht,[1] and Gerard A. Maugin[2]

[1] Centre for Nonlinear Studies, Institute of Cybernetics at Tallinn University of Technology, Akadeemia tee 21, 12618 Tallinn, Estonia.
[2] Laboratoire de Modélisation en Mécanique, Université Pierre et Marie Curie, UMR 7607, Tour 65-55, 4 Place Jussieu, Case 162, 75252, Paris Cédex 05, France.
[3] To whom correspondence should be addressed. Arkadi Berezovski, Centre for Nonlinear Studies, Institute of Cybernetics at Tallinn University of Technology, Akadeemia tee 21, 12618 Tallinn, Estonia; e-mail: Arkadi.Berezovski@cs.ioc.ee.

Abstract

A thermodynamically consistent finite-volume numerical algorithm for thermoelastic phase-transition front propagation is described. A simple mathematical model of martensitic phase transition front propagation is considered. The phase transition front is viewed as an ideal mathematical discontinuity surface. The problem remains nonlinear even in this simplified description that requires a numerical solution. A nonequilibrium description of the process is provided by means of nonequilibrium jump relations at the moving phase boundary, which are formulated in terms of contact quantities. The same contact quantities are used in the construction of a finite-volume numerical scheme. The additional constitutive information is introduced by a certain assumption about the entropy production at the phase boundary. Results of numerical simulations show that the proposed approach allows us to capture experimental observations in agreement with theoretical predictions in spite of the idealization of the process.

Keywords: Finite volume methods, martensitic phase transformations, moving phase boundary, thermomechanical modeling

1. Introduction

The propagation of waves and phase-transition fronts in thermoelastic media is governed by the same field equations and equations of state (at least in the integral formulation). However, although these equations are sufficient for the description of thermoelastic waves, that is not the case for the phase transition fronts. It is well known that initial boundary value problems, formulated according to the usual principles of continuum mechanics, can suffer from a lack of uniqueness of the solution when the body is composed of a multiphase material.[1] The solution in this case involves a propagating phase boundary that separates the austenite from the martensite; the speed V_N of this interface remains undetermined by the usual continuum theory.

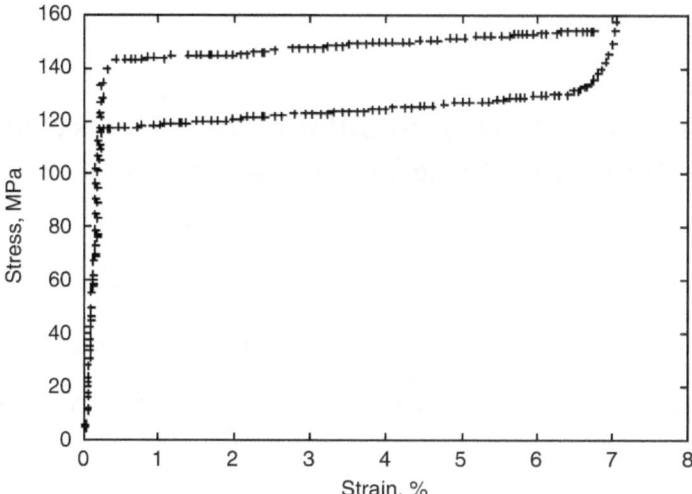

Fig. 13.1. Experimental stress–strain relation for Cu–Zn–Al shape-memory alloy.[4]

The propagation of phase interfaces in shape-memory alloys under applied stress is an experimentally observed phenomenon.[2,3] It is also connected with a superelastic effect. Originally in the austenitic phase, martensite is formed, upon loading, beyond a certain stress level, resulting in the stress plateau shown in Figure 13.1. The cause of stability of martensite at sufficiently high temperature is the applied stress, and therefore upon unloading martensite becomes unstable and reverts to its parent phase gaining its undeformed shape. This effect, which causes the material to be extremely elastic, is known as pseudoelasticity or superelasticity. Therefore, the propagation of phase interfaces results in a nonclassical nonlinear behavior of shape-memory alloys.

The simplest possible formulation of the stress-induced phase transition front propagation problem is given by Abeyaratne et al.[5] in the case of an isothermal uniaxial motion of a slab in small-strain approximation. The phase front is represented by a jump discontinuity separating the different austenite and martensite branches of the N-shaped local stress–strain curve. A shift of the martensitic branch of the curve is provided by the incorporation of a transformation strain, which is considered as an experimentally determined material constant.

From a thermodynamic point of view, a phase transition is a nonequilibrium process; entropy is produced at the moving phase boundary at a rate $f_S V_N$.[6] The entity f_S is called the driving force and may be expressed in terms of the limiting stress, deformation gradient, and free-energy on the two sides of the interface.[1,5–8] The uniqueness of the solution is provided by the introduction of two additional constitutive relations: a kinetic law for a driving force that establishes the speed of the transformation front

$$V_N = \phi(f_S), \qquad (13.1)$$

where a constitutive function ϕ provides the continuum theory with a suitable description of the lattice transformation mechanism, and a nucleation criterion.[1,6,9]

The prescription of the kinetic relation, of the nucleation criterion, and of the transformation strain means that the material behavior is completely known, and the numerical simulation is needed only for adjusting the values of coefficients of the model. In the considered model, the local equilibrium approximation is exploited in spite of the irreversibility of the phase transformation process. Moreover, to perform simulations of practical examples we need to move to a numerical approximation. In this case, we face a nonequilibrium behavior of finite-size discrete elements or computational cells. It is clear that the local equilibrium approximation is not sufficient to describe such a behavior.

Therefore a nonequilibrium description of the stress-induced phase-transition front propagation is preferable. To do this we need to choose an appropriate nonequilibrium theory. Our choice is influenced by numerical aspects of the modeling. This means that we need to have not only the nonequilibrium description of states of (finite volume) computational elements, but also the description of their interactions. In our opinion, the best possibility is provided by the thermodynamics of discrete systems.[10] In this theory, in addition to usual local equilibrium quantities, so-called contact quantities are introduced to provide the description of interactions between the systems. Therefore, the thermodynamic state space is extended.

The next step is to establish the nonequilibrium jump conditions at the phase interface. Each model of the stress-induced martensitic phase-transition front propagation uses its own jump relations.[11-15] All of them differ from the classical equilibrium jump relations, which consist in the case of thermoelastic solids in the continuity of temperature and chemical potential and the continuity of the normal Cauchy traction at the phase boundary.[16,17]

We apply the nonequilibrium jump relations,[18] which should be fulfilled for each pair of adjacent discrete elements. Supplementary constitutive information is introduced by means of certain assumptions about the entropy production at the phase boundary.

In order to include the nonequilibrium jump relations in the simulation, we apply a procedure which is similar to that proposed in,[19] but with a completely different numerical algorithm, based on the wave-propagation method.[20,21] However, we have made certain essential improvements to be able to apply it in the case of moving phase boundaries, for example. In effect, we reformulate the algorithm in terms of contact quantities and nonequilibrium jump relations. The nonequilibrium jump relations are different for processes with and without entropy production.[22,23] This gives us the possibility to apply distinct nonequilibrium jump relations in the bulk (for the wave propagation without the entropy production) and at the phase boundary (where entropy is produced, because the phase transition is dissipative). The latter plays the role of a kinetic relation without an explicit specification. A thermodynamic criterion of initiation of the phase transition process follows from the simultaneous satisfaction of both distinct nonequilibrium jump relations at the phase boundary.

The chapter is organized as follows. The governing equations and jump relations for the simplest problem of a uniaxial phase transition front propagation in a slab are given in Section 2. A discrete representation of the formulated problem is presented in Section 3. Nonequilibrium jump relations at the phase boundary are introduced in Section 4. The finite volume numerical scheme is discussed in Section 5. The algorithm is presented in terms of contact quantities. We describe in detail how the contact quantities can be computed in bulk and at the phase boundary. Results of numerical simulations and a comparison with available experimental data are given in Section 6. Finally, main conclusions are presented in Section 7.

2.   Simple Example: Uniaxial Motion of a Slab

In order to explain some of the key ideas with minimal mathematical complexity, it is convenient to work in an essentially one-dimensional setting. Following Abeyaratne et al.,[5] we consider a slab, which in an unstressed reference configuration occupies the region $0 < x_1 < L$, $-\infty < x_2, x_3 < \infty$, and assume an uniaxial motion of the form

$$u_i = u_i(x, t), \quad x = x_1, \tag{13.2}$$

where t is time, x_i are the spatial coordinates, and u_i are the components of the displacement vector. In this case, we have only three nonvanishing components of the strain tensor

$$\varepsilon_{11} = \frac{\partial u_1}{\partial x}, \quad \varepsilon_{12} = \varepsilon_{21} = \frac{1}{2}\frac{\partial u_2}{\partial x}, \quad \varepsilon_{13} = \varepsilon_{31} = \frac{1}{2}\frac{\partial u_3}{\partial x}. \tag{13.3}$$

Particle velocities associated with Eq. (13.2) are

$$v_i(x, t) = \frac{\partial u_i}{\partial t}. \tag{13.4}$$

Without loss of generality, we can set $\varepsilon_{13} = 0$, $v_3 = 0$ because of zero initial and boundary conditions for these components. Then we obtain uncoupled systems of equations for longitudinal and shear components that express the balance of linear momentum and the time derivative of the Duhamel–Neumann thermoelastic constitutive equation, respectively,[22,23]

$$\frac{\partial(\rho_0(x)v_1)}{\partial t} - \frac{\partial\sigma_{11}}{\partial x} = 0, \quad \frac{\partial}{\partial t}\left(\frac{\sigma_{11}}{\lambda(x) + 2\mu(x)}\right) - \frac{\partial v_1}{\partial x} = m(x)\frac{\partial\theta}{\partial t}, \tag{13.5}$$

and

$$\frac{\partial(\rho_0(x)v_2)}{\partial t} - \frac{\partial\sigma_{12}}{\partial x} = 0, \quad \frac{\partial}{\partial t}\left(\frac{\sigma_{12}}{\mu(x)}\right) - \frac{\partial v_2}{\partial x} = 0, \tag{13.6}$$

which are complemented by the heat conduction equation

$$C(x)\frac{\partial\theta}{\partial t} = \frac{\partial}{\partial x}\left(k(x)\frac{\partial\theta}{\partial x}\right). \tag{13.7}$$

Here σ_{ij} is the Cauchy stress tensor, ρ_0 is the density, θ is temperature, and C is the heat capacity per unit volume for a fixed deformation. The dilatation coefficient α is related to the thermoelastic coefficient m, and the Lamé coefficients λ and μ by $m = -\alpha(3\lambda + 2\mu)$. The indicated explicit dependence on the point x implies that the body is materially inhomogeneous in general.

The above description is well known and these systems of equations can be solved separately. We focus our attention on the system of equations for shear components [Eq. (13.6)] because the martensitic phase transformation is expected to be induced by shear.

2.1 Jump Relations

To consider the possible irreversible transformation of a phase into another one, the separation between the two phases is idealized as a sharp discontinuity surface S across which most of the fields undergo finite jumps. Let $[A]$ and $< A >$ denote the jump and mean value of a discontinuous field A across S, the unit normal to S being oriented from the "minus" to the "plus" side:

$$[A] := A^+ - A^-, \qquad < A > := \frac{1}{2}(A^+ + A^-). \qquad (13.8)$$

Let $\tilde{\mathbf{V}}$ be the material velocity of the geometrical points of S. The material velocity $\mathbf{V}$ is defined by means of the inverse mapping $X = \chi^{-1}(x, t)$, where X denotes the material points[24]

$$\mathbf{V} := \left.\frac{\partial \chi^{-1}}{\partial t}\right|_x. \qquad (13.9)$$

The phase transition fronts considered are *homothermal* (no jump in temperature; the two phases coexist at the same temperature) and *coherent* (they present no defects such as dislocations). Consequently, we have the following continuity conditions.[25,26]

$$[\mathbf{V}] = 0, \qquad [\theta] = 0 \qquad \text{at } S. \qquad (13.10)$$

Jump relations associated with the conservation laws in the bulk are formulated according to the theory of *weak solutions* of hyperbolic systems. Thus the jump relations associated with the balance of linear momentum and balance of entropy read[25,26]

$$\tilde{V}_N[\rho_0 v_2] + [\sigma_{12}] = 0, \qquad \tilde{V}_N[S] + \left[\frac{k}{\theta}\frac{\partial\theta}{\partial x}\right] = \sigma_S \geq 0, \qquad (13.11)$$

where S is entropy, $\tilde{V}_N$ is the normal component of the material velocity of the points of S, and σ_S is the entropy production at the interface. As shown in References[25,26] the entropy production can be expressed in terms of the driving force f_S such that the dissipation at the interface reads

$$f_S \tilde{V}_N = \theta_S \sigma_S \geq 0, \qquad (13.12)$$

where θ_S is the temperature at S. In addition, the *balance of material forces* at the interface between phases is found in the form[25,26]

$$f_S = -[W] + < \sigma_{ij} > [\varepsilon_{ij}], \qquad (13.13)$$

where W is the free energy per unit volume.

2.2 Dynamic Loading

In a dynamic problem we look for piecewise smooth velocity and stress fields $v_2(x, t)$, $\sigma_{12}(x, t)$ for inhomogeneous thermoelastic materials, which obey the following initial and boundary conditions,

$$\sigma_{12}(x, 0) = v_2(x, 0) = 0, \qquad \text{for} \quad 0 < x < L, \tag{13.14}$$

$$v_2(0, t) = v_0(t), \quad \sigma_{12}(L, t) = 0, \quad \text{for} \quad t > 0, \tag{13.15}$$

and satisfy the following field equations

$$\frac{\partial(\rho_0(x)v_2)}{\partial t} - \frac{\partial \sigma_{12}}{\partial x} = 0, \qquad \frac{\partial}{\partial t}\left(\frac{\sigma_{12}}{\mu(x)}\right) - \frac{\partial v_2}{\partial x} = 0, \tag{13.16}$$

and jump conditions

$$\tilde{V}_N[\rho_0 v_2] + [\sigma_{12}] = 0, \quad [\mathbf{V}] = 0, \quad [\theta] = 0 \quad \text{at} \ \mathcal{S}, \tag{13.17}$$

$$f_{\mathcal{S}} = -[W] + <\sigma_{ij}> [\varepsilon_{ij}], \quad f_{\mathcal{S}}\tilde{V}_N \geq 0. \tag{13.18}$$

It should be noted that Eqs. (13.17) and (13.18) are useless unless we can determine the value of the velocity of the phase boundary. A possible solution is the introduction of an additional constitutive relation between the material velocity at the interface and the driving force in the form of a kinetic relation.[1,6,9] Because the nonlinearity of the formulated problem due to the moving phase boundary requires a numerical solution, we postpone the introduction of the supplementary constitutive information to the numerical approximation.

3. Discrete Representation

3.1 Integral Balance Laws for Discrete Elements

Following the main ideas of finite volume numerical methods,[21] we divide the body in a finite number of identical elements of elementary volume Δx. Integration over the finite volume element of Eq. (13.16) yields the following set of integral forms.

$$\frac{\partial}{\partial t}\int_{\Delta x}\rho_0 v_2 dx = (\sigma_{12})^{right} - (\sigma_{12})^{left}, \tag{13.19}$$

$$\frac{\partial}{\partial t}\int_{\Delta x}\sigma_{12} dx = (\mu v_2)^{right} - (\mu v_2)^{left}. \tag{13.20}$$

3.2 Averaged Quantities and Fluxes

Introducing averaged quantities at each time step

$$\bar{v}_2 = \frac{1}{\Delta x}\int_{\Delta x} v_2 dx, \qquad \bar{\sigma}_{12} = \frac{1}{\Delta x}\int_{\Delta x} \sigma_{12} dx, \tag{13.21}$$

and numerical fluxes at the boundaries of each element

$$F \approx \frac{1}{\Delta t}\int_{t_l}^{t_{l+1}} \sigma_{12}\,dt, \quad G \approx \frac{1}{\Delta t}\int_{t_l}^{t_{l+1}} \mu v_2\,dt, \tag{13.22}$$

we are able to write a finite volume numerical scheme for Eqs. (13.19) and (13.20) for a uniform grid (n) in the form (l denotes time steps)

$$(\bar{v}_2)_n^{l+1} - (\bar{v}_2)_n^l = \frac{\Delta t}{\rho_n \Delta x} \left((F^{right})_n^l - (F^{left})_n^l \right),$$
(13.23)

$$(\bar{\sigma}_{12})_n^{l+1} - (\bar{\sigma}_{12})_n^l = \frac{\Delta t}{\Delta x} \left((G^{right})_n^l - (G^{left})_n^l \right).$$
(13.24)

The main difficulty in the construction of a numerical scheme is the proper determination of the numerical fluxes F, G.[21] In fact our discrete elements are not in equilibrium, especially in the presence of phase transformation. Even if we can associate the averaged quantities with local equilibrium parameters, we still need to have a description of the nonequilibrium states of discrete elements. Moreover, we need also a description of interaction between these nonequilibrium elements, because classical equilibrium conditions are not valid in the case of fast propagation of sharp phase interfaces through the material during a stress-induced martensitic phase transformation.

4. Nonequilibrium Jump Conditions at the Phase Boundary

We start with the classical equilibrium conditions at the phase boundary. The classical equilibrium conditions at the phase boundary consist, for single-component fluidlike systems, in the equality of temperatures, pressures, and chemical potentials in the two phases; that is,

$$[\theta] = 0 \quad \text{or} \quad \left[\left(\frac{\partial U}{\partial S} \right)_{V,M} \right] = 0,$$
(13.25)

$$[p] = 0 \quad \text{or} \quad \left[\left(\frac{\partial U}{\partial V} \right)_{S,M} \right] = 0,$$
(13.26)

$$[\mu] = 0 \quad \text{or} \quad \left[\left(\frac{\partial U}{\partial M} \right)_{S,V} \right] = 0,$$
(13.27)

where U is the internal energy, M is mass, V is volume, p is pressure, and μ is the chemical potential.

In the considered homothermal case, the continuity of temperature at the phase boundary still holds, and the continuity of the chemical potential can be replaced by the expression for the nonzero driving force [Eq. (13.18)]. What we need is to change the equilibrium condition for pressure [Eq. (13.26)]. In nonequilibrium, we expect that the value of internal energy of an element differs from its equilibrium value[27]

$$U = U_{eq} + U_{ex},$$
(13.28)

where the excess energy U_{ex} is the difference between the nonequilibrium and equilibrium values. Therefore, we can make a direct generalization of classical equilibrium condition for pressure using the excess energy

$$\left[\left(\frac{\partial (U_{eq} + U_{ex})}{\partial V} \right)_{S,M} \right] = 0. \tag{13.29}$$

However, the obtained jump relation corresponds to a fixed entropy at the boundary. At the same time, it is well understood that the martensitic phase transformation is a dissipative process, which involves entropy change. Therefore, we propose to replace the jump relation [Eq. (13.29)] by another nonequilibrium jump relation. Our choice of the fixed variables is influenced by the stability conditions for single-component fluidlike systems[28]

$$\left[\left(\frac{\partial (U_{eq} + U_{ex})}{\partial V} \right)_{\theta,M} \right] = 0, \qquad \left[\left(\frac{\partial (U_{eq} + U_{ex})}{\partial V} \right)_{p,M} \right] = 0. \tag{13.30}$$

The last two jump relations differ from Eq. (13.29) only by fixing different variables in the corresponding thermodynamic derivatives.

To be able to exploit the jump relations, we need to have a more detailed description of nonequilibrium states than by only introducing the energy excess. The most convenient description of the nonequilibrium states may be obtained by means of the thermodynamics of discrete systems,[10] where the thermodynamic state space is extended by means of so-called *contact quantities*.

4.1 Contact Quantities

We still deal with single-component fluidlike systems. A *discrete system*[10] is considered as a domain separated from its equilibrium environment by a contact surface. In a Schottky system *per se*, the interaction between the system and the environment consists of heat, work, and mass exchanges. These exchange quantities allow us to define so-called *contact quantities*. For instance, considering the heat exchange $\dot{Q}$, the *contact temperature* Θ is defined by the inequality[10]

$$\dot{Q} \left(\frac{1}{\Theta} - \frac{1}{T^*} \right) \geq 0 \tag{13.31}$$

for vanishing work and mass exchange rates. Here T^* is the thermostatic temperature of the equilibrium environment. From Eq. (13.31) it follows that $\dot{Q}$ and the bracket always have the same sign. We now suppose that there exists exactly one equilibrium environment for each arbitrary discrete system for which the net heat exchange between them vanishes. Then Eq. (13.31) determines the contact temperature Θ of the system as the thermostatic temperature T^* of the system's environment for which this net exchange vanishes. The *dynamic pressure p* and *chemical potential, μ* are defined analogously:[10]

$$\dot{V}(p - p^*) \geq 0, \qquad \dot{M}(\mu^* - \mu) \geq 0, \tag{13.32}$$

where $\dot{V}$ is the time rate of volume, and $\dot{M}$ is the time rate of mass.

The contact quantities so defined together with common local equilibrium variables provide a complete thermodynamic description of nonequilibrium states of a separated discrete system.

In the required extension to the *thermoelastic case*, the state of each element is identified with the thermodynamic state of a discrete system associated with it, each element being assumed in local equilibrium. In thermoelasticity, in addition to Θ and Eq. (13.31), which governs heat exchange, we must define a *contact dynamic stress tensor* Σ_{ij}. Analogously to Eq. (13.31) that holds for $\dot{\varepsilon}_{ij} = 0$ we have

$$\frac{\partial \varepsilon_{ij}}{\partial t}(\Sigma_{ij} - \sigma_{ij}^*) \geq 0, \tag{13.33}$$

for vanishing heat and mass exchange rates. Here σ_{ij}^* is the Cauchy stress tensor in the environment.

In the thermoelastic case, the thermodynamic derivatives that we should exploit instead of $(\partial U / \partial V)_\theta$ and $(\partial U / \partial V)_p$ are:[28]

$$\left(\frac{\partial \bar{E}}{\partial \varepsilon_{ij}}\right)_\theta = -\bar{\theta}\left(\frac{\partial \bar{\sigma}_{ij}}{\partial \theta}\right)_\varepsilon + \bar{\sigma}_{ij}, \quad \left(\frac{\partial \bar{E}}{\partial \varepsilon_{ij}}\right)_\sigma = \bar{\theta}\left(\frac{\partial \bar{S}}{\partial \varepsilon_{ij}}\right)_\sigma + \bar{\sigma}_{ij}, \tag{13.34}$$

where E is the internal energy per unit volume and overbars denote the local equilibrium values.

Contact quantities are assumed to be connected with the excess energy in a similar way

$$\left(\frac{\partial E_{ex}}{\partial \varepsilon_{ij}}\right)_\theta = -\Theta\left(\frac{\partial \Sigma_{ij}}{\partial \theta}\right)_\varepsilon + \Sigma_{ij}, \quad \left(\frac{\partial E_{ex}}{\partial \varepsilon_{ij}}\right)_\sigma = \Theta\left(\frac{\partial S_{ex}}{\partial \varepsilon_{ij}}\right)_\sigma + \Sigma_{ij}, \tag{13.35}$$

where the interaction entropy S_{ex} is still undetermined. Using Eqs. (13.34) and (13.35) we obtain from Eq. (13.30) that the parameters of the adjacent nonequilibrium elements of a thermoelastic continuum should satisfy the thermodynamic consistency conditions, the first of which is valid for all processes with no entropy production

$$\left[-\bar{\theta}\left(\frac{\partial \bar{\sigma}_{ij}}{\partial \theta}\right)_\varepsilon + \bar{\sigma}_{ij} - \Theta\left(\frac{\partial \Sigma_{ij}}{\partial \theta}\right)_\varepsilon + \Sigma_{ij}\right] \cdot N_j = 0, \tag{13.36}$$

and the second one corresponds to any inhomogeneity accompanied by entropy production

$$\left[\bar{\theta}\left(\frac{\partial \bar{S}}{\partial \varepsilon_{ij}}\right)_\sigma + \bar{\sigma}_{ij} + \Theta\left(\frac{\partial S_{ex}}{\partial \varepsilon_{ij}}\right)_\sigma + \Sigma_{ij}\right] \cdot N_j = 0. \tag{13.37}$$

Here N_j are components of the unit normal at the boundary of a discrete element. Now we are able to describe the nonequilibrium states of discrete elements and to exploit the nonequilibrium jump relations, if we can determine the values of contact quantities, which can be done at least numerically.

5. Finite Volume Numerical Scheme

5.1 Contact Quantities in the Bulk

We now need to solve the system of equations [Eq. (13.16)]. First we apply Eq. (13.36) to determine the values of the contact quantities in the absence of phase transformation.

Because shear components of the stress tensor are independent of temperature, Eq. (13.36) reduces to

$$[\bar{\sigma}_{12} + \Sigma_{12}] = 0. \tag{13.38}$$

In the uniaxial case we have at the interface between elements $(n-1)$ and (n),

$$(\Sigma_{12}^+)_{n-1} - (\Sigma_{12}^-)_n = (\bar{\sigma}_{12})_n - (\bar{\sigma}_{12})_{n-1}. \tag{13.39}$$

This relation should be complemented by the kinematic condition between material and physical velocity,[24] which in the small-strain approximation become

$$[\mathbf{v} + \mathbf{V}] = 0. \tag{13.40}$$

Assuming that the jump of the contact velocity is determined by the second term of Eq. (13.40),

$$[\mathcal{V}] = [\mathbf{V}], \tag{13.41}$$

we obtain in the uniaxial case

$$(\mathcal{V}_2^+)_{n-1} - (\mathcal{V}_2^-)_n = (\bar{v}_2)_n - (\bar{v}_2)_{n-1}. \tag{13.42}$$

At this step we need to introduce constitutive relations between contact stresses and contact velocities. Our choice is motivated by the possible reduction to the wave-propagation algorithm. In fact, introducing the relations between contact stresses and contact velocities

$$(\mathcal{V}_2^-)_n = -\frac{(\Sigma_{12}^-)_n}{\rho_n c_n}, \qquad (\mathcal{V}_2^+)_{n-1} = \frac{(\Sigma_{12}^+)_{n-1}}{\rho_{n-1} c_{n-1}}, \qquad c = \sqrt{\frac{\mu}{\rho}}, \tag{13.43}$$

we then obtain a linear system of equations for the unknown contact velocities

$$(\mathcal{V}_2^+)_{n-1} - (\mathcal{V}_2^-)_n = (\bar{v}_2)_n - (\bar{v}_2)_{n-1}, \tag{13.44}$$

$$(\mathcal{V}_2^+)_{n-1}\rho_{n-1}c_{n-1} + (\mathcal{V}_2^-)_n\rho_n c_n = (\bar{\sigma}_{12})_n - (\bar{\sigma}_{12})_{n-1}. \tag{13.45}$$

The corresponding numerical scheme (13.23), (13.24) can be represented as

$$(\bar{\sigma}_{12})_n^{l+1} - (\bar{\sigma}_{12})_n^l = \frac{\Delta t}{\Delta x}\mu_n((\mathcal{V}_2^+)_n^l - (\mathcal{V}_2^-)_n^l), \tag{13.46}$$

$$(\bar{v}_2)_n^{l+1} - (\bar{v}_2)_n^l = \frac{\Delta t}{\Delta x}\frac{1}{\rho_n}((\Sigma_{12}^+)_n^l - (\Sigma_{12}^-)_n^l). \tag{13.47}$$

The two relations [Eqs. (13.44) and (13.45))] together express a characteristic property for the cell-centered numerical fluxes in the conservative wave-propagation algorithm,[29] whose advantages we can therefore exploit. However, phase transitions are always accompanied by the production of entropy. Hence we need to apply another nonequilibrium jump relation at the phase boundary.

5.2 Contact Quantities at the Phase Boundary

Suppose that the interface between two thermoelastic phases is placed between elements numbered $(p-1)$ and (p). For the left element adjacent to the phase boundary, the contact quantities $(\Sigma_{12}^-)_{p-1}$ at the left boundary of the element can be determined within the above-described numerical procedure. However, we need a more careful consideration for values of the contact stresses $(\Sigma_{12}^+)_{p-1}$ at the right side of the element which corresponds to the phase boundary. Similarly, for the right element adjacent to the phase boundary, we need to determine the values of $(\Sigma_{12}^-)_p$. The corresponding procedure is based on the nonequilibrium jump relation [Eq. (13.37)] that is specified in the isothermal uniaxial case to be

$$\left[\bar{\theta} \left(\frac{\partial \bar{S}}{\partial \varepsilon_{12}} \right)_\sigma + \bar{\sigma}_{12} + \Sigma_{12} \right] = 0. \tag{13.48}$$

Here we should make a certain assumption about the entropy production at the phase boundary. The simplest one is the continuity of contact stresses at the phase boundary

$$[\Sigma_{12}] = 0. \tag{13.49}$$

Another relation follows from the coherency conditions for the material velocity [Eq. (13.10)] which can be expressed in the small-strain approximation as follows,

$$[V_2] = 0. \tag{13.50}$$

In terms of the contact stresses, Eq. (13.50) yields

$$\frac{(\Sigma_{12}^+)_{p-1}}{\rho_{p-1} c_{p-1}} + \frac{(\Sigma_{12}^-)_p}{\rho_p c_p} = 0. \tag{13.51}$$

It follows from Eqs. (13.49) and (13.51) that the values of contact stresses vanish at the phase boundary

$$(\Sigma_{12}^+)_{p-1} = (\Sigma_{12}^-)_p = 0. \tag{13.52}$$

Now all the contact quantities at the phase boundary are determined, and we can update the state of the elements adjacent to the phase boundary.

The material velocity at the interface is determined by means of the jump relation for linear momentum [Eq. (13.17)][1]

$$V_N^2 = \frac{[\bar{\sigma}_{12}]}{< \rho_0 > [\bar{\varepsilon}_{12}]}. \tag{13.53}$$

The direction of the front propagation is determined by the positivity of the entropy production [Eq. (13.12)],

$$\sigma_S = \frac{f_S V_N}{\theta_S} \geq 0. \tag{13.54}$$

The obtained relations at the phase boundary are used in the described numerical scheme for the simulation of phase-transition front propagation.

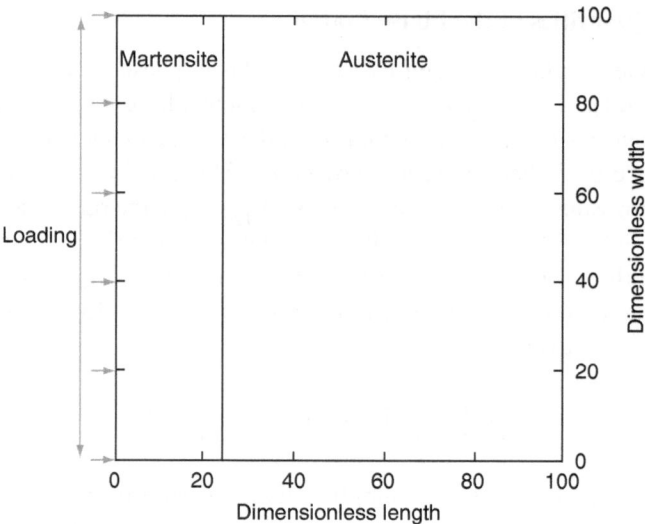

Fig. 13.2. Plane wave: geometry.

6. Numerical Simulations

6.1 Interaction of a Plane Wave with Phase Boundary

As a first example, we consider the interaction of a plane wave with a phase boundary to confirm the results of phase-transition front propagation in the one-dimensional case.[30–32] The geometry of the problem is shown in Figure 13.2. The wave is excited at the left boundary of the computation domain by prescribing a time variation of a component of the stress tensor. Upper and bottom boundaries are stress-free; the right boundary is assumed to be rigid. The time-history of loading is shown in Figure 13.3. If the magnitude of the wave is high enough, the phase transformation process is activated at the phase boundary. The maximal value of the Gaussian pulse is chosen as 0.7 GPa. Material properties correspond to Cu-14.44Al-4.19Ni shape-memory alloy[33] in austenitic phase: the density $\rho = 7100 \, \text{kg/m}^3$, the elastic modulus $E = 120 \, \text{GPa}$, the shear wave velocity $c_s = 1187 \, \text{m/s}$, the dilatation coefficient $\alpha = 6.75 \cdot 10^{-6} \, 1/\text{K}$.

It was recently reported[34] that elastic properties of the martensitic phase of Cu–Al–Ni shape-memory alloy after impact loading are very sensitive to the amplitude of loading. Therefore, for the martensitic phase we choose, respectively, $E = 60 \, \text{GPa}$, $c_s = 1055 \, \text{m/s}$, with the same density and dilatation coefficient as above. As a first result, the stress–strain relation is plotted in Figure 13.4 at a fixed point inside the computational domain which was initially in the austenitic state. As we can see in Figure 13.4, the stress–strain relation is at first linear corresponding to elastic austenite. Then the strain value jumps along a constant stress line to its value in the martensitic state due to phase transformation. Afterwards both loading and unloading correspond to elastic martensite. The value of the strain jump between straight lines, the slope of which is prescribed by material properties of austenite and martensite, respectively, is determined by the value of stress, which conforms to the critical value of the driving

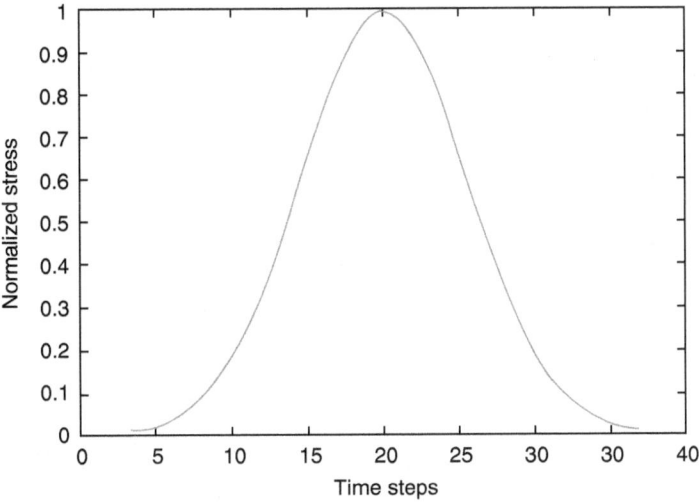

Fig. 13.3. Loading time history.

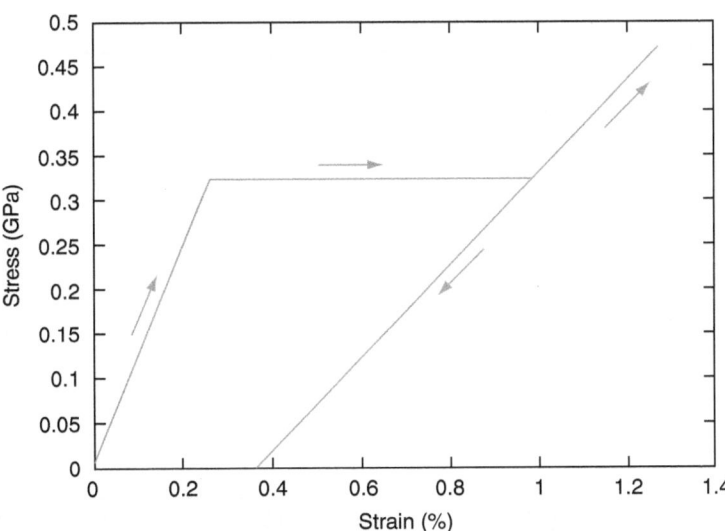

Fig. 13.4. Stress–strain behavior at a fixed point if the transformation strain is taken into account.

force, in agreement with the barrier of potential that we have to overcome to go from one phase to the other. Therefore, the stress value corresponding to the critical value of the driving force can be associated with the transformation stress, and the value of the strain jump corresponds to the transformation strain. We should then take into account that martensite can exist only in the deformed state; that is, the martensitic line should start from a nonzero value of the transformation strain. The result shown in the Figure 13.4, looks very much like the stress–strain dependence given in Reference 5.

The obtained stress–strain relation at any fixed point results in overall pseudoelastic response of a specimen. The overall stress–strain behavior can be compared with the dynamic experiment provided in Reference[34] after adjusting the applied pulse width

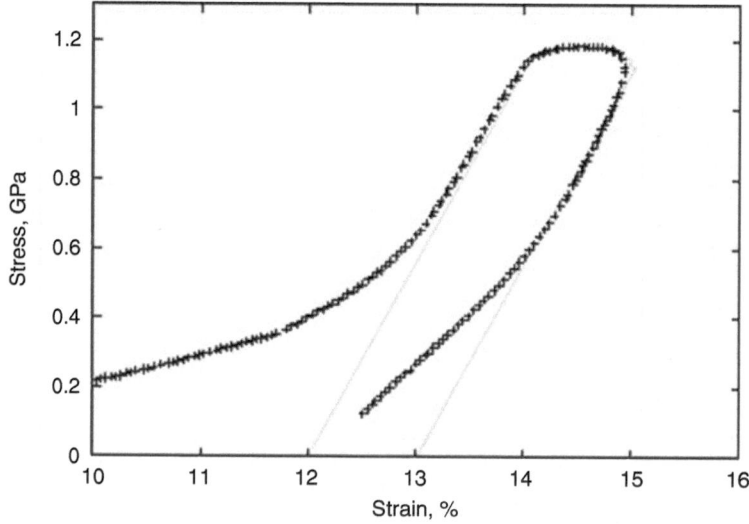

Fig. 13.5. Stress–strain relation: comparison with experimental data from Reference 34 (sample 1).

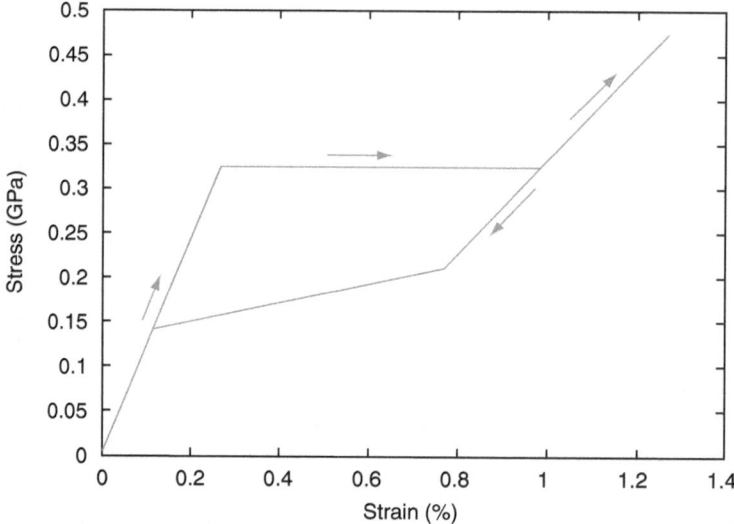

Fig. 13.6. Stress–strain behavior at a fixed point with full recovering of austenite.

and shape (see Figure 13.5) with an excellent agreement in the phase transformation region.

6.2 Hysteretic Behavior

Up to now it was supposed that austenite is not recovered after unloading which is not the case if the value of the reference temperature is above the onset of the reverse transformation temperature. The inverse phase transformation should occur

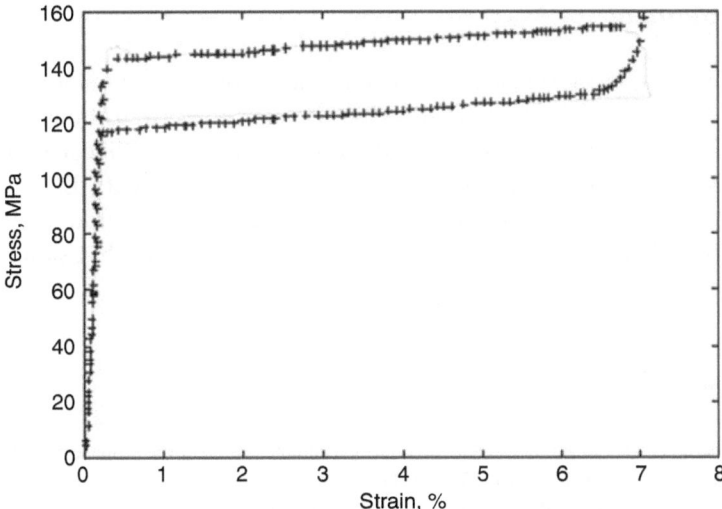

Fig. 13.7. Stress–strain relation at the phase boundary: comparison with experimental data from Reference 4.

immediately when the actual deformation of martensitic elements becomes less than the transformation strain. Because the inverse transformation is governed by a condition other than the direct transformation, we obtain a hysteretic stress–strain behavior (Figure 13.6). Again, the overall stress–strain dependence can be compared with experimental data. See Figure 13.7, where the experimental data of a quasistatic loading of a similar material with relatively high applied loading rate (1 MPA/s) from Reference[4] are given. The applied stress in this case was linearly increased and the duration of the impulse was chosen to fit the experimental data.

7. Conclusions

Attempts at numerical simulations of moving phase boundaries in solids meet the problems with constitutive modeling of the nucleation criterion and kinetic relation at the phase boundary, as well as with the construction of a proper numerical algorithm. In spite of the accuracy and stability of the wave propagation method for inhomogeneous media, its application to the phase-transition problems is impossible unless we can predict the values of numerical fluxes at the phase boundary. We have proposed to determine all the needed quantities by means of nonequilibrium jump relations at the phase boundary, which are presented by means of contact quantities derived from the thermodynamics of discrete systems. In this case the construction of the algorithm is complemented by the development of a thermodynamic model of phase-transition front propagation.

Results of numerical simulations show that the proposed approach allows us to reproduce experimental observations, in spite of the idealization of the process.

Acknowledgment

Support of the Estonian Science Foundation under contracts No. 4504, 5765 (A.B., J.E.), of the European Network TMR. 98-0229 on "Phase transitions in crystalline substances" (G.A.M.) and of ESF Scientific Programme on Nonlinear Acoustic Techniques for Micro-Scale Damaged Diagnostics (NATEMIS) (A.B., J.E.) is gratefully acknowledged. G.A.M. benefits from a Max Planck Award for International Cooperation (2001–2005). Research achieved within the Parrot French–Estonian Program (2003–2004).

References

[1] R. Abeyaratne and J.K. Knowles, Kinetic relations and the propagation of phase boundaries in solids, *Arch. Rat. Mech. Anal.* **114**, 119–154 (1991).

[2] J.A. Shaw and S. Kyriakydes, Thermomechanical aspects of NiTi, *J. Mech. Phys. Solids* **43**, 1243–1281 (1995).

[3] J.A. Shaw and S. Kyriakydes, On the nucleation and propagation of phase transformation fronts in a NiTi alloy, *Acta Mater.* **45**, 673–700 (1997).

[4] B.C. Goo and C. Lexcellent, Micromechnics-based modeling of two-way memory effect of a single-crystalline shape-memory alloy, *Acta Mater.* **45**, 727–737 (1997).

[5] R. Abeyaratne, K. Bhattacharya, and J.K. Knowles, Strain-energy functions with local minima: Modeling phase transformations using finite thermoelasticity, in: *Nonlinear Elasticity: Theory and Application*, edited by Y. Fu and R. W. Ogden (Cambridge University Press, Cambridge, UK, 2001), pp. 433–490.

[6] R. Abeyaratne and J.K. Knowles, On the driving traction acting on a surface of strain discontinuity in a continuum, *J. Mech. Phys. Solids* **38**, 345–360 (1990).

[7] L. Truskinovsky, Dynamics of nonequilibrium phase boundaries in a heat conducting nonlinear elastic medium, *J. Appl. Math. Mech.* (PMM) **51**, 777–784 (1987).

[8] G.A. Maugin and C. Trimarco, The dynamics of configurational forces at phase-transition fronts, *Meccanica* **30**, 605–619 (1995).

[9] J.K. Knowles, Stress-induced phase transitions in elastic solids, *Comp. Mech.* **22**, 429–436 (1999).

[10] W. Muschik, Fundamentals of nonequilibrium thermodynamics, in: *nonequilibrium Thermodynamics with Application to Solids*, edited by W. Muschik (Springer, Wien, 1993), pp. 1–63.

[11] Y.C. Chen and D.C. Lagoudas, Impact induced phase transformation in shape memory alloys, *J. Mech. Phys. Solids* **48**, 275–300 (2000).

[12] A. Bekker, J.C. Jimenez-Victory, P. Popov, and D.C. Lagoudas, Impact induced propagation of phase transformation in a shape memory alloy rod, *Int. J. Plasticity* **18**, 1447–1479 (2002).

[13] D.C. Lagoudas, K. Ravi-Chandar, K. Sarh, and P. Popov, Dynamic loading of polycrystalline shape memory alloy rods, *Mech. Mater.* **35**, 689–716 (2003).

[14] J.A. Shaw, A thermomechanical model for a 1-D shape memory alloy wire with propagating instabilities, *Int. J. Solids Struct.* **39**, 1275–1305 (2002).

[15] V. Stoilov and A. Bhattacharyya, A theoretical framework of one-dimensional sharp phase fronts in shape memory alloys, *Acta Mater.* **50**, 4939–4952 (2002).

[16] M.A. Grinfeld, *Thermodynamic Methods in the Theory of Heterogeneous Systems* (Longman, London, 1991).

[17] P. Cermelli and S. Sellers, Multi-phase equilibrium of crystalline solids, *J. Mech. Phys. Solids* **48**, 765–796 (2000).

[18] A. Berezovski and G. A. Maugin, On the thermodynamic conditions at moving phase-transition fronts in thermoelastic solids, *J. Non-Equilib. Thermodyn.* **29**, 37–51 (2004).

[19] X. Zhong, T.Y. Hou, and P.G. LeFloch, Computational methods for propagating phase boundaries, *J. Comp. Phys.* **124**, 192–216 (1996).

[20] R.J. LeVeque, Wave propagation algorithms for multidimensional hyperbolic systems, *J. Comp. Phys.* **131**, 327–353 (1997).

[21] R.J. LeVeque, *Finite Volume Methods for Hyperbolic Problems* (Cambridge University Press, Cambridge, UK, 2002).

[22] A. Berezovski and G.A. Maugin, Simulation of thermoelastic wave propagation by means of a composite wave-propagation algorithm, *J. Comp. Phys.* **168**, 249–264 (2001).

[23] A. Berezovski, J. Engelbrecht, and G.A. Maugin, Thermoelastic wave propagation in inhomogeneous media, *Arch. Appl. Mech.* **70**, 694–706 (2000).

[24] G.A. Maugin, *Material Inhomogeneities in Elasticity* (Chapman and Hall, London, 1993).

[25] G.A. Maugin, Thermomechanics of inhomogeneous - heterogeneous systems: application to the irreversible progress of two- and three-dimensional defects, *ARI* **50**, 41–56 (1997).

[26] G.A. Maugin, On shock waves and phase-transition fronts in continua, *ARI* **50**, 141–150 (1998).

[27] W. Muschik and A. Berezovski, Thermodynamic interaction between two discrete systems in non-equilibrium, *J. Non-Equilib. Thermodyn.* **29**, 237–255 (2004).

[28] H.B. Callen, *Thermodynamics* (Wiley & Sons, New York, 1960).

[29] D.S. Bale, R.J. LeVeque, S. Mitran, and J.A. Rossmanith, A wave propagation method for conservation laws and balance laws with spatially varying flux functions, *SIAM J. Sci. Comp.* **24**, 955–978 (2003).

[30] A. Berezovski and G.A. Maugin, Thermoelastic wave and front propagation, *J. Thermal Stresses* **25**, 719–743 (2002).

[31] A. Berezovski and G.A. Maugin, Thermodynamics of discrete systems and martensitic phase transition simulation, *Technische Mechanik* **22**, 118–131 (2002).

[32] A. Berezovski, J. Engelbrecht, and G.A. Maugin, A thermodynamic approach to modeling of stress-induced phase-transition front propagation in solids, in: *Mechanics of Martensitic Phase Transformation in Solids*, edited by Q.P. Sun (Kluwer, Dordrecht, 2002), pp. 19–26.

[33] J.C. Escobar and R.J. Clifton, On pressure-shear plate impact for studying the kinetics of stress-induced phase-transformations, *Mat. Sci. Eng.* **A170**, 125–142 (1993).

[34] Y. Emel'yanov, S. Golyandin, N.P. Kobelev, S. Kustov, S. Nikanorov, G. Pugachev, K. Sapozhnikov, A. Sinani, Ya.M. Soifer, J. Van Humbeeck, and R. De Batist, Detection of shock-wave-induced internal stresses in Cu-Al-Ni shape memory alloy by means of acoustic technique, *Scripta mater.* **43**, 1051–1057 (2000).

14

The Spectral Analysis of a PM Space Unit in the Context of the Classification of Nonlinear Phenomena in Ultrasonic Wave Propagation

Sigrun Hirsekorn

Fraunhofer Institut für zerstörungsfreie Prüfverfahren (IZFP), Universität, Geb. 37, 66123 Saarbrücken, Germany.
To whom correspondence should be addressed: Sigrun Hirsekorn, IZFP, Universität, Geb. 37, 66123 Saarbrücken, Germany; e-mail:sigrun.hirsekorn@izfp.fraunhofer.de.

Abstract
Experiments on diverse materials, such as rocks, soil, cement, concrete, and damaged metals, have revealed evidence for nonlinearity, hysteresis, and discrete memory in their elastic behavior. A variety of nonlinear effects in quasistatic as well as dynamic measurements was observed, for example, the resonance-frequency downwards shift with increasing excitation amplitude, the generation of higher harmonics, the so-called slow dynamics, and so on. For the simulation of these effects on the propagation of ultrasonic waves, various models have been proposed. They usually assume the presence of a large number of soft interstitial regions, which are taken to be responsible for the nonlinear and hysteretic behavior of the material specimen. In order to simplify the treatment, a so-called "PM-space" of pairs of preassigned interstice strain states and corresponding pressure values at which transitions from one state to the other are assumed to take place, is often considered. The relationship between the choice of the PM-space and the consequent nonlinearity is, however, inferred only phenomenologically. Starting with the case of only one interstice, the interdependence among the parameters of the model, the input excitation, and the spectral contents of the specimen's response are derived analytically. The results are related to the strains and restoring forces as present in thin bonded interfaces and discussed with regard to the inverse problem and the classification of defects and weak bonds.

Keywords: Classification, harmonics, hysteresis, interstice, phenomenology, PM-space, spectral analysis, theory, ultrasound, wave propagation

1. Introduction

Experimental investigations of diverse materials such as rocks, soil, cement, concrete, damaged metals, etc have revealed evidence for nonlinearity, hysteresis, and discrete memory in their elastic behavior. These discoveries suggest the existence of a nonlinear mesoscopic elasticity (NME) universality class, to which all the aforementioned materials, as well as many others, belong.[1] Hence the appearance of a variety of nonlinear effects in both quasistatic and dynamic experiments such as, for example, the resonance-frequency downwards shift with increasing excitation amplitude, the generation of higher harmonics, the so-called slow dynamics, and so on. To simulate these effects on the propagation of ultrasonic waves in nonlinear mesoscopic elastic materials[1-7] various models have been developed on the base of a statistical Preisach–Mayergoitz space.[8-10] These models assume the presence of a large number of soft interstices, which are taken to be responsible for the nonlinear and hysteretic behavior of the material.[2-5] In order to simplify the treatment, the so-called "PM-space" of pairs of preassigned interstice strain states and corresponding pressure values at which transitions from one state to the other are assumed to take place, is considered. The relationship between the choice of the PM-space and the consequent nonlinearity is, however, inferred only phenomenologically. The investigation of the binding forces in adherent joints, in which the interface between bonded elements is the primary source of the nonlinearity[11-13] and the general theoretical analysis of the background[14,15] may allow more detailed and realistic conclusions about the originating forces of nonclassical nonlinear (NCNL) effects and may lead to classification methods of weak bonds and defects.

Starting with the case of only one interstice described as a hysteretic mesoscopic elastic unit (HMEU) in a homogeneous linear elastic material, the interdependence among the parameters of the PM model, the input excitation, and the spectral contents of the specimen's response is derived analytically. The calculations were carried out for a rectangular[16] as well as rhombic HMEU. The results are related to the strains and restoring forces as present in thin bonded interfaces and discussed with regard to the inverse problem and the classification of defects and weak bonds.

2. The Spectral Response of a Single Rectangular Hysteretic Mesoscopic Elastic Unit (HMEU)

A sample of a linear elastic substrate material containing one adherent joint, the soft interstice which is the source of the nonlinearity, is considered. In the first instance, the bonded interface, contrary to References,[11-13] is described by a rectangular hysteretic mesoscopic elastic unit (HMEU) as defined, for example, in Reference[9] (Figure 14.1a). The interface width may have only two stable values l_c and l_o, $l_c < l_o$, which correspond to a so-called closed and open state, respectively. If the HMEU is initially in its closed state l_c and a decreasing external pressure (tension) is applied, the interface remains in the closed state until the pressure P_o is reached, at which point the interface width changes abruptly to its open state l_o. Further decrease of the applied pressure (increase in tension) no longer changes the interface width. If now the pressure is increased again, the HMEU remains open until the pressure $P_c \geq P_o$ is reached, where it jumps back into its closed state and remains there even if the pressure is further increased. The HMEU shows hysteresis if $P_c > P_o$; no hysteresis occurs if $P_o = P_c$. Because of the force balance, the external pressure is equal to the restoring forces in the interface referred to in References,[11-13] but of opposite sign. Corresponding considerations hold if the starting position of the interface is its open state. Both cases are described in detail in Reference.[16] Here, only the results with the closed state as

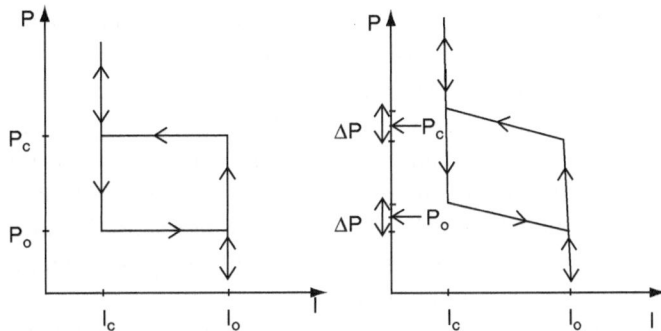

Fig. 14.1. Schematic sketch of the stress–displacement relation of a hysteretic mesoscopic elastic unit of (a) rectangular shape as used in the PM–space model[9] and of (b) rhombic shape similar to the model used in Reference,[17] respectively.

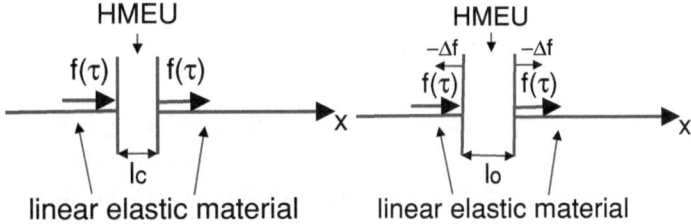

Fig. 14.2. One-dmensional model of a HMEU embedded in a linear elastic substrate material: (a) inital state, closed without external load; (b) open state because of external tensile load.

starting point are repeated briefly and used as a base of interpretation and discussion in the context of the classification of the elastic behavior of bonds.

Figure 14.2 shows a one-dimensional model of the HMEU with an initially closed state embedded in a linear elastic substrate. The HMEU is excited by an external sinusoidal force per area; the pressure is

$$P = f(\tau) = -f_0 \sin \tau. \qquad (14.1)$$

Here, $\tau = \omega t$ is the normalized time, and ω is the excitation angular frequency. The phase is chosen in agreement with the description in References,[11–15] that is, the excitation starts with a tension force to increase the interface width. The resulting restoring force in the interface is $F = -P$. The force amplitude f_0 of the excitation has to be larger than the absolute values of the opening and closing pressures P_o and P_c (i.e., $f_0 \geq |P_o|$, $f_0 \geq |P_c|$) to guarantee a continual alternation between the two states closed and open. The opening and closing pressures P_o and P_c are related to opening and closing times τ_o and τ_c with, for an initially closed state, $\tau_o \leq \tau_c$ within a vibration cycle:

$$P_o = -f_0 \sin \tau_o, \left. \frac{df(\tau)}{d\tau} \right|_{\tau = \tau_o} = -f_0 \cos \tau_o \leq 0, \qquad (14.2a)$$

$$P_c = -f_0 \sin \tau_c, \left. \frac{df(\tau)}{d\tau} \right|_{\tau = \tau_c} = -f_0 \cos \tau_c \geq 0; \qquad (14.2b)$$

that is, $0 \leq \tau_o \leq \pi/2, \pi/2 \leq \tau_c \leq 3\pi/2$ in the first cycle. Of course, opening and closure occur respectively for decreasing and increasing pressure.

The interstice is embedded in the linear elastic substrate. Hence the transfer from one state to the other, the jump in interstice width, acts against the elastic forces of the substrate. Thus, because of the balance of forces, the opening and closing pressures P_o and P_c cannot be independent of the substrate stiffness. For example, if the substrate is infinitely stiff, the opening tensile force (negative pressure) for an initially closed interstice will be infinite also; that is, it is not possible to cause the jump. This case is excluded in the following considerations.

The excitation of the HMEU generates elastic waves in the joined linear elastic matrix material. We consider the transmission into the positive x-direction. As in References,[13-16] the strain of the transmitted waves, the response, can be represented as a Fourier series:

$$\varepsilon_t(X, \tau) = \varepsilon_0 + \sum_{n=1}^{\infty} \varepsilon_n \sin(n\tau - nX + \varphi_n), \tag{14.3}$$

where $X = kx$ is the normalized length coordinate, $k = \omega/v_L$ is the wave number, and v_L the sound velocity in the substrate. The amplitudes ε_n and phases φ_n of the transmitted waves can be determined by the boundary conditions at the interface, as carried out in the following.

The HMEU is located at the position $X = 0$. The interstice width and the ultrasonic wavelength are assumed to be small compared to the length of the substrate; the jump in the interstice width must not cause a parallel translation of the whole substrate, and the stresses have to be continuous. As long as the interface distance is rigid, the sinusoidal force $f(\tau)$ is directly transferred into the elastic material. An abrupt change in strain of the amount $\Delta l/l_c$ occurs for an initially closed interface when a transition between the two interface states takes place, $\Delta l = l_o - l_c$ being the difference in the interface width between both states. Thus, in the PM-space unit the strain is either 0 or $\Delta l/l_c$. In the substrate the strain is caused either only by the external force $f(\tau)$ or by the external force and in addition by the force acting on the substrate because of the induced interstice strain. In the linear substrate the stress is related to the strain by Hooke's law; that is, the ratio is the elastic constant of the substrate c_{11}, and the stress in the substrate at the interface (i.e., at $X = 0$) during the first vibration cycle is

$$c_{11}^{\varepsilon_t}(X = 0, \tau) = \begin{array}{ll} f_0 \sin \tau & 0 \leq \tau \leq \tau_o, \tau_c \leq \tau \leq 2\pi \\ f_0 \sin \tau - c_{11}\Delta l/l_c & \text{if} \quad \tau_o \leq \tau \leq \tau_c \end{array}, \tag{14.4}$$

$c_{11} = \rho v_L^2$ is the elastic constant, and ρ is the density in the linear elastic substrate. From the spectral representation (14.3) of the strain at $X = 0$ and Eq. (14.4) the amplitudes ε_n and the phases φ_n of the response can be calculated using the orthogonality relations of the trigonometric functions $\sin(n\tau)$ and $\cos(n\tau)$. The results are

$$\varepsilon_0 = -\frac{\Delta l}{l_c} \frac{\tau_c - \tau_o}{2\pi}, \tag{14.5a}$$

$$\varepsilon_n \sin \varphi_n = -\frac{2\Delta l}{n\pi l_c} \cos \frac{n(\tau_c + \tau_o)}{2} \sin \frac{n(\tau_c - \tau_o)}{2}, \tag{14.5b}$$

$$\varepsilon_n \cos \varphi_n = \frac{f_0}{c_{11}} \delta_{n1} - \frac{2\Delta l}{n\pi l_c} \sin \frac{n(\tau_c + \tau_o)}{2} \sin \frac{n(\tau_c - \tau_o)}{2}. \tag{14.5c}$$

Equations (14.5) show that in general the response of a sinusoidally excited rectangular HMEU contains the incident frequency, all of its higher harmonics, and a static part. The amplitudes ε_n and phases φ_n of the transmitted waves contain the amplitude of the excitation as a parameter indirectly via the opening and closure times τ_o and τ_c. The parameters of the transmitted fundamental frequency additionally depend on the ratio of the excitation amplitude and the elastic constant c_{11} in the substrate. The static part ε_0 is below zero if the initial state of the interface is closed; that is, the static force pushes apart the surfaces forming the HMEU, and the mean interface distance increases during insonification, which causes a decrease in strain in the substrate. The amplitudes of the higher harmonics ε_n decrease as do the elements of a harmonic series with 1 over the order n. That is, a truncation of the Fourier series (3) considering only a finite number of higher harmonics cannot be a good approximation of the response of a rectangular HMEU.

If $P_c = P_o$ no hysteresis occurs, and Eqs. (14.5b) and (14.5c) yield, in agreement with the general result,[11–15] two possibilities for the phase of each of the transmitted waves:[16]

$$\varepsilon_{2n+1} \sin \varphi_{2n+1} = 0, n = 0, 1, 2, \ldots; \quad \text{i.e., } \cos \varphi_{2n+1} = \pm 1,$$
$$\varphi_{2n+1} = 0, \pi; \tag{14.6a}$$
$$\varepsilon_{2n} \cos \varphi_{2n} = 0, n = 1, 2, 3, \ldots; \quad \text{i.e., } \sin \varphi_{2n} = \pm 1,$$
$$\varphi_{2n} = \pi/2, 3\pi/2. \tag{14.6b}$$

The amplitudes of the higher harmonics in the response of a rectangular HMEU depend on the difference of opening and closure times within a vibration cycle, and the phases depend on the sum [Eqs. (14.5b,c)]. This entails that the information about hysteresis in the stress–strain relation is mainly contained in the phases, whereas the amplitudes inform about the general shape. This prediction may be confirmed by the calculation of the response of a rhombic, that is, a somehow "smoothed" rectangular, HMEU.

3. The Spectral Response of a Single Rhombic Hysteretic Mesoscopic Elastic Unit (HMEU)

The adherent joint embedded in the linear elastic substrate material, the soft interstice which is the source of the nonlinearity, is now described by a rhombic HMEU similar to the model used in Reference[17] (Figure 14.1b). The interface width still may have only two stable values l_c and $l_o, l_c < l_o$, corresponding to a so-called closed and open state, respectively. However, the transition between the two interface states no longer takes place abruptly, but linearly with the decrease or increase of the applied pressure within a certain pressure range ΔP. If the HMEU is initially in its closed state l_c and a decreasing external pressure (tension) is applied, the interface remains in the closed state until the pressure $P_o + \Delta P/2$ is reached. Then, the interface width changes linearly with the decrease in pressure and reaches its open state l_o at the pressure value $P_o - \Delta P/2$. A further decrease of the applied pressure (increase in tension) does not change the interface width anymore. If now the pressure is increased again, the HMEU remains open until the pressure $P_c - \Delta P/2$, $P_c \geq P_o$, is reached. Then, its width changes linearly with increasing pressure and reaches its closed state l_c again at the pressure value $P_c + \Delta P/2$. The interface remains in its closed state even if the pressure is further increased. The HMEU shows hysteresis if $P_c > P_o$, and no hysteresis if $P_o = P_c$. Corresponding considerations hold if the interface starts in its open state.

The HMEU is excited by an external sinusoidal force, the pressure P of Eq. (14.1). The force amplitude f_0 of the excitation now has to be larger than the absolute values of the end points of the opening and the closing process $P_o - \Delta P/2$ and $P_c + \Delta P/2$, respectively; that is, $f_0 \geq |P_o - \Delta P/2|$, $f_0 \geq |P_c + \Delta P/2|$, to guarantee a continual complete alternation between the two states l_c and l_o. Along the lines of Eqs. (14.2), the pressures describing the opening and the closing process are related to the corresponding times within a vibration cycle:

$$P_o = -f_0 \sin \tau_o, \quad P_o \pm \Delta P/2 = -f_0 \sin \tau_{o1,o2},$$
$$\left. \frac{df(\tau)}{d\tau} \right|_{\tau = \tau_{o1,o,o2}} = -f_0 \cos \tau_{o1,o,o2} \leq 0, \tag{14.7a}$$

$$P_c = -f_0 \sin \tau_c, \quad P_c \pm \Delta P/2 = -f_0 \sin \tau_{c2,c1},$$
$$\left. \frac{df(\tau)}{d\tau} \right|_{\tau = \tau_{c1,c,c2}} = -f_0 \cos \tau_{c1,c,c2} \geq 0, \tag{14.7b}$$

with $0 \leq \tau_{o1} \leq \tau_o \leq \tau_{o2} \leq \pi/2 \leq \tau_{c1} \leq \tau_c \leq \tau_{c2} \leq 3\pi/2$ in the first cycle if the initial state is closed.

As in the case of a rectangular HMEU, the strain of the waves transmitted into the joined linear elastic substrate material, the response to the excitation of the HMEU, can be represented as a Fourier series, as in Eq. (14.3). The boundary conditions at the interface, which determine the amplitudes ε_n and phases φ_n of the transmitted waves, have to be modified following the rhombic stress–strain relation. Again, as long as the interface distance is rigid, the sinusoidal force is directly transferred into the elastic material. The additional change in strain during the opening and closing process of the interface is linear with the change in the applied pressure and with the maximum strain present in an initially closed interface $\Delta l/l_c$, $\Delta l = l_o - l_c$. The strain in the substrate at the interface (at $X = 0$) during the first vibration cycle is

$$\varepsilon_t(X = 0, \tau) = \frac{f_0}{c_{11}} \sin \tau - \frac{\Delta l}{l_c} \begin{cases} 0 & 0 \leq \tau \leq \tau_{o1}, \tau_{c2} \leq \tau \leq 2\pi \\ \frac{f_0}{\Delta P}(\sin \tau - \sin \tau_{o1}) & \tau_{o1} \leq \tau \leq \tau_{o2} \\ 1 & \text{if} & \tau_{o2} \leq \tau \leq \tau_{c1} \\ \frac{f_0}{\Delta P}(\sin \tau - \sin \tau_{c2}) & \tau_{c1} \leq \tau \leq \tau_{c2} \end{cases}.$$
$$\tag{14.8}$$

Equations (14.3) and (14.8) and the orthogonality relations of the trigonometric functions $\sin(n\tau)$ and $\cos(n\tau)$ yield the amplitudes ε_n and the phases φ_n of the response:

$$\varepsilon_0 = -\frac{\Delta l}{2\pi l_c} \left\{ \frac{\tau_{c1} + \tau_{c2} - \tau_{o1} - \tau_{o2}}{2} + \frac{P_o}{\Delta P}(\tau_{o2} - \tau_{o1}) + \frac{P_c}{\Delta P}(\tau_{c2} - \tau_{c1}) \right. \quad (14.9a)$$

$$\left. + \frac{f_o}{\Delta P}(\cos \tau_{o1} - \cos \tau_{o2} + \cos \tau_{c1} - \cos \tau_{c2}) \right\}$$

$$\varepsilon_1 \sin \varphi_1 = \frac{\Delta l}{\pi l_c} \frac{P_c - P_o}{f_o}, \qquad (14.9b)$$

$$\varepsilon_1 \cos \varphi_1 = \frac{f_o}{c_{11}} + \frac{\Delta l}{\pi l_c} \left\{ \frac{P_o}{2\Delta P}(\cos \tau_{o2} - \cos \tau_{o1}) + \frac{P_c}{2\Delta P}(\cos \tau_{c2} - \cos \tau_{c1}) \right.$$

$$(14.9c)$$

$$\left. + \frac{\cos \tau_{c1} + \cos \tau_{c2} - \cos \tau_{o1} - \cos \tau_{o2}}{4} - \frac{f_o}{2\Delta P}(\tau_{o2} - \tau_{o1} + \tau_{c2} - \tau_{c1}) \right\},$$

$$\varepsilon_n \sin \varphi_n = \frac{\Delta l}{\pi (n-1)(n+1)l_c} \left\{ \frac{\sin n\tau_{c1} + \sin n\tau_{c2} - \sin n\tau_{o1} - \sin n\tau_{o2}}{2n} \right.$$

$$(14.9d)$$

$$+ \frac{P_o}{n\Delta P}(\sin n\tau_{o2} - \sin n\tau_{o1}) + \frac{P_c}{n\Delta P}(\sin n\tau_{c2} - \sin n\tau_{c1})$$

$$\left. - \frac{f_o}{\Delta P}(\cos n\tau_{o2} \cos \tau_{o2} - \cos n\tau_{o1} \cos \tau_{o1} + \cos n\tau_{c2} \cos \tau_{c2} - \cos n\tau_{c1} \cos \tau_{c1}) \right\},$$

$$\varepsilon_n \cos \varphi_n = \frac{\Delta l}{\pi (n-1)(n+1)l_c} \left\{ \frac{\cos n\tau_{o1} + \cos n\tau_{o2} - \cos n\tau_{c1} - \cos n\tau_{c2}}{2n} \right.$$

$$(14.9e)$$

$$- \frac{P_o}{n\Delta P}(\cos n\tau_{o2} - \cos n\tau_{o1}) - \frac{P_c}{n\Delta P}(\cos n\tau_{c2} - \cos n\tau_{c1})$$

$$\left. - \frac{f_o}{\Delta P}(\sin n\tau_{o2} \cos \tau_{o2} - \sin n\tau_{o1} \cos \tau_{o1} + \sin n\tau_{c2} \cos \tau_{c2} - \sin n\tau_{c1} \cos \tau_{c1}) \right\}.$$

Equations. (14.5) present a special case of Eqs. (14.9) and therefore can be obtained by a limiting procedure. One has to carry out $\tau_{c1} \rightarrow \tau_c, \tau_{c2} \rightarrow \tau_c, \tau_{o1} \rightarrow \tau_o$, and $\tau_{o2} \rightarrow \tau_o$, using P_o, P_c, and ΔP from Eqs. (14.7). Note, that there are two different equations for the ΔP related to P_o and P_c, respectively.

In general, the response of a sinusoidally excited rhombic HMEU also contains the incident frequency, all of its higher harmonics, and a static part changing the mean interface width during insonification. The amplitudes ε_n and phases φ_n of the transmitted waves depend on the excitation amplitude f_0 as well as on the parameters of the opening and closure process, P_o, P_c, ΔP, l_o, and l_c. In the limit condition $P_o = P_c$ (i.e., no hysteresis); Eqs. (14.9) yield the general results for the phases as given in Eqs. (14.6). In contrary to a rectangular HMEU, where the transitions between the two interface states take place abruptly, now the amplitudes of the higher harmonics ε_n decrease with $1/n^2$, where n is the harmonic order; that is, smoothing the interface stress–strain relation entails in the response a faster decrease of the higher harmonics amplitudes in dependence on their order.

The results, Eqs. (14.9), are too complex for a more detailed interpretation of parameter dependencies and additional analytical investigations of further smoothing effects on the output, but nevertheless clearly show that the spectral analysis of the response gives information of the interface behavior and can be used as a characterization tool. Especially, the significance of the phases in the spectral contents of the response, which contain different information than the amplitudes, is shown.

4. Strains and Restoring Forces in Thin Bonded Interfaces

For thin bonded interfaces in components, which can be approximately described by interaction forces only without explicitly taking into account the material properties of the adhesive, measured amplitudes and phases of the transmitted ultrasonic waves may be used to determine the force–distance curve in the interface[11–13], or the equivalent, its stress–strain relation. In the preceding sections, the opposite was carried out. The force–distance relation of a thin bonded interface in a linear elastic substrate material was described by the elastic behavior of a rectangular and of a rhombic HMEU. For both cases, the response to a monochromatic compressional excitation, the amplitudes and phases of the waves generated in transmission, were calculated.

The results given in Eqs. (14.5) and (14.9) show that even the rough description of the force–distance relation in the interface as a rectangular HMEU yields the general behavior of a nonlinear interface as the generation of higher harmonics and the change in the mean interface width during insonification. But contrary to our experiments on nonlinear ultrasonic transmission through thin bonded interfaces, where we have measured amplitudes of higher harmonics discernable from noise only up to the third order,[11–13] the amplitudes of the higher harmonics generated by a rectangular PM-space unit (Figure 14.1a), especially for high order, seem to be much too large to describe realistic interstices and decrease with $1/n$, n being the order, only [Eqs. (14.5)]. This overestimation of the higher harmonics is due to the unrealistic edges in the force–distance relation. The effect was already reduced by using a rhombic (Figure 14.1b) instead of a rectangular HMEU as the description for the elastic behavior of the bond, which is equivalent to smoothing the force–distance relation. This is in agreement with the fact that the PM-space model yields good results in the simulation of wave propagation in materials with a large number of nonlinear HMEUs and a convenient distribution of different pairs of opening and closure pressures and interface distances (e.g. Reference[7]). The integration over a large number of different HMEUs has a smoothing effect.

For low excitation amplitudes that are not capable of causing a transition out of the initial state, a rectangular as well as a rhombic HMEU do not reproduce the reflection of waves at interfaces with a linear elastic behavior different from that of the substrate, but behave as a perfect bond. Likewise, if the amplitude of the excitation force is very large so that it exceeds the maximum of the restoring force in the interface[11–13], a description by a HMEU such as presented in the models in Figure 14.1 can no longer be used, because those models do not include a possible breaking of the joint. The effect of those features on the spectral parameters of the output still has to be investigated.

5. Summary

The spectral contents of the response to a sinusoidal excitation of a rectangular and of a rhombic HMEU in a linear elastic material was derived analytically. The results were related to strains and restoring forces as present in thin bonded interfaces. It was shown that the amplitudes as well as the phases of the spectral components contained in a response give information about the elastic behavior of bonds and contacts and might be used for their characterization and evaluation. An important result is that the information contained in the amplitudes and in the phases are complementary; that is, only amplitude and phase measurements of the ultrasonic response may allow a complete characterization of defects and weak bonds.

Acknowledgment

This work has been supported by the ESF-PESC program NATEMIS and by the EC 6th Framework Programme, Priority 4: Aeronautics and Space, within the Specific Targeted Research Project FP6-502927, Health Monitoring of Aircraft by Nonlinear Elastic Wave Spectroscopy AERONEWS. The author also likes to thank P.P. Delsanto for stimulating and fruitful discussions.

References

[1] R.A. Guyer and P.A. Johnson, Nonlinear mesoscopic elasticity: evidence for a new class of materials, *Physics Today* **52**, 30–36 (1999).

[2] R.A. Guyer, J.A. Tercate, and P.A. Johnson, Hysteresis and the dynamic elasticity of consolidated granular materials, *Phys. Rev. Lett.* **82**, 3280–3283 (1999).

[3] J.A. Tencate, E. Smith, and R.A. Guyer, Universal slow dynamics in granular solids, *Phys. Rev. Lett.* **85**, 1020–1024 (2000).

[4] B. Capogrosso-Sansone and R.A. Guyer, Dynamic model of hysteretic elastic systems, *Phys. Rev. B* **66**, 224101–224112 (2002).

[5] M. Scalerandi, P.P. Delsanto, and P.A. Johnson, Stress induced conditioning and thermal relaxation in the simulation of quasi-static compression experiments, *J. Phys. D: Appl. Phys.* **36**, 288–293 (2003).

[6] M. Scalerandi, V. Agostini, P.P. Delsanto, K. Van Den Abeele, and P.A. Johnson, Local interaction simulation approach to modeling nonclassical nonlinear elastic behavior in solids, *Jour. Acoust. Soc. Am.* **113**, 3049–3059 (2003).

[7] P.P. Delsanto and M. Scalerandi, Modeling nonclassical nonlinearity, conditioning, and slow dynamics effects in mesoscopic elastic materials, *Phys, Rev. B* **68**, 064107–064115 (2003).

[8] D.J. Holcomb, Memory, relaxation and microfracturing in dilatant rock, *J. Geophys. Res.* **86**, 6235–6248 (1981).

[9] K.R. McCall and R.A. Guyer, Equation of state and wave propagation in hysteretic nonlinear elastic materials, *J. Geophys. Res.* **99**, 23887–23897 (1994).

[10] K.R. McCall and R.A. Guyer, A new theoretical paradigm to describe hysteresis, discrete memory and nonlinear elastic wave propagation in rock, *Nonlin. Proc. in Geophys.* **3**, 89–101 (1996).

[11] S. Hirsekorn, Nonlinear transfer of ultrasound by adhesive joints - a theoretical description, *Ultrasonics* **39**, 57–68 (2001).

[12] P.P. Delsanto, S. Hirsekorn, V. Agostini, R. Loparco, and A. Koka, Modeling the propagation of ultrasonic waves in the interface region between two bonded elements, *Ultrasonics* **40**, 605–610 (2002).

[13] S. Hirsekorn, A. Koka, and W. Arnold, Theoretical and Experimental Investigations of Interaction Forces in Thin Bonded Interfaces by Nonlinear Ultrasonic Transmission, 2nd Workshop "NDT in Progress", Prague, Czech Republic, Oct. 6–8, 2003, ISBN 80-214-2475-3, 99–106 (2003).

[14] P.P. Delsanto and S. Hirsekorn, A Unified Treatment of Nonclassical Nonlinear Effects in the Propagation of Ultrasound in Heterogeneous Media, *Ultrasonics* **42**, 1005–1010 (2004).

[15] S. Hirsekorn and P.P. Delsanto, On the universality of nonclassical nonlinear phenomena and their classification, *Appl. Phys. Lett.* Vol. **84**, Issue 8, 1413–1415 (2004).

[16] S. Hirsekorn, Spectral Analysis of a PM Space in the Simulation of Nonlinear Ultrasonic Wave Propagation, World Congress on Ultrasonics (WCU), Paris, France, Sep. 7-10, 2003, Conference Proceedings, Eds.: D. Cassereau, M. Deschaps, P. Laugier, and A. Zarembowitch, Société Française d'Acoustique (SFA), ISBN 2-9515619-8-9 (Vol. I) and 2-9515619-9-7 (Vol. II), 491–494 (2003).

[17] M. Scalerandi, E. Ruffino, P.P. Delsanto, P.A. Johnson, and K. Van den Abeele, Review of Progress in Quantitative Nondestructive Evaluation (QNDE) Vol. **19B**, edited by D.O. Thompson and D.E. Chimenti, Plenum Press, New York, 1393–1399 (2000).

15

Nonlinear Ultrasonic Transmission Through Thin Bonded Interfaces: Theoretical Background and Numerical Simulations

S. Hirsekorn,[1,3] M. Hirsekorn,[2] and P.P. Delsanto[2]

[1]Fraunhofer Institut für zerstörungsfreie Prüfverfahren (IZFP), Universität, Geb. 37, 66123 Saarbrücken, Germany.
[2]INFM – Politecnico di Torino, Dipartimento di Fisica, C.so Duca degli Abruzzi 24, 10129 Torino, Italy.
[3]To whom correspondence should be addressed: Sigrun Hirsekorn, Fraunhofer Institut für zerstörungsfreie Prüfverfahren (IZFP), Universität, Geb. 37, 66123 Saarbrücken, Germany, sigrun.hirsekorn@izfp.fraunhofer.de.

Abstract

Adhesive bonds in composite structures influence the mechanical behavior of components and limit their load capacity. Consequently, the investigation of the interaction forces in adhesive joints is an important task in nondestructive testing. For this purpose the nonlinear stress–strain relation of adhesive joints for high-amplitude excitation resulting in higher and/or subharmonics generation can be exploited. This contribution discusses ultrasonic transmission through samples consisting of two plates joined together by a thin adhesive layer. It also describes the calibration procedure of the input and output wave parameters in order to obtain absolute values of the forces acting within the interface . Numerical simulations assist in the interpretation and evaluation of the experimental data.

Keywords: Adhesive, bond, interface, measurement, numerical simulation, theory, transmission, ultrasound

1. Introduction

Bonded interfaces in composite materials significantly influence the mechanical behavior of components and limit their load capacity. Consequently, the investigation of interaction forces in adhesive joints and the development of techniques to evaluate the bond quality are important tasks in nondestructive testing. As in all materials, the nonlinear part of the stress–strain relation of adhesive joints becomes more and more important with increasing strain.[1] This property causes a nonlinear modulation of ultrasonic waves resulting in the generation of higher harmonics and maybe also subharmonics at the interface, both in reflection and transmission, provided the strain of the excitation is large enough. A large amount of research has been carried out with the objective of relating the generation of higher harmonics in ultrasonic transmission through bonded structures to the quality of the bonds.[2–9] Commonly used are the nonlinearity parameter β, a measure of the second harmonic generation,[2] and the distortion factor K, a measure of the total nonlinear contents in the response.[7] Thin bonded interfaces can be approximately described by binding forces without taking directly into account the material properties of the adhesive.[3,7–9] Then, the local interaction forces, which may induce damping and hysteretic effects, can be probed by the measured amplitudes and phases of the ultrasonic waves transmitted through the interface.

Potential errors in characterizing bonded interfaces by nonlinear transmission of ultrasound are due to the fact that nonlinear signals may be caused not only by the interface but also by the measuring system (transmitting probes, amplifiers, etc.), the coupling medium, the bonded components themselves, frequency-dependent ultrasonic damping in the sample, and so on. These contributions are often larger than the effect stemming from the nonlinear interaction force in the bond. Therefore, it is necessary to improve the significance of the experimental data by separating the nonlinear effects of the interface from the errors, reducing the error effects, or calibrating the experimental results by measuring data from similar samples without bonded interfaces. Computer simulations of the experiments[10] serve as a powerful tool for the evaluation of the error magnitude in the measured data caused by unwanted nonlinear effects, even if their origin is not known in detail.

In this chapter, the ultrasonic transmission through samples consisting of two aluminum plates joined together by a thin epoxy layer is discussed. Contrary to recent investigations on similar samples,[5,6] which are restricted to the generation of the second harmonics, the third harmonics and the phases of the transmitted waves are also considered. These parameters contain valuable information especially for high-amplitude ultrasonic loading. In order to obtain absolute values of the forces in the interface, the input and output wave parameters must be calibrated capacitatively or interferometrically.[11,12] The calibration procedure is described in detail in Section 3. The experiments are numerically simulated using the Local Interaction Simulation Approach LISA[13] (see Section 4). The results confirm that the measured nonlinearity originates from the bonded interface.

2. Binding Forces in Thin Bonded Interfaces

We consider a sample of two plates of the same material bound together by a thin adhesive layer of static equilibrium width a_s. It is assumed that the interface is so thin that it can be described by binding forces only, without explicitly taking into account the material properties of the adhesive layer. The binding force in the interface $F(a)$ is a nonlinear function of the interface width a, its static equilibrium value being $F(a_s)$. The elastic behavior of the two plates is assumed to be linear within the amplitude range covered by the ultrasonic waves in the experiments. A monochromatic compressional wave

$$\varepsilon_{in} = \varepsilon_I \sin(\omega t - kx) \tag{15.1}$$

of circular frequency ω, wave number k, and strain amplitude ε_I is injected on one side of the specimen and arrives perpendicularly to the interface, modulating its width sinusoidally with the same frequency. In conditions of dynamic equilibrium (i.e., after transient phenomena have relaxed), the resulting time-dependent interaction force can be represented as a sum of sinusoidal forces of the incident frequency and its higher harmonics.[3] The general form

$$F\left(a\left(t\right)\right) = F_0 + \sum_{n=1}^{\infty} F_n \sin\left(n\omega t + \varphi_n\right) \tag{15.2}$$

contains the force amplitudes F_n and phases φ_n as parameters. This is a special application of the general problem of nonclassical nonlinear response of a system to an arbitrary excitation (see Chapter 1 of this book and References[14,15]). Here, the "cause" is the sinusoidal incident ultrasonic wave (15.1), and the "effect" is the restoring force in the bonded interface (15.2). In the following, the general formalism is shortly recalled for application to the special case considered.

The sinusoidal forces $F_n \sin(n\omega t + \varphi_n)$ generate the transmitted (and reflected) ultrasonic waves, so that the phases φ_n are transferred directly to the transmitted waves of fundamental frequency and to its higher harmonics. Their strain amplitudes ε_n multiplied by the elastic constant C_{11} relevant for the propagation of the waves in the plates are equal to the force amplitudes F_n ($F_n = C_{11}\varepsilon_n$). These ultrasonic waves additionally modulate the interface width. The constant part F_0 causes a static distortion of the interface; that is, the mean interface width a_e during insonification is, in the case of nonlinear interaction forces, not equal to the static equilibrium interface width a_s. The modulation by higher harmonics and the change in mean width are effects of higher order provided that $F_n < F_1$. The linear interface modulation

$$a\left(t\right) = a_S\left(1 + \varepsilon_{BI}\sin\omega t\right) = a_S + a_0\sin\omega t \tag{15.3}$$

has the amplitude a_0 determined by the interface strain amplitude ε_{BI}. In general, the interface vibration is phase-shifted relative to the incident wave in Eq. (15.1), depending on the amplitudes and phases of the reflected and transmitted waves of fundamental frequency at the interface. This phase-shift is referred to later in the calibration analysis and the numerical simulations of the experiments [see Eq. (15.5c)]. Due to the nonlinearity of the interaction force $F(a)$, the relation between the strain amplitudes of the incident wave in the components ε_I and in the interface ε_{BI} is expected to be nonlinear.

During the cycle ν, the interface width reaches two times the value a_s (at $t^+ = 2\pi\nu$ for increasing and at $t^- = (2\nu+1)\pi$ for decreasing width). From Eq. (15.2), we obtain

$$F\left(a(t) = a_S\right) = F_0 \pm \sum_{n=0}^{\infty} F_{2n+1}\sin\varphi_{2n+1} + \sum_{n=1}^{\infty} F_{2n}\sin\varphi_{2n}, \tag{15.4}$$

and the average value of $F(a = a_s)$ within one cycle, the static equilibrium interstice force is

$$F\left(a_S\right)_{av} = F_0 + \sum_{n=1}^{\infty} F_{2n}\sin\varphi_{2n}. \tag{15.5}$$

Likewise, within the same cycle $a(t_{max / min}) = a_{max / min}$ for $t_{max} = (2\nu + \frac{1}{2})\pi$, $t_{min} = (2\nu + \frac{3}{2})\pi$, and

$$F(a_{max / min}) - F(a_S)_{av} = \pm \sum_{n=0}^{\infty} F_{2n+1}(-1)^n \cos \varphi_{2n+1} + \sum_{n=1}^{\infty} F_{2n}((-1)^n - 1) \sin \varphi_{2n}.$$
(15.6)

These formulae include hysteresis because of the different values of the interaction force at the static interface width a_s. The forces in the interface relative to the static equilibrium are

$$\Delta F(a(t)) = F(a(t)) - F(a_S)_{av} = \sum_{n=1}^{\infty} F_n \sin(n\omega t + \varphi_n) - \sum_{n=1}^{\infty} F_{2n} \sin \varphi_{2n}. \quad (15.7)$$

The results are independent of the static part F_0; that is, these relative binding forces can be determined by the measured amplitudes and phases of the transmitted ultrasonic waves of the incident frequency and its higher harmonics. No hysteresis occurs if the time-dependent force acting within the interface during insonification fulfills the symmetry conditions in time

$$F(a(t_{max} + \Delta t)) = F(a(t_{max} - \Delta t)) \quad \text{and} \quad F(a(t_{min} + \Delta t)) = F(a(t_{min} - \Delta t)).$$
(15.8)

This restricts the phases in Eq. (15.2) to two values each and thus reduces the two values of the forces at the static equilibrium interface width in Eq. (15.4) to one; that is, $F_{nh}(a_s) = F(a_s)_{av}$. The index nh indicates "no hysteresis." In the considered example of nonlinear ultrasonic transfer through a thin bonded interface, the input is the incident ultrasonic wave. If nonlinear interface modulations are neglected, the deviation Δa of the interface width from its static equilibrium value a_s can be considered as input as well [Eq. (15.3)]. The output is the restoring force in the interface [Eq. (15.7)], which has to tend to infinite repulsive values if the interface width approaches zero. This leads to a further restriction of the phases in the nonhysteretic nonlinear case, and we get

$$\varphi_{2n} = n\pi - \frac{\pi}{2} \quad \text{and} \quad \varphi_{2n} = n\pi, \quad \text{for } n = 1, 2, 3, \ldots. \quad (15.9)$$

These phases inserted into Eq. (15.7) yield the relative binding forces as function of time for a monochromatic ultrasonic excitation and a nonhysteretic response

$$\Delta F_{nh}(a(t)) = F_1 \sin \omega t + F_2(\cos 2\omega t - 1) - F_3 \sin 3\omega t + \cdots. \quad (15.10)$$

For the maximum interface width during a vibration cycle, this equation reduces to

$$\Delta F_{nh}(a_{max}) = F_{nh}(a_{max}) - F_{nh}(a_S) = F_1 - 2F_2 + F_3 + F_5 - 2F_6 \pm \cdots. \quad (15.11)$$

So far, this approximate equation has been used to estimate the local relative binding forces in thin bonded interfaces from measured amplitudes of transmitted ultrasonic waves.[3,8,9]

The relative binding forces in the general case [Eq. (15.7)] as well as in the non-hysteretic approximation [Eq. (15.10)] can easily be represented as a function of the normalized linear interface width modulation $\Delta a_N(t) = (a(t) - a_S)/a_0 = \sin \omega t$; that is, $-1 \le \Delta a_N \le 1$ [Eq. (15.3)]:

$$\Delta F(\Delta a_N) = F_1 \left(\Delta a_N \cos \varphi_1 \pm \sqrt{1 - \Delta a_N^2} \sin \varphi_1 \right)$$
$$+ F_2 \left(\pm 2 \Delta a_N \sqrt{1 - \Delta a_N^2} \cos \varphi_2 - 2 \Delta a_N^2 \sin \varphi_2 \right)$$
$$+ F_3 \left(\left(3 - 4 \Delta a_N^2 \right) \Delta a_N \cos \varphi_3 \right) \pm \left(1 - 4 \Delta a_N^2 \right) \sqrt{1 - \Delta a_N^2} + \cdots ,$$

$$(15.12)$$

$$\Delta F_{nh}(\Delta a_N) = F_1 \Delta a_N - 2 F_2 \Delta a_N^2 - F_3 \left(3 - 4 \Delta a_N^2 \right) + \cdots . \qquad (15.13)$$

We restrict ourselves to harmonics up to the third order because, up to now, in our experiments the measured amplitudes of harmonics beyond third order have always been below noise level. Positive and negative signs correspond to increasing and decreasing interface width, respectively. Of course, in the nonhysteretic approximation, there is no difference between the forces for increasing and decreasing interface width.

The maximum interface width during a vibration cycle is reached at $\Delta a_N = 1$. This value inserted into Eqs. (15.12) and (15.13) yields the corresponding relative binding force in the hysteretic and in the nonhysteretic cases, respectively,

$$\Delta F(a_{\max}) = F_1 \cos \varphi_1 - 2 F_2 \sin \varphi_2 - F_3 \cos \varphi_3 + \cdots , \qquad (15.14)$$
$$\Delta F_{nh}(a_{\max}) = F_1 - 2 F_2 + F_3 + \cdots . \qquad (15.15)$$

The result in the nonhysteretic case (15.15) is, of course, equal to Eq. (15.11).

3. Calibration of the Measurement Data

3.1 Calibration of the Transmitted Ultrasonic Amplitudes

For the calibration procedure[11, 12] we use a plate of the same material and thickness as the one at the receiver side of the sample under investigation. We inject monochromatic compressional waves of various amplitudes from the side of the plate, which in the composite sample is bonded to the second plate. The displacement amplitude at the backwall is measured interferometrically as a function of the incident wave amplitude (known as voltage amplitude at the sending probe). The measurements are carried out for the injected frequency and then repeated successively for the second and third harmonics. The same measurements at the same excitation amplitudes and frequencies are repeated using a piezoelectric transducer at the receiver side. This yields a relation between the absolute amplitude of the transmitted wave in case of a free backwall and the voltage this wave generates in the piezoelectric transducer. Due to the coupling of the receiver to the backwall of the plate the latter is no longer free, which requires further evaluation of the calibrated values.

Throughout the calibration measurements, a sending probe causes a sinusoidal force

$$F_{in,n}(t) = F_n \sin(n \omega t + \varphi_n), \quad \text{with } F_n = C_{11} \varepsilon_n, \qquad (15.16)$$

at the intromission side of the plate at $x = 0$ (x is the coordinate in the thickness direction). ε_n is the strain amplitude generated at the surface of the plate, and C_{11} is its elastic constant (here $C_{11} = C_{11,Al} = 107.8\,\mathrm{GPa}$). We assume that the bonded interface is located at $x = 0$ and Eq. (15.16) corresponds to the nth harmonic of the restoring force in the interface. Ultrasonic reflections at the plate surfaces cause forward and backward propagating waves,

$$\varepsilon_{pl,n}^{I,R}(t) = \varepsilon_n^{I,R+} \sin\left(n\omega t + \varphi_n^{I,R+} - k_n x\right) + \varepsilon_n^{I,R-} \sin\left(n\omega t + \varphi_n^{I,R-} + k_n x\right),$$
(15.17)

$$u_{pl,n}^{I,R}(t) = \frac{\varepsilon_n^{I,R+}}{k_n} \cos\left(n\omega t + \varphi_n^{I,R+} - k_n x\right) - \frac{\varepsilon_n^{I,R-}}{k_n} \cos\left(n\omega t + \varphi_n^{I,R-} + k_n x\right),$$
(15.18)

for the strain ε_{pl} and displacement u_{pl}, respectively. $k_n = n\omega/v_L$ is the wave number of the nth harmonic of the insonified frequency, and v_L is the compressional sound velocity in the plate. The indices I and R indicate amplitudes and phases in the case of interferometric and piezoelectric receiver probe measurements, respectively. At $x = 0$ we have the boundary condition

$$\varepsilon_n^{I,R+} \sin\left(n\omega t + \varphi_n^{I,R+}\right) + \varepsilon_n^{I,R-} \sin\left(n\omega t + \varphi_n^{I,R-}\right) = \varepsilon_n \sin\left(n\omega t + \varphi_n\right). \quad (15.19)$$

During the interferometric measurements the backwall of the plate at $x = D$ (D is the plate thickness) is free and fulfills the boundary condition

$$\varepsilon_n^{I+} \sin\left(n\omega t + \varphi_n^{I+} - k_n D\right) + \varepsilon_n^{I-} \sin\left(n\omega t + \varphi_n^{I-} + k_n D\right) = 0. \quad (15.20)$$

Eqs. (15.17), (15.18), and (15.20) lead to

$$\varepsilon_n^{I+} = \varepsilon_n^{I-}, \quad \varphi_n^{I+} = \varphi_n^{I-} + \pi, \quad \text{and} \quad u_{pl}^{I}(x = D) = \frac{2\varepsilon_n^{I+}}{k_n}. \quad (15.21)$$

The displacement amplitude $u_{pl}^{I}(x = D)$ at the free backwall is measured interferometrically. Consequently, the two measurement series described above yield a relation between the strain amplitude ε_n^{I+} of the forward propagating wave in the case of a free sample backwall and the voltage this wave generates at the oil coupled receiver probe. The relation between the measured strain ε_n^{I+} and the strain ε_n from Eq. (15.16) results from Eqs. (15.19) to (15.21):

$$\varepsilon_n = 2\,|\sin\left(k_n D\right)|\,\varepsilon_n^{I+}. \quad (15.22)$$

In our case (two aluminum plates of 4 mm thickness joined together by a 30μm thick epoxy layer, fundamental frequency $f = 2.25\,\mathrm{MHz}$, sound velocity in aluminum $v_L = 6318.7\,\mathrm{m/s}$), the calibrated measured amplitudes ε_n^{I+} have to be multiplied by $2|\sin k_n D| = 0.915$, 1.628, and 1.979 for the transmitted fundamental frequency and its second and third harmonics, respectively, in order to yield the force amplitudes $F_n = C_{11}\varepsilon_n$ acting within the interface.

3.2 The Phases of the Transmitted Ultrasonic Waves

The measured signal in the experiments on the composite samples is the AC-voltage generated at the receiving probe. Its Fourier transformation yields the amplitude and phase spectra. The force amplitudes in the interface follow from the voltage amplitudes by the calibration measurements described in the preceding section. A relation between the phases of the waves of different frequencies at the transducer $\varphi_{t,n}$ to those of the force components in the bonded interface φ_n can be derived as follows.

When the transmitted wave is detected by a receiver probe, the backwall of the sample is no longer free. In the coupling layer of thickness d both forward and backward propagating waves are generated, whereas in the transducer only the former occur.

$$
\varepsilon_{c,n}(t) = \varepsilon_{c,n}^+ \sin\left(n\omega t + \varphi_{c,n}^+ - k_{c,n}(x-D)\right)
$$
$$
+ \varepsilon_{c,n}^- \sin\left(n\omega t + \varphi_{c,n}^- + k_{c,n}(x-D)\right), \tag{15.23}
$$

$$
u_{c,n}(t) = \frac{\varepsilon_{c,n}^+}{k_{c,n}} \cos\left(n\omega t + \varphi_{c,n}^+ - k_{c,n}(x-D)\right)
$$
$$
- \frac{\varepsilon_{c,n}^-}{k_{c,n}} \cos\left(n\omega t + \varphi_{c,n}^- + k_{c,n}(x-D)\right), \tag{15.24}
$$

$$
\varepsilon_{t,n}(t) = \varepsilon_{t,n} \sin\left(n\omega t + \varphi_{t,n} - k_{t,n}(x-D-d)\right), \tag{15.25}
$$

$$
u_{t,n}(t) = \frac{\varepsilon_{t,n}}{k_{t,n}} \cos\left(n\omega t + \varphi_{t,n} - k_{t,n}(x-D-d)\right) \tag{15.26}
$$

are the corresponding strains and displacements, the indices c and t stand for coupling medium and transducer, respectively. Instead of (15.20) we obtain the boundary conditions

$$
C_{11,Al}\lfloor \varepsilon_n^{R+} \sin\left(n\omega t + \varphi_n^{R+} - k_n D\right) + \varepsilon_n^{R-} \sin\left(n\omega t + \varphi_n^{R-} + k_n D\right)\rfloor
$$
$$
= C_{11,c}\lfloor \varepsilon_{c,n}^+ \sin\left(n\omega t + \varphi_{c,n}^+\right) + \varepsilon_{c,n}^- \sin\left(n\omega t + \varphi_{c,n}^-\right)\rfloor, \tag{15.27}
$$

$$
\frac{\varepsilon_n^{R+}}{k_n} \cos\left(n\omega t + \varphi_n^{R+} - k_n D\right) - \frac{\varepsilon_n^{R-}}{k_n} \cos\left(n\omega t + \varphi_n^{R-} + k_n D\right)
$$
$$
= \frac{\varepsilon_{c,n}^+}{k_{c,n}} \cos\left(n\omega t + \varphi_{c,n}^+\right) - \frac{\varepsilon_{c,n}^-}{k_{c,n}} \cos\left(n\omega t + \varphi_{c,n}^-\right), \tag{15.28}
$$

$$
C_{11,c}\lfloor \varepsilon_{c,n}^+ \sin\left(n\omega t + \varphi_{c,n}^+ - k_{c,n}d\right) + \varepsilon_{c,n}^- \sin\left(n\omega t + \varphi_{c,n}^- + k_{c,n}d\right)\rfloor
$$
$$
= C_{11,t}\varepsilon_{t,n} \sin\left(n\omega t + \varphi_{t,n}\right), \tag{15.29}
$$

$$
\frac{\varepsilon_{c,n}^+}{k_{c,n}} \sin\left(n\omega t + \varphi_{c,n}^+ - k_{c,n}d\right) - \frac{\varepsilon_{c,n}^-}{k_{c,n}} \sin\left(n\omega t + \varphi_{c,n}^- + k_{c,n}d\right) = \frac{\varepsilon_{t,n}}{k_{t,n}} \sin\left(n\omega t + \varphi_{t,n}\right).
$$
$$
\tag{15.30}
$$

The boundary condition (15.19) is, of course, still valid. From these equations a relation between the phases φ_n of the forces at the bonded interface and the phases $\varphi_{t,n}$ measured at the receiver can be derived:

$$\tan \varphi_n = \frac{\sin \left(\varphi_{c,n}^+ - \varphi_{nD}^{R+}\right) \sin \left(\varphi_{nD}^{R-} - k_n D\right) - A \sin \left(\varphi_{c,n}^+ - \varphi_{nD}^{R-}\right) \sin \left(\varphi_{nD}^{R+} + k_n D\right)}{\sin \left(\varphi_{c,n}^+ - \varphi_{nD}^{R+}\right) \cos \left(\varphi_{nD}^{R-} - k_n D\right) - A \sin \left(\varphi_{c,n}^+ - \varphi_{nD}^{R-}\right) \cos \left(\varphi_{nD}^{R+} + k_n D\right)},$$

(15.31)

$$\sin \left(\varphi_{c,n}^+ - \varphi_{nD}^{R\pm}\right) = \cos \left(k_{c,n} d\right) \frac{\tan \varphi_{t,n}^+ + \tan \left(k_{c,n} d\right) - \left(1 - \tan \varphi_{t,n}^+ \tan \left(k_{c,n} d\right)\right) \tan \varphi_{nD}^{R\pm}}{\sqrt{1 + \tan^2 \varphi_{t,n}^+} \sqrt{1 + \tan^2 \varphi_{nD}^{R\pm}}},$$

(15.32)

$$\sin \left(\varphi_{nD}^{R\pm} \pm k_n D\right) = \cos \left(k_n D\right) \frac{\tan \varphi_{nD}^{R\pm} \pm \tan \left(k_n D\right)}{\sqrt{1 + \tan^2 \varphi_{nD}^{R\pm}}},$$

(15.33)

$$\cos \left(\varphi_{nD}^{R\pm} \pm k_n D\right) = \cos \left(k_n D\right) \frac{1 \mp \tan \varphi_{nD}^{R\pm} \tan \left(k_n D\right)}{\sqrt{1 + \tan^2 \varphi_{nD}^{R\pm}}},$$

(15.34)

$$\tan \varphi_{nD}^{R\pm} = \frac{B^\pm \tan \varphi_{t,n} + \tan \left(k_{c,n} d\right)}{B^\pm - \tan \varphi_{t,n} \tan \left(k_{c,n} d\right)},$$

(15.35)

$$\varphi_{nD}^{R\pm} = \varphi_n^{R\pm} \mp k_n D,$$

(15.36)

$$A = \frac{I_{pl} - I_c}{I_{pl} + I_c}, \qquad B^\pm = \frac{\left(I_{pl} \pm I_t\right) I_c}{I_{pl} I_t + I_c^2}.$$

(15.37)

Here, $I_{pl} = \rho v_L$, $I_c = \rho_c v_{L,c}$, and $I_t = \rho_t v_{L,t}$ are the acoustic impedances, ρ, ρ_c, and ρ_t are the densities, and v_L, $v_{L,c}$, and $v_{L,t}$ are the compressional sound velocities of the aluminum substrate, the coupling medium, and the receiver, respectively. The results show that, in general, the phases measured at the receiver are not the same as those at the interface.

In some limit cases the Eqs. (15.31)–(15.37) simplify remarkably. If $B^\pm \ll \tan(k_{c,n} d)$, Eq. (15.35) becomes

$$\tan \varphi_{nD}^{R\pm} \approx \frac{1}{\tan \varphi_{t,n}},$$

(15.38)

and, with the additional assumption $I_c \ll I_{pl}$, Eq. (15.31) yields

$$\tan \varphi_n \approx \tan \varphi_{t,n}.$$

(15.39)

For very thin coupling layers ($d \approx 0$), we may get $B^\pm \gg \tan(k_{c,n} d)$ (depending on the numerical values of $B^\pm$), and Eq. (15.35) becomes

$$\tan \varphi_{nD}^{R\pm} \approx \tan \varphi_{t,n}.$$

(15.40)

Then, the additional assumption $I_c \ll I_{pl}$ leads to

$$\tan \varphi_n \approx -\frac{1}{\tan \varphi_{t,n}} = \tan \left(\varphi_{t,n} - \tfrac{\pi}{2}\right).$$

(15.41)

In our experiments, the thickness of the coupling layer (about $2\,\mu\text{m}$) yields $\tan(k_{c,n} d) = 0.0195, 0.0390$, and 0.0585 for the fundamental frequency and its second and third harmonics, respectively. The numerical evaluation was carried out with the density $\rho_c = 1.17\,\text{g/cm}^3$, sound velocity $v_{L,c} = 1450\,\text{m/s}$, and the resulting acoustic impedance $I_c = 1.697 \cdot 10^6\,\text{kg/m}^2\text{s}$ (material parameters of the oil). For the numerical evaluation of B^+ and B^- from Eq. (15.37), the material parameters of aluminum (density $\rho = 2.7\,\text{g/cm}^3$, compressional sound velocity $v_L = 6318.7\,\text{m/s}$, resulting acoustic impedance $I_{pl} = 17.06 \cdot 10^6\,\text{kg/m}^2\text{s}$) and the acoustic impedance of the receiver probe (ceramic) of $I_t \approx 31 \cdot 10^6\,\text{kg/m}^2\,\text{s}$ were used yielding $B^+ \approx 0.153$ and $B^- \approx -0.0449$; that is, neither of the considered limiting cases apply.

3.3 Calibrated Ultrasonic Input Strain Amplitudes

In order to calibrate absolutely the ultrasonic amplitudes incident onto the bonded interface, we perform similar measurements as described in Section 3.1 for a plate of the same material and thickness as the one at the intromission side of the composite sample under investigation. Again, the ultrasonic backwall vibration amplitudes are measured interferometrically. However, because only the fundamental frequency is insonified into the plate, measurements at higher harmonics are not necessary. The excitation voltage at the sending probe causes the sinusoidal force

$$F_i\,(t) = C_{11}\varepsilon_i \sin\,(\omega t) \tag{15.42}$$

at the intromission side of the plate. Here, ε_i is the strain amplitude generated at the surface of the plate. The phase of the force (15.42) is assumed to be zero, which can be obtained by a convenient tuning of the excitation voltage. The relation between the phase of the excitation voltage φ_t and the phase of the sinusoidal force at the intromission side of the plate (here equal to zero) can be derived from the conditions of continuous displacements and stresses at the boundaries between sending probe, coupling medium, and plate:

$$\tan \varphi_t = \frac{I_c}{I_t} \tan\,(k_c d)\,. \tag{15.43}$$

I_c and I_t are the acoustic impedances of the coupling medium and sending probe, respectively; $k_c = \omega/v_{L,c}$ is the wave number, $v_{L,c}$ is the sound velocity in the coupling layer, and d is its thickness. For the same data as before, we get $\tan \varphi_t = 0.00107$; that is, the phase of the excitation voltage is in a very good approximation equal to the phase of the exciting force at the intromission side of the sample.

Due to ultrasonic reflections at both surfaces of the plate, the excitation force $F_i(t)$ causes forward and backward propagating waves in the plate with strain and displacement

$$\varepsilon_i^{I,R}\,(x,t) = \varepsilon_i^{I,R+} \sin\,\left(\omega t + \varphi_i^{I,R+} - kx\right) + \varepsilon_i^{I,R-} \sin\,\left(\omega t + \varphi_i^{I,R-} + kx\right), \tag{15.44}$$

$$u_i^{I,R}\,(x,t) = \frac{\varepsilon_i^{I,R+}}{k} \cos\,\left(\omega t + \varphi_i^{I,R+} - kx\right) - \frac{\varepsilon_i^{I,R-}}{k} \cos\,\left(\omega t + \varphi_i^{I,R-} + kx\right). \tag{15.45}$$

$\varepsilon_i^{I,R+}, \varphi_i^{I,R+}$ and $\varepsilon_i^{I,R-}, \varphi_i^{I,R-}$ are the strain amplitudes and phases of the forward and backward propagating waves, respectively. At the intromission surface, we have

$$\varepsilon_i^{I,R+} \sin\left(\omega t + \varphi_i^{I,R+}\right) + \varepsilon_i^{I,R-} \sin\left(\omega t + \varphi_i^{I,R-}\right) = \varepsilon_i \sin \omega t. \tag{15.46}$$

During the calibration measurements, the boundary condition at the free backwall is

$$\varepsilon_i^{I+} \sin\left(\omega t + \varphi_i^{I+} - kD_i\right) + \varepsilon_i^{I-} \sin\left(\omega t + \varphi_i^{I-} + kD_i\right) = 0, \tag{15.47}$$

where D_i is the thickness of the plate. In the composite sample, the boundary condition at the backwall of the intromission plate is determined by the forces at the interface; that is,

$$\varepsilon_i^{R+} \sin\left(\omega t + \varphi_i^{R+} - kD_i\right) + \varepsilon_i^{R-} \sin\left(\omega t + \varphi_i^{R-} + kD_i\right) = \varepsilon_1 \sin(\omega t + \varphi) \tag{15.48}$$

for the excitation frequency. In general, the phase φ in Eq. (15.48) is not equal to the phase φ_1 defined in Eq. (15.31), because φ is related to the phases φ_i^{R+} and φ_i^{R-} of the incoming wave, and φ_1 is related to the phases $\varphi_{t,n}^+$ of the transmitted waves measured at the receiver probe. However, by tuning the phase of the excitation voltage we can adjust the phase of the excitation force, so that it is the same in all the measurements. Then $\varphi = \varphi_1$. In the following we consider this case.

The interferometric calibration measurements yield the amplitude of the free plate backwall vibration, that is, the amplitude $u_i^I(D_i, t)$, and consequently a relation between the strain amplitude ε_i^{I+} and the excitation voltage at the sending probe. Eqs. (15.46) and (15.47) relate the amplitudes and phases of the calibration signal and of the signal at the intromission surface of the plate,

$$2\,|\sin(kD_i)|\,\varepsilon_i^{I+} = \varepsilon_i, \tag{15.49}$$

$$\tan \varphi_i^{I+} = -\cot(kD_i) = \tan\left(kD_i + \tfrac{\pi}{2}\right). \tag{15.50}$$

In the experiments on the composite sample, Eqs. (15.46) and (15.48) lead to

$$4\sin^2(kD_i)\left(\varepsilon_i^{R+}\right)^2 = \varepsilon_1^2 + \varepsilon_i^2 - 2\varepsilon_1\varepsilon_i\cos(\varphi_1 - kD_i), \tag{15.51}$$

$$\tan\left(\varphi_i^{R+}\right) = -\frac{\varepsilon_1\cos\varphi_1 - \varepsilon_1\varepsilon_i\cos(kD_i)}{\varepsilon_1\sin\varphi_1 - \varepsilon_1\varepsilon_i\sin(kD_i)}. \tag{15.52}$$

In the limiting case of very weak bonding, the backwall of the intromission plate is almost free, because the amplitude of the transmitted wave ε_1 is small compared to the amplitude ε_i generated at the sample surface. Then, approximately, the strain amplitude of the wave incident onto the bonded interface is equal to the amplitude measured by the interferometric calibration, and the phase of the transmitted wave is given by the plate thickness,

$$\varepsilon_I = \varepsilon_i^{R+} \approx \varepsilon_i^{I+}, \quad \tan \varphi_1 = \tan(kD_i). \tag{15.53}$$

For a strong joint, when the insonified ultrasonic wave is transmitted almost completely through the bonded interface, that is, the amplitude ε_i^{R-} of the reflected beam is small compared to the amplitude ε_i^{R+} incident onto the bonded interface, we get approximately

$$\varepsilon_I = \varepsilon_i^{R+} \approx \varepsilon_1 \approx \varepsilon_i = 2 \left| \sin \left(k D_i \right) \right| \varepsilon_i^{I+}, \quad \tan \varphi_1 = - \tan \left(k D_i \right) . \tag{15.54}$$

In our measurements, neither of the limiting cases was a good approximation, hence the amplitudes ε_i^{R+} are determined by Eq. (15.51). We define α as the ratio of the transmitted amplitude ε_1 and the calibrated measured amplitude ε_i^{I+}, and β as the ratio of the strain amplitude incident onto the bonded interface in the experiment $\varepsilon_I = \varepsilon_i^{R+}$ and the calibrated measured amplitude ε_i^{I+},

$$\varepsilon_1 = \alpha \varepsilon_i^{I+}, \quad \varepsilon_I = \varepsilon_i^{R+} = \beta \varepsilon_i^{I+} . \tag{15.55}$$

Equation (15.51) yields

$$\beta^2 = 1 + \frac{\alpha^2}{4 \sin^2 \left(k D_i \right)} - \frac{\alpha}{\left| \sin \left(k D_i \right) \right| \cos \left(\varphi_1 - k D_i \right)} . \tag{15.56}$$

The ratio α can be calculated from the experimental data. Equations (15.39) for $\alpha = 0$ is the limit (15.53); that is, the amplitudes measured in the calibration are approximately equal to the amplitudes incident onto the bonded interface in the sample ($\beta = 1$). For $\alpha = 2 \left| \sin(k D_i) \right|$, the limit (15.54) is obtained; that is, $\beta = 2 \left| \sin(k D_i) \right|$, which leads to $\varepsilon_I = \varepsilon_i^{R+} \approx \varepsilon_1 \approx \varepsilon_i$. Equations (15.39) allow the determination of the ratio β from measured data.

3.4 The Phase of the Interface Vibration

The interface vibration is sinusoidal if interface distortions by generated higher harmonics and the static part are neglected. In general, the interface vibration is phase-shifted relative to the excitation (15.30):

$$\Delta a \left(t \right) = a \left(t \right) - a_S = a_S \varepsilon_{BI} \sin \left(\omega t + \varphi_0 \right) . \tag{15.57}$$

In addition, a phase-shift between the interface vibration and the transmitted ultrasonic wave of fundamental frequency may occur, which results in hysteretic interaction forces even in the linear case (e.g., caused by damping). The interface modulation $\Delta a(t)$ is the difference of the ultrasonic displacements of the two surfaces of the bonded aluminum plates

$$\Delta a \left(t \right) = u_{pl,1}^R \left(x = 0, t \right) - u_i^R \left(x = D_i, t \right) . \tag{15.58}$$

The displacements $u_{pl,1}^R \left(x = 0, t \right)$ and $u_i^R \left(x = D_i, t \right)$ follow from Eqs. (15.18) and (15.45), respectively. Using the boundary conditions for the strains at the intromission surface (15.46) and at the interface (15.19) and (15.48) as well as Eq. (15.35), the displacements $u_i^R \left(x = D_i, t \right)$ and $u_{pl,1}^R \left(x = 0, t \right)$ are expressed in terms of measured data from the calibration and/or experiments by the equations

$$u_i^R \left(x = D_i, t \right) = \frac{\varepsilon_i - \varepsilon_1 \cos \varphi_1 \cos \left(k D_i \right) \sin \omega t - \varepsilon_1 \sin \varphi_1 \cos \left(k D_i \right) \cos \omega t}{k \sin \left(k D_i \right)},$$
$$(15.59)$$

$$u_{pl,1}^R \left(x = 0, t \right) = \frac{\varepsilon_1^{R+}}{k} \cos \left(\omega t + \varphi_1^{R+} \right) - \frac{\varepsilon_1^{R-}}{k} \cos \left(\omega t + \varphi_1^{R-} \right)$$
$$= \frac{\varepsilon_1^{R+} \cos \varphi_1^{R+} - \varepsilon_1^{R-} \cos \varphi_1^{R-}}{k} \cos \omega t - \frac{\varepsilon_1^{R+} \sin \varphi_1^{R+} - \varepsilon_1^{R-} \sin \varphi_1^{R-}}{k} \sin \omega t,$$
$$(15.60)$$

$$\varepsilon_1^{R+} \cos \varphi_1^{R+} - \varepsilon_1^{R-} \cos \varphi_1^{R-} = \frac{\tan \varphi_1^{R-} + \tan \varphi_1^{R+} - 2 \tan \varphi_1}{\tan \varphi_1^{R-} - \tan \varphi_1^{R+}} \varepsilon_1 \cos \varphi_1, \quad (15.61)$$

$$\varepsilon_1^{R+} \sin \varphi_1^{R+} - \varepsilon_1^{R-} \sin \varphi_1^{R-} = \frac{\cot \varphi_1^{R-} + \cot \varphi_1^{R+} - 2 \cot \varphi_1}{\cot \varphi_1^{R-} - \cot \varphi_1^{R+}} \varepsilon_1 \sin \varphi_1. \quad (15.62)$$

Equations (15.58)–(15.62) yield the interface vibration $\Delta a(t)$ in the form

$$\Delta a \left(t \right) = C_1 \cos \omega t + C_2 \sin \omega t. \quad (15.63)$$

The constants C_1 and C_2 can be calculated from the measurement data. Comparison of Eqs. (15.63) and (15.57) yields a relation of these constants to the phase of the interface vibration, which no longer contains the unknown static equilibrium interface distance a_s and the measure ε_{BI} of the strain amplitude in the interface during vibration:

$$a_s \varepsilon_{BI} \cos \varphi_0 = C_2 \quad \text{and} \quad a_s \varepsilon_{BI} \sin \varphi_0 = C_1 \Rightarrow \tan \varphi_0 = \frac{C_1}{C_2}. \quad (15.64)$$

This relation allows us to relate the phase of the interface vibration in ultrasonic transmission experiments to the phase of the excitation voltage, and thus, with the results from Section 3.2, the phase shifts of the transmitted ultrasonic waves compared to the interface vibration $\varphi_n - n\varphi_0, n = 1, 2, 3, \cdots$. The important role of these phases is discussed in Section 3.2.

The calibration and evaluation procedure of the experimental data allows for a simple simulation of the experiments, such as the one described in the following section. The numerical simulations do not have to take into account the finite thickness of the two plates and, as a consequence, forward and backward propagating waves.

4.   A One-Dimensional Simulation Model

Due to its symmetry in the direction normal to the wave propagation, the geometry of the experiment suggests the application of a 1-D model for its numerical simulation. The 1-D approach is applicable if we focus on the interstice deformation in an area at the center of the specimen, whose transversal section is much smaller than the specimen length. Because we simulate linear elastic wave propagation except for a very localized nonlinearity (the interstice) we use the Local Interaction Simulation Approach (LISA).[13] Its local character makes it particularly suitable for our experiment. We only have to replace the linear relation between stress and strain at the bonding by a nonlinear one.

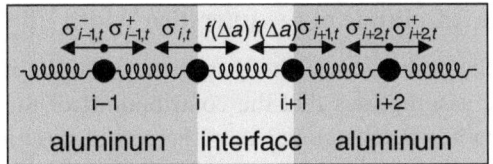

Fig. 15.1. Discretization of the specimen for the LISA simulations.

4.1 Derivation of the Iteration Equations for Numerical Simulations

The LISA code is similar to the formalism of Finite Difference Equations (FDE). The specimen is divided into a number of equally sized homogeneous cells (of length δ and cross-section α), whose masses are concentrated in a node, labeled with an index j. The bonding is located at $j = i$ (see Figure 15.1). The elastic behavior of the material is described by springs connecting neighboring nodes. The equation of motion for a node j is given by

$$m_j \ddot{u}_j = \sigma_j^+ - \sigma_j^-, \tag{15.65}$$

where $m_j = \rho_j \delta$ is the mass of the material cell per its cross-section α, u_j is its displacement from equilibrium, σ_j^- and σ_j^+ are the stresses in the contiguous cells to the left and to the right of node j, respectively. According to Hooke's law the stresses within each aluminum plate are

$$\sigma_j^\pm = C_{11} \left. \frac{\partial u_j}{\partial x} \right|_\pm . \tag{15.66}$$

The x-derivative of u_j is calculated within the material cell to the left $(-)$ or to the right $(+)$ of the node. Using a finite difference scheme, the discretized equation of motion becomes

$$\rho_j \delta \ddot{u}_j (t) = \rho_j \delta \frac{u_{j,t+1} - 2u_{j,t} + u_{j,t-1}}{\tau^2} = -\frac{C_{11}}{\delta} \left(u_{j,t} - u_{j-1,t} \right)$$

$$+ \frac{C_{11}}{\delta} \left(u_{j+1,t} - u_{j,t} \right) = \sigma_j^+ - \sigma_j^-. \tag{15.67}$$

Within the adhesive layer (i.e., between the nodes i and $i + 1$), the linear stress–strain relation is no longer valid. Therefore we replace the linear stresses between the nodes i and $i + 1$ by a nonlinear function $f(\Delta a_t)$, where $\Delta a_t = u_{i+1,t} - u_{i,t}$ is the deviation of the interface width from its static equilibrium at the time t. Under the assumptions that the contribution of higher harmonics to the interface oscillation is negligible [Eq. (15.57)], that the forces are nonlinear but not hysteretic, and that no conditioning occurs, the interstice stress as a function of $\Delta a / a_0$ is given by Eq. (15.13). The amplitude of the interstice width oscillation a_0 depends on the strain amplitudes and phases of the injected, reflected, and transmitted wave. As discussed below, these assumptions are valid up to a certain threshold of the excitation. A further increase of the input amplitude requires explicitly taking into account the phases of the transmitted waves and using the interstice stress function given by Eq. (15.12) for numerical simulations of the experiments.

4.2 Derivation of the Amplitude of the Interstice Width Vibration

In order to simplify the derivation of the amplitude a_0 of the interstice width vibration we again use the assumptions that the contributions of the higher harmonics to the interstice vibration are much smaller than the one of the fundamental frequency. Thus, the interstice vibration can be considered as almost sinusoidal. We express the incoming u_i, reflected u_r, and transmitted u_t wave as monochromatic waves of the input frequency (2.25 MHz in our experiment) with phase angles relative to the phase ϕ_0 of the interstice vibration. The origin $x = 0$ is set to the center of the interstice. Hence, for finite interstice width, the displacement oscillation at the interstice borders are phase shifted by $\frac{1}{2}k_i a_s$ with respect to ϕ_0. k_i is the wave number correspondent to the input frequency in the adhesive, and a_s is the static equilibrium width of the interstice. The obtained wave equations are

$$u_i(x, t) = a_i \cos\left(\omega t - k\left(x + \tfrac{1}{2}a_s\right) + \varphi_0 + \varphi_i\right), \quad \text{for } x \leq -\tfrac{1}{2}a_s, \quad (15.68a)$$

$$u_r(x, t) = a_r \cos\left(\omega t + k\left(x + \tfrac{1}{2}a_s\right) + \varphi_0 + \varphi_r\right), \quad \text{for } x \leq -\tfrac{1}{2}a_s, \quad (15.68b)$$

$$u_t(x, t) = a_t \cos\left(\omega t - k\left(x - \tfrac{1}{2}a_s\right) + \varphi_0 + \varphi_t\right), \quad \text{for } x \geq +\tfrac{1}{2}a_s, \quad (15.68c)$$

where k is the wave number corresponding to the input frequency in the sample material. The stress waves within the sample material are obtained from the displacement waves by deriving by x and multiplying with the elastic constant C_{11}. The stress within the interstice $F(t)$ is assumed sinusoidal as well, but may have another phase

$$F(t) = F_1 \sin\left(\omega t - k_i x + \varphi_0 + \varphi_1\right), \quad \text{for} -\tfrac{1}{2}a_s \leq x \leq +\tfrac{1}{2}a_s. \quad (15.69)$$

Because the stresses must be continuous at the boundaries between the aluminum slab and the bonding, we obtain for the right border ($x = \tfrac{1}{2}a_s$),

$$C_{11}k a_t \sin(\omega t + \varphi_0 + \varphi_t) = F_1 \sin\left(\omega t - \tfrac{1}{2}k_i a_s + \varphi_0 + \varphi_1\right), \quad (15.70)$$

which yields

$$F_1 = C_{11}k a_t, \quad (15.71)$$

$$\varphi_t = \varphi_1 - \tfrac{1}{2}k_i a_s. \quad (15.72)$$

The continuity of stresses on the left border ($x = -\tfrac{1}{2}a_s$) yields

$$C_{11}k a_i \sin(\omega t + \varphi_0 + \varphi_i) - C_{11}k a_r \sin(\omega t + \varphi_0 + \varphi_r)$$
$$= F_1 \sin\left(\omega t + \tfrac{1}{2}k_i a_s + \varphi_0 + \varphi_1\right), \quad (15.73)$$

and thus after time derivation and using Eqs (15.71) and (15.72)

$$a_i \cos(\omega t + \varphi_0 + \varphi_i) - a_r \cos(\omega t + \varphi_0 + \varphi_r) = a_t \cos\left(\omega t + \varphi_0 + \varphi_1 + \tfrac{1}{2}k_i a_s\right). \quad (15.74)$$

The deformation of the interstice is then calculated as the difference between the displacements at the right border and at the left border of the bonding

$$\Delta a\left(t\right) = a_t \cos\left(\omega t + \varphi_0 + \varphi_t\right) + a_t \cos\left(\omega t + \varphi_0 + \varphi_1 + \tfrac{1}{2}k_i a_S\right)$$
$$-2a_i \cos\left(\omega t + \varphi_0 + \varphi_i\right) = a_t \cos\left(\tfrac{1}{2}k_i a_S\right)\cos\left(\omega t + \varphi_0 + \varphi_1\right)$$
$$-2a_i \cos\left(\omega t + \varphi_0 + \varphi_i\right). \tag{15.75}$$

Using Eq. (15.57) and the orthogonality of sine and cosine we obtain

$$0 = a_0 \cos\left(-\tfrac{\pi}{2}\right) = 2a_t \cos\left(\tfrac{1}{2}k_i a_S\right)\cos\left(\varphi_1\right) - 2a_i \cos\left(\varphi_i\right) \tag{15.76}$$
$$-a_0 = a_0 \sin\left(-\tfrac{\pi}{2}\right) = 2a_t \cos\left(\tfrac{1}{2}k_i a_S\right)\sin\left(\varphi_1\right) - 2a_i \sin\left(\varphi_i\right). \tag{15.77}$$

The amplitude of the interstice vibration is then given by

$$a_0 = 2\sqrt{a_i^2 - a_t^2 \cos^2\left(\tfrac{1}{2}k_i a_S\right)\cos^2\left(\varphi_1\right)} - 2a_t \cos\left(\tfrac{1}{2}k_i a_S\right)\sin\left(\varphi_1\right), \tag{15.78}$$

or expressed in terms of the strain amplitudes $\varepsilon_I = a_i/k$ and $\varepsilon_1 = a_t/k$ in the sample material

$$a_0 = \frac{2}{k}\left(\sqrt{\varepsilon_I^2 - \varepsilon_1^2 \cos^2\left(\tfrac{1}{2}k_i a_S\right)\cos^2\left(\varphi_1\right)} - \varepsilon_1 \cos\left(\tfrac{1}{2}k_i a_S\right)\sin\left(\varphi_1\right)\right). \tag{15.79}$$

For very thin interstices ($\tfrac{1}{2}k_i a_S \ll 1$), Eq. (15.79) simplifies to

$$a_0 = \frac{2}{k}\left(\sqrt{\varepsilon_I^2 - \varepsilon_1^2 \cos^2\left(\varphi_1\right)} - \varepsilon_1 \sin\left(\varphi_1\right)\right). \tag{15.80}$$

If there is no phase shift between the interstice deformation and the resulting stress and vice versa (i. e., instantaneous reaction of the material on applied forces), a_0 is calculated by

$$a_0 = \frac{2}{k}\sqrt{\varepsilon_I^2 - \varepsilon_1^2 \cos^2\left(\tfrac{1}{2}k_i a_S\right)} \xrightarrow{a_S \to 0} \frac{2}{k}\sqrt{\varepsilon_I^2 - \varepsilon_1^2}. \tag{15.81}$$

This formula is a good approximation to experiments in which no linear hysteresis occurs. The results presented in the following section are obtained using a_0 of Eq. (15.81).

It is interesting to check the limit when the sample does not contain a bonded interface, that is, when this cell is equal to all other cells. Then, the complete incident wave is transmitted, and we obtain the displacement amplitude in aluminum. In fact, in this case it is

$$a_0 = \frac{2\varepsilon_I}{k}\sin\left(\tfrac{1}{2}k_i a_S\right) \overset{k_i a_S \ll 2}{\approx} \varepsilon_I a_S. \tag{15.82}$$

4.3 Results of the LISA Simulations

The calibrated measurements have been carried out for a sample consisting of two aluminum plates of 4 mm thickness bound together by an epoxy layer of less than 0.03 mm thickness.[11, 12] The calibration procedure transforms the measured data from the receiver and the input transducer into the strain amplitudes and the phases of the fundamental and higher harmonics of the incoming and transmitted waves. Waves reflected from the borders of the aluminum slabs are eliminated. Therefore, in our LISA simulations we assume infinite aluminum slabs at both sides of the bonding. This is realized using a Perfectly Matching Layer (PML) of absorbing boundaries at both ends of the specimen.[16] We choose the discretization step δ equal to the static equilibrium width of the bonding a_s (approximately 0.03 mm). A monochromatic compressional wave with a frequency of 2.25 MHz is injected at the left end of the specimen. Its wavelength in aluminum is about 100 times the discretization step, ensuring a sufficiently smooth description of the wave and that the effects due to finite propagation time of the wave through the interstice are negligible.

The values F_1, F_2, and F_3 in Eq. (15.13) are derived from the experimental data at the largest forcing amplitude for which hysteresis is negligible ($\varepsilon_{in} = 109 \cdot 10^{-6}$ for the first and $\varepsilon_{in} = 130 \cdot 10^{-6}$ for the second sample)[10–12], that is, in both cases from one measuring point only. We calculate numerically the transmitted wave for input strain amplitudes below and above the reference measuring point. Figure 15.2 shows the numerical results as well as the experimental data[10] of the first, second, and third harmonic as a function of the input strain. Both simulations yield a very good agreement up to the threshold when hysteresis sets in. For larger excitation amplitudes the second and third harmonics show large discrepancies, suggesting that the aforementioned assumptions are no longer valid. Nonclassical effects, such as hysteresis of the higher harmonics, which are not included in the model, are responsible for these differences. Because these effects generally occur in the presence of defects in the bonding, the discrepancies may give a measure of the bonding quality. The agreement between theoretical and experimental results up to the nonclassicity threshold is a good indicator of the quality of the calibration procedure and shows that the measured nonlinearity actually stems from the adhesive interface. In fact, simulations performed from uncalibrated measurements (data not shown for brevity) yield a much poorer agreement with the experimental data.

5. Discussion

Our measurements were carried out for samples of two aluminum plates joined together by thin adhesive epoxy layers, which are typical bonds in the aircraft industry. They are good examples of our theoretical description of thin bonded interfaces by interaction forces without taking into account explicitly the material properties of the adhesive. The ultrasonic waves transmitted through the bonded interface show a threshold behavior. Up to a certain point, their amplitudes depend on the excitation following the power series expansion of a quasistatic interaction force curve (Figure 15.2), and the phases vary little (within the measurement accuracy).[11, 12] Then, another regime sets in. The amplitudes of the second and third harmonic grow stronger than the second and third power of the excitation strain, respectively, and the phases of the transmitted waves change dramatically.[11, 12] From a slightly higher threshold on ($\varepsilon_{in} = 160 \cdot 10^{-6}$ in both samples), the amplitude of the transmitted fundamental

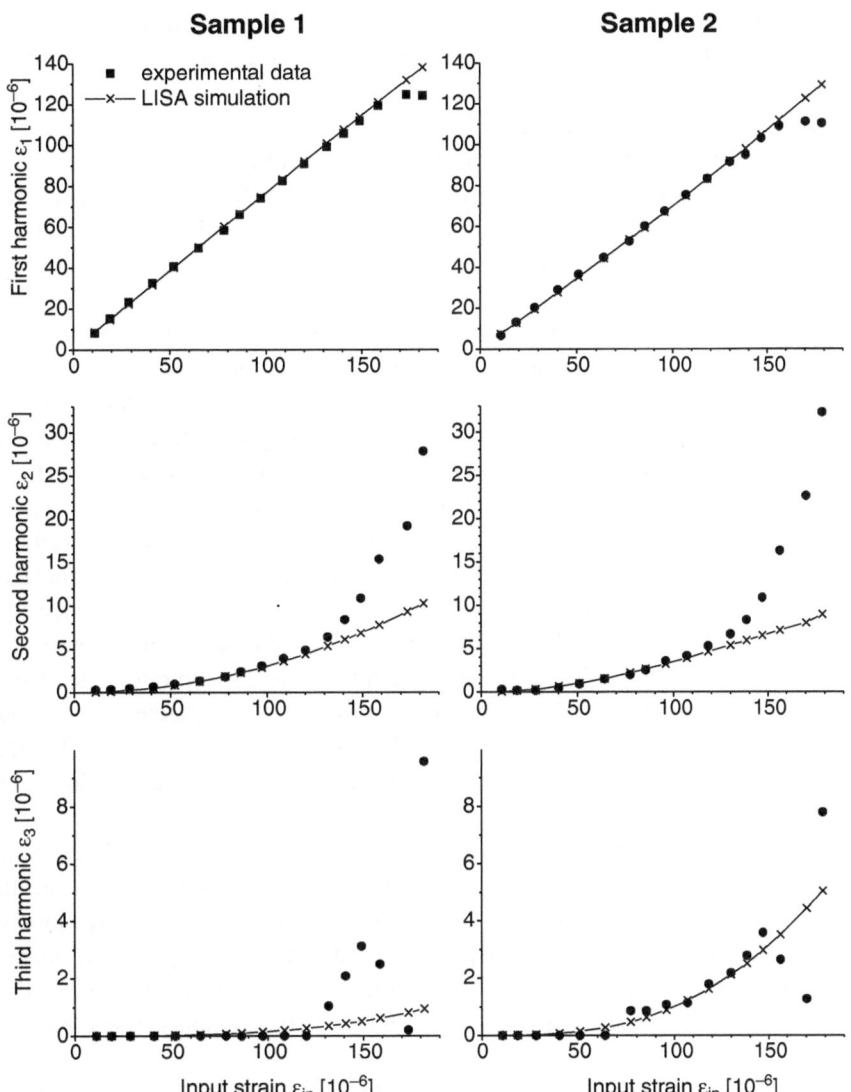

Fig. 15.2. Experimental and numerical strain amplitudes of the first three harmonics of the transmitted wave.

frequency (Figure 15.2) as a function of the excitation amplitude changes from linear to a smaller slope. Conversely, the increase of the strain amplitude in the interface becomes larger than linear. These thresholds, which are characteristic quantities of the bonding and might be used for its quality evaluation, indicate a complete change in its dynamic behavior. Above these thresholds, the approximation "no hysteresis and no conditioning" is no longer valid. The evaluation of the transmitted waves to yield the restoring forces in the interface and the simulation of the experiments have to take into account explicitly the phases of the waves. The measurement data at the first threshold yield an interface vibration amplitude of $a_0 \approx 0.2\,\mu$m. With an interface thickness of about 30μm, this corresponds to a strain of about 10^{-2}. At these strain values the transition from elastic to viscoelastic/plastic behavior of epoxy in quasistatic tension experiments occurs.[1]

Contrary to the experimental and simulation results presented here, in several papers[17–20] a quadratic behavior of the third harmonic with the increase of the excitation amplitude was considered. This behavior refers to an effect which is evidently not dominant in our experiments. Equation (8) in Reference[17] and Eq. (1) in Reference[18] clearly show that in the cited publications and related work a theory is developed to describe wave propagation in overall nonlinear bulk materials. This theory does not describe a localized nonlinearity in a linear bulk material as in the case of two components of linear elastic materials bound together by a more or less weak adhesive interface. References[19] and[20] describe resonance experiments in a rock bar. For this purpose, the theory of References[17] and[18] is applied to the one-dimensional resonance case with standing waves. This allows the assumption that the dominant terms in the equation of motion are those that feed back into the driving frequency.[19] Therefore, in a first-order approximation, terms proportional to β (the second-order nonlinearity) and $\pm\alpha\partial u/\partial x$ (hysteresis) are negligible. With these assumptions, the quadratic amplitude dependence of the third harmonic is predicted [Eq. (3) in Reference[20]]. That is, the quadratic behavior of the third harmonic is caused by end point memory, which corresponds to the contribution to the strain amplitude and strain rate dependent modulus [Eq. (1b) of Reference[20]] that remains if hysteresis is neglected. The numerical results plotted in Figure 15 of Reference[18] show for strains larger than 10^{-7} that the quadratic behavior of the third harmonic is not valid in general, even not in the case of bulk nonlinearity.

So far, phase measurements have not been very accurate due to a large signal-to-noise ratio. Nevertheless, the results[11,12] show that, at least below a certain threshold, hysteresis is only a small effect in our experiments. This was also observed in experiments on rocks. The numerically determined PM-space densities in Barea sandstone show that the hysteretic mechanical elements (those off the diagonal in the PM-space) have a much lower density than the nonhysteretic elements (those on the diagonal).[21,22]

Samples like ours were investigated with a different method by the research groups of Prof. S. I. Rokhlin, Ohio State University, Columbus, Ohio, U.S.A., and Prof. L. Adler, Adler Consultants, Columbus, Ohio, U.S.A. Their ultrasonic spectroscopy does not determine binding forces, but the effective interface stiffness, which in our method corresponds to the slope of the interaction force curve if no hysteresis occurs. Recently, nonlinearity was included into ultrasonic spectroscopy by means of a pressure dependent interface stiffness.[23]

The relation between the amplitudes and phases of the transmitted ultrasonic waves and the forces acting in the interface, which we use for the evaluation of our experimental data, is only valid for thin bonded interfaces, which do not require a description by an adhesive layer of finite thickness with all its material properties. Thick bonded interfaces can be modeled by interaction forces between the surfaces of the two plates and the surfaces of the adhesive layer, whereas the latter is described as a more or less viscous medium. The description may become rather complicated and may entail difficulties when trying to extract parameters of interest in NDE problems from measured transmitted or reflected ultrasonic waves. In this case as well, LISA simulations may provide an important tool for the validation of such an extended model and in the development of practical application methods and techniques.

Acknowledgments

This work has been supported by the ESF-PESC program NATEMIS, the EC project FP6-502927 AERONEWS, the INFM Parallel Computing Initiative, and MURST (prot. 2002022571, Italy). The authors wish to thank the CINECA supercomputing laboratory (Bologna, Italy) and the Bioindustry Park Canavese for providing its multiprocessing Linux cluster, and W. Arnold, A. Koka, and M. Scalerandi for fruitful discussions.

References

[1] See for example: G. Habenicht, *Kleben*, 3rd edition (Springer, 1997).

[2] D.C. Hurley and C.M. Fortunko, Determination of the nonlinear ultrasonic parameter ß using a Michelson interferometer, *Meas. Sci. Technol.* **8**, 634–542 (1997).

[3] S.U. Faßbender and W. Arnold, Measurement of adhesion strength of bonds using nonlinear acoustics, in: *Rev. Progr. QNDE 1995* **15**, edited by D.O. Thompson and D.E. Chimenti, Plenum Press, New York, 1321–1328 (1996).

[4] O. Buck, W.L. Morris, and J.M. Richardson, Acoustic harmonic generation at unbonded interfaces and fatigue cracks, *Appl. Phys. Lett.* **33**(5), 371–373 (1978).

[5] M. Rothenfußer, M. Mayr, and J. Baumann, Acoustic nonlinearities in adhesive joints, *Ultrasonics* **38**, 332–326 (2000).

[6] C. Bockenheimer, D. Fata, W. Possart, M. Rothenfußer, U. Netzelmann, and H. Schäfer, The method of non-linear ultrasound as a tool for the non-destructive inspection of structural epoxy - metal bonds – a résumé, *International Journal of Adhesion & Adhesives* **22**, 227–233 (2002).

[7] e.g., S. Hirsekorn, A. Koka, A. Wegner, and W. Arnold, Quality assessment of bond interfaces by nonlinear ultrasonic transmission, in: *Rev. Progr. QNDE 1999* **19B**, edited by D.O. Thompson and D.E. Chimenti, Plenum Press, New York, 1367–1374 (2000).

[8] A. Wegner, A. Koka, J. Janser, U. Netzelmann, S. Hirsekorn, and W. Arnold, Assessment of the adhesion quality of fusion-welded silicon wafers with nonlinear ultrasound, *Ultrasonics* **38**, 316–321 (2000).

[9] S. Hirsekorn, Nonlinear transfer of ultrasound by adhesive joints - a theoretical description, *Ultrasonics* **39**, 57–68 (2001).

[10] P.P. Delsanto, C. Camagna, M. Hiresekorn, and S. Hirsekorn, Nonlinear Effects in the Propagation of Ultrasonic Waves Through Thin Interfaces, 2nd Workshop "NDT in Progress", Prague, Czech Republic, Oct. 6–8, 2003, ISBN 80-214-2475-3, 69–74 (2003).

[11] S. Hirsekorn, A. Koka, and W. Arnold, Theoretical and Experimental Investigations of Interaction Forces in Thin Bonded Interfaces by Nonlinear Ultrasonic Transmission, 2nd Workshop "NDT in Progress", Prague, Czech Republic, Oct. 6–8, 2003, ISBN 80-214-2475-3, 99–106 (2003).

[12] S. Hirsekorn, A. Koka, S. Kurzenhäuser, and W. Arnold, Calibration and evaluation of nonlinear ultrasonic transmission measurements on thin bonded interfaces, in: *Adhesion – Current Research and Applications*, edited by Wulff Possart, WILEY-VCH, Weinheim, Berlin, 403–419 (2005).

[13] P.P. Delsanto, R.B. Mignogna, M. Scalerandi, and R.S. Schechter, Simulation of ultrasonic pulses propagation in complex media, in: *New Perspectives on Problem in Classical and Quantum Physics*, Vol. 2, edited by P.P. Delsanto and A.W. Saenz, Gordon and Breach, New Delhi, 51–74 (1998).

[14] S. Hirsekorn and P.P Delsanto, On the universality of nonclassical nonlinear phenomena and their classification, *Appl. Phys. Lett.* Vol. **84**, Issue 8, 1413–1415 (2004).

[15] P.P. Delsanto and S. Hirsekorn, A Unified Treatment of Nonclassical Nonlinear Effects in the Propagation of Ultrasound in Heterogeneous Media, *Ultrasonics* **42**, 1005–1010 (2004).

[16] A.J. Safjan, Highly accurate non-reflecting boundary conditions for finite element simulations of transient acoustics problems, *Comput. Methods Appl. Mech. Engrg.* **152**, 175–193 (1998).

[17] K.R. McCall and R.A. Guyer, Equation of state and wave propagation in hysteretic nonlinear elastic materials, *J. Geophys. Res.* **99**, 23887–23897 (1994).

[18] K.E.A. Van Den Abeele, P.A. Johnson, R.A. Guyer, and K.R. McCall, On the quasi-analytic treatment of hysteretic nonlinear response in elastic wave propagation, *J. Acoust. Soc. Am.* **101**, 1885–1898 (1997).

[19] R.A. Guyer, K.R. Mccall, and K.E.A. Van Den Abeele, Slow elastic dynamics in a resonant bar of rock, *Geophys. Res. Lett.* **25**, 1585–1588 (1998).

[20] K.E.A. Van Den Abeele, A. Sutin, J. Carmeliet, and P.A. Johnson, Micro-damage diagnostics using nonlinear elastic wave spectroscopy, *NDT&E International*, **34** (4), 239–248 (2001).

[21] R.A. Guyer, K.R. McCall, and G.N. Boinott, Hysteresis, discrete memory, and nonlinear wave propagation in rock, *Phys. Rev. Lett.* **74**(17), 3491–3494 (1995).

[22] R.A. Guyer, K.R. McCall, G.N. Boinott, L.B. Hilbert Jr, and T.J. Plona, Quantitative use of Preisach-Mayergoyz space to find static and dynamic elastic moduli in rock, *J. Geophys. Res.* **102**(B3), 5281–5293 (1997).

[23] S.I. Rokhlin, L. Wang, B. Xie, V.A. Yakovlev, and L. Adler, Modulated angle beam ultrasonic spectroscopy for evaluation of imperfect interfaces and adhesive bonds, *Ultrasonics* **42**, 1037–1047 (2004).

16

A LISA Model of the Nonlinear and Hysteretic Response of Interstitial Regions to Applied Stresses

Antonio S. Gliozzi,[1] Matteo Nobili, Marco Scalerandi

INFM, Physics Department, Politecnico di Torino, C.so Duca degli Abruzzi 24, 10129, Torino, Italy.
[1]To whom correspondence should be addressed: antonio.gliozzi@polito.it.

Abstract

Several models have been proposed to describe the propagation of elastic waves in nonlinear hysteretic media, most of them based on an implementation of the Preisach–Mayergoyz approach. Here, we present a phenomenological model, in which interstitial regions between material grains or at adhesive bondings, are treated as bi-stable harmonic oscillators. Different PM-space implementations are proposed. The model, which may be solved analytically, is used to predict the elastic response of an interstitial region under different conditions and to different applied stresses.

Keywords: Elastic behavior, harmonic oscillator, hysteretic media, interfaces, PM models, virtual experiments

1. Introduction

In the last decades, several experiments have shown evidence of nonlinear hysteretic elastic response of many different materials to an external loading. Among them, rocks [1–3], concrete [4], structural materials [5], ceramics [6], composites [7], and so on. Nonlinear effects in the acoustic wave propagation are one of the first indications of damage progression in such materials [8], with the resulting wealth of applications in the field of quantitative nondestructive evaluation.

For this purpose, an accurate comprehension of the basic properties of the materials employed in the industrial processes or products is required and their behavior under changes of environmental conditions, component interactions, and the like must be understood. Unfortunately, the understanding of the physical origin of elastic nonlinearity is still an open question, in particular when it cannot be described within the framework of the traditional Landau theory [9]. In recent years, a tremendous amount of research work has been carried out in order to study the microscopic structure and dynamics of various materials, using molecular dynamics techniques [10], including ab initio calculations, pseudopotential energy approaches [11], atomistic models, and so on. A purely microscopic description is, however, not sufficient to predict the response of a macroscopic system, because in general, it neglects the collective behavior emerging from global interactions.

An alternative approach is offered by modeling and numerical simulations at the mesoscopic level [12], which provide a natural and powerful approach to exploiting the advances of molecular dynamics for applications in nondestructive evaluation, material science, seismic studies, and engineering. As a result, several phenomenological models have been developed in the last years to describe hysteretic elasticity in materials [13–22].

Such different approaches reflect the richness and variety of phenomenological observations, ranging from the formation of higher-order harmonics with anomalous rates [23], anomalous elastic behavior under temperature changes [24], relaxation and slow dynamics phenomena [25], modulation and self-induced transparency [26], anomalous resonance-frequency shifts [27], plastic behavior under fatigue cycling [28], and so on. It is, however, surprising that not much attention has been given to emphasizing the common features shared by the various effects, which suggest the possibility to speak of a "universal behavior" for some classes of nonlinear elastic materials [23, 29].

The search for such a universality cannot set aside the fact that elastic nonlinearity, in its various forms, always stems from the complexity of the system. Here, complexity is not intended simply as "complicated," but rather, in the sense given by Holland. It conveys the meaning that damage, its dynamics and interaction with elastic waves, are emergent phenomena, deriving from interactions of a large number of simple constitutive units on a lower space scale. It is therefore important to identify the basic features shared by such systems, which make them complex from an elastic point of view: multiscaling, feedback, fractality, irreversibility, sensibility to the initial conditions, and so on.

In this contribution, we propose a model based on the Local Interaction Simulation Approach (LISA) [30], applied in conjunction with a spring model [31], of nonlinear hysteretic elasticity, using the Preisach–Mayergoyz (PM) space [32] formalism. After a brief review of the existing models, we present our approach in Section 3. Then, in Section 4, we present a few results that are obtained analytically. A more extensive review of the applications to experimental observations are given in Chapter 17 of this book [33].

2. Models of Elastic Nonlinear Systems

As remarked in the introduction, a large number of phenomenological models were proposed in the past years to describe elastic hysteresis under dynamic conditions. This section contains a brief overview on a few continuous models and a more detailed description of discrete approaches based on a Preisach–Mayergoyz formalism.

As far as continuous models are concerned, several authors have approached the problem of the nonlinearity arising at the interface between two surfaces in contact when damage is present in the bonding region. Delsanto et al. [34] derived the interface forces between two aluminum slabs agglutinated with epoxy, starting from a spectral analysis of the experimentally measured transmitted signal. Kim et al. [35] proposed an elastoplastic micromechanical model, based on noninteracting single asperity contacts, using different springs during the loading and unloading part of a stress cycle. The approach was used to describe the plasticity-induced hysteresis in the ultrasonically measured interfacial stiffness during loading–unloading cycles. Rough surfaces in contacts were studied [36] using normal and transverse stiffness for different incident angles and considering both elastic and adhesive forces [37].

Van den Abeele et al. [15] introduced a continuum formulation (based on a sort of mean-field approach) with modulus discontinuities, in which the dynamic modulus differs between increasing and decreasing strains. A similar approach was proposed by Nazarov et al. [38], who adopted a stress–strain relation, which corresponds qualitatively to the hysteresis of the modified Granato–Lucke model. More recently, Vakhnenko et al. [19] proposed a model in which fast subsystems, identified with the field of rapid longitudinal displacements, are interacting with slow subsystems, representing, for example, the concentration of defects in intergrain contact bonds. Their approach also allows the description of relaxation phenomena, such as slow dynamics.

In the Preisach–Mayergoyz approach [32] the specimen is considered as a system of many simple parts, either with (Hysteretic Elastic Elements—HEE) [17, 18, 20] or without (Hysteretic Mesoscopic Units—HMU) [15, 16, 21, 22] elastic properties. The macroscopic behavior results from the collective behavior of a large number of HMU/HEEs. The relevant material properties (e.g., the modulus) are extracted from their statistical behavior.

The approaches based on a PM-space have in common that each unit may be in two states (called "open" and "closed" in the following), characterized by different equations of state. The transition between the states is driven by an external variable P, which is usually chosen as stress or strain. For each unit a pair of parameters (P_c, P_o) is defined such that at $P < P_c$ the unit is in the open state up to $P = P_c$, after which it switches to the closed state. For decreasing P, the unit is in the closed state for $P > P_o$. At $P = P_o$, it switches back to the open state. The state transition protocol is schematized in Figure 16.1. It follows that the unit is always in the closed (open) state if $P > P_c (P < P_o)$, whereas in the intermediate pressure range ($P_o < P < P_c$) the unit may be in one state or the other depending on the previous stress history. It is then possible to consider, in the latter region, thermally activated random transitions (with transition rates q_1 and q_2) between the two states [18, 20]. The parameter pairs are

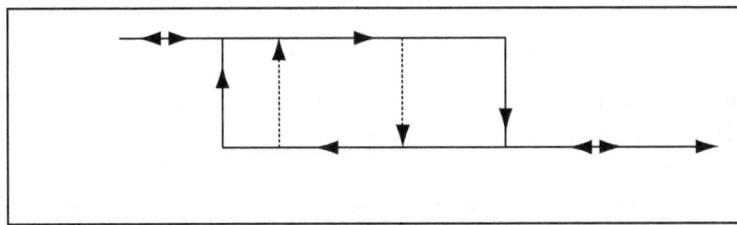

Fig. 16.1. Bi-state representation of a traditional PM protocol, as discussed in the text.

distributed statistically in the (P_c, P_o)-plane (the PM-space), accounting for nonlinear hysteretic elasticity and memory effects.

A basic difference between the various PM approaches is whether the two states have different elastic properties (HEE) or not (HMU). Van den Abeele et al. [21] introduced a multiscale approach, in which the HMUs may be in two different states with different strains. The macroscopic modulus is calculated from the total strain change due to the elastic deformation of the material and the HMU state transitions, induced by an infinitesimal stress change. A PM-space implementation was also proposed by Gusev et al. [16], who assumed that the stress depends not only on the strain, but also on the variation of a parameter responsible for the hysteretic behavior of the medium. Likewise, Capogrosso Sansone and Guyer [22] recently proposed an approach based on a chain of bi-stable springs, in which the two states differ in the spring rest-length.

Parallel to these approaches, models based on HEEs (i.e., considering explicit different elastic behaviors) were proposed [17, 18, 39]. In particular, we proposed a model, in which linear elastic portions are alternated with nonlinear ones. The latter may be either in a rigid or in a poroelastic state [17]. In our approach fast dynamics observations (e.g., the resonance frequency shift) are a consequence of conditioning [18, 20], that is, of the change in the material structure induced by the propagating wave, which increases with the excitation amplitude. This assumption seems to be in agreement with recent experimental observations [40]. As a consequence, an approach based on HEEs rather than HMUs is, to our knowledge, the only PM-based model capable of describing slow dynamics [18]. Our approach is discussed partly in the following sections and more in detail in Chapter 17 [33].

3. A Bi-Stable Oscillator Model for Interstitial Regions

3.1 From the Equations of State to the Equation of Motion

We consider the interstitial regions (called HE in the following and representing regions between grains in granular materials, bonding regions between surfaces in contact, cracked regions in extensively damaged materials, etc.) as poroelastic viscous media, composed of a matrix of fluid and a solid portion. Then, we define phenomenologically (with some analogy to Biot theory: see, for example, Eqs. 2.1 and 2.2 in [41]) the total stress tensor τ as

$$\tau = \sigma + p_f, \tag{16.1}$$

where

$$\sigma = K\varepsilon + \alpha d\varepsilon/dt \tag{16.2}$$

is the usual viscoelastic stress tensor of the solid portion (Kelvin–Voigt constitutive equation). Here K is the material elastic constant, α the dynamic viscosity tensor, and ε the interstice strain.

The term p_f in the stress equation corresponds to the contribution to the volumetric pressure due to the fluid, which is proportional to the applied pressure P : $p_f = -\gamma P$. The constant γ depends, among others, on the saturation and porosity density and the applied pressure P is due to the external forces $F^\pm$ applied to the right and left tips delimiting the HE (labeled with the index $\pm$): $P = -(F^+ - F^-)/2\Sigma$ Note that we assume the forces to be orthogonal to the interstice cross-section Σ (which is correct when symmetries allow a 1-D treatment). In the following $P > 0$ indicates compression.

It follows that the HEs behave as damped linear springs (of rest-length δ_0 equal to the interstice thickness and elastic constant $\beta = K\Sigma/\delta_0$) with a point mass $m_\pm$ at each end of the spring. In the following, we assume the two masses to be equal: $m = m_+ = m_-$. The term p_f in Eq. (16.1) can be in fact included in an effective external forcing, acting on each of the tips:

$$\overline{F}^\pm = F^\pm \mp \frac{\gamma}{2}(F^+ - F^-). \tag{16.3}$$

The equation of motion of the point masses at the ends of the springs is then:

$$m\frac{d^2 x^\pm}{dt^2} = \overline{F}^\pm \mp K\Sigma\frac{x^+ - x^-}{\varepsilon_0} - \alpha\Sigma\delta_0\frac{\omega^2}{v_0^2}\frac{dx^\pm}{dt}, \tag{16.4}$$

where $x^\pm$ are the positions of the point masses and $(x^+ - x^-)/\delta_0$ is the HE deformation (strain) δ. Here, ω and v_0 are the wave frequency and speed. The attenuation term [last on the right of Eq. (16.4)] has been derived assuming the approximation of a sinusoidal behavior of the displacement. It follows that the second space derivative of the displacement is equal to $k^2 x^\pm$, where $k = \omega/v_0$ is the wave number.

Subtracting the two equations and introducing the HE volume $V = \Sigma\delta_0$, we obtain the equation of motion for δ as

$$m\frac{d^2\delta}{dt^2} + \eta\frac{d\delta}{dt} = \frac{(1-\gamma)}{\delta_0}\left(F^+ - F^-\right) - 2\beta\delta, \tag{16.5}$$

where $\eta = \alpha V\omega^2/v_0^2$. From now on, η is considered as the damping parameter characterizing the HE.

3.2 A PM Representation for a Bi-State System

The parameters $\{\gamma, \beta, \eta\}$ completely define the elastic behavior of the HE. Following a PM-space representation, as discussed in the previous section, we assume that different states (also called phases) are allowed at different pressure ranges. In other words, pressure is considered as the control parameter driving the transition between the states, even though other state variables (e.g., strain or temperature) may be chosen without changing the validity of our approach.

Besides the general poroelastic state (labeled in the following PE) defined by arbitrary values of the parameters, we can define three limit states with a specific physical interpretation:

- State R: $\gamma = 1.0$, $\beta = 0$. We call this state "rigid," because the interstice deformation is independent of the applied pressure [see Eq. (16.5)], although the strain may evolve with time to a steady value, which no longer contributes to the interstice stress τ [see Eq. (16.5)]. Once equilibrium is reached, the interstice is no longer deformed, hence the interstice does not play any role in the propagation of an elastic wave and an incident wave is completely transmitted through the interstice without time delay.

- State B: $\gamma = 0, \beta = 0, \eta = 0$. This is called the "broken state." In fact, the interstice stress is zero [see Eqs. (16.1) and (16.2)] and the two edges of the interstice move independently from each other [see Eq. (16.4)].

- State VE: $\gamma = 0, \beta > 0$. The interstice behaves as a solid viscoelastic medium described by a Young's modulus K.

Note that other states may also be introduced but are not of interest here.

Likewise, the protocols describing transitions between the two states may be defined in various ways, following the traditional PM-space description discussed in the previous section. Here, we propose four alternative protocols, which are discussed in the next subsection and applied to specific stress protocols in the following section.

1. The simplest protocol accounts for random transitions between the states; that is, it assumes that both states are allowed for any value of the state variables (strain–stress–temperature) of the system. As we show, however, such a protocol, called in the following a *random protocol*, does not introduce any nonlinearity (see Figure 16.2a).

2. The transition occurs at a fixed value of pressure. Introducing for each state a chemical potential μ, which is a function of the pressure applied to the interstice, the temperature and other state variables (e.g., the strain), transitions occur if the chemical potentials of the two phases have the same values (see Figure 16.2b). Such a protocol, called a *deterministic protocol*, accounts for nonlinearity but not for hysteresis.

3. The third alternative is the traditional PM approach (*hysteretic protocol*); see Figure 16.2c. Here, the HE is in the open state if $P < P_o$ and in the closed state if $P > P_c$. In the intermediate region, the state depends on the previous stress history. Such a protocol is obtained if the chemical potential depends explicitly on the strain (e.g., as proposed in Reference [42] for a surface chemical potential that depends on the surface curvature) or if different chemical potentials are defined for increasing and decreasing state variable. A review of results using this protocol is reported in [17, 33, 43].

4. A similar protocol (*randomized protocol*) can be introduced by allowing metastable states. As in the deterministic case, the transition takes place for a given value of the relevant state variable, that is, when the chemical potentials of the two phases are equal. However, one state may remain "stable" beyond the transition point, even if its chemical potential is larger (i.e., the free energy of the system is not minimized: see Figure 16.2d). If the HMU is in a metastable state, the transition occurs randomly due to thermal fluctuations. The transition rate depends on the difference between the chemical potentials.

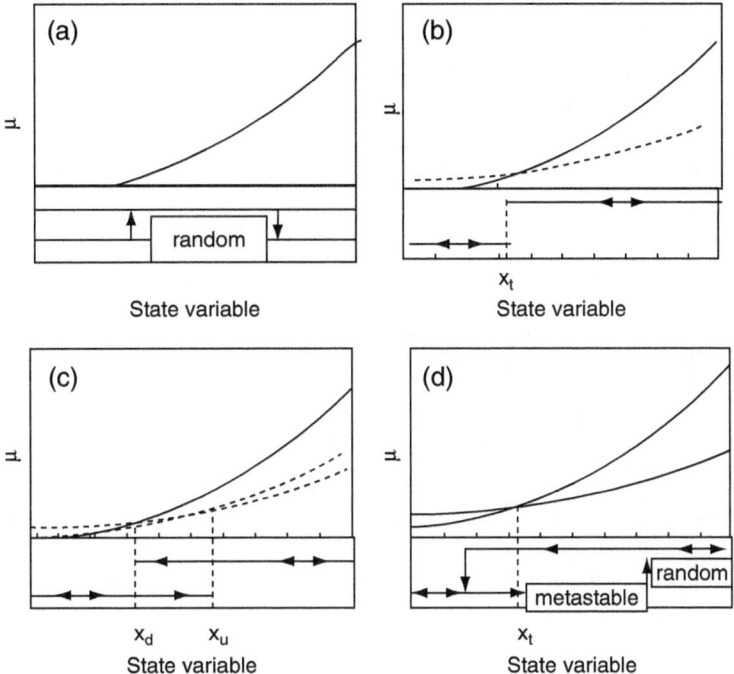

Fig. 16.2. Representation of different transition protocols as discussed in the text. a) Random protocol: chemical potentials μ are the same for the two states and transitions between them occur randomly. b) Deterministic protocol: chemical potentials are different for open (solid line) and closed (dots) states. Transition occurs at the intersection of the chemical potentials. c) Hysteretic protocol: in addition to b) chemical potentials for the open state are different when the state variable increases (dashed line) or decreases (solid line). d) Randomised protocol: in addition to b), transitions occur randomly around the point of intersection between the chemical potentials.

3.3 Behavior of the Transition Protocols

To discuss the significance of the four protocols, we consider a numerical experiment, in which a sinusoidal forcing with unit amplitude (in arbitrary units) and frequency 1 MHz is applied to a single HE. The chemical potentials for the various phases should of course be defined starting from physical considerations. Here, due to the lack of a realistic description of the mechanisms responsible for elastic hysteresis, we choose them arbitrarily. For protocol 3, we have chosen $P_c = 0.1$, $P_o = 0.01$. For protocols 2 and 4, we have selected chemical potentials that depend linearly on the external pressure only (i.e., strain dependencies are neglected): $\mu_{o,c} = a + b_{o,c}(P - P_c)$ for the open/closed states. The numerical value of $b_{o,c}$ has been set to 0.05. Furthermore, random transition probabilities from the open to the closed state (and vice versa) are $p_{o-c} = p_{c-o} = 0.001$ events/s (for the random protocol) and

$$p_{o-c} = p_1 \frac{1}{1 + e^{-(b_o - b_c)(P - P_c)/d}}$$

$$p_{c-o} = p_2 \frac{1}{1 + e^{(b_o - b_c)(P - P_c)/d}}$$

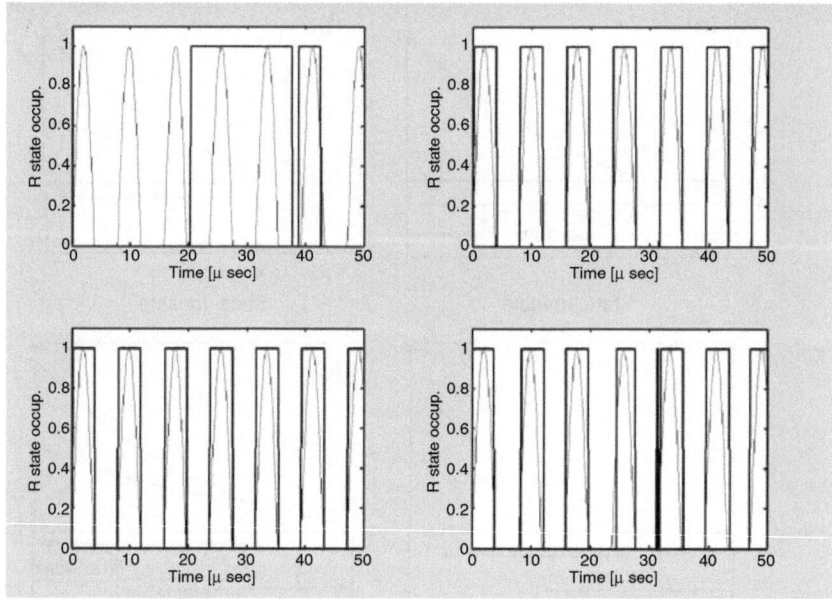

Fig. 16.3. Occupancy of the rigid state as a function of time for a sinusoidal forcing: a) random protocol; b) hysteretic protocol; c) deterministic protocol; d) randomised protocol.

(for the randomized protocol), with $p_1 = p_2 = 0.001$ and $d = 0.01(b_o - b_c)$. These formulas have been introduced in order to account for a dependence of the transition rates from the chemical potential differences between the two states. d plays the role of a generalized temperature. Note that the values of a, b_o, and b_c are not relevant for our simulations.

In Figure 16.3, the state of the HE as a function of time is reported for the four protocols (the applied forcing is reported for reference as a thin solid line). Except for the random protocol, transitions between the states occur always close to P_c and P_o. Because the difference between P_o and P_c is very small compared to the forcing amplitude, in Figure 16.3c hysteresis is not visible (although it is evident in Figure 16.4). Conversely, the effects of metastable states in the randomized protocol are observable in Figure 16.3d: the transitions at $t = 15$ and 25 s are delayed with respect to the time at which $P = P_{c,o}$ and repeated closed–open transitions occur at about 32 s.

We then calculate the average occupancy of the closed state, distinguishing between occupancy during loading and unloading. For this purpose, we sample the loading part of each period T of the applied pressure in $J (J = 50)$ intervals of duration $\Delta t = T/2J$. (The period of the loading phase is half the period of applied pressure). Then, we consider $M = 500$ sinusoidal cycles, resulting in $C = JM$ intervals, covering a time interval CT. To each interval (labeled with an index $j = 1 \ldots C$) we associate the variable r_j, which assumes the value 1 if, in the corresponding time interval, the HE was in the closed state (and 0 otherwise). The relative time T_{closed} spent by the HE in the PE state is then given as

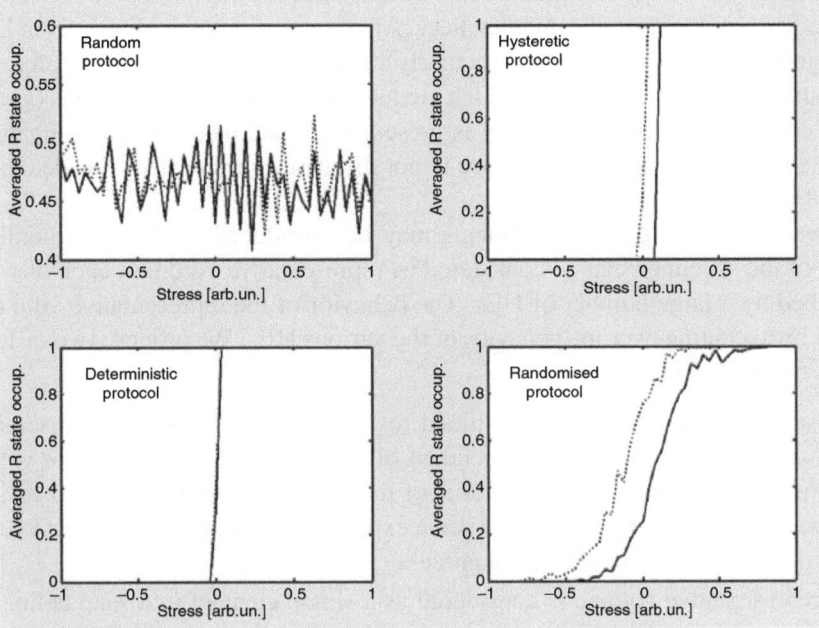

Fig. 16.4. Average occupancy of the rigid state for a sinusoidal applied forcing: random (a), deterministic (b), hysteretic (c) and randomised (d) protocol. Solid (dotted) lines represent occupancy during loading (unloading) phases.

$$T_{\text{closed}} = \frac{\sum_{j=1}^{C} r_j \Delta t}{T}.$$

The same is done to calculate the average occupancy during unloading. The results for the four protocols are reported in Figure 16.4. Additional details of the protocol implementation are reported in Section 4.1.

3.4 Combinations of Hysteretic Elements

As already remarked, the macroscopic behavior results from the collective behavior of a large number of HEs. In fact, a single HE is nonlinear only in correspondence with the state transition and hysteretic only in a small pressure range around the transition values. Therefore it cannot account sufficiently for the nonlinearity of the system. For each of the HE, the chemical potentials of the two states and the other relevant parameters are defined independently. State transitions occur at different pressures for the various HEs, providing very large statistics, from which the macroscopically observed nonlinear response is obtained. Different arrangements of HEs (or eventually combinations) can then be chosen to describe extended (e.g., in multigrain or extensively damaged materials) or localized (e.g., defects, interfaces, etc.) nonlinearities.

In previous papers [18, 20, 33], serial arrangements of HEs have been used by alternating interstices consisting of a single HE only and linear material grains. Such an approach introduces a specific spatial scale for the nonlinear region, because N HE occupy the space of N grains plus N interstices. Also, separation of nonlinear from linear zones is implicit. Such a choice is particularly suitable if the nonlinearity is present in a large number of small interstices, which are located between elastic regions (grains). It is not convenient if the nonlinearity is present in a zone that is small compared to the total specimen dimension. This case is not discussed further here (for details see Chapter 17 of this book).

However, more complex arrangements may be introduced. In fact, the nonlinear portions of the specimen can be considered as representative volumes, each of which is described by a large number of HEs. The behavior of the representative volume is obtained by averaging over the behavior of the various HEs. We propose two different methods for averaging.

(a) The same external pressure is assumed to act on each HE in the representative volume, but their strains are independent of each other. The strain of the volume is defined as the average of the strains of the HEs. This approach yields residual strains at zero stress (as observed in some experiments) due to the HEs that become rigid during the loading/unloading process.

(b) The representative volume is considered as a single element in a state defined by parameters $\{\gamma, \beta, \eta\}$, given by the average of the parameters calculated for a large number of virtual HEs subject to the external pressure applied to the interstice. This approach is adopted in the case where one of the two states is broken. In fact, strain cannot be defined for a broken HE. Hence, averaging over strains is meaningless.

4. Results and Discussions

In this section we give a few numerical examples of application of the proposed model, limiting ourselves to consider parallel arrangements of HEs for different choices of the applied forcing and different choices of the allowed states.

Equation (16.5) is solved analytically for a given external force F. For simplicity we assume $F^+ = F^- = F$. Depending on the values of γ, β, and η, the different solutions (under, over, and critically damped) of the harmonic oscillator are found. The solutions are valid for each of the N HEs in the interstice. For each HE the solution has to be calculated at intervals of constant state, imposing continuity of the strain δ and velocity $d\delta/dt$ when a state change occurs. Furthermore, in each time interval, different HEs may be in different states, hence described by different values of the parameters. Their arrangement is then treated as discussed in the previous subsection.

The following cases are considered.

- Case 1: Stress loading–unloading allowing a poroelastic (open) and a rigid (closed) state
- Case 2: Sinusoidal stress allowing a poroelastic and a broken state
- Case 3: Stress loading allowing two poroelastic states with different elastic constants

The applied stresses as a function of time are reported in Figure 16.5, where the stress is $P = F/\Sigma$. Here we assume positive stresses corresponding to compression.

4.1 Case 1: Transitions Between Poroelastic (PE) and Rigid (R) States

The response of a system containing 200 HE to repeated loading–unloading cycles (see Figure 16.5a) is reported in Figure 16.6. The parameters of the poroelastic state are $K = 90$ Gpa, $\eta = 2$ GPa $\cdot$ s, and $\gamma = 0.8$. Four different state transition protocols are examined.

(a) Random protocol: Transitions from one state to the other occur at a rate of $p = 0.001$ events/s.

(b) Hysteretic protocol: P_o and P_c are distributed statistically between -0.12 and 0.12 GPa.

(c) Deterministic protocols: The chemical potentials for the PE (R) states are chosen as linearly dependent from the applied pressure P:

$$\mu_{o,c} = x_{o,c} + y_{o,c} P.$$

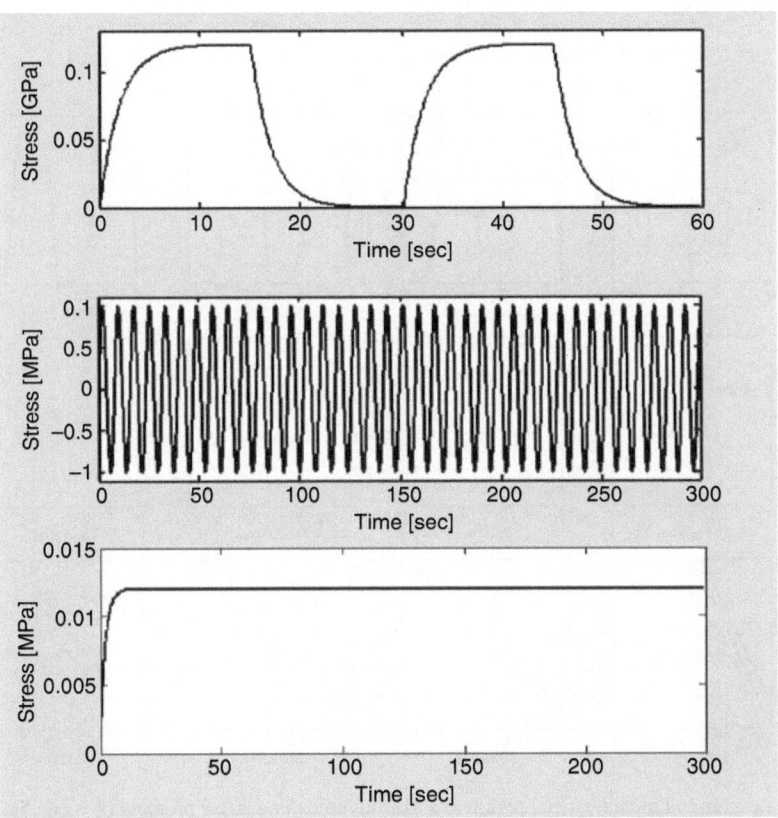

Fig. 16.5. Applied stress protocols (stress vs. time). (a): repeated loading-unloading cycles. (b): sinusoidal stress. (c): loading applied stress.

From $\mu_o = \mu_c$ follows $P_{\text{trans}} = P_c = P_o = (x_o - x_c)/(y_c - y_o)$. The distribution of the random parameters $x_{o,c}$ and $y_{o,c}$ is chosen to ensure P_c lies between -0.12 and 0.12 GPa.

(d) Randomized protocols: the same chemical potentials are adopted as in (c). Furthermore, transitions from PE to R (or from R to PE) occur with the stress-dependent probabilities given in Section 3.3.

The strain and the number of HEs in the rigid state as a function of time, and the stress as a function of strain are shown in Figure 16.6. The solid lines represent the linear response, that is, if no transitions to the rigid state are allowed. For the four protocols, the strain follows the applied stress with a delay due to attenuation. The consequence is the formation of a loop in the stress–strain curve already in the linear case. At zero stress, the strain returns to zero only in the deterministic case. In the other cases (only slightly visible for the randomized case), units with P_c or P_o smaller than zero cause the appearance of residual strains, which slowly disappear with time (as discussed in Reference [17]).

The two cases with random terms (random and randomized protocols in the first and last column, respectively) show a remarkable increase of the delay, particularly during the unloading phase. The stress–strain loops are highly deformed and wider (indicating larger attenuation than in the linear case). In the hysteretic/deterministic

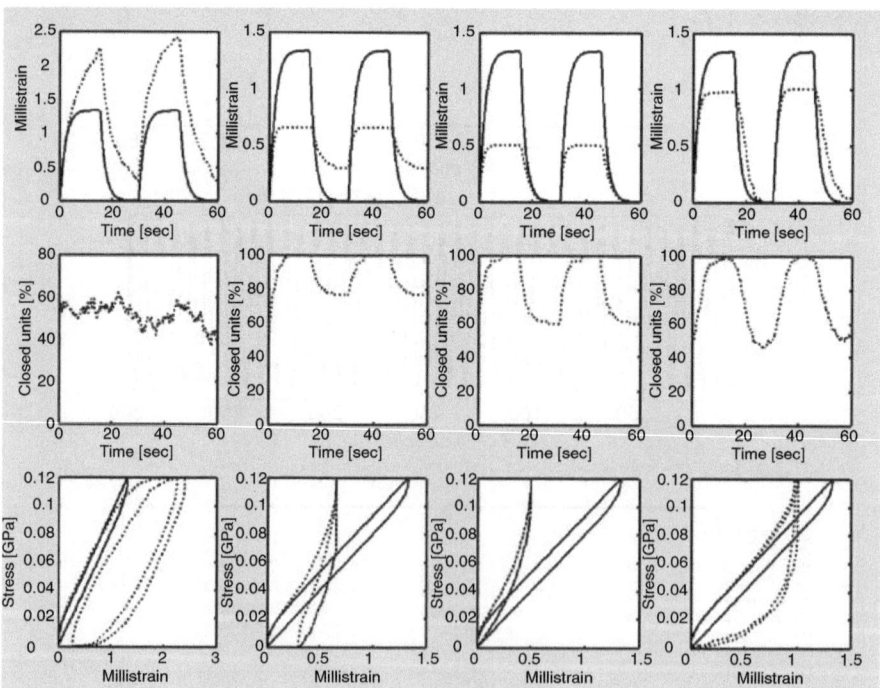

Fig. 16.6. Response of the intersticial region to a loading/unloading stress protocol (see Fig. 5a). Strain vs time (first row), number of units in the rigid state vs. time (second row), and stress vs. strain (third row) are shown for random, hysteretic, deterministic and randomised protocols in the four columns, respectively. Solid lines represent the linear case in which all HE are constrained in the PE state.

(second and third column) cases, the increase in delay with respect to the linear case is less evident. More important is that the strain does not reach the same maximum as in the linear case, due to the fact that most HEs (about 50%) became rigid already at about $t = 5$ s (see second row), that is, at stresses of the order of 0.09 GPa. In both cases, hardening occurs if the applied stress increases (as shown by the increase in the slope of the stress–strain curve with respect to the linear case), but only in the hysteretic case an increased attenuation is observed, by means of the widening of the gap between up-going and down-going curves.

In both hysteretic and randomized protocols, the stress–strain curves exhibit different slopes during loading and unloading at any given stress value. This provides evidence of the presence of hysteresis, manifested in the different elastic properties during loading/unloading. No similar effect is generated in the random and the deterministic protocol. Furthermore, the random protocol does not induce distortion of the stress–strain curve, only rotation and widening. Hence the random protocol does not account for nonlinearity, as is also demonstrated in Figure 16.7.

4.2 Case 2: Transitions Between Poroelastic (PE) and Broken (B) States

Here we consider an interstice containing $N = 200$ HE, with a PE state described by the same parameters as in Subsection 4.1. and a B state. We consider the same

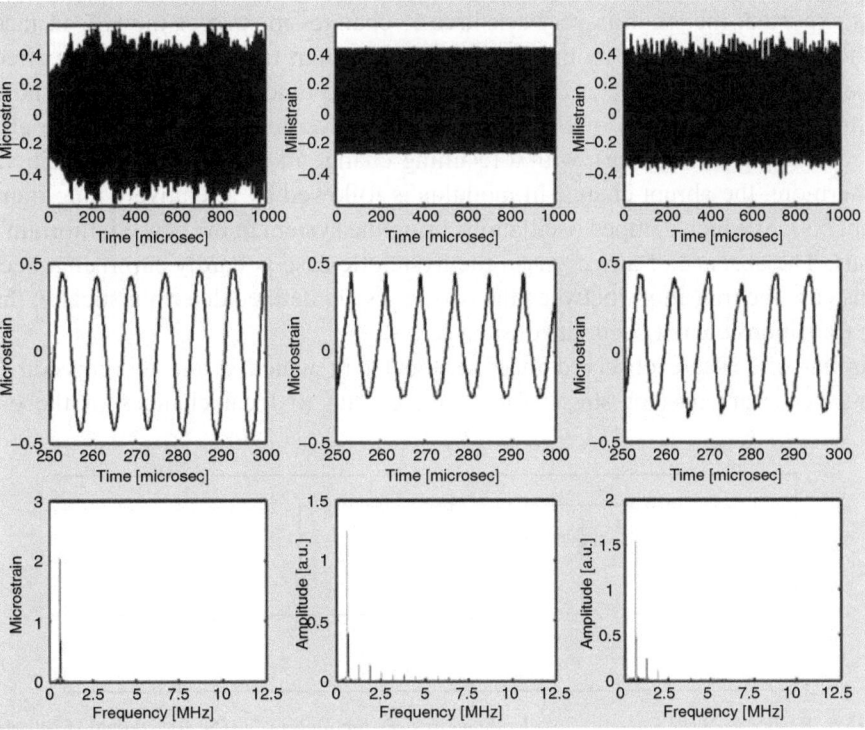

Fig. 16.7. Response of the interstice to sinusoidal forcing in the case of random (left), hysteretic (middle) and randomised (right column) protocols. First row: strain vs. time; second row: strain evolution is zoomed to a shorter time-window; third row: Fast Fourier Transform of the strain signal.

protocols for the state transitions as discussed above and the sinusoidal stress protocol represented in Figure 16.5b.

The response of the system is reported in Figure 16.7 for random, hysteretic, and randomized protocols, respectively, in the three columns. The strain as a function of time is reported in the first row for the full signal and zoomed to a shorter time window in the second row. In the random protocol, the signal amplitude is randomly modulated, but not distorted. The interstitial region is linear elastic, as is confirmed by the absence of higher-order harmonics in the fast Fourier transform spectrum (third row).

Indications of nonlinearity and hysteretic effects are visible in the other two cases. Both show a strong deformation of the signal, with formation of a typical triangular profile (indication of hysteresis), which is reflected in a considerable content of higher-order harmonics in the FFT spectrum. Also, in both hysteretic and randomized protocols the mean of the signal is larger than zero (corresponding to a net contraction of the interstice).

4.3 Case 3: Acoustic Emission

As a final example, we consider an interstice composed of a single HE with two PE states (differing only in the parameter $K = 90$ GPa and 70 GPa for the two states). A loading is applied to the interstice and kept constant (see Figure 16.5c). Figure 16.8 shows the strain response (upper row), the state of the interstice (second row: 1 denotes the state with $K = 70$ GPa) and the displacement velocity response (third row).

As expected, the strain is very sensitive to changes in state of the HE. In fact, after the stress has reached its maximum value, random transitions in either direction between the two PE states, as occur for random and randomized protocols (shown in the first and last columns of Figure 16.8) cause a variation of the interstice elasticity (softening or hardening), with a resulting change in the equilibrium strain value. Furthermore, the abrupt change in modulus is followed by a relatively long transient (about 5 s), in which damped oscillations bring the system to the new equilibrium conditions. The absence of any event in the hysteretic case is hardly surprising, because in this case the transitions between the two states are defined deterministically; that is, state transitions cannot occur at constant stress.

Similar effects are observed in experiments, in which Acoustic Emission (AE) events occur at constant stress, for example, due to local changes of the elastic

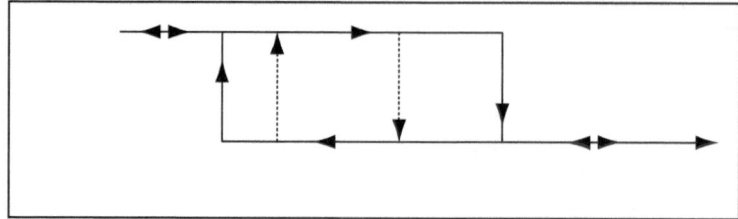

Fig. 16.8. Response of the interstice to a loading stress as reported in Fig. 5c for random (first column), hysteretic (second) and randomised (third) protocols. The plots show strain (first row), state (second row) and displacement velocity (third row) as a function of time. State 1 corresponds to the softer state. Vertical lines indicate transitions.

properties (in general associated with damaging). In these measurements, the change in strain is normally hardly measurable, because the AE events occur only in a very small portion of the specimen. Therefore, the strain velocity is usually used for characterizing and counting the number of events. The signal simulated here (last row of Figure 16.8, where a proper time window has been selected) is qualitatively similar to signals observed in experiments [44]. Note that the time length of the signal reproduced in the present simulations is not comparable with experimental data, because its frequency and its duration are determined by the properties of the interstitial region (i.e., its modulus, damping, contact quality parameter, and length). Here the topic is not discussed further, because it is still the subject of investigation [45].

5. Conclusions

We have proposed a simple bi-state model to describe interstitial regions as superpositions of linear damped harmonic oscillators, allowed to jump between different elastic states. Introducing different transition protocols, we have shown that we can describe random phenomena, deterministic transitions (leading to nonhysteretic nonlinearity), and hysteretic protocols (both the traditional Preisach–Mayergoyz approach and its randomized variation).

The proposed approach allows us to obtain an analytical solution for the response of the interstice to different kinds of applied stresses. Besides reproducing features typical of traditional models (e.g., hysteresis loops, generation of harmonics, etc.), it allows us also to model additional phenomena, such as acoustic emission [45] or damage progression [47].

The validation of the model is given in Chapter 17 of this book [33].

Acknowledgments

We thank Professor P.P. Delsanto and Doctor M. Hirsekorn (Politecnico di Torino) for useful comments and suggestions. This work has been supported by the ESF (grant NATEMIS) and by the EC (AERONEWS grant nr. FP6-502927).

References

[1] J. Tencate, E. Smith, and R.A. Guyer, Universal slow dynamics in solids, *Phys. Rev. Lett.* **85**, 1024 (2000); Chapter 4, in this book.

[2] B. Zinszner, P.A. Johnson, and P.N.J. Rasolofosaon, Influence of change in physical state on elastic nonlinear response in rock: effects of confining pressure and saturation, *J. Geoph. Res.* **102**, 8105–20 (1997).

[3] Chapter 25, in this book.

[4] J.C. Lacouture, P. Johnson, and F. Cohen-Tenoudji, Study of critical behavior in concrete during curing by application of dynamic linear and nonlinear means, *J. Acoust. Soc. Am.* **113**, 1325–1332 (2003); A. Carpinteri, G. Ferro, and G. Ventura, Size effects on flexural response of reinforced concrete elements with a nonlinear matrix, *Eng. Fracture. Mech.* **70**, 995 (2003).

[5] N. Pugno and A. Carpinteri, Tubular adhesive joints under axial load, *J. Appl. Mech.* **70**, 832–839 (2003).

[6] K. van den Abeele and K. van den Helde, Correlation between dynamic nonlinearity and static mechanical properties of corroded E-glass reinforce polyester composites, in: *Review of Progress in Quantitative Nondestructive Evaluation* (AIP Conf. Proceedings, 2000), pp. 1359.

[7] R. El-Guerjouma, Nondestructive evaluation of damage and failure of fibre reinforced polymer composites using ultrasonic waves and acoustic emission, *Adv. Eng. Mater.* **3**, 601 (2001); Chapter 24, in this book.

[8] K.Van den Abeele et al., Non linear elastic wave spectroscopy techniques to discern material damage, Part II, *Res. Nondestruct. Eval.* **12**, 31–42 (2000); Micro-damage diagnostics using Non Linear Elastic Wave Spectroscopy, *NDT & E Int.* **34**, 329 (2001).

[9] L.D. Landau and E.M. Lifsits, *Theory of elasticity*, (Pergamon Press, Oxford, 1981).

[10] F. Cleri, Representation of mechanical loads in molecular dynamics simulations, *Phys. Rev. B.* **65**, 141071 (2002).

[11] Chapter 11, in this book.

[12] P. Van, C. Papenfuss, and W. Muschik, Mescoscopic dynamics of microcracks, *Phys. Rev. E* **62**, 6206 (2000).

[13] K. McCall and R.A. Guyer, Equation of state and wave propagation in hysteretic nonlinear elastic materials, *J. Geoph. Res.* **99**, 23887 (1994).

[14] R.A. Guyer et al., Quantitative use of Preisach-Mayergoyz space to find static and dynamic elastic moduli in rock, *J. Geoph. Res.* **102**, 5281–93 (1996).

[15] K. van den Abeele et al., On the quasi-analytic treatment of hysteretic nonlinear response in elastic wave propagation, *J. Ac. Soc. Am.* **101**, 1885 (1997).

[16] V. Gusev and V. Aleshin, Strain wave evolution equation for nonlinear propagation in materials with mesoscopic mechanical elements, *J. Ac. Soc. Am.* **112**, 2666–79 (2002).

[17] M. Scalerandi, P.P. Delsanto, V. Agostini, K. Van Den Abeele, and P.A. Johnson. Local interaction simulation approach to modeling nonclassical, nonlinear elastic behavior in solids, *J. Ac. Soc. Am.* **113**, 3049–3059 (2003); M. Scalerandi, P.P. Delsanto, and P.A. Johnson, Stress induced conditioning and thermal relaxation in the simulation of quasi-static compression experiments, *J. Phys. D: Appl. Phys.* **36**, 288 (2003).

[18] P.P. Delsanto and M. Scalerandi, Modeling nonclassical nonlinearity, conditioning and slow dynamics effects in mesoscopic elastic materials, *Phys. Rev. B* **68**, 064107 (2003).

[19] O.O. Vakhnenko et al., Strain-induced kinetics of intergrain defects as the mechanism of slow dynamics in the nonlinear resonant response of humid sandstone bars, *Phys. Rev. E* **70**, 015602(R) (2004).

[20] M. Nobili and M. Scalerandi, Temperature effects on the elastic properties of hysteretic elastic media: modeling and simulations, *Phys. Rev. B* **69**, 104105 (2004).

[21] Chapter 12, in this book.

[22] B. Capogrosso Sansone, and R.A. Guyer, Dynamic model of hysteretic elastic systems, *Phys. Rev. B* **66**, 2241011 (2002).

[23] R.A. Guyer and P.A. Johnson, Non linear mesoscopic elasticity: Evidence for a new class of materials, *Phys. Today* **52**, 30 (1999).

[24] T.J. Ulrich and T.W. Darling, Observation of anomalous elastic behavior in rock at low temperatures, *Geoph. Res. Lett.* **28**, 2293 (2001).

[25] R.A. Guyer et al., Slow elastic dynamics in a resonant bar of rock, *Geoph. Res. Lett.* **25**, 1585 (1998).

[26] V. Aleshin, V. Gusev, and V. Zaitsev, Propagation of initially bi-harmonic sound waves in a 1D semi-infinite medium with hysteretic non-linearity, *Ultrasonics* **42**, 1053–9 (2004).

[27] L. Ostrowsky and P.A. Johnson, Dynamic nonlinear elasticity in geomaterials, *Riv. del Nuovo Cimento* **24**, 1–46 (2001).

[28] M. Mizuno et al. Stress, strain and elastic modulus behavior of SiC/SiC Composites during creep and cyclic loading, *J. Europ. Ceramic Soc.* **18**, 1869 (1998).

[29] Chapter 1, in this book.

[30] P.P. Delsanto, R. Mignogna, M. Scalerandi, and R. Schechter, Simulation of Ultrasonic Pulse Propagation in Complex Media, in: *New Perspectives on Problems in Classical and Quantum Physics*, edited by P.P. Delsanto and A.W. Saenz, (Gordon & Breach New Delhi, 1998), vol. 2, pp. 51–74.

[31] P.P. Delsanto and M. Scalerandi, A spring model for the simulation of wave propagation across imperfect interfaces, *J. Acoust. Soc. Am.* **104**, 2584–2591 (1998).

[32] D.J. Holcomb, *J. Geoph. Res.* **B86**, 6235 (1981); R.A. Guyer et al., Hysteresis, discrete memory, and nonlinear wave propagation in rock: A new paradigm, *Phys. Rev. Lett.* **74**, 3491 (1995).

[33] Chapter 17, in this book.

[34] P.P. Delsanto, S. Hirsekorn, V. Agostini, R. Loparco, and A. Koka, Modeling the propagation of ultrasonic waves in the interface region between two bonded elements, *Ultrasonics* **40**, 605 (2002); Chapter 15, in this book.

[35] J.-Y. Kim, A. Baltazar, and S.I. Rokhlin, Ultrasonic assessment of rough surface contact between solids from elastoplastic loading–unloading hysteresis cycle, *J. Mech. Phys. Solids* **52**, 1911–1934 (2004).

[36] A. Baltazar, S.I. Rokhlin, C. Pecorari, On the relationship between ultrasonic and micromechanical properties of contacting rough surfaces, *J. Mech. Phys. Solids* **50**, 1397–1416 (2002).

[37] C. Pecorari, Adhesion and nonlinear scattering by rough surfaces in contact: Beyond the phenomenology of the Preisach–Mayergoyz framework, *J. Ac. Soc. Am.* **116**, 1938 (2004); Chapter 19, in this book.

[38] V.E. Nazarov and A.B. Kolpakov, Experimental investigations of nonlinear acoustic phenomena in polycrystalline zinc, *J. Ac. Soc. Am.* **107**, 1916 (2000).

[39] O. Bou Matar et al., Pseudo Spectral simulation of 1-D nonlinear propagation in heterogeneous elastic media, to be submitted.

[40] J.A. Tencate et al., Nonlinear and nonequilibrium dynamics in geomaterials, *Phys. Rev. Lett.* **93**, 065501 (2004); Chapter 26, in this book.

[41] M.A. Biot, Mechanics of deformation and acoustic propagation in porous media, *J. Appl. Phys.* **33**, 1482 (1956); M.D. Collins et al., Wave propagation in poro-acustic media, *Wave Motion* **25**, 265 (1997).

[42] B. Yang et al., Local strain-mediated chemical potential control of quantum dot self-organization in heteroepitaxy, *Phys. Rev. Lett.* **92**, 025502 (2004).

[43] A.S. Gliozzi, M. Nobili, and M. Scalerandi, A two-states damped harmonic oscillator as a basic block to describe elastic non linearity, submitted to *Phys. Rev. B* (2005).

[44] A. Carpinteri, G. Lacidogna, and N. Pugno, Acoustic emission to evaluate the energy dissipation during the compression of hereogeneous materials, *Proc. of the Tenth Int.Congress on Sound and Vibration*, 1187 (2003).

[45] F. Bosia, A. Gliozzi, M. Griffa, M. Nobili, and M. Scalerandi,Acoustic emission and nonlinear elasticity in fatigue induced damage, *Proc. of the Twelfth Int. Congress on Sound and Vibration*, Paper Nr. 870 (2005).

[46] M. Scalerandi et al., A PM approach to fatigue induced irreversibility, *Phys. Rev. B.* **73**, 092103 (2006).

17

Numerical Analysis of the Anomalous Elastic Behavior of Hysteretic Media: Quasistatic, Dynamic, and Relaxation Experiments

Marco Scalerandi,[1] Matteo Nobili, Michele Griffa, Antonio S. Gliozzi, Federico Bosia

INFM, Physics Department, Politecnico di Torino, C.so Duca degli Abruzzi 24, 10129, Torino, Italy.
[1]To whom correspondence should be addressed: marco.scalerandi@infm.polito.it.

Abstract

Models at a mesoscopic level may be useful tools to support data interpretation and the design of experimental setups. They may even help to provide a better understanding of the physical mechanisms responsible for hysteresis in damaged elastic media. Of course, they should prove capable of reproducing at least most of the observed phenomenology in which the anomalous elastic behavior of various materials under applied stresses manifests itself. In this contribution, we briefly describe a phenomenological model based on a Preisach–Mayergoytz space approach and we review its application to simulate quasistatic, fast dynamic, and slow dynamic experiments.

Keywords: Elastic behavior, frequency shift, hysteretic media, numerical simulation, relaxation, virtual experiments

1. Introduction

Evidence has been shown in the past decade of the importance of the nonlinear elastic response of materials to an external loading, in particular to predict the mechanical integrity of materials such as rocks [1], concrete [2], structural materials [3], ceramics [4], composites [5], and so on. Nonlinear features are indeed the first manifestation of damage progression [6], for example, crack propagation [7]. On the other hand, the material intrinsic nonlinearity may be seen as the cause of fatigue and, eventually, lead to the healing of specimens subjected to cyclic loading, for example, during operation. As a consequence, a detailed understanding of the physical origin of elastic nonlinearity and the consequent modeling of the nonlinear response to applied stresses is a prerequisite for the prediction of mechanical failure of materials.

The huge amount of experimental data gathered on the subject shows that nonlinearity emerges particularly when mesoscopic features (i.e., on a space scale of the order of few microns) are present in the material, even though the origin of the response of such elements has ultimately to be traced back to microscopic details. Furthermore, a wide phenomenology is observed, such as the formation of temperature gradients [8], relaxation phenomena [9], creep [10], and so on. All these features suggest that it is possible to speak of a "universal behavior" of nonlinear elastic materials [11, 12].

Among the various observations, an anomalous elastic behavior has been observed in several materials (especially rocks, sandstone, concrete, and damaged materials) and has been traced back to the presence of elastic hysteresis. A very first peculiarity of the phenomenon is that a huge nonlinearity manifests itself over a wide range of strain levels, and is remarkable already at low strains (about 10^{-8}).

In the high-strain regime, quasistatic experiments have been conducted and measurements have shown the existence of hysteretic loops in the stress–strain curves with end point memory [13]. In the low-strain regime (10^{-7} to 10^{-5}), resonance frequency shift (with anomalous dependence on the strain amplitude) [14], strong dependence on humidity/temperature [15], and generation of higher-order harmonics (with nonclassical rates) [16] have been observed, together with anomalous modulation effects, both as far as sideband generation [17] and attenuation [18] are concerned.

Recent experimental data have shown that most of the effects listed above (classified under the term "fast dynamics") are due to the simultaneous presence of nonlinearity and nonequilibrium dynamics [19]. The latter manifests itself in the so-called conditioning effect, that is, the memory effect due to the "temporary" modification of the elastic properties of the material induced by the excitation wave. This effect, known to be due to large amplitude disturbances [20], has been shown not to be separable from fast dynamics even at low strains [19]. Also, very recently it has been proved that conditioning, at least in materials such as damaged ceramics, is not instantaneous, rather occurs on a time scale of the order of a few seconds [21].

As remarked, such "temporary" conditioning is fully reversible, in the sense that the elastic properties of the material slowly recover their unperturbed values [1, 22]. This phenomenon is usually called slow dynamics and takes place on a time scale ranging from a few hours to several days.

Several models have been developed to attempt an understanding of this phenomenology [23–30]. Most of them are based on a Preisach–Mayergoyz (PM) approach [31]. For a detailed description of the latter, see Chapter 16 in the present book [32].

In particular, in past years we have proposed a model, based on the Local Interaction Simulation Approach (LISA) [33], applied in conjunction with the spring model [34] and a PM representation [31]. We have shown that several of the features observed in nonlinear hysteretic and nonclassical systems may be well reproduced by introducing a bi-state constitutive equation for the nonlinear features, with adequately driven transitions between two states with different elastic properties [24, 25].

In this contribution, after a brief discussion of the model, we review most of the results obtained to show how our approach is capable of capturing a good part of the phenomenology observed in either quasistatic [24], fast dynamics [25, 35], and relaxation [25, 36] experiments.

2. The Model

Let us consider a multigrained material specimen. If it consists of a thin bar (of length L and cross section $\Sigma \ll L^2$), as is the case in most experimental setups, we can simplify the problem with a 1-D schematization, in which N grains alternate with $N - 1$ interstices. Grain portions are much larger than interstices, of the order of 10–20 μm versus 1 μm, respectively. For simplicity, we assume that at equilibrium (i.e., in the absence of external forces and at some given temperature T_0) all grains have the same rest length λ_0, the same elastic constant K and the same mass $m = \rho \, \Sigma \lambda_0$, where ρ is the mass density. We consider as negligible both the interstice rest length δ_0 and its mass.

Under an external perturbation $F(t)$, the length of each grain (labeled with an index i) and each interstice [$\lambda_i(t)$ and $\delta_i(t)$, respectively] are time dependent and can be determined once the equation of state for both grains and interstices are given. The index i and the time dependence are omitted in the following, whenever no confusion can arise.

As discussed elsewhere [24, 25], we assume the hysteretic nonlinearity to be confined to the interstice region. Therefore grains follow the usual thermoelastic theory [36,37] and their constitutive equations (i.e., the relation between stress and strain) at a given temperature T are given as

$$\sigma(t) = K\varepsilon(t) - K\alpha\,(T - T_0)\,, \tag{17.1}$$

where α is the thermal expansion coefficient of the specimen and $\varepsilon(t) = \partial u/\partial x \approx (\lambda - \lambda_0)/\lambda_0$, u being the displacement.

In addition to Eq. (17.1), the state of the grain is determined by the 1-D thermal conduction equation:

$$C\frac{\partial T}{\partial t} = k\frac{\partial^2 T}{\partial t^2} - \alpha K\frac{\partial}{\partial t}\frac{\partial \varepsilon}{\partial x}\,, \tag{17.2}$$

where k is the thermal conductivity and C is the specific heat.

The constitutive equation for interstices should take into account nonclassical nonlinear effects and has already been discussed in another chapter of this same book [32]. We assume the stress τ on the interstice to be given as

$$\tau(t) = 2a_1 P(t) + a_2\eta(t) + a_3\,\dot{\eta}(t) + a_4\Delta T\,, \tag{17.3}$$

where a dot denotes a time derivative, $P(t)$ is the pressure applied on the interstice under consideration and $\eta(t)$ corresponds to the intersticial region strain. ΔT is the temperature difference between the interstice tips. For simplicity, we assume here that the temperature is uniform along the interstice. Hence $\Delta T = 0$. Such an assumption is a reasonable first approximation (due to the short length of the interstice and the 1-D treatment adopted), even though an explicit thermoelastic equation similar to Eq. (17.2) for the interstice itself may help to capture additional experimentally observed features.

We remark again that Eq. (17.3) is only a phenomenological equation which, as we show, captures most of the phenomenology, but should be modified according to the eventually emerging knowledge of interaction mechanisms taking place at the molecular or mesoscopic level, for example, dislocations theory [38], Biot theory [39], glass transitions [40], and so on.

To introduce nonlinearity in the linear system of Eqs. (17.1)–(17.3), following a Preisach–Mayergoyz approach [31], we assume that the interstice may be in two different linear states, the nonlinearity being only due to sudden transitions from one state to the other. In the first state, labeled $r = 1$, the interstice is rigid ($a_1 = 0.5$, $a_2 = 0, a_3 = a_{3R}$) and any disturbance travels across the interstice without further straining it; the other ($r = 2$) is a linear elastic state, where the values given to the parameters $a_n(n = 1, 2, 3)$ are constants and their ratio to the rigid state values define the level of hysteretic nonlinearity. Other states are also possible, as discussed in References [32, 41].

The rules for the transition from the rigid to the elastic state (and vice versa) are defined as follows: we define for each interstice a parameter pair (P_c, P_o) with $P_o < P_c$ and starting for any given interstice at a given pressure $P < P_c$, we assume that the interstice length varies elastically up to $P = P_c$, at which point it becomes rigid. Conversely, when P decreases, the interstice remains rigid up to the value $P = P_o$, where it becomes elastic again.

It follows that the interstice is always in the rigid (elastic) state if $P > P_c(P < P_o)$, whereas in the intermediate pressure range ($P_o < P < P_c$) the two states coexist and the interstice is in one state or the other depending on the previous stress history. It is then natural to consider, in the latter pressure region, thermally activated random transitions (with rates q_c and q_o) between the two linear states. Because it is reasonable to assume that the rigid state is more stable than the elastic one (the rigid–elastic transition implies the rupture of a sort of "static bond"), we require for the thermally activated transition rates that $q_c > q_o$. Clearly, these hopping transition rates increase with temperature, but in the present context this dependence is not explicitly included. Likewise, any other dependence of the rates on, for example, the applied pressure P is neglected.

It is well known that hysteretic nonlinearity depends strongly on the initial state of the specimen; that is, it depends on the previous stress history. Hence, the definition of the initial conditions is, at least in the presented approach, a key point. We assume therefore that before starting the "virtual experiment" the specimen is completely relaxed; that is, a long time has passed since the last perturbation of the specimen, either thermal or mechanical. This can amount to several hours in real experiments. Also, we assume that the specimen is initially kept at atmospheric pressure (the pressure scale is chosen so that atmospheric pressure is scaled to zero) and room temperature. It follows that the state of each interstice is defined as follows: elastic state ($r = 2$) for all interstices with $P_o > 0$ and rigid state ($r = 1$) for interstices with $P_c < 0$. The state of the remaining interstices cannot be deterministically defined, because both states are possible. It follows that they are statistically distributed according to the thermal relaxation probabilities $q_{c/o}$ defined above, where the probabilities of being in an elastic/rigid state are given by $q_{c/o}/(q_c + q_o)$.

In addition, the distribution of the (P_c, P_o) pairs for the various interstices is very important and should be derived from inversion of experimental data [42]. Here, for simplicity, we always adopt a uniform distribution in the $P_c - P_o$ space, provided $P_o < P_c$.

Once the constitutive laws are given, the equations of motion can be easily obtained. Denoting with $u_i^{\pm}$ the displacements of the right and left tips of the ith interstice, Newton's law becomes

$$\frac{\rho}{2} \frac{d^2 u_i^{\pm}}{dt^2} = \sigma_i^{\pm} \mp \tau_i - \gamma \frac{d u_i^{\pm}}{dt}, \tag{17.4}$$

where γ accounts for attenuation and $\sigma_i^{\pm}$ are the stresses supported by the grains adjacent to the ith interstice, which, in turn, supports a strain τ_i, as defined in Eqs. (17.1) and (17.3). Equation (17.4) can be solved numerically (e.g., through finite differences) together with Eq. (17.2) to give the response to an external loading of the specimen as a function of time.

3. Results and Discussion

In the course of the experiments, several macroscopic effects have been observed, as discussed in the introduction, for very different strain ranges and different excitation types. In this section, we show how the proposed approach is capable of reproducing most of the observations reported in the literature. In particular we discuss:

(a) The existence of *hysteretic loops and discrete memory* in the stress–strain relation, deriving from *quasi-static experiments*, that is, when the forcing applies a large slowly varying strain to the specimen, so that the system is considered to be in equilibrium at all times. This effect is called "quasistatic hysteresis".

(b) Existence of "mechanical" interactions which induce *softening of the effective elastic moduli* of the specimen, shown mostly in dynamic experiments where a rapidly varying sinusoidal forcing is applied. We call this effect "fast dynamics".

(c) Existence of *stress-history effects*, leading to variations in the dynamic moduli which do not immediately disappear when the disturbance (both mechanical and thermal) is removed. In other words, the specimen is "conditioned", that is, the elastic properties at a fixed amplitude depend on the history of the specimen.

(d) Existence of *relaxation processes*, which render the conditioning fully reversible; that is, after the change in the elastic properties due to a large amplitude disturbance, the latter slowly recover their equilibrium state. The effect is generally called "slow dynamics".

In all the experiments reported in the following we have chosen a bar of length $L = 61.02$ mm and cross section $\Sigma = 10$ mm^2, with a damaged region extending along its entire length. Grain and interstice lengths are 0.75 mm and 1.5 μm, respectively. We have assumed for the specimen a bulk modulus $K = 23210$ MPa and a linear Q factor of 600.

3.1 Quasistatic Experiments

We consider first quasistatic experiments, in which the external stress $\sigma_{ext}(t) = F(t)/\Sigma$, acting on both edges of the bar, is varied with time according to a certain protocol, but always leaving sufficient time between successive steps for the stress to distribute itself homogeneously through the specimen. Thus, the system can be assumed to be in equilibrium at all times, all time derivatives can be set to zero, and the results are independent of damping. Quasistatic conditions can be easily achieved, because the time for an ultrasonic pulse to cross the entire bar (whose length is assumed to be $< 1\,\text{m}$) is of the order of $10^{-4}\,\text{s}$, whereas several seconds (or even minutes) elapse between successive loading steps.

In the following, we consider virtual compression experiments performed with a certain protocol, in which the applied stress varies from 0 to a given σ_{max} and back to zero. At each time step, because the system is in equilibrium, the known applied stress is uniformly distributed along the bar and Eqs. (17.1)–(17.3) provide the strain at each node and interstice. Note that for rigid interstices the strains at a given time are unchanged from the values at the preceding time step.

A sinusoidal protocol with exponentially increasing amplitude, as reported in the inset, is considered in Figure 17.1, where a stress–strain plot is reported. The stress cycles generate stable and closed hysteretic loops (note that conditioning effects during quasistatic experiments, not shown here, may also be described by our model as reported in [24]). It is interesting to remark that the curvature is always facing upwards; that is, the stiffness increases with the applied stress, although less so in the up-going branch, in agreement with experimental observations in rocks. Also, it is noticeable

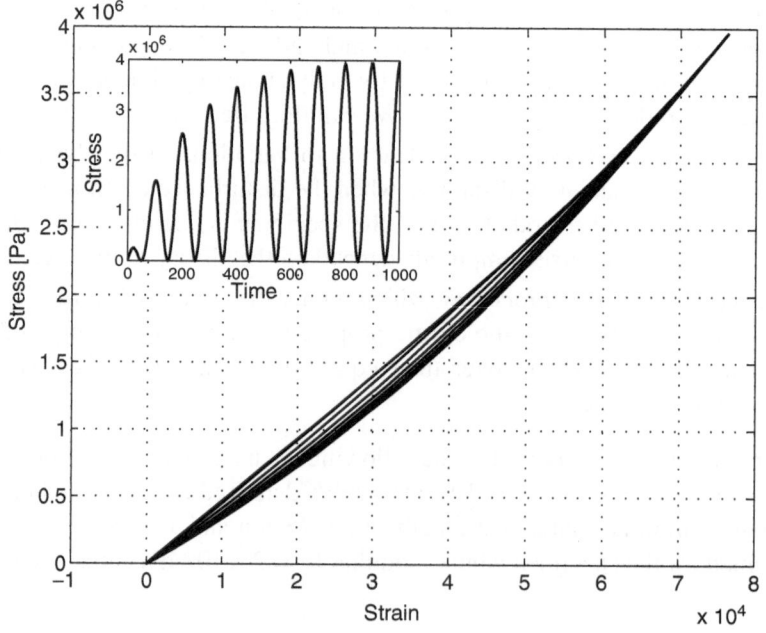

Fig. 17.1. Stress–strain curves for the applied stress protocol reported in the inset.

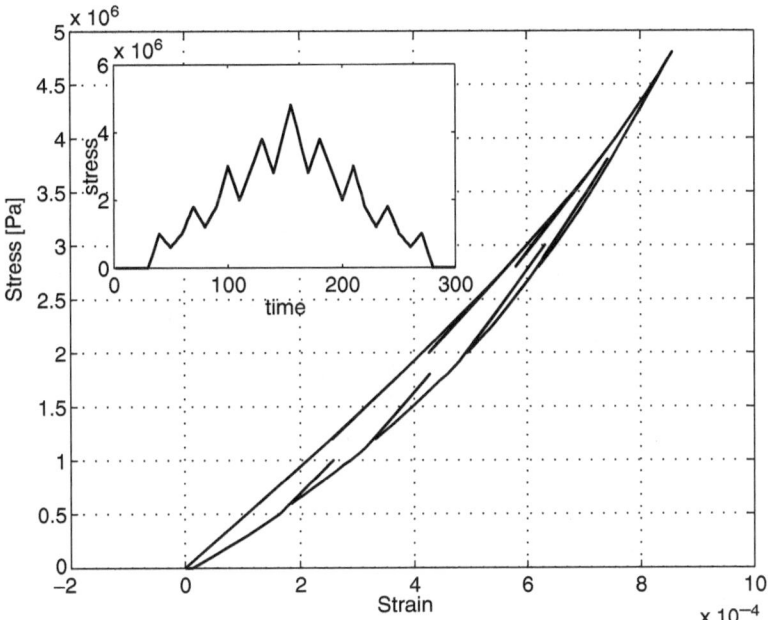

Fig. 17.2. Stress–strain curves for the applied stress protocol reported in the inset.

that the up-going branch is unchanged through successive cycles, whereas the down-going branch is affected by the maximum previously achieved strain amplitude. Finally, a typical feature of our model is the change in the average slope as a function of the maximum amplitude. Indeed it can be noted that, increasing the applied maximum stress, the average elastic constant increases (hardening with increasing amplitude).

Figure 17.2 shows the stress–strain curve for a triangular protocol with nested small amplitude loops, as shown in the inset. The curvature is the same as that in Figure 17.1, pointing to the same hardening effect. The plot also illustrates the discrete or end point memory effect: during loading each of the smaller loops terminates on the same loading curve; that is, the stress–strain curve always follows the same path and during each loading cycle the specimen "remembers" its past history. The same phenomenon also takes place during the unloading phase. Additionally, it is interesting to note that smaller loops occuring at increasing applied stresses correspond to exciting a more rigid specimen; that is, the specimen hardens at higher confining pressures, in agreement with experiments [14].

3.2 Fast Dynamics Experiments

To simulate a dynamic experiment, we consider a bar with a forced edge and free boundaries on the opposite tip. In particular, we assume that monochromatic waves of driving amplitude F_0 and varying frequency ω are injected in a rod-shaped specimen by a transducer attached to one end of the specimen. The signal is recorded by an accelerometer attached to the other end. At any given excitation level, the frequency is swept through the fundamental resonance mode ω_R of the specimen and the

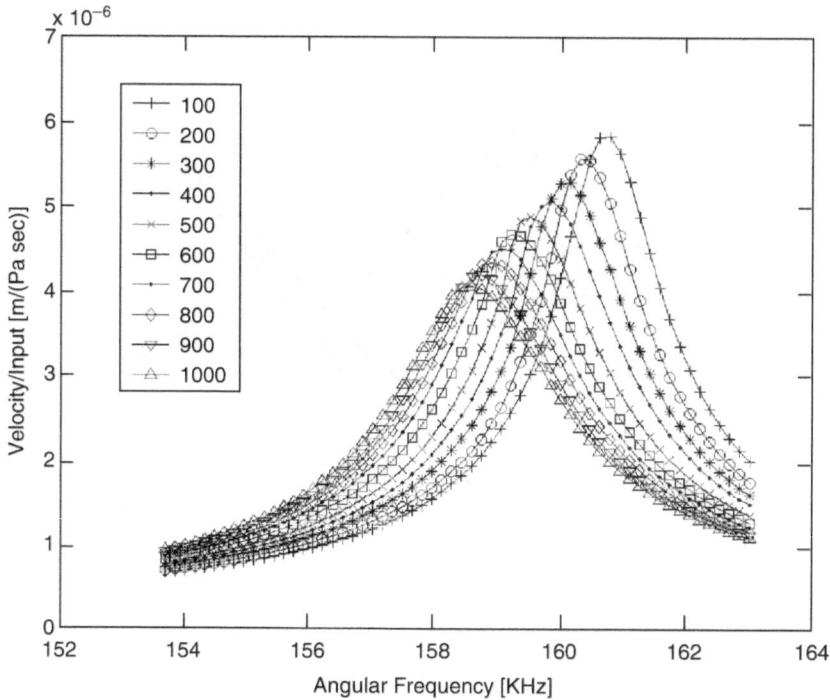

Fig. 17.3. Resonance curves for different driving amplitudes. The resonance frequency (i.e., the frequency corresponding to the peak amplitude) shifts downward and attenuation increases with increasing driving amplitudes. Here $a_1 = 0.495$, $a_2 = 6.5 \times 10^{-5}$ and $a_3 = 7$.

time-averaged acceleration amplitude A (in stationary conditions) is recorded. This procedure of resonance curve tracking is repeated for several different levels of excitation or under different environmental conditions.

Here the applied stress is $\sigma_{ext}(t) = \sigma_0 \cos(\omega t) = F_0 \cos(\omega t)/\Sigma$.

In particular, the experiments reported in this subsection are presented to support the claim that several mechanisms may equivalently cause a larger level of nonlinearity in the specimen, manifested in an increasing softening of the material. In particular we consider here softening of the material due to the increase in driving amplitude (Figure 17.3), the increase in the damage level (Figure 17.4), and abrupt temperature changes (Figure 17.5).

The resonance frequency shift is analyzed in Figure 17.3, where the average velocity recorded on the free edge of the bar is plotted versus frequency for several values of the driving amplitude. Velocities are normalized to the respective input amplitude. In agreement with experimental results, the resonance frequency ω_R is shifted downwards for increasing driving amplitudes. Note also a nonlinear attenuation effect due to hysteretic loops: the peak amplitude is not proportional to the driving amplitude F_o and the width of the resonance curve becomes larger with F_o. The resonance frequency shift is linear with the peak amplitude, in agreement with observations.

A very similar behavior is obtained when the driving amplitude is kept fixed but one of the nonlinear indicators a_i is varied. In agreement with experimental results,

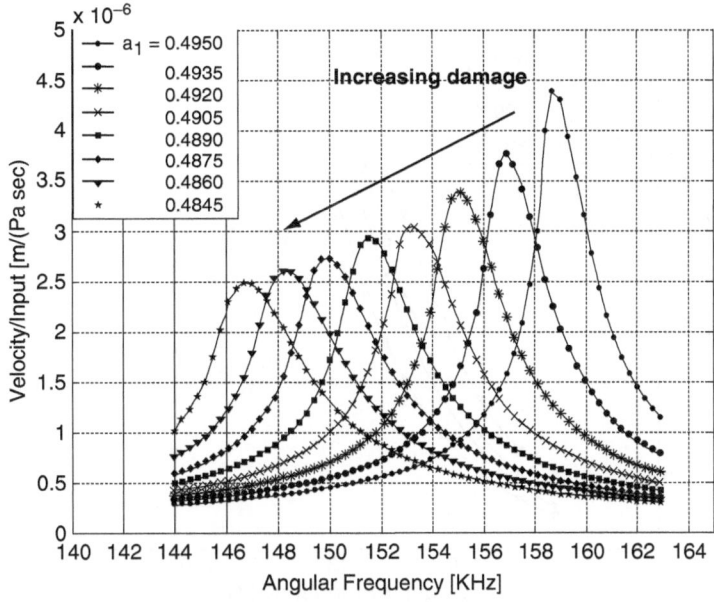

Fig. 17.4. Resonance curves for different damaged states and a forcing amplitude $\sigma_0 = 700$ Pa. Here $a_2 = 6.5 \times 10^{-5}$ and $a_3 = 7$.

a larger nonlinearity (e.g., smaller a_1 in Figure 17.4, farther from the rigid value of 0.5) corresponds to a larger resonance frequency shift and a stronger attenuation. It is interesting to notice the sensitivity of the resonance frequency on the parameter a_1: a change of approximately 2% in the parameter causes an 8% change in the resonance frequency.

As a last example of a dynamic experiment, we report the resonance curves at a fixed amplitude for decreasing temperature values, as shown in Figure 17.5. Here the experiment has been conducted by modifying the temperature after each resonance curve measurement (i.e., without allowing the system to relax). Temperature changes are reported in the plot legend. As discussed in [36, 43], the stresses generated in the material by the temperature change induce a softening which is shown in Figure 17.5, and also a large contribution to attenuation. As a result, the shift with decreasing temperature is downward, with an anomalous behavior: in fact, in a linear case, materials are expected to become stiffer with decreasing temperature. Indeed, a right shift is observed if conditioning is removed by letting the system relax after each temperature change before measuring the resonance curve (not reported for brevity).

3.3 Conditioning

As previously mentioned, conditioning consists of a temporary variation of the elastic constants after a large amplitude perturbation has been applied. To show the effect, the following virtual experiment has been performed.

- The resonance curve is measured at a small excitation amplitude ($\sigma_0 = F_0/\Sigma = 235$ Pa).

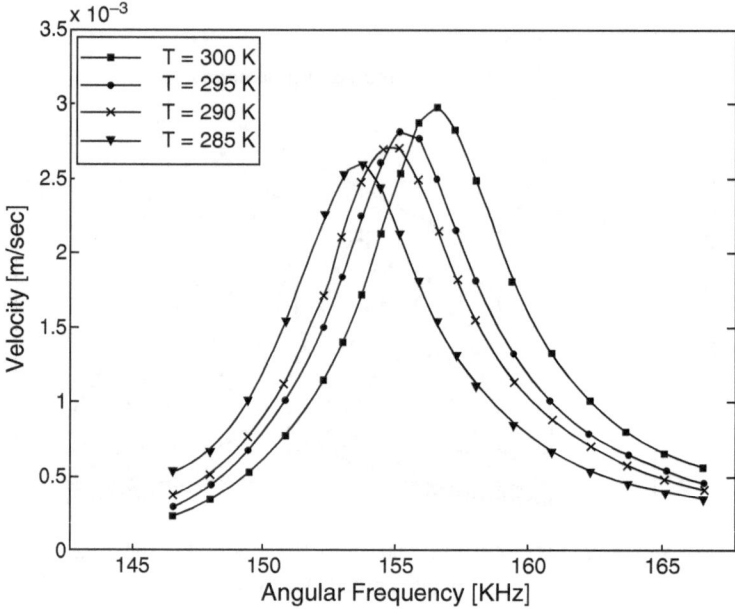

Fig. 17.5. Resonance curves for different temperatures and a forcing amplitude $\sigma_0 = 700$ Pa. Here $a_1 = 0.495$, $a_2 = 6.5 \times 10^{-5}$ and $a_3 = 7$. Note that the output velocity here is not normalized to the input amplitude.

- The measurement is repeated with a large excitation amplitude ($\sigma_0 = F_0/\Sigma = 1730$ Pa).
- A further measurement is performed at the same large amplitude.
- The resonance curve is measured again at the lower excitation amplitude.

Results in the aforementioned four cases are reported in Figure 17.6, where the following can be noted.

- The resonance angular frequency ω_1 for the first large amplitude measurement is shifted downward (as expected from the observations reported in the previous subsection) with respect to the low-amplitude resonance frequency ω_0.
- Both the resonance frequency and distortion of the curves measured at large amplitude are slightly different ($\omega_2 \neq \omega_1$); in particular, the material is softer during the repetition of the measurement at large amplitude, evidence of further conditioning (in the softening direction) of the material. Something similar is reported in Reference [21].
- When the large amplitude perturbation is removed, the resonance frequency of the material remains close to ω_2, that is, the material elastic properties have temporarily been changed even though the forcing amplitude is small. Similar experimental results are reported in Reference [22].

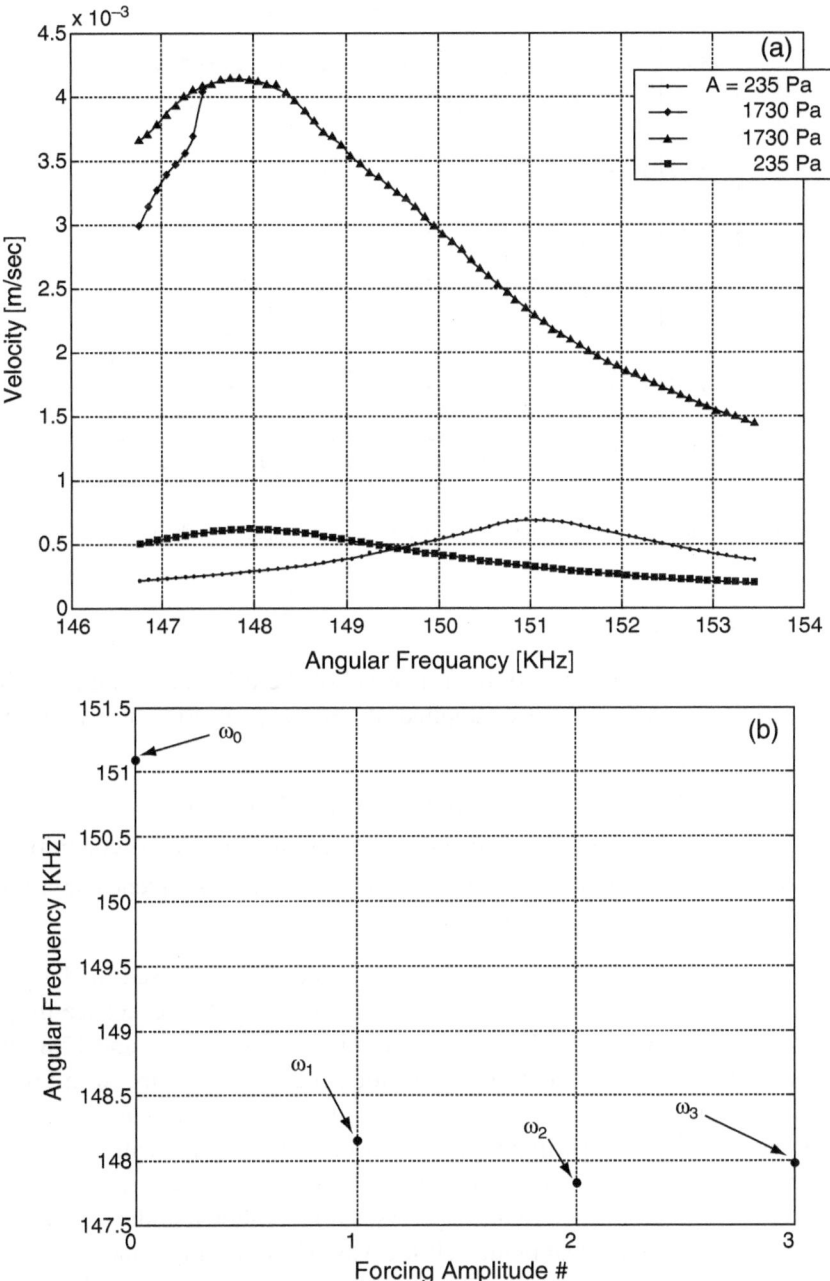

Fig. 17.6. (a) Resonance curves corresponding to two different levels of forcing amplitude; each amplitude is applied twice. (b) Angular resonance frequencies obtained from Figure 17.6a using a Lorentz-fit: ω_0 corresponds to the first curve at $\sigma_0 = 235$ Pa; ω_1 and ω_2 are obtained after conditioning at $\sigma_0 = 1730$ Pa, and ω_3 is obtained after recovery at $\sigma_0 = 235$ Pa.

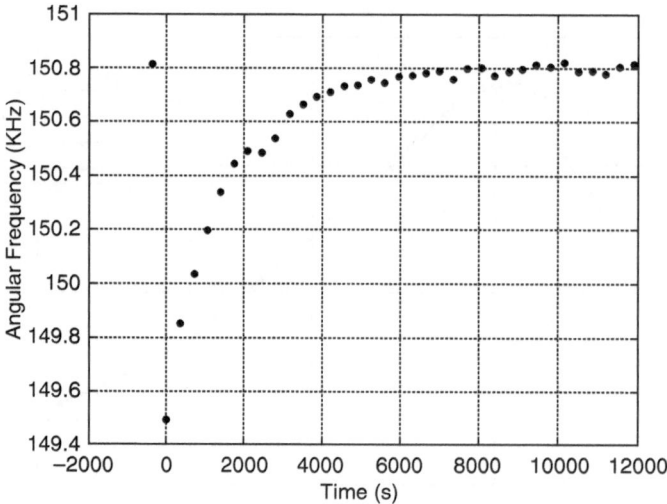

Fig. 17.7. Resonance frequency recovery after a mechanical perturbation at $t = 0$.

3.4 Relaxation Experiments

Relaxation consists in the recovery by a material of the original elastic condition after large amplitude driving has been applied. A recovery experiment is performed as follows.

- The resonance frequency is measured at a low amplitude level.
- Large amplitude loading is applied and as a result, the material is conditioned.
- After having removed the perturbation, the resonance frequency at fixed low driving amplitude is tracked in time.

Results are presented here to support the claim that the proposed model predicts, in agreement with experimental data, a log time recovery of the resonance frequency. Also, we wish to show that mechanical perturbations (Figures 17.7 and 17.8) or thermal shocks (Figure 17.9) produce similar results. Indeed, as discussed in [36, 43], a thermal shock causes mechanical deformations in the specimen that are responsible for a nonlinear softening superimposed to the usual softening/hardening due to variations of the elastic constants with increasing/decreasing temperature [14].

In Figure 17.7, the resonance frequency versus time is shown in the case when conditioning is caused by a large amplitude elastic wave (forcing amplitude $\sigma_0 = F_0/\Sigma = 800$ Pa) propagating in the specimen. The angular resonance frequency before conditioning (about 150.8 KHz) suddenly drops as soon as the large amplitude driving is applied to 149.5 KHz. The recovery process takes place in about three hours, a reasonable time scale if compared to experiments [1].

The same plot is shown in Figure 17.8, using a log scale for time. Here, the log time recovery is quite evident at least up to 7000 s. At later times, saturation inevitably occurs.

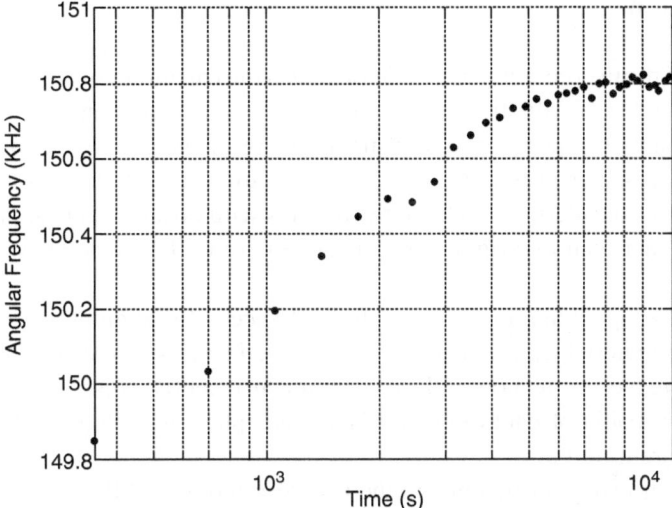

Fig. 17.8. Same as Figure 17.7, but with a log scale on the x-axes.

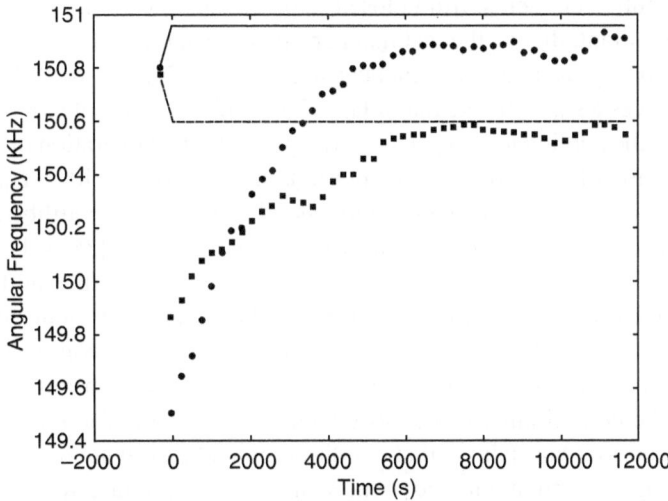

Fig. 17.9. Resonance frequency recovery after a thermal perturbation at time $t = 0$. Circles: $\Delta T = -5°$K; squares $\Delta T = +5°$K.

An altogether similar behavior is observed when a change in temperature is applied, as shown in Figure 17.9. Here, the resonance frequency at ambient temperature ($T = 300°$K) is about 150.8 KHz. When a temperature variation of $\Delta T = \pm 5°$ K is introduced, the resonance frequency suddenly drops, more so in the case of cooling. At subsequent time steps, the resonance frequency relaxes back to the value expected in the linear case, with a slightly larger value than the original case when cooling occurs (due to the hardening of the elastic constant with a temperature decrease) and smaller value when heating occurs. For more details see [36,43] and experimental data in [17, 22].

4. Discussion

The review of results presented here demonstrates the capability of the proposed model to reproduce, at least qualitatively, most of the observed phenomena in quasistatic, fast, and slow dynamics experiments. The main feature of the PM-space implementation adopted herein is indeed the possibility of simulating the whole set of experiments. This is done by considering two intersticial states with different elastic properties [24, 25, 36, 43], rather than two rigid states, as done in other PM-space-based approaches [23, 27–29].

The model adopted herein has been the first successful attempt to describe conditioning of the material [25] induced by low applied strains as recently observed [19]. To our knowledge, up to now only the approach recently presented by Vakhnenko et al. [30] is capable of capturing slow-dynamics phenomenology, although starting assumptions are different.

As a result of using elastic units in the PM description, a different interpretation of the origin of nonlinearity is suggested here. Indeed, fast-dynamics effects (in particular the resonance frequency shift) are not only due to continuous changes of state in time of some constitutive elements, from closed to open and vice versa. Rather, conditioning is more important, that is, the fact that perturbations induce changes in the material structure. In our approach, any mechanical stress with $-P_{max} < \sigma < P_{max}$, whatever its cause, produces a state change that relaxes back only because of slow dynamics, independently of the applied loading. This corresponds to the generation of a nonequilibrium configuration of some of the interstices, namely the ones with $-P_{max} < P_o < 0$ and $0 < P_c < P_{max}$. These interstices only slowly recover the equilibrium condition once the perturbation is removed, giving rise to slow dynamics [36, 43].

As we have shown in Section 3.2, the change in elastic properties at the scale level of the units used to implement the PM space, may be equivalently achieved by varying driving amplitude or temperature, or by directly changing the nonlinear parameters. In other words, the frequency shift due to increasing excitation amplitude is mediated through the conditioning of the specimen caused by the forcing itself, in a sort of feedback process. Our claim is that conditioning and fast dynamics are not only always present together, but are linked to the same mechanism, in agreement with recent experimental observations [19].

Furthermore, other features, which agree with experimental observations, are typical of a bi-elastic equation of state, and are not reproduced by classical PM-space approaches as reported in the literature. Among these is the change, with respect to the rigid (intact) case, of the resonance frequency and attenuation factor in the zero amplitude limit. In other words, the limit of the resonance frequency when the forcing amplitude goes to zero is different from the frequency of the undamaged sample.

Of course, the model presented so far is still limited by being a phenomenological model and a connection between the so-called "damage indicator parameters" $\{a_i\}$ and the structural integrity of the specimen is needed, requiring a deep understanding of the physical origin of nonlinearity [44]. Specifically, it would be desirable to link the nonlinear parameters of our approach with the output of existing damage progression models [45].

Likewise, efforts are currently being made to implement additional features into our approach, which are typical of the complexity and intrinsic to the structure of the material under investigation. Among them, the following aspects are currently being studied.

(a) *Multiscaling*: We introduce a sort of hierarchy from an upper (macroscopic) to a lower level. Each intermediate level can be regarded as a statistical ensemble of basic units and the behavior of each of these is estimated as the collective or averaged behavior of a statistical ensemble defined on the lower hierarchical level. In this approach, we are considering the implementation of transitions to irreversible states, typical of plastic transitions or mechanical failure or fracture [46]. This would enable us to consider Acoustic Emission (AE) effects [47] and, in particular, to explore the eventual correlation between AE and slow dynamics [48].

(b) *Thermodynamics of the system*: Implementation of the transition between states as stochastic events, for example, driven by the minimization of the free energy of the system, is in progress [32]. Such an approach will allow the definition of the "heat" released during the "phase transition" and its influence on the thermal equilibrium (feedback).

(c) *Frequency dependence*: Transitions driven not only by the magnitude of the applied pressure (as in the approach so far developed), but also by the rate of the pressure variation and other features will be considered in the future. In particular, to account for the fractal microstructure of the media analyzed, power law distributions (with noninteger exponents) will be chosen for both the transition rates and their correlation [47].

Acknowledgments

We thank Professor P.P. Delsanto (Politecnico di Torino) for comments and suggestions and acknowledge support from ESF (grant NATEMIS) and EC (AERONEWS grant nr. FP6-502927).

References

[1] J. Tencate, E. Smith and R.A. Guyer, Universal slow dynamics in solids, *Phys. Rev. Lett.* **85**, 1024 (2000).

[2] K. van den Abeele and J. De Vissche, Damage assessment in reinforced concrete using spectral and temporal nonlinear vibration techniques, *Cement Concrete Res.* **30**, 1453 (2000); A. Carpinteri, G. Ferro, and G. Ventura, Size effects on flexural response of reinforced concrete elements with a nonlinear matrix, *Eng. Fracture. Mech.* **70**, 995 (2003).

[3] N. Pugno and A. Carpinteri, Tubular adhesive joints under axial load, *J. Appl. Mechanics* **70**, 832–839 (2003).

[4] K. van den Abeele and K. van den Helde, Correlation between dynamic nonlinearity and static mechanical properties of corroded E-glass reinforce polyester composites, in: *Review of Progress in Quantitative Nondestructive Evaluation* (AIP Conf. Proceedings, 2000), pp. 1359.

[5] Chapter 24, in this book.

[6] K. van den Abeele et al., Non linear elastic wave spectroscopy techniques to discern material dam-age, Part II, *Res. Nondestruct. Eval.* **12**, 31–42 (2000); Micro-damage diagnostics using non linear elastic wave spectroscopy, *NDT & E Int.* **34**, 329 (2001).

[7] A. Carpinteri, B. Chiaia and P. Cornetti, On the mechanics of quasi-brittle materials with a fractal microstructure, *Eng. Fracture. Mech.* **70**, 2321 (2003).

[8] V. Zaitsev, V. Gusev and B. Castagnede, Luxemburg-Gorky effect retooled for elastic waves: A mechanism and experimental evidence, *Phys. Rev. Lett.* **89**, 105502 (2002).

[9] A.E. Romanov and J.S. Speck, Stress relaxation in mismatched layers due to threading dislocation inclination, *Appl. Phys. Lett.* **83**, 2569 (2003).

[10] P.V. Yasnii et al., Modeling of material creep damage process with a superimposed high-frequency cyclic component, *Strength of Materials* **35**, 31 (2003).

[11] Chapter 4, in this book.

[12] Chapter 1, in this book.

[13] N.G. Cook and K. Hogdson, Some detailed stress-strain curves for rock, *J. Geoph. Res.* **70**, 2883 (1965).

[14] T.J. Ulrich and T.W. Darling, Observation of anomalous elastic behavior in rock at low temperatures, *Geoph. Res. Lett.* **28**, 2293 (2001); B. Zinszner, P.A. Johnson, P.N.J. Rasolofosaon, Influence of change in physical state on elastic nonlinear response in rock: effects of confining pressure and saturation, *J. Geoph. Res.* **102**, 8105–20 (11997).

[15] P.A. Johnson, B. Zinszner and P.N.J. Rasolofosaon, Resonance and nonlinear elastic phenomena in rock, *J. Geoph. Res.* **101**, 11553 (1997).

[16] L. Ostrowsky and P.A. Johnson, Dynamic nonlinear elasticity in geomaterials, *Riv. del Nuovo Cimento* **24**, 1–46 (2001).

[17] Chapter 23, in this book.

[18] V. Aleshin, V. Gusev and V. Zaitsev, Propagation of initially bi-harmonic sound waves in a 1D semi-infinite medium with hysteretic non-linearity, *Ultrasonics* **42**, 1053–9 (2004).

[19] Chapter 25, in this book.

[20] J.A. TenCate and T.J. Shankland, Slow dynamics in the non linear elastic response of Berea sand-stone, *Geophys. Res. Lett.* **23**, 3019 (1996).

[21] M. Bentahar and R. El Guerjouma, Experimental investigations in nonlinear acoustics for damage characterisation of composite materials, submitted to *Ultrasonics* (2005).

[22] R.A. Guyer et al., Slow elastic dynamics in a resonant bar of rock, *Geoph. Res. Lett.* **25**, 1585 (1998).

[23] K. McCall and R.A. Guyer, Equation of state and wave propagation in hysteretic nonlinear elastic materials, *J. Geoph. Res.* **99**, 23887 (1994).

[24] M. Scalerandi, P.P. Delsanto, V. Agostini, K. Van Den Abeele and P.A. Johnson. Local interaction simulation approach to modeling nonclassical, nonlinear elastic behavior in solids, *J. Ac. Soc. Am.* **113**, 3049–3059 (2003); M. Scalerandi, P.P. Delsanto and P.A. Johnson, Stress induced conditioning and thermal relaxation in the simulation of quasi-static compression experiments, *J. Phys. D: Appl. Phys.* **36**, 288 (2003).

[25] P.P. Delsanto and M. Scalerandi, Modeling nonclassical nonlinearity, conditioning and slow dynam-ics effects in mesoscopic elastic materials, *Phys. Rev. B* **68**, 064107 (2003).

[26] Chapter 15, in this book.

[27] V. Gusev and V. Yu. Zaitsev, Theory of non-collinear interactions of acoustic waves in an isotropic material with quadratic non linearity, *J. Acoust. Soc. Am.* **111**, 80 (2002).

[28] B. Capogrosso Sansone and R.A. Guyer, Dynamic model of hysteretic elastic systems, *Phys. Rev. B* **66**, 2241011 (2002).

[29] Chapter 12, in this book.

[30] O.O. Vakhnenko et al., Strain-induced kinetics of intergrain defects as the mechanism of slow dy-namics in the nonlinear resonant response of humid sandstone bars, *Phys. Rev. E* **70**, 015602(R) (2004).

[31] D.J. Holcomb, *J. Geoph. Res.* **B86**, 6235 (1981); R.A. Guyer et al., Hysteresis, discrete memory and nonlinear wave propagation in rock, *Phys. Rev. Lett.* **74**, 3491 (1995).

[32] Chapter 16, in this book.

[33] P.P. Delsanto, R. Mignogna, M. Scalerandi and R. Schechter, Simulation of ultrasonic pulse propagation in complex media, in: *New Perspectives on Problems in Classical and Quantum Physics*, edited by P.P. Delsanto and A.W. Saenz, (Gordon & Breach, New Delhi, 1998), vol. 2, pp. 51–74.

[34] P.P. Delsanto and M. Scalerandi, A spring model for the simulation of wave propagation across imperfect interfaces, *J. Acoust. Soc. Am.* **104**, 2584–2591 (1998).

[35] A.S. Gliozzi, M. Nobili and M. Scalerandi, Modeling localised nonlinear damage and analysis of its influence on resonance frequencies, submitted to *J. Acoust. Soc. Am.* (2005).

[36] M. Nobili and M. Scalerandi, Temperature effects on the elastic properties of hysteretic elastic media: Modeling and simulations, *Phys. Rev. B* **69**, 104105 (2004).

[37] L.D. Landau and E.M. Lifshits, *Statistical Physics*, (Pergamon, Tarrytown, NY, 1979).

[38] A.E. Romanov and J.S. Speck, Stress relaxation in mismatched layers due to threading dislocation inclination, *Appl. Phys. Lett.* **83**, 2569 (2003).

[39] M.D. Collins et al., Wave propagation in poro-acustic media, *Wave Motion* **25**, 265 (1997).

[40] X. Xia and P.G. Wolynes, Microscopic theory of heterogeneity and nonexponential relaxations in supercooled liquids, *Phys. Rev. Lett.* **86**, 5526–8 (2001); V. Lunchenko and P.G. Woylnes, Intrinsic quantum excitations of low temperature glasses, *Phys. Rev. Lett.* **87**, 195901 (2001).

[41] M. Scalerandi, M. Nobili, M. Griffa, A.S. Gliozzi, and F. Boisa, Preisach-Mayergoyz approach to fatigue-induced irreversibility, *Phys. Rev. B* **73**, 092103 (2006).

[42] R.A. Guyer et al., Quantitative use of Preisach–Mayergoyz space to find static and dynamic elastic moduli in rock, *J. Geoph. Res.* **102**, 5281–93 (1996).

[43] P.P. Delsanto, M. Nobili and M. Scalerandi, Temperature dependence of the elastic properties of hysteretic materials, *Mater. Sci. Forum* **480–481**, 573–578 (2004). (special issue "Cross-Disciplinary Applied Research in Materials Science and Technology")

[44] Chapter 19, in this book.

[45] A. Carpinteri, B. Chiaia and P. Cornetti, A mesoscopic theory of damage and fracture in heterogeneous materials, *Theor. Appl. Fract. Mech.* **41**, 43 (2004).

[46] M. Scalerandi et al, A Preisach–Mayergoytz approach to fatigue induced irreversibility, submitted to *Phys. Rev. B* (2005).

[47] F. Bosia, A.S. Gliozzi, M. Griffa, M. Nobili, M. Scalerandi, Acoustic emission and nonlinear hysteretic elasticity: a combined model, in: Proc. of 12th International Congress on Sound and Vibration, Lisbon, Portugal (2005).

[48] M. Bentahar, private communication.

A 2-D Spring Model for the Simulation of Nonlinear Hysteretic Elasticity

Martin Hirsekorn,[1,3] Antonio Gliozzi,[1] Matteo Nobili,[1] and Koen Van Den Abeele[2]

[1]INFM - Dip. Fisica, Politecnico di Torino, C.so Duca degli Abruzzi 24, 10129 Torino, Italy.
[2]Interdisciplinary Research Center, Katholieke Universiteit Leuven Campus Kortrijk, Etienne Sabbelaan 53, 8500 Kortrijk, Belgium.
[3]To whom correspondence should be addressed: e-mail: martin.hirsekorn@polito.it.

Abstract
A technique is proposed to expand existing 1-D spring models for the simulation of nonlinear hysteresis attributed to interstitial regions between linearly elastic material grains to 2-D. Special emphasis is given to the explicit formulation of the approach in order to provide the tools for the development of specialized models and for the implementation on numerical simulation codes.

Keywords: Grain interstice model, hysteresis, nonlinear elasticity, PM-space, two dimensions

1. Introduction

For the application to NonDestructive Evaluation (NDE) of structures that are used in such diverse artifacts as aircraft fuselage and historic buildings, the development of 2-D and 3-D models is indispensable. In this chapter we describe in detail a recently proposed expansion to 2-D [1] of a 1-D spring model that uses the PM concept for the description of NCNL effects of small interstices between much larger grains with classical elastic behavior [2]. Special emphasis is placed on the explicit formulation of interstice and grain stresses and strains in the applied discretization scheme. A technique for numerical simulation of elastodynamics that is particularly suited for the implementation of local effects is the Local Interaction Simulation Approach (LISA) [3]. It has been developed for linear elasticity in 1-D [4], 2-D [5], and 3-D [6], and then expanded to simulate classical nonlinearity [7].

In Section 2 we summarize briefly the 1-D interstice model for heterogeneous materials, using the physical parameters relevant for an approach in 2-D and 3-D. The passage from 1-D to 2-D is performed in Section 3. In order to simplify the model, we express the formalism in terms of eigenstresses and eigenstrains (Section 3.3). Under the assumption of an orthotropic interstices the number of parameters reduces significantly. Then, the PM protocol of the 1-D model [2, 8] can be easily included

(Section 5). Finally, we exemplify the potential of the 2-D model by means of some simple simulations (Section 6).

2.   A 1-D Interstice Model

The 1-D model that is used as basis for the development of the 2-D model is described in [8] and [9]. There, a stress–strain relation of the interstice is proposed that includes an elastic term depending on the interstice deformation, a damping term accounting for viscoelastic effects and a pressure term, which depends on the forces applied to the interstice by the surrounding grains. The latter term originates from a "contact quality parameter" introduced in the model in order to account for imperfect contacts between material grains [2,10]. The reasons for the introduction of this term are purely practical. Basically, it only rescales the elastic constant and the damping term of the interstice. However, it allows for an easy simulation of viscoelastic and rigid (i.e., of infinite elastic modulus) interstices within the same approach. The interstice stress is hence given by

$$\sigma = -aP + K\epsilon + \eta\dot{\epsilon}, \tag{18.1}$$

where K is the elastic constant and η the dynamic viscosity parameter of the interstice. a is the parameter identified with the contact quality of the grains contiguous to the interstice.

Recently, Gliozzi et al. [8] discovered that the stress–strain relation (18.1) resembles the stress–strain relation of liquid saturated porous media derived from the theory of Biot [11]. They called it therefore the "poroelastic" interstice model. It is not yet clear under which conditions a stress–strain relation in the form of Eq. (18.1) really can be derived from Biot's theory of poroelasticity, but because NCNL elasiticy has been found in particular in porous media [12] this approach seems to be promising.

2.1 External Stress

In the intended 2-D treatment, it is convenient to use external stresses rather than pressures. The total pressure on the interstice is equal to the sum of the external forces that act on the interstice divided by the total area, on which they act. We call $\bar{F}^-$ and $\bar{F}^+$ the forces acting on the subnodes at the left and right side of the interstice, respectively. The forces are defined to be positive if they act in the positive x direction. Hence, the interstice pressure is given by

$$P = \frac{\bar{F}^- - \bar{F}^+}{2\varepsilon_y \varepsilon_z}. \tag{18.2}$$

In order to eliminate the arbitrary sizes of a material cell (ε_y and ε_z) in the directions that are not considered in the 1-D case (i.e., y and z), we use in the following force densities,

$$\bar{F} = \varepsilon_x \varepsilon_y \varepsilon_z F, \tag{18.3}$$

where $\varepsilon_i, i \in \{x, y, z\}$ is the size of a material cell in the i direction. The external forces are due to stresses in the material cells contiguous to the interstice. Stress is defined as force divided by the area, on which the force acts. Note that positive stresses in the left material cell (i.e., if it is under tension) cause a force in the negative x direction on the left subnode, and vice versa. Hence,

$$\pm \sigma^\pm = \frac{1}{\varepsilon_y \varepsilon_z} \bar{F}^\pm = \varepsilon_x F^\pm, \tag{18.4}$$

where $\sigma^\pm$ are the stresses in the material cells contiguous to the interstice in the positive $(+)$ and negative $(-)$ x direction. From this we obtain for the external interstice pressure

$$P = -\frac{\sigma^+ + \sigma^-}{2}. \tag{18.5}$$

2.2 Time Evolution of the Interstice Strain

The equation of motion on each subnode is obtained from Newton's second law

$$m^\pm \ddot{u}^\pm = \bar{F}^\pm + \bar{f}^\pm, \tag{18.6}$$

where $\bar{f}^\pm$ are the internal forces due to the interstice stress. The internal forces divided by the area, on which they act, yield the internal interstice stress σ.

$$\mp \sigma = \frac{1}{\varepsilon_y \varepsilon_z} \bar{f}^\pm = \varepsilon_x f^\pm, \tag{18.7}$$

in analogy with the definition of external force densities. We assume the interstice to be of zero mass, which is a good approximation for interstices that are much thinner than the material cells (interstice width $\delta_x \ll \varepsilon_x$). Then the mass of each subnode is equal to half the mass of the contiguous material cell; that is,

$$m^\pm = \frac{1}{2} \varepsilon_x \varepsilon_y \varepsilon_z \rho^\pm, \tag{18.8}$$

where $\rho^\pm$ are the densities of the material cells to the right $(+)$ and to the left $(-)$ of the interstice.

The interstice strain is given by the difference of the subnode displacement divided by its equilibrium width δ_x.

$$\dot{\bar{\epsilon}} = \frac{\ddot{u}^+ - \ddot{u}^-}{\delta_x} = \frac{2}{\delta_x \varepsilon_x} \left(\frac{\sigma^+}{\rho^+} + \frac{\sigma^-}{\rho^-} - \frac{\sigma}{\rho^+} - \frac{\sigma}{\rho^-} \right). \tag{18.9}$$

We simplify this relation defining the reduced and average densities

$$\frac{1}{\rho_{red}} = \frac{1}{2} \left(\frac{1}{\rho^+} + \frac{1}{\rho^-} \right) \qquad \rho_{av} = \frac{1}{2} (\rho^+ + \rho^-), \tag{18.10}$$

and the external stress

$$\sigma^{ext} := \frac{1}{2} \rho_{red} \left(\frac{\sigma^+}{\rho^+} + \frac{\sigma^-}{\rho^-} \right) = \frac{\rho^- \sigma^+ + \rho^+ \sigma^-}{2 \rho_{av}}, \tag{18.11}$$

and obtain

$$\ddot{\epsilon} = \frac{4}{\delta_x \varepsilon_x \rho_{red}} (\sigma^{ext} - \sigma). \qquad (18.12)$$

Note that in homogeneous media, the external stress is equal to the mean of the stresses in the contiguous material cells. For inhomogeneous media, we define the external stress differently, in order to obtain a driven damped harmonic oscillator relation for the interstice strain as in the case of homogeneous media [Eq. (18.1)]. Interpreting the external pressure as the negative external stress yields the following differential equation for the time dependence of the interstice strain,

$$\ddot{\epsilon} = \frac{4}{\delta_x \varepsilon_x \rho_{red}} \left((1 - a)\sigma^{ext} - K\epsilon - \eta\dot{\epsilon} \right). \qquad (18.13)$$

2.3 Rigid Interstice

We define as rigid an interstice whose strain is constant with time and thus does not depend on the external stress. To fulfill this, we have to set the parameter a in Eq. (18.13) to 1. A constant strain rate $\dot{\epsilon}$ is guaranteed by a zero elastic parameter K. If both conditions are satisfied, the interstice is permanently rigid. In a nonlinear hysteretic model (see Section 5), an interstice is not permanently rigid, but may be subject to state transitions from, for example, viscoporoelastic to rigid. In this case, $\dot{\epsilon}$ is in general not zero when the interstice becomes rigid. Then the parameter η in Eq. (18.13) is related to the time τ needed by the interstice to become rigid by

$$\eta = \frac{\delta_x \varepsilon_x \rho_{red}}{4\tau}. \qquad (18.14)$$

2.4 Equation of Motion of the Interstice Mass Center

The displacement of the interstice mass center is obtained from the average of the weighted displacement of the subnodes

$$u^{mc} = \frac{m^+ u^+ + m^- u^-}{m^+ + m^-} = \frac{\rho^+ u^+ + \rho^- u^-}{2\rho_{av}}. \qquad (18.15)$$

Its acceleration is then calculated using Eq. (18.6),

$$\ddot{u}^{mc} = \frac{\sigma^+ - \sigma^-}{\varepsilon_x \rho_{av}}. \qquad (18.16)$$

This is equivalent to the equation of motion for a completely linear material as used for the simulation of linear wave propagation [4]. However, because the stresses in the linear material cells are calculated from the displacements of the contiguous subnodes (not of the mass center), the NCNL effects created in the interstice propagate through the material.

3. From 1-D to 2-D

In a 2-D model, we split the specimen into rectangular cells of linear elastic material and rectangular NCNL interstices, which connect the linear material cells at their corners (see Figure 18.1a). The interstice is now represented by the four subnodes at its

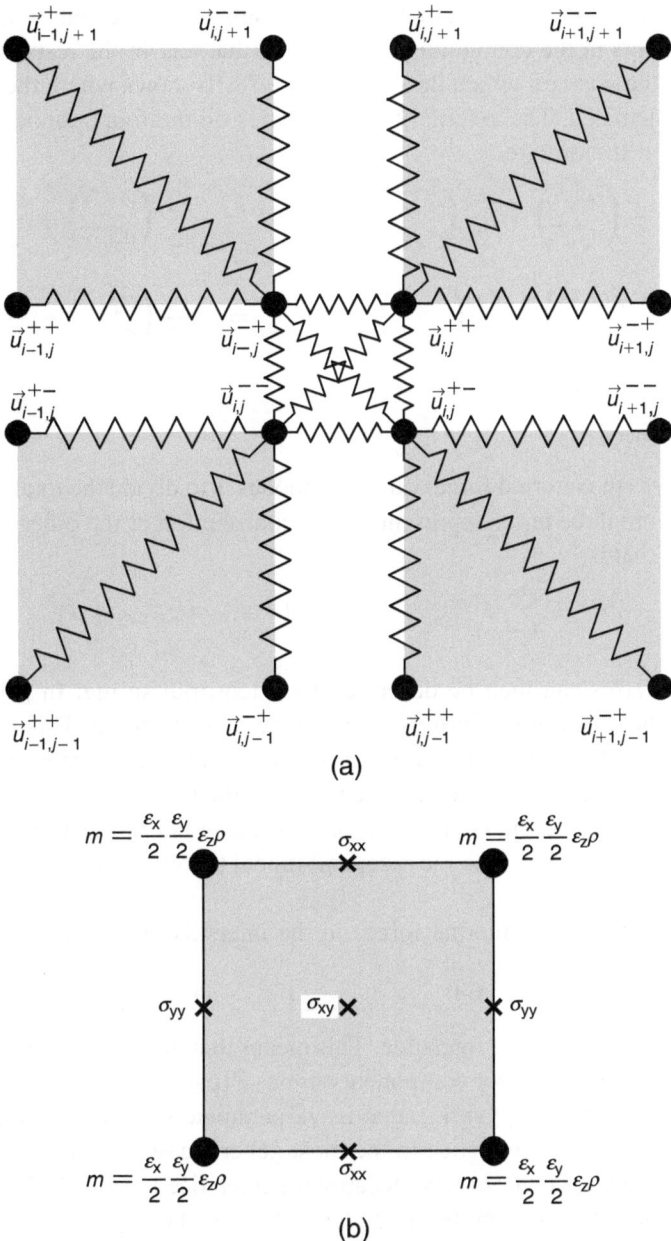

(a)

(b)

Fig. 18.1. (a): 1-D schematization of the nodes (black dots) and the force terms (springs) of four material cells (gray areas) and the connecting interstice. (b): Locations where the elastic stresses in a material cell are calculated. The mass of the cell is shared by the four contiguous subnodes.

corners (displacement vectors $\vec{u}^{++}, \vec{u}^{-+}, \vec{u}^{--}$, and $\vec{u}^{+-}$). Each subnode is also part of one of the contiguous material cells. Therefore, the total force that acts on a subnode is due to the interstice stress and the stress in the respective material cell.

3.1 External Forces

The components of the total external force on a subnode are given by the sum of all stress components in the contiguous material cell that tear in the respective direction multiplied by the area, on which they act. Figure 18.1b shows where the material cell stresses are calculated. The external force densities on the four subnodes in terms of the material cell stresses are

$$
\vec{F}^{++} = \frac{\varepsilon_x}{2} \begin{pmatrix} \sigma_{xx}^{++} \\ \sigma_{xy}^{++} \end{pmatrix} + \frac{\varepsilon_y}{2} \begin{pmatrix} \sigma_{xy}^{++} \\ \sigma_{yy}^{++} \end{pmatrix} \qquad
\vec{F}^{-+} = -\frac{\varepsilon_x}{2} \begin{pmatrix} \sigma_{xx}^{-+} \\ \sigma_{xy}^{-+} \end{pmatrix} + \frac{\varepsilon_y}{2} \begin{pmatrix} \sigma_{xy}^{-+} \\ \sigma_{yy}^{-+} \end{pmatrix}
$$

$$
\vec{F}^{--} = -\frac{\varepsilon_x}{2} \begin{pmatrix} \sigma_{xx}^{--} \\ \sigma_{xy}^{--} \end{pmatrix} - \frac{\varepsilon_y}{2} \begin{pmatrix} \sigma_{xy}^{--} \\ \sigma_{yy}^{--} \end{pmatrix} \qquad
\vec{F}^{+-} = \frac{\varepsilon_x}{2} \begin{pmatrix} \sigma_{xx}^{+-} \\ \sigma_{xy}^{+-} \end{pmatrix} - \frac{\varepsilon_y}{2} \begin{pmatrix} \sigma_{xy}^{+-} \\ \sigma_{yy}^{+-} \end{pmatrix}.
$$

$$(18.17)$$

3.2 Internal Forces

A way to express the internal forces on the subnodes is to divide the total internal force on a subnode into three terms representing the contribution of the other subnodes (see Figure 18.1a); that is,

$$
\vec{f}^{(n)} = \sum_k \vec{f}^{(nk)}, \qquad n, k \in \{++, -+, --, +-\}. \tag{18.18}
$$

Each of these terms can then be described by a tensorial spring. In [2], the tensors describing the behavior of the springs are assumed to be diagonal. This means that the differences in the displacement components in Cartesian coordinates are uncoupled; that is, for example, a relative displacement in y of the $(++)$ node with respect to the $(-+)$ node would move the $(-+)$ node only in the y but not in the x direction. In general, this cannot be true. The most general model therefore has to assume tensorial springs.

Because the sum of the internal forces in the interstice must be equal to zero, we obtain

$$
f^{(nk)} = -f^{(kn)}, \qquad n, k \in \{++, -+, --, +-\} \tag{18.19}
$$

for the tensorial springs of the interstice. This means that in this model we have to consider 6 (springs) times 4 (tensor components) times 3 [parameters of the viscoporoelastic stress–strain relation, Eq. (18.1)], that is, 72 parameters for a complete description of a 2-D viscoporoelastic interstice in the most general case. These parameters, however, are not all independent. In fact, because the interstice is described completely by the four subnodes, its effect on the subnodes is described completely by 4 (subnodes) times 2 (components), that is, 8 independent forces. In Cartesian coordinates, however, a single subnode and also the displacement difference of two subnodes cannot be expressed independently of the other subnodes. It is therefore convenient to regroup the subnode displacements into physically meaningful interstice deformations.

3.3 Interstice Eigenstrains

In the 1-D model, the static interstice is completely described by the displacement of the two subnodes. Therefore, apart from the displacement of the mass center, there is only one independent variable: the interstice strain. In the 2-D model, however, we have 4 subnodes with 2 independent displacement components each. Therefore, we have 8 independent variables: 2 mass center displacement components, and 6 independent strain components. We choose three of them to be the volumetric, deviatoric, and shear eigenstrain of the interstice (see Section 4), calculated at the center of the interstice. The other three strains are chosen such that the displacement difference between two arbitrary subnodes can be expressed in a unambiguous way in terms of the interstice strains. They can be interpreted as eigenstrain differences, that is, the change of the eigenstrains if calculated at different locations in the interstice (see Appendix A.1),

$$
\epsilon_V = \frac{1}{2(\delta_x + \delta_y)}(u_x^{++} - u_x^{-+} - u_x^{--} + u_x^{+-} + u_y^{++} + u_y^{-+} - u_y^{--} - u_y^{+-})
$$

$$
\tilde{\epsilon}_V = \frac{1}{2(\delta_x + \delta_y)}(-u_x^{++} + u_x^{-+} - u_x^{--} + u_x^{+-} - u_y^{++} + u_y^{-+} - u_y^{--} + u_y^{+-})
$$

$$
\epsilon_D = \frac{1}{2(\delta_x + \delta_y)}(u_x^{++} - u_x^{-+} - u_x^{--} + u_x^{+-} - u_y^{++} - u_y^{-+} + u_y^{--} + u_y^{+-})
$$

$$
\tilde{\epsilon}_D = \frac{1}{2(\delta_x + \delta_y)}(-u_x^{++} + u_x^{-+} - u_x^{--} + u_x^{+-} + u_y^{++} - u_y^{-+} + u_y^{--} - u_y^{+-})
$$

$$
\epsilon_S = \frac{1}{2(\delta_x + \delta_y)}(u_x^{++} + u_x^{-+} - u_x^{--} - u_x^{+-} + u_y^{++} - u_y^{-+} - u_y^{--} + u_y^{+-})
$$

$$
\tilde{\epsilon}_S = \frac{1}{2(\delta_x + \delta_y)}(u_x^{++} + u_x^{-+} - u_x^{--} - u_x^{+-} - u_y^{++} + u_y^{-+} + u_y^{--} - u_y^{+-}).
$$

$$(18.20)$$

Hence, the displacement of each subnode can be expressed in terms of these 6 strains and the displacement of the interstice's mass center (see Appendix A.1). In order to analyze the time dependence of these interstice strain components, we proceed as in Section 2.2 for the 1-D case, and obtain equivalent relations

$$
\ddot{\epsilon}_i = \frac{16}{(\delta_x + \delta_y)(\varepsilon_x + \varepsilon_y)\rho_{red}}(\sigma_i^{ext} - \sigma_i)
$$

$$
\ddot{\tilde{\epsilon}}_i = \frac{16}{(\delta_x + \delta_y)(\varepsilon_x + \varepsilon_y)\rho_{red}}(\tilde{\sigma}_i^{ext} - \tilde{\sigma}_i),
$$

$$(18.21)$$

where $i \in \{V, D, S\}$ indicates the eigenstrain and eigenstress components as defined in Eq. (18.20) and Appendix A.2. A detailed derivation of these equations and the definitions of the external and internal stresses can be found in Appendix A.2. The inverse relations are also derived, which permit us to express the total internal and total external forces on a single subnode in terms of these internal and external stresses.

3.4 Equation of Motion of the Interstice Mass Center in 2-D

The mass center displacement of the 2-D interstice is given by

$$\vec{u}^{mc} = \sum_n \frac{m^{(n)}\vec{u}^{(n)}}{m} = \sum_n \frac{\rho^{(n)}\vec{u}^{(n)}}{\rho_{av}}, \qquad n, \in \{++, -+, --, +-\}. \qquad (18.22)$$

Hence, its acceleration in terms of external and internal force densities is

$$\ddot{\vec{u}}^{mc} = \sum_n \frac{\vec{F}^{(n)}}{\rho_{av}} + \sum_n \frac{\vec{f}^{(n)}}{\rho_{av}}, \qquad n, \in \{++, -+, --, +-\}. \qquad (18.23)$$

The total sum of the internal forces on all subnodes must be equal to zero. Thus, the interstice acceleration is given by the sum of the external force densities divided by the average density of the contiguous material cells. The motion of the mass center is completely independent of what happens in the interstice.

4. A 2-D Interstice Model

With Eqs. (18.21) we are able to compute the interstice strain dependence, if we know the external and internal force densities. The external force densities are easily obtained from the stresses in the surrounding linearly elastic material cells, whereas the internal force densities depend on the assumptions of the physical properties of the interstice that we introduce in our model. In this work, we assume the interstice stress–strain relation of Eq. (18.1) [8,13]; that is, the internal interstice stresses depend on the external stresses, the interstice strain, and strain rate. In the most general case, any internal force component of any subnode may depend on any component of the external force on that subnode, and any component of displacement and displacement velocity of the other subnodes relative to the considered subnode.

We assume that the interstice is orthotropic in its internal behavior. In this case the eigenvectors of the stress–strain relation are the volumetric ϵ_V, deviatoric ϵ_D, and shear ϵ_S strain, as defined in Eq. (18.20). These eigenvectors are independent of each other; that is, in these coordinates the matrix parameters become diagonal. Initially we assume that the eigenstrain differences $\tilde{\epsilon}_V, \tilde{\epsilon}_D$, and $\tilde{\epsilon}_S$ as defined in Eq. (18.20) are also independent of each other and independent of the eigenstrains. With these assumptions, the number of independent parameters of the model is reduced to 18. The viscoporoelastic relations for the internal stresses are then given, in analogy to the 1-D case, by

$$\sigma_i = a_i \sigma_i^{ext} + K_i \epsilon_i + \eta_i \dot{\epsilon}_i \qquad \tilde{\sigma}_i = \tilde{a}_i \tilde{\sigma}_i^{ext} + \tilde{K}_i \tilde{\epsilon}_i + \tilde{\eta}_i \dot{\tilde{\epsilon}}_i, \qquad (18.24)$$

with $i \in \{V, D, S\}$. From Eq. (18.21) we obtain the differential equations for the time dependence of the interstice strains

$$\ddot{\epsilon}_i = \frac{16}{(\delta_x + \delta_y)(\varepsilon_x + \varepsilon_y)\rho_{red}} \left((1 - a_i)\sigma_i^{ext} - K_i \epsilon_i - \eta_i \dot{\epsilon}_i\right)$$

$$\ddot{\tilde{\epsilon}}_i = \frac{16}{(\delta_x + \delta_y)(\varepsilon_x + \varepsilon_y)\rho_{red}} \left((1 - \tilde{a}_i)\tilde{\sigma}_i^{ext} - \tilde{K}_i \tilde{\epsilon}_i - \tilde{\eta}_i \dot{\tilde{\epsilon}}_i\right). \qquad (18.25)$$

Note that for each eigenstrain we have obtained the relation of a driven damped harmonic oscillator, as in the 1-D case for the interstice strain [Eq. (18.13)].

In our model we have assumed that the interstice is much thinner than the material cells. Therefore, in a good approximation, the interstice eigenstrain differences $\tilde{\epsilon}_V$, $\tilde{\epsilon}_D$, and $\tilde{\epsilon}_S$ are zero (see Appendix A.1). This condition is realized if we impose the condition for a rigid interstice in 1-D (Section 2.3) on the internal eigenstress differences and set the interstice strain differences and its first derivative to zero at the beginning of the simulations. For the interstice parameters, this means imposing the values $\tilde{a}_i = 1$ and $\tilde{K}_i = 0$ for all $i \in \{V, D, S\}$. Because the rigid condition is imposed permanently, the damping parameters $\tilde{\eta}_V$, $\tilde{\eta}_D$, and $\tilde{\eta}_S$ have no influence on the interstice behavior and can be eliminated. Thus, the behavior of a 2-D viscoporoelastic interstice is described by nine independent parameters. If state transitions are allowed, the damping parameter of an interstice that becomes rigid within a time τ, is obtained, analogously to the 1-D case, by

$$\eta_i = \frac{(\delta_x + \delta_y)(\varepsilon_x + \varepsilon_y)\rho_{red}}{16\tau}.$$ (18.26)

5. A Two-State Hysteretic Protocol

Nonclassical Nonlinear (NCNL) effects, such as hysteresis, end point memory, slow dynamics effects, and so on, may be introduced into the interstice behavior by defining a minimum of two different states for the interstices (here, we choose the simplest case of two different states [9, 14]). The nonlinearity of the interstice emerges as a result of phase transitions between the two states. In the 1-D model, the external stress on the interstice has been chosen as driving parameter for the state transitions. Other possible choices are the internal stress or the interstice strain.

For distinction we call the two states "open" and "closed" [8]. The names suggest the necessary conditions for the driving parameter being that the interstice is in a certain state. The closed state is the state that is reached if the external stress falls under a certain "closure stress" σ^c (negative stress corresponds to compression of the interstice, therefore closed state). The open state is reached if the stress rises over a certain "opening stress" σ^o (tension of the interstice). Hysteresis occurs, if the opening stress is larger than the closure stress. In this case, if the external stress σ^{ext} on the interstice is larger than σ^o, the interstice is always open. If $\sigma^{ext} < \sigma^c$ the interstice is always closed. If the external stress lies between opening and closure stress $\sigma^c < \sigma^{ext} < \sigma^o$, the state of the interstice is determined by the history of the external stress; that is, it remains in the state that has most recently been imposed. Figure 18.2a shows a schematization of the state transition protocol. The opening and closure stresses can be identified as coordinates in the so-called PM-space (Preisach–Mayergoyz space) [8]. In our case of a viscoporoelastic interstice, the two coordinates of the PM-space are the opening and the closure stress, defining the state transitions of the interstice (see Figure 18.2b). In a NCNL elastic material, there are a large number of these hysteretic interstices, whose opening and closure stresses are statistically distributed over the PM-space. Relaxation phenomena, such as creep and the so called "slow dynamics,"

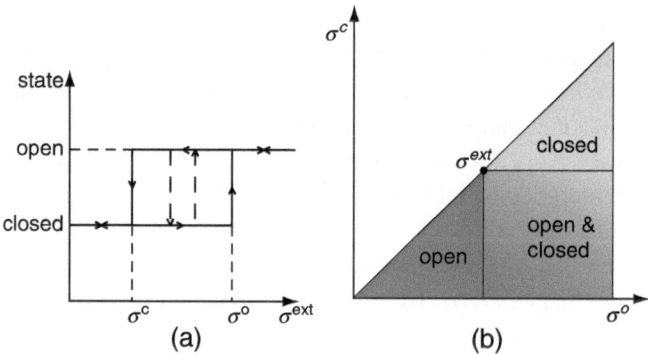

Fig. 18.2. (a): Illustration of the state transition protocol. Dashed arrows indicate random transitions that may occur at any external stress σ^{ext} with $\sigma^c < \sigma^{ext} < \sigma^o$. (b): Sample distribution of open and closed states in a PM-space. The elements that have a closure stress $\sigma^c > \sigma^{ext}$ are closed, the elements with opening stress $\sigma^o < \sigma^{ext}$ are open. The state of the elements with $\sigma^c < \sigma^{ext} < \sigma^o$ depends on the history of σ^{ext}.

are taken into account if we permit thermal state transitions [8]. These thermal transitions occur randomly if the external stress lies between opening and closure stress.

Because in contrast to 1-D, in 2-D there is more than one independent stress–strain relation (three in our model), the choice of an appropriate driving parameter and its associated state transitions is not straightforward. An intuitive choice is to assume that the state transitions of the eigenstress–eigenstrain relations [Eq. (18.24)] are independent of each other and driven by the respective eigenstress. For example, if the external volumetric stress σ_V^{ext} rises over a certain volumetric opening stress σ_V^o, only the parameters of the interstice associated to the volumetric stress–strain behavior (i.e., a_V, K_V, and η_V) change. The deviatoric and shear parameters are not affected. Another possibility is to assume that the state transitions of every eigenstress–eigenstrain relation are strictly connected. In this case, changes in the elastic behavior of the interstice under volumetric, deviatoric, and shear stresses are driven by the same driving parameter (e.g., σ_V) and occur therefore always simultaneously. The advantage of this description is the great reduction of independent parameters to only three. These assumptions describe a very special case of NCNL elasticity, but facilitate the analysis of the model predictions. The simulations presented in the following section are carried out using this simplified hysteretic protocol.

Detailed experimental investigations have to show whether one of these assumptions correctly describes the behavior of NCNL materials. Most probably, neither of them will be correct for anisotropic media. In isotropic media, however, the analysis of a potential coupling between applied stresses and stress–strain relations of other eigenstates (e.g., a change in the shear modulus due to a volumetric stress) may be of great relevance for the analysis and characterization of different defect types.

In the following, we define the closed state as "rigid" in the sense of Section 4. This choice is arbitrary and corresponds to the simplest case of a hysteretic protocol. Other choices may be two elastic states with different parameters [15], two rigid states with different strains; that is, strain jumps occur at the state transitions [16]. Broken states may be introduced easily in this model [8]. In the 2-D case, however, we have to distinguish if the interstice behaves as a broken interstice with respect to volumetric, deviatoric, shear, or any kind of external stress.

6. Numerical Simulations

In order to illustrate the model, we perform a set of 2-D simulations on a linearly elastic material specimen with an NCNL inclusion. As material properties we use the values of aluminum (Lamé constants $\lambda = 58.5\,\mathrm{GPa}$, $\mu = 26.0\,\mathrm{GPa}$, material density $\rho = 2700\,\mathrm{kg/m^3}$). The specimen size is 12 mm in x and 6 mm in y. A plane wave is injected in the x-direction at the border of the specimen. In order to account for a perfectly plane incident wave, we implement periodic boundary conditions in y; that is, the simulated specimen is a section of a material slab, which is infinitely extended in y where in the simulated section is periodically repeated. The boundary condition at the incidence side is given by the forcing that produces the injected wave. The boundary at the opposite side is free (zero stress). The specimen is discretized using a spatial grid of square cells with a side length of $\varepsilon_x = \varepsilon_y = 0.02\,\mathrm{mm}$. The temporal discretization step is $\tau = 2.8\,\mathrm{ns}$, which fulfills the condition of numerical stability $\frac{\varepsilon}{\tau} > c$, where c is the maximum wave propagation velocity of the materials present in the simulation.

The linearly elastic part of the specimen is simulated by imposing rigid contact to the interstices between the material grains (identified with the grid cells). Approximately in the center of the specimen (centered at $x = 5.5\,\mathrm{mm}$ and $y = 3.0\,\mathrm{mm}$), we include a squared zone of NCNL elasticity (side length 3 mm = 150 grid cells), where state transitions of the interstices are allowed from rigid to elastic and back again. The dimensions of the interstices are $\delta_x = \delta_y = 10^{-5}\,\mathrm{m}$. The contact quality parameter of the elastic state is $a_i = 0.95$, $i \in \{V, D, S\}$, the elastic moduli K_i 0.01 times the elastic moduli of the material grains, and the damping parameters $\eta_i = 0.95\,\mathrm{Pa\,s}$, which corresponds to a relaxation time of 2τ [see Eq. (18.26)]. These parameters yield strong nonlinear behavior without causing numerical instability problems during the simulations. The interstice opening and closure are driven by the external volumetric stress. The state transition values are distributed homogeneously in a PM-space of minimum stress $\sigma_{\min} = 10\,\mathrm{kPa}$ and maximum stress $\sigma_{\max} = 10\,\mathrm{kPa}$. This corresponds to about 68% coverage of the PM-space at maximum amplitude (5.5μm) in the simulations described in the following.

In the first simulation we inject a plane sinusoidal wave of frequency $f = 0.3\,\mathrm{MHz}$ and displacement amplitude $A = 5\,\mu$m in x. The displacement as a function of time is recorded at different locations at the right boundary (at $y = 3.0\,\mathrm{mm}$, 1.5 mm, 1.0 mm, and 0.6 mm from the boundary) over 30,000 time steps (corresponding to 84 μs). The Fourier transform of the signal is compared to the linear case (all interstices in the specimen fixed to the rigid state), in which the elastic wave propagates unaffectedly. Figure 18.3 shows the differences between the displacement spectrum of the specimen with the nonlinear inclusion and the spectrum without inclusion, normalized by the longitudinal displacement spectrum of the specimen without inclusion. In the longitudinal spectrum (left graph of Figure 18.3), the amplitude of the fundamental frequency

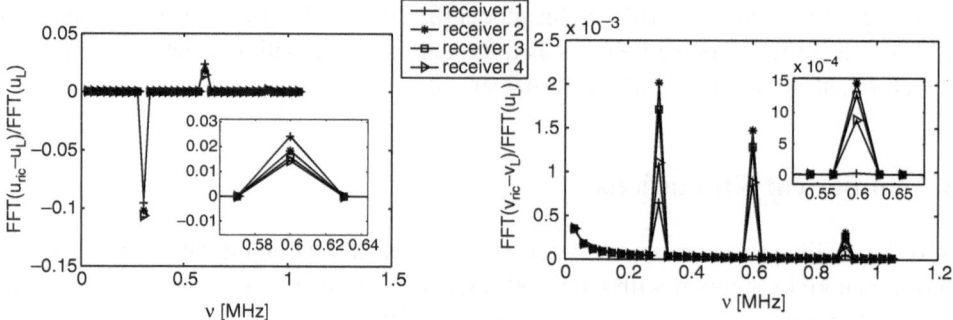

Fig. 18.3. Results of the Fast Fourier Transform (FFT) of the displacement time signal recorded at $y = 3.0\,\mathrm{mm}$ (receiver 1), $1.5\,\mathrm{mm}$ (2), $1.0\,\mathrm{mm}$ (3), and $0.6\,\mathrm{mm}$ (4). The difference between the spectrum obtained from the specimen with NCNL inclusion and the linear specimen without inclusion are normalized by the longitudinal linear spectrum. The left graph shows the comparison for the longitudinal, the right graph for the shear displacement component.

is reduced with respect to the linear specimen. This is partially due to reflection of the wave from the defect, and partially due to energy conversion from the fundamental frequency to higher harmonics, which can be clearly observed in the spectrum. In the inset, the peaks of the second harmonic are shown more in detail.

At the central receiver (at $y = 3.0\,\mathrm{mm}$), the shear displacement is much smaller than at the other receivers (right graph of Figure 18.3). This is due to the geometric symmetry of the experiment. In a deterministic experiment, the symmetry should elimininate completely the shear displacement component along the symmetry axis. However, in the NCNL inclusion, the rigid and elastic states are distributed statistically, causing a slight asymmetry. In material characterization, such symmetries can be of great importance, because they it may hide NCNL effects even if they are present in the material.

In order to analyze the dependence of the creation of higher harmonics on the amplitude of the incident wave, we repeated the simulation for different excitation amplitudes A (from $0.5\,\mu\mathrm{m}$ to $5.5\,\mu\mathrm{m}$ in steps of $0.5\,\mu\mathrm{m}$). The displacement is recorded at $y = 1.0\,\mathrm{mm}$, because there the symmetry effects are minimized. The time dependence of the longitudinal and the shear wave component are shown in Figures 18.4a and b, respectively. The longitudinal wave is dominated by the fundamental frequency component. Hence, its shape seems to be always sinusoidal. In the shear wave, the contribution of higher harmonics is much stronger compared to the fundamental frequency component, causing a visibly distorted wave form. The Fourier transforms of the signal are shown in Figures 18.4c and d. The amplitude of the higher harmonics increases with increasing A. From the spectra it can be seen that its increase is much larger than the increase of the amplitude of the fundamental component, which is almost linear with A. This is due to the fact that a higher amplitude causes state transitions of more interstices in the PM-space. As a consequence, the nonlinearity increases.

Figure 18.5 shows the amplitudes of the second and third harmonics as a function of the amplitude of the fundamental frequency of the transmitted wave in a double-logarithmic plot. The fitted lines show that the dependence follows a power law with

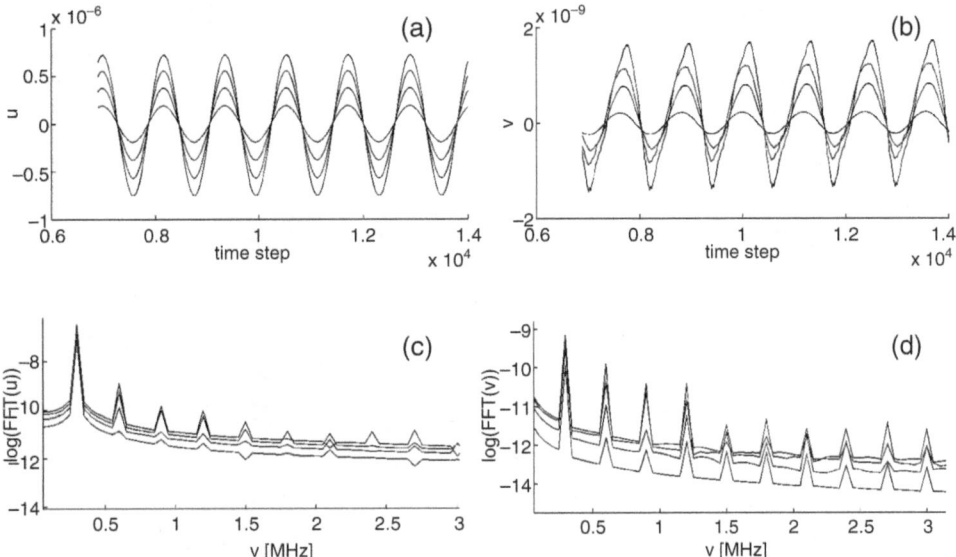

Fig. 18.4. Dependence of the transmitted wave displacement (in m) on the incident wave amplitude A. The time signal is shown in (a) longitudinal and (b) shear. The logarithm of the Fourier transform is shown in (c) londitudinal and (d) shear.

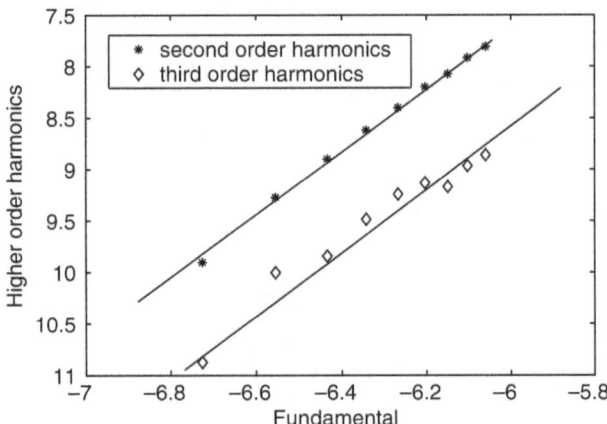

Fig. 18.5. Dependence of the second and third harmonics' amplitude on the amplitude of the fundamental frequency component of the longitudinal transmitted wave. The amplitudes are shown as logarithm of the displacement in units of m.

an exponent between 2 and 3. The lines are almost parallel; that is, the second and third harmonic follow the same power law, which is a nonclassical effect.

Finally, in order to visualize the 2-D effects of the simulations, we injected a transient Gaussian pulse of displacement amplitude $A = 5\,\mu$m in x and full time width at half maximum $\Delta t = 84$ ns. Figure 18.6 shows the displacement contour of the specimen at different time steps for a square and a circular (diameter 3 mm) NCNL inclusion. The wave is slowed down within the NCNL zone, indicating a softening of

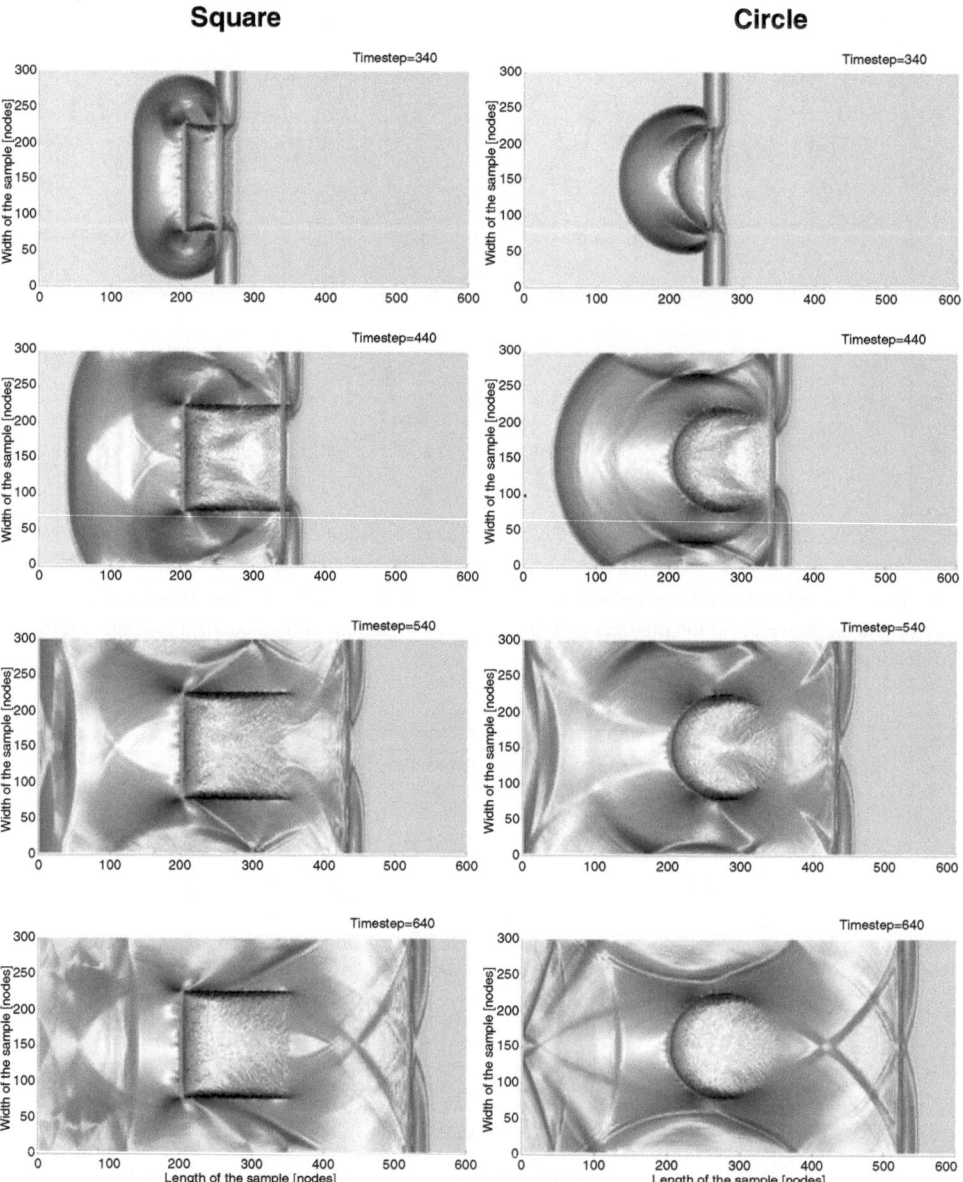

Fig. 18.6. Displacement amplitude in the specimen at different time steps during the passage of a Gaussian pulse. The maximum amplitude is represented by the white color.

the material. In the reflected pulse, two maxima can be distinguished clearly. This indicates a mode conversion from longitudinal to shear wave components, which occurs also in the case of a circular inclusion. This mode conversion is due to the difference between the effective modulus of the linear specimen and the inclusion (indicated by the lower pulse propagation velocity in the inclusion). Because the interstices in the NCNL zone are in different states (depending on its opening and closure stresses), the inclusion is not homogeneous. This causes multiple reflections within the inclusion and is responsible for the grainy shape of the pulse within the inclusion and of the transmitted and reflected pulses.

7. Conclusions

The approach presented in this chapter gives a basic idea of how to generally develop a 2-D expansion of the 1-D interstice model proposed in [2]. The utilization of eigenstress–eigenstrain constitutive relations allows for an easy implementation of the phenomenology of the 1-D model in a 2-D or 3-D context under conservation of the material symmetries. Here, the approach is presented for orthotropic materials, because the derivation of eigenstress–eigenstrain relations is straightforward and the expressions are simple. Analogous relations can be derived for any kind of anisotropic materials.

The case of isotropic materials is not as simple as it seems. In 2-D there are only two, in 3-D only three different eigenvalues of the constitutive equation. The corresponding eigenvectors (eigenstrains and eigenstresses) are degenerate (e.g., in 2-D, the deviatoric and the shear strain are degenerate). Hence, a change in the shear stress–strain relation (that implies a change in the shear modulus) should also change the deviatoric stress–strain relation. This coupling between degenerate eigenstates has not yet been included in the model.

Moreover, it is not yet clear, whether the stress–strain relations associated with different eigenvalues are independent of each other. In Section 5 we presented two different approaches: one with independent eigenstress–eigenstrain relations and one with fully coupled relations. Experimental investigations have to be carried out in order to examine possible coupling. If present, such a coupling may also cause a change of the eigenstates. It then has to be clarified how to interpret the stress history associated with the eigenstresses.

A similar procedure for the development of a 2-D model for NCNL elasticity is presented in [16]. The fundamental difference between the approaches is the use of two rigid states with a constant strain jump at state transitions compared to one elastic and one rigid state in the present model. Such a protocol yields a symmetric hysteresis in the stress–strain protocol, which implies the generation of only odd harmonics. However, material conditioning is not considered for brevity. In fact, in order to account for material conditioning in a simple PM space model, at least two states of different elastic behavior have to be permitted. Quasistatic and dynamic experimental investigations have to show which approach is best suited for which kind of materials. A possible intermediate approach would be a PM protocol with two states of different elastic behavior (e.g., one rigid and one elastic state), including strain jumps at the state transitions.

At present, damping is accounted for by a term proportional to the displacement velocity. Such a term with a constant damping factor yields a quality factor that decreases inversely proportional to the wave frequency. In most materials, however, a quality factor almost constant with frequency has been observed over wide frequency ranges [17]. The implementation of viscoelasticity proposed by Robertsson et al. [18]

is therefore preferable. Bou Matar et al. [19] have implemented this viscoelasticity approach in their model for the NCNL elasticity simulation. They also use, as well as de Van Den Abeele et al. [16], a continuous PM-space implementation rather than a statistic one. The continuous PM-space overcomes the shortcomings due to limited statistics if only a finite number of HE are used. Such an approach is particularly useful if the zone of NCNL elasticity is very localized. It allows us to put the effects of a whole PM-space with an infinite number of elements into a single interstice.

The most urgent future work should aim at an extensive experimental study of NCNL materials with set-ups that are particularly designed for the analysis of 2-D effects. In collaboration between the groups that developed the cited models, the respective advantages will be combined, and the models will be adapted to simulations of particular applications.

Acknowledgments

The work has been supported by the European Science Foundation—Physical and Engineering Sciences (ESF-PESC) Program NATEMIS. The authors wish to thank the CINECA supercomputing laboratory (Bologna, Italy) and the Bioindustry Park Canavese for providing their multiprocessing Linux cluster, and Professor P. P. Delsanto and Doctor M. Scalerandi from the Politecnico di Torino, Italy for fruitful discussions and continuous support.

A. Appendix

A.1 Interstice Strains

In this appendix we illustrate the definition of interstice strains in Section 3.3. We start from the Cartesian strain components, which are defined as

$$\epsilon_{xx} = \frac{\partial u_x}{\partial x} \qquad \epsilon_{yy} = \frac{\partial u_y}{\partial y} \qquad \epsilon_{xy} = \frac{1}{2}\left(\frac{\partial u_x}{\partial y} + \frac{\partial u_y}{\partial x}\right). \tag{A.1}$$

We have two possibilities to calculate the diagonal components of the strain tensor from the subnode displacements,

$$\epsilon_{xx}^+ = \frac{u_x^{++} - u_x^{-+}}{\delta_x} \qquad \text{and} \qquad \epsilon_{xx}^- = \frac{u_x^{+-} - u_x^{--}}{\delta_x}, \tag{A.2}$$

$$\epsilon_{yy}^+ = \frac{u_x^{++} - u_x^{+-}}{\delta_y} \qquad \text{and} \qquad \epsilon_{yy}^- = \frac{u_x^{-+} - u_x^{--}}{\delta_y}. \tag{A.3}$$

The diagonal strain tensor components at the center of the interstice are given by the mean of these two strains,

$$
\epsilon_{xx} = \frac{u_x^{++} - u_x^{-+} + u_x^{+-} - u_x^{--}}{2\delta_x} \qquad \epsilon_{yy} = \frac{u_y^{++} - u_y^{+-} + u_y^{-+} - u_y^{--}}{2\delta_y}. \qquad (A.4)
$$

The differences between the strains are

$$
\tilde{\epsilon}_{xx} = \frac{-u_x^{++} + u_x^{-+} + u_x^{+-} - u_x^{--}}{2\delta_x} \qquad \tilde{\epsilon}_{yy} = \frac{-u_y^{++} + u_y^{+-} + u_y^{-+} - u_y^{--}}{2\delta_y}. \qquad (A.5)
$$

These strains tend to zero for infinitesimally thin interstices, where the approximation of uniform stress throughout the interstice is valid.

The volumetric and deviatoric eigenstrains of the interstice are calculated from these strain components as follows.

$$
\epsilon_V = \frac{\delta_x \epsilon_{xx} + \delta_y \epsilon_{yy}}{\delta_x + \delta_y} \qquad\qquad \epsilon_D = \frac{\delta_x \epsilon_{xx} - \delta_y \epsilon_{yy}}{\delta_x + \delta_y}. \qquad (A.6)
$$

We also define the volumetric and deviatoric strain differences:

$$
\tilde{\epsilon}_V = \frac{\delta_x \tilde{\epsilon}_{xx} + \delta_y \tilde{\epsilon}_{yy}}{\delta_x + \delta_y} \qquad\qquad \tilde{\epsilon}_D = \frac{\delta_x \tilde{\epsilon}_{xx} - \delta_y \tilde{\epsilon}_{yy}}{\delta_x + \delta_y}. \qquad (A.7)
$$

From these equations we obtain the interstice strains as defined in Eq. (18.20).

The shear strain in terms of the displacement of the subnodes states

$$
\epsilon_S = \frac{u_x^{++} + u_x^{-+} - u_x^{--} - u_x^{+-} + u_y^{++} - u_y^{-+} - u_y^{--} + u_y^{+-}}{2(\delta_x + \delta_y)}. \qquad (A.8)
$$

The shear strain difference

$$
\tilde{\epsilon}_S = \frac{u_x^{++} + u_x^{-+} - u_x^{--} - u_x^{+-} - u_y^{++} + u_y^{-+} + u_y^{--} - u_y^{+-}}{2(\delta_x + \delta_y)} \qquad (A.9)
$$

corresponds to

$$
\epsilon_{xy} = \frac{1}{2}\left(\frac{\partial u_x}{\partial y} - \frac{\partial u_y}{\partial x}\right), \qquad (A.10)
$$

which in an infinitesimal interstice becomes zero.

Inversion of these strain relations yields well-defined expressions for the subnode displacement components

$$u_x^{++} = u_x^{(mc)} + \frac{\delta_x + \delta_y}{8\rho_{av}}\Big[+ (\rho^{-+} + \rho^{--})\epsilon_V - (\rho^{-+} + \rho^{+-})\tilde{\epsilon}_V + (\rho^{-+} + \rho^{--})\epsilon_D$$
$$- (\rho^{-+} + \rho^{+-})\tilde{\epsilon}_D + (\rho^{--} + \rho^{+-})\epsilon_S + (\rho^{--} + \rho^{+-})\tilde{\epsilon}_S\Big]$$

$$u_x^{-+} = u_x^{(mc)} + \frac{\delta_x + \delta_y}{8\rho_{av}}\Big[- (\rho^{++} + \rho^{+-})\epsilon_V + (\rho^{++} + \rho^{--})\tilde{\epsilon}_V - (\rho^{++} + \rho^{+-})\epsilon_D$$
$$+ (\rho^{++} + \rho^{--})\tilde{\epsilon}_D + (\rho^{--} + \rho^{+-})\epsilon_S + (\rho^{--} + \rho^{+-})\tilde{\epsilon}_S\Big]$$

$$u_x^{--} = u_x^{(mc)} + \frac{\delta_x + \delta_y}{8\rho_{av}}\Big[- (\rho^{++} + \rho^{+-})\epsilon_V - (\rho^{-+} + \rho^{+-})\tilde{\epsilon}_V - (\rho^{++} + \rho^{+-})\epsilon_D$$
$$- (\rho^{-+} + \rho^{+-})\tilde{\epsilon}_D - (\rho^{++} + \rho^{-+})\epsilon_S - (\rho^{++} + \rho^{-+})\tilde{\epsilon}_S\Big]$$

$$u_x^{+-} = u_x^{(mc)} + \frac{\delta_x + \delta_y}{8\rho_{av}}\Big[+ (\rho^{-+} + \rho^{--})\epsilon_V + (\rho^{++} + \rho^{--})\tilde{\epsilon}_V + (\rho^{-+} + \rho^{--})\epsilon_D$$
$$+ (\rho^{++} + \rho^{--})\tilde{\epsilon}_D - (\rho^{++} + \rho^{-+})\epsilon_S - (\rho^{++} + \rho^{-+})\tilde{\epsilon}_S\Big]$$

$$u_y^{++} = u_y^{(mc)} + \frac{\delta_x + \delta_y}{8\rho_{av}}\Big[+ (\rho^{--} + \rho^{+-})\epsilon_V - (\rho^{-+} + \rho^{+-})\tilde{\epsilon}_V - (\rho^{--} + \rho^{+-})\epsilon_D$$
$$+ (\rho^{-+} + \rho^{+-})\tilde{\epsilon}_D + (\rho^{-+} + \rho^{--})\epsilon_S - (\rho^{-+} + \rho^{--})\tilde{\epsilon}_S\Big]$$

$$u_y^{-+} = u_y^{(mc)} + \frac{\delta_x + \delta_y}{8\rho_{av}}\Big[+ (\rho^{--} + \rho^{+-})\epsilon_V + (\rho^{++} + \rho^{--})\tilde{\epsilon}_V - (\rho^{--} + \rho^{+-})\epsilon_D$$
$$- (\rho^{++} + \rho^{--})\tilde{\epsilon}_D - (\rho^{++} + \rho^{+-})\epsilon_S + (\rho^{++} + \rho^{+-})\tilde{\epsilon}_S\Big],$$

$$u_y^{--} = u_y^{(mc)} + \frac{\delta_x + \delta_y}{8\rho_{av}}\Big[- (\rho^{++} + \rho^{-+})\epsilon_V - (\rho^{-+} + \rho^{+-})\tilde{\epsilon}_V + (\rho^{++} + \rho^{-+})\epsilon_D$$
$$+ (\rho^{-+} + \rho^{+-})\tilde{\epsilon}_D - (\rho^{++} + \rho^{+-})\epsilon_S + (\rho^{++} + \rho^{+-})\tilde{\epsilon}_S\Big]$$

$$u_y^{+-} = u_y^{(mc)} + \frac{\delta_x + \delta_y}{8\rho_{av}}\Big[- (\rho^{++} + \rho^{-+})\epsilon_V + (\rho^{++} + \rho^{--})\tilde{\epsilon}_V + (\rho^{++} + \rho^{-+})\epsilon_D$$
$$- (\rho^{++} + \rho^{--})\tilde{\epsilon}_D + (\rho^{-+} + \rho^{--})\epsilon_S - (\rho^{-+} + \rho^{--})\tilde{\epsilon}_S\Big],$$
$$\tag{A.11}$$

where $\vec{u}^{(mc)}$ is the mass center displacement of the interstice, given by

$$4\rho_{av}u_x^{(mc)} = \rho_1 u_x^{++} + \rho_2 u_x^{-+} + \rho_3 u_x^{--} + \rho_4 u_x^{+-}$$
$$4\rho_{av}u_y^{(mc)} = \rho_1 u_y^{++} + \rho_2 u_y^{-+} + \rho_3 u_y^{--} + \rho_4 u_y^{+-}. \tag{A.12}$$

A.2 Interstice Strain Time Dependence in 2-D

A.2.1 Internal and External Stresses

In Section 3.3 and Appendix A.1 we have defined six independent interstice strains. We now derive their time dependence in a manner analogous to the 1-D case (Section 2.2). Then, we apply Newton's second law to the second time derivative of Eq. (18.20) in order to express the subnode accelerations in terms of the applied forces. We call $\vec{F}^{(n)}$ the external force density due to the stresses in the material cell, which is represented by the subnode (n) (see Section 3.1), and $\vec{f}^{(n)}$ the total internal force density on the subnode (n) due to the viscoporoelastic interstice. The mass of the subnode (n) is equal to a quarter of the mass of the material cell contiguous to (n),

$$m^{(n)} = \frac{\varepsilon_x}{2}\frac{\varepsilon_y}{2}\varepsilon_z \rho^{(n)}, \quad n, \in \{++, -+, --, +-\}. \tag{A.13}$$

The acceleration of the subnodes is then given by

$$\ddot{\vec{u}}^{(n)} = 4\frac{\vec{F}^{(n)} + \vec{f}^{(n)}}{\rho^{(n)}}, \quad n, \in \{++, -+, --, +-\}. \tag{A.14}$$

Inserting these expressions into the second time derivative of Eq. (18.20) yields the strain accelerations in terms of external and internal force densities. The combinations of terms containing internal force density components correspond to the internal eigen-stresses and eigenstress differences with analogous definitions as for the eigenstrains and eigenstrain differences in Appendix A.1. Thus, we obtain six independent internal stresses in place of the single internal stress in the 1-D case.

$$\sigma_V = \frac{\varepsilon_x + \varepsilon_y}{8} \rho_{red} \left(-\frac{f_x^{++}}{\rho^{++}} + \frac{f_x^{-+}}{\rho^{-+}} + \frac{f_x^{--}}{\rho^{--}} - \frac{f_x^{+-}}{\rho^{+-}} - \frac{f_y^{++}}{\rho^{++}} - \frac{f_y^{-+}}{\rho^{-+}} + \frac{f_y^{--}}{\rho^{--}} + \frac{f_y^{+-}}{\rho^{+-}} \right)$$

$$\tilde{\sigma}_V = \frac{\varepsilon_x + \varepsilon_y}{8} \rho_{red} \left(+\frac{f_x^{++}}{\rho^{++}} - \frac{f_x^{-+}}{\rho^{-+}} + \frac{f_x^{--}}{\rho^{--}} - \frac{f_x^{+-}}{\rho^{+-}} + \frac{f_y^{++}}{\rho^{++}} - \frac{f_y^{-+}}{\rho^{-+}} + \frac{f_y^{--}}{\rho^{--}} - \frac{f_y^{+-}}{\rho^{+-}} \right)$$

$$\text{(A.15)}$$

$$\sigma_D = \frac{\varepsilon_x + \varepsilon_y}{8} \rho_{red} \left(-\frac{f_x^{++}}{\rho^{++}} + \frac{f_x^{-+}}{\rho^{-+}} + \frac{f_x^{--}}{\rho^{--}} - \frac{f_x^{+-}}{\rho^{+-}} + \frac{f_y^{++}}{\rho^{++}} + \frac{f_y^{-+}}{\rho^{-+}} - \frac{f_y^{--}}{\rho^{--}} - \frac{f_y^{+-}}{\rho^{+-}} \right)$$

$$\tilde{\sigma}_D = \frac{\varepsilon_x + \varepsilon_y}{8} \rho_{red} \left(+\frac{f_x^{++}}{\rho^{++}} - \frac{f_x^{-+}}{\rho^{-+}} + \frac{f_x^{--}}{\rho^{--}} - \frac{f_x^{+-}}{\rho^{+-}} - \frac{f_y^{++}}{\rho^{++}} + \frac{f_y^{-+}}{\rho^{-+}} - \frac{f_y^{--}}{\rho^{--}} + \frac{f_y^{+-}}{\rho^{+-}} \right)$$

$$\text{(A.16)}$$

$$\sigma_S = \frac{\varepsilon_x + \varepsilon_y}{8} \rho_{red} \left(-\frac{f_x^{++}}{\rho^{++}} - \frac{f_x^{-+}}{\rho^{-+}} + \frac{f_x^{--}}{\rho^{--}} + \frac{f_x^{+-}}{\rho^{+-}} - \frac{f_y^{++}}{\rho^{++}} + \frac{f_y^{-+}}{\rho^{-+}} + \frac{f_y^{--}}{\rho^{--}} - \frac{f_y^{+-}}{\rho^{+-}} \right)$$

$$\tilde{\sigma}_S = \frac{\varepsilon_x + \varepsilon_y}{8} \rho_{red} \left(-\frac{f_x^{++}}{\rho^{++}} - \frac{f_x^{-+}}{\rho^{-+}} + \frac{f_x^{--}}{\rho^{--}} + \frac{f_x^{+-}}{\rho^{+-}} - \frac{f_y^{++}}{\rho^{++}} - \frac{f_y^{-+}}{\rho^{-+}} - \frac{f_y^{--}}{\rho^{--}} + \frac{f_y^{+-}}{\rho^{+-}} \right).$$

$$\text{(A.17)}$$

For a better understanding of the constant factor and the signs in these equations, we derive the volumetric internal stress at the center of the interstice for homogeneous material (constant ρ) and quadratic material cells ($\varepsilon_x = \varepsilon_y$). The σ_{xx} component of the internal stress tensor is obtained by the sum of all internal forces that act in the x-direction divided by the total area on which they act. The sign of the force terms is given by the direction of the normal vector of the area. If it points in the positive x-direction, the x-component of the force causes a positive contribution to σ_{xx}. The normal vectors of the area on which internal forces act are always directed towards the interstice. Hence,

$$\sigma_{xx} = \frac{\varepsilon^3(-f_x^{++} + f_x^{-+} + f_x^{--} - f_x^{+-})}{2\varepsilon^2}. \tag{A.18}$$

Analogously the σ_{yy} component states

$$\sigma_{yy} = \frac{\varepsilon^3(-f_y^{++} - f_y^{-+} + f_y^{--} + f_y^{+-})}{2\varepsilon^2}. \tag{A.19}$$

The volumetric stress is then given by

$$\sigma_V = \frac{\sigma_{xx} + \sigma_{yy}}{2}, \tag{A.20}$$

which is in terms of the internal force densities,

$$\sigma_V = \tfrac{\varepsilon + \varepsilon}{8}(-f_x^{++} + f_x^{-+} + f_x^{--} - f_x^{+-} - f_y^{++} - f_y^{-+} + f_y^{--} + f_y^{+-}). \tag{A.21}$$

In analogy to the 1-D case we define the external stresses in the same way. The expressions in terms of external force densities have the same form as Eqs. (A.15)–(A.17), but with opposite signs in the force terms and after replacing the internal force density by the external force density on the same subnode. With these stresses, we obtain Eqs. (18.21).

Note that external "eigenstresses" on the interstice do not correspond to the eigenstresses in the material cells. It is therefore erroneous to identify the external volumetric stress on the interstice (i.e., the part of the external stress that provokes a volumetric deformation of the interstice) with the mean of the volumetric stresses in the material cell. Quite the contrary, if in the material cells only volumetric stresses are present, but they differ between the cells that are in contact with the interstice, in general also deviatoric and shear deformations of the interstice occur due to these stresses.

A.2.2 Inverse Relations

In the same way as the subnode displacement has been expressed in terms of the mass center displacement and the interstice strains, we can express the total internal and external force densities on a subnode in terms of the derived stress relations. For the internal force densities we obtain, for example,

$$\frac{f_x^{++}}{\rho^{++}} = \frac{1}{2(\varepsilon_x + \varepsilon_y)\rho_{red}}\left[+ \frac{\rho^{-+} + \rho^{--}}{\rho_{av}}\sigma_V - \frac{\rho^{-+} + \rho^{+-}}{\rho_{av}}\tilde{\sigma}_V + \frac{\rho^{-+} + \rho^{--}}{\rho_{av}}\sigma_D \right.$$
$$\left. - \frac{\rho^{-+} + \rho^{+-}}{\rho_{av}}\tilde{\sigma}_D + \frac{\rho^{--} + \rho^{+-}}{\rho_{av}}\sigma_S + \frac{\rho^{--} + \rho^{+-}}{\rho_{av}}\tilde{\sigma}_S \right]. \tag{A.22}$$

The relations for the other internal force density components are omitted for brevity. They may be easily derived from Eq. (A.11) by substituting strains by stresses and changing the constant factor in the same way as has to be done to obtain Eq. (A.22) from the expression for u_x^{++} in Eq. (A.11).

The expressions for the external force densities have the same form, but with opposite signs in the stress terms and an additive term of the form $(\frac{\vec{F}_i^{mc}}{\rho_{av}})(i \in \{x, y\})$, where

$$F^{mc} = F^{++} + F^{-+} + F^{--} + F^{+-} \tag{A.23}$$

is the sum of the external force densities which is equal to the total force density on the mass center of the interstice (see Section 3.4), in analogy with the terms $u_i^{(mc)}$ in Eq. (A.11).

References

[1] P. Delsanto, A.S. Gliozzi, M. Hirsekorn, and M. Nobili, "A 2D spring model for the simulation of ultrasonic wave propagation in nonlinear hysteretic media," *Ultrasonics* in press (2006).

[2] P.P. Delsanto and M. Scalerandi, "A spring model for the simulation of the propagation of ultrasonic pulses through imperfect contact interfaces," *J. Acoust. Soc. Am.* **104**(5), 2584–2591 (1998).

[3] P.P. Delsanto, R.B. Mignogna, M. Scalerandi, and R.S. Schechter, "Simulation of ultrasonic pulse propagation in complex media," in *New Perspectives on Problems in Classical and Quantum Physics*, edited by P. P. Delsanto and A. W. Saenz (Gordon and Breach Science Publishers, New Delhi, 1997), pp. 51–74.

[4] P.P. Delsanto, T. Whitcombe, H.H. Chaskelis, and R.B. Mignogna, "Connection machine simulation of ultrasonic wave propagation in materials. I: The one-dimensional case," *Wave Motion* **16**, 65–80 (1992).

[5] P.P. Delsanto, R.S. Schechter, H.H. Chaskelis, R.B. Mignogna, and R. Kline, "Connection machine simulation of ultrasonic wave propagation in materials. II: The two-dimensional case," *Wave Motion* **20**, 295–314 (1994).

[6] P.P. Delsanto, R.S. Schechter, and R.B. Mignogna, "Connection machine simulation of ultrasonic wave propagation in materials. III: The three-dimensional case," *Wave Motion* **26**, 329–339 (1997).

[7] M. Scalerandi, P. P. Delsanto, C. Chiroiu, and V. Chiroiu, "Numerical simulation of pulse propagation in nonlinear 1-D media," *J. Acoust. Soc. Am.* **106**(5), 2424–2430 (1999).

[8] Chapter 16 of this book.

[9] P.P. Delsanto and M. Scalerandi, "Modeling nonclassical nonlinearity, conditioning and slow dynamics effects in mesoscopic elastic materials," *Phys. Rev.* B **68**(6), 064107 (2003).

[10] M. Scalerandi, P. Delsanto, and P. Johnson, "Stress induced conditioning and thermal relaxation in the simulation of quasi-static compression experiments," *J. Phys. D: Appl. Phys.* **36**, 288–293 (2003).

[11] T. Yamamoto, "Acoustic propagation in the ocean with a poro-elastic bottom," *J. Acoust. Soc. Am.* **73**(5), 1587–1596 (1983).

[12] L. A. Ostrovsky and P. A. Johnson, "Dynamic nonlinear elasticity in geomaterials," *Rivista del Nuovo Cimento C* **24**(7), 1–46 (2001).

[13] Chapter 17 of this book.

[14] M. Scalerandi, E. Ruffino, P.P. Delsanto, P.A. Johnson, and K.E.-A. Van Den Abeele, "Non-linear techniques for ultrasonic micro-damage diagnostics: A simulation approach," in *Review of Progress in Quantitative Nondestructive Evaluation, Vol. 19B*, edited by D.O. Thompson and D.E. Chimenti (1999), pp. 1393–1399.

[15] M. Scalerandi, P.P. Delsanto, V. Agostini, K. Van Den Abeele, and P.A. Johnson, "Local interaction simulation approach to modeling nonclassical, nonlinear elastic behavior in solids," *J. Acoust. Soc. Am.* **113**(6), 3049–3059 (2003).

[16] Chapter 12 of this book.

[17] J.W. Spencer, "Stress relaxations at low frequencies in fluid-saturated rocks: Attenuation and modulus dispersion," *J. Geophys. Res.* **86**, 1803–1812 (1981).

[18] J.O.A. Robertsson, J.O. Blanch, and W.W. Symes, "Viscoelastic finite difference modeling," *Geophysics* **59**(9), 1444–1456 (1994).

[19] O. Bou Matar, S. Dos Santos, J. Fortineau, L. Haumesser, and F. Van Der Meulen, "Pseudo-spectral simulation of 1D nonlinear propagation in heterogeneous elastic media," in press (2006).

19

Nonclassical Nonlinear Dynamics of Solid Surfaces in Partial Contact for NDE Applications

Claudio Pecorari[1,3] and Igor Solodov[2]

[1]Marcus Wallenberg Laboratory, Royal Institute of Technology, 10044 Stockholm, Sweden, e-mail: pecorari@kth.se.
[2] Faculty of Physics, M.V. Lomonosov Moscow State University, 119992 Moscow, Russia and Institute for Polymer Testing and Polymer Science (IKP) –Non-destructive Testing- (ZFP), University of Stuttgart, Pfaffenwaldring 32, D-70569 Stuttgart, Germany, e-mail: solodov@ikp.uni-stuttgart.de.
[3]To whom correspondence should be addressed.

Abstract
In this chapter, theoretical analysis and experimental results on nonlinear acoustic wave interaction at the interface between solid surfaces in partial contact are presented. In the first section, the nonlinear macroscopic dynamic behavior of the interface is studied on the basis of three distinct mechanisms, controlling the interaction of asperities in contact at the microscopic level. They include Hertzian nonlinearity of two spheres in contact in response to the normal force applied to them, the nonlinear dynamics of two spheres in contact subjected to an oscillating tangential force, and the effect of adhesion forces of Van der Waals type. In each of the aforementioned cases, the solution of the nonlinear scattering problem is presented and the nonlinear responses of the interface to longitudinal and shear waves are calculated. In the latter two cases the results predict a nonclassical nonlinear behavior of the higher harmonics caused by elastic hysteresis and end point memory. Phenomenology and experimental verification of nonlinear dynamics of an interface with intermittent contact are considered in the second section. In this case, the two basic mechanisms of contact nonlinearity that are analyzed are associated with surface clapping and friction force nonlinearity. Nonclassical features of higher harmonic spectra in both cases are demonstrated by calculations and experimentally verified. A variety of mixed nonlinear signatures, which we propose for application to nonlinear nondestructive evaluation (NDE), is observed in realistic materials with flaws.

Keywords: Adhesion, clapping, contact, friction, higher harmonics, hysteresis, interface, nondestructive evaluation, nonlinearity

1. Introduction

In dealing with the characterization of bond quality, a major problem that the NDE community is confronted with is the similarity of the acoustic response of partial and kissing bonds. A possible criterion to distinguish the two types of bond was proposed by Nagy,[1] who suggested using the ratio between the interface reflection coefficients for shear and longitudinal waves. In fact, for an interface between identical materials, this ratio is close to one for kissing bonds, and it approaches two if the bond is partial. Nagy's criterion is useful when the effect of the interface imperfections on the linear acoustic response of the interface itself is easily measurable.

On the contrary, the quality of a bond between materials with very dissimilar acoustic properties is very difficult to assess. Equally difficult is the assessment of the interface integrity when the closure of a kissing bond makes such an interface between identical materials nearly transparent to an inspecting wave. Another example in which linear measurements may fail to detect and identify the nature of a defect appears when a partially closed crack, possibly of small dimensions, is surrounded by a medium with coarse microstructure. In this case, in fact, the linear response of the crack may be hidden by the incoherent field scattered by the microstructure, or may be confused with coherent signals scattered by surrounding geometrical irregularities of the complex structure of the system.

To overcome such problems, researchers have lately focused their efforts on exploiting nonlinear phenomena arising from the interaction of ultrasonic waves with imperfect interfaces, and, in particular, with interfaces between rough surfaces in contact. The interest in the latter system stems from their role as prototypical models of real kissing bonds. It has been experimentally shown that the spectrum of a wave scattered by such a discontinuity is characterized by higher harmonic components and other features that may be considered as the fingerprints of cracklike discontinuities. Nonlinear scattering phenomena, therefore, appear to offer new tools for inspecting materials and engineering components in search of cracks.[2–7]

In spite of the great wealth of laboratory results, the theoretical understanding of the mechanisms responsible for the generation of nonlinear effects has lagged behind the progress marked on the experimental front. In this chapter, a review of some recent results concerning the nonlinear acoustic properties of rough surfaces in contact is presented. The subsequent material presented is divided in two main sections. The first one considers the phenomena involving an interface between two rough surfaces in constant partial contact; that is to say, there are always portions of the surfaces in contact with each other. For this to happen, the amplitude of the total acoustic field to which the interface is subject is assumed to be sufficiently small compared to the roughness of the two surfaces in contact. Thus, perturbation approaches may be adopted to treat the scattering phenomena of interest. Micromechanical models of the characteristic dynamics of these systems are also presented. They stem from well-known results developed in the field of contact mechanics, the validity of which has been extensively confirmed by independent experiments. The second part deals with phenomena that appear as the amplitude of the external excitation increases to the point that the stresses generated at the interface by the total field overcome the compressive stress field, if any, which keeps the interface partially closed. The detailed mathematical description of such nonlinear vibrations presents a conspicuous challenge that is still waiting for a satisfactory solution. In this chapter, therefore, a phenomenological treatment is followed, which provides an intuitive framework within which the presented experimental observations are considered.

2. Nonlinear Scattering by Interfaces: Permanent Contact

In the framework of the spring model the macroscopic mechanical properties of an interface and, among these, its nonlinearity, are accounted for by suitable distributions of normal and tangential springs having stiffness constants K_N and K_T. The stiffness constants link the relevant components of the total stress field, σ_{33} and σ_{31}, to the components of the extra-opening displacement of the interface, Δu_1 and Δu_3, as follows:[8]

$$\sigma_{33}(x_3 = 0^+) = K_N(\Delta u_3)\Delta u_3, \tag{19.1}$$

$$\sigma_{31}(x_3 = 0^+) = K_T(\Delta u_1, \Delta u_3)\Delta u_1. \tag{19.2}$$

In Eqs. (19.1) and (19.2), the interface is assumed to lie in the plane $x_3 = 0$, and the stiffness constants K_N and K_T to be functions of the variation of the interface opening displacement $\Delta \vec{u} = (\Delta u_1, \Delta u_3)$. Such dependence must be derived from micromechanical models of asperities in contact.

In the following, three possible macroscopic dynamic behaviors are investigated, which aim at describing the nonlinear macromechanics of a kissing bond. The first mechanism is described by the Hertz law, which relates the relative displacement between the centers of two spheres in contact with the normal force applied to them.[8] The second mechanism concerns the dynamics of two spheres in contact, which are subject to a tangential oscillating force.[8] Finally, the third mechanism includes the effect of adhesion forces of the Van der Waals type.[9] In each of the aforementioned cases, the solution of the scattering problem is presented, along with some details concerning the adopted methodology.

2.1 Elastic Interface

The normal stiffness K_N of an interface formed by two rough elastic surfaces in contact is given by[8]

$$K_N = n \left\langle \frac{E}{1-\nu^2} \right\rangle \left\langle R^{1/2} \right\rangle \int_0^\delta (\delta - z)^{1/2}\, \varphi(z; N)\, dz. \tag{19.3}$$

In Eq. (19.3), n is the number of contacts per unit area, E and ν are the Young modulus and the Poisson ratio of the material, respectively, R is the radius of curvature of the asperities, and φ is the height distribution of the asperities of the composite surface. The latter is defined by a linear combination of the profiles of the two surfaces, which maps the actual contacts of the interface onto the asperities of the composite surface. Following Brown and Scholz,[11] this function is properly modeled by an inverted chi-squared distribution that depends on an integer parameter, $N \geq 2$, known as the "number of degrees of freedom." The integration variable z is defined by the transformation $z = Z_0 - z'$, where Z_0 is the coordinate of the highest asperities of the composite surface, and z' is the actual coordinate of the asperity measured from the surface mean plane. Finally, δ is the relative approach between the mean planes of the rough surfaces; it is null when no pressure is applied to the interface.

The nonlinear nature of the dependence of K_N on the relative approach δ can be accounted for by considering the expansion of K_N in powers of the variation $\Delta\delta$ in which only the first-order term is retained,

$$K_N(\delta + \Delta\delta) = K_N(\delta) + \frac{\partial K_N}{\partial \delta}\Delta\delta = K_{N,0} + K_{N,1}\Delta\delta. \tag{19.4}$$

In Eq. (19.4), the constant $K_{N,0} = K_N(\delta)$ can be evaluated by means of Eq. (19.3), and $K_{N,1}$ is given by

$$K_{N,1} = \frac{\partial K_N}{\partial \delta} = \frac{n}{2}\left\langle\frac{E}{1-v^2}\right\rangle\langle R^{1/2}\rangle\int_0^\delta (\delta - z)^{-1/2}\varphi(z; N)dz. \tag{19.5}$$

The boundary conditions that control the scattering of a longitudinal plane wave at normal incidence on a interface with nonlinear stiffness can be conveniently cast in nondimensional form as follows.

$$\frac{\partial U^+}{\partial X} = \frac{\bar{K}_{N,0}}{\kappa^2}(\Delta U - \varepsilon_N \Delta U^2), \tag{19.6a}$$

$$\frac{\partial U^+}{\partial X} = \frac{\partial U^-}{\partial X}. \tag{19.6b}$$

The superscripts $+$ and $-$ refer to the half-spaces for which the X-coordinate is positive or negative, respectively. To obtain these equations, the particle displacement is normalized with respect to the amplitude of the incident wave, A_{in}, $u = A_{in}U$. Similarly, new nondimensional space and time variables are introduced by means of the following definitions: $X = k_T x$, where k_T is the wavenumber of the shear wave, and $\tau = \omega t$, where ω is the angular frequency of the incident wave. The normalized stiffness is defined by $\bar{K}_{N,0} = K_{N,0}/Z_T\omega$, where Z_T is the shear acoustic impedance of the medium, and the nonlinear parameter $\varepsilon_N = A_{in}K_{N,1}/K_{N,0}$. Except for interfaces that are nearly open, for which K_N approaches 0, ε_N is always much smaller than 1, and decreases as the interface stiffens. Therefore, a perturbation approach to find the solution of the normalized boundary value problem can be employed.

The solutions of the zeroth-order system describe waves with the same angular frequency as the incident wave, and complex amplitudes that are proportional to the well-known complex reflection R and transmission T coefficients for an imperfect linear interface.[8] The solutions of the first-order system include a term that is proportional to $\exp(j2\tau)$, and their complex amplitude is given by

$$A(2\omega) = -\frac{\varepsilon_N}{4}\frac{j\frac{\bar{K}_{N,0}}{\kappa}}{1 - j\frac{\bar{K}_{N,0}}{\kappa}}(T - 1 + R)^2. \tag{19.7}$$

Figure 19.1 illustrates the dependence of the second harmonic amplitude $A(2\omega)$ on the normalized interfacial stiffness. The interface is formed by two steel rough surfaces with parameters given in Table 19.1, where $M = \frac{2}{3}n\langle E/(1-v^2)\rangle\langle R^{1/2}\rangle$. As expected, after reaching a maximum value in the neighborhood of $\bar{K}_{N,0} = 1$, the

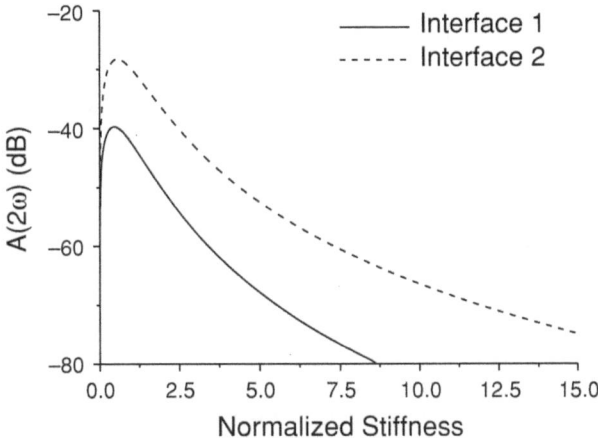

Fig. 19.1. Amplitude of the second harmonic components, $A(2\,\omega)$, versus the normalized interfacial stiffness for the two interfaces characterized by the parameter of Table 19.1.

Table 19.1. Statistical parameters of the interfaces

	Roughness (μm)	M (GPa/(μm$^{3/2}$))	Degrees of Freedom
Interface 1	0.68	5.4	3
Interface 2	0.23	76.8	5

nonlinear response of the interface is drastically reduced as the interface becomes stiffer. For the interface with smaller roughness, $A(2\,\omega)$ reaches values that are only 30 dB below that of the incident wave.

The tangential stiffness of two rough surfaces held together by an external pressure can be derived using the model by Mindlin and Deresiewicz[8] for two elastic spheres in contact and subjected to an oscillating tangential force, F_{tan}. This model describes a hysteretic loop, the origin of which rests in the relative partial slipping of the contacting spheres. Within the framework of the Greenwood and Williamson approach, the normalized nonlinear boundary conditions enforced at the interface when a shear wave impinges upon it at normal incidence are

$$\frac{\partial U^+}{\partial X_3} = \bar{K}_T \left[\Delta U - \frac{\varepsilon_T}{2} \left[\text{sgn}\left(\frac{\partial \Delta U}{\partial \tau}\right)(\Delta U^2 - \Delta U_{max}^2) + \Delta U\,\Delta U_{max} \right] \right], \quad (19.8a)$$

$$\frac{\partial U^+}{\partial X_3} = \frac{\partial U^-}{\partial X_3}. \quad (19.8b)$$

Here again, the perturbation parameter $\varepsilon_T = A_{in}\,K_{T,1}/K_{T,0}$, where $K_{T,0}$, and $K_{T,1}$ are defined by expressions similar to those obtained for longitudinal incidence.[8] The solution is sought by following again the same perturbation approach used earlier.

Figure 19.2 presents plots of the first and higher harmonics generated by the smoother interface of Table 19.1 versus the normalized transverse stiffness. Only

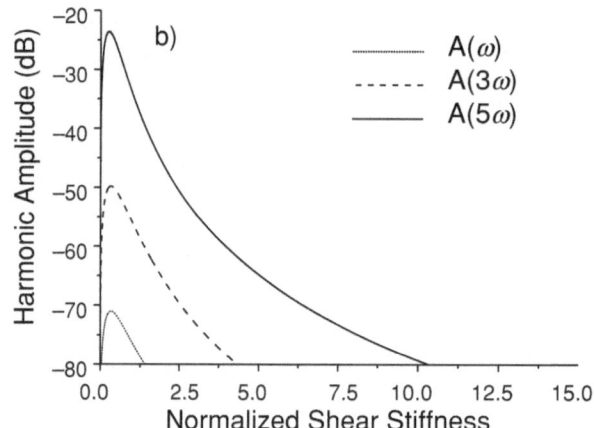

Fig. 19.2. Amplitude of the first three odd harmonic components, $A(n\,\omega)$, $n = 1, 3, 5$, versus the normalized interfacial shear stiffness for Interface 2 of Table 19.1.

harmonics of odd order are generated. The reduced nonlinear response of this kind of interface to a shear excitation, compared to the response to a longitudinal wave, can be partly explained by the manifestation of the first-order nonlinearity in the third and higher odd-harmonics, and partly by the magnitude of the coefficient ε_T compared to ε_N. These results indicate that the magnitude of the nonlinear response of interfaces formed by rough surfaces in contact with a longitudinal wave exceeds that of a shear wave by about 20 dB.

2.2 Interface with Adhesion

In this section, the role of adhesion on the scattering of a longitudinal wave insonifying an interface at normal incidence is examined. The force law controlling the contact between two spheres is described by the Greenwood–Johnson model,[10] and Figure 19.3 offers a typical example.

The relationship between the applied pressure P and the relative approach, Δ, between the mean planes of the rough surfaces in contact can be again written as

$$P(\Delta) = 2\pi \, n \, R \, \Delta\gamma \int_0^\Delta F\left(\Delta - t \,|\, \mu\right) \varphi(t) \, dt. \tag{19.9}$$

In Eq. (19.9), F is the force law between asperities, which depends on the normalized relative approach, Δ, as well as on the Tabor parameter. Following Fuller and Tabor,[12] all the contacts are assumed to have the same composite radius of curvature, R. Having brought the two surfaces to a maximum normalized approach $\Delta_{\max}$ at the end of the

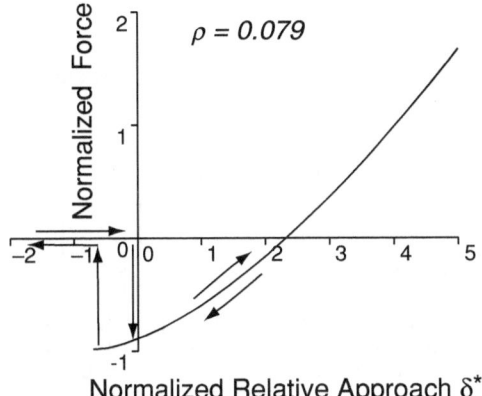

Fig. 19.3. Normalized force-displacement relationship for a contact between Perspex and steel spheres according to the Greenwood and Johnson model. The arrows indicate the path followed by the force as the relative approach varies.[9]

loading phase of the first cycle, the relationship between the applied pressure P and the normalized relative approach Δ during unloading is given by

$$P(\Delta) = 2\pi N R \Delta\gamma \int_0^{\Delta+D} F\left(\Delta - t \mid \mu\right) \varphi(t)dt, \qquad (19.10)$$

where $D = \Delta_{max} - \Delta$, if $0 < \Delta_{max} - \Delta < \bar{\delta}$ and $D = \bar{\delta}$, if $\Delta_{max} - \Delta > \bar{\delta}$. The inclusion of D in the upper limit of integration accounts for the stretching of the peaks that have been formed last during the preceding loading phase of the cycle.

The main interest of this investigation is in the dynamic behavior of the interface when it is subject to a cyclic loading. Thus, if Δ_{min} is the relative approach at the end of the unloading phase of the cycle, the pressure–approach relationship during all the following loading phases is given by

$$P(\Delta) = 2\pi N R \Delta\gamma \int_0^{\Delta+D} F\left(\Delta - t \mid \mu\right) \varphi(t)dt, \qquad (19.11)$$

where $D = \bar{\delta} - \Delta + \Delta_{min}$, if $0 < \Delta - \Delta_{min} < \bar{\delta}$ and $D = 0$, if $\Delta - \Delta_{min} > \bar{\delta}$. The upper limit of integration accounts for the effect of contacts that are under tension at the end of the unloading cycle, and are now progressively set under increasing compression again during the current compressive phase of the cycle.

The effect of adhesion is to introduce hysteresis with end point memory in the relationship between pressure and relative approach.

The interaction between a longitudinal wave with amplitude A_{in} and an interface of the type just discussed is described by the following boundary conditions.

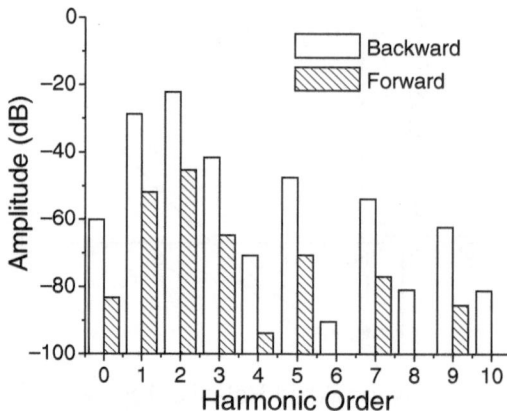

Fig. 19.4. Normalized spectrum of nonlinear waves that are scattered forward and backward by a Plexiglas–steel interface, which is characterized by a normalized linear component of the interface stiffness $K_0/(Z^-\omega) = 1.29$.

$$k^-(\lambda^+ + 2\,\mu^+)\frac{\partial U^+}{\partial\eta} = K_0\Delta U - K_1\,A_{in}\Delta U^2 - K_2'\left[H\left(\text{sgn}\left(\frac{\partial\,\Delta U}{\partial\tau}\right)\right)\right.$$

$$\left.H(\Delta U + \Delta U_{\max} - \bar{\Delta}) + \theta H\left(\text{sgn}\left(-\frac{\partial\Delta U}{\partial\tau}\right)\right)H(\Delta U_{\max} - \bar{\Delta} - \Delta U)\right]\Delta U,$$

$$(19.12\text{a})$$

$$(\lambda^+ + 2\,\mu^+)\frac{\partial U^+}{\partial\eta} = (\lambda^- + 2\,\mu^-)\frac{\partial U^-}{\partial\eta}. \qquad (19.12\text{b})$$

in which $U^{-,+}(z,t) = u^{-,+}(z,t)/A_{in}$, $\eta = k^- z$, where k^- is the longitudinal wave number in the negative half-space, and $\tau = \omega t$. Furthermore, $\bar{\Delta}$ has been redefined as $\bar{\Delta} = \bar{\Delta}/A_{in}, \theta = K_2''/K_2'$, and the Lamé constants λ^- and μ^-, and λ^+ and μ^+ refer to the negative and positive half-space, respectively.

The solutions are sought in the form of a perturbation series in two small parameters, $\varepsilon_1 = K_1 A_{in}/K_0$ and $\varepsilon_2 = K_2' A_{in}/K_0$. Figure 19.4 illustrates an example of normalized spectrum of the higher harmonics generated upon scattering of an incident wave with amplitude $A_{in} = 2\,\text{nm}$, frequency $f = 1\,\text{MHz}$, and propagating in the Perspex half-space. The material of the second half-space is steel. The amplitude of the incident wave is used as normalization constant. The Perspex–steel interface has a composite roughness $\sigma = 30\,\text{nm}$, and a probability distribution density with a number of degrees of freedom $N = 10$. The normalized stiffness is equal to 1.29. To the first-order approximation, and when hysteresis is activated, the amplitude of all the higher harmonics is a function of A_{in}^2, as experimentally verified in damaged materials, at least for the third harmonic component.[13]

It is important to point out the remarkable resemblance between the spectra in Figure 19.4 and those predicted by Van Den Abeele et al.[14] in their theoretical investigation on nonlinear propagation of acoustic waves in nonlinear media, in which the P-M model was used to characterize the type and degree of nonlinearity. Meegan et al.[15] reported measured spectra of waves propagating in sandstone, which bear a strong resemblance to those of Figure 19.4. In particular, the slow decay of the amplitude of the higher harmonics together with the dominant presence of the odd components seem to constitute the acoustic signature of an hysteretic interface with end point memory.

3. Nonlinear Scattering by Interfaces: Intermittent Contact

3.1 "Clapping" Nonlinearity

Let us assume now that the amplitude of acoustic excitation is high enough to produce alternating stresses that exceed static stress of the originally closed interface. For normal traction developed at normal incidence of the longitudinal incident acoustic wave, it results in clapping of the surfaces in contact. To find out the basic features of such a "clapping" nonlinearity we consider a simple model of an acoustically driven clapping interface between flat surfaces. To this end, the nonlinearity comes from asymmetrical dynamics of the contact stiffness: the latter is, apparently, higher in a compression phase than for tensile stress when the contact (crack) is assumed to be supported only by the edge-stresses. Such a behavior of a clapping interface can be approximated by a piecewise stress (σ)-strain (ε) relation[3] (Figure 19.5):

$$\sigma = C^{II}[1 - H(\varepsilon)(\Delta C/C^{II})]\varepsilon, \tag{19.13}$$

where $H(\varepsilon)$ is the Heaviside unit step function, $\Delta C = [C^{II} - (d\sigma/d\varepsilon)_{\varepsilon > 0}]$, and C^{II} is the intact material second-order (linear) stiffness.

For $\varepsilon = \varepsilon(t) - \varepsilon^0$, where $\varepsilon(t) = \varepsilon_0 \cos v_0 t$ and ε^0 is the static contact strain, the stiffness modulation $H(\varepsilon_0 \cos v_0 t - \varepsilon^0)(\Delta C/C^{II}) = \Delta C(t)$ is a pulse-type function of period $T = 2\pi/v_0$ (Figure 19.5). The spectrum of the nonlinear part in Eq. (19.13) $\sigma^{NL}(t) = \Delta C(t)[\varepsilon(t) - \varepsilon^0]$ can be easily found using the modulation theorem and the stress amplitude of the nth harmonic ($n = 0, 1, 2, 3\ldots$) then takes the form:

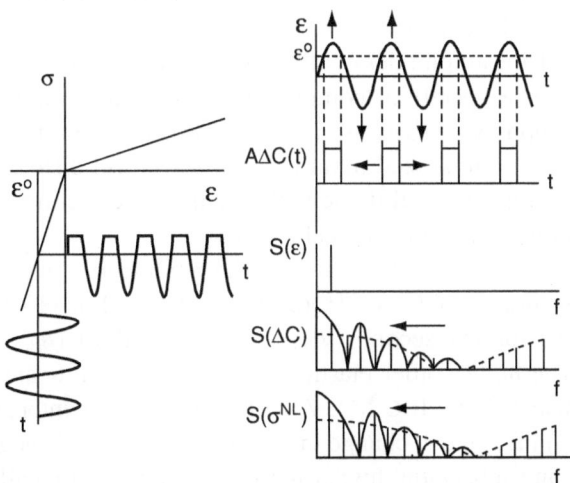

Fig. 19.5. Phenomenological model of clapping interface: piecewise stress-strain relation (left) and formation of higher harmonic spectrum (right).

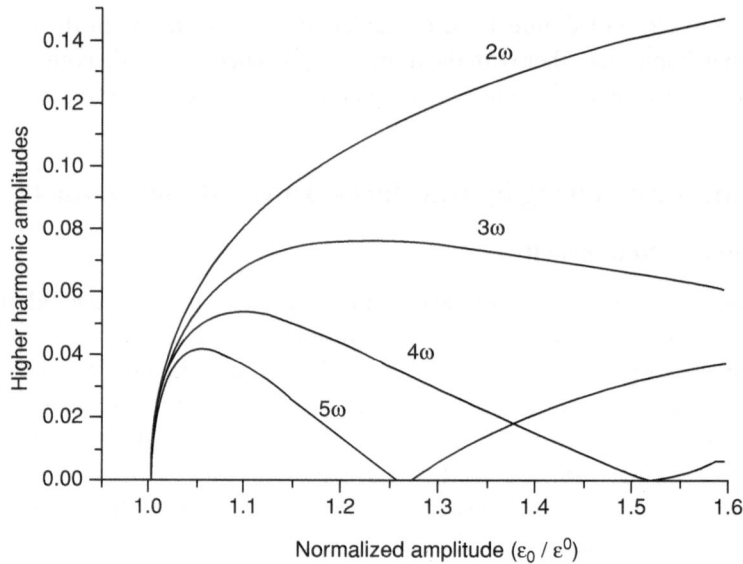

Fig. 19.6. Dynamic characteristics of higher harmonic generation by clapping interface.

$$A_n = \Delta C \Delta \tau \varepsilon_0 [sinc((n+1)\Delta\tau) - 2\cos(\pi\,\Delta\tau)sinc(n\Delta\tau) + sinc((n-1)\Delta\tau)], \tag{19.14}$$

where $\Delta\tau = \tau/T (\tau = (T/\pi)\text{Arc}\cos(\varepsilon^0/\varepsilon_0)$ is the normalized modulation pulse length.

The spectrum of the nonlinear oscillations (19.14) is illustrated in Figure 19.5 and contains a number of both odd and even higher harmonics arising simultaneously as soon as $\varepsilon > \varepsilon^0$ (threshold of clapping). Due to the pulse-type stiffness variation, the harmonic amplitudes are always modulated by the *sinc* envelope function. Its argument depends on τ: as the wave amplitude ε_0 increases, τ grows from 0 to $T/2$ accompanied by corresponding "compression" of the envelope function (shown by arrows in Figure 19.5). This affects dynamic characteristics of the higher harmonics (Figure 19.6): beyond the threshold, the amplitudes of all harmonics increase monotonically followed by oscillations due to the spectrum "compression" effect, unless finally ($\varepsilon_0 >> \varepsilon^0$) all odd harmonics are suppressed (because $\tau = T/2$). Such spectral features of the "clapping" nonlinearity are summarized in an unusual waveform distortion (Figure 19.5): the interface acts as a mechanical diode with a half-period rectified output instead of the sawtoothlike profile found in classical materials.

For further comparison with classical material nonlinearity, we assume $\Delta C(t) = (C^{III}/2C^{II})\varepsilon_0 \cos \omega_0 t + (C^{IV}/6C^{II})(\varepsilon_0 \cos \omega_0 t)^2$ in Eq. (19.13) to obtain traditional power laws for the second ($u_{2\omega} \sim \varepsilon_0^2$) and third ($u_{3\omega} \sim \varepsilon_0^3$) harmonics, respectively. The higher-order elastic moduli vary insignificantly for all materials available, so that: $C^N \sim 10C^{N-1}$. As a result, even for intense acoustic waves ($\varepsilon_0 \sim 10^{-5} - 10^{-4}$) the relative change in stiffness ($\Delta C/C^{II}$) is negligibly small and one cannot expect any substantial higher harmonics generated locally in the classical case.

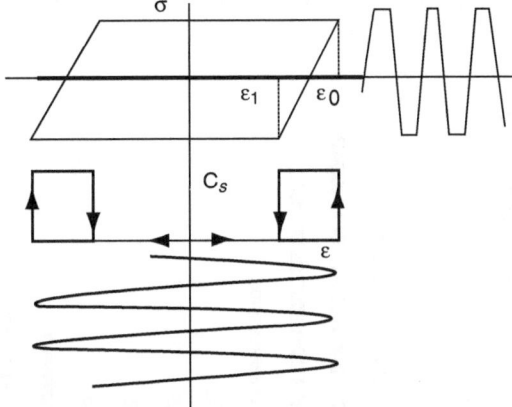

Fig. 19.7. Stress-strain relation and stiffness variation for sliding friction coupled surfaces in contact.

3.2 Friction Force Nonlinearity

Consider now the nonbonded interface between two friction coupled surfaces subjected to an oscillating tangential traction (shear wave scattering) strong enough to cause their sliding.[16] For a harmonic shear wave strain ε, transition between static and kinematic friction makes such a nonbonded contact to follow the hysteretic stress–strain relation shown in Figure 19.7. The contact stiffness changes between C_s (for a stick phase) and zero (slide phase) twice over the input strain period according to the relation:

$$C(t) = (C_S/2)\{1 - \text{sign}(\dot{\varepsilon})\text{sign}[\varepsilon + \text{sign}(\dot{\varepsilon})\varepsilon_1]\}. \quad (19.15)$$

Unlike asymmetrical dynamics of the stiffness for the "clapping" mechanism, the friction force nonlinearity causes a symmetrical variation of tangential interface stiffness as a function of time (Figure 19.7). By integrating Eq. (19.15) on strain one obtains σ as a double valued linear function of ε with the integration constant providing its continuity:

$$\sigma(t) = C(t)\varepsilon(t) + C(t)\varepsilon_1\text{sign}(\dot{\varepsilon}) + (C_s/2)(\varepsilon_0 - \varepsilon_1)\text{sign}(\dot{\varepsilon}). \quad (19.16)$$

Because $C(t)$ is a $2\nu_0$-function, none of the terms in Eq. (19.16) contain even-order higher harmonics. The lack of even harmonics is also substantiated by the fully rectified output waveform observed in this case (Figure 19.7). The amplitudes of the odd harmonics are found from the spectrum of Eq. (19.16) as follows.

$$B_N = C_s\varepsilon_0[\Delta\tau_f(sinc(N-1)\Delta\tau_f + sinc(N+1)\Delta\tau_f$$
$$-2\Delta\varepsilon sinc\, N\Delta\tau_f) - 0.5(1 - \Delta\varepsilon)\,|\,sinc\, N/2|], \quad (19.17)$$

where $\Delta\varepsilon = \varepsilon_1/\varepsilon_0$, $\Delta\tau_f = (1/2\pi)\text{Arc}\cos(\Delta\varepsilon)$ and $N = 2n + 1$ ($n = 0, 1, 2, 3\ldots$). Spectrum (19.17) is shown in Figure 19.8 and features obvious evidence for the friction-dependent $sinc$-modulation. From Eq. (19.17), no friction force nonlinearity is observed for the surfaces with very low ($\Delta\varepsilon \to 1$) and high friction ($\Delta\varepsilon \to -1$). In the latter case, the interface is in a permanent contact and its nonlinearity is due to the mechanism of nonlinear tangential stiffness described in Section 2.1.

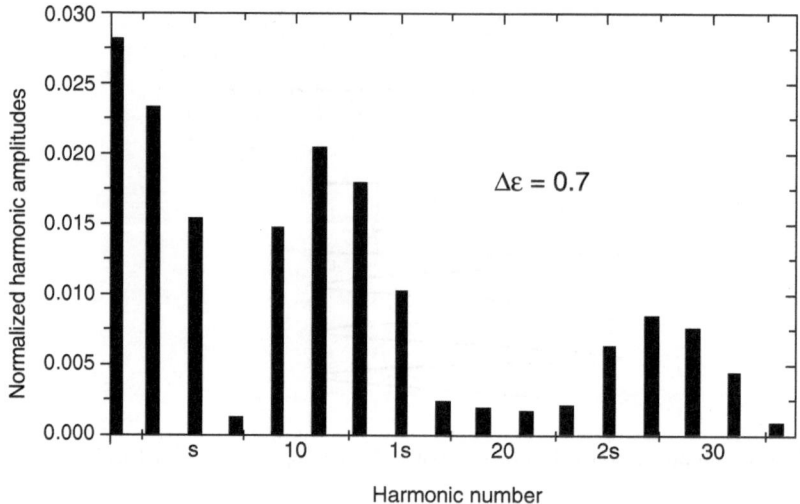

Fig. 19.8. Typical higher harmonic spectrum for friction force nonlinearity.

3.3 Experimental Verification and Opportunities for Nonlinear NDE

The nonclassical higher harmonic generation described above is expected to accompany nonlinear scattering of acoustic waves in imperfect materials with internal interfaces either intrinsic to the material structure or formed by fractured defects (cracks, delaminations, impacts, etc.). Because diverse internal boundaries are supposed to display different nonlinear behavior, some important NDE applications can be developed by solving the inverse problem of material (or defect) characterization by its nonlinear signature. However, in realistic materials with flaws, the latter, apparently, are not confined to the two basic nonlinear responses shown above to be characteristic of the intermittent contact. A few nonlinear interface signatures given below for specific experimental situations illustrate a variety of modified modes one should be aware of to interpret the results of the nonlinear NDE.

Figures 19.9a,b show the spectrum and the vibration pattern in the delamination area of a carbon fibre-reinforced composite plate (CFRP) subject to 20 kHz 10 W-acoustic excitation. The *sinc*-modulation of the higher harmonics (both odd and even orders) and half-period rectified waveform clearly reveal the above-predicted features of the "clapping" nonlinearity. An intact CFRP is a quite linear material so that the average level of the higher harmonics observed is an indicator of overall damage in the sample.

In polymers and natural composites, such as wood, acoustic wave propagation is accompanied by a viscous deformation associated with molecular or fiber internal friction. Figure 19.10a shows a typical single-spot nonlinear vibration spectrum measured in an intact soft wood material (spruce). An evident odd harmonic domination reflects the symmetry in nonlinear stiffness of the intact material caused by partial slip and adhesion mechanisms similar to those discussed in Sections 2.1 and 2.2. The situation is different, however, for local nonlinearity of compliant areas caused by flaws in wood (e.g., cracks, delaminations, knots, etc.). A weak bonding in such areas results in micro- or even macroclapping of internal boundaries in defects that brings about the higher harmonics of even orders (Figure 19.10b). The change from symmetrical to

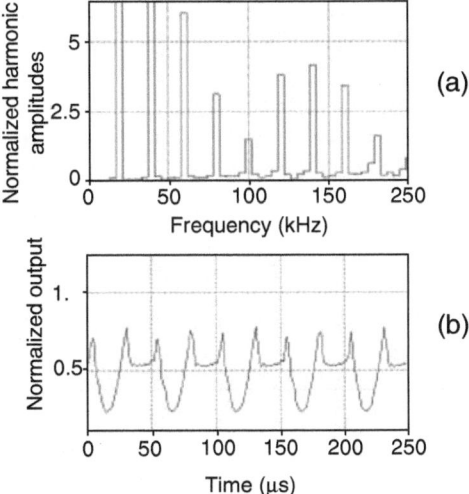

Figs. 19.9 a, b. Higher harmonic spectrum (a) and vibration pattern (b) measured in delaminated area of carbon fibre-reinforced composite.

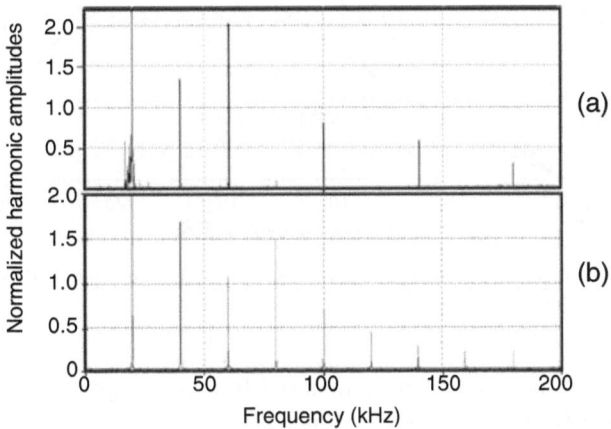

Figs. 19.10 a, b. Measured higher harmonic vibration spectra in wood: intact specimen (a) and sample with delamination (b).

asymmetrical nonlinearity enables us therefore, to use the even-order harmonic signature for discerning and imaging flaws in wood and wood composites.[17]

Another example of symmetrical nonlinearity is shown in Figure 19.11 for cellulose fiber-reinforced gypsum compound. An exclusive odd harmonic spectrum measured for a 20 kHz acoustic excitation (Figure 19.11) confirms symmetrical stiffness variation of the intact material. A sample of this porous material was then subject to a static tensile load of about 3 MPa superimposed on a low frequency (5 Hz) cyclic stress of 0.1–0.2 MPa amplitude. As the number of cycles increased, the material nonlinearity changed from symmetrical to asymmetrical with even harmonics of the low-frequency stress growth shown in Figure 19.12.[18] Similar to the above, such a transition may be associated with the clapping mechanism due to internal microcracking induced by the low-frequency fatigue of the material. Instructively, the higher harmonics reached their

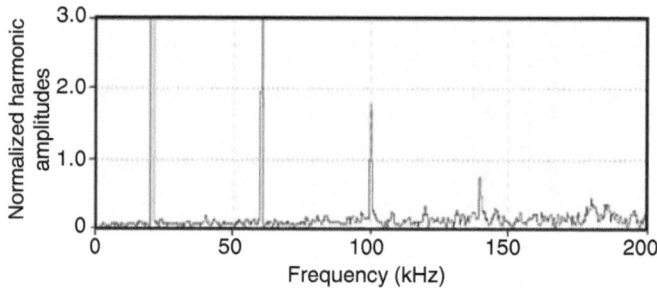

Fig. 19.11. Higher harmonic vibration spectrum observed in cellulose fibre-reinforced gypsum compound.

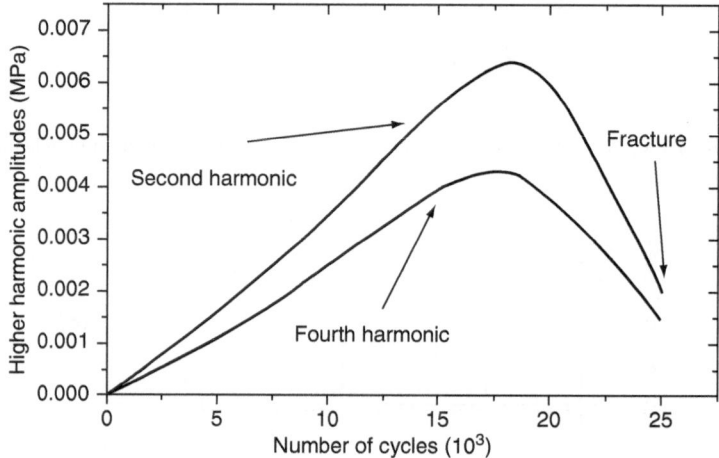

Fig. 19.12. Development of second- and fourth harmonics cellulose fibre-reinforced gypsum compound in low-frequency cyclic fatigue test.

maxima a few thousand cycles before material fracture, which shows another opportunity of nonlinear NDE in evaluation of material endurance.

Besides the higher harmonic generation, the asymmetry of the "clapping" nonlinearity also manifests in the DC-nonlinear response.[19] The lower stiffness of the interface in the tensile phase makes the contact extension dominate, resulting in acoustically induced expansion of the average contact gap against the pressure keeping the surfaces in contact. For a pulsed acoustic excitation, the gap expansion will cause transient "DC"-acoustic compression pulses in both media across the interface. Similarly, closing of the gap by the trailing edge of the acoustic excitation will produce tensile strain pulses in the adjacent materials. Such a bipolar "DC"-nonlinear response of the glass–YZ–LiNbO$_3$ interface, supported by a static contact pressure, is shown in Figure 19.13. The calibration of the receiver confirmed that the negative polarity of the signal in Figure 19.13 complied with the gap expansion, that is, the "DC"-acoustic displacement directly opposite to the static contact force.

The asymmetric stiffness variation was assumed to exist for acoustic stress normal to the interface whereas the tangential stress was supposed to produce symmetrical stiffness change. The latter was based on the symmetry in the friction force response due to

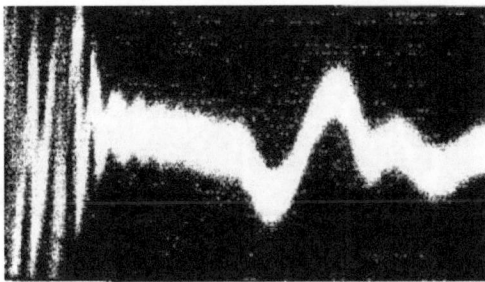

Fig. 19.13. "DC"-acoustic compression pulses observed for glass-YZ-LiNbO$_3$ interface driven by 15-MHz acoustic excitation.

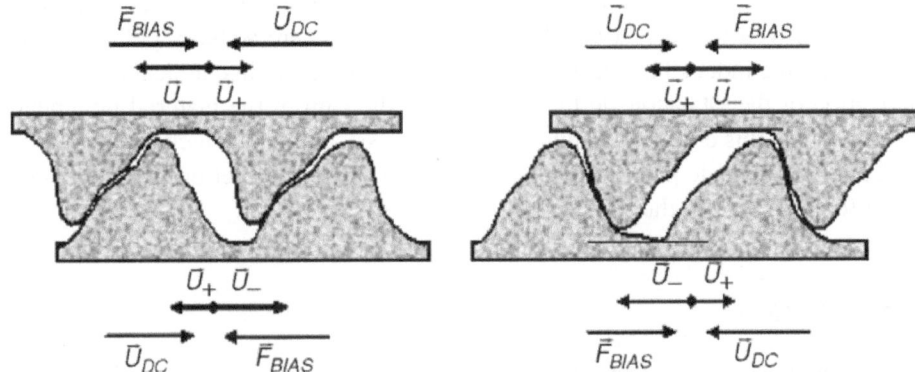

Fig. 19.14. Positions of the neighboring asperities on the rough interface for two opposite directions of tangential biasing external force applied to the interface.

statistically homogeneous distribution of the asperities along the interface. However, such a generally reasonable assumption may not always be valid. This can be seen from Figure 19.14, which sketches the positions of the neighboring asperities on the rough interface for two opposite directions of static tangential biasing force ($\vec{F}_{BIAS}$) applied to the interface. It can be easily observed that for the biased contact, the friction force and tangential stiffness are asymmetrical: the contact is stiffer when the driving shear traction has the same orientation as the biasing force and softer in the opposite phase. As a result, the friction force mechanism acquires the features of the asymmetrical nonlinearity and, similarly to the clapping mechanism, can produce the higher harmonics of even order (in particular, zero-order or DC-field).

The latter is illustrated in Figure 19.14: due to the stiffness asymmetry the interface shear displacement against the biasing force ($\vec{U}_-$) dominates and thus $\vec{U}_- > \vec{U}_+$. Similarly to the clapping interface, this forces the DC-shear displacement ($\vec{U}_{DC}$) to be always in the direction opposite to the bias applied, and hence in the same direction as the static friction force it triggers. In pulse mode, the onset and release of a harmonic driving stress will produce a rectified shear acoustic pulse with polarization dependent on the direction of the contact biasing force. The experimentally measured "DC"-shear responses of the tangentially biased glass–YZ–LiNbO$_3$ interface are shown in Figure 19.15; the follow-up behind the first rectified pulse in the pictures is due to

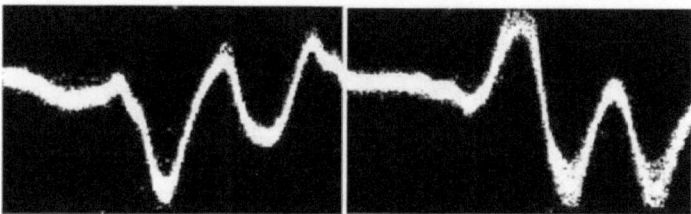

Fig. 19.15. Shear wave "DC"-nonlinear response of glass-YZ-LiNbO$_3$ interface driven by 15-MHz acoustic excitation for two opposite orientations of tangential biasing force.

"ringing" of the receiving transducer. It can be seen that the polarity of the rectification response changes to the opposite as the interface bias inverses. The transducer calibration proved that the "DC"-shear pulses observed are actually polarized along the friction force induced by biasing. It is worth recalling that in classical materials the polarization of the DC-acoustic field is fixed by the sign of the material nonlinearity parameter.[20] To this end, the nonlinearity of nonclassical media which comprise rough interfaces can be made positive or negative by varying the direction of the tangential biasing force applied to the interface.

4. Summary

Interfaces with no strength but finite stiffness may respond to an interrogating acoustic wave in a nonlinear fashion. Three mechanisms have been considered here, which determine such a behavior. They all involve micromechanics of individual contacts that is determined by well-known interaction laws between asperities. In particular, it is shown that an interface between asperities in Hertzian contact is characterized by a response in which the second harmonic component dominates the spectrum of the nonlinear scattered waves. On the other hand, mechanisms such as partial slip and adhesion produce a cascade of higher-order harmonics, with the odd harmonics dominating over the even ones. Indeed, when an interface between rough surfaces in contact is excited by a shear wave at normal incidence, only odd harmonics are generated by its nonlinear response. When adhesion becomes significant, the amplitude distribution of the nonlinear components in the spectrum of the scattered waves closely resembles that found in geomaterials and materials with distributed damage. In all cases, the amplitude of the scattered waves has been shown to decrease with increasing roughness.

The strength of the interface bonding is a crucial factor for development of nonlinear acoustic phenomena in imperfect materials. A weakly bonded interface driven by an intense acoustic wave displays a specific nonlinear dynamics of the intermittent contact, associated with either asymmetrical or symmetrical stiffness variation when driven by normal or tangential tractions, respectively. As a result, a full-scale integer multiple higher harmonic spectrum in the former case of clapping interface is changed by the odd harmonic domination for the friction force nonlinearity. Both nonlinear spectra exhibit nonclassical features such as threshold onset, unconventional waveform distortion, *sinc*-envelope modulation, and nonclassical dynamic characteristics.

In realistic materials with flaws, the nonlinear interface signatures are not confined to the two distinctive nonlinear responses characteristic of intermittent contacts. In particular, fractured defects or mechanical bias, with a static load in the friction force-driven nonlinear materials, result in the transition to asymmetrical nonlinearity, and thus the even-order harmonic signature induced can be used for material characterization and nonlinear NDE of flaws. Another implication of the asymmetrical interface stiffness variation is manifested in the rectification effect accompanied by the DC-elastic response. For mechanically biased contacts, the interface nonlinearity always works against the biasing stress and provides the DC-displacements on the interface polarized opposite to the biasing force applied. As compared with molecular nonlinearity of classical materials, nonclassical media comprising internal interfaces exhibit a unique degree of flexibility in shaping their nonlinear signatures by mechanical impact.

Acknowledgments

C. Pecorari acknowledges the support of the Swedish Inspectorate for Nuclear Power, the Swedish Council of Research and the European Science Foundation. The results of Section 2 and the associated figures have been reproduced here with the kind permission of the American Institute of Physics.

References

[1] P.B. Nagy, Ultrasonic classification of imperfect interfaces, *J. Nondestructive Eval.* **11**, 127–140 (1992).

[2] I.Y. Solodov, Ultrasonics of non-linear contacts: propagation, reflection and NDE applications, *Ultrasonics* **36**, 383–390 (1998).

[3] I.Y. Solodov, N. Krohn, and G. Busse, CAN: an example of non-classical acoustic nonlinearity in solids, *Ultrasonics* **40**, 621–625 (2002).

[4] I.Y. Solodov and B.A. Korshak, Instability, chaos, and 'memory' in acoustic wave – crack interaction, *Phys. Rev. Lett.* **88**, 014303 (2002).

[5] V.Y. Zaistev, V. Gusev, and B. Castagnede, Observation of the 'Luxenburg-Gorky' effect for elastic waves, *Ultrasonics* **40**, 627–631 (2002).

[6] V.Y. Zaistev, V. Gusev, and B. Castagnede, Luxenburg-Gorky effect retooled for elastic waves: a mechanism and experimental evidence, *Phys. Rev. Lett.* **89**: 105502 (2002).

[7] D. Donskoy, A. Sutin, and A. Ekimov, Nonlinear acoustic interaction on contact interfaces and its use for nondestructive testing, *NDT&E Int.* **34**, 231–238 (2001).

[8] C. Pecorari, Nonlinear interaction of plane ultrasonic waves with an interface between rough surfaces in contact, *J. Acoust. Soc. Am.*, **113**, 3065–3072 (2003).

[9] C. Pecorari, Adhesion and nonlinear scattering by rough surfaces in contact: Beyond the phenomenology of the Preisach–Mayergoyz framework, *J. Acoust. Soc. Am.*, **116**, 1938–1947 (2004).

[10] J.A. Greenwood and J.B.P. Williamson, Contact of nominally flat surfaces, *Proc. R. Soc. London A* **295**, 300–319 (1966).

[11] S.R. Brown, and C.H. Scholz, Closure of random elastic surfaces in contact" *J. Geophys. Res.* **90**, 5531–5545 (1985).

[12] K.N.G. Fuller and D. Tabor, The effect of surface roughness on the adhesion of elastic solids, *Proc. R. Soc. of London A* **345**, 327–342 (1975).

[13] K.E.-A. Van Den Abeele, P.A. Johnson, and A. Sutin, Nonlinear elastic wave spectroscopy (NEWS) techniques to discern material damage, Part I: Nonlinear wave modulation spectroscopy, *Res. Nondestr. Eval.* **12**, 17–30 (2000).

[14] K.E.-A. Van Den Abeele, P.A. Johnson, R.A. Guyer, and K.R. McCall, On the quasi-analytic treatment of hysteretic nonlinear response in elastic wave propagation, *J. Acoust. Soc. Am.* **101**, 1885–1898 (1997).

[15] D.G. Meegan, Jr., P.A. Johnson, R.A. Guyer, and K.R. McCall, Observations of nonlinear elastic wave behavior in sandstone, *J. Acoust. Soc. of Am.* **94**, 3387–3391 (1993).

[16] E.M. Ballad, B.A. Korshak, I.Yu. Solodov, N. Krohn, and G. Busse, Local nonlinear and parametric effects for non-bonded contacts in solids, *Nonlinear Acoustics at the Beginning of the 21st Century*, Eds. O. Rudenko and O. Sapozhnikov (MSU, Moscow) 727–734 (2002).

[17] I. Yu. Solodov, K. Pfleiderer, and G. Busse, Nondestructive characterization of wood by monitoring of local elastic anisotropy and dynamic nonlinearity, *Holzforschung*, **58**, 504–510 (2004).

[18] H. Gerhard and U. Lampater, Nonlinear approach to fracture tests with miniature tensile machine, Joint project of IKP-ZFP & IFF Stuttgart University, (2004) (unpublished).

[19] B.A. Korshak, I.Yu. Solodov, and E.M. Ballad, DC-effects, sub-harmonics, stochasticity and "memory" for contact acoustic non-linearity, *Ultrasonics*, **40**, 707–713, (2002).

[20] W.T. Yost and J.H. Cantrell, Jr., Acoustic-radiation stress in solids. II. Experiment, *Phys. Rev. B*, 1984, **30**, 3221–3227 (1984).

20

Nonlinear Dynamics in Granular Materials

Claes Hedberg[1,3] and Philippe Martinet[2]

[1]Blekinge Tekniska Högskola, Gräsvik, 371 30 Karlskrona, Sweden, claes.hedberg@bth.se.
[2]Kungliga Tekniska Högskolan, 100 44 Stockholm, Sweden, martinet@kth.se.
[3]To whom correspondence should be addressed: Claes Hedberg: Blekinge Tekniska Högskola, Gräsvik, 371 30 Karlskrona, Sweden email: claes.hedberg@bth.se.

Abstract

This short review is about nonlinear behavior in granular materials. We have selected particular behaviors relevant to the main topic of this book which is the nonlinearity relation between the microscale and macroscale properties of a material. At the mesoscale, granular material present intricate physical processes such as dynamical arches formations.

The conductivity in the proximity of the conductivity threshold depends exponentially on a universal criterion that only depends on the dimensionality of the problem. We present a model based on 1-D flow equations whose solution is a universal attractor representing an exponentially increasing stress with the solidity field at a critical value about 0.5.

Granular avalanches present segregation mechanisms making large grains move to the front of the avalanche, and the smaller grains to the tail. One can also observe nonlinear upwards-propagating dispersed shock waves.

We give a stress–strain relationship typical for granular media with mesoscopic inhomogeneities. From the vibration of two grains the solution is generalized to a distributed fluid-saturated grainy medium. When the wave velocity is equal to the speed of sound, a wave resonance phenomenon can be located in a layer at a particular depth.

Keywords: Fluid-saturated granular media, grain fluidization, grain vibration, granular avalanches, inertia nonlinearity, multiscale nonlinearity, universal criteria

1. Introduction

The global behavior of a large number of grains that only interact through collisions and friction is typically nonlinear and different from any other standard and familiar forms of matter as solids, liquids, and gases. Vibrated granular materials are perfect examples of such multiscale nonlinearity. Density waves may have intricate patterns that oscillate between one symmetrical pattern to its inverse in resonancelike states. Granular materials are often referred as a new form of matter on their own. Granular materials present different scales of interest. At the microscale (i.e., at the scale of a few grains) a granular material behaves as an inelastic solid.

At the macroscale, granular material behaves very much as does a liquid or a gas depending of the density of grains (e.g., [1]). But it is at the mesoscale of 5–7 grain sizes, that granular materials show intricate mechanisms. The existence of such mesoscale was estimated by P. Martinet in the context of internal erosion of earth dams [2]. The same critical value of 5 grains was more precisely recalculated by Koenders et al. [3]–[5] who also demonstrated the reality and existence of the mesoscale range in granular media.

Biot [6] in 1955 developed a theory of dynamic deformation of media with a solid and a fluid phase. He stated that there exist two pressure waves, and one displacement wave. In the laboratory Paterson in 1955 [7] measured two pressure waves in marine sands. He found a fluid wave, and one frame wave which he believed to be the pressure wave of second type of Biot, its velocity increases with pressure, and velocity decreases with fluid increase. The Biot-based models correspond relatively well with measurements [8].

Marine sands are homogeneous in grain sizes, sorted by natural processes. In experiments an ultrasound pulse separates in propagation into two parts: one high frequency with high velocity (fast wave), and one low frequency with low velocity (slow wave) [9]. The fast wave has the same frequency as the excitation, and the slow wave-frequencies depend on the nonlinear interaction in the sand internal structure. The slow longitudinal wave was confirmed in water-saturated glass beads by Plona [10]. The slow wave exists also when the fluid is air [11]–[13]. A mathematical model of this phenomena was developed by Nikolaevskii [14], [15] which reduced the problem to Korteweg-de-Vries, Burgers, or a fourth-order equation. In sands with low fluid saturation (less than approximately 15%) only the slow wave with velocity 200–300 m/s was found. For fully fluid-saturated sands only the fast wave with velocities of 1600 m/s was found. In sands with 17–20% saturation, both kinds of waves appeared. As the velocity of the slow wave (the frame wave) increases with pressure, it might reach the same velocity as the fast wave at higher depths. A peculiar behavior is that the dissipative coefficient of an acoustic wave decreases with increasing amplitude [16]. This dissipative coefficient is sensitively dependent on the moisture content. An increase in the static pressure results in high-frequency pulses in wave velocity increases, and amplitude increases because the attenuation decreases [17].

A perfect saturation is difficult to attain as even a few tenths of a percent of air change the sound velocity from 1800 m/s (for fully saturated) to 200 m/s [18]. The cause is the same that stands as a fundament in the phenomena in this book: very small soft partitions are introduced between the harder bulk part of the material leading to a major change in the macroscopic behavior.

The elastic waves that propagate through the frame part of a fluid–grain material are well described by the grain contact mechanics [18], such as the Hertz contact description. The idea to replace the real sandlike medium by a model of packed spherical grains was used many years ago to describe the properties of seismic waves propagating in geological structures. An exhaustive review of these works is given by White [19]. More recent studies devoted to nonlinear properties of such models are reviewed in Reference [20].

2. Vibrated Granular Materials

Under vertical vibrations granular materials behave following two different processes: a fast motion due to the collision of the grains with the vibrated plane and a slow convective motion located in the bulk of the material.

For shallow granular beds that are confined inside a regular shape, surface waves may appear and, typically, produce square patterns. These waves correspond to variations in the local density and are also called density waves. In general, the waves oscillate between two different patterns at a frequency $f/2$ where, f is the frequency of the vertical vibration [21]. See Figure 20.1.

3. Solidification and Fluidization

Solidification and fluidization are two mechanical processes that have both tremendous industrial and scientific implications. The flow equations for highly compacted granular materials that are externally stressed present strong non-linearities. One example is non-linear consolidation. A first-order quantitative estimate was done by Koenders [4]. The result gives a clue on how semistable filtration takes place. The model is based on 1-D flow equations that are continuity equations for the solid and liquid phases and the force equilibrium equations. Two critical values for the velocity are obtained and, very reasonably, agree with measured values. The presence of waves in the medium is required by the presence of exponentially increasing solutions of the characteristic equation for the first-order term of the linearized system of equations. A universal attractor is found and the solutions represent an exponentially increasing stress with the solidity field at a critical value about 0.5. If the granular flow appears in narrow pipes

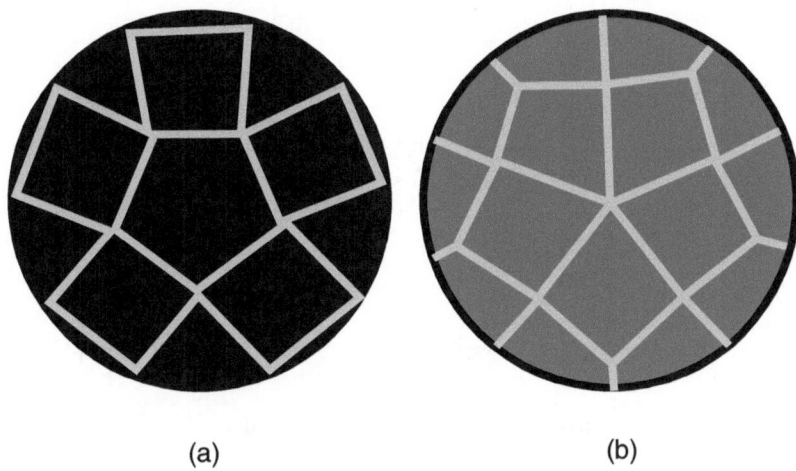

(a)   (b)

Fig. 20.1. Snapshots of fivefold symmetry in resonancelike states of shallow granular beds vibrating vertically (a) and the inverse symmetry after one period (b). (a) Conductivity σ as a function of the probability p to have a grain that is a conductor. (b) Infinite conductive cluster in a granular material with a proportion of p conductive grains and (1–p) isolated grains.

periodical clogging and declogging appear spontaneously [22]. In the case of a sand-clock this effect give rise to a "tic tac"-like sound that is a combination of fluidization and solidification of the thin colon of sand located along the narrow canal.

4. Granular Avalanches

If the slope of a sandpile is less than the angle of repose, the friction between the grains is larger than the gravitational forces and the granular material behaves as a solid. If the slope is higher than the angle of repose, a fluidlike motion appears that we called granular flow. In contrary to a classical fluid, the motion is limited to the top layer. For simple binary granular materials that contain only two sorts of grains with two different sizes, observations show granular avalanches with two different waves: typical roll waves and upwards-propagating dispersed shock waves [23] (see Figure 20.2). Segregation (i.e., size separation) is one of the first observed behaviors of granular materials and is often referred to as the Brazil nut effect. It is a simple mechanical process that inhabits the large grains that are obliged to move upward. In avalanches, segregation results in a front of large grains moving downward with the front of the avalanche.

5. Universal Criteria

If a granular material is compressed, a nonuniform stress field composed of "back-bones" appears [24], [25]. The density threshold between the rigid and nonrigid phase is a first-order transition because there is discontinuity in the backbone density. With $\epsilon = p - p_c$ being the distance to the critical point representing the the density threshold p_c, the correlation length ξ diverges as ϵ goes to zero as

$$\xi \approx \epsilon^{-\nu},$$

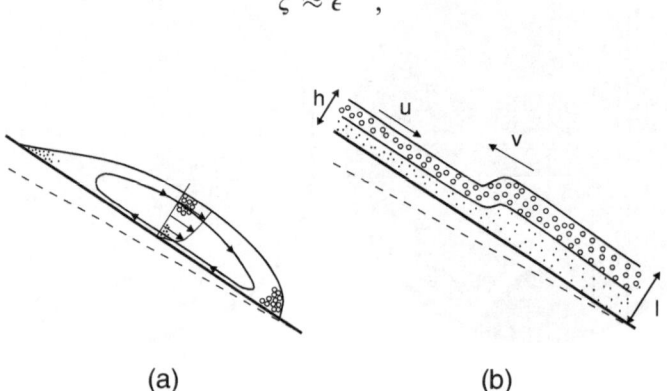

(a)   (b)

Fig. 20.2. (a) Typical roll wave in granular avalanche. The large grains (white) overlie the small grains (dark). (b) Upwards-propagating shock wave. The grains below the shock are almost at rest whereas the grains above are flowing rapidly downslope.

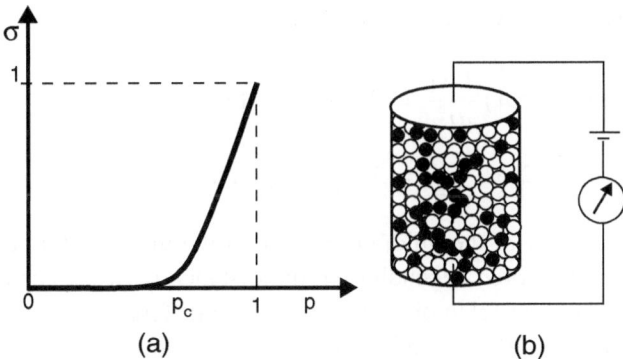

Fig. 20.3. (a) Conductivity σ as a function of the probability p to have a grain that is a conductor. (b) Infinite conductive cluster in a granular material with a proportion of p conductive grains and $(1-p)$ isolated grains.

where ν is a universal correlation length exponent. Inside regions that have a size smaller than the correlation length ξ, mean values depend on the size of the system and homogeneous approaches cannot be applied.

An example of such critical behavior is given by the classical experiment of a mixture of isolated and conductive mono-size grains. At a critical density p_c of conductive grains the entire medium becomes conductive ($\sigma = 1$) because an infinite conductive cluster appears in the medium. The value of the critical density is a universal parameter that only depends on the dimensionality of the problem. See Figure 20.3

6. Vibration of Grains

The typical nonlinear responses of grainy media are based on nonlinear stress–strain relationships typical for solids containing mesoscopic inhomogeneities or defects in their structure (see, e.g., chapters in this book and the reviews [20]– [26]). However, there exists a different type of nonlinearity that manifests itself due to inertial forces between grains, appearing if a system of interacting particles is placed into a vibrating fluid. Structural internal forces are caused by a nonuniform mass distribution. In the following, a model of grainy medium is developed that first deals with two grains, and then with an ensemble of grains, immersed into a vibrating fluid. The inertial attractive forces have a hydrodynamic origin, and the repulsive forces are caused by deformation of colliding grains.

6.1 Vibration of a Pair of Grains

Let us consider two bodies immersed in an ideal fluid, being in contact during part of the period of an acoustic wave. We consider here two spherical bodies with different densities ρ_1, ρ_2 and volumes V_1, V_2. The coordinates x_1, x_2 indicating their center of mass satisfy the coupled system of equations

$$(\rho_0 + 2\rho_1)\frac{d^2 x_1}{dt^2} = 3\rho_0\frac{dv}{dt} - \frac{2}{V_1}[F(x_2 - x_1) - F_0] \tag{20.1}$$

$$(\rho_0 + 2\rho_2)\frac{d^2 x_2}{dt^2} = 3\rho_0\frac{dv}{dt} + \frac{2}{V_2}[F(x_2 - x_1) - F_0]. \tag{20.2}$$

Here $\bar{v}$ is the velocity of the fluid, $m_0 = \rho_0 V$ is the mass of fluid displaced by the body, F is a repulsive force depending on the distance between centers, and F_0 is a constant external holding (clamping) force. For the difference $\xi = x_2 - x_1$ one can derive from (20.1) and (20.2) the inhomogeneneous nonlinear equation [27],

$$\frac{d^2 \xi}{dt^2} - a[F(\xi) - F_0] = b\frac{dv}{dt}, \tag{20.3}$$

where

$$a = 2\frac{V_1(\rho_0 + 2\rho_1) + V_2(\rho_0 + 2\rho_2)}{V_1 V_2(\rho_0 + 2\rho_1)(\rho_0 + 2\rho_2)}, \qquad b = 6\rho_0\frac{\rho_1 - \rho_2}{(\rho_0 + 2\rho_1)(\rho_0 + 2\rho_2)}. \tag{20.4}$$

According to Hertz's theory, and for a harmonic fluid vibration caused by a mono-chromatic acoustic wave, a nondimensional equation is obtained [27]:

$$\frac{d^2 y}{d\tau^2} + \beta^2[y^{3/2}\Theta(y) - 1] = D \cdot \cos(\tau). \tag{20.5}$$

Here $y = h/h_0, \tau = \omega t, \beta^2 = \alpha^2/\omega^2\sqrt{h_0}, D = bv_0/\omega h_0, h = R_1 + R_2 - (x_2 - x_1), \alpha^2 = aE\sqrt{R_1 R_2/R_1 + R_2}$, and Θ is the Heaviside step-function. The constants v_0 and ω are the amplitude and the frequency of vibrational velocity of the fluid, and $h = h_0$ is the equilibrium position of the spheres defined by the equation

$$\alpha^2 h_0^{3/2} = aF_0 . \tag{20.6}$$

In Figure 20.4 the position, velocity, and the amplitude of the frequency content of the velocity are shown for $\beta = 1$ and $D = 2$. The usual nonlinear generation of higher harmonics takes place, and the appearance of low-frequency spectra is well pronounced. These low frequencies do not depend on the driving frequency of the acoustic source. The transformation of energy to low frequencies is connected with the high amplitude of negative displacement and the long time spent by particles in free motion between collisions. After each short collision the particles move slowly away from their neighbors and then back, leading to a very broad spectrum. Such spectral broadening for a harmonic input signal during its propagation through grainy media is a well-known phenomenon.

7. A Distributed Fluid-Saturated Grainy Medium

We consider a more general distributed one-dimensional nonlinear model. The struc-ture has periodically located nonlinear elements that form an infinite discrete chain.

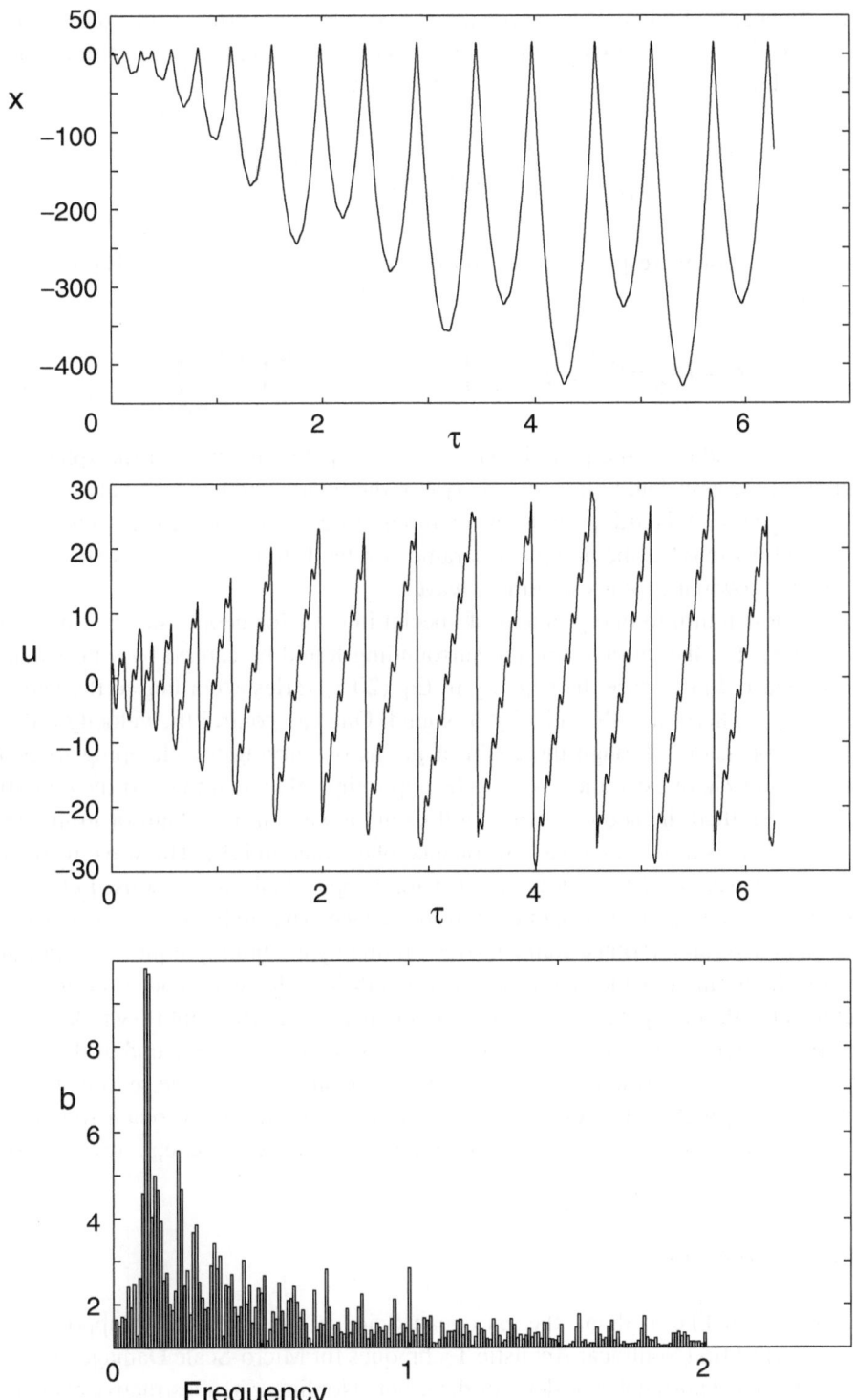

Fig. 20.4. The position x, the velocity u, and the frequency content of the velocity u for the parameter values $\beta = 1$ and $D = 2$.

Each element of this chain is an aggregate of two spherical particles, like those studied in the preceding section. One can pass now to the continuum leading to an inhomogeneous nonlinear wave equation in the displacement ξ [28]:

$$\frac{\partial^2 \xi}{\partial t^2} = c^2 \left(1 - 2\epsilon \frac{\partial \xi}{\partial x}\right) \frac{\partial^2 \xi}{\partial x^2} + \frac{3m_0}{2m} \frac{\partial v}{\partial t} \ . \tag{20.7}$$

The velocity c of wave propagation and the nonlinear parameter ϵ in Eq. (20.7) are given by

$$c = L(\frac{3}{2m})^{1/2} \left(\frac{R}{2} E^2 F_0\right)^{1/6} , \qquad \epsilon = \frac{L}{4} \left(\frac{RE^2}{2F_0^2}\right)^{1/3} . \tag{20.8}$$

Here R is the radius of the particle, $m = \rho V + m_0/2$ is the mass of the sphere with account for the associated mass, $m_0 = \rho_0 V$ of the liquid, and v_n is the velocity of liquid flowing around the nth particle. At clamping force F_0 tending to zero the velocity c also tends to zero but the nonlinear parameter ϵ tends to infinity. So, at small F_0 we deal with a slow but strongly nonlinear wave.

Let us now return to one problem of special interest for geophysics and nonlinear wave theory. Let the vibration of the surrounding liquid be caused by a propagating acoustic wave. In this case the velocity in Eq. (20.7) varies according to the law $v = v(t - x/c_0)$, where c_0 is the velocity of sound. One can control the velocity c (20.8) of wave propagation through the chain of grains by varying the clamping force F_0. By increasing F_0, one can increase c to be approximately equal to c_0. If the condition $c \approx c_0$ is fulfilled the acoustic wave will excite a wave in the chain of grains most efficiently; this is a so-called wave resonance phenomenon [29]. The wave resonance phenomenon described here is of interest for geophysical applications. Let a wave propagate in water along a horizontal bottom surface. The subbottom layers consist of water-saturated sand or other grains moved around by a vibrating liquid. The pressure pressing the grains together increases with depth into the sediment. Therefore, at a specific depth the clamping force—the pressure increasing with depth—will lead to the equality of velocity of wave propagation (20.8) and the speed of sound at that depth. Consequently, wave resonance occurs at this depth, and waves in the sediment will be excited with high efficiency only in this layer. The existence of resonant frequencies of vibration of grains was discovered more than 20 years ago (see, e.g., review [30]).

Acknowledgments

Hedberg would like to thank the European Science Foundation for support of the project NATEMIS (Nonlinear Acoustic Techniques for Micro-Scale Damage Diagnostics), and Vetenskapsrådet, Sweden, for the grant "Nonlinear nondestructive evaluation of material conditions—resonance and pulse techniques."

References

[1] R.P. Behringer, E. Clément, J. Geng, D. Howell, L. Kondic, G. Metcalfe, C. O'Hern, G. Reydellet, S. Tennakoon, L. Vanel, and C. Veje, Science in the sandbox: fluctuations, friction and instabilities, in *Coherent Structures in Complex Systems*, (Ed. by D. Reguera, L.L. Bonilla, J.M. Rubí, Lecture Notes in Physics, Springer, New York), vol. 567, 351–391,(2001).

[2] P. Martinet, Flow and clogging mechanisms in porous media with application to dams, doctoral thesis, The Royal Institute of Technology, Stockholm, Sweden, TRITA-AMI PHD 1017, 1998.

[3] M.A.Koenders and A.F. Williams, Flow equations of particle fluid mixtures, *Acta Mech.*, **92**, 91–116, 1992.

[4] M.A.Koenders, Invited review article, A first order constitutive model for a particulate suspension of spherical particles, *Acta Mech.*, **122**, 1–19, 1997.

[5] M.A.Koenders, J.R.F Arthur, and T. Dunstan, The behavior of granular materials at peak stress, yielding, damage and failure of anisotropic solids, EGF5, 805–818, 1990.

[6] M.A. Biot, Theory of propagation of elastic waves in a fluid-saturated porous solid, *J. Acoust. Soc. Am.* **28**, 168–191 (1956).

[7] N.R. Paterson, Seismic wave propagation in porous granular media, *Geophysics* **VXXI**(3), 691–714 (1955).

[8] N.P. Chotiros, An inversion for Biot parameters in water-saturated sand, *J. Acoust. Soc. Am.* **112**(5), 1853–1868 (2002).

[9] N.A. Vilchinska and Y.L Dzilna, Resonance frequencies in sands, In: *Field Measurements in Geomechanics*. Edited by Geraldine Sørum (Balkema, Rotterdam, 1991), pp. 927–935.

[10] T.J. Plona, Observation of a second bulk compressional wave in a porous medium at ultrasonic frequencies, *Appl. Phys. Lett.* **36**(4), 259–261 (1980).

[11] P.B. Nagy, L. Adler, and B.P. Bonner, Slow wave propagation in air- filled porous materials and natural rocks, *Appl. Phys. Lett.* **56**, 2504–2506 (1990).

[12] R.G. Stearns, Measurement of a multicomponent granular system using acoustic slow waves, *J. Appl. Phys.* **71**(2), 606–611 (1992).

[13] A. Moussatov, B. Castagnède, and V. Gusev, Observation of nonlinear interaction of acoustic waves in granular materials: Demoulation process, *Phys. Lett. A* **283**, 216–223 (2001).

[14] V.N. Nikolaevskii, *Soviet Science Academic Doklady* **283**(3), 1321–1324 (1985) (in Russian).

[15] V.N. Nikolaevskiy, Nonlinear models of waves in granulated partially wet media. In: *Nonlinear Acoustics at the Beginning of the 21st Century*. Edited by O.V.Rudenko and O.A.Sapozhnikov, Faculty of Physics, MSU, Moscow, 2002, Vol.1, 245–250.

[16] V.E. Nazarov, A.V. Radostin, and Y.A. Stepanyants, Influence of water content in river sand on the self-brightening of acoustic waves, *Appl. Acoust.* **62**, 1347–1358, (2001).

[17] V.Yu. Zaitsev, A.B. Kolpakov, and V.E. Nazarov, Detection of acoustic pulses in river sand: experiment, *Acoust. Phys.* **45**, 235–241, (1999).

[18] F.D. Shields, and J.M. Sabatier, The effect of moisture on compressional and shear wave speeds in unconsolidated granular material, *J. Acoust. Soc. Am.* **108**(5), 1998–2004 (2000).

[19] J.E. White, *Underground Sound. Application of Seismic Waves*, Elsevier, Amstrdam, 1983.

[20] V.G. Bykov *Nonlinear Wave Processes in Geological Media*, Vladivostok, Dalnauka, 2000 (in Russian).

[21] T. Metcalf, J.B. Knight, and H.M. Jaeger, Surface patterns in shallow beds of vibrated granular material, *Physica A*, **236**, 202 (1997).

[22] Density waves in narrow pipe by Thorsten Pöschel [http://www.edpsciences.org/articles/jp1/abs/1994/04/jp1v4p499/jp1v4p499.html].

[23] Pattern formation in granular avalanches [http://www.ma.man.ac.uk/ sngray/Papers/CMT_9_1997.pdf]}.

[24] S.N. Coppersmith, C.-H. Liu, S. Majumdar, O. Narayan, and T.A. Witten, Model of force fluctuations in bead packs, *Phys. Rev. E*, **53**, 4673 (1996).

[25] D.M. Mueth, H.M. Jaeger, and S.R. Nagel, Force distribution in a granular medium, *Phys. Rev. E* **57**, 3164 (1998).

[26] O.V. Rudenko, Nonlinear methods in acoustic diagnostics, *Russian J. Nondestructive Testing*, **29**(8), 583–588 (1993).

[27] C.M.Hedberg, Nonlinear dynamics of grains in a liquid-saturated soil, In: *Nonlinear Acoustics at the Beginning of the 21st Century*. Edited by O.V.Rudenko and O.A.Sapozhnikov, Faculty of Physics,MSU, Moscow, 2002, Vol.2, 735–738.

[28] O.V. Rudenko and C.M. Hedberg Nonlinear dynamics of grains in a liquid-saturated soil, *Nonlinear Dynamics*, **35**, 2004, 187–200.

[29] M.B. Vinogradova, O.V.Rudenko, A.P.Sukhorukov, *Theory of Waves*, Moscow, Nauka, 1990 (2nd edition, in Russian).

[30] N.N. Kochina, P.Ya. Kochina, V.N. Nikolaevskii, *The World of Underground Liquids*, Earth Physics Institute, Moscow, 1994 (in Russian).

21

Thermally Induced Rate-Dependence of Hysteresis in Nonclassical Nonlinear Acoustics

Vitalyi Gusev[1] and Vincent Tournat

Université du Maine, Avenue Olivier Massaien, 72085 Le Mans, Cedex 09, France.
e-mail: vitali.goussev@univ-lemans.fr; vincent.tournat@univ-lemans.fr.
[1]To whom correspondence should be addressed.

Abstract

Contribution of hysteretic mechanical elements to the stress/strain relationship of microinhomogeneous material is analyzed within the framework of the Preisach–Arrhenius model where the transitions between the different mechanical states of the individual elements in addition to acoustic loading can be induced by thermal fluctuations. The model provides an explanation of why with increasing wave amplitude a transition from a behavior, which is quasi-independent of wave amplitude, to another, characterized by the dominance of nonclassical hysteretic quadratic nonlinearity, takes place in microinhomogeneous materials. Analytical evaluation of the Preisach–Arrhenius model for the acoustic hysteresis confirms the expectation that thermal relaxation effects are capable of recovering the dependence of the nonlinear acoustic properties of the material on acoustic wave frequency. The theory predicts the boundaries for an intermediate interval of frequencies where hysteretic quadratic nonlinearity dominates in microinhomogeneous materials. Outside this interval (at sufficiently low or sufficiently high frequencies) the nonlinearity significantly diminishes. However the width of the frequency interval for the hysteretic quadratic nonlinearity depends on the acoustic wave amplitude and increases with the increasing wave amplitude. The low-frequency cutoff of the interval diminishes with increasing wave amplitude and the high-frequency cutoff increases. As a result, if the system manifests linearity or quasinonhysteretic nonlinearity at sufficiently low acoustic amplitudes, sooner or later with increasing wave amplitude it will manifest hysteretic quadratic nonlinearity.

Keywords: Dispersion of nonlinearity, hysteretic nonlinearity, microinhomogeneous materials, nonclassical nonlinearity, Preisach–Arrhenius model, rate-dependent hysteresis, thermal relaxation

1. Introduction

The objective of nonlinear acoustics is the evaluation of material nonlinearity, that is to say, of a deviation of the material mechanical behavior from the Hooke's law, by application of low-amplitude (acoustic) strain waves. Typical amplitude values of periodic strain waves do not exceed 10^{-5} and the nonlinear contribution to the material stress/strain relationship is small. Currently there exists a consensus, according to which the nonlinear mechanical properties of microinhomogeneous materials (such as rocks, polycrystalline metals, and ceramics, e.g.) are dominated by nonclassical hysteretic nonlinearity,[1–4] as opposed to the nonlinearity of the interatomic interactions and the kinematic nonlinearity.[5,6] Hysteretic nonlinearity is understood phenomenologically in terms of the nonlinear hysteretic motion of the mesoscopic mechanical elements such as dislocations, intergrain contacts, or defects, for example, with the dimensions exceeding interatomic distances but significantly smaller than the acoustic wavelength.[1,2] As a mathematical tool for the description of hysteresis in nonlinear mechanical properties, the Preisach–Mayergoyz (PM) model of hysteresis[7–10] can be applied. Even in its simplest formulation the PM model explains what is, perhaps, the best known and the most common manifestation of the hysteretic nonlinearity, that is to say, the shift of the resonance frequency of a solid microinhomogeneous bar proportional to the wave amplitude in the bar.[1–4,11] However, the PM model does not explain either experimentally observed dependence of the nonlinear phenomena on frequency[12,13] or the absence of the hysteretic quadratic nonlinearity at very low amplitudes of the acoustic loading.[14–16]

We note here that the Preisach (Preisach–Mayergoyz) formalism[7–10] attributes hysteresis in the nonlinear stress/strain relationship to combined behavior of individual bistable (two-level) hysteretic mechanical units, sometimes referred to as hysterons.[17,18] The transitions (Barkhausen jumps[10]) between two possible states (i.e., energy levels) are assumed to take place instantaneously and exactly at some critical levels of varying stress (strain). For different individual mechanical elements, the levels are different. This model of the hysteretic nonlinearity is essentially dispersionless, that is to say, it is frequency-independent, because there are no characteristic scales of either time or length in the model. The PM model does not take into account that hysteresis is always a dynamic phenomenon. If thermal fluctuations are taken into account in the description of the mesoscopic elements, then there will be no hysteresis in the static limit because the thermal fluctuations are always pushing the system to a unique equilibrium state. In quasistatic experiments, hysteresis will appear at frequencies for which thermal fluctuations have insufficient time to force the system during a wave period in a state having free energy at its absolute minimum value. Instead, the system will be in a state in which its free energy is in a local minimum, that is to say, in a metastable state. Consequently, the nonlinear mesoscopic mechanical elements, in reality, are nonhysteretic in the static limit and hysteretic only in their dynamic behavior. In the theory of magnetism, the Preisach–Mayergoyz model is considered as a zero-temperature limit for rate-independent hysteresis,[10] because the thermal fluctuations are not included and because the stress/strain relation depends only on the sign of the strain rate but not on its magnitude.

The Preisach–Arrhenius (PA) model for the description of thermally activated relaxation, or "after-effect," in magnetic materials[10,18] takes into account that the transitions between the energy levels of the system can be thermally activated and that the probability of the transition is controlled by the Boltzmann factor $\exp(-\Delta E/k_B T)$, where ΔE is the difference in energy levels, or some activation energy, k_B is the Boltzmann constant, and T is the absolute temperature. The thermally controlled transition is not instantaneous statistically. Rather, there is a characteristic time scale for each individual mechanical element that can be estimated by $\tau_0 \exp(\Delta E/k_B T)$ as defined by the Arrhenius formula for the transition time, where τ_0 is some characteristic attempt time associated with the Barkhausen jump between the energy levels.[10] Consequently, dispersion in the acoustic nonlinearity is expected in the Preisach–Arrhenius model. In the rate-independent approximation assumed by the PM model the external action on the system remains nearly unchanged during the time needed to complete the Barkhausen jump. Thus, the external field creates the conditions for the system instability and spontaneous (or thermally initiated) Barkhausen jumps from one local energy minimum to the next. Therefore, when appreciable variations of an external action take place during individual Barkhausen jumps, then rate-independence no longer applies. At high frequency and weak acoustic wave amplitude the characteristic time for a thermally stimulated transition to occur can significantly exceed the acoustic wave period. Thus, the individual elements have insufficient time to modify their state even when loading makes it for some time allowed by energy considerations.[10]

The acoustic wave affects the system through the modulation of the difference ΔE between the energy levels, and in doing so, renders the thermally activated relaxation processes amplitude dependent. Qualitatively speaking, the Preisach–Arrhenius model describes nonlinear temperature-dependent relaxation of the hysteretic mechanical elements. Consequently it might be expected that the nonlinearity of the system is due not only to the intrinsic nonlinearity of the bistable hysteretic elements but also due to the nonlinearity of the relaxation process.

2. Preisach–Arrhenius Model for Acoustic Response of Microinhomogeneous Media

There exists a consensus that microinhomogeneous materials may contain some mechanical elements that are mesoscopic (with the dimensions exceeding the atomic scale but significantly smaller than the acoustic wavelength) and hysteretic (such as reversible Griffith cracks[8] or contacts with adhesion, e.g.). The hysteresis in the behavior of an individual mechanical element might be imagined in the simplest way as being related to the possibility for the element to be in different states under the same mechanical loading. In which state the mechanical element is actually a function of the acoustic loading history. Both in the Preisach–Mayergoyz[7–10] and the Preisach–Arrhenius[10,18,19] models it is assumed that the mechanical elements have two states (two energy levels) and that the contribution σ' of an element to the macroscopic stress in material depends on its state. This phenomenological description assumes that the free-energy of the material, which possesses multiple local minima reflecting the complexity of the mutual interactions among the system's components, can be represented as a linear superposition of two-level bi-stable contributions.[10] In the PM theory the transition of an element from state 1 to state 2 takes place when $\partial\varepsilon/\partial t > 0$, $\varepsilon = \varepsilon_2$, and the inverse transition takes place when $\partial\varepsilon/\partial t < 0$, $\varepsilon = \varepsilon_1 < \varepsilon_2$ (Figure 21.1). The difference between the critical switching strains ε_2 and ε_1 ($\varepsilon_2 \neq \varepsilon_1$) is at the origin of the hysteretic nature of these elements. If the notation $f(\varepsilon_1, \varepsilon_2)$ is introduced

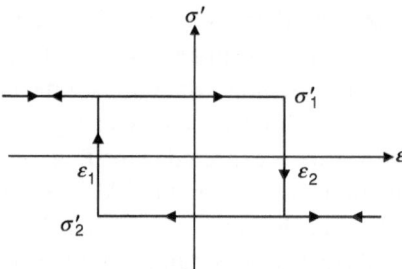

Fig. 21.1. Contribution σ' of an individual mechanical element to stress in the framework of the Preisach–Mayergoyz model. Arrowheads indicate direction of strain variation in time.

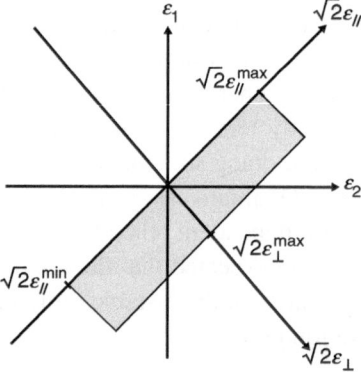

Fig. 21.2. Presentation of mechanical element distribution at Preisach–Mayergoyz plane $(\varepsilon_2, \varepsilon_1)$, where ε_2 and ε_1 are the critical strain values for switching the elements between the levels. A distribution, limited in PM plane by $\varepsilon_\perp \equiv (\varepsilon_2 - \varepsilon_1)/2 \le \varepsilon_\perp^{max}$ and $\varepsilon_{//}^{min} \le \varepsilon_{//} \equiv (\varepsilon_2 + \varepsilon_1)/2 \le \varepsilon_{//}^{max}$, is presented in gray as an example.

to represent the distribution function of the elements in the plane $(\varepsilon_2, \varepsilon_1)$ then the contribution of all the elements to the stress is given as

$$\sigma = \int_{-\infty}^{\varepsilon_2} d\varepsilon_1 \int_{\varepsilon_1}^{\infty} d\varepsilon_2 \sigma'(\varepsilon_1, \varepsilon_2, \varepsilon) f(\varepsilon_1, \varepsilon_2). \qquad (21.1)$$

Here $f(\varepsilon_1, \varepsilon_2) d\varepsilon_1 d\varepsilon_2$ is the number of the elements with critical strains belonging to the intervals $(\varepsilon_1, \varepsilon_1 + d\varepsilon_1)$ and $(\varepsilon_2, \varepsilon_2 + d\varepsilon_2)$ of the PM plane $(\varepsilon_2, \varepsilon_1)$. Due to the assumed condition $\varepsilon_2 > \varepsilon_1$ the integration in the PM plane is in the sector to the right of the diagonal $\varepsilon_2 = \varepsilon_1$ (Figure 21.2). The arguments of the function $\sigma'(\varepsilon_1, \varepsilon_2, \varepsilon)$ indicate that, in general, the contribution of an element to the total stress depends on its position in the PM plane and the loading history as it is presented in Figure 21.1. An important feature of the PM model is that hysteresis in the mechanical behavior of the individual elements exists independently of the magnitude of the strain rate, because the transitions at critical levels ε_2 and ε_1 are assumed to be instantaneous. It is assumed that the transition $1 \Rightarrow 2$ always happens when the strain $\varepsilon (\partial\varepsilon/\partial t > 0)$ exceeds ε_2 independently of how fast ε returns back to the region $\varepsilon < \varepsilon_2$. From a physics point of view, in the PM model the acoustic loading not only creates the conditions for the transition but also induces the change of the state.

The physical nature of $\sigma'(\varepsilon_1, \varepsilon_2, \varepsilon)$ in the Preisach–Arrhenius model is very different. In fact, the acoustic field is no longer the only physical actor that can induce transitions between the states 1 and 2. There are also thermal fluctuations that statistically can cause the same transitions. In the PA model, the transition from state 1 to state 2, for example, can occur during a finite time interval and at values of ε that do not strictly satisfy the conditions $\varepsilon = \varepsilon_2$ $(\partial \varepsilon / \partial t > 0)$. At finite values of the temperature, the elements can overcome the energy barrier by thermal activation at lower strains as long as there is a second (local) energy minimum in which they can jump. Qualitatively speaking, thermal fluctuations accelerate the transitions below the critical level of strain ε_2. At the same time, above the critical strain ε_2, thermal fluctuations induce inverse transitions (from state 2 to state 1), which are completely forbidden in the zero-temperature model. The picture of the inverse transitions $2 \rightarrow 1$ near the critical strain ε_1 is similar.

In the Arrhenius model for thermally initiated transitions, the transition time τ_{12} from level 1 to level 2 is equal to $\tau_{12} = \tau_0 \exp[d(\varepsilon_2 - \varepsilon)/k_B T]$, where d measures the variation of energy difference ΔE_{12} between states 1 and 2 caused by an applied unit strain (deformation potential). Accordingly the transition time τ_{12} diminishes exponentially with increasing strain when the applied strain exceeds the critical level ε_2. Similarly, the time τ_{21} of the inverse transition is $\tau_{21} = \tau_0 \exp[d(\varepsilon - \varepsilon_1)/k_B T]$. The transition times τ_{12} and τ_{21} control the probabilities W_1 and W_2 to find the element in states 1 and 2, respectively,

$$\partial W_1 / \partial t = -W_1 / \tau_{12} + W_2 / \tau_{21},$$
$$\partial W_2 / \partial t = W_1 / \tau_{12} - W_2 / \tau_{21}, \qquad (21.2)$$
$$W_1 + W_2 = 1.$$

These equations are sufficient to describe the variation of stress in response to acoustical loading. Actually the average level of $\sigma'(\varepsilon_1, \varepsilon_2, \varepsilon)$ in the absence of the acoustic wave does not contribute to dynamic stress in Eq. (21.1). Thus it is useful to evaluate the variations of $\sigma'(\varepsilon_1, \varepsilon_2, \varepsilon)$ relative to the average level $(\sigma_1' + \sigma_2')/2$, where σ_1' and σ_2' are the contributions to stress when the element is in positions 1 and 2, respectively. Then the contributions of states 1 and 2 to stress that can be modified by acoustic excitation are described as $(\sigma_1' - \sigma_2')/2 = \Delta\sigma'(\varepsilon_1, \varepsilon_2)$ and $(\sigma_2' - \sigma_1')/2 = -\Delta\sigma'(\varepsilon_1, \varepsilon_2)$, respectively. Taking into account the probabilities of finding the element in the corresponding states, the wave-dependent contribution $\sigma''(\varepsilon_1, \varepsilon_2, \varepsilon)$ to $\sigma'(\varepsilon_1, \varepsilon_2, \varepsilon)$ can be presented as

$$\sigma''(\varepsilon_1, \varepsilon_2, \varepsilon) = \Delta\sigma'(\varepsilon_1, \varepsilon_2)W_1 - \Delta\sigma'(\varepsilon_1, \varepsilon_2)W_2$$
$$= \Delta\sigma'(\varepsilon_1, \varepsilon_2)(W_1 - W_2) \equiv \Delta\sigma'(\varepsilon_1, \varepsilon_2)Q. \qquad (21.3)$$

The relations (21.2) lead to a single equation describing the dynamics of the function Q, which has been introduced in Eq. (21.3) to characterize the asymmetry of the element distribution between the two levels,

$$\partial Q / \partial t + (1/\tau_{21} + 1/\tau_{12})Q = (1/\tau_{21} - 1/\tau_{12}). \qquad (21.4)$$

An obvious but important conclusion based on Eq. (21.4) is the absence of the hysteresis in the contribution of an element to stress under the static conditions. For $\partial/\partial t \to 0$ (zero frequency of the acoustic action) the solution of Eq. (21.4) is

$$Q_0 = -\tanh\left[d\left(\varepsilon - \frac{\varepsilon_1 + \varepsilon_2}{2}\right)/k_B T\right]. \tag{21.5}$$

Thus, in contrast to the PM model the hysteresis in the PA model is a dynamic phenomenon due to the finite rate of acoustic loading (compare the solutions in Figure 21.1 and in Figure 21.3).

For the following analysis the characteristic strain $\varepsilon_0 = k_B T/d$, which provides a scale for the amplitude of acoustic loading necessary for significant (e times) modification of the relaxation times τ_{12} and τ_{21}, is introduced. All the strains are normalized to this level ($\varepsilon/\varepsilon_0 \equiv \varepsilon$, $\varepsilon_{1,2}/\varepsilon_0 \equiv \varepsilon_{1,2}$). Two new variables $\varepsilon_{//} = (\varepsilon_2 + \varepsilon_1)/2$ and $\varepsilon_\perp = (\varepsilon_2 - \varepsilon_1)/2$ are then introduced. Qualitatively speaking $|\varepsilon_{//}|$ characterizes the average energy of the mechanical element (from the acoustics point of view), and $\varepsilon_\perp$ characterizes the separation of the energy levels 1 and 2 in the absence of acoustic loading. On the other hand, $\varepsilon_{//}$ and $\varepsilon_\perp$ have a clear geometrical interpretation: with reference to the diagonal $\varepsilon_2 = \varepsilon_1$ in the PM plane, they are proportional to the co-ordinates measured along that line and the direction orthogonal to it, respectively[10] (Figure 21.2).

In the introduced notations, Eq. (21.4) takes the form

$$\partial Q/\partial\theta + (2/F)\exp(-\varepsilon_\perp)\cosh(\varepsilon(t) - \varepsilon_{//})Q = -(2/F)\exp(-\varepsilon_\perp)\sinh(\varepsilon(t) - \varepsilon_{//}).$$

$$\tag{21.6}$$

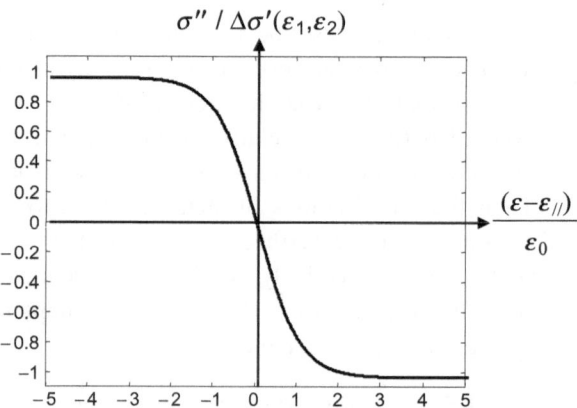

Fig. 21.3. Contribution σ'' of an individual mechanical element to stress in the framework of the Preisach–Arrhenius model in the case of infinitely low frequency of acoustic action. In accordance with Eqs. (21.3) and (21.5) the element behaves in response to strain variation as a two-level but a nonhysteretic unit.

Here the time is normalized to the period T_A of acoustic loading ($\theta = t/T_A$), and the parameter $F = \tau_0/T_A$ is the normalized frequency of the acoustic action. The integral relation (21.1) for the evaluation of the stress becomes

$$\sigma = -\varepsilon_0^2 \int\limits_0^\infty d\varepsilon_\perp \int\limits_{-\infty}^\infty d\varepsilon_{//} \Delta\sigma'(\varepsilon_\perp, \varepsilon_{//}) f(\varepsilon_\perp, \varepsilon_{//}) Q(\varepsilon_\perp, \varepsilon_{//}, \varepsilon(t)). \qquad (21.7)$$

To investigate the acoustic properties of the Preisach–Arrhenius model, Eq. (21.6) is integrated. The exact solution subjected to the conditions of periodicity ($Q(\theta + 1) = Q(\theta)$) is

$$Q = -\frac{\int_\theta^{\theta+1} d\theta' g_s(\theta') \exp\left[-\int_{\theta'}^{\theta+1} g_c(\theta'') d\theta'' \right]}{1 - \exp\left[-\int_\theta^{\theta+1} g_c(\theta'') d\theta'' \right]}, \qquad (21.8)$$

where $g_s = (2/F) \exp(-\varepsilon_\perp) \sinh(\varepsilon(\theta) - \varepsilon_{//})$, $g_c = (2/F) \exp(-\varepsilon_\perp) \cosh(\varepsilon(\theta) - \varepsilon_{//})$.

The formulae (21.7) and (21.8) with an appropriate modeling of the distributions $\Delta\sigma'(\varepsilon_\perp, \varepsilon_{//})$ and $f(\varepsilon_\perp, \varepsilon_{//})$ are sufficient for the description of the acoustic response of materials in the frame of the PA model. Here the results of the analysis are presented for the simplest variation of $\Delta\sigma'(\varepsilon_\perp, \varepsilon_{//})$ and $f(\varepsilon_\perp, \varepsilon_{//})$ in the PM plane ($\varepsilon_\perp, \varepsilon_{//}$). For this purpose the product $\Delta\sigma'(\varepsilon_\perp, \varepsilon_{//}) f(\varepsilon_\perp, \varepsilon_{//})$ is estimated by its characteristic value $(\Delta\sigma' f)_0$ and the extent of the element distribution in the PM plane is assumed to be limited by the boundaries $0 \le \varepsilon_\perp \le \varepsilon_\perp^{max}$, $\varepsilon_{//}^{min} \le \varepsilon_{//} \le \varepsilon_{//}^{max}$ ($\varepsilon_{//}^{min} < 0$, $\varepsilon_{//}^{max} > 0$) (Figure 21.2). It is worth mentioning that the assumption $\Delta\sigma'(\varepsilon_\perp, \varepsilon_{//}) f(\varepsilon_\perp, \varepsilon_{//}) \approx const$ is rather common in the applications of the Preisach–Mayergoyz model to acoustics, because only a small area of the PM plane with the dimensions $\propto \varepsilon_A \varepsilon_A/2$ (where ε_A is the amplitude of the acoustic wave) interacts with sound in the PM model.[2,3,9] In this case the details of the $\Delta\sigma' f$ distribution outside this small area play no role. In the Preisach–Arrhenius model the situation is different because the acoustic wave perturbs the relaxation of all the elements of the PM plane and it may appear of considerable relevance (in particular, for the case of low-frequency acoustic loading) that the distribution of the elements is somehow limited (i.e., $\left| \Delta\sigma' f \right|$ diminishes when $\varepsilon_\perp \to \infty$ and $\left| \varepsilon_{//} \right| \to \infty$).

In Figure 21.4 the results of the numerical evaluation of the hysteresis stress/strain loops predicted by Eq. (21.7) and Eq. (21.8) are presented[19] for the particular case of a sinusoidal strain variation and a homogeneous element distribution inside the rectangular area $\varepsilon_\perp \le \varepsilon_\perp^{max} = 10$, $-10 = \varepsilon_{//}^{min} \le \varepsilon_{//} \le \varepsilon_{//}^{max} = 10$. Modification of the hysteresis loop with increasing wave amplitude at intermediate nondimensional frequency $F = 1$ is demonstrated in Figure 21.4a. The transformation of an elliptical loop, which is typical for linear hysteresis in a stress/strain relationship, to a nonelliptical loop, which is typical of nonlinear hysteresis, with increasing wave amplitude, is clearly seen. Figure 21.4b demonstrates the opening of the hysteresis loop with increasing frequency, indicating the dynamic nature of hysteresis phenomena captured by the Preisach–Arrhenius model.

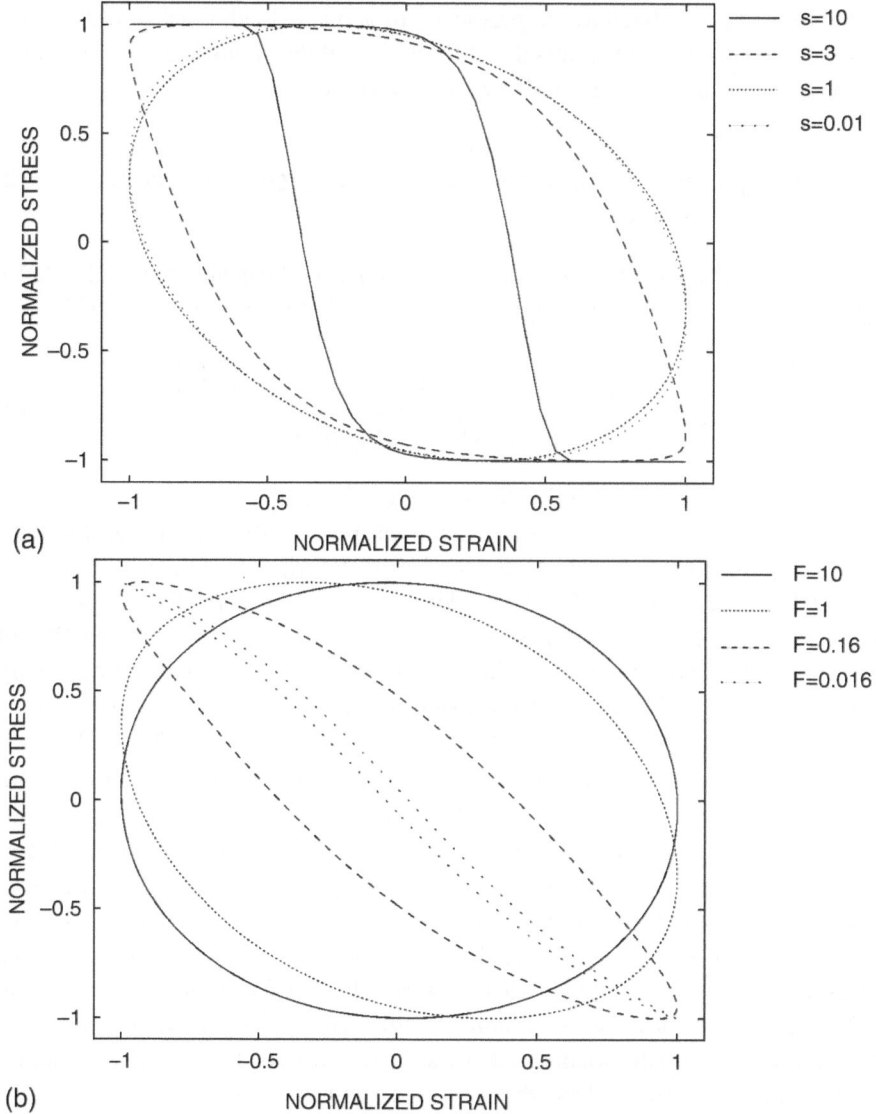

Fig. 21.4. Numerically obtained normalized stress/strain hysteretic dependences in the case of homogeneous element distribution inside the rectangular $\varepsilon_\perp \leq 10$, $-10 \leq \varepsilon_{//} \leq 10$. The path of the material state variation is directed clockwise along the loops. Modification of the hysteresis loop with increasing wave amplitude at fixed frequency $F = 1$ (a). Modification of the hysteresis loop with increasing frequency for the fixed wave amplitude $\varepsilon_A = 1$ (b).

3. Transition from Rate-Dependent to Rate-Independent Hysteresis

From the qualitative analysis of the validity limits of the Preisach–Mayergoyz model (presented in Section 1) it could be concluded that the PM regime should be located between the quasiequilibrium and the quasifrozen limits of the Preisach–Arrhenius model. From physical considerations, the PM regime is absent at very low frequencies, because there are nearly no hysteresis phenomena. In fact, an element has sufficient time both during loading and unloading to assume the same equilibrium configuration (see Figure 21.4b). At very high frequencies, the role of hysteresis is expected to be nearly negligible because the elements have no time to switch from one level to another. The numerical analysis of Section II has also confirmed that the transition from linear to nonlinear hysteresis tends to occur with increasing wave amplitude (see Figure 21.4a). These qualitative arguments are supported by the analytical estimates of the nonlinear contribution to the elastic modulus, which can be obtained[19] in the frame of the mathematical formalism presented in Eqs. (21.7) and (21.8). The so-called secant modulus[20] $\langle E \rangle \equiv \sigma (\varepsilon = \varepsilon_A)/\varepsilon_A$, which is one of the possible forms of presenting the modulus defect, was estimated analytically under the assumption of the infinite extension of the homogeneous distribution of the elements in the PM plane (in other words, $\varepsilon_\perp^{max} \to \infty$, $\varepsilon_{//}^{max} \to \infty$, $\varepsilon_{//}^{min} \to -\infty$), and by approximating the sinusoidal strain variation in the acoustic wave by a sawtooth profile.

The analysis has demonstrated that the linear decrease of the modulus defect with the acoustic wave amplitude $\langle E \rangle \propto -\varepsilon_A$, which is characteristic of rate-independent hysteresis in the frame of the PM model, can be realized only at high amplitudes of the acoustic loading ($\varepsilon_A \gg 1$). However, the latter should be in the region of the homogeneity of the elements' distribution (formally $\varepsilon_\perp^{max} \to \infty$, $\varepsilon_{//}^{max} \to \infty$, $\varepsilon_{//}^{min} \to -\infty$, when $\varepsilon_A \gg 1$). Three different frequency regimes can be identified.

In the high-frequency regime, determined by the inequality $F \gg F_H \equiv \exp(2\varepsilon_A)/(4\varepsilon_A)$, the contribution to the modulus (which, in the following, is normalized by $(\Delta\sigma' f)_0 \varepsilon_0^2$) is very small

$$|\langle E \rangle| \approx \left[1/(4F\varepsilon_A^2) \right] [\ln(F/F_H)/(F/F_H)] \ll 1. \tag{21.9}$$

The significant values of $\langle E \rangle$ with the dominant contribution, which is linear in strain, have been found only in the intermediate frequency regimes $\exp(\varepsilon_A)/(4\varepsilon_A) \equiv F_I \ll F \ll F_H \equiv \exp(2\varepsilon_A)/(4\varepsilon_A)$ and $\exp(-\varepsilon_A/2)/(4\varepsilon_A) \equiv F_L \ll F \ll F_I \equiv \exp(\varepsilon_A)/(4\varepsilon_A)$, where the secant modulus varies as $\langle E \rangle \approx -4\varepsilon_A + [\ln(4F\varepsilon_A)]^2/\varepsilon_A$ and $\langle E \rangle \approx -\varepsilon_A + 2\ln(4F\varepsilon_A)$, respectively. Linear dependence of the modulus on strain amplitude disappears in the low-frequency regime defined by the inequality $F \ll F_L \equiv \exp(-\varepsilon_A/2)/(4\varepsilon_A)$, where the dependence of the modulus on the strain amplitude is logarithmically weak

$$\langle E \rangle \approx -2\ln \left[1/(4F\varepsilon_A) \right]. \tag{21.10}$$

The obtained estimates correlate with the qualitative expectations. First, the Preisach–Mayergoyz regime, in which $\langle E \rangle \propto -\varepsilon_A$, has been recovered as a particular case of the Preisach–Arrhenius model. It is predicted that the PM regime can be obtained for $\varepsilon_A \gg 1$ in a wide frequency interval

$$\exp(-\varepsilon_A/2)/(4\varepsilon_A) \equiv F_L \ll F \ll F_H \equiv \exp(2\varepsilon_A)/(4\varepsilon_A). \tag{21.11}$$

Note that for $\varepsilon_A \gg 1$ we have $F_L \ll 1$, whereas $F_H \gg 1$. The theory predicts that acoustic nonlinearity grows with increasing frequency of a high-amplitude excitation ($\varepsilon_A \gg 1$) in the low-frequency domain $F \ll F_L$, that it weakly (logarithmically) depends on the frequency in the intermediate domain $F_L \ll F \ll F_H$ of quadratic hysteretic nonlinearity, and that it falls in the high-frequency domain $F \gg F_H$.

Second, in accordance with the derived formulae in transition from the low-frequency regime $F \ll F_L$ to the intermediate-frequency regime $F_L \ll F \ll F_H$, the dominant contribution to the secant modulus changes from being nearly strain independent to having a linear dependence on the strain. So, for a material loaded by high-amplitude acoustic waves, the critical frequency F_L can be identified as a transition frequency from the regime where its elements behave as if they were in quasiequilibrium (Figure 21.3), to the regime where they behave as bi-stable units (Figure 21.1).

Third, in accordance with the derived formulae, in the transition from the intermediate-frequency regime $F_L \ll F \ll F_H$ to the high-frequency regime $F \gg F_H$, there is a significant diminishing in secant modulus magnitude that is accompanied by the disappearance of the contribution which is linear in strain amplitude. Consequently, the critical frequency F_H can be identified as a transition frequency from the regime where the mesoscopic mechanical elements behave as bi-stable units, to the regime where they behave as quasifrozen ones.

In accordance with the obtained results, if the dominant contribution to the modulus defect in experiment is linear in wave amplitude, this necessitates the strong inequality $s_A \gg 1$. In other words, the dimensional acoustic strain amplitude should significantly exceed the characteristic strain $s_0 = k_B T / d$ of the material. In this limiting case, the theory predicts that the dispersion of the nonlinearity, which is accompanied by the deviation from the $\langle E \rangle \propto -\varepsilon_A$ law, might be expected in the frequency ranges $F \leq F_L$ and $F \geq F_H$.

It should be also noted that the obtained results correlate well with the experimentally observed dependence of the modulus defect on the wave amplitude.[14] For the comparison it should be taken into account that in the high-amplitude regime the dependence of the critical frequencies on the wave amplitude is exponentially strong [see Eq. (21.11)]. For example, if for the initial amplitude of the acoustic excitation with $\varepsilon_A \gg 1$ the system is in the low-frequency regime $F \ll F_L$, then with increasing ε_A the characteristic frequency $F_L \equiv \exp(-\varepsilon_A/2)/(4\varepsilon_A)$ diminishes and sooner or later the opposite condition $F_L \ll F$ will be fulfilled. This corresponds to the transition of the system with increasing wave amplitude from the low-frequency quasilinear regime (21.10) to the intermediate-frequency regime characterized by $\langle E \rangle \propto -\varepsilon_A$ typical of PM model.

If for the initial amplitude of the acoustic excitation with $\varepsilon_A \gg 1$ the system is in the high-frequency regime $F \gg F_H$, then with increasing ε_A the characteristic frequency $F_H \equiv \exp(2\varepsilon_A)/(4\varepsilon_A)$ increases and sooner or later the opposite condition $F \ll F_H$ will be fulfilled. This corresponds to the transition of the system with increasing wave amplitude from the high-frequency quasifrozen regime (21.9) to the intermediate frequency regime characterized by $\langle E \rangle \propto -\varepsilon_A$ typical of PM model.

Taking into account that the PA model naturally describes quasilinear behavior of the microinhomogeneous material at weak amplitudes of acoustic loading ($\varepsilon_A \ll 1$; see Figure 21.4a), it can be also concluded that the developed theory predicts the transition from the amplitude-independent modulus defect to the law $\langle E \rangle \propto -\varepsilon_A$ (typical of hysteretic quadratic nonlinearity) with acoustic amplitude increasing from $\varepsilon_A \ll 1$ to $\varepsilon_A \gg 1$.

4. Discussion

Prior to drawing any conclusion from the work presented above, it should be clearly stated that the thermal relaxation Preisach–Arrhenius model does not include all the effects producing rate-dependence of the hysteresis. See, for comparison, the description of rate-dependent hysteretic phenomena in magnetism.[10] The rate-dependence may also appear due to the fact that the acoustic field cannot, in principle, transform the state of a mechanical element infinitely fast.[10,21] In other words, an individual mechanical element cannot change its configuration instantaneously either due to direct effect of the acoustic field or due to thermal fluctuations. In the Preisach–Arrhenius model, the finite transition time appears only statistically in averaging over all the elements, whereas each of the elements still exhibits instantaneous transitions as in the zero-temperature (PM) model. To introduce finite transition times for the individual elements, either a micromechanical model of the transition between the different states should be formulated,[10,22] or the finite transition times could be introduced phenomenologically as a temperature-independent relaxation process.[21] Surely, the generalized theoretical model of hysteresis should include a correct description of the time evolution of both the transitions caused by thermal fluctuations and of those directly induced by the acoustic forces. The development of a generalized model would be highly desirable for the quantitative interpretation of the experiments,[12,13,15,16,23] where the dependence of the acoustic nonlinearity of the microinhomogeneous materials on frequency has been observed.

5. Conclusions

The evaluation of the Preisach–Arrhenius model for the acoustic hysteresis demonstrates that thermal effects are capable of inducing a dependence on wave frequency of the nonlinear acoustic properties of microinhomogeneous materials. Thermal effects can also lead to an amplitude-dependent behavior of the material which differs from that predicted by the Preisach–Mayergoyz model in several important aspects. The Preisach–Arrhenius model of rate-dependent acoustic hysteresis also explains the possible transition in acoustic behavior of microinhomogeneous materials with increasing wave amplitude from a linear one to another characterized by dominance of the hysteretic quadratic nonlinearity. From the physics point of view this is due to the fact that the higher the amplitude of the material mechanical loading, the more difficult for the thermal fluctuations to retain the system in a unique quasiequilibrium state.

References

[1] V.E. Nazarov, L.A. Ostrovsky, I.A. Soustova, and A.M. Sutin, Nonlinear acoustics of microinhomogeneous media, *Phys. Earth Planet. Inter.* **50**(1), 65–73 (1988).

[2] R.A. Guyer and P.A. Johnson, Nonlinear mesoscopic elasticity: Evidence for a new class of materials, *Phys. Today* **52**(April), 30–36 (1999).

[3] L.A. Ostrovsky and P.A. Johnson, Dynamic nonlinear elasticity of geomaterials, *La Rivista del Nuovo Cimento* **24**(7), 1–46 (2001).

[4] P.A. Johnson, Nonequilibrium nonlinear dynamics in solids: State of the art, This book.

[5] L.D. Landau and E.M. Lifshitz, *Theory of Elasticity* (Pergamon, London, 1959).

[6] L.K. Zarembo and V.A. Krasilnikov, Nonlinear phenomena in the propagation of elastic waves in solids, *Sov. Phys. Uspekhi* **13**(6), 778–797 (1971).

[7] I.D. Mayergoyz, *Mathematical Models of Hysteresis* (Springer-Verlag, Berlin, 1992).

[8] D.J. Holcomb, Memory, relaxation, and microfracturing of dilatant rock, *J. Geophys. Res.* **86**(B7), 6235–6248 (1981).

[9] R.A. Guyer, K.R. McCall, and G.N. Boitnott, Hysteresis, discrete memory, and nonlinear wave propagation in rock: a new paradigm, *Phys. Rev. Lett.* **74**(17), 3491–3494 (1995).

[10] G. Bertotti, *Hysteresis in Magnetism* (Academic Press, San Diego, 1998).

[11] J.A. Ten Cate, T.J. Shankland, and P.A. Johnson, Nonlinear elastic wave experiments: Learning about the behaviour of rocks and geomaterials, This book.

[12] J.K. Na and M.A. Breazeale, Ultrasonic nonlinear properties of lead zirconate–titanate ceramics, *J. Acoust. Soc. Am.* **95**(6), 3213–3221 (1994).

[13] V.E. Nazarov, Amplitude dependence of internal friction in zinc, *Acoust. Phys.* **46**(2), 186–190 (2000).

[14] A.S. Nowick, Variation of amplitude-dependent internal friction in single crystals of copper with frequency and temperature, *Phys. Rev.* **80**(2), 249–257 (1950).

[15] J.A. TenCate, D. Pasqualini, S. Habib, K. Heitmann, D. Higdon, and P.A. Johnson, Nonlinear and nonequilibrium dynamics in geomaterials, *Phys. Rev. Lett.* **93**(6), 065501 (2004).

[16] D. Pasqualini, Intrinsic nonlinearity of geomaterials: Elastic properties of rocks at low strain, This book.

[17] M.A. Krasnosel'skii and A.V. Pokrovskii, *Systems with Hysteresis* (Springer-Verlag, Berlin, 1989).

[18] E. Della Torre, *Magnetic Hysteresis* (IEEE Press, New York, 1999).

[19] V. Gusev and V. Tournat, Amplitude- and frequency-dependent nonlinearities in the presence of themally-induced transitions in the Preisach model of acoustic hysteresis, *Phys. Rev. B* **72**(5), 054104 (2005).

[20] A.B. Lebedev, Amplitude-dependent elastic-modulus defect in main dislocation-hysteresis models, *Phys. Solid State* **41**(7), 1105–1111 (1999).

[21] V. Gusev, W. Lauriks, and J. Thoen, Dispersion of nonlinearity, nonlinear dispersion and frequency-dependent nonlinear absorption of sound in micro-inhomogeneous materials with relaxation, *J. Acoust. Soc. Am.*, **103**(6), 3216–3226 (1998).

[22] B. Capogrosso-Sansone and R. A. Guyer, Dynamic model of hysteretic elastic systems, *Phys. Rev. B* **66**(22), 224101 (2002).

[23] V.E. Nazarov, A.V. Radostin, and I.A. Soustova, Effect of an intense sound wave on the acoustic properties of a sandstone bar resonator. *Experiment. Acoust. Phys.* **48**(1), 85–90 (2002).

Inverse Problems and Genetic Algorithms

Applications in NDE/NDT & Ultrasonics

Silvia Delsanto,[1,3,4] Michele Griffa,[2,3] and Lia Morra,[1,3]

[1]Dept. of Computer Engineering, Politecnico of Torino, Corso Duca degli Abruzzi 24, 10129, Torino, Italy.
[2]Dept. of Physics, Politecnico of Torino, Corso Duca degli Abruzzi 24, 10129, Torino, Italy.
[3]Bioinformatics and High Performance Computing Labs, Bioindustry Park of Canavese, via Ribes 5, 10010, Colleretto Giacosa (TO), Italy.
[4]silvia.delsanto@polito.it

Abstract

Inverse problems are omnipresent in natural and engineering sciences, for example, in material characterization. Impressive results have been obtained by applying analytical–numerical techniques to their solution; however, in many practical cases these methods present drawbacks, which impede their application. In this scenario, Genetic Algorithms (GAs) arise as interesting alternatives, especially for the solution of complicated inverse problems, such as those resulting from the modeling and characterization of complex nonlinear systems, such as in particular materials with nonlinear elastic behavior. In this chapter, we present a brief introduction to inverse problem solution, highlighting the difficulties inherent in the application of traditional analytical–numerical techniques, and illustrating how genetic algorithms may in part obviate these problems.

Keywords: Elastic wave propagation, genetic algorithms, inverse problems, materials characterization, model fitting

1. Introduction to Inverse Problems

When seeking to model a physical system or phenomenon, one wishes to determine mathematical models describing the relationship between the excitation introduced in the system (input) and its related response (output). A *Forward Problem* (herefore denoted as FP) consists in the prediction of the response of the system, once the excitation and/or the internal properties are known, whereas an *Inverse Problem* (herefore denoted as IP) aims at reconstructing the excitation and/or internal structure, starting from the response. In various important applications, such as tomography, model fitting, image analysis, geophysics, and many others, the input of the system is not directly accessible, but can be somehow recovered by measuring its output.

According to the original definition, solving an IP corresponds to reconstructing a vector of values, named *image $\vec{i}$*, given a data vector $\vec{d}$. The discrete or continuous nature of the image and data spaces determines the type of IP and particularly of the forward operator $\hat{A}$, which maps a point of the image space into a point of the data space:

$$\hat{A}(\vec{i}) = \vec{d}. \tag{22.1}$$

The associated IP can thus be solved by determining the vector $\vec{i}$ for which Eq. (22.1) is satisfied, knowing the data $\vec{d}$ and the operator $\hat{A}$. In many important applications, however, the image $\vec{i}$ is actually known, whereas the input includes boundary conditions and/or the parameters describing the transfer function $\hat{A}$, which can only be indirectly determined starting from the knowledge of $\vec{i}$ and $\vec{d}$. These two types of problems, however, are closely related; in the following, we adopt a general definition of IP, denoting as "input" any kind of information one wishes to determine, unless otherwise stated.

In this section, classical theory about IPs is briefly revised, and the concept of ill-posedness and ill-conditioning introduced. Dealing with ill-posedness and noise is an important aspect because most IPs of practical interest are not well-posed. However we show that, even in this case, useful information may still be extracted. An extensive literature on IPs theory is available; for more detailed information; see for instance the works of Kirsh,[1] Sabatier,[2] Tarantola,[3] and Tan and Fox.[4]

1.1 Ill-Posedness and Approximated Solutions

Given the FP, solving the IP involves finding the image $\vec{i}$ for given data $\vec{d}$. Independently of whether $\vec{i}$ and $\vec{d}$ are continuous or discrete in nature, the IP is termed *well-posed*[5] if it satisfies the following conditions.

1. Existence: A solution exists for any data $\vec{d}$ in data space.

2. Uniqueness: The solution is unique in image space.

3. Continuity: The inverse mapping $\vec{d} \to \vec{i}$ is continuous.

The first two requirements simply state that the operator $\hat{A}$ should have a well-defined inverse $\hat{A}^{-1}$, with codomain equal to the entire data space, whereas the third is a necessary, yet not sufficient, condition for the stability of the solution.

A solution can be considered stable if a small deviation $\Delta\vec{d}$ in the data vector results in a small deviation $\Delta\vec{i}$ of the corresponding image point. An important quantity for characterizing the stability of an IP is the condition number, *cond* $(\hat{A})$, which can be defined as

$$cond(\hat{A}) = \|\hat{A}\|\|\hat{A}^{-1}\|, \tag{22.2}$$

where $\|\hat{A}\|$ means the norm of the operator $\hat{A}$ and $\hat{A}^{-1}$ indicates the inverse (or pseudoinverse in the case the inverse doesn't exist) of the same operator. It can be shown that

$$\frac{\|\Delta\vec{i}\|}{\|\vec{i}\|} \leq cond(\hat{A})\frac{\|\Delta\vec{d}\|}{\|\vec{d}\|}, \tag{22.3}$$

where $\| \Delta \vec{d} \|$ is the variation of $\vec{d}$ and $\| \Delta \vec{i} \|$ the corresponding variation of $\vec{i}$. Equation (22.2) entails that the condition number controls relative error propagation from the data to the solution, so that the IP admits stable solutions only if it is also *well-conditioned*, that is, the condition number is not too large. It is clear that the definition of ill-conditioned problems is rather vague, compared to that of ill-posed ones. However, it should be noted that ill-conditioned problems can show properties very similar to those of ill-posed ones, in terms of sensitivity to noise and high-frequency perturbations.

In practical applications, data $\vec{d}$ are collected through measurements, and thus are affected by noise. Usually, measured data can be represented as the superposition of the "true" data vector $\vec{\tilde{d}}$, which can be obtained through the forward process as formulated in Eq. (22.1), and a stochastic variable $\vec{n}$ representing the noise process; the IP thus becomes:

$$\hat{A}(\vec{i}) = \vec{d} = \vec{\tilde{d}} + \vec{n}. \tag{22.4}$$

However, after adding random noise, this equation may no longer admit a solution: the IP must therefore be reformulated as an *optimization problem*, where the quantity to be minimized is the *misfit* $C(\vec{i})$ between the measured data $\vec{d}$, and the data calculated from a given image $\vec{i}$. Thus, an approximated solution can be found by minimizing the following function,

$$C(\vec{i}) = \| \vec{d} - \hat{A}(\vec{i}) \|. \tag{22.5}$$

In the presence of noise and ill-conditioned problems, the invertibility of the operator $\hat{A}$ turns out to be an issue of relatively little interest: even if the problem can be exactly solved from a mathematical point of view, the effects of noise amplification can be disruptive to the point that the solution is actually determined by the noise itself, rather than by relevant measurement information. Due to the uncertainty introduced by noise, the global minimum of $C(\vec{i})$ could be not the optimal solution, whereas better results can be obtained by considering a feasible set of solutions [specifically those satisfying the condition $C(\vec{i}) < C_0$, with C_0 depending on the level of noise] which can be considered consistent with the observed data.

Even if experimental data were noise-free, the IP could still admit multiple feasible solutions because of its indetermination, either due to the lack of available experimental data or because the forward operator, failing conditions 1 and 2, is not exactly invertible. When multiple potential solutions are available, and minimizing the misfit function may lead to instability, a compromise between stability and accuracy of the solution can be reached by including a priori information.

In fact, one usually has an idea of what a good solution should "look like", that is, of which properties it should reasonably possess. In the IP field, techniques employed to take into account such a priori information are known as *regularization techniques*. Common methods include Tikhonov's, Levenberg's, and Levenberg–Marquardt's regularization techniques.[6] Although an exhaustive discussion is beyond the scope of this chapter, a brief explanation of at least the most common and widely known technique, *Tikhonov's regularization*, is deemed essential.

Let $\vec{i}_0$ be the default solution for a given IP and let $\hat{L}$ be a linear operator. For instance, $\vec{i}_0$ could be determined according to a priori information, when available, or can be simply set equal to the null solution $\vec{0}$. The Tikhonov's regularization scheme consists in minimizing, instead of the quantity given in Eq. (22.5), the following function,

$$\Lambda(\vec{i}) = \lambda^2 \cdot \Omega(\vec{i}) + C(\vec{i}) = \lambda^2 \|\hat{L}(\vec{i} - \vec{i}_0)\|^2 + \|\vec{d} - \hat{A}(\vec{i})\|^2. \tag{22.6}$$

Two competing terms are thus jointly minimized: the former is the misfit function, and the latter penalizes solutions "distant" from the default solution, according to the operator $\hat{L}$. In the simplest case (i.e., with $\vec{i}_0 = \vec{0}$ and $\hat{L}$ equal to the identity operator), this term simply reduces to the norm of the solution. The weight parameter λ controls the amount of regularization of the solution: by adjusting its value, one can regulate the sensitivity of the solution to measured data, and therefore to the noise therein, in order to counterbalance the effect of perturbations. It is thus clear that the optimal value for λ is noise dependent. A soft computing method for the choice of λ is briefly illustrated in Section 6.1.

1.2 Classical Methods: A Brief Overview and Related Difficulties

Many techniques have been proposed for solving IPs, which are often domain-specific and exploit the peculiarities of a given problem. However, even in this variegated scenario some common characteristics can be identified.

In some problems, the unknown image and the data can be related by an invertible, although generally nonlinear, operator, so that Eq. (22.1) can be exactly solved. This approach is suitable only for restricted classes of IPs, because these methods are unable to deal with ill-posedness, ill-conditioning, data uncertainty, and underdetermination.

As previously mentioned, IPs are usually reformulated as optimization problems. Supposing that an analytical representation of the forward operator is available, a formal solution can often be readily found. In the simplest case, $\hat{A}$ is a linear operator, and the problem is reduced to zeroing the derivatives of the misfit function $C(\vec{i})$ with respect to $\vec{i}$ and solving an (often large) system of coupled linear equations. In the most general case, however, that is, when the operator is nonlinear, an analytical solution of this system may not be available. Unless the problem is somehow simplified (e.g., by linearization), we have to resort to iterative methods for multidimensional, nonlinear optimization, such as the steepest descent algorithm (belonging to the set of conjugate gradient methods) or the Gauss–Newton method. Such methods are based on the exploration of the "search space," starting from an initial guess for the solution and then moving towards a local minimum based on the values of the derivates in the current point. These optimization techniques represent a valid method in the solution of IPs. However, they may suffer from various drawbacks. In particular, they tend to be computationally intensive and liable to the presence of local minima, issues that may be particularly critical when the error landscape tends to present many local optima. Furthermore, these techniques require an analytical formulation of the objective function to be minimized, for example, $C(\vec{i})$, which is generally not possible.

Novel techniques such as Genetic Algorithms (GAs) emerge in this scenario as an interesting alternative, due to the fact that they do not require an analytical formulation of the objective function, and are intrinsically parallel in conception, and thus naturally adequate to parallel implementations, which may allow a viable solution of otherwise impractically burdensome problems from a computational point of view. Last but not least, they are relatively robust to a poor initialization and in general to the presence of local minima.

2. Genetic Algorithms and Inverse Problems

Genetic Algorithms (GAs) are a popular Soft Computing (SC) technique, a term which denotes an ensemble of methodologies that "differ from conventional (hard) computing in that, unlike hard computing, they are tolerant of imprecision, uncertainty and partial truth" in order to "achieve tractability, robustness and low solution cost."[7] Some of these methodologies can be collected in the so-called "biologically inspired computing" class: borrowing features and abilities of biosystems, which seem to be particularly adept at solving precisely those classes of computationally hard problems that do not seem to lend themselves well to classical algorithmic approaches.

Genetic algorithms in particular mimic the evolutionary process by creating a population of solutions, which evolve through principles inspired by the natural selection criterion. They are very effective in search and optimization when the underlying search (parameters) space is large, multidimensional, and characterized by complicated and unknown landscapes, often characterized by the presence of local optima. Excellent references in the field are the books by Goldberg[8] and Michalewicz.[9]

In this section, we briefly review the basic principles of GAs in order to illustrate their application and their advantages in solving IPs. Examples of interest in ultrasonics and NDE/NDT are presented in the following section.

The first step in a genetic algorithm consists in the definition of an initial "population" of solutions. In our specific case, each solution corresponds to a determined set of values of the input parameters of the system, that is, the components of the vector $\vec{i}$ as introduced in Section 1 (see Figure 22.1). Each member of the population is termed chromosome or genome and is composed of genes. The correspondance between genes and parameters may be biunivocal, with each real parameter encoded by a single real-valued gene (direct genotype-to-phenotype mapping), or else each parameter may be associated with a string of genes assuming values from a finite alphabet (typically a binary alphabet), whose members are termed alleles (Figure 22.2). The initialization of the population may be random, although the use of supplementary information may be quite useful in accelerating convergence.

Once the population is defined, a process of selection is enacted, in order to guarantee the survival of the genetic heritage of the fittest individuals. The definition of fitness is related to the value of the objective function corresponding to the individual. It is worth noting that the definition of the objective function is a fundamental step in the implementation of a GA because it is at this point that the generic problem is reinterpreted as an optimization one. Furthermore, the effectiveness and the efficiency

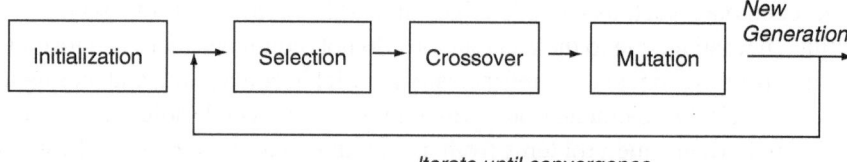

Fig. 22.1. Flow chart for a basic genetic algorithm.

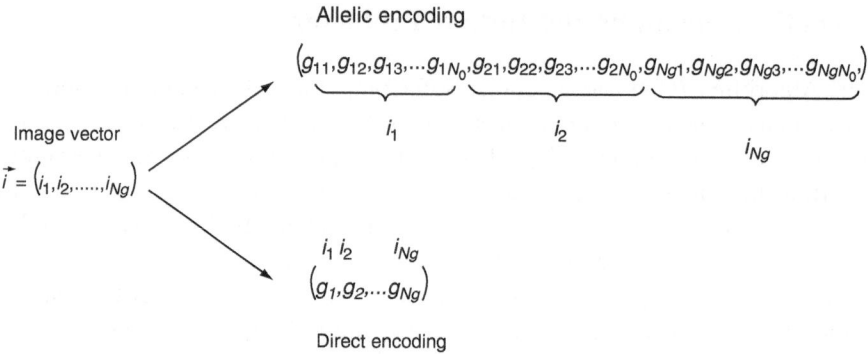

Fig. 22.2. GA encoding.

of the algorithm are strongly related to the measure in which the objective function quantifies the fitness of the solutions. In our particular case, the degree of fitness is reflected by the error between the observed data and the computed output of the system characterized by the parameters associated to the individual.

Having defined the fitness function, it is now necessary to explicate how the selection is implemented. Different schemes are proposed in the literature, but the most common are the roulette-wheel selection, the ranking selection, and the stochastic binary tournament selection (Figure 22.3). In the roulette-wheel selection, each individual is selected with a probability proportional to its degree of fitness. There are many kinds of ranking selection; perhaps the most common selects n copies of the N/n fittest individuals as the successive generation, where N is the number of elements in the population. The value of n determines the greediness of the algorithm; high values of n ensure a rapid convergence, but may likewise imply an inadequate exploration of the parameter space. Finally, in the stochastic binary tournament selection, pairs of individuals are randomly chosen among the population and the winner (i.e., the fittest individual in the pair) is then selected. This technique apparently converges faster than both the roulette-wheel and ranking schemes.

After the selection process has been completed, the chosen individuals undergo a process, whose objective is to create new hybrid individuals, which combines desirable properties of the "parent" individuals. This process is named *crossover* (Figure 22.3), and consists in randomly choosing two individuals and substituting them with probability p_{cross} with their children, that is, two new individuals obtained by crossing the

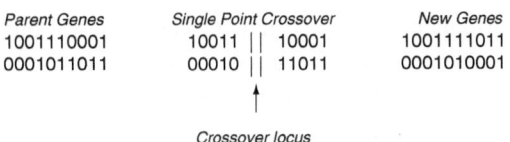

Parent Genes	Single Point Crossover	New Genes
1001110001	10011 || 10001	1001111011
0001011011	00010 || 11011	0001010001

Crossover locus

Fig. 22.3. Single-point crossover.

genes of the parents. This process is particularly important in the search for the optimal solutions, as the new individuals represent new solutions, which combine properties of individuals known to possess a relatively high degree of fitness. Specifically, in the most common variant of crossover known as single-point crossover, the chromosomes are divided in two parts at a specific point termed the *locus*, and the corresponding halves of the two parents are switched to create the children. Other variants of the crossover scheme exist, whose aim is to to allow a greater versatility in the change, for example, avoiding that genes on different ends are always separated during a crossover. For example, in uniform probability crossover, each gene is exchanged among parents with uniform probability.

Finally, the last operation applied to the population is *mutation*. Mutation is a technique applied in order to ensure a certain degree of variety in the genetic heritage of the population of solutions and thus avoid premature convergence of the population. It consists in the perturbation with probability p_{mut} of each gene of each individual of the population.

The genes that are the result of the selection–crossover–mutation process, often termed offspring, may completely or in part replace the original population. The new population thus derived is consequently termed the new generation and constitutes the successive "guess" on the input of the system. The succession selection–crossover–mutation–replacement is iterated until a threshold on the number of iterations, on the value of the objective function, or on the convergence of the genetic heritage is attained.

Though intuitively the concepts underlying the GAs' efficacy are evident, a formal treatment of the methodology capable of explaining and predicting their behavior is still largely incomplete. Existing theories developed in particular for binary-coded, roulette-wheel selection, single-point crossover variants are based principally on the concept of *schemata*, that is, similarity templates. In particular a schema is a set of chromosomes that share identical values of corresponding genes. For example, the schema **1*0*** corresponds to the set of (8 bit, binary-coded) chromosomes, that have the third gene set to one and the fifth gene set to zero and don't-cares in all the other positions. Two important results in the theory of schemata are the schema theorem, which affirms that short, low-order schemata corresponding to genes with above-average objective function values will receive exponentially increasing representation in the successive generations, especially when the defined bits are contiguous, and the implicit parallelism theorem, which states that the GA treats $O(N^3)$ schemata in each generation. Upon the first result is built the so-called *building block hypothesis*, which states that a GA seeks near-optimal performance through the juxtaposition of short, low-order, high-performance schemata, hence referred to as "building blocks." Under this hypothesis, the importance of the gene encoding process becomes clear, and in fact it is advisable to encode the chromosomes in such a way as to guarantee as far as possible that similar individuals have similar error objective function values. On the other hand, the second theorem indicates that doubling the population may more than halve the runtime, so that if the objective function evaluation is not excessively computationally intensive, the use of a large population is recommended.

Theoretical considerations aside, a certain number of practical indications have been developed due to experience gathered by an intensive use of GAs. In particular, typical values for crossover and mutation rates in binary-coded GAs commonly found in the literature[10] are respectively $0.6 < p_{cross} < 0.9$ and $0.001 < P_{mut} < 0.01$. This mutation probability corresponds to the mutation of at most a couple of alleles per chromosome and a couple of chromosomes for generation. In general, in any case, real-valued parameters are better managed with real-valued encoding. Furthermore, stochastic binary tournament selection works faster than roulette-wheel and ranking selection and has fewer problems of convergence.

3. Inverse Problems in NDE/NDT

In this section, we present a brief overview of different examples of interest in ultra-sonics NonDestructive Evaluation (NDE), which require the solution of IPs in order to characterize the elastic properties of a specimen and to discern its internal composition. The respective IPs have been tackled using both classical methods and genetic algorithms.

See also the works of Windels and Van Den Abeele[11] and of Prevorovsky et al.[12] in this book for other examples related to the ones shown here. Other IPs in the NDE/NDT domain include, for example, acoustic emission sources localization and identification[11] and the description and modeling of nonclassical nonlinear elastic materials via a Preisach–Mayergoyz approach to elastic hysteresis.[12, 14]

3.1 Measurements of the Elastic Constants of a Solid Specimen

3.1.1 Resonant Ultrasound Spectroscopy

The determination of the elements of the second-order elastic constants tensor, C_{ijkl}, of a 3-D finite body (under the approximation of linear viscoelastic behavior) is a type of IP very well known in ultrasonics and NDE/NDT.[15] It can be solved in different ways, depending on the choice of the specific model for the associated FP.

One of the most common formulations of this problem is based on the measurements of the resonant frequencies of a body, which vibrates in accordance to specific eigenmodes. The resonant responses of viscoelastic bodies depend on their shape, elastic constants, crystallographic orientation, mass density, and anelastic behavior (dissipation). Thus, the measurement of eigenfrequencies may yield information on these parameters.

Continuous-Wave (CW) methods[15] are based on the excitation of eigenmodes of vibration through the injection of plane elastic waves. The use of plane waves yields several experimental difficulties, among which we mention the necessity of large samples in order to obtain quasiplane traveling waves and to avoid diffraction and scattering, the use of transducers with a diameter much greater than the ultrasound wavelength, but much smaller than the linear sample dimension, which implies a reduction in the amplitude and an increase in wave distorsion of the signals from/to the electronic measurement chain.

The excitation of eigenmodes of vibration through plane waves leads to a simple formulation of the FP. After choosing a particular direction of propagation related to the crystallographic orientation of the specimen, a plane wave solution of the initial/boundary value conditions (Cauchy–Dirichlet) problem for the elastodynamic wave equation leads to a set of algebraic linear equations with variables given by a subset of the elastic constants tensor. The squares of phase velocities are known (Christoffel eq.s)[16], considering that the phase velocities are exactly related to the particular modes (i.e., *eigenfrequencies*) of vibration, so they can be estimated from the resonance spectrum after mode identification.[16] This procedure involves many measurement sessions, using distinctly prepared specimens, in order to analyze wave propagation along the different directions: the number of sessions must be at least equal to the number of independent components of C_{ijkl} in order to obtain a complete estimation of the elastic constants tensor, implying that the estimation of elastic constants of low-symmetry materials[15] may be a cumbersome procedure.

A second methodology, called *Resonant Ultrasound Spectroscopy* (RUS),[17, 18] tries to extract more information from a single resonance spectrum. Instead of repeatedly injecting plane waves to excite eigenmodes of different types in the specimens, the experimental protocol of RUS involves only one measurement. The specimen (usually with a parallelepipedlike shape) is held lightly at two of its corners between two transducers, in order to excite eigenmodes with low-amplitude contact forces. Under this protocol, the FP can be formulated with good accuracy as a free-boundary value problem with the elastodynamic wave equation.[19] It has been shown that the Cartesian components of the displacement vector field minimize the Lagrangian of the system.[20] The problem of recovering the displacement vector field becomes then a variational problem that can be solved through the Rayleigh–Ritz method.[21] The three displacement components are substituted by their expansion in a suitable basis of a functional space, of which we wish to determine the coefficients. Minimizing the Lagrangian reduces to solving a generalized eigenvalue problem, where the eigenvalues are proportional to the squares of the resonance frequencies and the matrices involved depend on the parameters, and in particular on the density and the elastic constants.

In this case, the formulation of the forward operator $\hat{A}$ consists in a complicated set of mathematical procedures leading to the generalized eigenvalue problem. This is precisely an example of an inverse operator that cannot be analytically determined. However, the solution to the corresponding IP can be obtained in many cases through a conjugate gradient approach to the least squares optimization problem (model fitting), as described in Section 1.2.[17, 22]

3.1.2 Elastic Constants Estimation for a Monoclinic Crystal Through a Dedicated GA

The estimation of the second- and third-order stiffness tensors of a crystal of caesium di-hydrogen phosphate (CeH_2PO_4) obtained by Chiroiu et al.[23] is reviewed here as an example of the solution of an IP by the use of a GA.

CeH_2PO_4 is a crystalline material with ferroelectric behavior. Ferroelectric crystals are used in the fabrication of memory devices in the microelectronic and nanoelectronic industry.[24] The detection of internal defects through elastic waves has increasingly obtained great importance. The estimation of both second- and third-order elastic constants is fundamental for the determination of the size, shape, and localization of defects through the analysis of classical nonlinear effects such as high-order harmonics generation.[25,26]

In their work,[23] Chiroiu et al. derived the linear algebraic equations relating the squares of phase velocities, based on experimental data, and the linear combinations of elastic constants, as discussed in the continous wave approach presented in Section 3.1.1.

In order to take into account the nonlinear generation of harmonics effect, the solution for the Cauchy–Dirichlet problem for the elastodynamic equation was formulated as a sum of plane wave (along a fixed direction) displacement fields, each of which was characterized either by the fundamental frequency or by one of its integer multiples. This formulation and the use of a nonlinear stress–strain equation involving the third-order stiffness tensor leads to the corresponding Christoffel equations with eigenvalues proportional to the squared phase velocities and matrix elements depending on linear combinations of both second- and third-order elastic constants.

CeH_2PO_4 has a monoclinic symmetry, so the total number K of independent elastic constants is 45 (13 of second order and 32 of third order). For each fixed direction of plane wave propagation, a FP of the previously cited type can be formulated and three possible solutions can be obtained, one quasilongitudinal (qP) and two quasishear (qS) with respective phase velocities. The total number N of experimental velocity measurements must be greater than K in order to avoid problems of numerical instability during the inversion, so the total number of selected directions of propagation is very high, as typical for CW methods.[27]

The inverse problem was formulated as an optimization problem (specifically a model identification in the least squares sense), where

$$\varepsilon(C) = \sum_{l=1}^{N} (v_l^m - v_l^c(C))^2 + \delta^2 \qquad (22.7)$$

is the objective function to be minimized, C is a collective symbol identificating the set of all the independent elastic constants, v_l^m is the measured phase velocity, $\forall l = 1, \ldots, N$, and v_l^c is the corresponding one calculated by resolving the forward problem using a guess for C. Finally, δ is a correction connected to the choice of the specific boundary value conditions.

Instead of using conjugate gradient methods as reported in Section 2.1, in this case the authors exploited a GA approach for the determination of the value of the elastic constants corresponding to the minimum of $\varepsilon(C)$. The fitness function (which has to be maximized) was expressed as follows,

$$F = \frac{\chi^2}{\varepsilon(C)}, \tag{22.8}$$

where χ^2 is the sum of the squared measured phase velocities.

The subsequent fundamental step was the coding of the parameter space: each of the 45 elastic constants was considered as a real-valued parameter belonging to the interval $[-100; +100]$ GPa; that interval was subdivided into $3 * 10^6$ subintervals in order to guarantee an estimation error of about 10^{-4} GPa. Each parameter was then coded as a binary string $(b_{21}^j b_{20}^j \ldots b_2^j b_1^j b_0^j)(\forall j = 1, \ldots, 45)$ of 22 bits representing an integer number between 1 and $3 * 10^6$ corresponding to each subinterval. The concatenation of the gene strings thus constituted the individual chromosome $(b_{21}^1 b_{20}^1 \ldots b_2^1 b_1^1 b_0^1 b_{21}^2 b_{20}^2 \ldots b_2^2 b_1^2 b_0^2 \ldots b_{21}^{45} b_{20}^{45} \ldots b_2^{45} b_1^{45} b_0^{45})$. In this way, the 45-dimensional parameter space was mapped onto a $45 * 22$-dimensional Boolean space.

The GA was then implemented as described in Section 2, using a roulette-wheel selection operator. An intermediate new step in the algorithm was moreover introduced after the application of the mutation operator: a fluctuation operator was used, whose aim was the improvement of the search in the neighborhood of the current fittest genome. The iterations were stopped when the fitness function F achieved a stationary value.

The total number N of chosen experimental data was 57, with 22 velocity values for qP waves and 35 for qS ones; the directions for which the measured phase velocities were very low were not considered, nor did the increase in the number of input values lead to a significant improvement in the accuracy of the estimated elastic constants. The GA achieved convergence after about 400 iterations. The authors showed that it was able to estimate the elastic constants with a good accuracy and with less computing time with respect to conjugate gradient methods.

3.2 Acoustic Tomography

3.2.1 Introduction

Tomography is a typical example of IP that, thanks to its vast range of potential applications, has always raised a great interest in the scientific community. Generally, tomography refers to the cross-sectional imaging of an object obtained from signals detected in various locations. Depending on the type of source (e.g., X-ray, magnetic radiofrequency pulses, ultrasounds), one can obtain highly accurate maps of different physical parameters of the object being imaged.

From a mathematical point of view, classic tomographic techniques deal with the reconstruction of an image from its projections, which can be rigorously defined as the integrals of the image in the directions specified by different angles. For instance, in X-ray imaging, the attenuation experienced by an X-ray traveling through the object is proportional to the line integral of the object density along that path. Various reconstruction algorithms have been defined for recovering the image from these integrals.

In the case of nondiffracting and nonrefracting sources (i.e., when energy source rays travel along straight rays), well-established algorithms, such as filtered backprojection or direct Fourier transform methods, are available. These methods are based on the assumption that integrals are calculated along straight rays, and exploit the properties of transforms (such as the Radon or Fourier transforms) to obtain fast and accurate reconstruction.

Ultrasonic Tomography (UT) employs N transmitters and M receivers, located along the border of the object being imaged, to record the *Times of Flights* (TOFs) or amplitudes of an ultrasonic pulse from each transmitter to each receiver, and exploits this information to recover the physical properties of the object (e.g., attenuation or wave speed in transmission tomography and acoustic impedance mismatch in reflection tomography, respectively). It constitutes a viable alternative to other tomographic techniques, such as X-ray tomography, particularly when avoiding the use of ionizing radiation is desirable, or when a propagation velocity map is more meaningful than an attenuation map.

However, in the UT case, the assumption that ultrasonic waves propagate along straight lines does not hold, due to the refraction of acoustic paths. When inhomogeneities in the object cause significant scattering of the wave field, one must take diffraction into account and resort to specific reconstruction algorithms for diffracting and/or refracting sources.

Iterative procedures, capable of taking into account irregular sampling geometry and ray bending, include Algebraic Reconstruction Technique (ART),[28] and Simultaneous Iterative Reconstruction Technique (SIRT).[29] These techniques are very demanding from a computational point of view, compared to transform-based methods. Although in principle capable of solving general tomographic problems, they often cannot lead to a good and reliable solution in a reasonable time.

Basically, these algorithms iteratively adjust the estimated slowness values in order to provide the best match between computed t_p and measured T_p times of arrival.[30,31] Such iterative methods thus require an initial estimate, which is usually set to a uniformly gray image, that is, to a constant stiffness (although other techniques have also been reported in the literature).

For instance, the ART method seeks to estimate the optimal slowness value for each ray p (i.e., for each receiver–transmitter pair). Let us suppose that the image is discretized in grid cells with slowness value $s_{ij} = 1/c_{ij}$. Then, at each iteration, the ART algorithm calculates the time of arrival along each path, by summing the contributions of all cells lying in the path itself, and compares this sum to the measured time. The difference is calculated, and the adjustment is distributed among the cells belonging to that path. The procedure is iteratively repeated for all paths until convergence is reached (according to the established convergence criterion). This iterative method applies to nondiffracting, nonrefracting sources, as well as diffracting and/or refracting ones. Note that, to account for ray bending, paths can no longer be assumed to be straight lines, but instead a suitable ray tracing method must be used to determine the actual path between each transmitter–receiver pair. The SIRT method is based on the same principle, but differs from ART in that it uses an average correction for all rays applied to each tomography cell, instead of working on a ray-by-ray basis.

3.2.2 Applications of GAs to Ultrasonic Tomography

Tomographic techniques, and specifically ultrasonic tomography, have been introduced in the previous section as an important class of IPs, with a wide range of potential applications. As therein mentioned, although iterative methods for solving general UT problems are available, their burdensome computational requirements often make their use impractical as the number of transmitters and/or receivers increases. In this scenario, GAs emerge as an interesting alternative to other IP-solving techniques, especially when some a priori information about the specimen is available. For instance, Delsanto et al.[32] have addressed a tomographic problem in which the specimen is assumed to be homogeneous, except for a homogeneous inclusion of arbitrary shape, of which they wish to predict the geometry. The elastic properties of both regions are assumed to be known.

Although these assumptions are rather restrictive, there are many important applications for which the nature of the inclusion is well known. Moreover, the approach can be easily extended to take into account a higher number of inclusions with different characteristics, or even to predict the physical properties of the entire specimen (after a suitable discretization), albeit at the expense of a higher computational load. However, the intrinsic parallelism of GAs can at least partially compensate for this increase in computation requirements by allowing a straightforward implementation on parallel hardware architectures.

To design the GA, a model for the propagation of ultrasonic pulses is required. In order to allow an easy and fair comparison between GAs and other techniques, a very simple model based on Snell's law has been considered, rather than a full-field simulation. In any case, a more realistic model of ultrasonic pulse propagation is not expected to strongly affect the convergence properties of the GA-based inversion technique.

Simply stated, times of flights are estimated by determining the ray path first, and then calculating the TOF t_{ij} for each transmitter i and receiver j. The agreement between estimated and experimental TOFs is then evaluated through the misfit function Φ, defined as

$$\Phi = \sum_{i=1}^{N} \sum_{j=1}^{M} \frac{(t_{ij} - \tilde{t}_{ij})^2}{t_{ij}\tilde{t}_{ij}}, \tag{22.9}$$

where $\tilde{t}_{ij}$ represent the experimentally determined TOFs.

Each genome encodes a discretized $m \times n$ grid representing the specimen as a 2-D array. Thanks to the simplifying assumption of homogeneous specimen and inclusion, a binary encoding can be used. Nevertheless, the number of possible configurations is still very large, and increases exponentially with the grid dimension. Because the size of the initial population required for convergence depends on the dimensionality of the search space, for nontrivial grid dimensions the increase in computational requirements would make this approach infeasible in practice, unless a method for reducing the number of possible configurations is available.

The number of configurations, as well as population size, is related to the number of schemata. Due to the rather compact and homogeneous nature of both specimen and inclusion, solving the problem on a coarser grid should give the large-scale structure of the inclusion, suggesting a schemata to which all the strings in the finer grid must belong. A multistep, multiresolution GA approach was thus followed, solving the problem on a very coarse grid first and then doubling the dimension of the grid at each step, until the desired resolution was achieved. At each step, each cell completely surrounded by similar cells is frozen, so that the number of possible configurations is strongly reduced. The size of the population is still dependent on the grid dimension; a careful experimental investigation has shown that, for a 160×160 grid, the size of the initial population can be reduced to 60–100 elements, thanks to the adopted multistep approach.

An important issue in the design of a GA is the selection of suitable operators. A "good" operator for a specific problem should, in general, favor the identification of valuable schemata, while avoiding the disruption of fit individuals. For this problem, an extension of the standard two-point crossover for 2-D arrays was selected, in which the two children are generated by swapping a submatrix.

On the contrary, usual mutation operators, which normally operate on isolated genes, are almost useless for this specific problem. In order to favor even further compact configurations, a variant of this operator was thus devised, assigning a randomly selected cell, along with its nearest neighbors, to be either all inclusion or specimen (according to the majority of the cells in the block).

Finally, it is worth noticing that, although crossover and mutation points are usually chosen at random, sometimes other approaches can be more efficient. In this case, random selection has proven to be very efficient in finding the bulk structure of the inclusion. On the contrary, to obtain precise border definition, it is convenient to force crossovers and mutations including cells on the border. Numerical experiments have shown that the best results are obtained if random and controlled selection iterations are alternated.

This method has been tested on a variety of synthetic data, showing that the GA-based method is much more efficient than ART, both in terms of quality of reconstruction and CPU time. In fact, it gives satisfactory results even for nontrivial configurations, including noncompact, concave, nonsimply connected inclusions, whereas ART requires, in such cases, a strong increase in the number of transmitters, leading to CPU times that are unaffordable with a PC. The higher performance of the GA technique demonstrates its superior ability to exploit available a priori information.

4. Genetic Algorithms in the Solution of Ill-Posed Inverse Problems

In the previous sections we have briefly explained the principles underlying GAs and illustrated a few examples of GAs applied to the NDE/NDT field. We have not, however, considered specifically how GAs may be applied to the solution of ill-posed IPs. We show in this section that GAs may prove useful both as aid to classical techniques for the treatment of ill-posed IPs,[33] as well as independently when used to resolve indetermined IPs, especially in the model-fitting domain.

In Section 1.1 ill-posed problems are formally defined and we have showed how Tikhonov's regularization may help obviate the problem. However, a question that is still open is the choice of the regularization parameter λ. An interesting approach to this problem that exploits GAs has been recently proposed by Mera et al.[34,35] In this work, the authors propose a multipopulation GA, where each population resolves the considered IP for a different value of λ chosen within a possible set of λ_j values, $\forall j = 1, \ldots N_{pop}$ (where N_{pop} is the total number of populations), yielding the solution $\vec{i}_{\lambda_j}$. The solution is then given by the $\vec{i}_{\lambda_j}$ with minimum j such that

$$\| \tilde{\vec{d}} - \hat{A}\left(\vec{i}_{\lambda_j}\right) \| \le c\,\delta; \tag{22.10}$$

according to Morozov's discrepancy principle.[36] The populations may be allowed to evolve independently or else they may share some individuals through a process termed *migration*, which involves the sharing of the best individuals among neighboring populations (in this case populations' neighbors are given by the populations with values of λ_j immediately greater and lesser). However, in the work it is demonstrated that a cooperative effort is rewarded with a faster convergence, with values of fitness in the independent evolution case still lesser after 1000 generations than the corresponding fitness obtained in 100 generations by the joint evolution strategy.

This example is interesting, as it shows how GAs may be combined with regularization schemes in the solution of ill-posed IPs. Regularization techniques, however, are in general appropriate only in those cases in which, either through previous knowledge of the system or through physical considerations, a priori information is possessed, as explained in Section 1.1. In many cases, and especially when the information one wishes to obtain by resolving the IP is the value of the parameters regulating a system (i.e., in model fitting), this may not be the case. Model-fitting problems, furthermore, often are ill-posed in the sense that they are undetermined, as many parameters are known to have an impact on system evolution, yet little information is known as output. Even in this case, though, GAs have often been shown to be useful, as an attentive study of the convergence rate of the parameters can yield insight into the relative impact on the evolution of the system, and thus help understand the behavior of the solutions.

In an interesting study[37] focused on the determination of the velocity and dispersivity parameters in a groundwater contaminant transport model, the authors generated pseudoexperimental data with a Monte Carlo-based simulation model, assuming specific values for the parameters, and successively tried to reconstruct them through a GA. Because the values of the constants were known, the authors were able to assess exactly the genetic algorithm's capability in their determination. Specifically, although the data computed with the reconstructed parameters always converged to the observed data, only four out of six parameters converged correctly to the true value of the psuedoexperimental data, thus demonstrating that multiple solutions were compatible with the data. The authors also noted that the parameters which converged first tended to be those with the strongest influence on the system behavior. Hence, a temptative sensitivity analysis of the system based on the rapidity of convergence of the different parameters was proposed.

Also based on this approach was a study on tumor growth proposed by Delsanto et al.[38] In this study, the authors tried to determine the relative impact of eight different parameters on the growth of multicellular tumor spheroids. The true value of the parameters was not determinable, as some of these constants jointly modeled the influence of different biological variables, and others were simply very difficult to determine experimentally. However, the relative convergence rate of the different parameters did seem to reflect the different impact of the parameters on the system; for example, the death rate of the cells had very little impact on the solution, a conclusion that was perfectly compatible with the modeling hypotheses, which did not take explicitly into account the tumor spheroid composition (the percentage of dead, quiescent, and proliferant cells).

It is worth noting that the imposition of additional conditions may allow discrimination between the different solutions. The difficulty in this case is due to the fact that in practice this coincides with a multiobjective optimization, and it is not always clear how the different objectives should be combined to perform the optimization. The treatment of multiobjective optimization is beyond the scope of this chapter; however, one of the most interesting branches of research in GAs is focused specifically on this subject; the interested reader is referred to the work of Fonseca and Fleming[39] for a comprehensive overview of the topic.

Acknowledgments

This work has been supported by the ESF NATEMIS Project through the Short Visit Grants #221-227-228. We are very grateful to Professor Z. Prevorovsky, Doctor M. Chlada, and Doctor J. Vodicka for many enlightening discussions on the use of soft computing techniques in the solution of inverse problems of interest in NDE/NDT.

References

[1] Kirsch, A., *An Introduction to the Mathematical Theory of Inverse Problems* (Springer-Verlag, New York, 1996).

[2] Sabatier, P.C., *Applied Inverse Problems* (Springer, Berlin, 1978).

[3] Tarantola, A., *Inverse Problem Theory: Methods for Data Fitting and Model Parameter Estimation* (Elsevier, New York, 1987).

[4] Tan, S.M., and Fox, C., *Inverse Problems Lecture Notes*; http://www.phy.auckland.ac.nz/Staff/smt/453707SC.html.

[5] Hadamard J., *Lectures on Cauchy Problem in Linear Partial Differential Equation* (Yale University Press, New Haven, CT, 1923).

[6] Engl, H.W., Hanke, M., and Neubauer, A., *Regularization of Inverse Problems* (Kluwer Academic, Dordrecht, 1996).

[7] Zadeh, L.A., 1994, private communication; http://wwwbisc.cs.berkeley.edu/BISCProgram/History.htm

[8] Goldberg, D.E., *Genetic Algorithms in Search, Optimization and Machine Learning*, 1st ed., (Addison-Wesley, Reading, MA, 1989).

[9] Michalewicz, Z., *Genetic Algorithms + Data Structures = Evolution Programs*, 3rd ed. (Springer-Verlag, Berlin, 1996).

[10] Weile, D.S., and Michielssen, E., Genetic algorithm optimization applied to electromagnetics: A review, *IEEE Trans. Antennas and Propagation*, **45**(3): 343–353 (1997).

[11] Windels, F. and Van Den Abeele, K., in: *The Universality of Nonclassical Nonlinearity, with Applications to NDE and Ultrasonics*, edited by P.P. Delsanto (Springer, New York, 2006), Chapter 23.

[12] Prevorovsky, Z., Chlada, M., and Vodicka, J., in: *The Universality of Nonclassical Nonlinearity, with Applications to NDE and Ultrasonics*, edited by P.P. Delsanto (Springer, New York, 2006), Chapter 32.

[13] Guyer, R.A., McCall, K.R., Boitnott, G.N., Hilbert Jr., L.B., and Plon, J.J., Quantitative implementation of Preisach–Mayergoyz space to find static and dynamic elastic modulii in rock, *J. Geophys. Res.* **102**: 5281–5294 (1997).

[14] Aleshin, V., Van Den Abeele, K., Desadeleer, W., and Sibai, M., 2005, AERONEWS internal communication.

[15] McSkimmin, H. J., in: *Physical Acoustics*, edited by W. P. Mason (Academic, New York, 1964), Chapter 4.

[16] Pollard, H.F., *Sound Waves in Solids*, 1st ed. (Pion, London, 1977).

[17] Leisure, R.G. and Willis, F.A., Resonant ultrasound spectroscopy, *J. Phys.-Condens. Mat.* **9**: 6001–6029 (1997).

[18] Migliori, A., and Sarrao, J.L., *Resonant Ultrasound Spectroscopy: Applications to Physics, Materials Measurements and Non-Destructive Evaluation*, 1st ed. (Wiley, New York, 1997).

[19] Achenbach, J.D., *Wave Propagation in Elastic Solids*, (North-Holland, Amsterdam,1976).

[20] Visscher, W.M., Migliori, A., Bell, T.M., and Reinert, R.A., On the normal modes of free vibration of inhomogeneous and anisotropic elastic objects, *J. Acoust. Soc. Am.* **90**:2154–2162 (1991).

[21] Wan, F.Y.M., *Introduction to the Calculus of Variations and its Applications*, 1st ed. (Chapman & Hall, New York, 1995).

[22] Zadler, B., Le Rosseau J.H.L., Scales, J.A., and Smith, M.L., Resonant ultrasound spectroscopy: Theory and application, *Geophys. J. Inter.* **156**, 154–169 (2004).

[23] Chiroiu, C., Munteanu, L., Chiroiu, V., Delsanto, P.P., and Scalerandi, M., A genetic algorithm for the determination of the elastic constants of a monoclinic crystal, *Inv. Probl.* **16**: 121–132 (2000).

[24] Auciello, O., Scott, J.F., and Ramesh, R., The physics of ferroelectric memories, *Phys. Today* **51**(7): 22–27 (1998).

[25] Teodosiu, C., *Elastic Models of Crystal Defects* (Springer, Berlin 1982).

[26] Green, R.E., Jr., *Treatise on Materials Science and Technology*, vol. 3, *Ultrasonic Investigation of Mechanical Properties* (Academic, New York, 1973).

[27] Prawer, S., Smith, T.F., and Finlayson, T.R., The room temperature elastic behaviour of CsH_2PO_4, *Aust. J.* **36**(1): 85–92 (1985).

[28] Gordon, R., A tutorial on ART, *IEEE Trans. Nucl. Sci.* **21**:78–93 (1974).

[29] Gilbert, P., Iterative methods for the three-dimensional reconstruction of an object from projections, *J. Theor. Biol.* **29**:105 (1972).

[30] Schubert, F., Basic principles of acoustic emission tomography, *European Conference on Acoustic Emission Testing*, Berlin, **2**: 693–708 (2004).

[31] Schechter, R.S., Mignogna, R.B., and Delsanto, P.P., Ultrasonic tomography using curved ray paths obtained by wave propagation simulations on a massively parallel computer, *J. Acoust. Soc. Am.* **100**: 2103–2111 (1996).

[32] Delsanto, P.P., Romano, A., and Scalerandi M., Application of genetic algorithms to ultrasonic tomography, *J. Acoust. Soc. Am.* **104**(3):1374–1381 (1998).

[33] Karr, C.L., Yakushin, I., and Nicolosi, K., Solving inverse initial-value, boundary-value problems via genetic algorithm, *Eng. Appl. Artif. Intell.* **13**: 625–633 (2000).

[34] Mera, N.S., Elliot, L., and Ingham, D.B., On the use of genetic algorithms for solving ill-posed problems, *Inv. Probl. Eng.* **11**:105–121 (2003).

[35] Mera, N.S., Elliott, L., and Ingham, D.B., A multi-population genetic algorithm approach for solving ill-posed problems, *Comput. Mech.* **33**:254–262 (2004).

[36] Morozov, V.A., On the solution of functional equations by the method of regularization, *Soviet. Math. Dokl.* **7**: 414–417 (1966).

[37] Giacobbo, F., Marseguerra, M., and Zio, E., Solving the inverse problem of parameter estimation by genetic algorithms: The case of the groundwater contaminant transport model, *Ann. of Nucl. En.* **29**: 967–981 (2002).

[38] Delsanto, S., Morra, L., Griffa, M., and Demartini, C., A genetic algorithm approach to the exploration of parameter space in mesoscopic multicellular tumor spheroid models, *Proceedings of the IEEE Engineering Sciences in Medicine and Biology Society Annual International Meeting*, San Francisco, Vol. 1, 675–678 (2004).

[39] Fonseca, C.M., and Fleming, P.J., An overview of evolutionary algorithms in multiobjective optimization, *Evolution. Comput.* **3**: 1–16 (1995).

Part III
Experimental Results and Applications

Part III

Characterization and Imaging of Microdamage Using Nonlinear Resonance Ultrasound Spectroscopy (NRUS): An Analytical Model

Koen Van Den Abeele[1] and Filip Windels

Katholieke Universiteit Leuven, Interdisciplinair Research Center (IRC), KULAK, E. Sabbelaan 53, B-8500 Kortrijk, Belgium.
[1]To whom correspondence should be addressed: Koen Van Den Abeele, Katholieke Universiteit Leuven, Interdisciplinair Research Center (IRC), KULAK, E. Sabbelaan 53, B-8500 Kortrijk, Belgium; e-mail: koen.vandenabeele@kuleuven-kortrijk.be.

Abstract

We present a nonlinear version of the Resonance Ultrasound Spectroscopy (RUS) theory by extending the formalism to the treatment of damage in the form of nonlinearity. General analytical equations are derived for the one-dimensional case (a 1-D bar), describing the excitation amplitude-dependent shift in the resonance frequency and the harmonic interaction between bar modes due to the presence of either localized or volumetrically distributed nonlinearity. The solutions are obtained for the case in which the damage area is represented by a cubic nonlinearity, as well as for the more interesting case in which hysteretic nonlinearity ought to be considered. The results are compared to numerical calculations from a multiscale model, described in detail in Chapter 12 of this book, showing excellent agreement. Finally the obtained formulae are exploited to infer critical information about the damage position, the degree of nonlinearity and the width of the damage zone either from the shifts in resonance frequency occurring at different excitation modes, or from the shift and the harmonics predicted at a single mode. Unlike other techniques, the NRUS method does not require a spatial scan to locate the defect, as it lets different excitation modes, with different vibration patterns, probe the structure.

Keywords: Hysteresis, inverse method, microdamage, nonlinear resonant ultrasound spectroscopy, nonlinearity

1. Introduction

"Can we 'hear' the shape of a drum?" is one of the most famous questions in physics that was posed by the mathematician Mark Kac in 1966 [Kac, 1966]. The answer is yes. If perfect, the drum will have a family of vibration modes, also known as resonances, and upon analyzing these modes, we are able to describe the physical parameters of the drum.

Now, suppose we make a hole in the drumhead using a hole punch. "Can we then 'hear' the location of the hole?" The answer is again yes. The hole is a localized perturbation in the drumhead, and causes a shift in some of the resonance frequencies depending on the corresponding motion of the drumhead at the location of the hole. If the hole is located in a place where a particular mode has a node, that mode will not show a frequency shift. If it is situated in a place at an antinode, that mode will show a frequency shift relative to the perfect drumhead. The information about the location of the hole is thus encoded in the frequency shifts of the spectral modes that are affected by the local perturbation. Standard tomographic inversion routines can be used to invert this information for the location of the hole.

Suppose we replace the hole in the drumhead with a very small defect such as a tear or a crack that is only visible with the aid of a magnifying glass. "Can we hear the location of the microdefect?" This is exactly the question that we would like to answer in this chapter, at least for the simple case of a localized defect in a one-dimensional (1-D) bar. The methodology to do this is based on an extension of the formalism of (linear) Resonance Ultrasound Spectroscopy (RUS) to Nonlinear Resonance Ultrasound Spectroscopy (NRUS).

Resonance Ultrasound Spectroscopy (RUS) [Visscher et al., 1991] is a bench-top measurement technique used to determine the full linear elastic tensor of a sample from the combined information contained in its resonance frequencies, its geometry, and its density. A self-consistent explanation of the fundamental details of linear RUS can be found in Chapter 22. RUS is very accurate when it is applied to samples having a well-defined geometry and homogeneously distributed linear elastic constants. Nevertheless, it has been successfully applied to determine the elastic constants of anisotropic media, to study thermoelectric materials, rocks, and the like. [Keppens et al. 1998; Ulrich et al. 2002; Ogi et al. 2002; Nakamura et al. 2004; Ichitsubo et al. 2002].

Microcracks caused by incipient damage will affect the resonance spectrum of a sample only very slightly. In many cases the effect is masked by the resolution of the frequency spectrum. Therefore, a standard RUS analysis may not be able to determine the presence of damage at early stages. However, this does not mean that we are not able to hear damage at all. We simply need to operate in a slightly different manner. It is known that damage produces a nonlinear relation between stress and strain and that the nonlinearity can be put into evidence by analyzing the response of the system at increasing excitation amplitudes. The more damage, the larger is the level of nonlinearity, and the sooner it can be picked up in the analysis. Several studies have shown that the sensitivity of the variation of the nonlinearity with increasing damage is far better than what can be obtained from the evolution of the linear material parameters [Nagy, 1998; Van Den Abeele et al. 2000a,b].

The above considerations underline the need for a nonlinear version of RUS: Nonlinear RUS or NRUS. Rather than only limiting the analysis to finding the location of resonances in the spectrum and comparing them to the spectrum of an intact sample, NRUS investigates and analyzes the amplitude dependence of certain resonance frequencies and uses this information to quantify the location and degrees of nonlinearity.

Several experimental techniques have been developed that exploit the principle of nonlinearity. Some are based on the nonlinear analysis of resonance modes, others are using the amplitude-dependent interaction of two-component signals. Examples are

Single Mode Nonlinear Resonant Ultrasound Spectroscopy (SIMONRUS) [Nazarov et al. 1988; Nazarov and Sutin 1989; Zimenkov and Nazarov 1995; Johnson et al. 1996; Van Den Abeele et al. 2000b, 2001; Van Den Abeele and De Visscher 2000], Nonlinear Wave Modulation Spectroscopy (NWMS) [Antonets et al. 1986; Sutin and Donskoy 1998; Van Den Abeele et al. 2000a], nonlinear wave propagation [Morris et al. 1979; Cantrell and Yost 1994; Krohn et al. 2002], nonlinear time of flight spectroscopy [Kazakov et al. 2002], slow dynamics [TenCate and Shankland 1996, TenCate and thers. 2000], and others. Additional references can be found in several other chapters included in this book related to nonlinear NDT application. Among them we mention Chapters 04, 12, 15, 24, 27–29 and 31.

In this chapter we focus on the theoretical support of SimonRUS and present a means to use this technique for the characterization and localization of damage. This development works in a two-way direction.

First, on the level of the direct problem, we study and predict the nonlinear signatures of the resonances from the given nonlinear elastic constants inside the sample. Over the last years, many researchers have developed numerical models to predict these effects using numerical methods such as the multiscale model or the local simulation approach (e.g., Chapters 12, 17, and 18 of this book and their references). However, in order to preserve the computational simplicity of linear RUS where the resonances can be directly determined as matrix eigenvalues [Visscher et al. 1991], we should avoid the use of numerical models, and return to analytical formulations. So from the viewpoint of computational speed and physical insight, a nonlinear variant of the analytical theory behind RUS would be highly desirable. Of course this can only be realized under certain limiting conditions.

Secondly, a simple solution of the direct problem always makes it easier to solve the inverse problem. From this perspective, it is clear that an analytical version of the NRUS model for the direct problem will be far more advantageous than its numerical counterpart in terms of inverse characterization of nonlinearity because of the calculation speed and the transparency of the formulas.

In this chapter, we limit ourselves to the derivation of analytical NRUS formulas for the case of a one-dimensional bar with distributed damage features. The chapter is built up as follows. In Section 2, we recall the semianalytical version of the nonlinear wave equation for nonlinear and nonunique equations of state, and comment on the general solution procedure in terms of normal modes. In Section 3, we treat the particular case in which the damage can be represented by a classical nonlinear perturbation in the local stress–strain relation, and we derive the solutions for the resonant frequency shift, and the harmonic amplitudes as a function of the nonlinear characteristics of defect and its position. In Section 4, we repeat this for a nonclassical representation of damage using a hysteretic stress-strain perturbation. In the fifth section, we address the inverse problem of damage characterisation and location using NRUS. Our study shows that the different vibration patterns of different modes probe different parts of the structure giving rise to mode-dependent nonlinear signatures/quantification, which can be used to solve the inverse problem. This has many advantages: the use of information from different modes eliminates the need for a laborious scanning apparatus as is used in traditional (linear and nonlinear) damage localization techniques [Kazakov et al. 2002;

Krohn et al. 2002; Stoessel et al. 2002; Ballad et al. 2004]. The detector and excitation source can therefore remain fixed (as in linear RUS), and the modes themselves do the scanning job.

2. General Equation and Solution

2.1 General Nonlinear 1-D Equation

In this section, we consider the wave equation for a one-dimensional bar in the presence of damage that is represented by a nonlinear and nonunique stress–strain equation. As a first approximation, classical nonlinearity and hysteresis in the stress–strain relation can be accounted for by assuming the following stress–strain relation [Landau and Lifshitz, 1969; Van Den Abeele et al. 1997; Guyer et al. 1998],

$$
\sigma = K\,(1+\beta\,\varepsilon+\delta\,\varepsilon^2+\cdots)\,\varepsilon + K\frac{\alpha}{2}\,\Big[\,sign(\partial_t\varepsilon)\,((\Delta\varepsilon)^2-\varepsilon^2)-2(\Delta\varepsilon)\varepsilon\Big], \quad (23.1)
$$

where K is the linear stiffness constant (modulus), $\varepsilon = \partial_x u$ the strain, and $\Delta\varepsilon$ the strain amplitude. The parameters β and δ are combinations of third- and fourth-order elastic constants representing the atomic anharmonicity or acoustoelasticity. The parameter α is the strength of the hysteresis. This latter parameter quantifies the opening of the stress–strain loop and is therefore also related to extra damping. The loop area is $4K\alpha(\Delta\varepsilon)^3/3$.

We focus on the description of "forced" wave resonances in a one-dimensional bar of length L, with stress-free boundaries. The excitation is supplied by a sinusoidal force with amplitude $\tilde{F}$ and circular frequency Ω at one end of the bar. The response is measured at the other end. Attenuation is accounted for by considering a damping term proportional to the velocity, containing a frequency-independent quality factor. However, other models are also possible. With the above representation of the stress–strain relation, the 1-D nonlinear equation for the displacement $u(x,t)$ as a function of space and time coordinates, including attenuation and external sinusoidal excitation, then becomes (ρ denotes the density, and $\delta_{x,0}$ is the continuous Kronecker symbol indicating that the force is located at at $x = 0$):

$$
\begin{aligned}
\rho\,\partial_{tt}^2 u = \partial_x\,\Big\{&K\,\partial_x u(1+\beta\,\partial_x u+\delta\,(\partial_x u)^2+\cdots)\\
&+ K\frac{\alpha}{2}\Big[sign(\partial_t\partial_x u)((\Delta\partial_x u)^2-(\partial_x u)^2)-2(\Delta\partial_x u)\partial_x u\Big]\Big\}\\
&-\rho\,\frac{\Omega}{Q}\,\partial_t u + \tilde{F}\cos(\Omega t)\,\delta_{x,0}.
\end{aligned}
\quad (23.2)
$$

2.2 General Solution

We seek the solution of Eq. (23.2) by decomposing the physical field u into a sum of products, separating the variables x and t [Pohit et al. 1999].

$$
u\,(x,t) = \sum_i \psi_i\,(x)\,z_i\,(t), \quad (23.3)
$$

where $\{\psi_i(x)\}$, $i = -\infty.. + \infty$, is a set of chosen—and hence known—*spatial* functions [satisfying the boundary conditions; i.e, $\psi_n(x) = \cos(n\pi x/L]$ for free boundaries), whereas $\{z_i(t)\}$, $i = -\infty..+\infty$, is a set of temporal functions that represent the new unknowns. The modal shape functions $\{\psi_i(x)\}$ have the property of being mutually orthogonal, both for the displacement and the strain.

To find the general solution to the problem, we now substitute the proposed "normal mode" solution given by Eq. (23.2) into Eq. (23.3), multiply both sides of the equation by $\psi_n(x)$, and integrate the result over the spatial coordinate x from 0 to L. Due to the hysteretic contribution, the resulting equation is a fairly complicated equation for the unknown spatial functions $z_n(t)$ in the normal mode expansion. We can easily simplify this equation using two assumptions: (1) the excitation is performed at a frequency Ω in the neighborhood of the resonance frequency of the mth mode, and (2) we only consider $A_m \cos(m\pi x/L) \cos(\Omega t + \phi_m)$ as the dominant contribution in the displacement field for the hysteretic nonlinear terms containing the strain amplitude (in other words, we assume $\Delta \partial_x u = m(\pi/L) A_m$).

After some tedious calculations we arrive at the following set of coupled differential relations (adopting the Einstein summation convention, however, no sum over n).

$$\partial_{tt}^2 z_n + \frac{\Omega}{Q} \partial_t z_n + \omega_n^2 z_n = F \cos(\Omega t) - B_{njk} z_j z_k - D_{njkl} z_j z_k z_l$$
$$- H_{n\,m} \left[sign(\partial_t z_m) (A_m^2 - z_m^2) - 2A_m z_m \right], \quad (23.4)$$

where $F = (2\tilde{F})/(\rho L)$, and ω_n the frequencies of the linear (low amplitude) resonances [Visscher et al. 1991]; that is, $\omega_n = n(\pi c/L)$ with c the linear bar velocity. The advantage of using normal mode shape functions is that the mode-coupling occurs solely by way of the nonlinear interaction. There is no coupling at the linear level. The coupling constants appearing in the second, third, and fourth terms on the right-hand side of Eq.(23.4) correspond to the components of higher-order tensors B, D, and H, and can be expressed as integrals over x:

$$B_{njk} = \frac{2K}{\rho L} \int_0^L dx\ \beta\ \partial_x \psi_n\ \partial_x \psi_j\ \partial_x \psi_k \quad (23.5a)$$

$$D_{njkl} = \frac{2K}{\rho L} \int_0^L dx\ \delta\ \partial_x \psi_n\ \partial_x \psi_j\ \partial_x \psi_k\ \partial_x \psi_l \quad (23.5b)$$

$$H_{n\,m} = \frac{K}{\rho L} \int_0^L dx\ \alpha\ \partial_x \psi_n\ |\partial_x \psi_m|\ \partial_x \psi_m. \quad (23.5c)$$

Thus, once nonlinearity is considered, all existing linear modes will interact with each other at the nonlinear level, provided they all have a nonzero strain level at those places where the nonlinearity is present. This is important for the inverse mapping of the damage location using different excitation modes.

In a SIMONRUS, one analyzes the dependence of a resonance curve at increasing forcing amplitude. In order to obtain a resonance curve (at a constant forcing amplitude F), the sinusoidal source frequency Ω applied to the sample is stepped up discretely over a small range encompassing a certain resonance frequency, for instance, around ω_m for mode m. At each discrete frequency during the scan, the steady-state amplitude response $|z_n(\Omega; F)|$ after $5Q$ periods is monitored, for $n = m$, $2m, 3m, 4m, 5m$, and so on. Let us first recall the solution of the linear problem.

2.3 Linear Solution

If everything were linear, and we would investigate the response of the bar at source frequencies Ω in the neighborhood ω_m, the only mode that would be receptive to this excitation is exactly the mode m. All other modes are not activated because the driving frequency is too far away from their corresponding resonance frequencies. The solution of Eq. (23.4) in the absence of nonlinearity ($\beta = \delta = \alpha = 0$) is of the form $A_m \cos(\Omega t + \phi_m)$. Basic differential calculus tells us that the linear response amplitude A_m and phase ϕ_m satisfy following equations,

$$A_m = |z_m(\Omega; F)| = F/\sqrt{(\Omega^2 - \omega_m^2)^2 + (\Omega^2/Q)^2}, \qquad (23.6a)$$

$$\tan \phi_m = -\Omega^2/(Q(\Omega^2 - \omega_m^2)). \qquad (23.6b)$$

Equation (23.6b) clearly shows that the phase lag for an excitation in the neighborhood of any mode is independent of the mode number. which implies that the attenuation depends linearly on the frequency. The true resonance frequency corresponds to the maximum amplitude response (solution of $dA/d\Omega = 0$), and is given by

$$\Omega_{res} = \omega_m/\sqrt{1 + 1/Q^2}. \qquad (23.6c)$$

This is the linear response. However, when nonlinearity is present in the system, one can expect that some signatures of the resonance spectrum will change accordingly. In general, the position of the resonance peaks will become amplitude dependent, harmonics start to appear, attenuation may become nonlinear, and so on. The observation of such nonlinear behavior betrays the presence of the nonlinear zone inside the sample. The deduction of analytical relations for the expected shifts and the harmonics can help in the inverse procedure of damage localization from nonlinear observations. This is exactly the subject of the rest of the chapter. We consider two cases: classical cubic nonlinearity and nonclassical hysteretic nonlinearity.

3.   Solution for Classical–Cubic Nonlinearity

For the purpose of illustrating the calculation procedure, we now only consider the nonlinear term containing the cubic "δ" nonlinearity in Eqs. (23.1), (23.2), and (23.4). We investigate the resulting effect of a "δ"-nonlinearity on the resonance frequency and on the generation of harmonics.

3.1 Shift of the Resonance Frequency for Sinusoidal Excitation near Mode "m"

Equation (23.4) for $\beta = \alpha = 0$ states that mode n can be influenced due to the cubic nonlinear interaction of the modes j, k, and l (summation over j, k, and l). If we concentrate on the excitation of the bar around the frequency ω_m, and investigate the dominant nonlinear effect affecting mode m due to this excitation, we can safely say that the first-order perturbation on the behavior of this mode will only be due to the combination of modes j, k and l with $j = k = l = m$. (All other responses are zero in the linear case.) Therefore the amplitude dependence of mode m is basically governed by the following equation (no sum over m):

$$\partial_{tt} z_m + \frac{\Omega}{Q} \partial_t z_m + \omega_m^2 z_m = F \cos(\Omega t) - D_{mmmm} z_m^3, \tag{23.7a}$$

where:

$$D_{mmmm} = \frac{2}{\rho L} \int_0^L dx \, K\delta \, (\partial_x \psi_m)^4 = \frac{2K}{\rho L} \left(m\frac{\pi}{L} \right)^4 \int_0^L dx \, \delta \, \sin^4 \left(m\frac{\pi}{L}x \right). \tag{23.7b}$$

One can solve Eq. (23.7) using rigorous perturbation methods or comprehensive multiple timescale methods, and so on [Landau and Lifshitz 1969; Nayfeh 1973; Pohit et al. 1999]. However, in this case, it suffices to simply substitute $z_m = A_m \cos(\Omega t + \phi_m)$ into Eq. (23.7a) and to equate the corresponding terms in $\cos(\Omega t)$ and $\sin(\Omega t)$ (neglecting the force term in $\cos(\Omega t)$ arising as a secondary part of the nonlinear contribution). As a result, we find that the amplitude A_m satisfies the following equations.

$$A_m = \frac{F}{\sqrt{\left(\Omega^2 - \omega_m^2 - \frac{3}{4} D_{mmmm} A_m^2 \right)^2 + \left(\frac{\Omega^2}{Q} \right)^2}}. \tag{23.8}$$

For a fixed forcing amplitude F, Eq. (23.8) gives an implicit relation between the response amplitude A_m and the driving frequency Ω. Because the nonlinear contribution to the amplitude change is small, we can consider the term in the denominator containing the square of A_m as being constant. Consequently, the resonance frequency, for which the response amplitude is maximal, expressed in terms of the maximal strain amplitude $\varepsilon_m = \Delta\varepsilon_{\max} = A_{m,\max}(m\pi/L)$, is approximately given by

$$\Omega_{res}(\varepsilon_m) \approx \Omega_{res}(0) \left[1 + \frac{3}{4L} \varepsilon_m^2 \int_0^L dx \, \delta \, \sin^4 \left(m\frac{\pi}{L}x \right) \right]. \tag{23.9}$$

The relative frequency shift is quadratic in the strain, which agrees with the literature on nonlinear harmonic oscillators [Landau and Lifshitz 1969]. It is important to notice that the quadratic dependence on strain is irrespective of the setting of the cubic nonlinearity: global or localized. For a localized damage centered at $x = x_d$, extending from $[x_d - d/2, x_d + d/2]$ (with $d \ll L$) and represented by a constant cubic nonlinearity δ (Figure 23.1a); we obtain:

$$\Omega_{res,Local}(\varepsilon_m) \approx \Omega_{res}(0) \left[1 + \frac{3}{4} \frac{\delta \, d}{L} \sin^4 \left(m\frac{\pi}{L}x_d \right) \varepsilon_m^2 \right] = \Omega_{res}(0) \left[1 + C_1 \varepsilon_m^2 \right].$$

$$\tag{23.10}$$

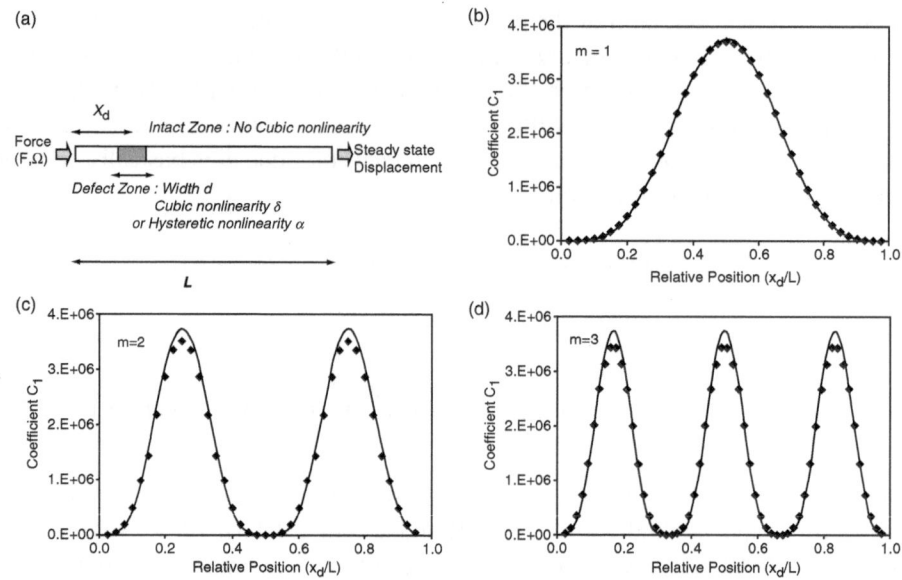

Fig. 23.1. Investigation of the nonlinear signature C_1 (proportionality coefficient in the strain amplitude dependence of the resonance frequency shift) from a cubic nonlinear defect as function of its position x_d in a resonant bar. (a) Schematic geometry for the resonant bar simulations. (b) Signature C_1 for excitation near the fundamental mode ($m = 1$). (c) Signature C_1 for the second bar mode excitation ($m = 2$). (d) Signature C_1 for the third bar mode excitation ($m = 3$). Lines represent the analytical formulae; diamonds correspond to the numerical simulations. The width of the defect zone is $d = L/20$.

For a global constant nonlinearity extending all along the bar, the shift becomes

$$\Omega_{res,Global}(\varepsilon_m) \approx \Omega_{res}(0) \left[1 + \frac{9}{32} \delta \varepsilon_m^2 \right]. \tag{23.11}$$

In most cases δ is considered to be negative (softening), implying that the resonance frequency reduces quadratically with amplitude. Note that the above expressions [Eqs. (23.10 and (23.11)] are just special cases to illustrate the effect of localized and global damage on the frequency shift. The proposed formalism is far more widely applicable because the integral in (23.9) can be calculated for an arbitrary distribution of the nonlinearity. The (sine)^4 dependence on the position x_d of the damage in Eq. (23.10) is quite understandable keeping in mind that modes will contribute to the nonlinearity to an amount proportional to their strain-field amplitude at the defect position. Because the shift is essentially the result of a threefold nonlinear interaction of mode m thereby affecting the same mode m, we indeed arrive at the fourth-order dependence. Another important observation is that the shift depends on the product of the nonlinearity strength δ and the relative width d/L, causing the parameter $\delta d/L$ to be an "effective" nonlinear strength parameter.

We have compared the analytical prediction of the shift for localized damage [Eq. (23.10)] to the shift calculated from the numerical multiscale model discussed in Chapter 12 of this book, ignoring hysteretic nonlinearity. Figure 23.1b shows the numerical and analytical predictions of the proportionality coefficient C_1 in Eq. (23.10) for the

Table 23.1. Parameters used in the simulations

General Parameters	NRUS Parameters	Numerical Simulation (NS) Parameters
$K = 10\,\text{GPa}$ $\rho = 2600\,\text{kg/m}^3$	**Cubic Nonlinearity** $\delta_{\text{NRUS}} = -10^8$	**Cubic Nonlinearity** $\delta_{\text{NS}} = -3\ 10^{-12}$ which yields $\delta_{\text{NRUS}} = (\delta_{\text{NS}} K^2)/3 = -10^8$
$L = 0.25\,\text{m}$ $d = L/20$ $Q = 80$	**Hysteretic Nonlinearity** $\alpha_{\text{NRUS}} = 200$	**Hysteretic Nonlinearity** $\gamma = 10^{-3}$ which yields $\alpha = \gamma K^2/5 \cdot 10^{14} = 200$ **NS Discretisation Parameters** (m = mode number; Ω = circular frequency) $\Delta x = L/(120m)$ $\Delta t = 2\pi/(\Omega * 384 * m)$

fundamental mode ($m = 1$) and for a fixed value of d and δ. The parameter values used in the simulations are given in Table 23.1. The agreement between the analytical predictions and the numerical simulation results is excellent and validates the NRUS predictions. Similar agreement can be found for higher order modes: $m = 2$ in Figure 23.1c and $m = 3$ in Figure 23.1d.

3.2 Amplitude of the Harmonics for Sinusoidal Excitation near Mode "m"

To find the first-order amplitude of the second and third harmonics, given a sinusoidal excitation near mode m, we need to solve the following equations.

$$\partial_{tt}^2 z_{2m} + \frac{\Omega}{Q}\partial_t z_{2m} + \omega_{2m}^2 z_{2m} = -D_{2m\,m\,m\,m} z_m^3. \tag{23.12}$$

$$\partial_{tt}^2 z_{3m} + \frac{\Omega}{Q}\partial_t z_{3m} + \omega_{3m}^2 z_{3m} = -D_{3m\,m\,m\,m} z_m^3. \tag{23.13}$$

By substituting the solution for z_m into Eqs. (23.12) and (23.13), the right-hand terms will be composed of source terms containing $\cos(\Omega t)$ and $\cos(3\Omega t)$. Neither has a frequency near ω_{2m}, thus there will be no activation of the second harmonic and therefore z_{2m} will be zero all the time. On the other hand, the 3Ω component in the right-hand side of Eq. (23.13) will produce a nonzero amplitude for the third harmonic when Ω is swept around ω_m. Representing z_{3m} by $A_{3m}\cos(3\Omega t + \phi_{3m})$, the amplitude A_{3m} can be easily calculated and reads:

$$A_{3m} = \frac{|D_{3m\,m\,m\,m}|}{4}\frac{A_m^3}{\sqrt{((3\Omega)^2 - \omega_{3m}^2)^2 + \dfrac{\Omega^2(3\Omega)^2}{Q^2}}}. \tag{23.14}$$

The amplitude of the strain field of the third harmonic for an excitation near the resonance frequency ω_m of mode m consequently becomes

$$\varepsilon_{3m} = \left| \frac{3Q}{2L}\int_0^L \delta\,\sin\left(\frac{3m\pi}{L}x\right)\,\sin^3\left(\frac{m\pi}{L}x\right)\,dx \right|\varepsilon_m^3. \tag{23.15}$$

The integral can be calculated for an arbitrary distribution of the cubic nonlinearity over the bar. Irrespective of the distribution, we notice that the third harmonic is proportional to the third power in the strain, and that the quality factor plays an important role in the efficiency of the harmonic generation. Higher Q-values (lower attenuation) will produce higher harmonics. For the two special cases of a localized damage zone and a globally distributed cubic nonlinearity we respectively get:

$$\varepsilon_{3m,Local} = \frac{3}{2} Q \frac{|\delta| d}{L} \left| \sin\left(\frac{3m\pi}{L} x_d\right) \sin^3\left(\frac{m\pi}{L} x_d\right)\right| \varepsilon_m^3 = C_3 \varepsilon_m^3 \quad (23.16)$$

$$\varepsilon_{3m,Global} = \frac{3}{16} Q |\delta| \varepsilon_m^3. \quad (23.17)$$

For the higher-order even harmonics, one can repeat the reasoning applied for the second harmonic. The cubic nonlinearity never produces a source contribution that contains an even multiple of the source frequency near ω_{2km}, for any integer k-value. All even harmonics thus remain zero. For the odd harmonics, it is always possible to identify combinations of source terms that lead to a source contribution with frequency $(2k + 1)\Omega$. For the fifth harmonic there are three terms that yield such a source term, giving rise to the following equation,

$$\partial_{tt}^2 z_{5m} + \frac{\Omega}{Q} \partial_t z_{5m} + \omega_{5m}^2 z_{5m} = -D_{5m\,3m\,m\,m}\, z_{3m} z_m^2 - D_{5m\,m\,3m\,m}\, z_{3m} z_m^2$$

$$- D_{5m\,m\,m\,3m}\, z_{3m} z_m^2. \quad (23.18)$$

After some calculus and in analogy with the third harmonic, we obtain the general expression for the strain amplitude of the fifth harmonic for an excitation near the resonance frequency ω_m of mode m:

$$\varepsilon_{5m} = \frac{45 Q^2}{4L^2} \left| \int_0^L \delta \sin\left(\frac{3m\pi}{L} x\right) \sin^3\left(\frac{m\pi}{L} x\right) dx \right|$$

$$\times \left| \int_0^L \delta \sin\left(\frac{5m\pi}{L} x\right) \sin\left(\frac{3m\pi}{L} x\right) \sin^2\left(\frac{m\pi}{L} x\right) dx \right| \varepsilon_m^5. \quad (23.19)$$

The fifth harmonic is proportional to the fifth power of the fundamental strain amplitude, irrespective of the damage distribution. For the two special cases of a localized damage zone and a globally distributed cubic nonlinearity, we respectively get:

$$\varepsilon_{5m,Local} = \frac{45}{5} \left[Q \frac{\delta d}{L} \right]^2 \left| \sin\left(\frac{5m\pi}{L} x_d\right) \sin\left(\frac{3m\pi}{L} x_d\right) \sin^2\left(\frac{m\pi}{L} x_d\right)\right| \varepsilon_m^5$$

$$= C_5 \varepsilon_m^5. \quad (23.20)$$

$$\varepsilon_{5m,Global} = \frac{45}{256} Q^2 \delta^2 \varepsilon_m^5. \quad (23.21)$$

Because the fifth harmonic is a second-order harmonic generation, its efficiency is quadratic in the nonlinearity strength and in the quality factor.

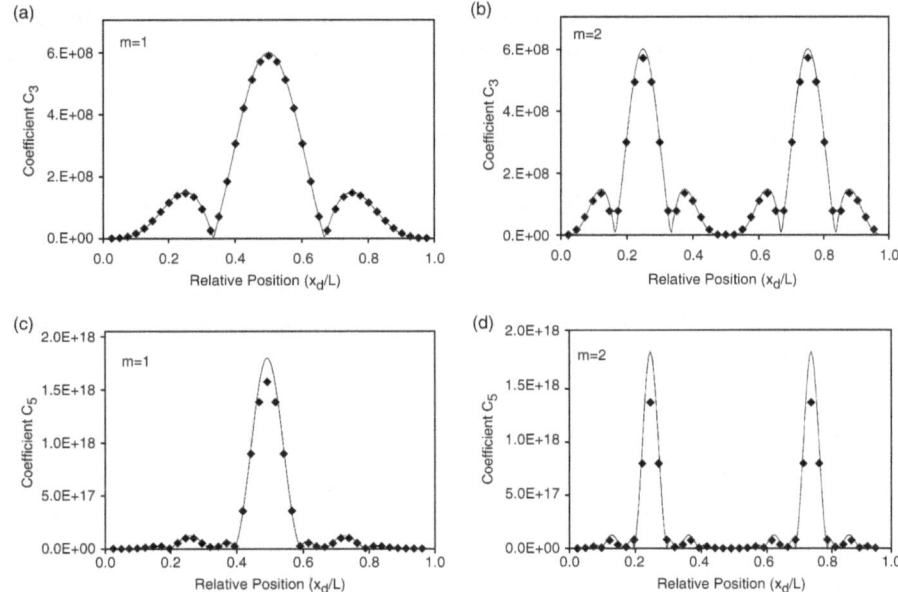

Fig. 23.2. Investigation of the nonlinear signature C_3 and C_5 (proportionality coefficient in the strain amplitude dependence of the third and fifth harmonics) from a cubic nonlinear defect as function of its position x_d in a resonant bar. (a) Signature C_3 for fundamental mode excitation ($m = 1$). (b) Signature C_3 for the second bar mode excitation ($m = 2$). (c) Signature C_5 for the fundamental mode excitation ($m = 1$), and (d) signature C_5 for the second bar mode excitation ($m = 2$). Lines represent the analytical formulae; diamonds correspond to the numerical simulations. The width of the defect zone is $d = L/20$.

The particular position-dependent behavior of the coefficients C_3 and C_5 in Eqs. (23.16) and (23.20) in the case of a localized damage zone with cubic nonlinearity is checked by comparing it to the results of the multiscale model (Chapter 12 of this book). Figure 23.3 illustrates the good agreement between the analytical and numerical results for the proportionality coefficients of the third and fifth harmonics taken at the resonance frequency of the fundamental of mode $m = 1$ and mode $m = 2$. When comparing the results for the resonance frequency shift in Figure 1 to the harmonic generation efficiency in Figure 23.2, we notice that a defect may show a reasonable resonance frequency shift without a sign of harmonics being generated, for instance, when the defect is located at $L/3$ for $m = 1$, or at $L/6$ for $m = 2$. This is entirely due to the influence of the mode shapes on the efficiency of the harmonic generation.

4. Nonclassical Nonlinearity

In this section, we focus on the influence of the nonuniqueness of the nonlinear stress–strain equation. Therefore, we set all classical nonlinearity to be zero ($\beta = \delta = 0$), and consider only the effect of the parameter α in Eqs. (23.1), (23.2), and (23.4).

The trick for finding the amplitudes of the fundamental and higher harmonics is to identify the oscillating terms in the right-hand side of Eq. (23.4) with circular frequencies equal to $(n/m)\Omega \approx \omega_n$. Only those terms will contribute to the amplitude of $z_n(t)$. To do this, we are forced to express the sign-function into a Fourier series. Using

$$sign(\partial_t z_m) = sign(-\Omega A_m \sin(\Omega t + \phi_m))$$

$$= -\sum_{k=0}^{\infty} \frac{4}{(2k+1)\pi} \sin((2k+1)(\Omega t + \phi_m)), \qquad (23.22)$$

it becomes clear that only the odd harmonics will be generated with nonzero amplitudes, in as much as there will be no oscillating contributions in the right-hand side source terms with even multiples of Ω.

4.1 Shift of the Resonance Frequency for Sinusoidal Excitation near Mode "m"

For $n = m$, we basically have to find the solution of

$$\partial_{tt}^2 z_m + \frac{\Omega}{Q} \partial_t z_m + \omega_n^2 z_m$$

$$= F\cos(\Omega t) - H_{m\,m}\left[\frac{-8}{3\pi}A_m^2 \sin(\Omega t + \phi_m) - 2A_m^2 \cos(\Omega t + \phi_m) + \cdots\right],$$

$$(23.23)$$

where the three dots represent terms containing higher-order oscillations. In a similar manner as in the case of cubic nonlinearity, we now obtain that

$$A_m = \frac{F}{\sqrt{(\Omega^2 - \omega_m^2 + 2H_{m\,m}A_m)^2 + \left(\frac{\Omega^2}{Q} + \frac{8}{3\pi}H_{m\,m}A_m\right)^2}}. \qquad (23.24)$$

Equation (23.24) gives an implicit relation between the response amplitude A_m and the driving frequency Ω, for a fixed forcing amplitude F. Under the assumption of small nonlinear contributions, the resonance frequency can be expressed, in terms of the maximal strain response amplitude $\varepsilon_m = A_{m,\max}(m\pi/L)$, as follows.

$$\Omega_{res}(\varepsilon_m) \approx \Omega_{res}(0)\left(1 - \left(1 + \frac{4}{3\pi Q}\right)\frac{c\,H_{m\,m}\varepsilon_m}{\omega_m^3}\right)$$

$$= \Omega_{res}(0)\left[1 - \left(1 + \frac{4}{3\pi Q}\right)\frac{\varepsilon_m}{L}\int_0^L dx\,\alpha\left|\sin^3\left(m\frac{\pi}{L}x\right)\right|\right]. \qquad (23.25)$$

The relative frequency shift for hysteretic nonlinearity changes linearly with the strain, and again this agrees with literature on lumped nonlinear harmonic oscillators [Guyer et al. 1998]. It is important to notice that the linear dependence on strain is irrespective of the status of the hysteretic nonlinearity: global or localized. For a localized damage centered at $x = x_d$, extending from $[x_d - d/2, x_d + d/2]$ (with $d \ll L$) and represented by a constant hysteretic nonlinearity α (Figure 23.1a), we obtain:

$$\Omega_{res,Local}(\varepsilon_m) \approx \Omega_{res}(0) \left[1 - \frac{\alpha\, d}{L} \left(1 + \frac{4}{3\pi\, Q}\right) \left|\sin^3\left(m\frac{\pi}{L}x_d\right)\right| \varepsilon_m\right]$$

$$= \Omega_{res}(0) \left[1 - X_1\, \varepsilon_m\right]. \tag{23.26}$$

For a global constant hysteretic nonlinearity extending all along the bar, the shift becomes

$$\Omega_{res,Global}(\varepsilon_m) \approx \Omega_{res}(0) \left[1 - \frac{4\alpha}{3\pi} \left(1 + \frac{4}{3\pi\, Q}\right) \varepsilon_m\right]. \tag{23.27}$$

Because α represents the hysteretic strength (and by consequence a positive parameter), the change in resonance frequency will always be directed towards lower values. The most general expression for the shift contains a weighted integral over an arbitrary distribution of the nonlinearity. Again, we can interpret the (sine)^3 dependence on the position x_d of the damage in Eqs. (23.25) and (23.26) keeping in mind that modes will contribute to the nonlinearity to an amount proportional to their strain-field amplitude at the defect position. This time, however, the shift is essentially the result of a twofold nonlinear interaction of mode m that affects the same mode m. The absolute value operation is the consequence of the nonuniqueness of the stress–strain relation. Furthermore, we notice in Eq. (23.26) that the shift for a localized damage feature depends on the "effective" nonlinear strength parameter $\alpha d/L$.

Besides quantifying the relative frequency shift at resonance, Eq. (23.24) reveals another interesting relation connected with the hysteresis contribution to attenuation. If the (linear) quality factor is not too small ($Q > 10$), the hysteretic nonlinearity will be responsible for an amplitude-dependent change in the loss factor, which is mainly caused by a change in the Q-factor satisfying:

$$\frac{1}{Q_{NL}} = \frac{1}{Q_L} + \frac{8}{3\pi\, \omega_m^2} H_{m\, m} A_m = \frac{1}{Q_L} + \frac{8}{3\pi} \frac{c\, H_{m\, m}\varepsilon_m}{\omega_m^3}. \tag{23.28}$$

The amplitude dependent Q-factor is typical for hysteretic systems only. For instance, Eq. (23.8) for classical cubic nonlinearity does not show any amplitude dependence in the second term of the denominator, and therefore no nonlinear damping is accounted for.

Recalling that $\varepsilon_m \approx m\pi\, F Q/(L\omega_m^2)$ at resonance, we can compare the relative decrease in strain amplitude at resonance with the relative decrease of the resonance frequency obtained in Eq. (23.25). We then find the following constant relation,

$$\frac{\dfrac{\Omega_{res,0} - \Omega_{res}(A)}{\Omega_{res,0}}}{\dfrac{\varepsilon_{res,0} - \varepsilon_{res}(A)}{\varepsilon_{res,0}}} = \frac{\dfrac{\Delta\Omega_{res}(A)}{\Omega_{res,0}}}{Q_{NL}\left(\dfrac{1}{Q_{NL}} - \dfrac{1}{Q_L}\right)} \approx \frac{3\pi}{8Q}\left(1 + \frac{4}{3\pi\, Q}\right) \approx \frac{3\pi}{8Q} = \frac{1.178}{Q},$$

$$\tag{23.29}$$

which implies that the relative amplitude decrease due to hysteretic nonlinearity is approximately a factor Q larger than the relative frequency shift.

We again take advantage of the multiscale model discussed in Chapter 12 of this book to validate the above analytical formulae in the limiting case of uniform modulus, uniform quality factor, and localized hysteretic nonlinearity. The Preisach–Mayergoysz (PM) spaces (see also Chapters 11, 12, 14, 16–19, 21) used to model the

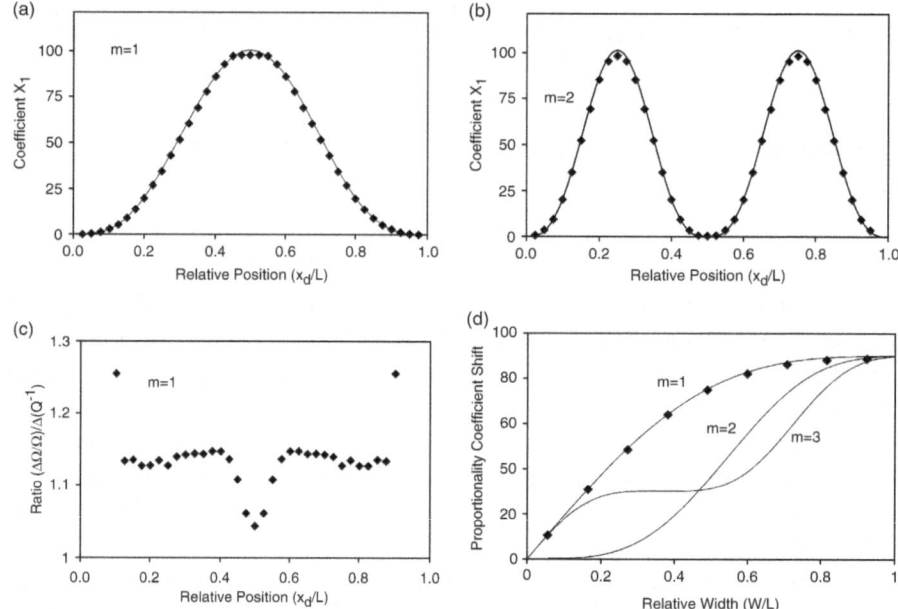

Fig. 23.3. Investigation of the nonlinear signature X_1 (proportionality coefficient in the strain amplitude dependence of the resonance frequency shift) from a hysteretic nonlinear defect ($d = L/20$) as a function of its position in a resonant bar. (a) Signature X_1 for excitation around the fundamental mode ($m = 1$). (b) Signature X_1 for the second bar mode excitation. (c) Ratio of relative frequency shift to absolute increase in Q^{-1} for the fundamental mode ($m = 1$) derived from the numerical simulations. (d) Investigation of the proportionality coefficient in the strain amplitude dependence of the resonance frequency shift from a hysteretic nonlinear defect as a function of its width (W). The defect is located in the middle of the bar ($x_d = L/2$): analytical predictions for the first three bar modes, and numerical results for the fundamental mode excitation ($m = 1$). Lines represent the analytical formulae; diamonds correspond to the numerical simulations.

hysteretic nonlinearity in the numerical simulations consist of uniformly distributed units over a range of -5 MP to 5 MPa. Figure 23.3a shows the numerical and analytical predictions of the proportionality coefficient X_1 in Eq. (23.26) for the fundamental mode ($m = 1$) and for a fixed value of d and α. The parameter values used in the simulations are listed in Table 23.1. A similar good agreement can be found for higher order modes: $m = 2$ in Figure 23.3b.

Figure 23.3c illustrates the ratio of relative frequency reduction to the absolute Q^{-1} reduction obtained with the numerical model. The average value is indeed close to the analytically predicted value of 1.178, except in the center of the bar, where second-order effects may play a role, and at the ends of the bar where the effect of the nonlinearity is highly tempered by the mode shape.

In addition, we investigated the influence of the width W of the defect zone for a fixed centered location of the defect ($x_d = L/2$) with a fixed nonlinearity α extending from $(L - W)/2$ to $(L + W)/2$. The analytical prediction and numerical results of the strain proportionality coefficient in Eq. (23.25) are illustrated in Figure 23.3d, and both results agree very well for $m = 1$. Initially, the shift increases linearly with the width

[Eq. (23.26) if $W = d \ll L$]. However, for larger values, a saturation is obtained due to the particular shape of the resonance mode that is weighing the contribution (strain-field is a sine function). The asymptotic behavior is the value of the proportionality coefficient for a complete volumetrically nonlinear bar with hysteretic nonlinearity α [Eq. (23.27)]. The predictions for higher modes are also shown in Figure 23.3d and can be readily understood from the particular mode shapes of the second and third resonance.

While performing the simulations as function of the width of the defect zone, we always observed the predicted constant value of 1.178 for the ratio of relative frequency shift to absolute attenuation increase. This leads to the conclusion that this ratio is independent of the defect characteristics.

4.2 Amplitude of the Harmonics for Sinusoidal Excitation near Mode "m"

We already stated that no even harmonics will be generated for a given sinusoidal excitation near mode m, because there will be no oscillating contributions in the right-hand side source terms of Eq. (23.4) with even multiples of Ω when $\beta = 0$. We can find the first-order amplitude of the third harmonics by solving the following equation.

$$\partial_{tt}^2 z_{3m} + \frac{\Omega}{Q}\partial_t z_{3m} + \omega_{3m}^2 z_{3m} = -H_{3m\,m}\frac{8}{15\pi}A_m^2 \sin(3(\Omega t + \phi_m)). \tag{23.30}$$

Representing z_{3m} by $A_{3m} \cos(3\Omega t + \phi_{3m})$, the strain amplitude for an excitation near the resonance frequency ω_m of mode m consequently reads:

$$\varepsilon_{3m} = \frac{8Q}{5\pi L}\left| \int_0^L \alpha \, \sin\left(\frac{3m\pi}{L}x\right) \left| \sin\left(\frac{m\pi}{L}x\right) \right| \sin\left(\frac{m\pi}{L}x\right) \, dx \right| \varepsilon_m^2 = X_3 \, \varepsilon_m^2. \tag{23.31}$$

Similarly the fifth strain amplitude can be calculated, and yields:

$$\varepsilon_{5m} = \frac{8Q}{21\pi L}\left| \int_0^L \alpha \, \sin\left(\frac{5m\pi}{L}x\right) \left| \sin\left(\frac{m\pi}{L}x\right) \right| \sin\left(\frac{m\pi}{L}x\right) \, dx \right| \varepsilon_m^2 = X_5 \, \varepsilon_m^2. \tag{23.32}$$

Both the third and fifth harmonic (and by extension all odd harmonics) are proportional to the second power of the fundamental strain amplitude, irrespective of the damage distribution. They all are generated by a second-order interaction between two fundamental components, which originated in the sign-function contribution. From Eqs. (23.31) and (23.32), one can easily obtain the expressions for the two special cases of a localized damage zone and a globally distributed hysteretic nonlinearity.

Again, we have checked the particular position-dependent behavior of the coefficients X_3 and X_5 in Eqs. (23.31) and (23.32) in the case of a localized damage zone with hysteretic nonlinearity by comparing it to the results of the multiscale model. Figure 23.4 illustrates the good agreement between the analytical and numerical results for the proportionality coefficients of the third and fifth harmonics taken at the resonance frequency of the fundamental of mode $m = 1$ and mode $m = 2$.

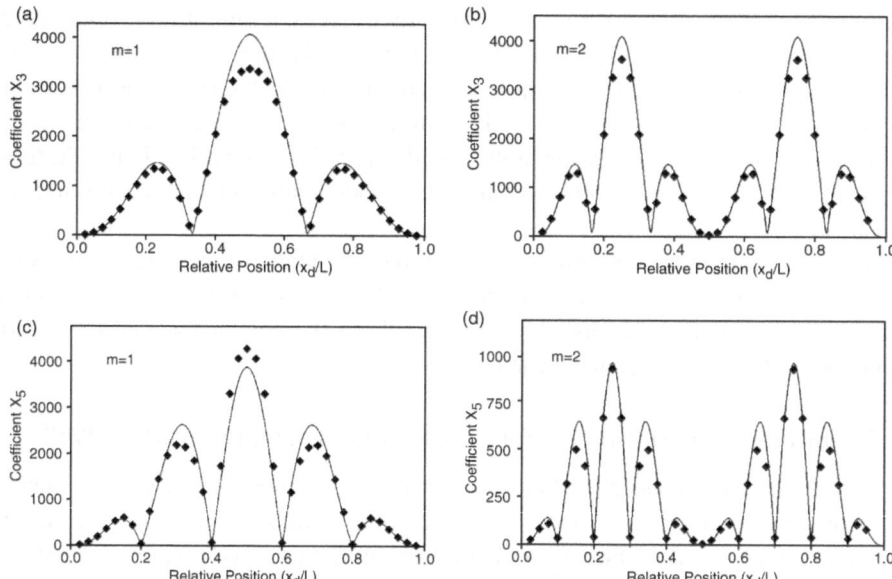

Fig. 23.4. Investigation of the nonlinear signatures X_3 and X_5 (proportionality coefficient in the strain amplitude dependence of the third and fifth harmonics) from a hysteretic nonlinear defect as a function of its position in a resonant bar. (a) Signature X_3 for the fundamental mode excitation ($m = 1$). (b) Signature X_3 for the second bar mode excitation ($m = 2$). (c) Signature X_5 for the fundamental mode excitation ($m = 1$), and (d) signature X_5 for the second bar mode excitation ($m = 2$). Lines represent the analytical formulae; diamonds correspond to the numerical simulations. The defect zone is $d = L/20$.

5. Inverse Modeling

As the NRUS predictions of frequency shifts and harmonic content are analytical in nature, they offer opportunities to solve the inverse problem of defect characterization and localization. Traditionally, scanning techniques are used to localize damage in a material. Zones with defects are spotted by analyzing the changing properties of reflected or transmitted signals. In most cases the scanning requires a mechanical operation and a coupling with the medium. A consistent coupling is indeed very important when nonlinear signatures are sought.

The degree to which nonlinearity can be observed highly depends on the value of the strain at the location of the nonlinearity. If the strain values are low the nonlinearity will not be activated. By using multiple mode information and combining harmonic and frequency shift observations, it is possible to interpret those pieces of information in a consistent manner and to invert them for the damage characteristics and location. The fact that global resonances can be used for the inversion is an important advantage over traditional techniques. Unlike other techniques, the NRUS method does not require a spatial scan to locate the defect, as them lets different excitation modes, with different vibration patterns, probe the structure.

In the next sections we give some possibilities on how to combine several of the above deduced formulae to infer the location of the defect, its degree of nonlinearity, and width, based on the obtained nonlinear signatures of the resonances.

5.1 Classical Cubic Nonlinearity

If the medium contains a localized defect that can be represented by a zone with classical nonlinearity, we can infer the location and its effective nonlinear strength by simply considering the lowest two resonance modes. In both cases the resonance frequency shift is given by Eq. (23.10) for $m = 1$ and $m = 2$. After some algebraic calculations, we otain an expression for the relative location of the defect:

$$x_d/L = \frac{1}{\pi} A\cos\left[\frac{1}{2}\sqrt[4]{\frac{\varepsilon_1^2 \, \Delta\omega_2/\omega_2}{\varepsilon_2^2 \, \Delta\omega_1/\omega_1}}\right]. \tag{23.33}$$

Experimental values for the shifts can be determined by the SIMONRUS technique. It is clear, however, that the inversion procedure by Eq. (23.33) is nonunique as defects located symmetrically with respect to the center of the bar cannot be distinguished from each other.

For what concerns the degree and width of the defect, we find that

$$\delta\frac{d}{L} = \frac{64 \, (\Delta\omega_1/\omega_1/\varepsilon_1^2)^2}{3 \, \left(4\sqrt{\Delta\omega_1/\omega_1/\varepsilon_1^2} - \sqrt{\Delta\omega_2/\omega_2/\varepsilon_2^2}\right)^2}. \tag{23.34}$$

This expression confirms that the relative width of the defect and its strength cannot be uncoupled for localized defects, and hence $\delta d/L$ should indeed be considered as the effective damage parameter characterizing the defect zone.

A similar inverse calculation can be obtained from the knowledge of the frequency shift and the third harmonic, for instance. For instance, combining Eqs. (23.10) and (23.16) yields:

$$x_d/L = \frac{1}{\pi} A\sin\left[\sqrt{\frac{3}{4} - \frac{\varepsilon_3}{8\,\varepsilon_1\,Q\,\Delta\omega_1/\omega_1}}\right]; \quad \delta\frac{d}{L} = \frac{4 \, (\Delta\omega_1/\omega_1/\varepsilon_1^2)}{3 \, \left(\frac{3}{4} - \frac{\varepsilon_3}{8\,\varepsilon_1\,Q\,\Delta\omega_1/\omega_1}\right)^2}. \tag{23.35}$$

For nonlocalized defects (spread over a finite width comparable to $L/2$, for instance), it is necessary to use the general equations given in Eqs. (23.9) and (23.15) for the inversion.

5.2 Non-Classical Hysteretic Nonlinearity

The above-outlined inversion can also be performed in cases when the defect is represented by a hysteretic nonlinearity. Using Eq. (23.26) for two modes ($m = 1, 2$) we get:

$$x_d/L = \frac{1}{\pi} A\cos\left[\frac{1}{2}\sqrt[3]{\frac{\varepsilon_1 \, \Delta\omega_2/\omega_2}{\varepsilon_2 \, \Delta\omega_1/\omega_1}}\right];$$

$$\alpha\frac{d}{L} = \frac{(\Delta\omega_1/\omega_1/\varepsilon_1)}{\left[1 - \frac{1}{4}\left(\frac{\Delta\omega_2/\omega_2/\varepsilon_2}{\Delta\omega_1/\omega_1/\varepsilon_1}\right)^{2/3}\right]^{3/2} \left(1 + \frac{4}{3\pi Q}\right)}. \tag{23.36}$$

On the other hand, if we use the strain dependence of the resonance frequency shift and of the third harmonic at the fundamental mode [Eqs. (23.26) and (23.31) for $m = 1$], we find that

$$x_d/L = \frac{1}{\pi} A \sin \left[\sqrt{\frac{3}{4} - \frac{5\pi \ \varepsilon_3 \left(1 + \frac{4}{3\pi Q}\right)}{32 \ \varepsilon_1 \ Q \ \Delta\omega_1/\omega_1}} \right] ;$$

$$\alpha \frac{d}{L} = \frac{(\Delta\omega_1/\omega_1/\varepsilon_1)}{\left[\frac{3}{4} - \frac{5\pi \ \varepsilon_3(1 + 4/(3\pi \ Q))}{32 \ \varepsilon_1 \ Q \ \Delta\omega_1/\omega_1}\right]^{3/2} \left(1 + \frac{4}{3\pi Q}\right)}. \tag{23.37}$$

Again it is necessary to use the more general equations and more sophisticated optimization algorithms for the inversion in the case of nonlocalized defects.

6. Conclusions

An analytical treatment of the effect of damage on the resonance mode characteristics is not simple, and probably only possible under severe assumptions. Nevertheless, if analytical formulae are available they can be of great use to (1) check the correctness of the simulations performed by numerical models in limiting cases, and (2) provide quick methods for the inversion of the damage characteristics.

We have presented a nonlinear version of the Resonance Ultrasound Spectroscopy (RUS) theory by extending the traditional formalism to the treatment of damage in the form of nonlinearity. General equations were developed in the 1-D case, describing the interaction between the modes induced by the nonlinearity. These equations were solved following a perturbation approach. We considered two different nonlinear stress–strain signatures that represent the damage: a reversible cubic nonlinearity as well as the more interesting hysteretic nonlinearity.

The solutions provide analytical expressions for the nonlinear shift of the modal resonance frequency and the harmonic generation as a function of the strength of the nonlinearity, the width of the defect, and its position in the 1-D system. Each analytical formula can be readily explained in terms of the vibration patterns of the modes that produce the particular nonlinear interaction. The dependence on the strain level is different for both cases of nonlinearity, but always independent of the defect location.

We compared the results to numerical calculations from a multiscale model, showing excellent agreement, provided the discretization is defined adequately.

The analytical formulae were exploited to infer critical information about the damage position, the degree of nonlinearity, and the width of the damage zone either from the shifts in resonance frequency occurring at different excitation modes, or from the shift and the harmonics predicted at a single mode. The use of multiple mode information for the inversion of damage characteristics and localization has an important advantage over traditional scanning techniques. Unlike other techniques, the NRUS method does not require a spatial scan to locate the defect, as it lets different excitation modes, with different vibration patterns, probe the structure.

Acknowledgments

The authors gratefully acknowledge the support of the Flemish Fund for Scientific Research. (G.0206.02 and G.0257.02), the provisions of the European Science Foundation Programme NATEMIS, and the European FP5 and FP6 Grants DIAS (EVK4-CT-2002-00080)and AERONEWS (AST3-CT-2003-502927).

References

Antonets VA, Donskoy DM, Sutin AM (1986), Nonlinear Vibro-diagnostics of flaws in multilayered structures, *Mech. of Composites Materials* **15**: 934–937.

Ballad E, Vezirov S, Pfleiderer K, Solodov I, Busse G (2004), Nonlinear modulation technique for NDE with air-coupled ultrasound, *Ultrasonics* **42**(1–9): 1031–1036.

Cantrell JH, Yost WT (1994), Acoustic Harmonic-Generation From Fatigue-Induced Dislocation Dipoles, *Phil. Mag. A* **69**: 315–326.

Guyer RA, McCall KR, Van Den Abeele K (1998), Slow elastic dynamics in a resonant bar of rock, *Geophys Res Lett* **25**(10): 1585–1588.

Ichitsubo T, Ogi H, Hirao M, Tanaka K, Osawa M, Yokokawa T, Kobayashi T, Harada H (2002), Elastic constant measurement of Ni-base superalloy with the RUS and mode selective EMAR methods, *Ultrasonics* **40**, 211–215.

Johnson PA, Zinszner B, Rasolofosaon PNJ (1996), Resonance and elastic nonlinear phenomena in rock, *J Geophys Res* **101**: 11553–11564.

Kac, Mark (1966), Can One Hear the Shape of a Drum?, *Amer Math Monthly* **73**: 1–23.

Kazakov VV, Sutin AM, Johnson PA (2002), Sensitive imaging of an elastic nonliniear wave-scattering source in a solid, *Appl Phys Lett* **81**(4): 646–648.

Keppens V, Mandrus D, Sales BC, Chakoumakos BC, Dai P, Coldea R, Maple MB, Gajewski DA, Freeman EJ, Bennington S (1998), Localized vibrational modes in metallic solids, *Nature* **395**(6705): 876–878.

Krohn N, Stoessel R, Busse G (2002), Acoustic non-linearity for defect selective imaging, *Ultrasonics* **40**(1–8): 633–637.

Landau L, E Lifshitz E (1969) *Mechanics* (Oxford, Pergamon).

Morris WL, Buck O, Inman RV (1979), Acoustic Harmonic-Generation Due To Fatigue Damage In High-Strength Aluminum, *J Appl Phys*, **50**(11): 6737–6741.

Nagy PB (1998), Fatigue damage assessment by nonlinear ultrasonic materials characterization, *Ultrasonics* **36**(1–5): 375–381.

Nakamura N, Ogi H, Hirao M (2004), Resonance ultrasound spectroscopy with laser-Doppler interferometry for studying elastic properties of thin films, *Ultrasonics* **42**(1–9): 491–494.

Neyfeh A (1973) *Perturbation Methods* (John Wiley & Sons, New York).

Nazarov VE, Ostrovsky LA, Soustova I, Sutin AM (1988), Nonlinear Acoustics Of Micro-Inhomogeneous Media, *Phys Earth and Planet Interiors* **50**(1): 65–73.

Nazarov VE, Sutin AM (1989), Theory Of A Parametric Sound Receiver Using A Nonlinear Layer, *Sov Phys Acoust USSR* **35**: 510–512.

Ogi H, Sato K, Asada T, Hirao M (2002), Complete mode identification for resonance ultrasound spectroscopy, *J Acoust Soc Am* **112**(6): 2553–2557.

Pohit G, Mallik A, Venkatesan C (1999), Free out-of-plane vibrations of a rotating beam with non-linear elastomeric constraints, *J Sound Vib* **220**(1): 1–25.

Stoessel R, Krohn N, Pfleiderer K, Busse G (2002), Air-coupled ultrasound inspection of various materials, *Ultrasonics* **40**(1–8): 159–163.

Sutin, AM, Donskoy, DM (1998), Vibro-acoustic Modulation Nondestructive Evaluation Technique, *Proceedings of SPIE* **3397**, pp. 226–237.

Tencate JA, Shankland TJ (1996), Slow dynamics in the nonlinear elastic response of Berea sandstone, *Geophys Res Lett* **23**(21): 3019–3022.

Tencate JA, Smith E, Guyer RA (2000), Universal slow dynamics in granular solids, *Phys Rev Lett* **85**(5): 1020–1023.

Ulrich T, McCall KR, Guyer RA (2002), Determination of elastic moduli of rock samples using resonant ultrasound spectroscopy, *J Acoust Soc Am* **111**(4): 1667–1674.

Van Den Abeele K, Johnson PA, Guyer RA, McCall KR (1997), On the quasi-analytic treatment of hysteretic nonlinear response in elastic wave propagation, *J Acoust Soc Am* **101**(4): 1885–1898.

Van Den Abeele K, Johnson PA, Sutin AM (2000a), Nonlinear elastic wave spectroscopy (NEWS) techniques to discern material damage, part I: Nonlinear wave modulation spectroscopy (NWMS), *Res Nondestruct Eval* **12**(1): 17–30.

Van Den Abeele K, Carmeliet J, Tencate JA, Johnson PA (2000b), Nonlinear Elastic Wave Spectroscopy (NEWS) techniques to discern material damage. Part II: Single Mode Nonlinear Resonant Acoustic Spectroscopy (SIMONRAS), *Res Nondestruct Eval* **12**(1): 31–42.

Van Den Abeele K, De Visscher J (2000), Damage assessment in reinforeced concrete using spectral and temporal nonlinear vibration techniques, *Cement and Concrete Research* **30**(9): 1453–1464.

Van Den Abeele K, Van De Velde K, Carmeliet J (2001), Inferring the degradation of pultruded composites from dynamic nonlinear resonance measurements, *Polymer Composites* **22**(4): 555–567.

Visscher W, Migliori A, Bell T, Reinert R (1991), On The Normal-Modes Of Free-Vibration Of Inhomogeneous And Anisotropic Elastic Objects, *J Acoust Soc Am* **90**(4): 2154–2162.

Zimenkov S, Nazarov V (1995), Propagation of nonlinear acoustic waves in rocks, Izv. Acad. Sci. USSR *Phys. Solid Earth* **30**(5): 437–439.

Laboratory Experiments using Nonlinear Elastic Wave Spectroscopy (NEWS): a Precursor to Health Monitoring Applications in Aeronautics, Cultural Heritage, and Civil Engineering

Koen Van Den Abeele,[1,4] Tomasz Katkowski,[2] Nicolas Wilkie-Chancellier,[3] and Wendy Desadeleer[1]

[1] Katholieke Universiteit Leuven, Interdisciplinair Research Center (IRC), KULAK, E. Sabbelaan 53, B-8500 Kortrijk, Belgium.
[2] University of Gdańsk, Institute of Experimental Physics, ul. Wita Stowsza 57, 80-952 Gdańsk, Poland.
[3] Université de Cergy-Pontoise, Equipe Circuit Instrumentation et Modélisation Electronique, rue d'Eragny, Neuville sur Oise, 95031 Cergy-Pontoise Cedex, France.
[4] To whom correspondence should be addressed: Koen Van Den Abeele, Katholieke Universiteit Leuven, Interdisciplinair Research Center (IRC), KULAK, E. Sabbelaan 53, B-8500 Kortrijk, Belgium; e-mail: koen.vandenabeele@kuleuven-kortrijk.be.

Abstract

In the framework of various European and nationally funded projects, intact and damaged samples of Carbon Fiber Reinforced Polymer (CFRP) composites, layered aluminum plates, and natural building stones are being examined using both linear and nonlinear acoustic and ultrasonic diagnostic methods. Similar techniques are also applied to the monitoring of the early curing stages of concrete. The goal of these projects is to develop the necessary means for an innovative microdamage inspection system based on Nonlinear Elastic Wave Spectroscopy (NEWS). This chapter contains a report on some of our latest laboratory results using different NEWS techniques. We investigated the heat damage in composite laminates at moderate temperatures using Nonlinear Resonance Ultrasound Spectroscopy (NRUS). The NRUS technique, which evaluates the amplitude dependence of a single resonance mode, has been implemented in two variations: time domain analysis (evaluating the signal reverberation) and frequency domain analysis (evaluating the sweep characteristics). Both methods show consistent results and provide a considerable sensitivity for nonlinear signatures as a function of heat temperature and exposure time. Linear and nonlinear wave propagation methods have also been used to examine near-surface deterioration of natural building stones used in restoration projects throughout Europe. Samples are locally excited with a high-frequency sinusoidal wave (order of 100 kHz) producing a surface wave, and in some cases a coupling with a low-frequency signal is produced by an impact. Linear wave speed and attenuation measurements are supplemented with nonlinear acoustic measurements investigating the creation of harmonics and intermodulation frequencies. Consistent observations show that undamaged areas are essentially linear in their response, and the damaged zones become highly nonlinear. Finally, this chapter also illustrates the use of NEWS in the monitoring of different steps in the early curing process of concrete. We analyzed the evolution in the linear and nonlinear ultrasonic behavior of concrete as a function of the degree of hydration during the first three days of the curing process. The results show a good correlation between the phase changes in the concrete due to chemical reactions and mechanical setting seen in the temperature profile, and the linear and nonlinear acoustic properties.

Keywords: Acoustic spectroscopy, aeronautics, composites, concrete, curing process, damage diagnosis, heat damage, microdamage, natural building stone, nonlinear acoustics, nonlinear elastic wave spectroscopy, nonlinearity

1. Introduction

Recent advances in modern material technology require the development of Non-Destructive Evaluation (NDE) techniques that allow the quantification of microstructural damage in a wide variety of materials during their manufacture and life cycle, ensuring both their quality and durability. In aeronautics, for instance, the performance and behavior characteristics of airframe structures can be adversely affected by structural degradation, resulting from sustained use within normal flight envelopes, as well as from exposure to severe environmental conditions, and from damage due to impact. Similar examples can be found in civil engineering applications. For primary load-bearing structures, these factors can have serious consequences in terms of safety, cost, and operation. Consequently, the timely and accurate detection, characterization, and monitoring of the development of structural defects over time (e.g., cracking, corrosion, delamination, material degradation, and other flaws, defects, or damage) are of major concern to the operational environment. Traditional NDE techniques such as high-quality linear acoustic, electromagnetic, and visual inspection methods are generally not sufficiently sensitive to the presence and development of domains of incipient and progressive damage. For this purpose, we are currently developing and validating innovative microdamage inspection systems based on various NDT methods within the class of Nonlinear Elastic Wave Spectroscopy (NEWS).

NEWS techniques primarily deal with the investigation of the amplitude dependence of material parameters such as wavespeed, attenuation, spectral content, and the like. The degree to which these material properties depend on the applied dynamic amplitude can be quantified by various nonlinear parameters. Several NEWS techniques have been developed to probe for the existence of damage (e.g., delaminations, microcracks, or weak adhesive bonds) by investigating the generation of harmonics, subharmonics, and intermodulation of frequency components, the amplitude-dependent shift in resonance frequencies, the nonlinear contribution to attenuation properties, slow dynamic effects, and phase modulation. Details about these methods and their applications in NDT can be found in References [1–18] as well as in Chapters 4, 15, 25, 28, and 29 of this book. Laboratory tests performed on a wide variety of materials subjected to different microdamage mechanisms of mechanical, chemical, and thermal origin, have shown that the sensitivity of such nonlinear methods to the detection of microscale features is far greater than that obtained with linear acoustical methods.

In the framework of various European and national funded projects, several of the above-mentioned techniques are being fine-tuned to specific applications. As a first step in this process, we investigated the use and robustness of these methods for materials with a simple geometry. In this chapter we restrict ourselves to the discussion of three examples: the study of heat damage in CFRP composites used in aeronautics, the investigation of surface deterioration in natural building stones on historical monuments, and the monitoring of the early stages in the concrete curing processes using NEWS.

2. Nonlinear Resonance Spectroscopy on CFRP Composites with Heat Damage at Moderate Temperatures

2.1 Rationale

A review of the mechanisms of heat damage in composites and a state of the art of non-destructive evaluation techniques currently used to evaluate heat damage is available in [19]. The exposure to heat induces chemical and microstructural changes that affect the mechanical behavior of the composite laminate, even at moderate temperatures. Studies have shown that thermal degradation is typically matrix dominated because by the time fiber properties such as tensile strength and modulus are affected, all other mechanical integrity is lost. Compressive, shear, and flexural properties are considered to be the most sensitive mechanical properties for use in the early detection of thermal degradation. Most of the work reported in the literature dealing with NDE for heat damage in composites is based on the following five methods: thermal (IR), ultrasonics, acoustic emission, dielectric properties, and radiography. These methods, although readily available and generally well developed, are limited in their capabilities to detect and characterize the changes in composite material properties associated with heat damage. As an example, experimental work carried out at CSM, Sweden, on unidirectional CF/epoxy laminates illustrates the fact that the mechanical properties change well before the microstructural alterations can be detected with classical ultrasonic techniques [20]. Figure 24.1 gives an idea of the detectability threshold of heat damage in unidirectional AS4-8552 CFRP laminates using conventional ultrasonics. Amplitude C-scans are performed for three samples exposed to 285, 290, and 300° C

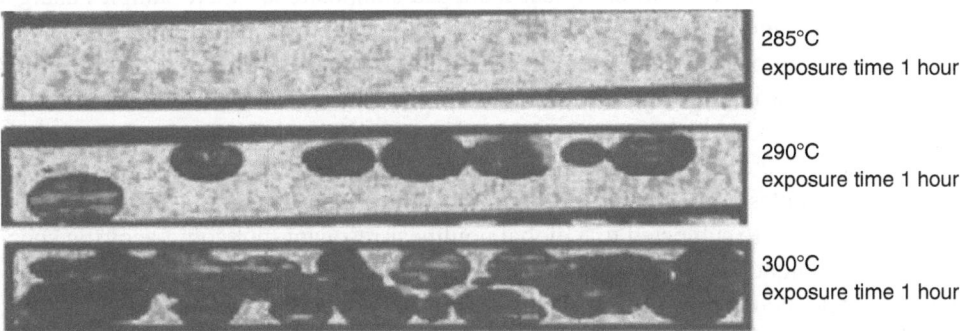

285°C
exposure time 1 hour

290°C
exposure time 1 hour

300°C
exposure time 1 hour

Fig. 24.1. Ultrasonic amplitude C-scan of unidirectional composite (CF/epoxy laminate) exposed for one hour to 285, 290, and 300°C.

for one hour. Delaminations clearly appear at 290°C whereas no sign of damage is seen at 285°C.

Nevertheless, the measured value of the interlaminar shear strength for the same type of samples changes from 121 MPa for nonexposed samples to 114 MPa when exposed at 200°C for the duration of one hour, to 84 MPa at 285°C and to 43 for samples at 300°C.

Most traditional NDE techniques are capable of detecting physical anomalies such as cracks and delaminations. However, to be effective for thermal degradation they must be capable of detecting initial heat damage that occurs on a molecular scale. Review of the literature from more recent years indicates that a vast number of NDE methods are currently under development and show various degrees of promise for characterizing heat damage in composites: thermal wave, vibrothermography, leaky lamb wave, ultrasonic backscatter, acoustoultrasonics, isotope radiation backscatter, embedded sensors, shearography, thermal imaging, backscattered X-rays, diffuse reflectance infrared Fourier transform, laser-induced fluorescence, and, of course, nonlinear elastic wave spectroscopy. More extensive information on the status of development of several of these NDE methods and their capabilities for detecting heat damage in composite laminates can be found in an extended state-of-the-art review available from NTIAC [21].

One of the goals of the European Union sponsored project, "Health Monitoring of Aircraft by Nonlinear Elastic Wave Spectroscopy" (AST-CT-2003-502927, AERONEWS, coordinated by Professor Koen Van Den Abeele [22]) is indeed to investigate and confirm the effectiveness of newly developed ultrasonic methods based on NEWS for detecting early damage in aeronautic components. Within AERONEWS, thermally loaded composites were considered to be good examples for the simulation of the initiation and progressive degree of microdamage as the result of extreme environmental conditions.

2.2 Experimental Set-Up and Methodology

To substantiate the potential of NEWS techniques to discern heat damage and test its postulated high sensitivity to incipient damage and micromechanical changes in the medium, we examined a set of heat-damaged composite laminate samples using a nonlinear resonant ultrasonic spectroscopy technique, and quantified their nonlinearity as a function of the heating temperature and exposure time. The set of 21 CFRP (AS4/8552 quasi-isotropic lay-up) samples consisted of one reference sample, which was left unexposed, and 20 samples exposed at five different temperatures (240, 250, 260, 270, and 300°C) for four different durations (15', 30', 45', and 60'). The nominal size of the samples was 120 mm × 20 mm × 4 mm (Figure 24.2a).

All samples were subjected to a global nonlinear resonant ultrasonic-acoustic spectroscopy technique which measures the amplitude dependence of the resonance behavior of a single mode of the samples. This technique has also been termed SIMONRUS: Single Mode Nonlinear Resonant Ultrasound Spectroscopy [11–15]. The mode under consideration in this study is the fundamental flexural mode of a beam, which has a stress concentration in the middle of the sample and displacement nodes at a distance

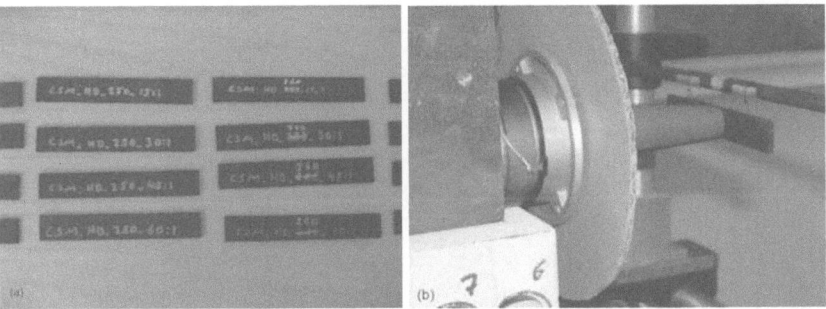

Fig. 24.2. (a) CFRP samples; (b) set-up for TD and FD SIMONRUS.

of 0.224 L from both edges, with L the length of the sample (120 mm). Two implementations of SIMONRUS were realized: an amplitude dependent analysis of the reverberation signal in the time domain (TD-SIMONRUS), and a study of the nonlinear resonance response in the frequency domain (FD-SIMONRUS). For both cases, we used the same set-up: a sample is supported by two nylon wires at the node lines, and is excited by a loudspeaker (diameter 32 mm, focused by a cone to 20 mm) centered in the middle of the sample (Figure 24.2b); The response is measured by a laser vibrometer near one of the edges. All equipment is computer controlled and operated through LabVIEW and GPIB. Acquisition of the signal is realized by a 5 MHz DAQ-card.

In the TD-SIMONRUS experiment we excite the sample with a 1000 period burst excitation at a given amplitude and with a frequency close to the fundamental flexural resonance frequency. We then record a total of 0.6 seconds (120,000 points at a sampling rate of 200 kHz) of the reverberation of the sample after the excitation is stopped. To achieve a high accuracy, we implemented a variable dynamic range acquisition procedure based on an automated feedback of the instantaneous amplitude response. Doing so, the dynamic range is adjusted each 4000 points, and the signals are averaged 10 times (see Figure 24.3a for a typical measurement result).

The analysis of the "composed" signals was done by fitting small moving time windows of the signal with an exponentially decaying sine function: $Ae^{-\alpha t} \sin(2\pi f t + \phi)$. The output of the TD-SIMONRUS analysis procedure yields the evolution of the frequency (f) and damping characteristic (α) as a function of the amplitude A in the decaying signal. If no nonlinearity is present we should obtain a constant frequency and a constant damping characteristic for all amplitudes. However, when nonlinearity is present and related to the thermally induced microdamage, we may expect a stronger dependence of the frequency and damping on the amplitude for increasing microdamage. This is indeed the case as shown in Figures 24.3b–d, in which we have gathered the results of frequency versus amplitude for the reference sample, a sample heated at 250°C for 45' and a sample exposed at 300°C for 60'.

Figure 24.4 illustrates the analyzed results of the amplitude dependence for the damping characteristic. Here we have plotted the Q factor ($Q = \pi f/\alpha$ inverse attenuation). Again we notice that this material parameter is amplitude dependent and that the nonlinearity is higher for samples with increased thermal damage.

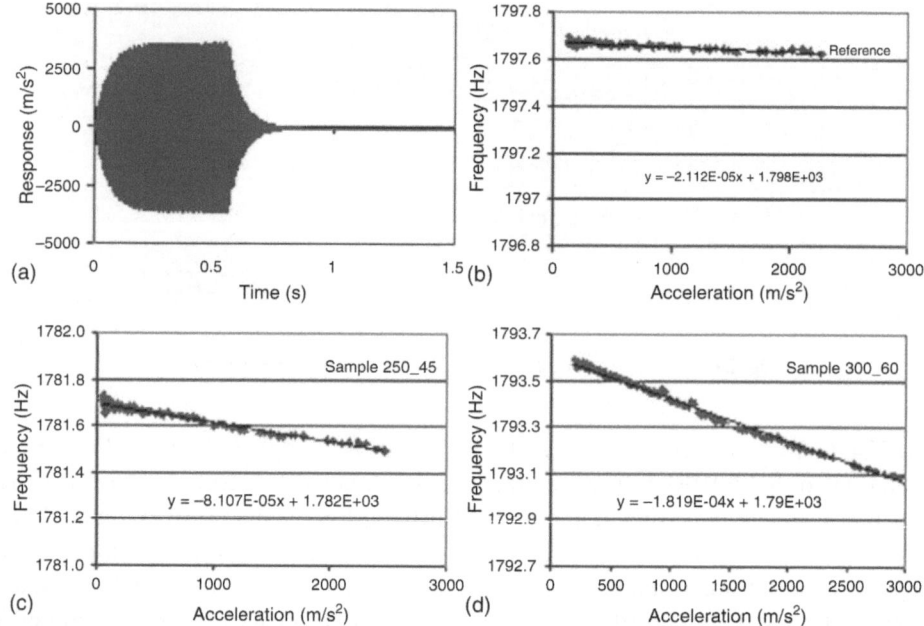

Fig. 24.3. TD SIMONRUS results: (a) typical recorded signal; (b) analyzed frequency versus acceleration amplitude for the reference sample showing almost no nonlinearity; (c),(d) for two samples at different heating temperature and exposure time: 250C for 45'(c) and 300C for 60' (d).

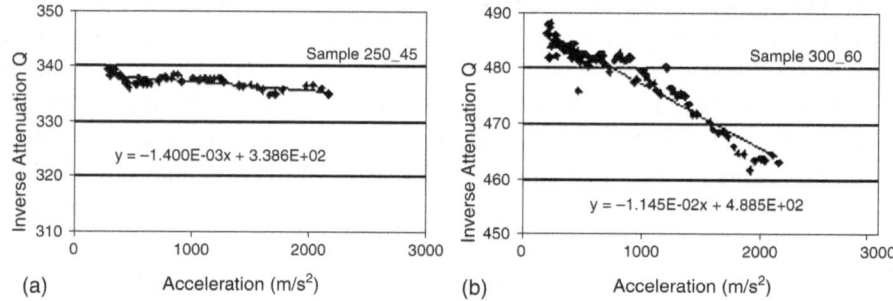

Fig. 24.4. TD SIMONRUS nonlinear Q-factor (inverse attenuation) analysis for two samples at different heating temperature and exposure time: 250°C for 45'(a) and 300°C for 60' (b).

In order to quantify the degree of nonlinearity, we calculate the linear proportionality coefficient γ between the relative resonance frequency shift and the strain amplitude, $\Delta f/f_o = \gamma\,\varepsilon$. The strain values ε were calculated from the measured acceleration values A, using the strain-acceleration conversion expression for beams: $\varepsilon = 0.225 \cdot d/(f \cdot L)^2 \cdot A$, with $d = 4\,\text{mm}$ and $L = 200\,\text{mm}$. It should be noted that, because of the global character of the applied NEWS method, γ only represents a global quantification of the nonlinearity, integrated over the whole sample. It contains no direct information on the localization of the defects.

To assure the reliability of the method, we verified that the obtained values of the nonlinearity γ were independent of the chosen initial excitation frequency and applied voltage. Figure 24.5 shows the robustness of the results for a composite laminate

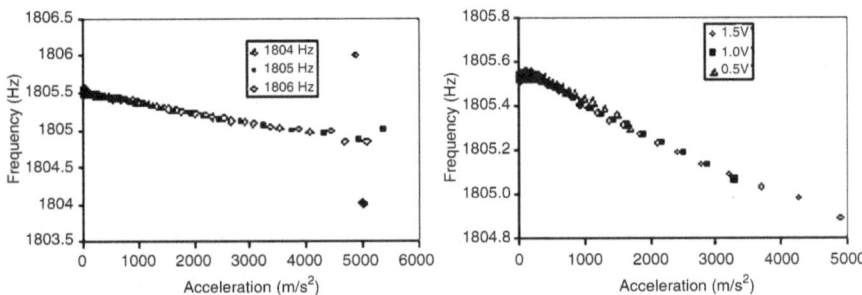

Fig. 24.5. Verification of the reliability of the TD SIMONRUS results for a single sample at various excitation frequencies (1804 Hz, 1805 Hz, 1806 Hz) for fixed amplitude (1.5 V) (left) and at various excitation amplitudes (0.5 V, 1 V, 1.5 V) for fixed excitation frequency (1805 Hz) (right).

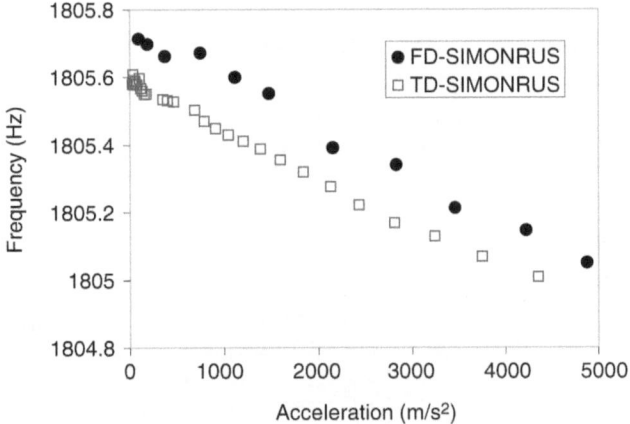

Fig. 24.6. Illustration of the consistency of TD SIMONRUS and FD-SIMONRUS results.

sample exposed at 300°C for 45'. In Figure 24.5a, the analyzed response at three different frequencies is illustrated for a fixed excitation amplitude. In Figure 24.5b, the response at fixed frequencies is illustrated for three different excitation amplitudes.

Finally, the TD-SIMONRUS results were also compared with results of FD-SIMONRUS experiments. In the latter method, which is well documented in earlier publications [11–15], we record and analyze the excitation dependence of the full resonance curve of the fundamental flexural mode while sweeping frequency. The experimental set-up is the same as for TD-SIMONRUS. As illustrated in Figure 24.6, the analysis of the resonance frequency as a function of the resonant response amplitude provides results that are remarkably consistent with the results from TD-SIMONRUS. The offset of both results is due to different environmental conditions.

2.3 Discussion of the Results in Comparison with Linear Measurements

The results for the global nonlinearity of all samples are summarized in the left part of Figure 24.7. The nonlinear parameter γ shows an overall increase with increasing exposure time and heat temperature, up to a factor 10 with respect to the reference value. A comparison with a traditional A-scan of the samples (Figure 24.8) reveals that

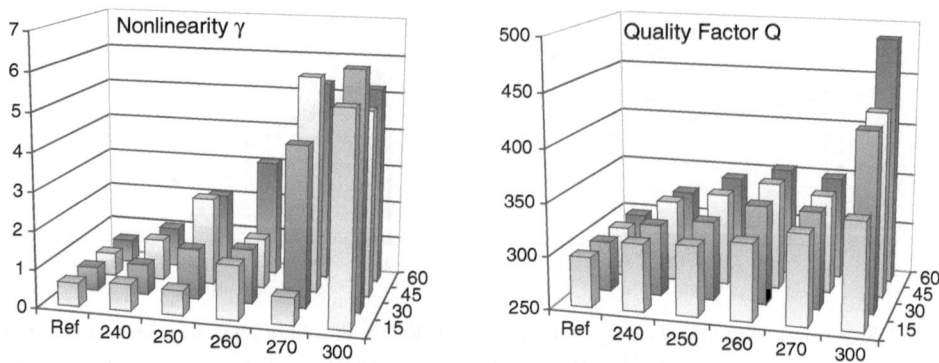

Fig. 24.7. Nonlinearity γ (left) and linear Q-factor (inverse attenuation, right) for all 21 samples as a function of heating temperature and exposure time.

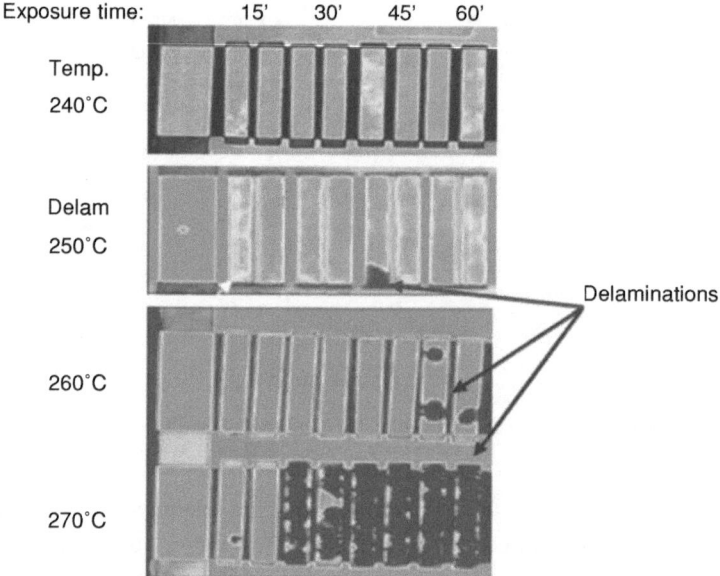

Fig. 24.8. Traditional amplitude A-scan for a set of composite laminates as function of heating temperature and exposure time.

the increase of nonlinearity (measured as a global property of the sample) is indeed an indication of the degradation of the mechanical properties of the sample due to the thermal loading. Samples with zones of delaminations clearly correspond to a higher global nonlinearity.

As a comparison with a measure of the linear characteristics of the sample (traditional ultrasound), we also examined the values of the linear (or low-amplitude) Q-factor of each of the samples. The results are displayed in Figure 24.7b. Apparently the linear attenuation for heat-treated samples reduces as a function of exposure time and heat temperature (Q-factor increases). If the linear measure of attenuation would be connected to damage, we would have expected the Q-factor to decrease instead of increase. Therefore, our conclusion is that the effect seen in the linear attenuation is not a measure of the microdamage, but is related to the chemical and physical change in linear material parameters due to the thermal loading (molecular changes in the resin matrix).

3. Nonlinear Wave Propagation and Wave Modulation Spectroscopy of Structural Damage in Natural Stones Used in Restoration of Historical Monuments

3.1 Rationale

The European Cultural Heritage deals with a great variety of structures of high intrinsic value. For centuries, historical monuments have been subjected to mechanical loads, and to various types of weathering (i.e., temperature variations, atmospheric pollution, freeze and thaw, moisture transport, etc.), which affect their quality and impose damage of a certain type and degree. The induced damage on natural building stones due to aggressive urban environment may lead to significant losses of the elasticity and strength properties of the outermost layers of stone. In time, this results in erosion features, cohesionless appearance, and disintegration. Damage diagnosis of NB stones is the first stage in the planning of the remedial steps to which the success of the restoration should be entrusted. An erroneous diagnosis may be very harmful to the cultural, structural, and economic outcome of the operation. Thus far, the classical processes for measuring the mechanical properties and effectiveness of the consolidation treatment of stones are observational, empirical, and destructive: a sample is taken from the structure—if it is permitted—it is cut and then standard rock mechanical tests are performed for characterization. Obviously, it would be much more appropriate if the knowledge of the mechanical properties and damage could be the result of nondestructive tests performed at the location itself. Up to now, however, there is no such validated "stone-friendly" technology, procedure, or apparatus available on the market that can do the in situ assessment of damage, verify the effectiveness of consolidation treatments of stones, and provide guidance in the choice of compatible quarry stones for repair.

This is exactly the goal of the European Union sponsored project "Integrated tool for in situ characterization of effectiveness and durability of conservation techniques in historical structures" (EVK4-CT-2002-00080, DIAS, coordinated by Professor George Exadactylos, Technical University of Crete, Greece [23]). The DIAS project will contribute to the evolution of EU policies by developing an integrated portable _Drilling-Indentation-Acoustics of Stones (DIAS)_ device that will furnish an easy-to-use and quasinondestructive technique to evaluate the in situ mechanical characteristics of a NB stone (elasticity and strength) and quantify and characterize its damage compared to the virgin material. The study presented in this article deals solely with the acoustical part of the project goal to quantify near-surface stone deterioration.

Among other nondestructive evaluation techniques, nonlinear wave propagation and nonlinear wave modulation spectroscopy are being increasingly used in damage detection applications [1–10]. The former technique investigates the creation of second and higher harmonics within the material during the propagation of a continuous wave,

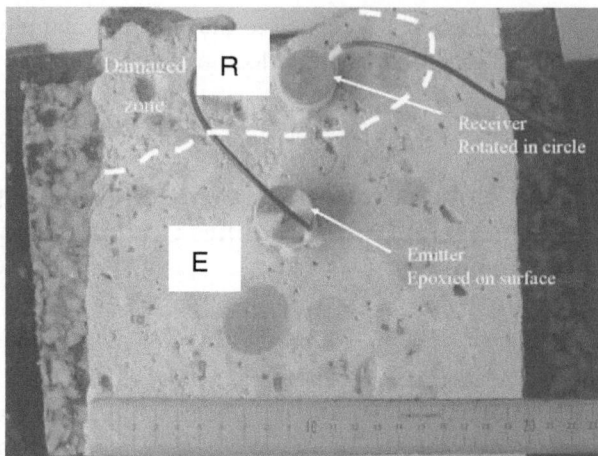

Fig. 24.9. Experimental set-up on the Balegem sample.

whereas the latter method consists of generating two signals of two separate frequencies in a sample, and inspecting the generation of sum and difference frequencies. In the following section we illustrate the use of both nonlinear techniques on an environmentally damaged block of Balegem in laboratory conditions.

3.2 Experimental Set-Up and Methodology

3.2.1 Description of the Balegem Stone Sample

Balegem is a natural stone that has been frequently used for the construction and restoration of historical buildings in Belgium and Holland over the past century. The Balegem stone is mainly composed of quartz grains in a calcite cement. In polluted city air the calcite is transformed into gypsum, causing a typical weathering.

For our laboratory experiments, we used a sample with an obvious peripheral zone that has deteriorated through external conditions (Figure 24.9). In order to investigate the damage quantification potential of the proposed acoustic techniques, the following set-up has been prepared. An emitting transducer is epoxied on the surface of the Balegem stone and a similar receiver is used and moved around to acquire signals in a circle around the emitter. Several positions are indicated on the ring (14 in total): the positions 13, 14, 1, and 2 are in the damaged zone; all the other positions are located on a supposedly intact surface.

3.2.2 Linear Acoustic Measurements

To get acquainted with the sample, we first performed time-of-flight measurements for pulsed Rayleigh waves propagating along the surface of the block. Figure 24.10 illustrates two typical temporal signals received at position 8 on the intact surface (Figure 24.10a) and at the diametrically opposite position (position 1) in the deteriorated zone (Figure 24.10b). We calculated the velocity of the surface wave from the arrival time, and plotted the obtained velocities at each position versus the positions on the

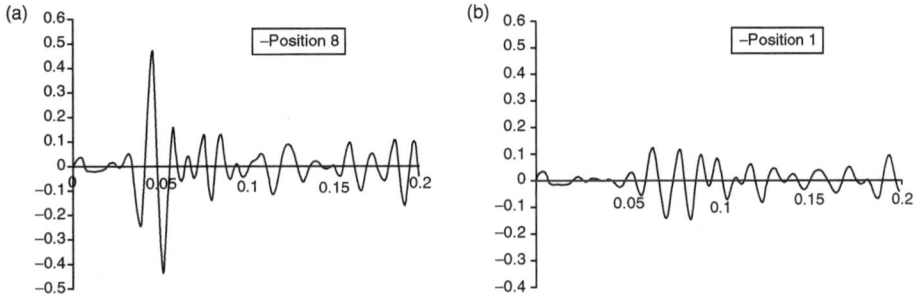

Fig. 24.10. Received temporal signals on the Balegem block: intact surface (a) and damaged surface (b).

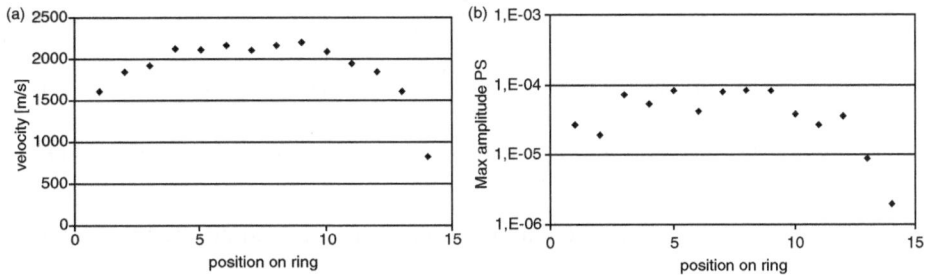

Fig. 24.11. (a) Wave velocity and (b) maximum power spectrum amplitude (arbitrary units) versus the positions on the ring.

ring (Figure 24.11a). The principal observation is that the wave velocity is considerably less in the deteriorated zone compared with the intact surface zones. We easily observe a 20 to 50% reduction in wave speed.

In addition, we also analyzed the amplitude characteristics of the signals. We calculated the Fourier transform of the entire temporal signal and identified the maximum amplitude in the power spectrum at each position. Figure 24.11b visualizes the relative amplitude of the received surface waves versus the position on the Balegem sample. As expected, we observed that the amplitude decreases when the surface is damaged (positions 1, 2, 13, and 14). A similar result has been found when integrating the power spectrum of the signals at each position (over a limited range in frequency), indicating that the "energy" is indeed decreasing in damaged zones. We observed a typical reduction of energy by a factor of 10 to 30.

3.2.3 Nonlinear Acoustic Measurements

In order to supplement the previous "linear" study and test the sensitivity of NEWS techniques, two types of nonlinear measurements are performed on the Balegem sample.

The first type of nonlinear measurements is based on the Nonlinear Wave Modulation Spectroscopy (NWMS) [7–10]. This technique consists of simultaneously exciting a sample with two autonomous waves, which frequency spectra are confined to two separate ranges, and inspecting the interference of the two waves in the

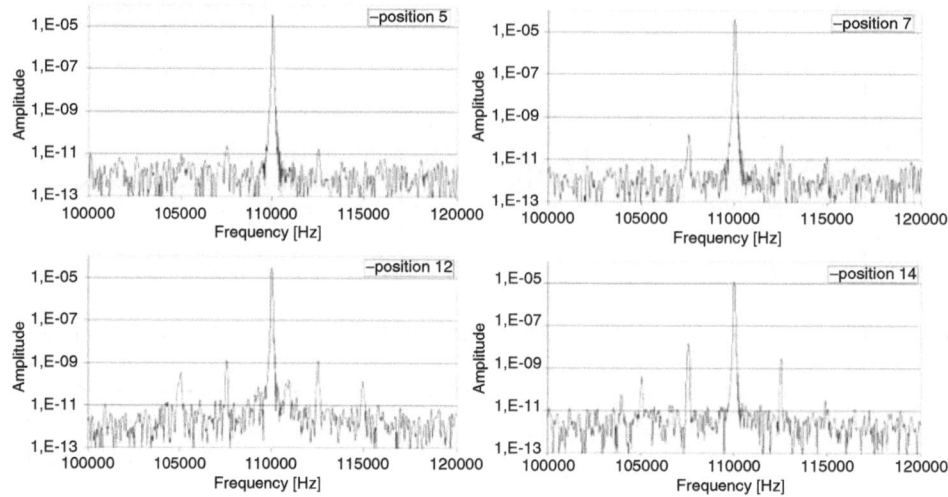

Fig. 24.12. Modulation phenomena for intact surface (5 and 7) and damaged zone (12 and 14).

material's response. We pay particular attention to the sum and difference frequencies (sidebands).

The study presented here, is performed using NWMS in "impact mode": the sample is excited by tapping it with an impact hammer, while a high-frequency continuous signal generated by a separate transducer is propagating through the sample. The interaction between the impact signal and the high frequency can be quantified and used in the assessment of damage within the sample. This method can be quickly applied and is ideally suited to applications where the question of damaged versus undamaged must be quickly addressed.

In this case, the high-frequency signal is generated by a piezoelectric transducer (E) epoxied to the Balegem sample. The emitted signal is fixed at 110 kHz with a constant voltage. The low-frequency signal, generated by the impact, is limited in time and is attenuated in the sample. The quite rich spectrum of the components corresponding to the hammer impact is limited in the frequency band to about 15 kHz. Typical intermodulation spectra are illustrated in Figure 24.12 for a frequency range centered around 110 kHz. On this figure, four subfigures are shown corresponding to four positions on the ring: two positions in the intact zone (positions 12 and 14, top figures) and two in the damaged zone (positions 5 and 7, bottom figures). Close observations of these spectra clearly reveal that the sideband amplitudes are much more important in the deteriorated zone. Apart from the increase of the first sideband, we can also note the manifestation of the second one.

For the quantification of the analysis, we have computed a simple parameter of non-linearity by calculating the energy included in the sidebands at each position (Figure 24.13). For the intact surface, the mean level of this energy is about 10^{-6}. When the wave has propagated through the environmental deteriorated zone, there is an increase in the nonlinearity parameter by a factor of 100 to 1000.

The second type of NEWS measurements we performed on the Balegem sample deals with the quantification of harmonic generation. A mono-frequency and quasi-

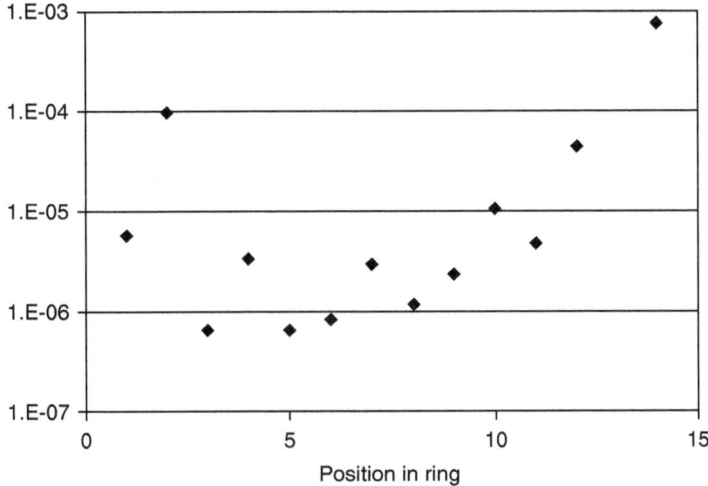

Fig. 24.13. Averaged sideband energy versus the positions on the ring.

continuous signal is generated by the transducer E, and response measurements are performed at transducer R, which is rotated on a circle around E. The response signals are analyzed at the fundamental excitation frequency as well as at its harmonic frequencies. In the current laboratory set-up, the frequency was fixed at 80 kHz and a 200 cycle burst signal was generated. The comparison of the harmonic generation [at 160 kHz (second harmonic) and 240 kHz (third harmonic)] in the spectra for an intact and a damaged zone is shown in Figure 24.14. We observe that the harmonic amplitudes are more important for the damaged zone (Figure 24.14b) than in the case of the intact surface (Figure 24.14a). The nature of the intermediate frequencies (appearing at odd multiples of 40 kHz) is not clear at the moment.

Another representation of the harmonic generation can be realized as follows: for each position, we store the amplitude of the fundamental frequency component together with those of the second and the third harmonics for different values of the amplitude of the excitation (starting from 100 mV to 1 V, increasing by 100 mV steps). By plotting the harmonics as a function of the fundamental, we get a typical view of the second and third harmonic dependence as illustrated in Figure 24.15 for position 5 on the ring. The difference between an intact surface and an environmental deteriorated zone is clearly shown in Figure 24.16 where the amplitude dependencies of the harmonics are represented for several positions on the ring. The nonlinearity is proportional to the shift of the curves. The larger the amplitude of the harmonic is for small amplitudes of the fundamental, the more the zone is nonlinear.

4. Monitoring the Curing of Concrete Using NEWS

4.1 Rationale

The durability of cement-based products and concrete structures is highly influenced by the early stages of hydration. The creation of an interfacial transition zone between

402   K.V.D. Abeele et al.

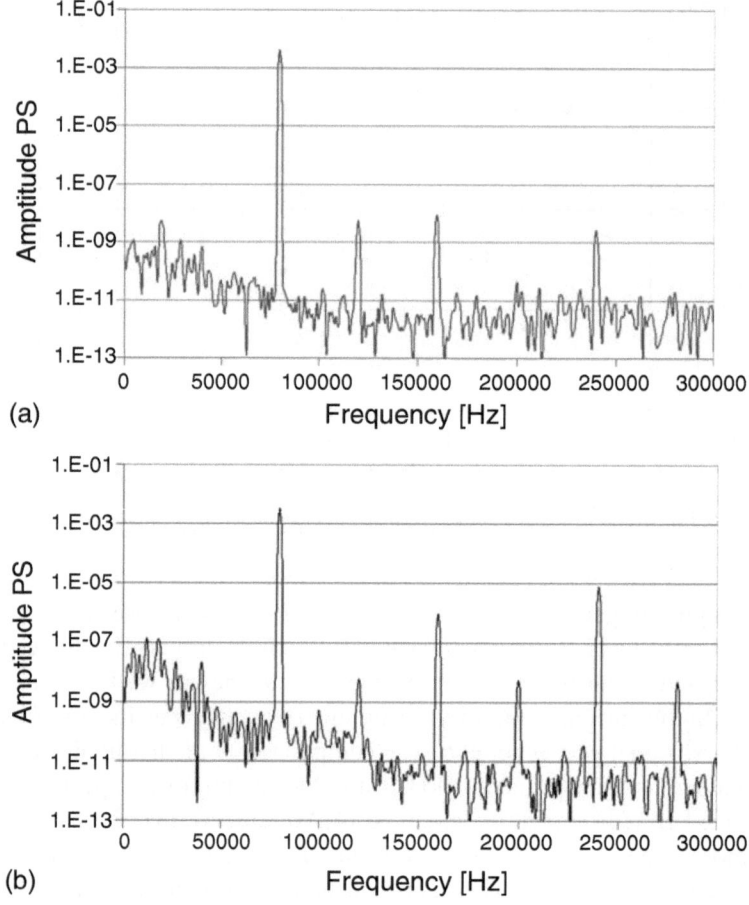

Fig. 24.14. Harmonic generation for intact surface (a) and deteriorated zone (b).

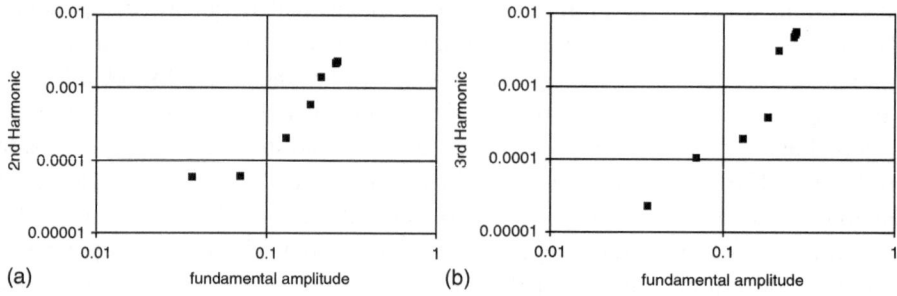

Fig. 24.15. Amplitude of the second (a) and the third harmonic (b) versus the fundamental amplitude at position 5.

the aggregates and the cement paste, with a thickness of up to 50 μ m, is considered to be the origin of primary defects in concrete leading to preferred paths for crack propagation and transport of aggressive agents threatening the durability of concrete [24]. A precise knowledge of the micromechanical properties during the successive

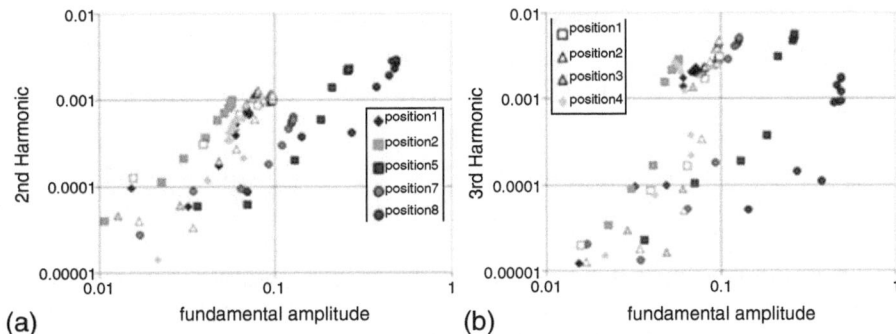

Fig. 24.16. Amplitude dependence of the second (a) and the third harmonic (b) for nine positions on the ring.

phases of the hydration process will provide information on the concrete resistance and allows assessment of its durability.

It is essential that nondestructive techniques be developed and applied in that respect. Boumiz [25,26] and Morin [27,28] used ultrasonic measurements of linear elastic coefficients in combination with volumetric shrinkage measurements to describe the evolution of the capillary network of a high-performance concrete during hardening. In general, however, lack of sensitivity and difficulties in interpretation of the linear ultrasonic measurements arise when validating the early stages of curing [29].

New developments in NDT, however, suggest that several alternative methods may be more appropriate in the evaluation of microstructural changes. In order to simultaneously probe the instantaneous microstructural activity and the micromechanical characteristics of freshly poured concrete during its curing process, we developed an integrated system of dynamic nondestructive techniques based on acoustic emission, linear ultrasonic wave propagation, and NEWS harmonic monitoring. With this study, we complement and extend the work of Lacouture et al. [30] who started the use of NEWS for monitoring the early chemical reaction phase in the curing process. Together with theoretical hydration models, this integrated approach should eventually lead to an improved prediction of the long-term behavior of concrete and its performance dependence on the curing processes.

4.2 Experimental Set-Up and Methodology

4.2.1 Materials and Sample Preparation

All material properties are measured inside a curing cell of $L \times W \times H = 200 \times 150 \times 100$ mm^3 (Figure 24.17). The cell has four circular openings containing four transducers: a transmitter and receiver for both compressional (P) and shear (S) waves. The compressional transducers (0.5 MHz central frequency) are positioned on the cross side and have a separation distance of 200 mm. The shear transducers (0.25 MHz central frequency), placed on the long side, have a mutual distance of 15 cm. A thin film closing the apertures prohibits the freshly poured concrete from leaking through. Springs in the transducer holders ensure perfect contact between the transducers and the concrete. In addition, two thermocouples, monitoring the evolution in

Fig. 24.17. Top view (left) and picture (middle) of the curing cell, containing AE sensors (AE), thermocouples (T), and compressional (P) and shear (S) transducers. The sample is placed in a climate chamber (right) and closed without active control, diminishing outside temperature and relative humidity fluctuations.

inside temperature, are fit in from the (open) top side. Lastly, two very sensitive sensors (0.375 MHz) attached on top of two protruding bars register the Acoustic Emission (AE) signals emerging from the microstructural activity in the concrete sample.

The concrete used in the experiment was composed as follows: $625 \, kg/m^3$ Sand 0/4, $1190 \, kg/m^3$ Aggregates 4/14, $450 \, kg/m^3$ CEM I 42.5R, Water $150 \, kg/m^3$, $4500 \, cc/m^3$ Plastifier Rheolmild 2000 PF, and corresponds to a concrete with a water:cement ratio of 33%.

4.2.2 Experimental Procedure

After pouring the fresh concrete in the cubic cell, the instrumented sample is placed in a climate chamber (Figure 24.17, right) to isolate it from large outside temperature and relative humidity fluctuations. The climate chamber is closed. However, climate conditions are not actively controlled in order to allow and to monitor variations of the concrete's inside temperature, which are characteristic for concrete curing. The (damped) temperature inside the climate chamber is also registered and is used to correct the thermal fluctuations in the concrete sample for day–night cycles.

The AE events are registered using a Digital Wave system, and are independent of the other measurements, except for the communication about the interruption of the registration at times when acoustic waves are transmitted through the medium. A LabVIEW script controls the active acoustic measurements, adjusting the function generator (Agilent 33250A) and the oscilloscope (LeCroy 9310AM) settings through GPIB. The acoustic responses, the temperature (three readings: inside temperatures $T_{in,1}$ and $T_{in,2}$, and outside temperature T_{out}) and relative humidity are logged on the same PC.

For each of the wave polarizations (P and S), two types of acoustic measurements are performed: a pulsed wave and a continuous wave train transmission. In the first experiment, we apply a unipolar pulse of $1 \, \mu s$ to the respective transducer. From the arrival time of the received pulses in the pulsed P and S wave transmission experiments, we determine the speed of the waves for longitudinal and shear polarization. In the continuous wave experiments, we typically use a sequence of 100 cycle bursts at 100 kHz and evaluate the spectral content (harmonics) at the receiver as a function of time.

We monitor the curing process during the first 72 hours. Sequences, consisting of registering condition parameters and acoustic responses, are repeated every 3 minutes in the first 12 hours of the experiment. Between 12 hours and 24 hours, sequences are repeated every 7 minutes. After one day, measurements are performed every 15 minutes. AE events are registered continuously, except at times when acoustic waves propagate through the medium.

4.3 Results and Analysis

4.3.1 Temperature Profile

The typical temperature change inside the concrete sample, calculated as the difference between inside and outside temperature (to correct for day–night cycle), is shown in the top figure of Figure 24.18, left. The temperature augmentation represents the internal accumulation of heat due to chemical reactions. The initiation of the first chemical reaction starts really early in the process, about 6 hours after the preparation of the concrete, and the increase lasts for about 12 hours. It is during this period that most clustering between the different particles is established, first between the smallest and later between the largest particles [23–26]. After reaching the hydration peak, the temperature decreases gradually back to room temperature during the subsequent mechanical setting phase.

4.3.2 Acoustic Emission Monitoring

The bottom figure in Figure 24.18 (left) illustrates the cumulative counts of the acoustic emission events, after automatic filtering out of bad readings. After an initial phase of activity, we observe a silent period, followed by an accelerated increase of counts that starts just prior to the main temperature change. At this point, it is believed that the largest particles in the concrete are becoming fully connected. The silence during the connection period of small particles is not easy to explain, but one of the reasons could be that the recording of high frequency pulses is hindered by the high attenuation in the slurry, and do not attain the threshold settings of the device. Figure 24.18 also illustrates that the accumulation of AE counts goes on well beyond the temperature peak. This reflects the mechanical setting (shrinkage) of the concrete during the curing process.

4.3.3 Ultrasonic Wave Velocity

Ultrasonic through transmission pulse mode experiments with separated pairs of P and S wave transducers were used to determine the speed of the waves for longitudinal and shear polarization. The arrival time of the received pulses was then calculated and transformed into longitudinal and shear wave velocities, as illustrated in Figure 24.18 (right). During the first part of the curing (earlier than six hours after the preparation) the attenuation was so high that it was impossible to transmit any sound wave through the sample. An alternative way to monitor the sound speed during this early period could be the monitoring of the reflection coefficient at the concrete–cell interface as was done by Lacouture et al. [30]. However, this method cannot be used for later

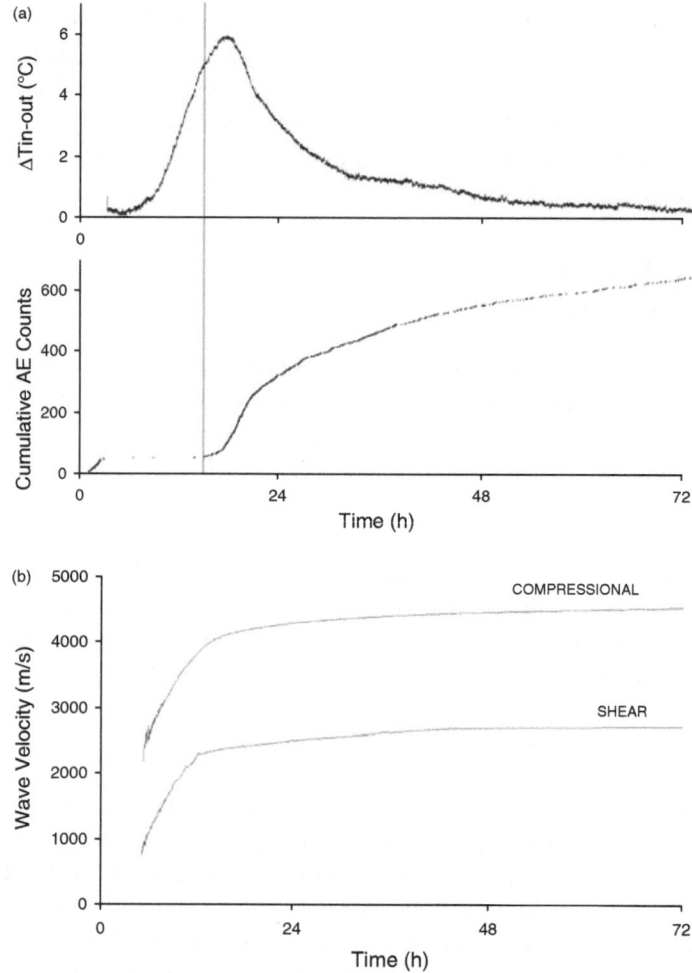

Fig. 24.18. Left: Cumulative counts of filtered AE events (bottom) in comparison with the variations in inside temperature (top); right: measured longitudinal and shear wave velocity during curing.

times because the shrinkage of the concrete causes a disbonding between the concrete and the lucite container, resulting in poor acoustic measurements. In our set-up the spring loading on the transducers assures a good permanent contact of transmitter and receiver to the concrete surface. Once the transmission of shear acoustic wave energy is possible, we observe a steep rise of the wave velocities which lasts almost up to the maximum temperature increase, and corresponds to the phase of connecting larger and larger particles. The steep rise is followed by a gradual increase to an asymptotic value. Actually, the inflection point in the longitudinal and shear velocity is situated just before the temperature peak, and indicates the transition between the grain connection phase and the pore-filling phase within the sample, controlled by diffusion of water and ions through the hydrate layers [23–26]. When transforming the wave velocities into Young's modulus and Poisson coefficient, we manifestly witness the complete transition between a fluidlike medium ($E = $ low and $v \approx 0.5$), and a real solid block of concrete ($E \approx 40\,\text{MPa}$ and $v \approx 0.2$) in the course of one day.

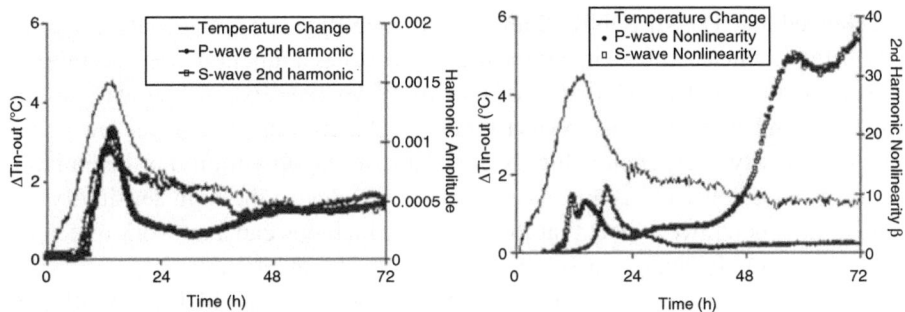

Fig. 24.19. (a) Evolution in second harmonic (compressional and shear) in comparison with the variations in inside temperature (top). (b) Evolution of the second harmonic longitudinal and shear wave nonlinearity coefficient.

4.3.4 Nonlinear Acoustic Analyses

Because of the large attenuation in the beginning of the curing process, we performed the continuous wave transmission experiments in a slightly modified curing cell with a smaller transmission distance (5 cm). Figure 24.19a shows the evolution of the second harmonic as the response to a 100 kHz compressional and shear burst signal in comparison with the variations in inside temperature. We observed a striking correspondence between the net temperature change due to the chemical and mechanical setting processes, and the harmonic content of the wave train transmission. After a fast build-up during the temperature rise (chemical phase), the harmonic generation reduces during the phase of mechanical setting, though keeping a sizeable level. The recordings of the third harmonic are not easy to interpret at this point.

In order to eliminate the effect of the fundamental amplitude on the level of the harmonics, we calculated the nonlinearity parameter β in the relation $A_2 = \beta(A_1)^2$, with A_2 the level of the second harmonic, and A_1 the fundamental amplitude. The evolution of β as function of curing time is shown in Figure 24.19b. We observed that the shear wave nonlinearity is dominant in the prepeak phase of the curing process, that is, during the formation of the connections between the particles. The level decreases together with the decreasing temperature, but rises again in the late mechanical setting. The longitudinal nonlinearity is delayed with respect to the shear nonlinearity and only exists in the pore-filling regime (early phase of the mechanical setting) during which hydration fills the capillary pores [23–26]. This analysis leads to the conjecture that the origin of the nonlinear components is connected to the micromechanical changes in the composition, both due to the initial chemical reactions (mostly shear nonlinearity) and the progressive mechanical setting of the sample (longitudinal and shear).

5. Conclusions

We have illustrated three laboratory applications of Nonlinear Elastic Wave Spectroscopy (NEWS), which could be used as precursors to real field applications in aeronautics and civil engineering.

We showed that TD-SIMONRUS and FD-SIMONRUS are both capable of discerning the effects of heat damage in composite laminates as a function of heat temperature and exposure time with a high degree of sensitivity. We have examined the reliability of the methods and compared the results to traditional measures of damage. The behavior of the nonlinearity coefficient is highly correlated to the observation of delaminations resulting from a linear ultrasonic A-scan. However, due to the high sensitivity of the nonlinear parameter, we expect that the nonlinear methods can also work to detect the more subtle property changes associated with modest temperatures.

Linear and nonlinear wave propagation techniques were applied to investigate the surface quality of intact and damaged natural building stones used in restoration projects of historical buildings. This has been realized by means of different measures of wave excitation: a pulsed signal generation for the time-of-flight and attenuation measurements, and two types of continuous wave generation, with and without a low-frequency impact, for the nonlinear acoustic measurements. In all cases, the results allow us to tell the difference between a damaged zone of a structure and an intact one. However, the sensitivity of the nonlinear methods is far superior.

Finally, NEWS techniques can also be applied to obtain complementary information about the different phases in curing processes. We have found that the origin of the nonlinearity is connected to the micromechanical changes in the composition due to the initial chemical reactions (mostly shear nonlinearity) and to the progressive mechanical setting of the sample (longitudinal and shear).

Acknowledgments

The authors gratefully acknowledge the support of the Flemish Fund for Scientific Research (G.0206.02 and G.0257.02), the provisions of the European Science Foundation Programme NATEMIS, and the European FP5 and FP6 Grants DIAS (EVK4-CT-2002-00080) and AERONEWS (AST3-CT-2003-502927).

References

[1] Buck O, Morris WL, Richardson JN (1978), Acoustic harmonic-generation at unbonded interfaces and fatigue cracks, *Appl Phys Letters* **33**(5): 371–373.

[2] Nazarov VE, Sutin AM (1997), Nonlinear elastic constants of solids with cracks, *J Acoust Soc Am* **102**: 3349–3354.

[3] Morris WL, Buck O, Inman RV (1979), Acoustic harmonic-generation due to fatigue damage in high-strength aluminum, *J Appl Phys* **50**(11): 6737–6741.

[4] Adler L, PB Nagy PB (1991), Second-order nonlinearities and their application in NDE, in DO Thompson, DE Chimenti (Eds), Review of Progress in quantitative Nondestructive Evaluation, Vol. 10B, Plenum, New York, pp. 1813–1820.

[5] Cantrell JH, Yost WT (1994), Acoustic harmonic-generation from fatigue-induced dislocation dipoles, *Phil Mag A* **69**: 315–326.

[6] Nazarov VE, Ostrovsky LA, Soustova I, Sutin AM (1988), Nonlinear acoustics of micro-inhomogeneous media, *Phys Earth and Planet Interiors* **50**(1): 65–73.

[7] Antonets VA, Donskoy DM, Sutin AM (1986), Nonlinear Vibro-diagnostics of flaws in multilayered structures, *Mech of Composites Materials* **15**: 934–937.

[8] Nagy PB (1998), Fatigue damage assessment by nonlinear ultrasonic materials characterization, *Ultrasonics* **36**(1–5): 375–381.

[9] Sutin, AM, Donskoy DM (1998), Vibro-acoustic Modulation Nondestructive Evaluation Technique, *Proceedings of SPIE* **3397**, pp. 226–237.

[10] Van Den Abeele K, Johnson PA, Sutin AM (2000), Nonlinear elastic wave spectroscopy (NEWS) techniques to discern material damage, part I: Nonlinear wave modulation spectroscopy (NWMS), *Res Nondest Eval* **12/1**: 17–30, 2000.

[11] Johnson PA, Zinszner B, Rasolofosaon PNJ (1996), Resonance and elastic nonlinear phenomena in rock, *J Geophys Res* **101**: 11553–11564.

[12] Van Den Abeele K, Carmeliet J, TenCate JA, Johnson PA (2000), Nonlinear Elastic Wave Spectroscopy (NEWS) techniques to discern material damage. Part II: Single Mode Nonlinear Resonant Acoustic Spectroscopy (SIMONARS), *Res Nondestr Eval* **12/1**: 31–42.

[13] Van Der Abeele K, Van De Velde K, Carmeliet J (2001), Inferring the degradation of pultruded composites from dynamic nonlinear resonance measurements, *Polymer Composites* **22**(4): 555–567.

[14] Van Den Abeele K, De Visscher J (2000), Damage assessment in reinforced concrete using spectral and temporal nonlinear vibration techniques, *Cement and Concrete Research* **30/9**: 1453–1464.

[15] Guyer RA, McCall KR, Van Den Abeele K (1998), Slow elastic dynamics in a resonant bar of rock, *Geophys Res Lett* **25**: 1585–1588.

[16] TenCate JA, Shankland TJ (1996), Slow dynamics in the nonlinear elastic response of Berea sandstone, *Geophys Res Lett* **23**(21): 3019–3022.

[17] TenCate JA, Smith E, Guyer RA (2000), Universal slow dynamics in granular solids, *Phys Rev Lett* **85**(5): 1020–1023.

[18] Vila M, Vander Meulen F, Dos Santos S, Haumesser L, Bou Matar O (2004), Contact phase modulation method for acoustic nonlinear parameter measurement in solid, *Ultrasonics* **42**: 1061–1065.

[19] GA Matzkannin, see http://wwwntiaccom/gamsoarhtml

[20] C Mattei, private communication.

[21] State of the art Review NTIAC-SR-98-02, see NTIAC publications at http://wwwntiaccom/

[22] For more information, visit: http://wwwkulakacbe/AERONEWS

[23] For more information, visit: http://minelabmredtucgr/dias/

[24] Ollivier JP, Masco JC, Bourdette B (1995), Interfacial Transition Zone in Concrete, *Advanced Cement Based Materials* **2**(1): 30–38

[25] Boumiz A, Vernet C, Cohen Tenoudji F (1996), Mechanical properties of cement pastes and mortars at early ages - Evolution with time and degree of hydration, *Advanced Cement Based Materials*, **3**(3-4): 94–106.

[26] Boumiz A (1995) Etude comparée des évolutions mécaniques et chimiques des pâtes de ciment et mortiers a très jeune age Développement des techniques acoustiques, Thèse de Doctorat de l'Université Paris 7, Paris.

[27] Morin V, Cohen Tenoudji F, Feylessoufi A, Richard P (2002), Evolution of the capillary network in a reactive powder concrete during hydration process, *Cement and Concrete Research* **32**(12): 1907–1914.

[28] Morin V, Cohen Tenoudji F, Vernet C, Feylessoufi A, Richard P (1998), Ultrasonic spectroscopic investigation of the structural and mechcanical evolutions of a reactive powder concrete. In: *Proc. Int. Symp. on High Performance Concrete and Reactive Powder Concrete* vol. 3, Sherbrooke University, Sherbrooke: 119–126.

[29] Sayers CM, Dahlin A (1993), Propagation of ultrasound through hydrating cement pastes at early times, *Advanced Cement Based Materials* **1**(1): 12–21.

[30] Lacouture JC, Johnson PA, Cohen Tenoudji F (2003), Study of critical behavior in concrete during curing by application of dynamic linear and nonlinear means, *J Acoust Soc Am* **113**: 1325–1332.

25

Nonlinear Elastic Wave Experiments: Learning About the Behavior of Rocks and Geomaterials

J.A. TenCate,[1] T.J. Shankland, and P.A. Johnson

[1]tencate@lanl.gov, MS D443, Earth and Environmental Sciences, Los Alamos National Laboratory, Los Alamos, NM USA 87545.

Abstract

Nonlinear acoustics was a huge topic of research in the 1960s and 1970s after the creation of the parametric array. (By mixing two high-frequency sound waves together in a nonlinear medium such as water, a very focused beam could be created.) Other applications were also suggested and research in the field exploded. In the mid-1980s a group at Los Alamos began exploring the nonlinearity of the earth with a mind to developing tools such as the parametric array for use in seismic imaging. Initial measurements showed rocks to be highly nonlinear. Yet, attempts at carefully quantifying the dynamic behavior of rocks were frustrating, as were attempts to model the physics. Rocks showed some extremely peculiar behavior, including memory effects (slow dynamics), hysteresis, and end point memory in addition to the expected Landau-type nonlinearity. This chapter traces (historically) the macroscopic experiments that led to our current understanding of the peculiar nonlinearity of not only rocks and geomaterials, but many other materials as well. Results from some recent microscopic measurements where neutron scattering is used to help ascertain the physical origin of the nonlinearity conclude the chapter.

Keywords: Geomaterials, hysteresis, memory effects, nonlinear acoustics, nonlinear elasticity, rocks, slow dynamics

1. Introduction

The study of nonlinearity in rocks was a rather natural outgrowth of the study of nonlinear waves in air and water. After the development of jet engines in the late 1940s, interest in loud (or finite amplitude rather than infinitesimal amplitude) sound boomed. With a rapid increase in submarine fleets and the advent of the Cold War, interest in nonlinear underwater acoustics also grew. Some of the first experiments to study the interaction of nonlinear waves occurred in the late 1950s. The development of the underwater parametric array in the early 1960s—which used the nonlinear mixing of two sound beams to form a narrow difference frequency beam (i.e., an acoustic spotlight)—really drove the field of nonlinear acoustics into a frenzy.

Nonlinearity in air and water can be manifested in various ways. Waves that propagate in a nonlinear fluid such as water distort and develop harmonics with distance. Even in (weakly nonlinear) air, if the wave is intense, distortion will develop and shocks will form, the sonic boom is probably the most well-known example. Wave-mixing effects in the form of intermodulation distortion and concomitant sideband generation may also be observed. Nonlinear effects such as the modulation (and suppression) of sound by sound have been known since the late 1950s.

In seismology, the "acoustic approximation" is frequently used. In fact much of the world's early seismic imaging is based on acoustic approximation. It was thus only natural to examine the nonlinearity of the earth and earth materials with nonlinear acoustics analogues. These nonlinear experiments were performed to explore and potentially develop techniques commonly used in nonlinear acoustics for seismic imaging applications. Early experiments at Los Alamos were carried out to create a parametric array in the earth (using an array of seismic sources). Although the results were inconclusive, the fact that earth materials were highly nonlinear was unmistakable. Nonlinear research in earth materials was thus scaled down to the laboratory (Johnson et al., 1987, 1989, 1991) and nearly ten years of research into the nonlinearity of earth materials resulted. Over this period, various types of rocks were studied and all were found to be highly nonlinear; notably, sedimentary rocks (which are oil and gas bearing) showed the largest and most interesting nonlinearities.

Two types of experiments in rocks and geomaterials are specifically discussed in this chapter. First, wave propagation experiments are carried out using (mostly) sedimentary solids (large blocks) and in long rods or rock cores. These experiments are analogous to acoustics experiments by D.T. Blackstock and his students at the University of Texas in Austin in the 1970s and 80s in air-filled ducts. Second are resonance experiments on long "thin" rods (or core samples). Although analogous (and mostly unremarkable) resonance experiments were attempted in air- and water-filled tubes, the much larger nonlinearity of rocks made resonance experiments and the effects of their nonlinearity much easier to study. Finally, a note about two-wave interaction experiments. Well-known two-wave nonlinear mixing experiments in water (e.g., modulation of sound by sound, scattering of sound by sound) were also performed in solids (Johnson et al., 1991) and lead somewhat naturally to techniques for nondestructive testing. Such experiments and techniques are discussed elsewhere in this book. This chapter concludes with a discussion of some very recent measurements using neutron scattering to learn about the microscopic behavior of the crystalline components of rocks and how these neutron scattering experiments relate to the peculiar nonlinear behavior that is now discussed.

2. Wave Propagation Experiments

Some of the first experiments on wave propagation in rocks (sandstone cores) were done at Los Alamos in the 1990s. Meegan et al. (1993) showed some of the very first wave propagation and harmonic measurements made in a long rock core of Berea sandstone. However, potential issues with receiver site effects due to bonding (a common problem in seismology), lack of strict environmental controls, and new modeling efforts led TenCate et al. (1996) to carefully repeat and expand Meegan's results in a more carefully controlled environment.

Both sets of experiments yielded data that strongly suggested rocks were more complex than expected. Quasistatic stress–strain loops on sandstones were known to be hysteretic and highly unusual since the 1900s (Adams and Coker, 1906) and quasistatic measurements by many others (see Guyer et al., 1995) suggested the need to include hysteresis in models developed at that time. Even so, propagation time waveforms didn't match simple expectations (Kadish et al., 1996) and the prediction of harmonic levels, even with ad hoc improvements (e.g., including hysteresis) in Landau theory (Van den Abeele et al., 1997), were not very encouraging. A new type of experiment was needed.

3. Resonance Experiments

The examination of a particular resonance mode at increasing excitation levels is a common experiment and often used to study nonlinear oscillators. Softening or hardening nonlinearity (with increasing drive amplitude) produces an easy-to-identify family of resonance curves. With a softening nonlinearity, the resonance frequency drops with increasing amplitude; with hardening, the resonance frequency rises. The Duffing oscillator (which includes an additional cubic nonlinearity in the spring constant) is perhaps one of the most well-known and frequently studied nonlinear oscillators. Noticeable peak shifts and jumps are possible and can be quite common. Experiments done in the early 1970s (Cruikshank, 1972) were performed to see if an air-filled tube showed any of the behavior typical of a nonlinear oscillator. Results were positive but at the same time disappointing. Air is simply not very nonlinear.

On the other hand, similar experiments on long (thin) core samples of various rocks produced dramatic sets of nonlinear resonance curves. Johnson et al. (1996) showed such results for a wide variety of rocks. Resonance frequencies of the samples they examined always softened with increasing drive level; in one particular Fontainebleau sandstone the frequency shift they observed was nearly 10%! Moreover, resonance curves obtained by sweeping frequencies upward while watching the sample's response differed from curves obtained sweeping downward. The results were highly reminiscent of curves one might obtain from a Duffing oscillator. As mentioned before, sedimentary rocks (especially clean sandstones) showed some of the most dramatic nonlinear peak shifts with increasing drive levels.

Efforts to describe the nonlinear resonance curves obtained on rocks with Duffing-like theoretical treatments failed, sometimes miserably, so additional experiments were performed. TenCate and Shankland (1996) discovered that the different up and down response curves obtained as a rock was swept through a resonance are repeatable, but only after the rock was "conditioned" first. Moreover, once the rock was given suitable time to rest or "recover," the whole resonance behavior was completely reproducible. In the case of one sandstone sample, the behavior of the rock was repeatable for hundreds of experiments; that is, the rock's macroscopic behavior was unchanged during these experiments. The authors dubbed this behavior "slow dynamics" (discussed in many places throughout this book). Fortuitously, the time scales of the slow dynamics in rocks were on the order of tens of minutes which made them very easy to study.

(Other materials of interest for nondestructive testing applications showed slow dynamics on much shorter time scales.) Finally, TenCate et al. (2000) discovered that the recovery back to the original state of many rocks went as the logarithm of time. One other notable recovery process can be described with a $\log(t)$ behavior, creep back to equilibrium. Slow dynamics, however, is not necessarily related to creep; slow dynamics is induced with an ac (acoustic) drive; creep is induced with a dc driving force. Rocks are peculiar solids.

During the above experiments, it was discovered that rocks (sedimentary rocks and concretes in particular) and their concomitant response were also highly susceptible to humidity and temperature; however, slow dynamics always remained an identifying feature of the rock's response unless the rock was fully saturated. Thus, great care and extreme measures were taken to be sure measurements were made in carefully controlled environments. As a consequence, an isolation chamber was built; careful measurements made in this chamber showed that there was a threshold above which slow dynamics became dominant; below that drive threshold, the rock behaved as a weakly nonlinear Duffing oscillator. It was also shown that different rocks have different thresholds. For more details, see Chapter 26 by D. Pasqualini in this book.

4. Microscopic Measurements—Neutron Scattering

Within the last few years, several neutron diffraction experiments were carried out by Darling et al. (2004a,b) on intact samples of rock. In these experiments the authors and their colleagues took simultaneous neutron diffraction data while performing quasistatic stress–strain loops, and while doing conditioning and recovery experiments. In this way, information on the atomic (crystalline lattice) scale was obtained at the same time as some of the classic nonlinear macroscopic measurements were made. In addition, neutron diffraction was recently used to determine how much of the rocks was amorphous and how much was crystalline; some fascinating hints at mechanisms for nonlinearity have been identified. Three sets of experiments are described in this final section.

Quasistatic stress–strain measurements on rocks show hysteresis loops as well as nonlinearity. As with many quasistatic stress–strain measurements there is an initial conditioning cycle followed by a repeatable banana-shaped loop. Neutron diffraction measurements (Darling et al., 2004), however, show that the crystalline lattice always behaves in a completely reversible and linear fashion. In fact, the authors estimate that only a few percent of the volume of the rock must contribute to the nonlinearity and hysteresis seen in the macroscopic measurements. They conclude that it is likely in the bond structure of the rock where all the peculiar nonlinearities occur. Placing the origin of the nonlinearity with the bond structure was not a new idea; however, these are the first compelling experiments that support that hypothesis.

Recent measurements (Page et al., 2004) show another interesting aspect of sandstones. When a pair distribution function technique is applied to the diffraction pattern obtained from a pure quartz sandstone (Fontainebleau sandstone in their particular case), it was found that there were an excess number of Si–O and O–O bonds not belonging to any long-range crystalline structure in the rock. The authors suggest there may be an amorphous phase (glass?) within the rock. The idea is appealing. Glassy dynamics is certainly reminiscent of many of the peculiar behaviors seen in rocks.

Finally, recent neutron diffraction measurements by TenCate et al. (2005) were taken while abruptly changing temperature and also while applying and removing a conditioning acoustic drive. Both sets of macroscopic measurements show abrupt changes in the state of the rock (i.e., initial drop of modulus) and then slow, log(time) recovery back to the original (or a new) equilibrium state. Neutron diffraction, on the other hand, suggests that the bulk of the crystalline material behaves as expected during the temperature changes (a dc "driving" force), with no unusual nonlinear behavior whatsoever. The acoustic (ac "driving" force) experiment has yet to be analyzed. Work on this topic is nearly complete and another publication is in preparation.

5. Summary

Rocks (especially sandstones and other sedimentary rocks) have been shown to have very peculiar nonlinearities. On the other hand, they are also easy to study, and their nonlinear properties have proven helpful for studies of a host of other materials that display rocklike behavior. Much has been learned but very careful measurements were necessary. Although it has long been suspected that most of the interesting nonlinear behavior seen in rocks lies in the way the rock is put together (the bond system), recent neutron measurements confirm what was long suspected. Applications of this work include better concretes, understanding more about the strength and durability of buildings made of stone, and numerous nondestructive testing applications.

Acknowledgments

This work is supported by Institutional Support (LDRD) at Los Alamos and by the U.S. Department of Energy Office of Basic Energy Science. Thanks to many colleagues for helpful discussions during this work. They include (in no particular order) Eric Smith, Tim Darling, Robert Guyer, Rick O'Connell, Abe Kadish, Brian Bonner, Koen Van den Abeele, Donatella Pasqualini, Marco Scalerandi, Pier Paulo Delsanto, Sven Vogel, Thomas Proffen, Salman Habib, and Alexander Sutin.

References

Adams, F.D., and E.G. Coker, "An investigation into the elastic constants of rocks, more especially with reference to cubic compressibility," *Publ. 46, Carnegie Inst. of Washington*, Washington, D.C. (1906).

Cruikshank, D.B., "Experimental investigation of finite-amplitude acoustic oscillations in a closed tube," *J. Acoust. Soc. Am.*, **52**, 1024–1036 (1972).

Darling, T.W., J.A. TenCate, D.W. Brown, B. Clausen, and S.C. Vogel, "Neutron diffraction study of the contribution of grain contacts to nonlinear stress–strain behavior," *Geophys. Res. Lett.*, **31**, L16604 (2004).

Guyer, R.A., K.R. McCall, and G.N. Boitnott, "Hysteresis, discrete memory, and nonlinear wave propagation in rock," *Phys. Rev. Lett.*, **74**, 3491–3494 (1995).

Johnson, P.A. and T.J. Shankland, "Nonlinear generation of elastic waves in crystalline rock and sandstone: continuous wave travel time observations," *J. Geophys. Res.*, **94**, 17729–17734 (1989).

Johnson, P.A., A. Migliori, and T.J. Shankland, "Continuous wave phase detection for probing nonlinear elastic wave interactions in rocks," *J. Acoust. Soc. Am.*, **89**, 598–603 (1991).

Johnson, P.A., T.J. Shankland, R.J. O'Connell, and J.N. Albright, "Nonlinear generation of elastic waves in crystalline rock,"*J. Geophys. Res.*, **92**, 3597–3602 (1987).

Johnson, P.A., B. Zinszner, P.N.J. Rasolofosaon, "Resonance and nonlinear elastic phenomena in rock," *J. Geophys. Res.*, **101**, 11553–11564 (1996).

Kadish, A., J.A. TenCate, and P.A. Johnson, "Frequency spectra of nonlinear elastic pulse-mode waves," *J. Acoust. Soc. Am.*, **100**, 1375–1382 (1996).

Meegan, G.D., P.A. Johnson, K.R. McCall, and R. Guyer, "Observation of nonlinear elastic wave behavior in standstone," *J. Acoust. Soc. Am.*, **94**, 3387–3391 (1993).

Page, K.L., Th. Proffen, S.E. McLain, T.W. Darling, and J.A. TenCate, "Local atomic structure of Fontainebleau sandstone: Evidence for an amorphous phase?" *Geophys. Res. Lett.*, **31**, L24606 (2004).

TenCate, J.A., and T.J. Shankland, "Slow dynamics in the nonlinear elastic response of Berea sandstone," *Geophys. Res. Lett.*, **23**, 3019–3022 (1996).

TenCate, J.A., T.W. Darling, S.C. Vogel, submitted to *Geophys. Res. Lett.*, (2006).

TenCate, J.A., E. Smith, and R.A. Guyer, "Universal slow dynamics in granular solids," *Phys. Rev. Lett.*, **85**, 1020–1023 (2000).

TenCate, J.A., K.E.-A. Van den Abeele, T.J. Shankland, and P.A. Johnson, "Laboratory study of linear and nonlinear elastic pulse propagation in sandstone," *J. Acoust. Soc. Am.*, **100**, 1383–1391 (1996).

Van den Abeele, K.E.-A., P.A. Johnson, R.A. Guyer, and K.R. McCall, "On the quasianalytic treatment of hysteretic nonlinear response in elastic wave propagation," *J. Acoust. Soc. Am.*, **101**, 1885–1898 (1997).

Intrinsic Nonlinearity in Geomaterials: Elastic Properties of Rocks at Low Strain

Donatella Pasqualini

EES-9, University of California, Los Alamos National Laboratory, Los Alamos, New Mexico 87545.
Correspondence: Donatella Pasqualini, Mail Stop D443, email: dondy@lanl.gov.

Abstract

The elastic properties of geomaterials are anomalous. Hysteresis with end point memory, slow dynamics, and linear variation of the resonance frequency with the strain are only some of these uncommon features. All these characteristics have been related to a nonclassical nonlinear elasticity. Chapter 1 introduces two strain regions where the experiments show different elastic behaviors. At low strain, rocks show their intrinsic nonlinearity until a strain material-dependent threshold, ϵ_{th}. A transition from linear to classical nonlinear behavior appears in this first region. For strains beyond ϵ_{th} the experimental data are contaminated by a complex nonequilibrium dynamics. Memory effects and conditioning complicate the characterization of the intrinsic nonlinearity of the sample and they do not allow a simple interpretation of the experimental data to prove the existence of anomalous nonclassical nonlinearity.

Keywords: Classical nonlinearity, conditioning, dynamical experiments, geomaterials, intrinsic nonlinearity, memory effects, nonequilibrium dynamics, nonlinearity

1. Introduction

Rocks are complex systems. Their extraordinary elastic properties are the manifestation of this complexity. The presence of hysteresis with end point memory (Cook and Hodgson, 1965, Gordon and Davis, 1967), long–time recovery relaxation phenomena (TenCate and Shankland, 1996, TenCate et al., 2000), and anomalous softening of the resonance frequency with strain (Johnson et al., 1996) are only some examples of the uncommon elastic properties of geomaterials. This experimental evidence led us to define rocks as anomalous nonlinear elastic materials. Several basic questions are still open about this nonlinearity.

In this chapter some recent experimental results are presented which help to better understand the nonlinearity of rocks. The focus of this chapter is to show the existence of a strain threshold, below which these materials show a classical nonlinear behavior, and beyond which complex memory and conditioning effects appear. These two regions are fundamental for understanding the nonlinear nature of rocks and to define the intrinsic nonlinear behavior of these materials. The key point is that beyond this threshold the experimental data cannot be simply used to prove the existence of nonclassical behavior due to the presence of nonequilibrium dynamics.

2. Intrinsic Nonlinearity and Conditioning

Most of the measurements of nonlinear effects are performed by resonance bar experiments. A detailed description of this experiment is presented by TenCate et al. in this book. In this type of experiment the resonance frequency (f_R) dependence on the external force is analyzed as an indicator of nonlinearity: the system is linear if the resonance frequency peak does not change with the external force, whereas a change in f_R indicates nonlinearity. There are many materials in nature that exhibit elastic nonlinearity. For these materials the nonlinearity is shown as a quadratic softening of the f_R increasing the drive amplitude. This nonlinearity, known as *classical nonlinearity*, has been described by Landau and Lifshitz (1998) using a Taylor expansion of the bar's displacement in the strain. Instead of a quadratic softening, rocks show a linear softening with the drive amplitude that was interpreted as an indicator of an anomalous nonlinearity (Guyer and Johnson, 1999) as shown by Johnson in this book.

In recent works (TenCate et al., 2004; Pasqualini et al., in press) it was shown that this shift has to be interpreted carefully because the measurements can be contaminated by the presence of a nonequilibrium dynamics. The external force can bring the rock into a nonequilibrium state complicating the dynamical behavior and the interpretation of the nonclassical behavior. Figure 26.1 shows the effect of the conditioning on the resonance frequency peak. The peak of the resonance curve is measured at different

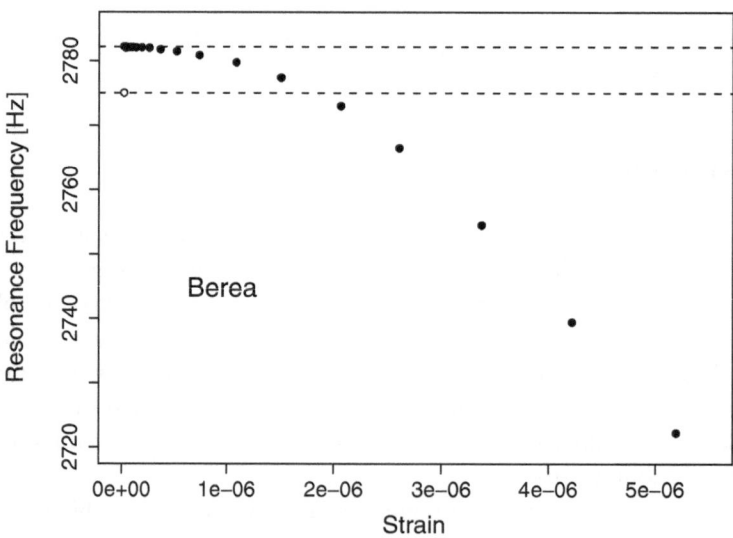

Fig. 26.1. Conditioning effect. The resonance frequency is plotted as a function of strain for Berea. The peak of the resonance curve is measured at different strains from low strain 10^{-9} up to 10^{-6}, the drive is then dropped of to the lowest strain. This last value (open dots) for f_R is different from the first one (full dot) as a consequence of conditioning.

strains from low strain 10^{-9} up to high strain 10^{-6}, and the external force is then reduced back to the lowest strain. The last value (open dots in Figure 26.1) for f_R is different from the first one (full dot): the reason for this difference is the conditioning.

The experimental data cannot be analyzed easily due to the presence of this nonequilibrium dynamics. In order to prevent this contamination, an ad hoc experimental strategy, named "zig–zag," was developed (see TenCate et al. in this book). This method consists of the systematical increase of the drive level passing through the up/down frequency sweeps and then releasing it back to the lowest strain to check if the f_R has changed.

3. Experimental Evidence

3.1 Two Regions

The samples analyzed are two sandstones, Berea and Fontainebleau. During the experiment the bar is driven by a frequency f and the acceleration of the bar end is measured. In order to compare samples with different length L the acceleration $\ddot{u}$ is converted into strain ϵ, using the convention $\epsilon = \ddot{u}/(4\pi L f^2)$. Different resonance curves are built at constant drive amplitude sweeping the frequency up/down (see Figure 26.2). In this experiment the frequency stability of the samples is $\simeq 0.1\,\mathrm{Hz}$ corresponding to a thermal stability of $10\,\mathrm{mK}$.

For each resonance curve the peak and the resonance frequency f_R are determined using a statistical analysis that was developed by D. Higdon of the Los Alamos National Laboratory. This analysis is based on a nonparametric Gaussian process to model the strain as a function of frequency (Banerjee et al., 2004). Using a Markov Chain Monte Carlo (MCMC) method, a Bayesian estimation for the peak and f_R are calculated together with their uncertainties.

The application of the zig–zag method reveals the presence of two strain regions. These two regions are divided by a strain threshold ϵ_{th}, below which there is no evidence of conditioning (first region) and beyond which the measures are contaminated by nonequilibrium dynamics (second region). The threshold ϵ_{th} is a function of the material and environmental quantities such as temperature, saturation, and so on. The value of the threshold for Fontainebleau is $\epsilon_{th} = 2 \cdot 10^{-7}$ and $\epsilon_{th} = 5 \cdot 10^{-7}$ for

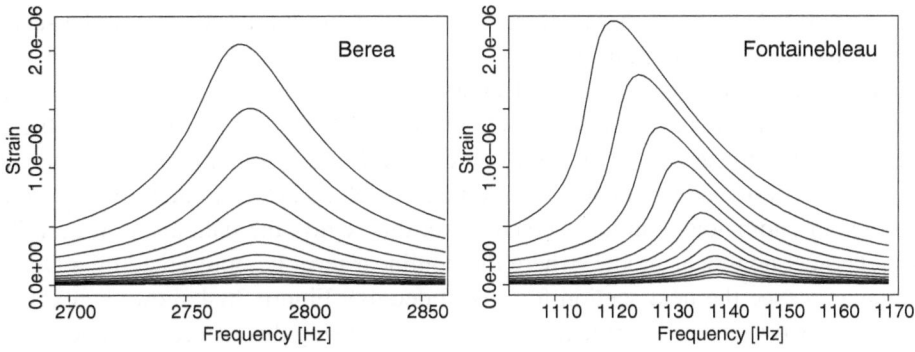

Fig. 26.2. Experimental resonance frequency curves for Berea and Fontainebleau.

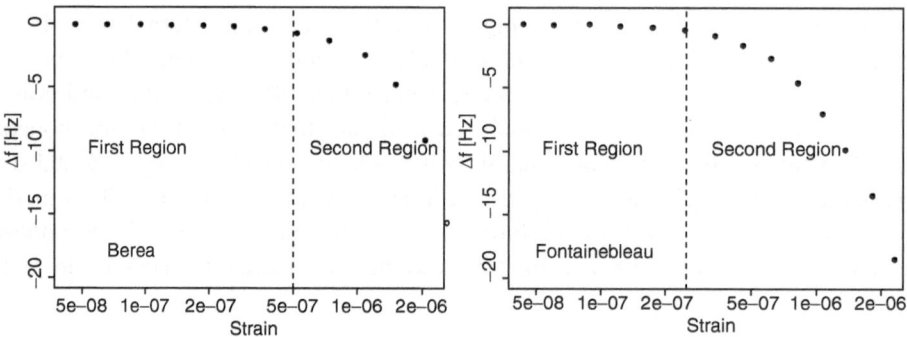

Fig. 26.3. Resonance frequency shift for both Berea and Fontainebleau. Dashed lines show the threshold.

Berea. In Figure 26.3, the shift of the resonance frequency for both samples is plotted as a function of strain. In the same figure the two regimes, nonconditioning and conditioning, are outlined.

In the first region, strain less than ϵ_{th}, there is no evidence of conditioning: Rocks show a reversible nonlinearity. The resonance frequency shift is repeatable: one can change how the experiment is carried out and the results do not change. In the absence of conditioning, the data show the intrinsic nonlinearity and consequently they can be simply analyzed and interpreted.

On the other hand, in the second region, strain bigger than ϵ_{th}, conditioning, and nonequilibrium dynamics are present. As a consequence, the experimental results are history dependent and not repeatable. Without a good understanding of the relationship between nonlinear dynamics and intrinsic nonlinearity, the data beyond that threshold cannot be simply interpreted to define intrinsic nonlinearity. Analysis of the data beyond ϵ_{th} without considering the nonequilibium contamination can only lead to erroneous conclusions.

3.2 First Region: Intrinsic Nonlinearity

Figure 26.4 shows the shift of the resonance frequency versus strain for the first region where the conditioning is not present. The strain range of this regime is $2 \cdot 10^{-9}$ to $2 \cdot 10^{-7}$ for Fontainebleau and $2 \cdot 10^{-9}$ to $5 \cdot 10^{-7}$ for Berea. The resonance frequency f_R and the respective strain are calculated using the MCMC method, which also computes the error bars. Note that the error bars for the strain are too small to be seen in Figure 26.4 for the strain range used.

The data analysis shows that f_R decreases quadratically as we increase the drive amplitude to the threshold ϵ_{th}. At very low strain, 10^{-8} to 10^{-7}, both rocks behave effectively as a linear elastic system: any change in the f_R can be seen in the error bars. There is no evidence of linear softening, which could lead one to believe in the presence of anomalous nonlinear behavior. In this region the data are well described by a classical nonlinear model, where the nonlinear term is represented by a Duffing nonlinearity (see the next section for model details). Figure 26.4 shows an excellent agreement between the experimental data and the fit using this theoretical model (solid lines). The quality factor Q is calculated as the ratio of the resonance frequency and

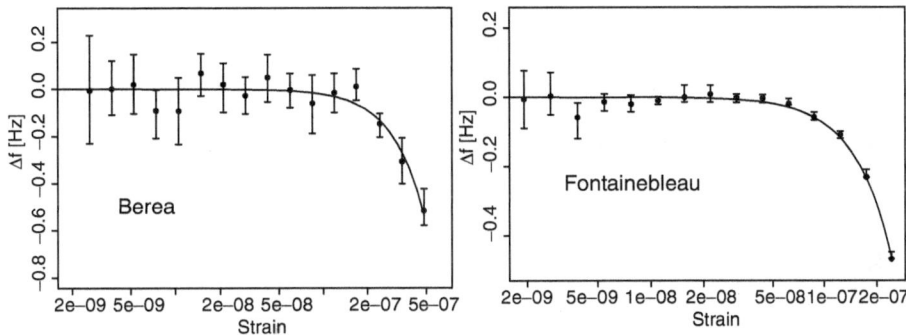

Fig. 26.4. First region: resonance frequency shift for Berea and Fontainebleau limited at the strain range where conditioning and memory effects do not occur. The solid lines are the theoretical fits: a classical Duffing nonlinear oscillator.

the width of the resonance curve at $1/\sqrt{2}$ of the maximum, Γ. It is important to point out that the uncertainties of the quality factor are larger than the ones of the resonance frequency peak. Analyzing the resonance curves, it is easy to see that Γ is constant within one percent. Therefore, the quality factor behaves as the resonance frequency, quadratically with the strain.

4. Model

As introduced in the previous section, in the low-strain regime the experimental data are accurately described by a simple phenomenological dynamical model. This model consists of a classical damped harmonic potential to which a quartic classical nonlinear term (Duffing) is added. The equation of motion for the displacement u can be written as follows.

$$\ddot{u} + \Omega_0^2 u + 2\mu\dot{u} + \gamma u^3 = F\sin(\omega t), \tag{26.1}$$

where Ω_0 is the linear resonance frequency, μ is the damping coefficient, and $\omega = 2\pi f$ is the angular frequency. $\gamma < 0$, the nonlinear parameter, leads to a softening nonlinearity as the experiments show. The amplitude of the driving force F is proportional to the amplitude of the voltage applied to the bar in the experiment. The derivation of an analytical approximation for the solution of Eq. (26.1) is given in detail in Nayfeh (1981) and leads to the following relation between the displacement amplitude, a, and the drive amplitude F.

$$\Omega_0^2 \mu^2 a^2 + a^2\left[(\omega - \Omega_0)\Omega_0 - \frac{3}{8}a^2\gamma\right]^2 = \frac{1}{4}F^2. \tag{26.2}$$

The peak of the resonance curve a_R, and the drive frequency $\omega_R = 2\pi f_R$, at which the peak occurs, can be easily calculated from Eq. (26.2):

$$\begin{cases} a_r = \dfrac{F}{2\mu\Omega_0} \\ \omega_R = \dfrac{3F^2\gamma}{32\mu^2\omega_0^3} + \Omega_0. \end{cases} \tag{26.3}$$

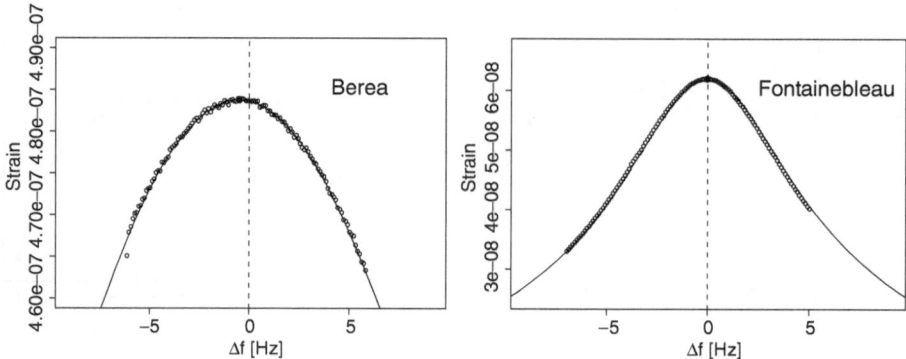

Fig. 26.5. First region: resonance frequency curve for Berea and Fontainebleau limited to the strain range where conditioning and memory effects do not occur. Solid lines are the theoretical model, open circles represent the experimental data, and full dots are the resonance frequency peaks calculated using MCMC.

The previous equation can be written in terms of the effective strain ϵ and the resonance frequency f_R as

$$f_R = \frac{3L^2\gamma}{16\pi^3\Omega_0}\epsilon^2 + \frac{\Omega_0}{2\pi}. \qquad (26.4)$$

Thus, in agreement with the measured data, the model predicts that the resonance frequency softens quadratically with the amplitude F, which is proportional to the strain amplitude. In Figure 26.4 the experimental data are fitted using the second equation in (26.3). The fitting parameters, Ω_0 and γ, are determined for both samples: for the Fontainebleau sample $\Omega_0 = 7262.8\,\text{rad/s}$, $\gamma = -7.6 \times 10^{19}\,\text{m}^{-2}\,\text{s}^{-2}$, and for the Berea sample, $\Omega_0 = 17375.7\,\text{rad/s}$, $\gamma = -5.3 \times 10^{19}\,\text{m}^{-2}\,\text{s}^{-2}$. Once the parameters Ω_0 and γ are fixed as just described, the resonance curves are reconstructed and compared with the experimental resonance curves in Figure 26.5. Both Figures 26.4 and 26.5 show an excellent agreement between theoretical prediction and measured data. The model also predicts that the width of the resonance curve is independent of the drive amplitude. Its theoretical value $\Gamma = 2\mu$ is in agreement with the experimental evidence.

5. Conclusions

The main idea presented in this chapter is that we need to be careful interpreting the experimental data to provide a basis for the existence of nonclassical behavior. Two strain regions have been delineated by a strain threshold, ϵ_{th}, which is material and environment dependent. The one at low strain is free of conditioning and memory effects and as a consequence the experiment is repeatable. In the second region the external force drives the sample into a new nonequilibrium state. Then the resonance frequency shift is not reversible anymore. Meanwhile in the first region the rock shows its intrinsic nonlinearity and the measured data can be simply interpreted; in the second region the data contaminated by the nonequilibrium dynamics do not have a simple interpretation.

Limiting the analysis to the first region, it has been found that rocks do not show a nonclassical nonlinearity as it was claimed in previous works (Smith and TenCate, 2000, Guyer et al., 1999). A classical Duffing nonlinearity is enough to capture the dynamical behavior in this region. A detailed explanation about the disagreement with previous works can be found in Pasqualini (in press). Anomalous features can be seen only in the region where the nonequilibrium dynamics contaminates the intrinsic nonlinearity and one cannot simply interpret this behavior as a sign of nonclassical nonlinearity in rocks.

As a consequence of this experimental evidence it is clear that we need a new theory that combines the intrinsic nonlinearity of the materials and nonequilibrium dynamics. We cannot speak about nonclassical nonlinearity in rocks until the fundamental relationship between intrinsic nonlinearity and nonequilibrium dynamics is understood.

Acknowledgments

The results presented in this chapter have been obtained in collaboration with several colleagues at Los Alamos National Laboratory. The theoretical part was done in collaboration with Salman Habib (T8) and Katrin Heitmann (ISR1), whereas the experimental part was done in collaboration with James A. TenCate (EES11). David Higdon (D1) developed the MCMC code.

This work was funded in part by the Institutional Support of Los Alamos through the Office of Basic Energy Science and Institute of Geophysics and Planetary Physics of Los Alamos National Laboratory (IGPP). I would like to thank Paolo Patelli for several discussions about the data analysis.

References

Banerjee, S., Carlin, B.P., and Gelfand, A.E., 2004, *Hierarchical Modeling and Analysis for Spatial Data*, Chapman and Hall/CRC Press, Boca Raton, FL.

Cook, N.G.W., Hodgson, K., 1965, Some detailed stressstrain curves for rock, *J. Geophys. Res.* **70**(12):2883–2888.

Gordon, R.B. and Davis, L.A., 1967, *J. Geophys. Res.* **73**:3917.

Guyer, R.A. and Johnson, P.A., 1999, Nonlinear mesoscopic elasticity, evidence for a new class of materials, *Phys. Today*, **52**:30–35.

Guyer, R.A., TenCate, J., and Johnson, P.A., 1999, Hysteresis and the dynamic elasticity of consolidated granular materials, *Phys. Rev. Lett.* **82**:3280–3283.

Johnson, P.A., this book.

Johnson, P.A., Zinszner, B., and Rasolofosaon, P.N.J., 1996, Resonance and elastic nonlinear phenomena in rock, *J. Geophys. Res.* **101**:11553–11564.

Landau, L.D., and Lifshitz, E.M., 1998, *Theory of Elasticity*, Butterworth-Heinemann, Boston.

Nayfeh, A.H., 1981, *Introduction to Perturbation Techniques*, Wiley, New York; Schmidt, G., and Tondl, A., 1986, *Non-Linear Vibrations*, Cambridge University Press, New York.

Pasqualini, D., Heitmann, K., TenCate, J.A., Habib S., Higdon, D., and Johnson, J.A., in press. *J. Geophys. Res.*

Smith, D.E. and TenCate, J., 2000, Sensitive determination of the nonlinear properties of Berea sandstone at low strains, *Geophys. Res. Lett.* **27**:1985–1988.

TenCate, J.A., et al. this book.

TenCate, J.A., Pasqualini, D., Habib, S., Heitmann, K., Higdon, D., and Johnson, P.A., 2004, Nonlinear and Nonequilibrium Dynamics in Geomaterials, *Phys. Rev. Lett.* **93**:065501-1–065501-4.

TenCate, J.A., Shankland, T.J., 1996, Slow dynamics in the nonlinear elastic response of Berea sandstone, *Geophys. Res. Lett.* **23**:3019–3022.

TenCate, J.A., Smith, E., Guyer, R. A., 2000, Universal slow dynamics in granular solids, *Phys. Rev. Lett.* **85**(5):1020–1023.

27

Nonlinear Acoustic Techniques for NDE of Materials with Variable Properties

Arvi Ravasoo[1] and Andres Braunbrück

Centre for Nonlinear Studies, Institute of Cybernetics at Tallinn University of Technology, Akadeemia tee 21, 12618 Tallinn, Estonia.
[1]Correspondence: Arvi Ravasoo, Institute of Cybernetics at Tallinn University of Technology, Akadeemia tee 21, 12618 Tallinn, Estonia; e-mail: arvi@ioc.ee.

Abstract

Acoustic techniques based on the interaction of two ultrasonic waves in materials with variable properties are proposed. Two types of materials are considered: (i) an inhomogeneously prestressed nonlinear elastic material undergoing two-parametric plane strain and (ii) a nonlinear elastic material with weakly inhomogeneous physical properties. The theoretical basis of the techniques is presented in detail. Analytical solutions to one-dimensional problems of nonlinear propagation and interaction of waves in the material (structural element) with two parallel traction-free boundaries are derived. The nonlinear part of boundary oscillations includes essential information about the predeformed state and the physical properties of the material. This enables us to propose (i) two techniques for qualitative and quantitative nondestructive evaluation (NDE) of the inhomogeneous predeformed state of the material and (ii) a resonance technique for NDE of weakly inhomogeneous physical properties of the nonlinear elastic material. Techniques are illustrated by numerical experiments.

Keywords: Acoustic NDE technique, inhomogeneity, nonlinearity, prestress, resonance, solid, wave interaction

1. Introduction

Nonlinear effects of wave propagation in materials with mechanical inhomogeneities (microscale damages, microstructure, etc.) are several orders of magnitude higher than in isotropic homogeneous materials. This fact encourages authors to elaborate techniques for NDE of the properties of inhomogeneous materials on the basis of nonlinear effects emerging in course of wave propagation and interaction.

The main aim of this work is to elaborate relatively simple nonlinear acoustic techniques for ultrasonic NDE of materials with variable properties, such as materials undergoing inhomogeneous prestress and materials with space-dependent physical properties. The amount of information necessary to be determined by NDE of inhomogeneous materials increases substantially in comparison with experiments with isotropic homogeneous materials. To achieve this goal, there are two principal possibilities for NDE: (i) to use more complicated techniques or (ii) to stay on the level of relatively simple tests. In this research the second approach is followed. The inverse problems of recovering the inhomogeneity of material properties are solved on the basis of the solutions to the direct problems of ultrasonic wave propagation. This approach makes it easier to use nonlinear effects that accompany multiwave propagation, reflection, nonlinear interaction, and so on, as the sources of information and to obtain relatively simple algorithms for NDE techniques.

Two types of inhomogeneous materials are investigated: (i) inhomogeneously pre-stressed nonlinear elastic materials and (ii) physically inhomogeneous elastic materials with weakly space-dependent density, and linear and nonlinear elastic properties. The prestressed state of the material is caused by the action of external forces and cor-responds to the plane strain. The problem (i) is considered as quasi one-dimensional and the problem (ii) as one-dimensional. In the quasi one-dimensional case, prior as-sumptions are made so that the wave process in two-dimensional prestressed materials is reduced to a one-dimensional problem. Equations of motion for both cases are de-rived from the theory of elasticity with quadratic nonlinearity. The variable material properties are described by polynomials.

The analytical solutions describing propagation, reflection, and interaction of two counterpropagating longitudinal waves with smooth arbitrary initial profiles in the in-homogeneous nonlinear elastic materials are derived. These solutions enable us to de-termine the dependencies of the wave characteristics on the parameters of the material inhomogeneity.

Harmonic wave propagation in inhomogeneous nonlinear elastic materials with two parallel boundaries is studied in detail. The wave process is excited in terms of par-ticle velocity and is recorded in terms of stress. Analytical expressions that describe the wave process are derived. Numerical simulation on the basis of the analytical solu-tions enables us to determine the distortion of wave profiles according to the material parameters and state.

The results of the analysis permit us to propose algorithms of qualitative and quan-titative techniques for NDE of (i) two-parametric prestressed state of the material and (ii) weakly inhomogeneous properties of the nonlinear elastic material.

2. Inhomogeneously Prestressed Material

2.1 Problem Formulation

The discovery of the piezoelectric effect by brothers Pierre and Jacques Curie in 1880 made it possible to use ultrasonic wave propagation data for NDE of material prop-erties. The first attempt in this sense was made in 1913.[1] After that an intensive the-oretical and experimental research was carried on, resulting in effective and versatile methods for evaluating different mechanical properties and states of materials.[2]

The detailed development of the mathematical basis and the experimental set-up of the ultrasonic NDE of materials with homogeneous prestress began much later. For example, Bergman and Shanbender in 1958[3] and Benson and Raelson in 1959[4] described the acoustoelastic effect, that is, the dependence of wave velocity on the value of initial stress in materials.

Inhomogeneity of prestress introduces additional difficulties in NDE, as the amount of information necessary to be determined increases dramatically. This problem is under intensive investigation nowadays.[5-7] In this work the efficiency of NDE is increased by taking advantage not only of the acoustoelastic effect, but also of the effects accompanying the nonlinear propagation and interaction of ultrasonic waves.[5,6]

The NDE problem for the inhomogeneously predeformed material is solved on the basis of two longitudinal waves that counterpropagate simultaneously in the material. An isotropic and homogeneous nonlinear elastic material is considered. Deformations of the material are described in the Lagrangian rectangular coordinates X_K, $K = 1, 2, 3$ on the basis of the nonlinear theory of elasticity. The physical and the geometrical nonlinearities are taken into account.[8]

Three different states of the isotropic homogeneous elastic material are distinguished. At the beginning, the material is in the initial, undeformed natural state. Then, the material is subject to the external forces and from now on is in the static (independent of time) prestressed state. At some instant two longitudinal waves are simultaneously excited at two boundaries, bringing the material in the final state.

In the final state, the components of the displacement vector $U_K^*(X_J, t)$ are expressed by the sum

$$U_K^*(X_J, t) = U_K^0(X_J) + U_K(X_J, t), \tag{27.1}$$

where $U_K^0(X_J)$ and $U_K(X_J, t)$ denote displacements evoked by prestress and wave motion, respectively and t denotes time.

Only two-dimensional deformations are considered, and therefore the components of the deformation vectors $U_3^0(X_J, t)$ and $U_3(X_J, t)$ are assumed to be zero.

In the case of plane strain, the equation of motion in terms of the displacement vector now takes the form of the following system of two equations[9]

$$[1 + k_1 U_{I,I}^* + k_2 U_{J,J}^*] U_{I,II}^* + [2 k_3 U_{I,J}^* + 2 k_4 U_{J,I}^*] U_{I,IJ}^*$$

$$+ [k_7 + k_3 U_{I,I}^* + k_3 U_{J,J}^*] U_{I,JJ}^* + [k_4 U_{I,J}^* + k_3 U_{J,I}^*] U_{J,II}^*$$

$$+ [k_3 U_{I,J}^* + k_4 U_{J,I}^*] U_{J,JJ}^* + [k_6 + k_5 U_{I,I}^* + k_5 U_{J,J}^*] U_{J,JI}^*$$

$$- c^{-2} U_{I,tt}^* = 0, \tag{27.2}$$

where the indices I, J, and t after a comma indicate differentiation with respect to X_I, X_J and time t. The indices I and J in Eq. (27.2) assume the values $I = 1$, $J = 2$ for the first equation and $I = 2$, $J = 1$ for the second. The coefficients

$$k_1 = 3 + 6 k (v_1 + v_2 + v_3), \quad k_2 = k (\lambda + 6 v_1 + 2 v_2),$$

$$k_3 = 1 + k (v_2 + 3 v_3/2), \quad k_4 = k (\mu + v_2 + 3 v_3/2),$$

$$k_5 = k [\lambda + \mu + 3 (2 v_1 + v_2 + v_3/2)],$$

$$k_6 = k (\lambda + \mu), \quad k_7 = k \mu, k = (\lambda + 2 \mu)^{-1}, c^{-2} = \rho_0 k \tag{27.3}$$

are functions of the Lamè constants λ and μ, the third-order elastic constants v_1, v_2 and v_3 and the material density ρ_0.

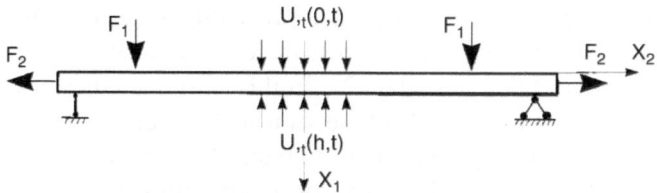

Fig. 27.1. Loading scheme. F_1, F_2 - external forces, U_t - boundary excitation.

The quasi one-dimensional problem of counterpropagation of two one-dimensional longitudinal waves in the material undergoing two-dimensional prestressed state is considered. The wave process is excited simultaneously on the surfaces $X_1 = 0$ and $X_1 = h$ of the material (Figure 27.1). The prestressed state and the ratio of the width of the excitation zone to the thickness of the material are assumed to be such that the spatial derivatives of the displacements due to the propagating waves are much larger in the direction of propagation X_1 than in the orthogonal direction X_2. Taking into account this assumption and the equilibrium of the material in the static prestressed state, the equation that describes longitudinal wave propagation in the material undergoing inhomogeneous plane strain yields:

$$[1 + k_1 U^0_{1,1} + k_2 U^0_{2,2}] U_{1,11} + [k_1 U^0_{1,11} + k_3 U^0_{1,22}$$

$$+ k_5 U^0_{2,12}] U_{1,1} + k_1 U_{1,11} U_{1,1} - c^{-2} U_{1,tt} = 0 \qquad (27.4)$$

The nonlinear second-order hyperbolic differential equation (27.4) with space-dependent coefficients can be solved provided that additional information about the prestressed state is available, that is, the coefficients in Eq. (27.4) are known. These coefficients are functions of the displacement U^0_K, that is, the solution of the equation of equilibrium of the material in the prestressed state

$$[1 + k_1 U^0_{I,I} + k_2 U^0_{J,J}] U^0_{I,II} + [2 k_3 U^0_{I,J} + 2 k_4 U^0_{J,I}] U^0_{I,IJ}$$

$$+ [k_7 + k_3 U^0_{I,I} + k_3 U^0_{J,J}] U^0_{I,JJ} + [k_4 U^0_{I,J} + k_3 U^0_{J,I}] U^0_{J,II} \qquad (27.5)$$

$$+ [k_3 U^0_{I,J} + k_4 U^0_{J,I}] U^0_{J,JJ} + [k_6 + k_5 U^0_{I,I} + k_5 U^0_{J,J}] U^0_{J,JI} = 0.$$

2.2 Counterpropagation of Longitudinal Waves

Our aim is to investigate how to use the nonlinear effects of wave propagation and interaction in NDE of the inhomogeneous prestressed state of a nonlinear elastic material. To achieve this goal, the quasi one-dimensional problem of counterpropagation of two longitudinal waves in the material is studied. This wave propagation process is described by Eq. (27.4). The two-dimensional prestressed state of the material that corresponds to plane strain is described by Eq. (27.5). The problem is solved under the assumption that the total deformations caused by prestress and wave motion are small but finite and remain elastic. The initial stage of distortion of the wave profile is considered and it is assumed that shock waves are not generated. These assumptions lead to the conclusion that in this problem the strain is small and the small parameter ε can be involved. Consequently, the problem may be solved by means of the perturbation technique. Following this procedure, the solution to Eq. (27.5) is sought in series

$$U_1 = \sum_{n=1}^{\infty} \varepsilon^n \, U_1^{(n)}, \tag{27.6}$$

and solution to Eq. (27.6) in series

$$U_K^0 = \sum_{m=1}^{\infty} \varepsilon^m \, U_K^{0\,(m)}, \tag{27.7}$$

where $\varepsilon \ll 1$ is a positive perturbation parameter and $K = 1, 2$.

In principle, the displacements caused by wave motion and prestress may be of different orders. The aim is to gather the maximum amount of information about the prestressed state from the wave propagation data. This problem was analyzed in Reference [18], reaching the conclusion that the amplitude of the excited wave is a critical aspect in ultrasonic NDE. The most informative wave propagation data for NDE can be obtained if displacements caused by the excited wave are of the same order as displacements caused by the prestress. This is the reason why the same small parameter is used in series (27.6) and (27.7).

Following the perturbation procedure the series (27.6) and (27.7) are introduced into Eq. (27.4) and a set of equations that determines the terms in series (27.6) is derived. The obtained equations are solved for the case of simultaneous propagation of two longitudinal waves with smooth arbitrary initial profiles $\varphi(t)$ and $\psi(t)$, that is, under the initial and boundary conditions

$$U_1(X_1, X_2, 0) = U_{1,t}(X_1, X_2, 0) = 0, \tag{27.8}$$
$$U_{1,t}(0, X_2, t) = \varepsilon a_0 \varphi(t) H(t), \tag{27.9}$$
$$U_{1,t}(h, X_2, t) = \varepsilon a_h \psi(t) H(t), \tag{27.10}$$

where $H(t)$ denotes the Heaviside unit step function, and a_0 and a_h are constants. The initial wave profiles satisfy the conditions $max \mid \varphi(t) \mid = 1$ and $max \mid \psi(t) \mid = 1$. The first term in series (27.6)

$$\begin{aligned} U_1^{(1)}(X_1, X_2, t) = \ & a_0 \, H(\xi) \int_0^\xi \varphi(\tau)d\tau + a_h \, H(\eta) \int_0^\eta \psi(\tau)d\tau \\ & - a_0 \, H(\theta) \int_0^\theta \varphi(\tau)d\tau - a_h \, H(\zeta) \int_0^\zeta \psi(\tau)d\tau \end{aligned} \tag{27.11}$$

$$\begin{array}{ll} \xi = t - X_1/c, & \eta = t - h/c + X_1/c, \\ \zeta = t - h/c - X_1/c, & \theta = t - 2\,h/c + X_1/c \end{array}$$

is the solution of the linear wave equation.

The second and the third terms in series (27.6) are solutions to the one-dimensional hyperbolic equations with constant coefficients where it is possible to separate the independent variables in the known right-hand sides. These equations are again solved exploiting the perturbation technique and the Laplace integral transform with respect to time.

The final expressions for the second and the third terms in series (27.6) are too cumbersome to be presented here. More details about these solutions are available in Reference [9]. The derived analytical solution (27.6) is valid in the time interval

$$0 \leq t < 2h/c. \tag{27.12}$$

As a result, the analytical solution (27.6) describing the counterpropagation and the interaction of two longitudinal waves with smooth arbitrary initial profiles in the inhomogeneously prestressed nonlinear elastic material is derived. The first term in series (27.6) describes simultaneous counterpropagation of two longitudinal waves in a homogeneous isotropic prestress-free linear elastic material. The subsequent terms correct the solution and take nonlinearity and prestress into account.

2.3 Prestressed State

Wave motion is described by solution (27.6) to Eq. (27.4) and is dependent on the prestressed state determined by the set of equations defining the equilibrium of the material (27.5).

The perturbation technique is used and the solution to this set of equations is sought in series (27.7) with the small parameter, deriving the sets of equations that determine the terms in series (27.7).

A two-dimensional material (structural element) with thickness h and length $2l$ is considered. The perturbation equations are solved for the special case of the prestressed state that corresponds to the plane strain (Figure 27.1). The surfaces $X_1 = 0$ and $X_1 = h$ are traction-free.

The aim is to solve the problem of NDE of the parameters of the predeformed state on the basis of wave propagation data. For this purpose it is convenient to describe the prestressed state by polynomials and define the polynomial boundary conditions in terms of the components $T^0_{KL}(X_1, X_2)$ of the symmetric Kirchhoff pseudostress tensor in the form

$$T^0_{11}(0, X_2) = T^0_{11}(h, X_2) = T^0_{12}(0, X_2) = T^0_{12}(h, X_2) = T^0_{21}(X_1, \pm l) = 0,$$

$$T^0_{22}(X_1, \pm l) = \varepsilon \sum_{n=0}^{5} w_n X_1^n, \tag{27.13}$$

where w_n is a constant. These boundary conditions may be expressed also in terms of displacement on the basis of the theory of elasticity.[8]

The polynomial solution to Eqs. (27.5) is sought. The linear set of equations has the following solution in terms of stress:

$$T^0_{11}(X_1, X_2) = T^0_{12}(X_1, X_2) = T^0_{21}(X_1, X_2) = 0,$$
$$T^0_{22}(X_1, X_2) = \varepsilon (w_0 + w_1 X_1),$$
$$T^0_{33}(X_1, X_2) = \varepsilon \nu (w_0 + w_1 X_1), \tag{27.14}$$

where ν denotes the Poisson ratio. Interestingly, the considered material is in equilibrium under boundary conditions (27.13) only when constants w_j, $j = 2, 3, \ldots$ are equal to zero.

Similarly, the second and the subsequent terms in series (27.7) are determined from the set of linear differential equations with known right-hand sides under the boundary conditions equal to zero. Expressions for these terms are omitted for brevity.

The result is that the prestressed state of the material is determined in terms of the prestress in leading orders: the component $T_{22}^0 = a + b\,X_1$ of the Kirchhoff pseudostress tensor, where the notation $a = \varepsilon\,w_0$, $b = \varepsilon\,w_1$ is used. The nonlinear correction to the predeformed state is rated to be of negligible magnitude.

Consequently, the considered two-dimensional prestressed state corresponds to the plane strain and can be considered as the two-parametric state. The parameter a characterizes the constant part of the prestress and the parameter b the linearly variable part.

2.4 Interaction of Sine Waves

In view of NDE sine waves with the same amplitude and frequency are excited on opposite surfaces of the material in terms of particle velocity. The boundary conditions (9) and (10) are transformed into

$$U_{1,t}(0, X_2, t) = \varepsilon\,a_0\,\sin\omega t\,H(t), \tag{27.15}$$

$$U_{1,t}(h, X_2, t) = \varepsilon\,a_h\,\sin\omega t\,H(t), \tag{27.16}$$

where ω denotes the radial frequency.

The evolution of the wave profile is recorded on the same surfaces in terms of stress. The distortion of the stress wave profile is analyzed on the basis of solution (27.6) to the equation of motion (27.4). This solution enables us to separate the linear wave propagation [first term in series (27.6)] from the nonlinear effects (second and subsequent terms) that accompany wave propagation, reflection, and interaction. In the case of sine waves nonlinear effects consist in the generation of higher harmonics and of nonlinear material–wave and wave–wave interaction.

The linear part of the solution describes the simultaneous propagation of two sine waves in the prestress-free physically linear material where wave interaction is determined by superposition of wave profiles.

The nonlinear effects of wave motion (Figure 27.2) that are analyzed here on the basis of the second term in series (27.6) are sensitive to the nonlinear physical properties of the material and to the prestress parameters. They are governed by the oscillation of double frequency (second harmonic) with respect to the frequency of excitation.

The nonlinear theory of elasticity[8] describes the stress as a function of the derivative of the particle displacement U with respect to the spatial coordinate X. This is the reason why, henceforth, the nonlinear effects are characterised by the function $U_{1,1}^{(2)}$. From the mathematical point of view, this function is described by the second term in the perturbation solution (27.6). The analytical expression for $U_{1,1}^{(2)} \equiv U_{,X}^{(2)}$ is derived

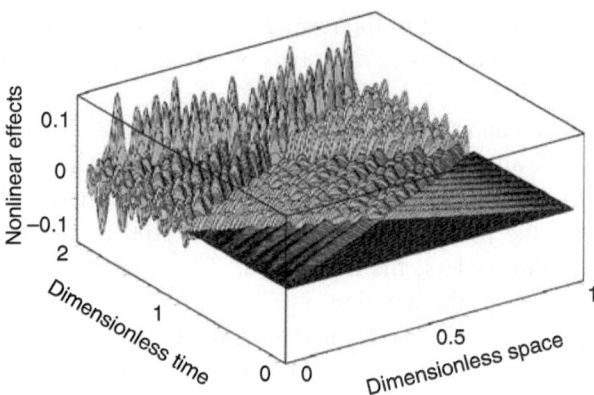

Fig. 27.2. Nonlinear effects evoked by two waves counterpropagation.

by means of an analytical computational software (Maple V), and assumes the form

$$U_{,X}^{(2)} = A_0^{(2)} + \sum_{j=1}^{m_1} A_{1j}^{(2)} \sin \omega \vartheta_j + \sum_{j=1}^{m_2} A_{2j}^{(2)} \cos \omega \vartheta_j$$

$$+ \sum_{j=1}^{m_3} A_{3j}^{(2)} \sin 2\omega \vartheta_j + \sum_{j=1}^{m_4} A_{4j}^{(2)} \cos 2\omega \vartheta_j, \qquad (27.17)$$

where $\vartheta_j = t + c_{1j} h/c + c_{2j} X_1/c$, and c_{1j}, c_{2j} are constants.

The expression (27.17) consists of nonperiodic term $A_0^{(2)}$ and periodic terms with arguments $\omega \vartheta_j$ and $2\omega \vartheta_j$. Analytical expressions for nonperiodic term $A_0^{(2)}$ and amplitudes $A_{ij}^{(2)}$, $i = 1, ..., 4$ are too cumbersome to be presented here.

From the physical point of view the function $U_{,X}^{(2)}$ describes the main part of nonlinear effects including the evolution of the second harmonic, influence of the prestress on the evolution of the first harmonic, nonlinear interaction between two first harmonics and influence of the material nonlinear physical properties on the wave propagation. Evolution of the third and the higher harmonics are neglected here along with the higher-order small phenomena. In the following, the dimensionless function $\varepsilon\, U_{,X}^{(2)}$ characterizes the ratio of the magnitude of nonlinear effects to the magnitude of the linear oscillation.

For demonstration purposes, let us consider a numerical experiment. Let the properties of the material (structural element) correspond to duralumin with density $\rho_0 = 2800$ kg/m^3, constants of elasticity $\lambda = 50$ GPa, $\mu = 27.6$ GPa, $\nu_1 = -136$ GPa, $\nu_2 = -197$ GPa, $\nu_3 = -38$ GPa, and dimensions $h = 0.1$ m and $l = 1$ m. The strain is characterized by the dimensionless constant ε that is set to $\varepsilon = 10^{-4}$. The sine wave amplitudes are determined by constants $a_0 = -a_h = c$ m/s in correspondence with the boundary conditions (27.15) and (27.16). The constant c is defined by Eq. (27.3). As a result, the amplitude of the excited particle velocity at the boundaries $X_1/h = 0$ and $X_1/h = 1$ has opposite sign and same absolute value $| \varepsilon\, a_0 | = 0.6130$ m/s. The wave excitation frequency ω is set to $\omega = 1.9 \cdot 10^6$ rad/s.

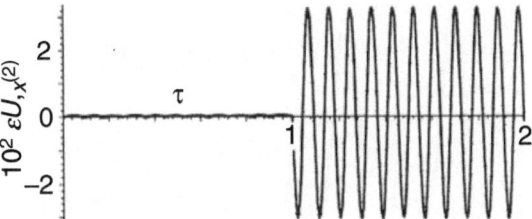

Fig. 27.3. Prestress-free material. Boundary oscillations.

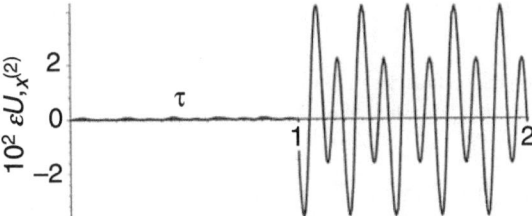

Fig. 27.4. Homogeneously prestressed material. Modulated oscillations on the boundaries.

The numerical simulation of the nonlinear part of boundary oscillations caused by the simultaneous propagation of two sine waves in the prestress-free physically non-linear elastic material is illustrated in Figure 27.3.

The oscillations on both boundaries, where $X_1/h = 0$ and $X_1/h = 1$, are similar. It is possible to distinguish two distinct intervals on the time axis: the propagation interval ($0 \leq \tau < 1$) and the interaction interval ($1 \leq \tau < 2$). Here X_1/h denotes the dimensionless spatial coordinate and $\tau = t\,c/h$ the dimensionless time.

The nonlinear wave interaction amplifies the boundary oscillation amplitude in the interaction interval about a hundred times. This phenomenon facilitates the usage of nonlinear effects of wave interaction in NDE of material properties and states.

2.5 NDE of Plane Strain

2.5.1 Qualitative NDE Technique

The nonlinear part of boundary oscillations in prestress-free materials are character-ized by the constant but different values of the amplitudes in the propagation and interaction intervals (see Figure 27.3). These amplitudes are sensitive to the physical properties of the material (density, elastic constants) but less sensitive to the value of the excitation frequency.

Homogeneous prestress ($T_{22}^0 = a$) modulates the boundary oscillation (Figure 27.4). The shape and the depth of modulation include information about the sign and the value of prestress. The oscillation profiles on both boundaries are similar.

Inhomogeneous prestress ($T_{22}^0 = a+b\,X_1$) modulates oscillation on different bound-aries in different way (see Figure 27.5).

The analysis of the influence of prestress on the boundary oscillations presented in Figures 27.3 to 27.5 leads to the conclusion that nonlinear effects of two sine wave counterpropagation data enable us to solve the problem of qualitative NDE of

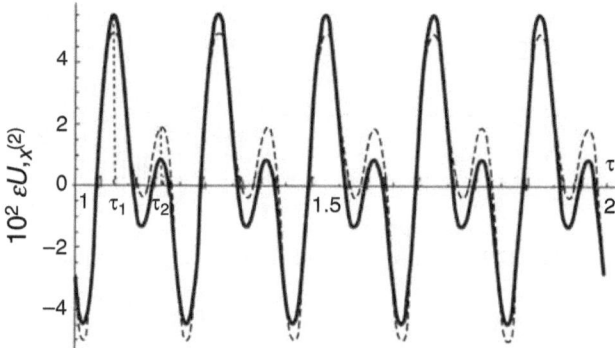

Fig. 27.5. Oscillations on the boundaries of inhomogeneously prestressed material recorded in the interval $1 \leq \tau < 2$ (solid line - $X_1/h = 0$, dashed line - $X_1/h = 1$).

prestress in materials. It is easy to distinguish qualitatively the presence and nature of prestress. The absence of modulation in boundary oscillations characterizes prestress-free materials. The similar modulation of oscillations on both boundaries is a sign of homogeneous prestress. The inhomogeneity in prestress causes disparity in oscillation profiles on different boundaries. Interestingly, in the case of pure bending when the stress T_{22}^0 has equal value but opposite sign on opposite boundaries, the boundary oscillation profiles coincide, but a phase shift occurs.

2.5.2 Quantitative NDE Technique

Quantitative NDE is based on the observation that the width of modulation of the nonlinear boundary oscillations, caused by the simultaneous propagation of two sine waves, ultimately depends on the inhomogeneous prestress.

For instance, let us consider the following model problem. It is assumed that the geometry and the physical properties (density, elastic constants of the second and third order) of the material are known. The preliminary inspection confirms the fact that the material is undergoing a predeformed state that corresponds to pure bending with tension or compression characterized by the constants a and b of the Kirchhoff pseudostress tensor component $T_{22}^0 = a + b \, X_1$. The purpose is to evaluate constants a and b starting from experimental boundary oscillation data.

In order to evaluate the unknown values of prestress parameters a and b for the real material, a suitable experiment is set up as follows. The recorded boundary oscillation profiles on opposite boundaries in the interval of interaction are plotted in Figure 27.5. The two instants τ_1 and τ_2 that correspond to the local maximums of the oscillation profile are fixed in this plot.

Using the analytical solution (27.6), boundary oscillation amplitudes at the instants τ_1 and τ_2 versus prestress parameters a and b are plotted, as shown in Figure 27.6.

The quantitative NDE of the prestress parameters for the case presented by Figure 27.5 includes the following steps. First, the values of oscillation amplitudes on opposite boundaries are determined for instants τ_1 and τ_2 (see Figure 27.5), and their difference is calculated for both time instants. Then, on the basis of the value of this difference for the instant τ_1 two possible values of parameter b are determined, as

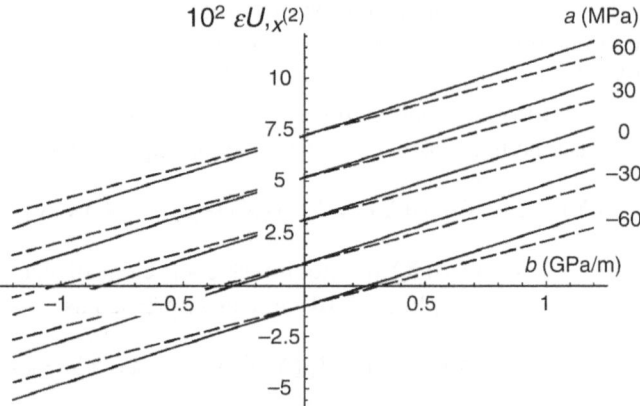

Fig. 27.6. Boundary oscillation amplitude versus prestress parameters a and b at the instant τ_1 (solid line - $X_1/h = 0$, dashed line - $X_1/h = 1$).

shown in Figure 27.6. Finally, parameter a is determined in Figure 27.6 according to the value of the oscillation amplitude in Figure 27.6 on one of the boundaries at the instant τ_1.

The corresponding plot for instant τ_2, not presented here, may be used to verify the values of the evaluated prestress parameters.

3. Weakly Inhomogeneous Material

Problem Formulation

Nonlinear effects arising from counterpropagation, interaction, and reflection of waves in different materials are intensively studied due to promising applications for nondestructive characterization of materials.[10] Effects of resonant wave–wave interaction,[11] amplitude amplification by interaction, and modulation of amplitudes by different external and internal effects[12] constitute only a short list of phenomena that may be exploited for practical purposes.

In this section the theoretical basis for counterpropagation of two harmonic waves in nonlinear weakly inhomogeneous elastic materials is outlined. The corresponding analytical solution is derived and analyzed numerically. The resonant values of interaction amplitudes as functions of the excitation frequency and amplitude are determined for various physically inhomogeneous nonlinear elastic materials. These resonant values are sensitive to the material properties and may be used for their nondestructive characterization.

The one-dimensional motion of an inhomogeneous nonlinear elastic material is described by the equation[12]

$$\left[1 + k_1(X)\, U_{,X}(X,t) \right] U_{,XX}(X,t) + k_2(X)\, U_{,X}(X,t) + k_3(X) \left[U_{,X}(X,t) \right]^2$$
$$- k_4(X)\, U_{,tt}(X,t) = 0. \tag{27.18}$$

Equation (27.18) is a nonlinear second-order partial differential equation with variable in space coefficients. These coefficients $k_i(X)$, $i = 1 \ldots 4$ are functions of the material variable physical properties, the density $\rho(X)$, the Lamé coefficients $\lambda(X)$ and $\mu(X)$ and the third-order coefficients of elasticity $\nu_1(X)$, $\nu_2(X)$, $\nu_3(X)$. In the one-dimensional case the five elastic coefficients are grouped together as follows:

$$\alpha(X) = \lambda(X) + 2\mu(X), \quad \beta(X) = 2 \left[\nu_1(X) + \nu_2(X) + \nu_3(X) \right], \tag{27.19}$$

where $\alpha(X)$ is the linear and $\beta(X)$ the nonlinear coefficient of elasticity. Now, the coefficients $k_i(X)$, $i = 1 \ldots 4$ in Eq. (27.18) may be expressed as

$$k_0(X) = [\alpha(X)]^{-1}, \quad k_1(X) = 3\left[1 + k_0(X)\beta(X)\right], \quad k_2(X) = k_0(X)\,\alpha_{,X}(X),$$

$$k_3(X) = 3\,k_0(X)\left[\alpha_{,X}(X) + \beta_{,X}(X)\right]/2, \quad k_4(X) = \rho_o(X)\,k_0(X). \tag{27.20}$$

It is assumed that the material has two parallel traction-free surfaces at $X = 0$ and $X = L$. Eq. (27.18) is solved under the initial and boundary conditions

$$U(X, 0) = U_{,t}(X, 0) = 0,$$

$$U_{,t}(0, t) = \varepsilon a_0 \varphi(t) H(t),$$

$$U_{,t}(L, t) = \varepsilon a_L \psi(t) H(t), \tag{27.21}$$

where the notation used is the same as in Eqs. (27.8)–(27.10).

The perturbation technique is again employed and the analytical solution to Eq. (27.18) is sought in the form of a series with a small parameter ε

$$U(X, t) = \sum_{n=1}^{\infty} \varepsilon^n \, U^{(n)}(X, t), 0 < \varepsilon \ll 1. \tag{27.22}$$

Henceforth, only the first three terms in series (27.22) are considered.

It is assumed that the inhomogeneity in the density $\rho(X)$, linear elastic coefficient $\alpha(X)$, and nonlinear elastic coefficient $\beta(X)$ is weak and may be described as a small deviation from the constant value of these properties by expression

$$\gamma(X) = \gamma^{(1)} + \varepsilon\gamma^{(2)}(X), \quad \gamma = \rho, \alpha, \beta, \tag{27.23}$$

where the function $\gamma^{(2)}(X)$ that describes the space-dependent properties of the material is formulated as a third-order polynomial

$$\gamma^{(2)}(X) = \gamma_{1\xi}X + \gamma_{2\xi}X^2 + \gamma_{3\xi}X^3,$$

$$\gamma^{(2)}(X) = \rho^{(2)}(X), \, \alpha^{(2)}(X), \, \beta^{(2)}(X),$$

$$\xi = \rho, \alpha, \beta. \tag{27.24}$$

Now, the inhomogeneous physical properties of a material are determined by nine constants $\gamma_{i\xi}$.

By introducing Eqs. (27.23) and (27.24) into expressions (27.20), the coefficient $k_0(X)$ may be expanded into the Taylor series

$$k_0(X) = \left[1 - \varepsilon \alpha^{(2)}(X)/\alpha^{(1)} + \left(\varepsilon \alpha^{(2)}(X)/\alpha^{(1)} \right)^2 - \ldots \right] / \alpha^{(1)}$$

and coefficients $k_i(X)$, $i = 1 \ldots 4$, after some algebraic calculations, are the following:

$$
\begin{aligned}
k_1(X) &= k_1^{(1)} + \varepsilon\, k_1^{(2)}(X) + \varepsilon^2 k_1^{(3)}(X), \\
k_2(X) &= \varepsilon\, k_2^{(2)}(X) + \varepsilon^2 k_2^{(3)}(X), \\
k_3(X) &= \varepsilon\, k_3^{(2)}(X) + \varepsilon^2 k_3^{(3)}(X), \\
k_4(X) &= k_4^{(1)} + \varepsilon\, k_4^{(2)}(X) + \varepsilon^2 k_4^{(3)}(X),
\end{aligned}
\tag{27.25}
$$

where functions $k_i^{(j)}(X)$, $i = 1, \ldots, 4$, $j = 2, 3$ are polynomials.

Finally, substituting the first three terms in series (27.22) and expressions (27.25) into Eq. (27.18) yields the governing equation. Following the perturbation procedure, (i.e., equating to zero the terms with the same power of ε and neglecting all terms higher than ε^3), a set of equations for the three first terms in series (27.22) follows.

3.1 Interaction of Harmonic Waves

From the point of view of practical applications, the counterpropagation of two harmonic waves in a physically nonlinear elastic material is investigated. The initial wave profiles are defined as sine functions

$$\varphi(t) = \psi(t) = \sin(\omega t), \tag{27.26}$$

where ω denotes the radial frequency.

Wave propagation and interaction is described by Eq. (27.22). The first term in series (27.22) is the solution to the linear wave equation. The latter has been derived under the initial and boundary conditions (27.21) and has the form

$$
\begin{aligned}
U_{,t}^{(1)}(X, t) &= a_0 H(\xi)\varphi(\xi) + a_L H(\eta)\psi(\eta) - a_0 H(\theta)\varphi(\theta) - a_L H(\zeta)\psi(\zeta), \\
\xi &= t - X/c, \eta = t - L/c - X/c, \theta = t - 2L/c - X/c, \zeta = t - X/c + L/c
\end{aligned}
\tag{27.27}
$$

where $c = (k_4^{(1)})^{(-1/2)}$ is the linear wave velocity.

The analytical expressions for the second and the third term in Eq (27.22) have been derived by means of the symbolic software Maple 9. Due to the their complexity, they are not presented here. The interested reader is referred to[12] for further detail.

Counterpropagation, reflection and interaction of two harmonic waves is illustrated on the basis of numerical simulations. The material is assumed to be duralumin with density $\rho^{(1)} = 3000$ kg/m^3 and coefficients of elasticity $\alpha^{(1)} = 100$ GPa and $\beta^{(1)} = -750$ GPa. The thickness of the specimen is $L = 0.1$ m. The wave process is excited by the values of constants $a_0 = -a_L = -c$ m/s and $\varepsilon = 10^{-4}$.

The wave process is analyzed in terms of $U_{,X}(X, t)$ that characterizes the stress distribution in the specimen. Function $U_{,X}(X, t)$ is derived from the solution (27.22).

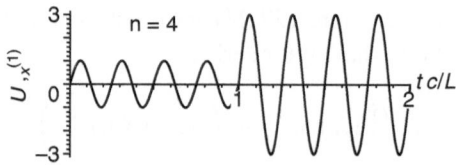

Fig. 27.7. Oscillation on the boundary of the material. Linear problem.

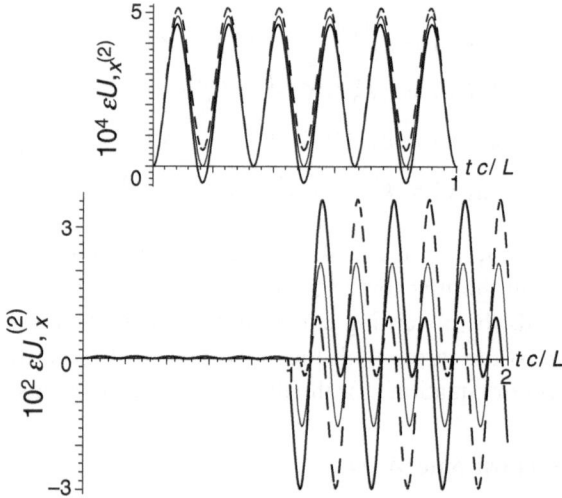

Fig. 27.8. Second-order effects of boundary oscillations. Thin solid line: homogeneous material, bold dashed line ($X = 0$) and bold solid line ($X = L$): inhomogeneous material.

The first term $U_{,X}^{(1)}(X, t)$ characterizes the stress distribution caused by the linear wave process in the homogeneous material and the subsequent terms $U_{,X}^{(i)}(X, t)$, $i = 2, 3$ take the nonlinearity and inhomogeneity of the problem into account. Linear wave propagation is studied. The excitation frequency is assumed to be $\omega = 1.53447 \cdot 10^6$ rad/s. On the boundaries of the material two different oscillation intervals may be distinguished: the wave propagation interval $0 \le t c/L < 1$ and the wave interaction interval $1 \le t c/L < 2$.

Amplification of the amplitude of the boundary oscillation in the wave interaction interval is dependent on the frequency ω. If the frequency is equal to $\omega = 2 \pi n c/L$, then the amplification is the highest: three times the initial amplitude, as demonstrated in Figure 27.7 for the case $n = 4$. In the special case when $\omega = 2 \pi (n + 0.5) c/L$ there is no amplification in the interval of wave interaction. For all other values of ω, the amplification lies between these two extremes.

The predominant factor of the nonlinear boundary oscillation $U_{,X}^{(2)}(X, t)$, plotted in Fig. 27.8 is governed by the double frequency i. e., the second harmonic. Here the excitation frequency satisfies the condition $\omega = 6 \pi c/L$ and two different nonlinear elastic materials are considered. The thin solid line corresponds to the nonlinear oscillation on the boundaries $X = 0$ and $X = L$ of the physically homogeneous material, and the bold dashed and bold solid lines describe oscillation in the inhomogeneous

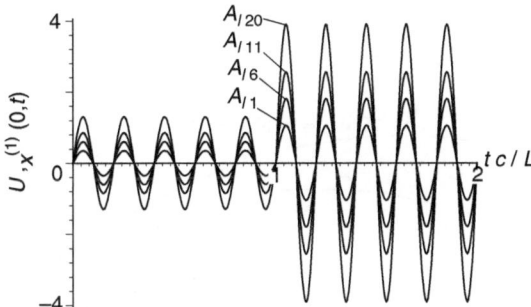

Fig. 27.9. Influence of the excitation amplitude on the boundary oscillations. Linear problem ($n = 5$).

material on boundaries $X = 0$ and $X = L$, respectively. Interestingly, in the interval of wave propagation, magnified in the upper part of Figure 27.8, the nonlinear effects are two orders of magnitude smaller that in the interval of wave interaction $1 \leq t c/L < 2$. This difference does not depend on the frequency, as it was in the case of linear wave interaction (Figure 27.7). It should also be noticed that that the oscillation on the boundaries of the homogeneous material has constant amplitude but the inhomogeneity in the material physical properties modulates it.

Qualitatively similar are the boundary oscillations described by the term $U_{,X}^{(3)}(X, t)$, except for the fact that these higher-order oscillations are modulated already on the boundaries of the homogeneous material.

3.2 Material Characterization by Nonlinear Resonance

This section focuses on exploring the feasibility of using wave interaction resonance caused by counterpropagation of two waves for nondestructive material characterization. The resonance (maximum) of the amplitude of the linear interaction (superposition) occurs if the excitation frequency satisfies the condition $\omega_l = 2 \pi n c/L$ (Figure 27.7; $n = 4$). The value of this resonance amplitude A_l depends on the excitation amplitude only. A simulation was performed for the following range of excitation amplitudes $-1.30 c \leq a_0 \leq -0.35 c, a_L = -a_0$ m/s and $\varepsilon = 10^{-4}$.

Some of the 20 computed values of the boundary oscillation amplitudes A_{li}, $i = 1, \ldots, 20$ are plotted in Figure 27.9 for the linear case. The maximum (resonant) amplitude in the interaction interval $1 < t c/L < 2$ occurs in all cases at the same frequency $\omega_l = 2 \pi c n/L$, where n is an integer.

As previously discussed, material inhomogeneity modulates the interaction amplitude (Figure 27.8). This is the reason why, henceforth, the first positive peak of the interaction amplitude will be considered. The value of this peak A is computed on the basis of the three first terms in solution (27.22) and takes nonlinearity and material inhomogeneity into account.

The following notation for the parameters of material inhomogeneity is adopted:

$$\gamma(X) = \gamma^{(1)}(1 + \delta_{i\xi}(X)), \quad \delta_{i\xi}(X) = \varepsilon \gamma_{i\xi} X^i / \gamma^{(1)}, \quad \delta_{i\xi}(L) \equiv \delta_{i\xi},$$
$$i = 1, 2, 3, \gamma, \quad \xi = \rho, \alpha, \beta. \tag{27.28}$$

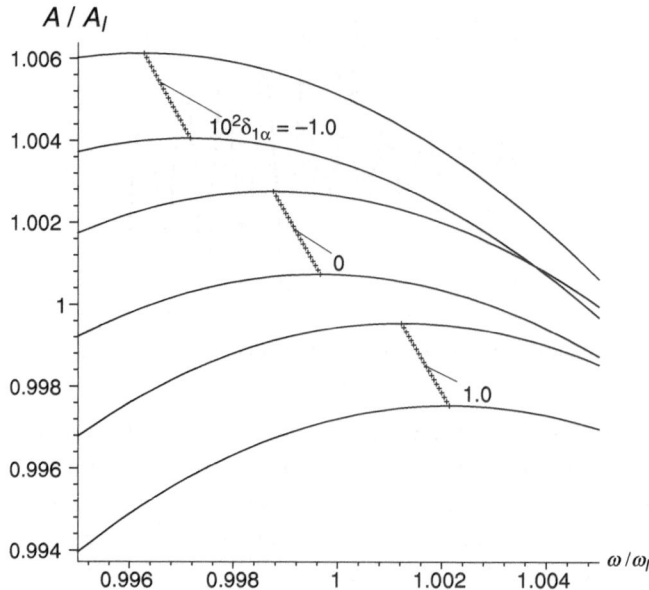

Fig. 27.10. Nonlinearity and inhomogeneity versus resonance frequency.

The relative value of the first positive peak of the interaction amplitude depends on the material inhomogeneity, on the excitation amplitude and frequency, as illustrated in Figure 27.10. Physical properties of the material are assumed to vary linearly along the X-axis and are described by Eq. (27.28) in the form

$$\gamma(X) = \gamma^{(1)}(1 + \delta_{1\xi}(X)), \quad \delta_{1\xi}(X) = \varepsilon\gamma_{1\xi}X/\gamma^{(1)}, \quad \gamma, \xi = \rho, \alpha, \beta. \quad (27.29)$$

The inhomogeneity of material properties is estimated by the value of the parameter of inhomogeneity $\delta_{1\xi}(L) \equiv \delta_{1\xi}, \quad \xi = \rho, \alpha, \beta$.

Three different materials with variable linear elastic property $\alpha(X)$ are considered. In the first case, the elastic coefficient $\alpha(X)$ varies linearly with a negative slope along the X-axes and the maximum deviation at $X = L$ from the basic value $\alpha^{(1)} = 100\,\text{GPa}$ is 1 %; that is, $10^2\delta_{1\alpha} = -1.0$. In the second case, the elastic coefficient $\alpha(X)$ is constant $\alpha^{(1)} = 100\,\text{GPa}$ (homogeneous material). In the third case, the elastic coefficient $\alpha(X)$ varies linearly with a positive slope along the X-axes ($10^2\delta_{1\alpha} = 1.0$).

By analyzing the curves presented in Figure 27.10, it can be easily seen that (i) the variation in the value of the linear elastic coefficient induces a shift in the value of the resonance frequency, (ii) the value of the resonance frequency is sensitive to the properties of the material and (iii) the resonance frequency may be used for qualitative and quantitative NDE of the properties of physically inhomogeneous nonlinear elastic materials.

The last conclusion is supported by the results of similar simulations for materials with variable density. In this case the shift of the resonant frequency occurred only in the horizontal direction (see Fig. 27.11).

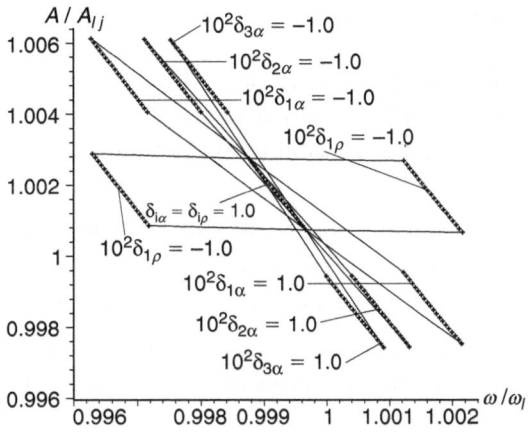

Fig. 27.11. Nonlinear resonance as function of excitation frequency and material properties ($j = 1$, ..., 20, $i = 1, 2, 3$).

The influence of the variation of the nonlinear elastic coefficient β on the resonance frequency is substantially weaker than that of the density and the linear elastic coefficient.

The results of the numerical simulation show cascades in different regions of $A/A_{li} - \omega/\omega_l$ plane (Figure 27.11). These cascades form different *corridors* for linear, quadratic and cubic inhomogeneities in material properties, as illustrated for $\alpha(X)$ in Figure 27.11. It is worth noting that inhomogeneity in density has an essentially different effect on wave interaction resonance than other material properties. This facilitates the use of the obtained results in NDE of the properties of inhomogeneous materials.

4. Conclusions

The key idea of exploiting nonlinear effects of ultrasonic wave propagation and interaction for elaboration of relatively simple methods of NDE of inhomogeneous materials is presented. In particular, two model problems are studied:

(i) Counterpropagation and interaction of two longitudinal waves in inhomogeneously prestressed nonlinear elastic materials (structural element).

(ii) The same wave propagation process in physically weakly inhomogeneous nonlinear elastic materials.

For those problems, a detailed theoretical basis is provided, including the derivation of analytical solutions.

Analysis of numerical simulations leads to the conclusion that the nonlinear effects of boundary oscillations evoked by the counterpropagation of two longitudinal waves in the material contain enough information for NDE techniques to evaluate the parameters of simple inhomogeneous materials.

Three NDE techniques are proposed:

(i) Qualitative NDE technique, on the basis of the profile of the nonlinear part of boundary oscillations, allows us to distinguish different states of plane strain in the material, including the possibility to discriminate between prestressed and prestress-free materials.

(ii) Quantitative NDE technique that enables us to evaluate the parameters of two-parametric prestressed state of the material on the basis of the values of two first peaks of the profile of nonlinear boundary oscillation to.

(iii) Resonance NDE technique, allowing us to distinguish qualitatively weak inhomogeneity in different physical properties of nonlinear elastic materials and to evaluate this weak inhomogeneity.

The proposed NDE techniques may be especially useful for nondestructive characterization of materials where nonlinear effects of wave propagation are several orders of magnitude higher than in the materials considered here, such as materials with microstructure, or affected by microscale damages, and so on.

Acknowledgments

The research was supported by the European Science Foundation under the NATEMIS program and by the Estonian Science Foundation under grant no. 4706.

References

[1] L.F. Richardson, Apparatus for warning a ship at sea of its nearness to large objects wholly or partially under water. *British Patent Specification*, 11125, March 27 (1913).

[2] J. Krautkrämer and H. Krautkrämer, *Ultrasonic Testing of Materials* (Springer-Verlag, Berlin, 1990).

[3] R.M. Bergman and R.A. Shanbender, Effect of statically applied stresses on the velocity of propagation of ultrasonic waves, *J. Appl. Phys.* **29**, 1736–1739 (1958).

[4] R.W. Benson and V.J. Raelson, Acoustoelasticity, *Product Engineering* **20**, 56–59 (1959).

[5] M. Sugiyama, Interaction between weak and short waves in a one-dimensional inhomogeneous nonlinear elastic material, *J. Acoust. Soc. Am.* **80**, 306–310 (1986).

[6] M. Kato, T.Sato, K.Kameyama, and H.Ninoyu, Estimation of the stress distribution in metal using nonlinear acoustoelasticity, *J. Acoust. Soc. Am.* **98**, 1496–1504 (1995).

[7] A. Ravasoo, Nonlinear longitudinal waves in inhomogeneously predeformed elastic media, *J. Acoust. Soc. Am.* **106**, 3143–3149 (1999).

[8] A.C. Eringen, *Nonlinear Theory of Continuous Media* (McGraw-Hill, New York, 1962).

[9] A. Ravasoo, and B. Lundberg, Nonlinear interaction of longitudinal waves in an inhomogeneously predeformed elastic medium, *Wave Motion* **34**, 225–237 (2001).

[10] D. Donskoy, A. Sutin, A. Ekimov, Nonlinear acoustic interaction on contact interfaces and its use for nondestructive testing, *NDT & E Int.* **34**, 231–238 (2001).

[11] L. Friedland, Autoresonant three-wave interactions, *Phys. Rev. Lett.* **69**(12), 1749–1752 (1992).

[12] A. Ravasoo, A. Braunbrück, Wave interaction for characterization of nonlinear elastic material, *Proc. Estonian Acad. Sci. Eng.* **6**(3), 171–185 (2000).

28

Nonlinear Elastic Behavior and Ultrasonic Fatigue of Metals

Cleofé Campos-Pozuelo,[3] Christian Vanhille,[2] and Juan A. Gallego-Juárez[1]

[1]Instituto de Acústica, CSIC, Serrano, 144, 28006 Madrid, Spain.
[2]ESCET, Universidad Rey Juan Carlos. Tulipán, s/n. 28933 Móstoles. Madrid, Spain.
[3]To whom correspondence should be addressed: C. Campos-Pozuelo,
[1]Instituto de Acústica, CSIC, Serrano, 144, 28006 Madrid, Spain; e-mail:
ccampos@ia.cetef.csic.es.

Abstract

Ultrasonic fatigue and nonlinear elastic behavior of metals are studied and their relation experimentally established. The high-cycle fatigue process is summarily described and the implication in the dynamical behavior of the material is outlined. In this framework, a new method and an experimental system for the study of the nonlinear behavior and fatigue failure of metals under high-intensity ultrasonic stresses has been developed and tested. The method is based on the measurement of the vibration velocity of bar-stepped samples for different excitation levels keeping the temperature of the sample constant. The experimental set-up is constituted by a driving system exciting the material samples at resonance and a nonintrusive data acquisition system. Results show the relation between the material nonlinearity parameter and the fatigue state. In this way the viability of nonlinear material characterization for fatigue damage assessment is established.

Keywords: Fatigue in metals, high-cycle fatigue, high-power ultrasonics, nonlinear characterization of metals, nonlinear vibrations

1. Introduction

The present study was originated in the framework of the design and construction of high-power ultrasonic transducers for use in air. There are numerous possibilities for airborne high-power ultrasound applications in industry, but they are generally limited by the maximum acoustic energy which is possible to generate. Two main causes introduce limitations: one is the nonlinear attenuation of pressure waves, leading to saturation and important energy losses by harmonic distortion, and the other is the limited power capacity of the transducers, due to fatigue of their metallic components. For the analysis of the second phenomenon a high-frequency fatigue experimental system has been designed, constructed, and tested as described in this chapter. Extensional and flexural vibrations to fatigue samples at about 20 kHz are applied and the effect is analyzed at different constant temperatures. The limiting strain, defined as the strain amplitude threshold beyond which fatigue can occur in a small number of cycles, is evaluated by using the experimental system and a theoretical procedure, which is here described. The limiting strain and the nonlinear behavior of attenuation (strain–amplitude dependence) are microscopically interpreted as changes occurring in the distribution and density of dislocations in relation with the initiation of the fatigue process. The detection of the fatigue process is fundamental in its first stages in order to avoid catastrophic damage of the material. To that purpose we have theoretically and experimentally established as described in this work the direct relation between the fatigue process and the nonlinear dynamical behavior of metallic materials extensionally and flexurally vibrating. Fatigue is one of the primary reasons for the failure of structural and particularly metallic components. Moreover, in many applications including aeronautic, energy industries, and, of course, high-power ultrasonic transducers, metallic components are expected to have a very long life ($> 10^9$ cycles) and often they are subjected to sonic or ultrasonic stressing. The study presented in this chapter concerns high-frequency (20 kHz) fatigue and more specifically the establishment of a relation between the high-cycle fatigue process and the evolution of the nonlinear elastic characteristics of metals.

A description of the fatigue process and a review of high-frequency fatigue experiments is presented in this section as a general introduction.

1.1 Fatigue Process in Metals

The term "fatigue" refers to the gradual accumulation of damage produced under alternating mechanical straining of a specimen. It is well known that a metal subjected to cyclic stressing will fracture at loads below the steady ultimate tensile strength. In fact, fatigue is one of the primary reasons for the failure of structural components. In general, there are many uncertainties in establishing fatigue reliability. Therefore, testing of the structures will always be necessary. The fatigue life contains two parts: initiation and propagation of the crack. Dislocations play a major role in the fatigue crack initiation phase. After a high number of loading cycles, dislocations pile up and form structures called persistent slip bands (PSB). PSBs are areas that rise above (extrusion) or fall below (intrusion) the surface. They leave tiny steps in the surface that serve as stress risers where fatigue cracks can initiate. The rupture surface presents a plane zone (where microcracks propagate) and a rough zone (where ductile deformation occurs).

At each stage of the complex fatigue process a certain phenomenon dominates. The fatigue process is usually represented by the Wöhler curve, where the applied stress is plotted versus the number of cycles to failure (Figure 28.1). The fatigue process has a statistical nature, such that an important dispersion of points in the Wöhler plot is expected. The relationship between strain amplitude and number of loading cycles S–N (Wöhler curve) depends on the material, its composition and previous history (heat treatment, mechanical treatment), as well as the conditions of the experiment (stress range, frequency of cycling, temperature, nature of the environment, etc.). The first part of the curve corresponds to low-cycle fatigue, in which high-stress amplitudes are applied and an important plastic deformation occurs. In the high-cycle region (high number of cycles and small stress amplitudes) the strain is almost completely elastic.

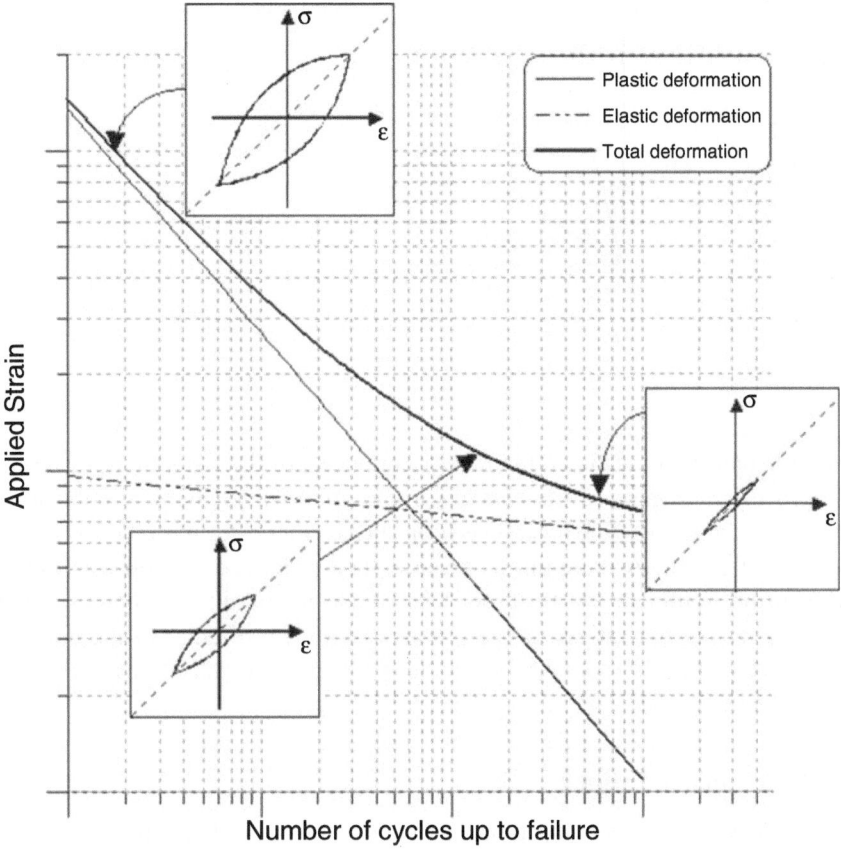

Fig. 28.1. The Wöhler curve. Relation between the mechanical properties and the fatigue process.

Fatigue research has often been focused on the study of the relation between fatigue damage accumulation and changes taking place on the surface during stressing (formation of slip lines, bands, extrusions, micro- and macrocracks). Further insight on fatigue life can be obtained from the study of changes in the mechanical properties of the stressed material.

The gradual changes in stress–strain curves as well as the changes in mechanical hysteresis loops as a function of the number of stress cycles were the object of early studies in the decades 1910–1920 (see Bratina, 1966 and references therein). The effect of fatigue stressing on the attenuation damping was also studied by different researchers during the fifties and sixties (Bratina, 1966; Mason, 1956, 1968).

Microscopic theories based on the movement and multiplication of dislocation loops explaining changes in attenuation with amplitude have been provided by Granato and Lücke (1956) and by Mason (1971). They proposed a quantitative theory of damping and modulus changes due to dislocation and they distinguished two kind of losses that are dependent on the frequency and on the amplitude, respectively. The amplitude-dependent loss is a hysteretical loss which has a strain-amplitude dependence. Figure 28.2 (from Granato and Lücke, 1956) outlines the stress–strain dependence obtained from the Granato–Lücke model for the second type of losses.

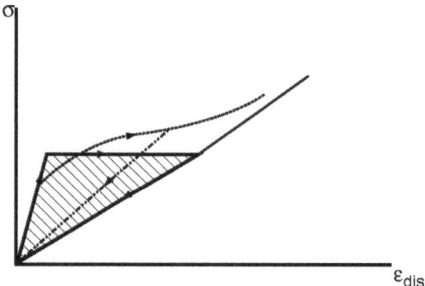

Fig. 28.2. Stress–strain ($\sigma - \varepsilon$) relation as proposed by Granato and Lücke (from Granato and Lücke, 1956).

In this chapter the focus is on the macroscopic description and quantification of the fatigue process by means of the changes in the nonlinear dynamical behavior of the material when fatiguing. The hysteresis loop surface increases with the strain and with the number of cycles. A damping increase is then produced as well as an important increase in the nonlinear elastic properties of the material. The increase of material damping during the fatigue process is a well-known effect. On the contrary, the changes produced in the nonlinear characteristics of the material have been scarcely evaluated (Nagy, 1998).

1.2 High-Frequency Fatigue

High-frequency mechanical vibrations in metals by means of high-intensity ultrasound can induce ultrasonic fatigue. Data about the behavior of materials at such loading-frequency range are useful for acoustic technology, power engineering, aircraft and rocket engineering, and so on.

One of the main advantages of ultrasonic fatigue testing is the important reduction of time, and therefore the possibility of analyzing a very high number of fatigue cycles (Puskár, 1982; Matikas, 2001). The test equipments are generally simple and give the possibility of testing the samples at different temperatures and under different environmental conditions.

High-frequency fatigue studies were initiated in the fifties (Puskár, 1982; and references therein) and nowadays the topic is still the object of numerous applied and theoretical investigations (Matikas, 2001; Wang et al., 2002; Holper et al., 2004; Marines et al., 2003; Papakyriacou et al., 2001; Mayer et al., 2001). In fact, due to the increasing requirements in fatigue strength and life of the components, high frequency fatigue analysis becomes the most appropriate method for material testing.

Puskár (1982) has reviewed experimental results on the fatigue limit obtained from low- and high-frequency fatigue tests in different materials and under different testing conditions. In general, experiments have been carried out by fatiguing the metal, and also by measuring attenuation variations with strain amplitude. Results show differences up to around 40% in the limiting strain for high- and low-frequency loading. In particular, for aluminium the limiting strain seems to be from 20 to 30% higher for high-frequency tests than for low-frequency tests. However, reports on iron indicate that the fatigue crack nucleates at higher values for low-frequency fatigue. Results for copper samples seem to be similar for high and low frequency. A number of contradictory results for steel samples are also reported in this reference. The complexity of the phenomenon is at the origin of such contradictions. In fact to be sure about the results, the material used and the conditions of the experiments have to be very carefully controlled. The grain size and the microstructure of the metal also have to be considered as an important feature in the limiting strain value.

Recent works (Marines et al., 2003; Wang et al., 2002) have described the fatigue behavior of different steels at high frequency. The main conclusion is that rupture can occur beyond 10^9 cycles and that the difference of fatigue strength between 10^6 cycles and 10^{10} cycles may be of about 200 MPa, where high-cycle tests present lower fatigue strength. Another conclusion is that the infinite fatigue life (i.e., the existence of a horizontal asymptote in the Wöhler curve) assumed for this kind of materials in classical fatigue analysis cannot be taken as a correct rule. For the materials tested, a direct effect of the loading frequency on the fatigue strength does not exist, but high frequencies seem to be more sensible to surface roughness. Other authors (Holper et al., 2004) have analyzed the fatigue process in aluminium alloys for low- and high-frequency fatigue, and their results show that in vacuum and under identical conditions, no influence of the frequency appears.

Other works also evaluate experimentally the effect of the loading frequency on fatigue strength for aluminium (Mayer et al., 2001), pure niobium, and tantalum and Ti 6Al 7Nb (Papakyriacou et al., 2001) under different environmental conditions.

In summary, the most recent works prove a stronger sensitivity of high-frequency fatigue tests to material characteristics and experimental conditions and explain in this way the contradictory results reported by early researchers.

Puskár (1989) described the different phases of high-cycle fatigue processes by means of a qualitative model (Puskár and Golovin, 1984). In this model, four stages are distinguished: fatigue process incubation (redistribution of dislocations, increase of dislocation density, microplastic deformation of surface grains), submicroscopic crack nucleation and propagation (appearance of persistent slip bands), crack propagation in high stress zones (surface striations), and final failure.

2. Method and Experimental Technique

A new method and an experimental system have been developed to study ultrasonic fatigue failure and its relation to the nonlinear behavior of metals under high intensity ultrasonic stresses. In this section, the theoretical hypothesis are summarized (Section 2.1), the experimental techniques and procedures are detailed (Section 2.2), and the method to quantify the damping in metals at high amplitudes (Section 2.3) as well as the procedure for nonlinear elastic characterization (Section 2.4) are described.

2.1 Nonlinear Elasticity Theories. Qualitative Description for a Phenomenological Analysis of the Nonlinear Elastodynamic Behavior of Fatigued Metals

The theory of high-cycle fatigue processes is related to the mechanical, and more specifically to the elastic properties of metals. The fatigue theory predicts that the surface of the hysteresis loop increases when the stress and/or the number of cycles increases. In fact, the microscopic theory proposed by Granato–Lücke, (1956) and Mason (1971) leads to a hysteretical stress–strain (σ–ε) dependence as plotted in Figure 28.2. It justifies the experimentally observed increase of damping both with stress amplitude and number of cycles.

To relate the nonlinear elastodynamic behavior of metals with the fatigue process, we show the relationship between the surface of the hysteresis loop and the nonlinearity parameter of the material. To this purpose, two nonlinear elasticity theories are presented briefly and applied to model nonlinear vibrations in a metallic rod vibrating in its first extensional resonant mode. First, the classical nonlinear elasticity is outlined and applied. Second, a simple hysteretical theory is considered.

The classical nonlinear elasticity theory may be used as a reference: the nonlinear elastic behavior of intact metallic samples may be well described by this approach. Nevertheless this classical model is not appropriate to describe the anomalous nonlinear elasticity observed in fatigued metals. A one-parameter hysteresis loop is proposed (Nazarov and Sutin, 1989). The parameter of the hysteresis loop is directly proportional to the surface of the loop, and the nonlinear elastodynamic behavior of metals is also shown to be directly proportional to this parameter.

Other more sophisticated models have been developed for the explanation of nonclassical nonlinear behavior of materials containing cracks (Maev and Solodov, 2000; Sutin and Nazarov, 1995; Donskoy et al., 2001; Van den Abeele et al., 1997; Scalerandi et al., 2003; and theoretical chapters of this book). Nevertheless the objective of this chapter is only to show in a simple and clear way the relation between the nonlinearity parameter and the physics of the fatigue process.

2.1.1 Classical Nonlinear Elasticity

The following constitutive equation is considered (Murnaghan, 1951),

$$\sigma = Y_0 \left(\frac{\partial u}{\partial x} \right) + Y_1 \left(\frac{\partial u}{\partial x} \right)^2 + Y_2 \left(\frac{\partial u}{\partial x} \right)^3, \qquad (28.1)$$

where σ is the Piola–Kirchhoff stress, u is the displacement, x is the spatial material coordinate and Y_0, Y_1, and Y_2 are, respectively, the second- third- and fourth-order elastic constants for extensional vibration. From the conservation of the momentum law, written in Lagrange coordinates, the following wave equation is obtained,

$$\rho_0 \frac{\partial^2 u}{\partial t^2} = Y_0 \frac{\partial^2 u}{\partial x^2} + Y_1 \frac{\partial}{\partial x} \left[\left(\frac{\partial u}{\partial x} \right)^2 \right] + Y_2 \frac{\partial}{\partial x} \left[\left(\frac{\partial u}{\partial x} \right)^3 \right], \qquad (28.2)$$

where ρ_0 is the density at the initial state and t is the time. A method of successive approximations is used to solve Eq. (28.2). The solution is assumed to be the addition of three terms: $u = u_l + u_2 + u_3$, where u_l is the first-order approximation and u_2 and u_3 are the second- and third-order perturbations. The boundary conditions are linear excitation at one end of the sample ($x = L$) and free vibration at the other ($x = 0$). Harmonic distortion and changes in frequency are then calculated. The well-known linear solution for u_l is easily obtained:

$$u_l(x, t) = -u_0 \cos \omega t \cos k_0 x \tag{28.3}$$

with u_0 being the displacement at the excitation point, ω the angular frequency, and k_0 the wavenumber. The second- and the third-order perturbations, calculated by a perturbative technique (Campos-Pozuelo et al., 2006), result in:

$$u_2(x, t) = \frac{Y_1 u_0^2}{Y_0}(A(x) + B(x) \cos 2\omega t) \tag{28.4}$$

$$u_3(x, t) = u_0^3 \left[F\left(x, \left(\frac{Y_1}{Y_0}\right)^2, \frac{Y_2}{Y_0}\right) \cos \omega t + G\left(x, \left(\frac{Y_1}{Y_0}\right)^2, \frac{Y_2}{Y_0}\right) \cos 3\omega t \right],$$
$$\tag{28.5}$$

where $A(x)$ and $B(x)$ are spatial functions,

$$F\left(x, \left(\frac{Y_1}{Y_0}\right)^2, \frac{Y_2}{Y_0}\right) \quad \text{and} \quad G\left(x, \left(\frac{Y_1}{Y_0}\right)^2, \frac{Y_2}{Y_0}\right)$$

are functions of space and also of the third- and fourth-order elastic constants. The change in the resonance frequency is thus:

$$k = k_0 + \Delta k, \Delta k = -\frac{k^3 u_0^2}{64}\left(21\left(\frac{Y_1}{Y_0}\right)^2 - 18\left(\frac{Y_2}{Y_0}\right)\right), \tag{28.6}$$

where k is the nonlinear wavenumber and k_0 is the low amplitude wavenumber (Campos-Pozuelo et al., 2006).

It must be noted that in this approach, the third harmonic component of the displacement is of the third order on $k u_0$. The change in frequency with amplitude $\Delta k / k$ is of the second order on the linear strain amplitude ($k u_0$).

2.1.2 Hysteretical Model

A hysteresis loop is assumed to describe the macroscopic relation between stress and strain. The following one-parameter hysteresis loop is proposed (Nazarov and Sutin, 1989),

$$\sigma = Y_0 \varepsilon + \begin{cases} -Y_0 \frac{\gamma}{2} \varepsilon^2 & \varepsilon > 0, \frac{\partial \varepsilon}{\partial t} > 0 \\ Y_0 \left(-\gamma \varepsilon_0 \varepsilon + \frac{\gamma}{2} \varepsilon^2\right) & \varepsilon > 0, \frac{\partial \varepsilon}{\partial t} < 0 \\ Y_0 \frac{\gamma}{2} \varepsilon^2 & \varepsilon < 0, \frac{\partial \varepsilon}{\partial t} < 0 \\ Y_0 \left(-\gamma \varepsilon_0 \varepsilon - \frac{\gamma}{2} \varepsilon^2\right) & \varepsilon < 0, \frac{\partial \varepsilon}{\partial t} > 0 \end{cases}, \tag{28.7}$$

where ε is the strain, Y_0 is the Young modulus, $\varepsilon_0(x)$ is the local strain amplitude, and γ is the parameter describing the opening of the hysteresis loop.

The same perturbation technique used in the classical elasticity theory is applied, but only up to the second order. From the conservation of the momentum law written in Lagrangian coordinates, a wave equation is obtained for every part of the cycle.

After applying an analytical Fourier transform and imposing the same boundary conditions as in the previous paragraph, the following expression is obtained for the displacement (Campos-Pozuelo et al., 2006),

$$u(x, t) = -u_0 \cos \omega t \cos kx + \gamma \, u_0^2 F(x) \sin \omega t \\ + \gamma \, u_0^2 G(x) \cos \omega t + \gamma \, u_0^2 H(x) \sin 3\omega t, \tag{28.8}$$

where $F(x)$, $G(x)$, and $H(x)$ are spatial functions. Changes in frequency are also calculated to be $\Delta k = -\gamma \, k u_0 / 3L$. The second-order perturbation from classical elasticity [Eq. (28.4)] should be added to this result to complete the solution.

It is important to note that the third harmonic is already present in this second-order model. Thus, second and third harmonic components are of the same order. This implies that the third harmonic amplitude shows a quadratic dependence on the excitation amplitude. On the other hand, the change of the frequency results in linear dependence on the amplitude, in contrast with the quadratic dependence obtained from the classical theory of elasticity.

2.2 Experimental Technique

2.2.1 High Frequency Fatigue System

The experimental system developed for the generation and analysis of ultrasonic fatigue in metallic samples is basically constituted by an excitation system to drive the samples at resonance and a noninvasive data acquisition system (Figure 28.3; Campos-Pozuelo et al., 2002).

Excitation system

The excitation system consists of an electronic generator designed ad hoc and a piezoelectric transducer (Gallego-Juarez et al., 1994). The electronic generator driving the transducer incorporates a feedback system in order to automatically adjust the excitation frequency to the resonance frequency of the transducer (Gallego-Juarez et al., 1994). The electronic generator also includes a switching circuit for the production of periodic interruptions in the driving signal in order to avoid the temperature of the sample increasing for long-term excitation (Kromp et al., 1973). The circuit establishes and counts the number and length of the tone bursts exciting the samples. In this way the number of applied cycles to the sample is controlled. Burst length and off-time can be varied between 0.1 s and 6000 s.

The driving transducer is an extensional resonant system at about 22 kHz constituted by two half-wave resonant elements: a piezoelectric sandwich and a stepped horn (Figure 28.3). The sandwich element consists of four piezoelectric ceramics placed between two metallic cylindrical rods. The stepped horn acts as a mechanical amplifier to achieve higher vibration amplitude at its thinner termination where the samples are attached.

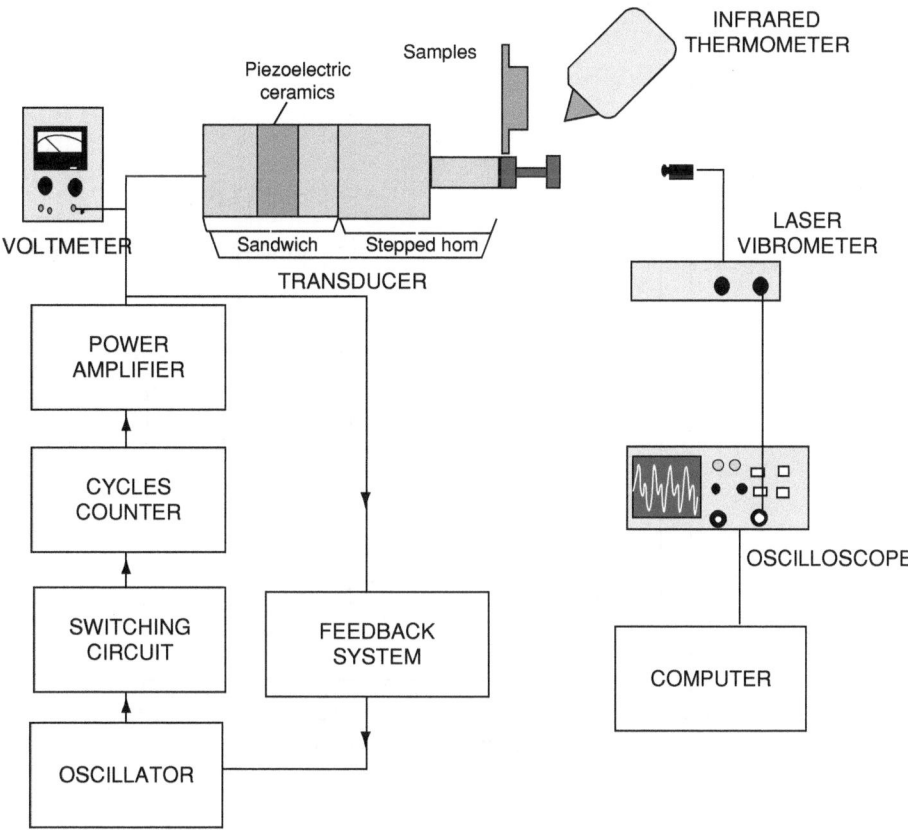

Fig. 28.3. Experimental set-up.

Samples

Two types of samples with stepped profile were designed and constructed to produce flexural and extensional standing waves. The samples were designed following two basic requirements: their first resonance mode has to match the transducer's resonance frequency and, in addition, they have to achieve high strains (enough for fatigue initiation and propagation) in certain sections, while they are driven at linear range strains in other sections. For extensional-wave tests, cylindrical stepped rods as shown in Figure 28.3 were used (Campos-Pozuelo et al., 2002). The theoretical strain distributions for two extensionally vibrating resonant samples of stepped and uniform shape are compared in Figure 28.4a (Campos-Pozuelo and Gallego-Juárez, 1995a). The strong increase of the strain in the central thinner section of the stepped sample allows a nonlinear behavior in this part, while keeping the excitation working in its linear range. In Figure 28.4b, we show the experimental validation of such assumptions. The direct measurement of the strain was made by experimental quantification of the transversal vibration of the sample and considering the relation of the classical and linear theory of the elasticity and the one-dimensional assumption. Uniform cylindrical samples were also used to determine the linear range of the transducer (Campos-Pozuelo and Gallego-Juárez, 1996).

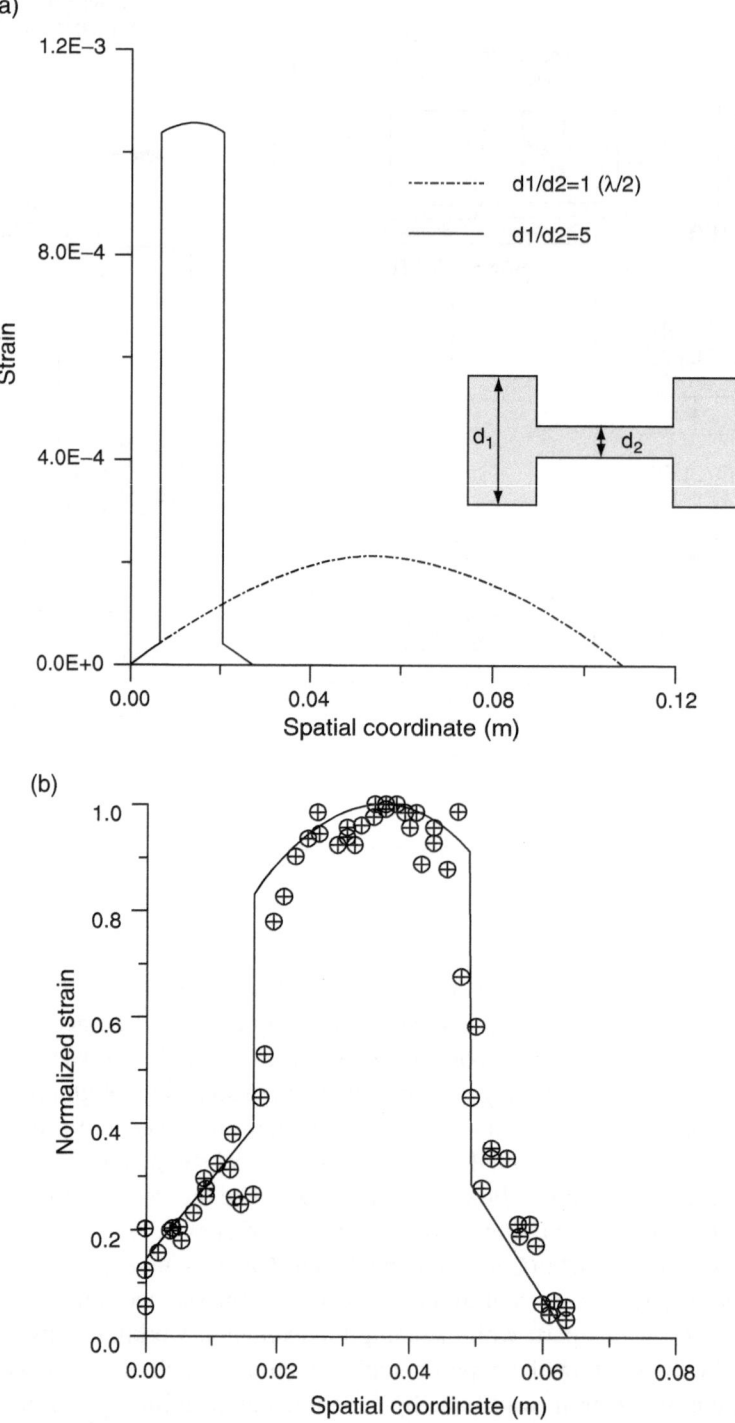

Fig. 28.4. (a) Strain distribution for two resonant samples extensionally vibration: a stepped rod with a diameters relationship $d_1/d_2 = 5$ (solid line) and $\lambda/2$ uniform rod (dash-dotted line). (b) Experimental–theoretical comparison of the strain distribution in a stepped sample extensionally vibration.

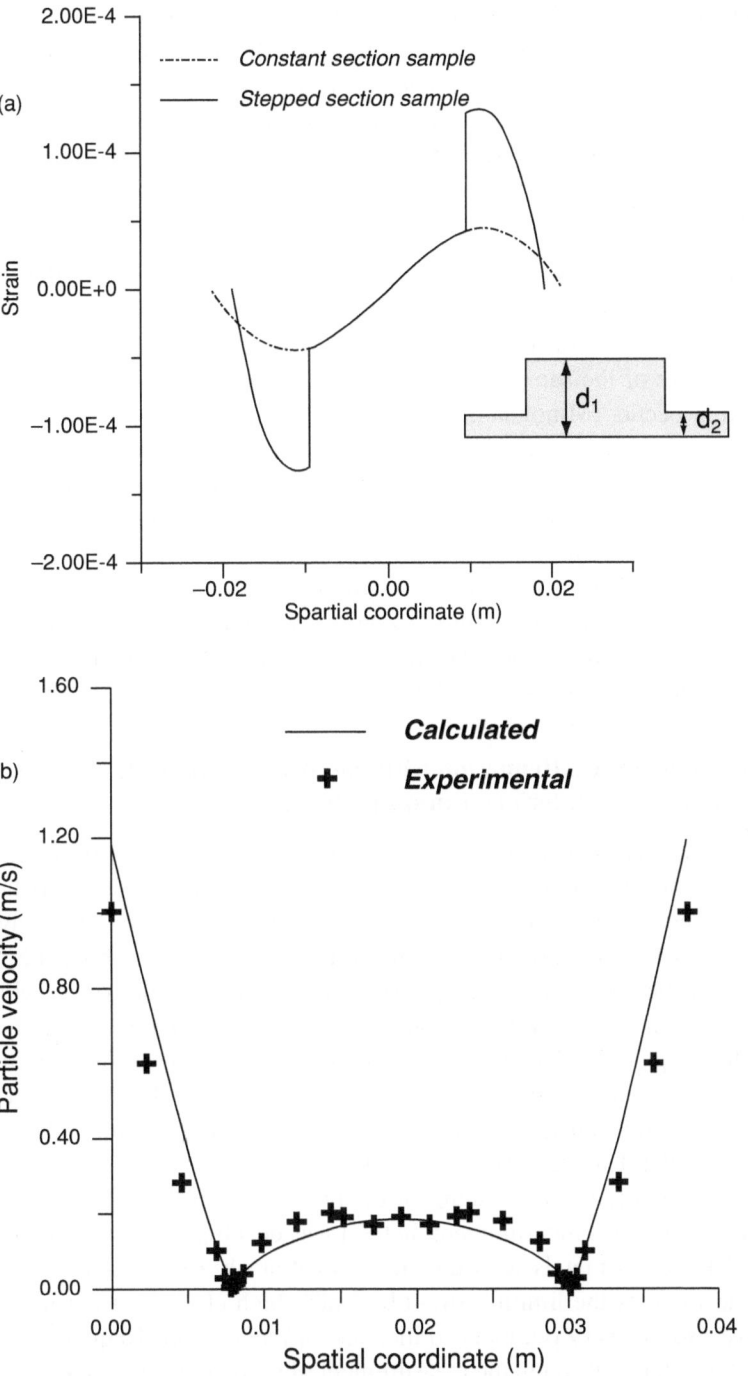

Fig. 28.5. (a) Strain distribution for two resonant samples flexurally vibrationg: a stepped sample with a step radio of $d_1/d_2 = 4$ (solid line) and a constant section bar (dash-dotted line). (b) Experimental–theoretical comparison of the strain distribution in a stepped sample flexurally vibration.

A similar idea was used to design samples for flexural vibration. Figure 28.5a shows the geometry of a stepped sample bar and its strain distribution compared with the distribution in a constant section prismatic bar. In Figure 28.5b the experimental validation of the assumed amplitude distribution is shown for a flexurally vibrating stepped sample (Campos-Pozuelo and Gallego-Juárez, 1995b).

Data acquisition

The data acquisition system is based on a He–Ne laser vibrometer (POLYTEC PFV-502) providing nonintrusive measurements of the particle velocity in the range of 10 microns/s up to 10 m/s for frequencies up to 1.5 MHz with an accuracy of 2 microns/s. The temperature of the sample was monitored at the nodal section, where the maximum heating occurs (Minogna et al., 1981), by using an infrared thermometer. The vibration signal was automatically acquired and treated by a PC.

To fatigue the samples, bursts of between 0.1 and 1 second with intervals of several seconds were applied to produce peak stress values high enough for crack nucleation (of the order of 300 MPa).

The system allows tests at different controlled temperatures by using a thermically isolated oven, provided with a control system to keep the inside temperature constant. The temperature may be measured by a thermocouple and regulated by a multifunction controller, connected to the electric heater.

2.2.2 Variation of the Attenuation with the Strain Amplitude for Intact (Nonfatigued) Materials: Limiting Strain

It has already been established that the internal friction increases with strain. Following the above-mentioned theory of Granato–Lücke, (1956), some authors (Mason, 1968, 1956; Mason and MacDonald, 1972; Nazarov, 1991; Kuz'menko, 1975) have assumed that the observed increase of attenuation for an ultrasonic wave above a critical strain amplitude can be associated with the formation of dislocation loops, which lead to fatigue crack. This critical value beyond which fatigue failure can easily be produced is called limiting strain. The increase of attenuation with the strain amplitude and its relationship with fatigue damage initiation has been analyzed by a number of authors and several experimental systems have been proposed (Mason, 1968, 1956; Mason and MacDonald, 1962; Kuz'menko, 1975; Puskár, 1977).

By using the experimental system for the production of the high-frequency fatigue previously described, a measurement method was developed to evaluate the variations of internal losses in metals with the strain level at ultrasonic frequency, and to determine the value of the limiting strain beyond which changes occur in the material, which give rise to a very pronounced increase of attenuation. To quantify the internal losses in the material, through the measurement of particle velocity and excitation voltage, a linear one-dimensional model of the assembly constituted by the transducer and the sample (Figure 28.6) has been developed (Campos-Pozuelo and Gallego-Juárez, 1996). This model is based on the following assumption: the piezoelectric sandwich may be approached by a piezoelectric ceramic rod having the same resonant frequency and mechanical impedance, and it is considered at one end to be free $x = 0$ and

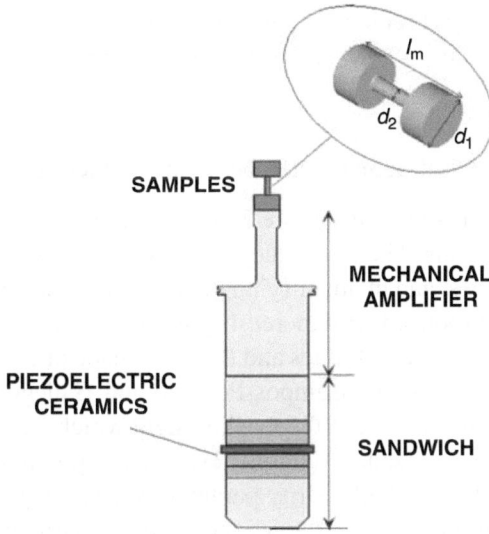

Fig. 28.6. Transducer with an extensional sample.

at the other to be coupled to the bigger section of the mechanical amplifier. That means that boundary conditions for the piezoelectric bar are $F(x = 0) = 0$ and $F(x = l_c) = Z_B \dot{u}$, Z_B being the termination impedance from the assembly constituted by the mechanical amplifier and the sample, F the force, $\dot{u}$ the particle velocity, and l_c the length of the piezoelectric rod. To calculate Z_B, energy loss is neglected in the mechanical amplifier, which is assumed to be resonant ($l = \lambda/2$). The input impedance of the sample (Z_m) is calculated by neglecting the energy dissipation at the external bigger sections, where the material vibrates under linear conditions. Also it is assumed that $\alpha_m l_m << k l_m$ (α_m is the attenuation of the sample material and l_m its length) and that the stepped samples are resonant ($\tan k \, l_m/4 = d_2/d_1$). To obtain the limit for linear behavior of the transducer, measurements were done with constant section rod samples, where no increase of the strain occurs.

With all these assumptions and by using the piezoelectric equations (Mason, 1968), the following relation is obtained between the applied voltage (V_a), and the vibration velocity at the end of the sample $\left(\dot{u}_f\right)$,

$$V_a = \dot{u}_f(K_t + K_m \alpha_m),\tag{28.9}$$

where

$$K_t = \left(\frac{D_2}{D_1}\right)^2 \frac{Z_{0c}\alpha_c l_c \pi D_1^2}{8\varphi}, \quad K_m = \left(\frac{D_1}{D_2}\right)^2 \frac{Z_{0m}\pi d_1^2 \alpha_c}{16\varphi} l_m,$$

D_1 and D_2 are the diameters of the two sections of the mechanical amplifier, Z_{0c} and Z_{0m} are the specific impedances of the piezoelectric element and of the sample, respectively, φ is the piezoelectric constant, and α_c the attenuation of the piezoelectric element. In the linear range of the transducer K_t is a constant.

From this equation the sample attenuation, α_m, can be expressed in the form

$$\alpha_m = \frac{\left(V_a/\dot{u}_f - K_t\right)}{K_m}.\tag{28.10}$$

In this way, from the measurement of the driving voltage V_a and the particle velocity at the end of the sample $\dot{u}_f$, the attenuation for different excitation levels may be obtained.

2.2.3 Analysis of the Nonlinear Elastic Behavior of Metals

The experimental set-up shown in Figure 3 is also valid for the nonlinear characterization of metallic materials. The technique basically consists of exciting the first resonant mode of the samples, and in quantifying the nonlinear harmonic distortion and the changes in frequency produced with increasing strain amplitudes. The vibration mode is driven at different voltage amplitudes and the waveshape of the vibration is analyzed by using standard FFT methods (Campos-Pozuelo and Gallego-Juarez, 1995a, b). It must be noted that by means of the feedback system which automatically adjusts the excitation frequency to the resonance frequency, it is also possible to quantify the nonlinear changes in frequency. Nonlinearity parameters may be quantified by measuring the slope of the straight line relating the second (or third) harmonic to the fundamental or the slope of the line relating changes in frequency with strain amplitude.

To analyze the changes in the nonlinearity parameter with the number of applied cycles during the fatigue process, the velocity response of the sample is captured and the signal at the turn-off of the tone burst exciting the sample is analyzed, as proposed by Van Den Abeele et al. (2002). In such a procedure the amplitude of the excitation signal is kept constant and the amplitude dependence is obtained by a time-windowed analysis of the reverberation signal. Figures 28.7 and 28.8 show examples of the different signals captured from the samples in the time and frequency domains, respectively. The reverberation technique, which is very useful for nonlinear characterization at constant excitation, presents the disadvantage of a high noise level at low amplitudes. This is because high- and low-amplitude measurements are made by using

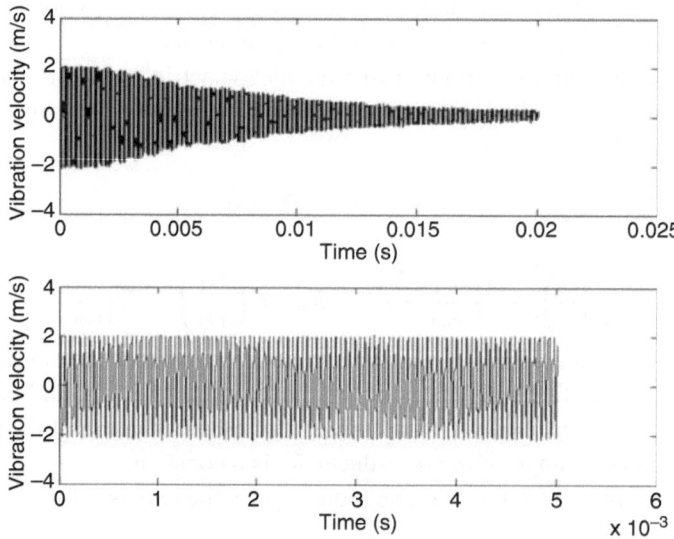

Fig. 28.7. Received signal from samples (time domain).

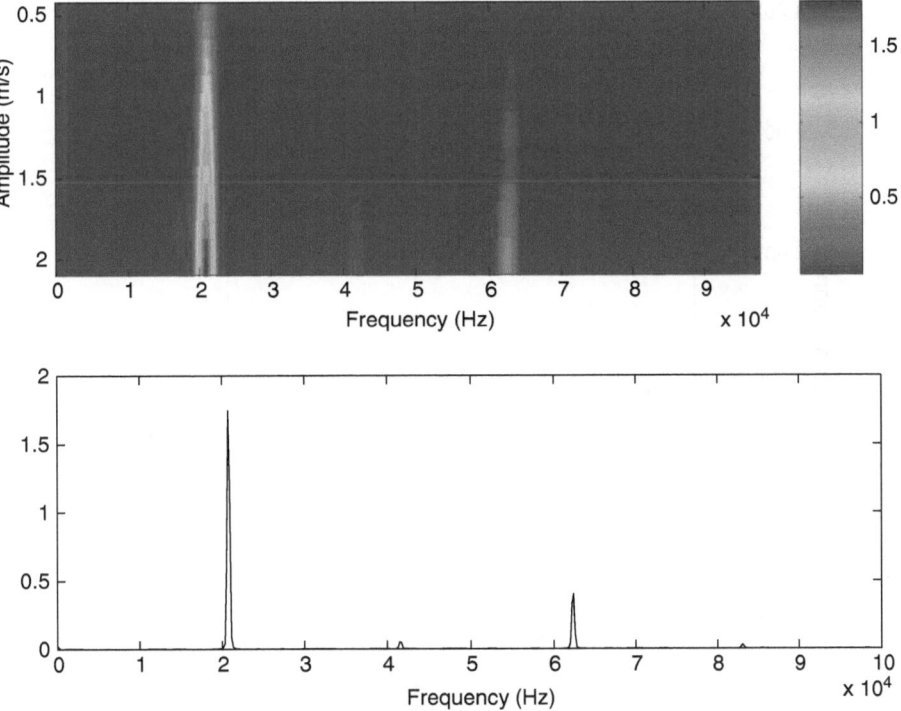

Fig. 28.8. Received signal from sample (frequency domain).

only one acquisition signal, and then the absolute vertical resolution is the same for both measurements. Therefore the relative vertical resolution for the low-level values is poorer than for the high-level values. For a more efficient use of this technique, more sophisticated signal treatment methods should be applied (Van Den Abeele et al., 2002).

3. Measurements

3.1 Limiting Strain of Metallic Alloys: Attenuation Versus Amplitude for Intact Samples

The procedure described in Section 2.2.2 has been applied to the determination of the limiting strain of some metallic alloys such as duraluminium and Ti 6Al 4V. The evolution of the internal friction with the strain amplitude has been obtained from the measurements of the applied voltage at the transducer and the vibration amplitude at the end of the samples. Results are shown in Figures 28.9 and 28.10. The threshold of amplitude beyond which the attenuation increases rapidly can be easily observed for duraluminum as well as for the titanium alloy (Ti 6%Al 4% V) in Figures 28.9 and 28.10, respectively. Such results have been obtained for a constant temperature at 23° C.

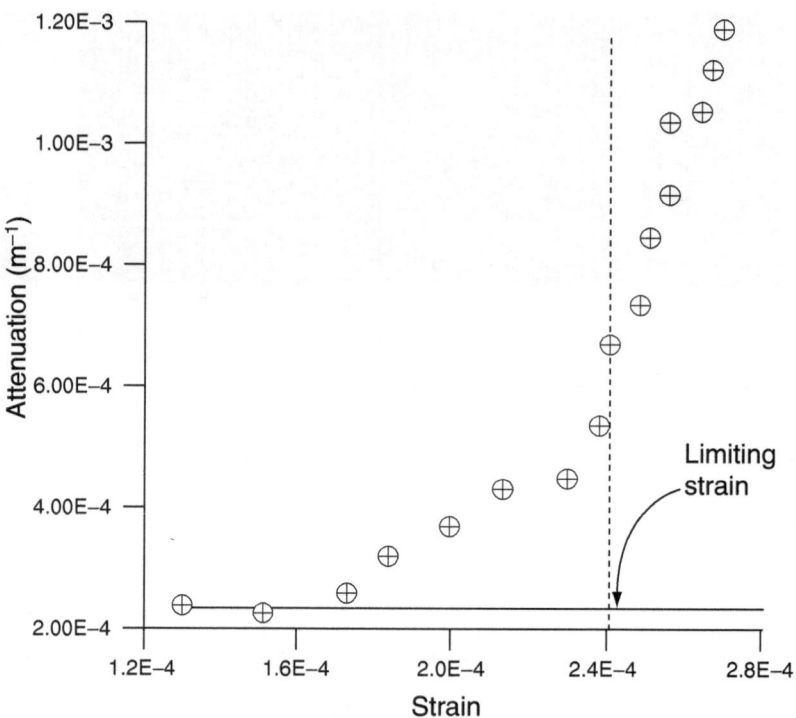

Fig. 28.9. Evolution of the attenuation with the strain samplitude in duraluminum.

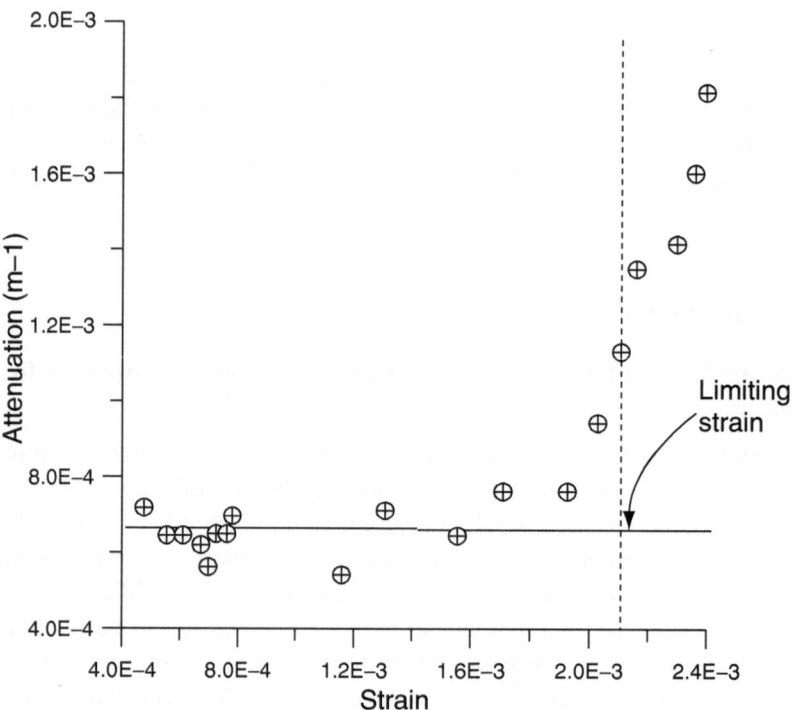

Fig. 28.10. Evolution of the attenuation with the strain amplitude in Ti6Al4V.

3.2 Comparative Analysis of the Nonlinear Behavior of Metallic Alloys Before and After Being Fatigued

Experimental tests about the nonlinear elastodynamic behavior of fatigued and intact samples have been carried out. The samples are fatigued by using the experimental set-up previously described. Two metallic alloys of current use in the aeronautics industry as well as in the construction of high-power ultrasonic transducers (duraluminium and Ti 6Al 4V) have been tested by using extensional and flexurally vibrating samples. The nonlinearity parameter was quantified from nonlinear harmonic distortion in the case of flexural vibrations and from the changes in frequency with strain amplitude for extensional vibrations. The nonlinear elastic behavior is compared for the same sample before and after fatigue crack nucleation.

In Figure 28.11 the third harmonic relative to the fundamental amplitude is plotted as a function of the fundamental amplitude, before and after fatiguing the sample, for titanium and aluminum alloys. The important increase of the nonlinear harmonic distortion after fatiguing the sample is evident for both materials. Figure 28.12 shows, for a Ti6Al4V sample, the comparison of the evolution of the second and third harmonics before and after fatiguing the material. The signals were picked up at the same measurement point. It is clear that the fatigue process produces a notable increase of the third harmonic, which becomes stronger than the second one. Such behavior cannot be

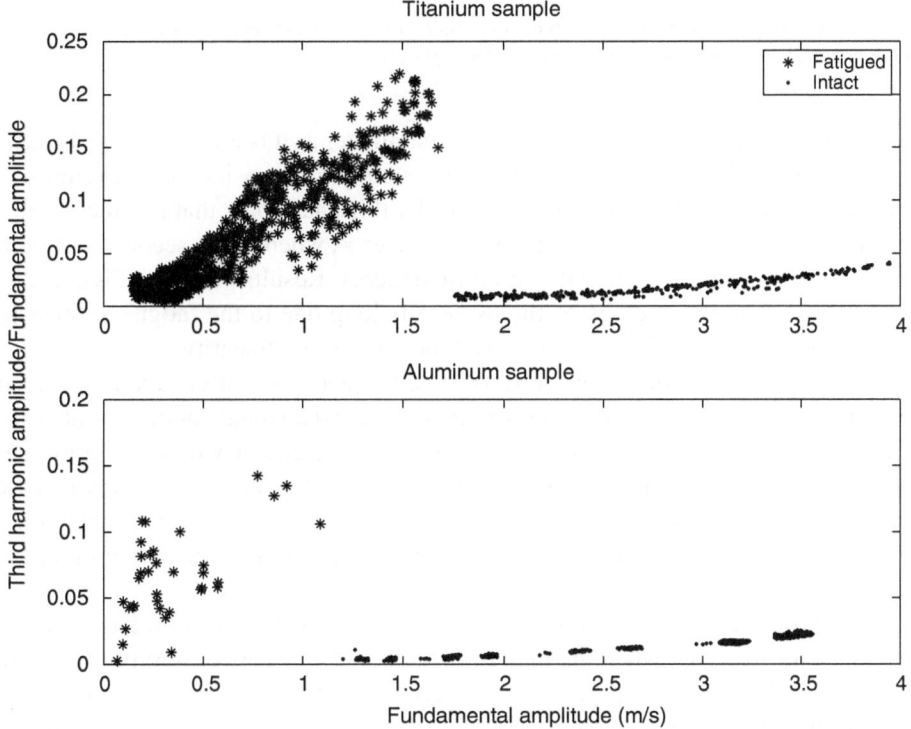

Fig. 28.11. Evolution with amplitude of the third harmonic normalized to the fundamental amplitude for intact and fatigued sample.

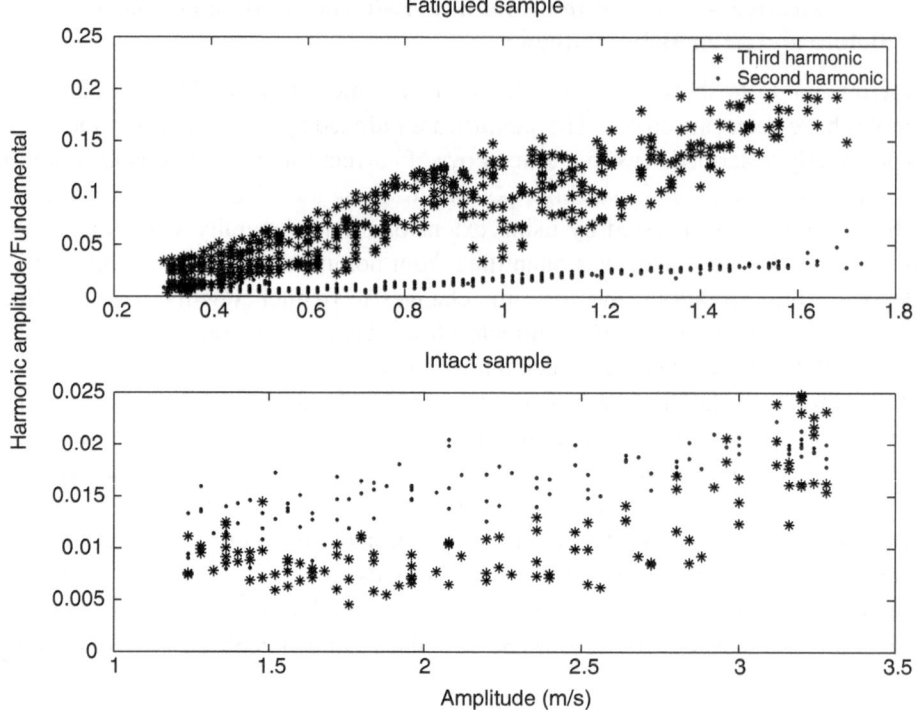

Fig. 28.12. Comparison of the evolution with amplitude of the third and second harmonics normalized to the fundamental amplitude for intact and fatigued sample.

interpreted as an increase of the nonlinearity parameter as it is understood in the classical theory of the elasticity (see Section 2.1.1). A simple theory including hysteresis, as proposed in Section 2.1, predicts a second-order third harmonic that is directly related to the hysteresis loop surface through the parameter γ, whereas the second harmonic is not affected by this parameter in this simplified theory. Results plotted in Figures 28.11 and 28.12 confirm the opening of the hysteresis loop due to the fatigue process (see Section 2.1.2) as the origin of the observed increase of nonlinearity.

The amplitude dependence of the resonance frequency of fatigued and intact samples on the two analyzed materials for their first extensional mode has also been experimentally quantified. A linear dependence of the frequency on the vibration amplitude, which is hard to justify from the classical nonlinear elasticity theory (see Section 2.1.1), is observed. Figure 28.13 shows the dependence of the frequency on the amplitude of the vibration for extensionally vibrating samples of duraluminum and Ti6Al4V.

In conclusion, it can be stated that the experimental results confirm the increase of the hysteresis loop area when the material is fatigued and the possibility of detecting such increase by a nonlinear elastic characterization of the material. The comparison between the different models and the experimental data shows that before fatiguing, the samples follow a classical nonlinear behavior whereas after crack nucleation, their nonlinear elastic behavior is better described by a hysteretical model.

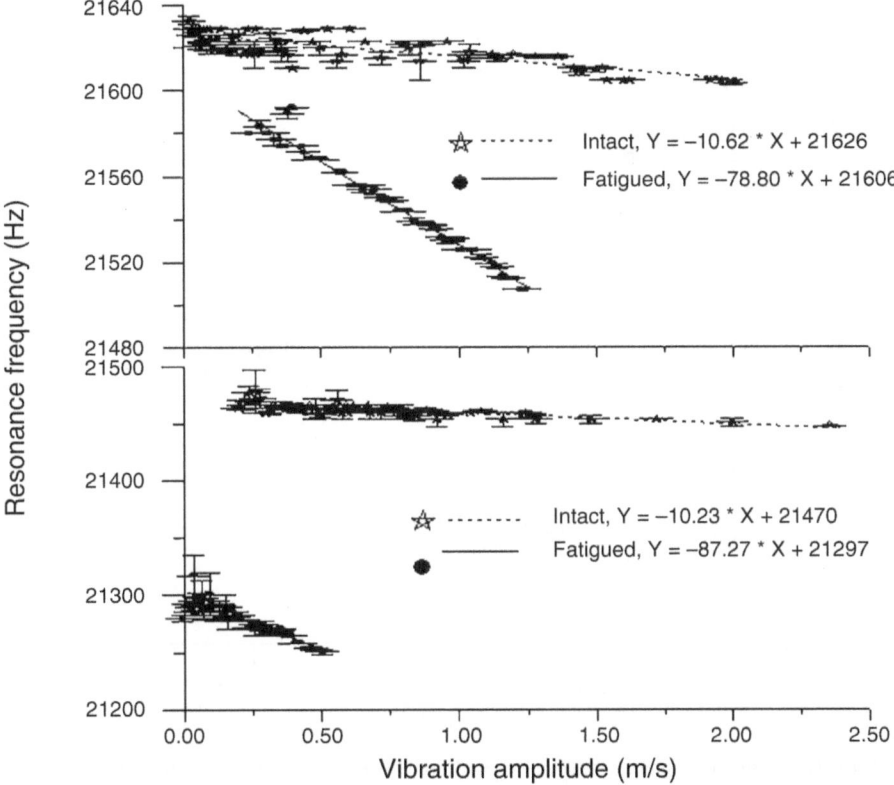

Fig. 28.13. Evolution of the fundamental frequency with amplitude for intact and fatigued sample.

3.3 Evolution of the Nonlinearity Parameters During the Fatigue Process

From the experimental results shown in Section 3.2 it seems clear that the nonlinear elastic characterization of metals can be used to monitor the fatigue process. The work of Nagy (1998) shows the viability of using nonlinear ultrasonic techniques for fatigue damage assessment. He measures the second-order acoustoelastic coefficient in a great variety of materials (including metals) during the fatigue process (classical, low-frequency fatigue). Van den Abeele et al. (2002) use nonlinear ultrasonic vibration to monitor the fatigue process. Both works conclude that nonlinear features are more sensible to fatigue damage than classical linear characterization. We propose here, on the basis of the fatigue system described in Section 2.2.1 and the nonlinear characterization methodology employed to compare intact and fatigued samples in Section 3.2, to use the high-amplitude ultrasonic vibrations fatiguing the samples for nonlinear characterization and monitoring of the linear and nonlinear elastic behavior of the metal during the fatigue process.

Samples are fatigued by applying a constant voltage to the driving transducer (see Section 2.2.1). When the fatigue crack nucleates, the sample attenuation increases. Then, according to the results of Section 2.2.2, the displacement of the sample decreases. Results will be always analyzed in terms of values normalized to the vibration amplitude. Bursts of length of 0.1 s at the resonance frequency are applied to the

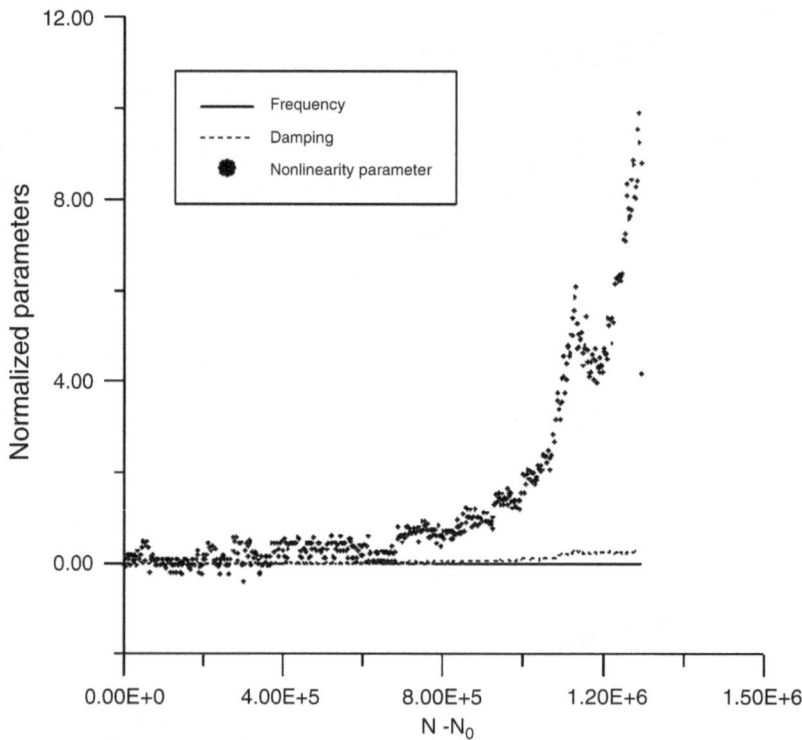

Fig. 28.14. Evolution of frequency, attenuation, and nonlinearity parameters with the number of applied cycles for an extensionally vibrating sample.

sample and the reverberation is automatically recorded for every burst. To characterize the linear and nonlinear elastic behavior of the samples during the fatigue process, frequency, attenuation, and nonlinearity parameters are evaluated for every acquisition. In Figure 28.14, results are plotted for a extensionally vibrating duraluminum sample. To construct this curve we consider the results of Figure 28.13 and define a nonlinearity parameter as the slope of the straight line fitting the frequency changes versus amplitude. To evaluate the attenuation, the envelope of the reverberation signal is calculated. All parameters are normalized by relating them to their initial values. Figure 28.15 shows results obtained for a flexurally vibrating Ti6Al4V sample. The harmonic distortion is represented by the ratio of the third harmonic to the fundamental amplitude for the maximum amplitude window in the reverberation signal. Both Figures 28.14 and 28.15 ratify the idea that nonlinear parameters are more sensitive than linear parameters to fatigue damage changes.

4. Conclusions

A study showing the relationship between ultrasonic fatigue and nonlinear elastic behavior of metals has been the object of the present chapter. An experimental system to produce and analyze high-frequency fatigue and a method for nonlinear

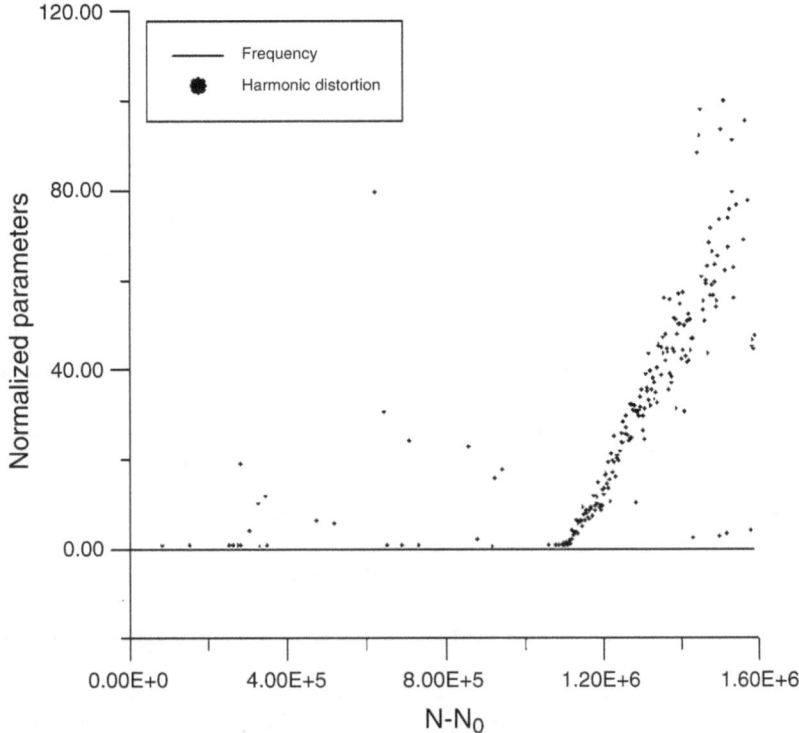

Fig. 28.15. Evolution of frequency, nonlinearity parameter, and harmonic distortion with the number of applied cycles for a flexurally vibrating sample.

characterization of metals by using high-amplitude ultrasonic vibrations have been described. Experimental results show that the nonlinearity parameters of the metal strongly increase with fatigue in the material. Moreover, it has been shown that the nature of the nonlinear elasticity exhibited by intact and fatigued specimens is different. Intact samples follow the predictions of the classical nonlinear elasticity theory whereas fatigued samples present a hysteretical behavior. Such results are in agreement with the description of the fatigue process outlined in the introduction of the chapter. The important changes in the nonlinear elastic behavior of metal samples (harmonic distortion, amplitude dependence of frequency and attenuation) compared to changes in linear elastic properties (frequency and attenuation) indicate that nonlinear characterization should be a good method for fatigue damage assessment in metals.

Acknowledgments

This work has been developed in the frame of the NATEMIS network of the European Science Foundation. It has been financed by the CAM project 07N/0115/02 and by the European Union project STREP FP6-502927 (AERONEWS).

References

Bratina, W.J., 1966, Internal friction and basic fatigue mechanisms in body-centered cubic metals, mainly iron and carbon steels, *Physical Acoustics* Vol.III , W. P. Mason, ed., Academic Press, New York and London, pp. 223–291.

Campos-Pozuelo, C. et al., 2002, Procedimiento y dispositivo para el estudio de la resistencia a fatiga de materiales metálicos a frecuencias ultrasónicas y temperatura constante, Patente española solicitud no. 200201700.

Campos-Pozuelo, C. and Gallego-Juarez, J.A., 1995a, Finite amplitude standing waves in metallic rods, *J. Acoust. Soc. Am.* **97:** 875–881.

Campos-Pozuelo, C. and Gallego-Juarez, J.A., 1995b, Finite amplitude flexural vibrations at ultrasonic frequencies in metallic bars *J. Acoust. Soc. Am.* **98:** 1742–1750.

Campos-Pozuelo, C. and Gallego-Juarez, J.A., 1996, Limiting strain of metlas subjected to high ultrasound, *Acustica-Acta Acustica* **82:** 823–828.

Campos-Pozuelo, C., Vanhille, C. and Gallego-Juárez, J.A., 2006, Comparative study of the nonlinear behaviour of fatigued and intact samples of metallic alloys, Under revision in *IEEE Transactions on Ultrasonics, Ferroelectrics, and Frequency Control.*, **53**: 175–184.

Donskoy, D., Sutin, A., and Ekimov, A., 2001, Nonlinear acoustic interaction on contact interfaces and its use for nondestructive testing. *NDT&E Int.* **34:** 231–238.

Gallego-Juarez, G., Rodriguez-Corral, J., San Emeterio, L. and Montoya-Vitini, F., 1994, European Patent EP 450,030 1991, US Patent 5,299,175.

Granato, A. and Lücke, K., 1956, Theory of mechanical damping due to dislocations, *J. Appl. Phys.* **27:** 583–593.

Holper, B., Mayer, H., Vasudevan, A.K., and Stanzl-Tschegg, S.E., 2004, Near threshold fatigue crack growth at positive load ratio in aluminium alloys at low and ultrasonic frequency: influences of strain rate, slip behaviour and air humidity, *Int. J. Fatigue* **26:** 27–38.

Kromp, W., Kromp, K., B-T, H., Langer, M., and Weiss, B., 1973, Techniques and equipment for ultrasonic fatigue testing. *Ultrasonic International 1973 Conference Proceedings.* Butterworth-Heinemann Ltd, Oxford 238–243.

Kuz'menko, V.A., 1975, Fatigue strength of structural materials at sonic and ultrasonic loading frequencies, *Ultrasonics* **13:** 21–30.

Maev, R.G. and Solodov, I.Y., 2000, Nonlinear acoustic spectroscopy of cracked flaws and disbonds: Fundamentals, techniques, and applications. *Review of Progress in Quantitative Nondestructive Evaluation.* Ed. D.O. Thompson and D.E. Chimenti. American Institute of Physics, 1409–1416.

Marines, I., Dominguez, G., Baudry, G., Vittori, J.-F., Rathery, S., Doucet, J.-P., and Bathias, C., 2003, Ultrasonic fatigue tests on bearing steel AISI-SAE 52100 at frequency of 20 and 30 kHz, *Int. J. Fatigue* **25:** 1037–1046.

Mason, W.P., 1956, Internal friction and fatigue in metals at large strain amplitude, *J. Acoust. Soc. Am.* **28:** 1207–1218.

Mason, W.P., 1968, Low and high-amplitude internal friction measurements in solids and their relation to imperfection motions, C.J. Mahon Jr., ed., Interscience, New York, pp. 288–364.

Mason, W.P., 1971, Internal friction at low frequencies due to dislocations: Applications to metals and rock mechanics, *Physical Acoustics* Vol. VIII, W. P. Mason and K.N. Thurston, ed., Academic Press, New York and London, pp. 347–371.

Mason, W.P. and Macdonald, D.E., 1972, The use of high-power ultrasonics (macrosonics) in studying fatigue in metals, *J. Acoust. Soc. Am.* **51:** 894–899.

Matikas, T.E., 2001, Specimen design for fatigue testing at very high frequencies, *Journal of Sound and Vibration* **247:** 673–681.

Mayer, H., Papakyriacou, M., Pippan, R., and Stanzl-Tschegg, S., 2001, Influence of loading frequency on the high cycle fatigue properties of AlZnMgCu1.5 aluminium alloy, *Mater. Sci. Eng. A* **314:** 48–54.

Murnaghan, F.D., 1951, *Finite Deformation of an Elastic Solid.* Ed. Dover, New York.

Minogna, R.B., Green, R.E. Jr., Duke, J.C. Jr., Henneke II, E.G., and Reifsnider, K.L. 1981, Thermographic investigation of high-power ultrasonic heating in materials, *Ultrasonics* **19:** 159–163.

Nagy, P.B., 1998, Fatigue damage assessment by nonlinear ultrasonic materials characterization, *Ultrasonics* **36**: 375–381.

Nazarov, V.E., 1991, Nonlinear acustic effects in annealed copper. *Sov. Phys. Acoust.* **37**: 75–78.

Nazarov, V.E. and Sutin, A.M., 1989, Harmonic generation in the propagation of elastic waves in nonlinear solid media, *Sov. Phys. Acoust.* **35**: 410–413.

Papakyriacou, M., Mayer, H., Pypen, C., Plenk, H. Jr., and Stanzl-Tschegg, S., 2001, Influence of loading frequency on high cycle fatigue properties of b.c.c. and h.c.p. metals, *Mater. Sci. Eng. A* **308**: 143–152.

Puskár, A., 1977, The thermal activation of cumulative fatigue damage at ultrasonic frequencies, *Ultrasonics* **15**: 124–128.

Puskár, A., 1982, *The Use of High-Intensity Ultrasonics,* Materials Science Monographs 13, Elsevier, Amsterdam-Oxford-New York.

Puskár, A., 1989, *Microplasticity and Failure of Metallic Materials,* Materials Science Monographs 56, Elsevier, Amsterdam-Oxford-New York-Tokyo.

Puskár, A. and Golovin, S.A., 1984, *Fatigue in Materials: Cumulative Damage Processes,* Elsevier, Amsterdam-Oxford-New York.

Scalerandi, M. et al., 2003, Local interaction simulation approach to modelling nonclassical, nonlinear elastic behaviour in solids *J. Acoust. Soc. Am.* **113**: 3049–3059.

Sutin, A.M. and Nazarov, V.E., 1995, Nonlinear acoustic methods of crack diagnostics. *Radiophysis. Quant. Electron.* **38**: 109–120.

Van den Abeele, K. et al., 1997, On the quasi-analytic treatment of hysteretic nonlinear response in elastic wave propagation, *J. Acoust. Soc. Am.* **101**: 1885–1898.

Van den Abeele, K. et al., 2002, Analysis of the nonlinear reverberation of titanium alloys fatigued at high amplitude ultrasonic vibration, *Revista de Acústica*, Vol XXXIII, CD-ROM (NON-02-002-IP).

Wang, Q.Y., Bathias, C., Kawagoishi, N., and Chen, Q., 2002, Effect of inclusion on subsurface crack initiation and gigacycle fatigue strength, *Int. J. Fatigue* **24**: 1269–1274.

Nonlinear Acoustic NDE: Inherent Potential of Complete Nonclassical Spectra

I. Solodov[1], K. Pfleiderer, and G. Busse

Institute for Polymer Testing and Polymer Science (IKP)–Nondestructive Testing–(ZFP), Stuttgart University, Stuttgart 70569, Germany
[1]To whom correspondence should be addressed: e-mail: solodov@ikp.uni-stuttgart.de

Abstract

The classical approach to nonlinear acoustic nondestructive evaluation (NDE) is based on the higher (ultra-)harmonic or mixed frequency response of an imperfect material. In the nonclassical case, beyond the well-known ultraharmonics and the modulation sidebands nonlinear spectra acquire a number of new spectral components: subharmonics, ultrasubharmonics, and ultrafrequency pairs. These nonclassical nonlinear modes demonstrate a high localization around defects and provide new opportunities for early detection and recognition of damaged areas. The chapter includes theoretical background and extensive experimental results on defect selective nonlinear imaging and NDE using complete multifrequency nonclassical spectra.

Keywords: Higher harmonics, hysteresis, imaging, instability, nonclassical nonlinearity, nondestructive evaluation (NDE), self-modulation, subharmonics, wave modulation

1. Introduction

A gradual pace of a 30-year history of nonlinear acoustics of solids has been disturbed by a dramatic turn over the last decade. In early 1960s, a classical field of investigations was aimed at homogeneous (flawless) crystals whose nonlinearity was associated with lattice anharmonicity. As a result, a unique means was created for experimental characterization of nonlinear behavior of interatomic forces in crystalline materials.[1] However, even in the first experimental studies of imperfect materials a substantial increase in nonlinearity was measured as soon as dislocations were induced in a single crystal of Al by a mechanical impact.[2] Further investigations confirmed an important role of internal boundaries and microinhomogeneities in enhancement of acoustic nonlinearity in fatigue materials with dislocations and in alloys containing internal interfaces between matrix and precipitate.[3]

A number of studies were then implemented to discover the mechanisms underlying boundary nonlinearity using surface and interface acoustic waves.[4,5] These results were supplemented by direct observations of efficient higher harmonic generation in bulk acoustic wave reflection from an interface between two nonlinear solids.[6] The experiments revealed an increase in acoustic nonlinearity by several orders of magnitude for both surface and bulk waves[7,8] in a weakly bonded contact. Besides the much higher efficiency, such a contact acoustic nonlinearity (CAN) was shown to exhibit a substantial qualitative departure from fundamental nonlinear effects of higher harmonics generation and acoustic wave interaction, which have been a predominant subject for most of the studies in classical nonlinear acoustics. The family of nonlinear contact phenomena included frequency transformations down on the spectrum, hysteresis, instabilities, stochastic effects, and so on[9-12] that are well known in other branches of nonlinear physics. Because the CAN closely simulates the nonlinear behavior of a crack, the nonlinear acoustics of the weakly bonded interface became a topical subject of numerous studies and applications concerned with NDE of cracked defects.

Another area where the deviations from classical nonlinear acoustics were found to be evident was acoustics of structurally inhomogeneous media and rocks in particular.[13] The grainy structure of rocks, apparently, comprises a number of nonideally bonded interfaces whose nonlinear response is the main source of nonlinearity in geomaterials. Different types of contact bonds in the interfaces between grains result in various mechanisms of structural nonlinearity[14] that are involved in interpreting nonclassical manifestations of acoustic nonlinearity in rocks. These mechanisms are complemented by the Preisach–Mayergoyz (P–M) formalism to include hysteresis of elastic properties of rocks.[15] The P–M scheme enables a phenomenological description of such nonclassical manifestations as elastic hysteresis and discrete memory observed in static and dynamic experiments. It also predicts unconventional features of quadratic dynamic characteristics of the higher harmonics and a linear amplitude dependence of the resonance frequency shift for acoustic waves in rocks.

The above approaches originated from quite different areas: nonlinear acoustics of interfaces and acoustics of structurally inhomogeneous media and rocks. However, ultimately they came to the common conclusion that these new nonlinear (nonclassical) phenomena may become a novel versatile tool for nonlinear material characterization and NDE, with a much broader area of applications than its classical predecessor. The latter fact is, mainly, due to the following circumstances. The stiffness of classical materials can be considered as a locally quasilinear characteristic because even for high acoustic strains $\approx 10^{-4}$ the contribution of nonlinear terms in the stiffness variation with the acoustic strain is usually below 10^{-3}. As a result, noticeable nonlinear effects are developed only because of the accumulation of the nonlinear response along distance and classical materials are said to display a distributed nonlinearity.

On the contrary, the acoustic wave interaction with a weakly bonded area in a solid is accompanied by a strong local stiffness variation: the stiffness of the interface (or a crack) can be substantially greater for compression than for tensile stress which is high enough to cause an intermittent contact between the crack surfaces.[16] In this case, the intact material outside the defect can be considered as a "linear carrier" of the acoustic wave and one can consider the localized nonlinearity of the imperfect solid. This nonclassical feature of CAN causes the effects of nonlinear reflection and scattering by cracked defects[8,17] and enables 2-D-imaging of the nonlinear excitations confined inside such defect areas.[18] Thus, nonlinear NDE of imperfect materials is inherently defect selective; that is, it distinctively responds to fractured flaws.

Fortunately, this group of flaws includes the most malignant defects for material strength: micro- and macrocracks, delaminations, debondings, impact and fatigue damages, and so on, which are discussed within the scope of this chapter.

Another implication of the localized nature of CAN is concerned with instability phenomena that can develop in the defect area.[19] The lower stiffness of the cracked area makes it behave as a localized oscillator which is, apparently, strongly nonlinear due to CAN. Therefore, for an intense acoustic excitation it can manifest such effects as subharmonic generation, instability, and transition to chaotic dynamics unconventional for distributed nonlinearity of classical materials but known for nonlinear resonators. As a result, the spectrum of local oscillations acquires a number of new nonlinear components and thus demonstrates a substantial departure from the classical multiple higher harmonic collection. Such nonclassical spectra open new opportunities for nonlinear NDE.

The latter fact is emphasized in this chapter: the physical effects that lay the basis for the emergence of nonclassical nonlinear spectra are considered through the prism of NDE applications. The chapter is organized in two main parts. The nonresonant manifestations of CAN are discussed in the first section, whereas the second one is concerned with instability phenomena. Each section opens with a phenomenological interpretation of nonlinear phenomena followed by case studies of nonlinear NDE applications.

2. Nonlinear NDE Using Nonresonant Effects of Acoustic Wave–Defect Interaction

2.1 Higher Harmonic Mode

If the amplitude of an acoustic wave exceeds the static stress of an originally closed interface it causes vibrations of intermittent contact between the defect fragments: clapping and/or rubbing of the microasperities provide a strongly nonlinear contact dynamics.[16] Clapping, apparently, results in asymmetrical modulation of the contact stiffness: it is higher for compression (C) and lower for the contact extension $(C - \Delta C)$. Such a bimodular behavior of a prestressed contact driven by a harmonic acoustic strain $\varepsilon(t) = \varepsilon_0 \cos \nu t$ is similar to a "mechanical diode" and results in a pulse-type modulation of its stiffness $\Delta C(t)$. It also provides an unconventional nonlinear wave-form distortion: a half-period rectified output instead of the sawtoothlike profile in classical materials. Because $\Delta C(t)$ is a pulse-type periodic function of the driving frequency ν, the spectrum of the stress induced in the damaged area $(\Delta C(t) \cdot \varepsilon(t))$ contains a number of its higher harmonics $n\nu$ (both odd and even orders) whose amplitudes are modulated by the *sinc*-envelope function. The depth of stiffness modulation $(\Delta C / C)$ can be as high as ~ 1 (for weakly stressed contacts) thus providing a very efficient ultraharmonic generation by clapping defects.[8]

The dynamics of the damaged area driven by shear traction results in friction-controlled rubbing between its microcontacts.[18] In this case, the stiffness modulation is caused by the transition between stick and slide phases: higher contact stiffness provided by the static friction in the stick phase drops substantially as the contact surfaces

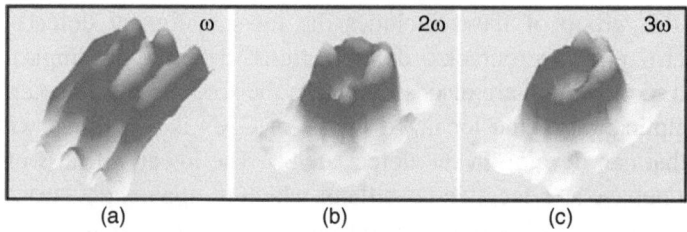

Fig. 29.1. Linear (50 kHz) (a), second (b) and third harmonic (c) images of oval delamination area in glass fiber-reinforced composite (GFRP).

start sliding. Such an abrupt transition, obviously, takes place whenever the harmonic driving force recovers from zero, that is, twice for the period of acoustic excitation (symmetric stiffness modulation). Therefore, the contact stiffness modulation $\Delta C(t)$ is a 2ν-pulse-type function that comprises its higher harmonics $2n\nu$. As a result, the spectrum of nonlinear shear vibrations of the defect ($\Delta C(t) \cdot \varepsilon(t)$) contains a number of odd harmonics of the driving frequency.

Apparently, the variety of the higher harmonic spectra in realistic materials is not confined to the two basic nonlinear responses shown above to be characteristic of the intermittent contacts. However, as shown below, a prevalence of any of the two mechanisms in the measured nonlinear response for a given acoustic excitation may cast light on the structure of the material and the type of defects it contains.

The experimental methodology used for the higher harmonic mode of nonlinear NDE includes an intense CW acoustic excitation in the kHz-frequency range combined with a fast and remote scanning laser vibrometry.[18] After a 2-D-scan and FFT of the signal received, the C-scan images of the sample area are obtained for any spectral line within the frequency bandwidth of 1 MHz.

Figures 29.1a–c show experimental imaging results for an oval delamination on top of a piezoactuator embedded into a glass fiber-reinforced composite (GFRP). The nonlinear images (Figures 29.1b,c) selectively reveal the boundary ring of the delamination where clapping and rubbing of the contact surfaces are, apparently, expected. Because the harmonics are generated locally within this area one would anticipate the source of nonlinearity to be primarily seen in the nonlinear vibration pattern. On the contrary, the driving frequency (50 kHz) image indicates a standing wave pattern over the whole area of the actuator. The strong localization of the nonlinear images is also facilitated by energy trapping for the higher harmonics generated inside the delamination area whose stiffness is substantially lower than that of the surrounding intact material.

The mechanism of friction nonlinearity was found to prevail in wood which is a natural fiber-reinforced composite.[20] In intact wood, the nonlinear spectrum averaged over the specimen surface exhibits an evident odd harmonic domination (see Figure 19.10, Chapter 19). However, due to the strong material inhomogeneity caused by the annual rings, a local nonlinear response of wood is also expected to be spatially inhomogeneous with a maximum nonlinear output in areas of the highest compliancy where peak strains are developed. A typical higher harmonic C-scan of the LR-plane (cut along the trunk) of a spruce specimen is shown in Figure 29.2. One can see the

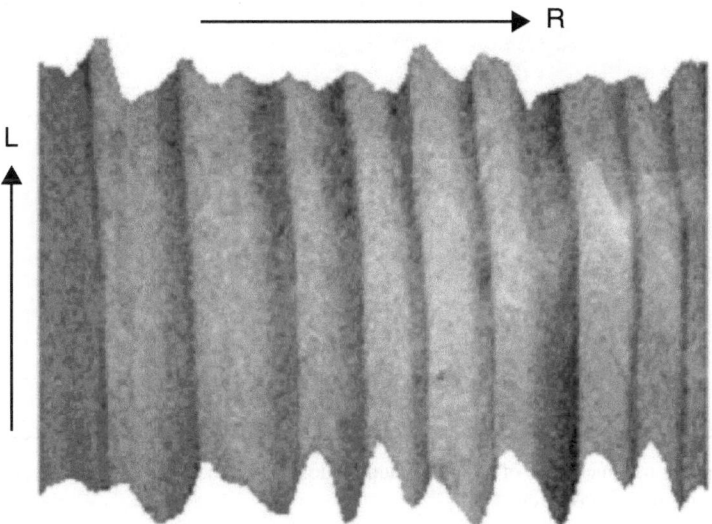

Fig. 29.2. Third harmonic C-scan of the LR-plate of spruce. Crests positions coincide with latewood/earlywood interface.

wavy distribution of local nonlinearity in the radial direction with maxima located in the earlywood area close to the latewood/earlywood transition interface. Therefore, the local nonlinear response indicates that the most load-vulnerable part of wood is formed early in the growing season when the thin-walled earlywood cells appear.

Because odd harmonic domination is a characteristic of intact wood, any deviation from such a model spectrum may indicate the presence of clapping defects responsible for even harmonic generation. Thus, the even harmonic distribution over the specimen enables us to localize and image the defects. Such "clapping-selective" imaging is shown in Figure 29.3 for simulated delaminations between a decorative oak veneer lamina and a particleboard substrate ($12 \times 6.5 \times 1$ cm). The delamination pattern was formed by $\cong 1.5$ cm wide periodic strips of unglued veneer areas. In Figure 29.3, the strips of the delaminated areas are clearly indicated by a sharp local increase in the fourth harmonic amplitude (dark strips) due to the clapping mechanism. It is worth noting that the alternative NDE technique of particleboard composites with air-coupled ultrasound normally fails due to the high damping for a thick specimen, whereas the nonlinear response is virtually independent of the specimen thickness.

2.2 Wave Modulation Mode

The stiffness modulation of a weakly bonded contact subjected to a two-wave acoustic excitation leads to an efficient mixing of the driving frequencies. The magnitudes of the modulation sidelobes are indicators of nonlinearity of the defect and were used for qualitative NDE of cracked flaws in metal parts, concrete, and composites.[21]

The flexibility and application area of the wave modulation technique can be expanded by combining the high sensitivity of the nonlinear approach with the benefits of noncontact ultrasonic excitation in a new air-coupled nonlinear modulation version.[22]

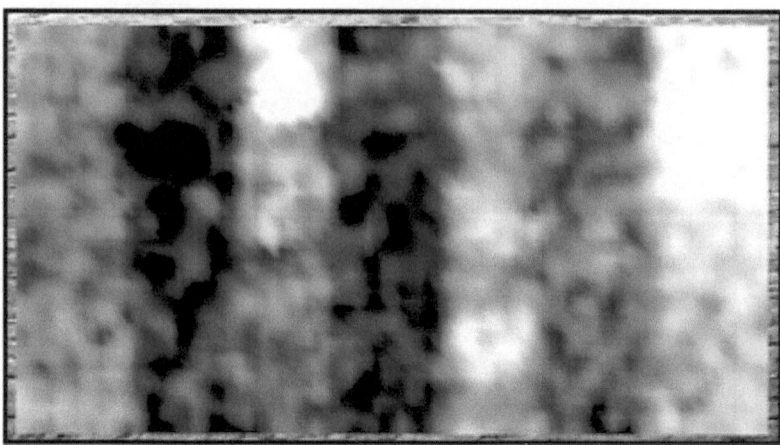

Fig. 29.3. Fourth-harmonic image of periodic delaminations between oak veneer lamina and particleboard plate. The specimen length is 12 cm.

It is based on the transmission of small-amplitude air-coupled ultrasound (frequency ω) through a cracked defect area in a sample subjected to low-frequency (LF) vibrations (Ω). The transmission coefficient T depends on the gap between the crack edges and can be assumed to be constant only for an infinitesimally small amplitude of LF vibrations (linear transmission mode): $V_{out}(t) = T \times V_{in}(\omega t)$. Otherwise, the open crack gap changes harmonically, first, and $T(t) \to (T_0 + T_\sim(\Omega t))$ causing linear modulation of the output signal: $V_{out}(t) = (T_0 + T_\sim(\Omega t)) \times V_{in}(\omega t)$. In addition to the probing wave of frequency ω, the spectrum acquires combination frequency components ($\omega \pm \Omega$) whose relative amplitudes are proportional to ($T_\sim/T_0$). A further increase in the pump amplitude causes clapping of the crack edges and results in a pulse-type modulation of the crack gap. Such a crack works as a modulator with a nonlinear transmission coefficient $T(t) \to (T_0 + \sum T_n(n\Omega t))$ and provides multiple sidebands around the fundamental frequency (nonlinear modulation).

Experimental evidence for the air-coupled modulation is shown in Figure 29.4. In the experiment, a focused beam of air-coupled ultrasound (frequency $\approx$ 452 kHz) is transmitted through an open surface cutting crack in a polymer specimen subject to a low-frequency (1 – 2 kHz) vibration. For an ultrasonic beam incident normally to the specimen surface, the transmission coefficient T_0 was relatively high, so that the modulation spectrum displays the complete multiple sideband pattern (Figure 29.4).

The air-coupled modulation develops locally within the cracked area and therefore can be used for locating and imaging defects. Scanning of a sample area with receiver tuned to a sidelobe frequency delivers information solely on the nonlinear cracked defects. The results, confirming the feasibility of defect-selective NDE in the wave modulation mode, are shown in Figures 29.5a–b for transmission through both a linear (a drop of water) and nonlinear (a crack) defect. The linear transmission image (Figure 29.5a) clearly reproduces both defects with a comparable negative contrast ($\Delta V/V \approx -0.6$) as one would expect due to wave damping and scattering. On the contrary, the sideband B-scans exhibit a strong rise in the nonlinear output ($\Delta V/V \approx 20$) in the crack area with zero contrast for the linear defect (Figure 29.5b).

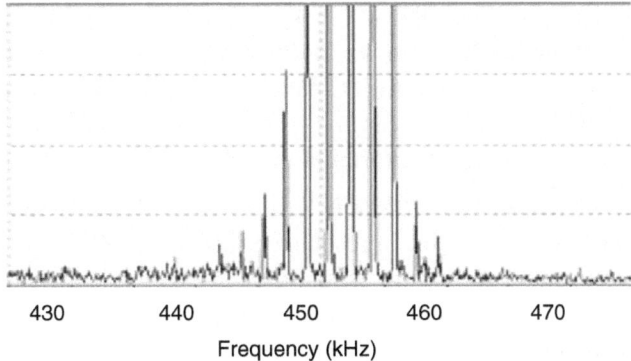

Fig. 29.4. Nonlinear modulation spectrum of air-coupled ultrasound: fundamental frequency is 452 kHz; modulation frequency ≈1.7 kHz.

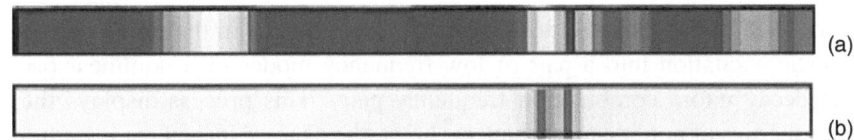

Fig. 29.5. Defect selective imaging with air-coupled ultrasound: fundamental frequency (a) and wave modulation (b) B-scans of the area with linear (drop of water, left) and nonlinear (crack, right) defects.

3. Nonlinear NDE Using Instability Modes

3.1 Subharmonics, Frequency Pairs and Self-Modulation

The higher harmonic and wave modulation modes do not require a frequency response of the defect to be taken into account. Some different scenarios of nonlinear dynamics of cracked defects that expand considerably nonclassical spectra are revealed in experiments[19,23] and are interpreted on the basis of nonlinear resonance. These scenarios exhibit forms of dynamic instability, that is, an abrupt change of the output for a slight variation of the input parameters.

To illustrate the feasibility of the new nonlinear vibration modes and ascertain their basic spectral patterns, we assume that the damaged area exhibits both resonance and nonlinear properties and can be identified as a nonlinear oscillator with s degrees of freedom.[19] The set of the s equations of motion for the oscillator driven by an external acoustic excitation in normal coordinates (Q_α) with quadratic nonlinearity approximation takes the form:[24]

$$\ddot{Q}_\alpha + \omega_\alpha^2 Q_\alpha = f_\alpha(t) + F_\alpha^{NL}, \qquad (29.1)$$

where $\alpha = 1, 2, \ldots, s$, ω_α- are the normal frequencies, and the nonlinear forces in the right-hand side are the quadratic forms of the normal velocities and accelerations.

In the first approximation of the perturbation theory: $Q_\alpha = Q_\alpha^{(1)} + Q_\alpha^{(2)} + \ldots$, we set $F_\alpha^{NL} = 0$ and for the harmonic excitation $f_\alpha(t) = f_0 \cos \nu t$. The solution to Eq. (29.1) then describes the independent oscillations in normal coordinates: $Q_\alpha^{(1)} = A_\alpha \cos \omega_\alpha t + B_\alpha \cos \nu t$, where A_α, B_α are constants and the phase factors are omitted.

The second-order equations $\ddot{Q}_\alpha^{(2)} + \omega_\alpha^2 Q_\alpha^{(2)} = F_\alpha^{NL}(Q^{(1)})$ take into account all self- and cross-interactions between the oscillations of normal and driven frequencies; by substituting the expression for $Q_\alpha^{(1)}$ we obtain for the nonlinear force: $F_\alpha^{NL} \sim \sum_{\beta\gamma} (A_\beta \cos \omega_\beta t + B_\beta \cos \nu t)(A_\gamma \cos \omega_\gamma t + B_\gamma \cos \nu t)$. Besides the second harmonics $2\omega_{\beta,\gamma}$, 2ν, and DC-terms, this formula includes the following combination frequency components: $\sum_{\beta\gamma} [F_{\beta\gamma}^{(2)} \cos(\omega_\beta \pm \omega_\gamma) + F_{\nu\beta}^{(2)} \cos(\nu \pm \omega_\beta)]$. It can be easily shown that the last term will cause a resonance increase in $Q_\alpha^{(2)}$ due to the external excitation if its frequency satisfies the condition $\nu \pm \omega_\beta \approx \omega_\alpha$, whereas a similar resonance behavior of $Q_\beta^{(2)}$ is obviously achieved when $\nu \pm \omega_\alpha \approx \omega_\beta$. Thus, the driving force provides a simultaneous resonance growth of the amplitudes for the pair of normal modes whose frequencies satisfy the condition:

$$\omega_\alpha + \omega_\beta \approx \nu. \tag{29.2}$$

Equation (29.2) shows that the nonlinear resonance results in the decay of the external acoustic excitation into a pair of low-frequency modes of a nonlinear oscillator (phonon decay into a combination frequency pair). This process displays the well-known properties of nonlinear resonance:[24] first, the exact values of $\omega_{\alpha,\beta}$ are functions of the driving amplitude so that Eq. (29.2) is an exact equality. Second, the resonance growth of the modes is possible only if the input excitation exceeds a certain threshold and is then affected by the amplitude and frequency instability and hysteresis. The resonance instability manifests in the avalanchelike amplitude growth of the nonlinear products beyond the input threshold. The reverse excursion of the driving amplitude results in bistability: the input amplitudes for the stepwise up and down transitions are different (amplitude hysteresis). Such a dynamics is totally different from the classical powerlaw dependences and is a distinctive signature of the nonlinear acoustic phenomena associated to nonlinear resonance.

It is worth noting that the particular case of Eq. (29.2) when $\omega_\alpha = \omega_\beta = \omega$ and hence $\nu \approx 2\omega$ corresponds to the subharmonic or the main parametric resonance. Thorough studies of the latter based on Hill's and Mathieu's equations[25,26] show a similar dynamic behavior (threshold, instability, hysteresis, etc.) for the subharmonics and frequency pairs[27] which, in light of the above, is not surprising because both phenomena stem from the family of nonlinear resonance effects. Instructively, both numerical calculations[28] and experimental simulations[9] demonstrate substantial broadening of the frequency bands where parametric resonances are observed as the input amplitude increases. Therefore, at high levels of acoustic excitation of the damaged area with a set of ω_α one can expect both subharmonics and frequency pairs to be generated virtually independently of the input frequency.

The resonance increase of the spectral components of combination frequency requires us to take into account the higher-order nonlinear terms in the driving force. It opens an opportunity for the combination-type resonance with a larger number of normal modes and also expands the nonlinear spectrum due to the interplay between existing resonant excitations $\omega_{\alpha,\beta}$ and ν from Eq. (29.2). In the latter case, after

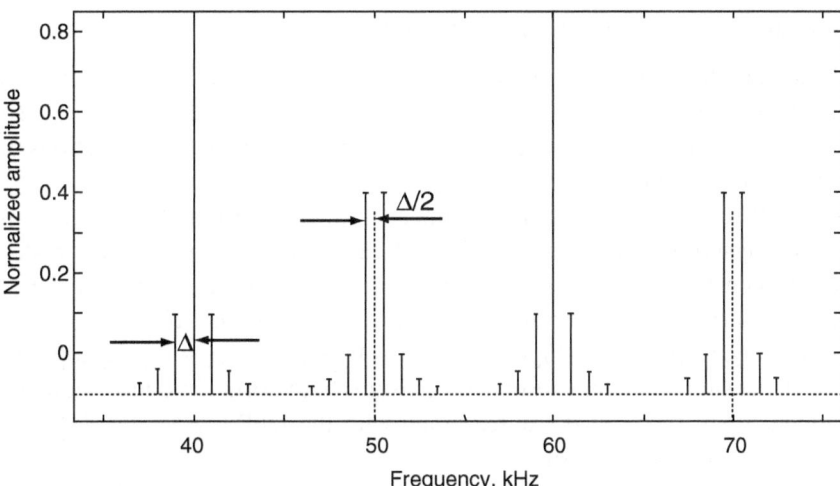

Fig. 29.6. Section of ultrasubharmonic and ultrafrequency pair spectrum calculated from Eq. (29.3): $\nu = 20\,kHz$; $\omega_\alpha = 9\,kHz$; $\omega_\beta = 11\,kHz$; $N = 8$.

accounting for the Nth-order terms, the spectrum includes the following frequency components,

$$F^{NL}(\omega) \sim \sum_{m,n,p} F_{mnp}(n\nu + m\omega_\alpha + p\omega_\beta), \qquad (29.3)$$

where $m + n + p = N - 1$.

Besides the ultraharmonics $n\nu$, the nonlinear spectrum in Eq. (29.3) comprises the ultrafrequency pairs (UFP) $(n\nu + m\omega_\alpha + p\omega_\beta)$ and $(n\nu + p\omega_\alpha + m\omega_\beta)$. They are separated by $|m - p|\,\Delta$ $(\Delta = \omega_\beta - \omega_\alpha)$, centered around $[n\nu + (m + p)\nu/2]$ and structured into two series:

1. A set centered around the ultrasubharmonics (USB) $(2a + 1)\nu/2$: in this case, $m + p = 2a + 1$ is odd as well as $m - p = [2(a - p) + 1]$. Thus, the first UFP components of the set are shifted by $\pm\Delta/2$ with respect to the USB. If $\omega_\alpha = \omega_\beta$, then the spectrum contains only the higher harmonics ($n\nu$) and USB $(2a + 1)\nu/2$.

2. A set centered around integer multiples of ν. In this case, $m + p = 2b$ is even as well as $(m - p)$ and the UFP series starts at $\pm\Delta$ from the harmonics.

These features of the nonlinear spectrum are shown in Figure 29.6 where the results of calculations based on Eq. (29.3) are given for the normalized amplitudes $F_{mnp} = 1$ and $N = 8$. The symmetrical positions of the UFP-side lobes around the higher harmonics and USB will result in an amplitude modulation of the nonlinear spectral components (see next section). Such a modulation is induced by nonlinear resonance in a multidegree-of-freedom system and, unlike the external wave modulation is called self-modulation.[23]

3.2 Nonclassical Spectra of Instability Modes

To substantiate the basic assumptions of the nonlinear resonance model and elucidate dynamical characteristics of the USB and UFP we studied experimentally the spectral responses of various fractured defects to a wide range of acoustic excitations beyond the threshold of instability.

Figures 29.1a–c demonstrate the nonlinear frequency responses of a crack in a 1-mm-thick polystyrene plate (commercial CD-case) measured in a sweep mode of the driving frequency ν. As ν decreases, successive resonance excitation of the subharmonics is observed for two normal modes $\omega_1 \approx 900\,\text{Hz}$ and $\omega_2 \approx 1100\,\text{Hz}$ when $\nu = 2\omega_1$, $2\omega_2$, respectively (Figures 29.7a,c). When $\nu \cong \omega_1 + \omega_2 \cong 2000\,\text{Hz}$ (Figure 29.7b), the sum-frequency resonance activates both normal modes simultaneously: the frequency pair (ω_1, ω_2) is excited along with the UFP ($\Delta \approx 200\,\text{Hz}$) due to the higher-order contact nonlinearity.

The dynamic properties of the nonlinear resonance modes are illustrated in Figures 29.8 and 29.9. Figure 29.8 shows the amplitude of $3\omega/2$-subharmonic wave generated in the reflection of 30-MHz acoustic waves from a crack in LiNbO$_3$ crystal,[23] as a function of the input voltage. One can clearly see the steplike thresholds $((V_{IN})_1 \geq 3V; \ (V_{IN})_2 \geq 4V)$ followed by the stable plateaus. The sharp amplitude increase at the thresholds confirms a transition into the instability region where an avalanchelike development of nonlinear oscillations takes place. The hysteresis of the curves in Figure 29.8 is evidence of bistability in the crack signature.

The experiments also reveal the staircase-like structure of the thresholds, shown in Figure 29.8, as functions of the driving amplitude. Such a structure is a possible indication of the successive excitation of the nonlinear oscillators associated with different

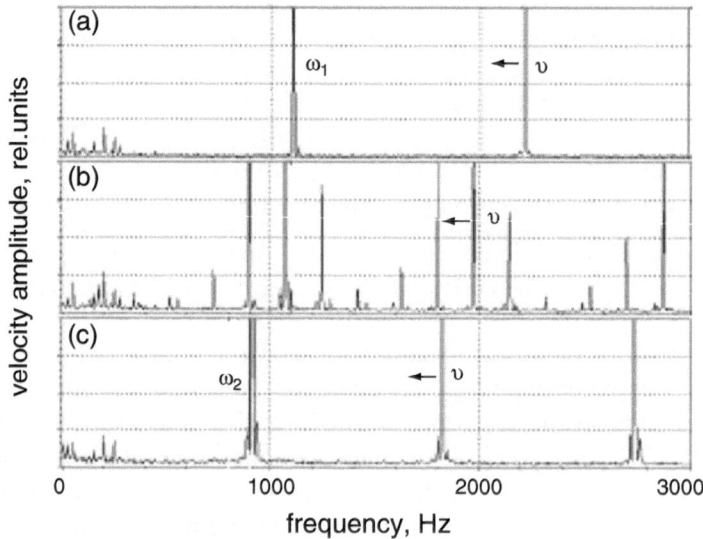

Fig. 29.7. Subharmonic (a, c) and UFP - (b) frequency responses of a crack in a sweep mode of driving frequency ν (frequency variation is shown by arrow).

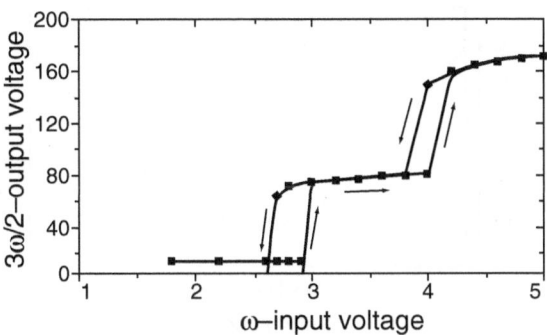

Fig. 29.8. Dynamic multi-stability for $3\omega/2$-ultra-subharmonic generated by a crack.

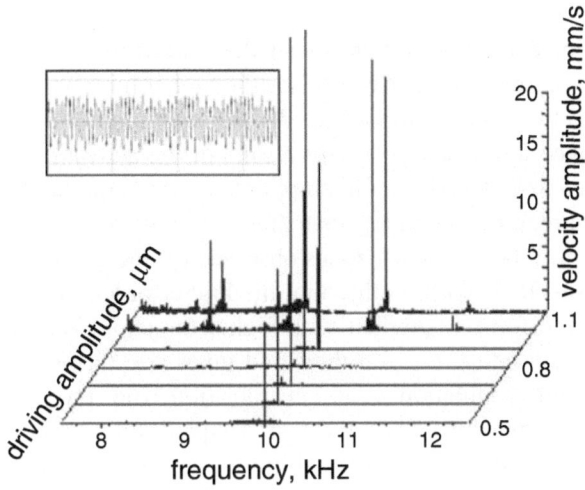

Fig. 29.9. Threshold decay of a subharmonic into ultra-frequency pairs in a delamination area of C/C-SiC-composite driven with 20 kHz acoustic excitation.

parts of the crack. Such a dynamics is consistent with the broadening of the frequency zones for the parametric resonance described above.

Figure 29.9 provides direct evidence of the spectral transformations beyond the subharmonic instability threshold for a delamination area in a C/C–SiC composite. As the driving amplitude of the 20-kHz excitation exceeds the threshold for the subharmonic mode ($\approx 0.5\ \mu$m in Figure 29.9), another instability threshold gives rise to an avalanchelike energy decay into the UFP-components (at $\approx 1\ \mu$m drive). The oscilloscope insert in Figure 29.9 illustrates the self-modulation of the output signal observed in the damaged area. A further increase of the input results in widening of the UFP-lines into quasicontinuous frequency bands which are the forerunners of the transition to chaos.

The experimental results on the frequency and dynamic nonlinear responses of the fractured flaws are summarized schematically in Figure 29.10 for a defect represented by a pair of coupled oscillators (normal frequencies ω_1 and ω_2). At low amplitude

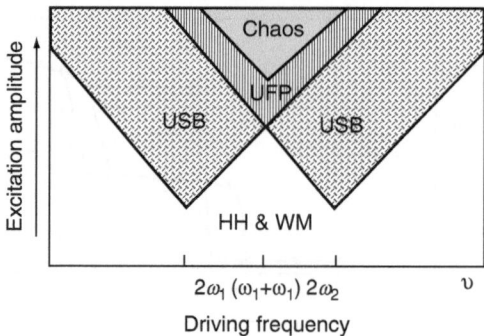

Fig. 29.10. Schematic diagram of frequency and amplitude nonlinear responses for a cracked defect conceived as a pair of coupled nonlinear oscillators.

(V_{IN}) of the driving excitation (frequency ν), the nonlinear spectrum follows the nonresonant scenario of Section 2 and comprises the higher harmonic (HH) and the wave modulation (WM) frequency components. As the input amplitude exceeds the threshold value, resonance instability generally results in the activation of the ultrasubharmonic components first. The threshold amplitude depends on the driving frequency: a minimal threshold requires frequency matching to the main subharmonic (parametric) resonance ($\nu = 2\omega$). The frequency zones observed for the USB expand readily as the excitation increases. It is worth noting that the higher-order fractional subharmonics can also be involved for $\nu = n\omega$ but with higher threshold values. Further increase of acoustic excitation above a given threshold gives rise to the ultrafrequency pairs accompanied by self-modulation. A direct transition from the nonresonance modes to UFP-instability is feasible when the sum–frequency resonance matching conditions are satisfied. Finally, temporal instability is developed and the self-modulation leads to chaotic beats and the quasicontinuous spectrum typical of noiselike excitations builds up.

3.3 Nonlinear NDE Using Subharmonic and Self-Modulation Spectra

In the previous section we have shown that the input acoustic power for the USB- and UFP-modes is basically about the same as that for regular nonlinear acoustic experiments. In practical terms, this requires an acoustic intensity of a few W/cm^2 in the high-MHz-frequency range and the driving amplitudes of μm-scale in the low-kHz range. A dramatic increase of the instability modes beyond the threshold, as a rule, leads to distinctive nonlinear spectra with multiple USB- and UFP-components which can provide abundant information on material properties and defects. A few examples of nonlinear NDE using these modes are given below to demonstrate their applicability and superior performance in the cases where the data obtained by nonresonant modes are insufficient.

As discussed above, because the subharmonic mode results from nonlinear resonance in acoustic wave–defect interaction, it may be sensitive to the input frequency, whereas the higher harmonics are normally invariant. This can lay the basis for a frequency-selective nonlinear NDE: for a given driving frequency only "resonant"

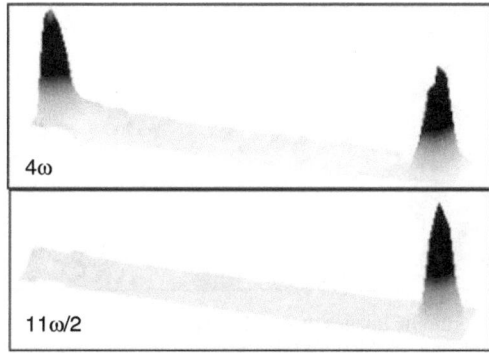

Fig. 29.11. Selective nonlinear NDE of an edge delamination (left) and a thermal damage (right) in CFRP-foam laminate: 4th harmonic (80 kHz) image (top); 11th subharmonic (110 kHz) image (bottom).

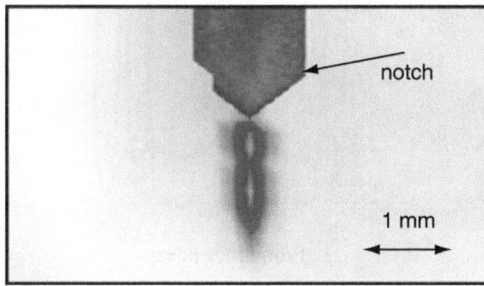

Fig. 29.12. Ultra-subharmonic (70 kHz) imaging of fatigue crack driven by 20 kHz input.

defects can be discerned in the subharmonic mode. Such a case is illustrated in Figure 29.11 for the multi-ply plate of carbon fiber-reinforced composite (CFRP) with two artificial defects, that is, edge delimination (left) and thermal impact (right). Both defects are detected in the higher harmonic mode (top) but the ultrasubharmonics are sensitive only to the resonance impact area (bottom).

Fatigue loads in metals (rotors, turbines, etc.) cause minute cracks of micrometer scale which gradually develop into major cracking and initiate an abrupt material fracture. The linear ultrasound is virtually unable to detect the fatigued crack at the early stage of its development. Examples of nonlinear imaging of fatigue induced microflaws and degradation of metal microstructure using the self-modulation and subharmonic modes are given in Figure 29.12, which shows fatigue cracking produced by cyclic loading in Ni-base super-alloys. Such a crack of less than 2 mm length, with average distance between the edges of only $\approx$ 5 μm, is clearly detected in the ultra-subharmonic ($7\nu/2$) image, whereas traditional linear NDE by using slanted ultrasonic reflection failed to work with such small cracks.

Fiber-reinforced composites constitute a class of hi-tech engineering materials whose application area is rapidly expanding in the aerospace and automotive industries. Fiber–metal laminates are new materials with excellent tolerance to impact, corrosion, and lightning stroke, low flammability, and low weight. The example in Figure 29.13 displays such an advanced material for aircraft industry: glass fiber-reinforced

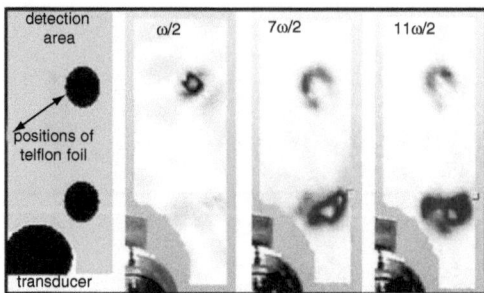

Fig. 29.13. Subharmonic images of simulated delaminations in a Glare® sample.

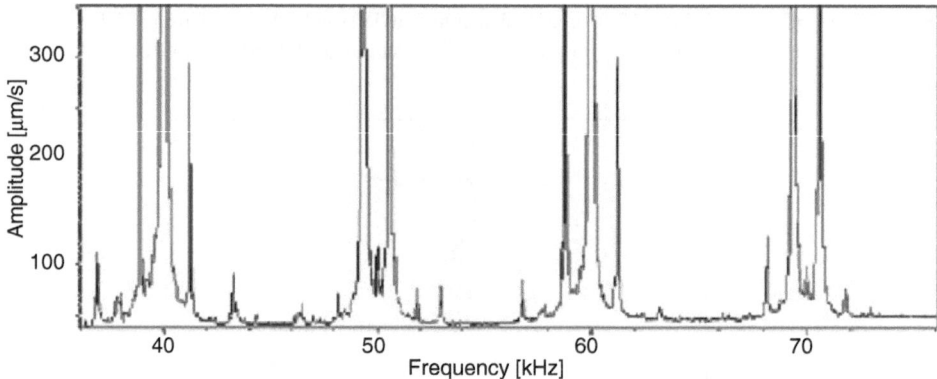

Fig. 29.14. UFP-spectrum in an impact damaged area of GFR-composite: $\nu = 20\,\text{kHz}$; $\Delta \approx 1.2\,\text{kHz}$.

aluminum laminate (Glare®). More specifically, it shows the ultrasubharmonic images of a Glare plate with two inserted circular Teflon foils to simulate local debonding. In the image, the defect can be recognized fairly well, and the quality is enhanced by the higher numbers of ultrasubharmonics. We believe that the latter is associated with the peculiar distribution of subharmonics over the delamination area and stronger acoustic dissipation outside the defect at higher frequencies.

Mechanical impacts in multi-ply composites produce a fracture that is a combination of matrix–fiber debonding, cracking, and delaminations. Such a combination of fractured defects makes the impact area strongly nonlinear and, normally, all nonlinear modes can be traced in the spectra observed. Figure 29.14 shows a section of the nonlinear spectrum measured in a 14-ply epoxy-based glass-fiber-reinforced composite (GFRP) with a 9.5 J impact damage for a 20-kHz excitation beyond the UFP-threshold. One can clearly identify the position of the higher harmonics, ultrasubharmonics, and ultrafrequency pairs ($\Delta \cong 1.2\,\text{kHz}$). The low amplitudes of the USB-components are due to their decay into the UFP above the threshold. It is worth noticing the striking similarity between the experimental spectrum in Figure 29.14 and the one calculated on the basis of Eq. (29.3) (Section 3.1, Figure 29.6), which substantiates the validity of the nonlinear resonance model developed above.

Similarly to all the nonlinear modes discussed, the frequency pair components generally display a strong spatial localization around the defects and are applicable for the detection of damage. The benefit of the UFP-mode is illustrated in Figures 29.15a–c

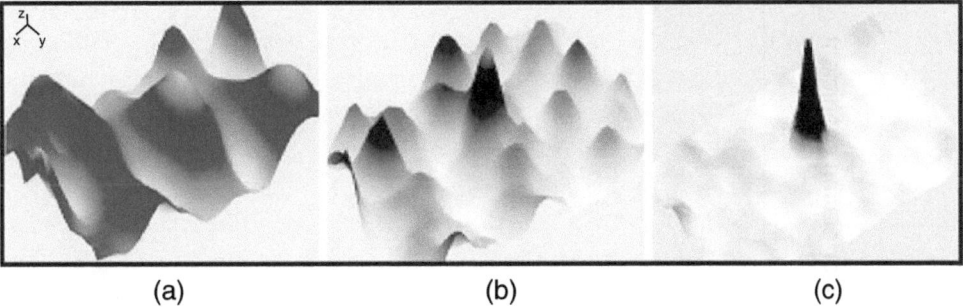

(a)   (b)   (c)

Fig. 29.15. Self-modulation imaging of the impact damaged area in the GFRP-composite: a) linear image (20 kHz); b) 4th harmonic image; c) image at the UFP-side-lobe around 10th harmonic (198.8 kHz).

for the GFRP-sample with the impact damage discussed above. The linear image taken at the driving frequency of 20 kHz reveals only a developed standing wave pattern over the whole sample (Figure 29.15a). The higher harmonic image is also corrupted by the standing wave pattern (b), whereas the image at the first UFP sidelobe of the tenth harmonic of the driving frequency (198.8 kHz) yields a very clear indication of the damaged area (Figure 29.15c).

4. Use of Nonlinear Spectra to Improve NDE Performance

The amplitude and the number of higher-order components in nonlinear spectra indicate the presence of fractured defects in the specimen. A simple and fast qualitative NDE of specimen integrity can be implemented on the basis of the "strength-and-width" of its nonlinear spectrum and used as a pass–fail test. A quantified version of such an algorithm, based on normalization of the higher harmonic and wave modulation spectra, was developed for quality assessment of thin-film coatings on composites.[29] The scanning approach described in this chapter enables precise measurement of the local vibration spectra and defects localization by applying a similar "strength-and-width" criterion to these local nonlinear spectra. In order to proceed with nonlinear imaging, a particular spectral line is chosen and the 2-D distribution of this frequency component calculated. The image obtained is only one of the many images intrinsic to the diverse nonclassical spectra shown in the preceding sections. Disregarding the spectral information provided by the other nonlinear modes reduces the reliability of nonlinear NDE and its efficiency in defect recognition and imaging. A straightforward approach based on a computer algorithm for defect identification and a tool for image processing are suggested below and applied to the problem of a comprehensive analysis of the nonlinear spectra.

4.1 Automated Identification of Defects

The feasibility of multifrequency recognition and imaging of defects by computer-based spectral data processing is evaluated using as an example an artificial delamination (of unknown shape and location) in a Glare specimen, subject to a dual-frequency

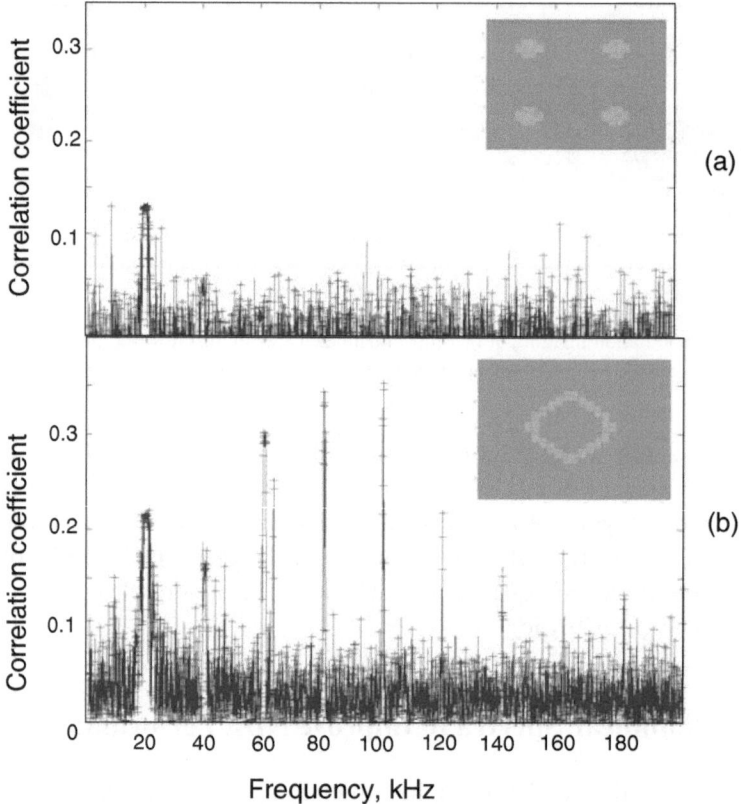

Fig. 29.16. Correlation spectra for two artificial reference images.

acoustic excitation at the 3.2 and 20 kHz frequencies. As expected, the output C-scan spectrum of nonlinear vibrations, averaged over the area of the specimen, revealed a number of ultraharmonics and combination frequency components. A routine analysis would normally comprise both the selection of specific frequencies to be used for computer imaging, and the analysis of obtained images by an operator for defects identification. Instead, a correlation approach can be applied for automated identification of the defect shape and location. It basically compares the defect of unknown form and position with a set of probing (reference) images. The latter can be generated as arbitrary point matrices or on the basis of some preliminary information about the defect. The values of the correlation factors between the reference matrix images and those obtained in the experiment enable us to identify the most probable shape and location of the nonlinear defect.

In a computer experiment, several artificial reference image matrices were generated and the correlation coefficient was calculated for the images obtained at each spectral component of the nonlinear spectrum. The examples in Figures 29.16a,b show the spectral distribution of the correlation factors calculated for the two reference matrices. From Figure 29.16b it is evident that the majority of the experimental images are similar to the second reference matrix, so that the shape of the nonlinear part of the defect is close to the ring located around the center of the scanning area. To further specify the defect in more detail, the procedure can be iterated after introducing smaller variations into the selected first-order reference image.

4.2 Image Quality Enhancement

As the shape and position of the defect are identified, one can proceed to improve image quality. First, the nonlinear spectral components with the highest correlation coefficients in Figure 29.16b are selected, so that the "raw" images of the delamination display all relevant features of the nonlinear response of the defect (Figure 29.17). These images are then reconstructed to remove background noise and to enhance contrast. The following image processing operations were applied for each pixel (p) after normalizing the amplitude (A) distributions of the spectral components in Figure 29.17.

First of all, amplitude averaging over the group of the n selected images could be performed, calculating the amplitude of each pixel in the final image (B) as: $B_p = \sum A_n^{(p)}/n$. As one would expect, averaging allows us to retain all relevant features in the original images, but at the expense of some increase in the background contrast outside the defect (Figure 29.18, left).

A straightforward way to improve the signal-to-noise ratio is to apply the multiplying operation to each pixel so that: $B_p = A_1^{(p)} \cdot A_2^{(p)} \cdot \ldots \cdot A_n^{(p)}$. This diminishes dramatically the low-amplitude parts of the image while the high-amplitude areas stay

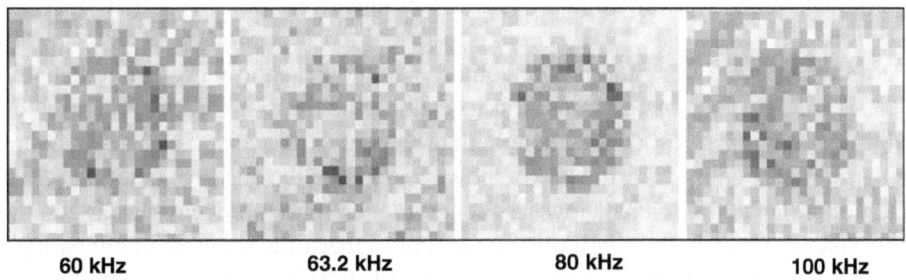

| 60 kHz | 63.2 kHz | 80 kHz | 100 kHz |

Fig. 29.17. Images of an oval delamination for a group of nonlinear spectral components with maximum correlation coefficient.

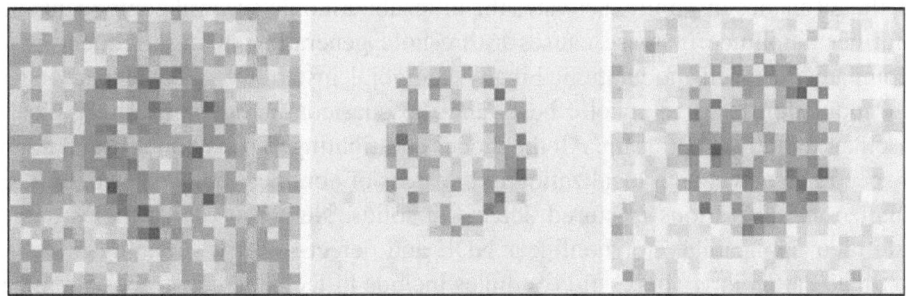

Fig. 29.18. Images of the oval delamination reconstructed by amplitude averaging (left), multiplying (middle) and squaring-and-averaging (right) operations.

basically unchanged. Figure 29.18 (middle) actually demonstrates substantial decrease in the background noise and thus, an apparent improvement of overall contrast. However, it may generally result in some loss of detail in the reconstructed image.

On the contrary, the operation of squaring-and-averaging ($B_p = \sum [A_n^{(p)}]^2/n$) is supposed to avoid such a destructive interference between the images, whereas the squaring is expected to increase the contrast of each image which then is averaged over the group. The results presented in Figure 29.18 (right) prove that a reasonably high contrast can be combined with a detailed image of the defect.

It is worth mentioning that the approach discussed above suggests only one of the opportunities in automated recognition and processing of the results of nonlinear NDE. Other advanced approaches that take advantage of the abundant information in the nonclassical spectra can be developed exploiting image analysis techniques presented in the literature.[30]

5. Summary

The nonlinear interaction of an acoustic wave with fractured defects is determined by nonlinear contact dynamics which strongly depends on the amplitude of the acoustic wave. At moderate driving amplitude, the contact acoustic nonlinearity suggests a fully deterministic scenario with higher harmonic generation and/or wave modulation. Unlike their classical counterparts, these effects feature much higher efficiency, specific dynamic characteristics, modulated spectra, and unconventional "rectified" waveform distortion.

At a higher level of excitation, the contact vibrations acquire a dynamic instability which is a forerunner of the transition to chaos. Such a dynamics is interpreted on the basis of nonlinear resonance phenomena for a defect conceived as a set of coupled oscillators. The nonlinear resonance is shown to result in the decay of external acoustic excitation either into a pair of low-frequency modes (phonon decay into a combination frequency pair) or a subharmonic mode. For higher-order contact nonlinearity, the nonlinear spectrum expands considerably to include the ultrasubharmonic and ultrafrequency pair (self-modulation) modes. Experiments show that even a moderate acoustic excitation of realistic cracked defects gives rise to instability vibration modes, which exhibit threshold behavior and distinctive hysteretic dynamics. Beyond the threshold, the resonance instability activates the ultrasubharmonic spectral components first. A further amplitude increase causes a threshold generation of ultrafrequency pairs accompanied by self-modulation. Finally, temporal instability is developed and the self-modulation leads to chaotic beats and the quasicontinuous spectrum typical of noiselike excitations builds up. All the modes contributing to such nonclassical nonlinear spectra display a high localization in the areas of nonlinear contacts and, thus, can visualize readily various fractured defects in solids. Numerous case studies demonstrate their applicability for nonlinear NDE and defect-selective imaging in various materials. Particularly successful examples include hi-tech and constructional materials: intact and damaged wood, impact damage and delaminations in fiber-reinforced plastics, fatigue microcracking in metals, and delaminations in fiber-reinforced metal laminates. Multiple frequency components comprising nonclassical nonlinear spectra in imperfect materials provide abundant information on properties and location of defects. Computerized automatic defect identification and application of image processing techniques enable a comprehensive analysis of the nonclassical spectra and improve reliability and quality of nonlinear NDE.

References

[1] M.A. Breazeale and J. Philip, Determination of third-order elastic constants from higher harmonic generation, in: *Physical Acoustics*, v. XVII, Ed. W.P. Mason, Academic Press, New York, 1965.

[2] A.A. Gedroitz and V.A. Krasilnikov, Elastic waves of finite amplitude and deviations from Hook's law, *Sov. Phys. JETP*, **16**, 1122–1131 (1963).

[3] J.H. Cantrell and W.T. Yost, Effect of precipitate coherency strains on acoustic harmonic generation, *J. Appl. Phys.*, **81**, 2957–2962 (1997).

[4] Y. Shui and I. Solodov, Nonlinear properties of Rayleigh and Stoneley waves in solids, *J. Appl. Phys.*, **64**, 6155–6165 (1988).

[5] M. Hamilton, D.J. Shull, Yu. A. Il'inskii, and E.A. Zabolotskaya, Harmonic generation in plane and cylindrical nonlinear Rayleigh waves, *J. Acoust. Soc. Am.*, **94**, 418–427 (1993).

[6] F.M.Severin, I.Yu.Solodov, and Yu.N.Shkulanov, Experimental observation of sound nonlinearity in the reflection from an interface between solids, *Moscow Univ. Bulletin, Phys.-Astron.*, **43**, 105–107 (1988).

[7] K.S. Len, F.M.Severin, and I.Yu.Solodov, Experimental observation of the influence of contact nonlinearity on the reflection of bulk acoustic waves and propagation of surface acoustic waves, *Sov. Phys. Acoust.*, **37**, 610–612 (1991).

[8] I.Y. Solodov, Ultrasonics of non-linear contacts: Propagation, reflection and NDE applications, *Ultrasonics*, **36**, 383–390 (1998).

[9] 9. E.M. Ballad, B.A. Korshak, I.Yu. Solodov, N. Krohn, and G. Busse, Local nonlinear and parametric effects for non-bonded contacts in solids, *Nonlinear Acoustics at the Beginning of the 21st Century*, Eds. O. Rudenko and O. Sapozhnikov, 727–734 (2002).

[10] A. Moussatov, V. Gusev, and B. Castagnede, Self-induced hysteresis for nonlinear acoustic wave in cracked material, *Phys. Rev. Lett.*, **90**, 124301 (2003).

[11] Yu. Solodov, N. Krohn, and G. Busse, CAN: An example of non-classical acoustic nonlinearity in solids, *Ultrasonics*, **40,** 621–625 (2002).

[12] B.A. Korshak, I.Yu. Solodov, and E.M. Ballad, DC-effects, sub-harmonics, stochasticity and "memory" for contact acoustic non-linearity, *Ultrasonics*, **40**, 707–713 (2002).

[13] R.G. Guyer and P.A. Johnson, Nonlinear mesoscopic elastisity: Evidence for a new class of materials, *Phys. Today*, 30–36 (April 1999).

[14] L.A. Ostrovsky and P.A. Johnson, Dynamic nonlinear elasticity in geomaterials, *Nuovo Cimento*, **24**, 1–46 (2001).

[15] R.A. Guyer and K.R. McCall, Hysteresis, discrete memory and nonlinear wave propagation in rock: A new paradigm, *Phys. Rev. Lett.*, **74**, 3491–3494 (1995).

[16] C. Pecorari and I. Solodov, Nonclassical nonlinear dynamics of solid interfaces in partial contact for NDE applications, Chapter 9, this book.

[17] V. Kazakov, A. Sutin, and P.A. Johnson, Sensitive imaging of an elastic nonlinear wave-scattering source in a solid, *Appl. Phys. Letts.*, **81**, 646–648 (2002).

[18] N. Krohn, K. Pfleiderer, I.Yu. Solodov, and G. Busse, Nonlinear vibro-acoustic imaging for non-destructive flaw detection, *Nonlinear Acoustics at the Beginning of the 21st Century*, Eds. O. Rudenko and O. Sapozhnikov, 779–786 (2002).

[19] I. Solodov, J. Wackerl, K. Pfleiderer, and G. Busse, Nonlinear self-modulation and subharmonic acoustic spectroscopy for damage detection and location, *Appl. Phys. Lett.*, **84**, 5386–5388 (2004).

[20] I. Solodov, K. Pfleiderer, and G. Busse, Nondestructive characterization of wood by monitoring of local elastic anisotropy and dynamic nonlinearity, *Holzforschung*, **58**, 504–510 (2004).

[21] K.E.-A. Van Den Abeele, P.A. Jonson, A. Sutin, Nonlinear elastic wave spectroscopy (NEWS) techniques to discern material damage, Part I: Nonlinear wave modulation spectroscopy (NWMS), *Res. Nondestr. Eval.*, **12**, 17–30 (2000).

[22] E.M. Ballad, S.Yu. Vesirov, K. Pfleiderer, I.Yu. Solodov, and G. Busse, Nonlinear modulation technique for NDE with air-coupled ultrasound, *Ultrasonics*, **42**, 1031–1036 (2004).

[23] I. Yu. Solodov and B.A. Korshak, Instability, chaos, and "memory" in acoustic wave-crack interaction, *Phys. Rev. Lett.*, **88**, 014303 (2002).

[24] L.D. Landau and E.M. Lifshitz, *Mechanics*, Pergamon Press, Oxford, 1960.

[25] N.W. McLachlan, *Theory and Application of Mathieu functions*, London, 1947.

[26] L. Adler and M.A. Breazeale, Generation of fractional harmonics in a resonant ultrasonic wave system, *J. Acoust. Soc. Am.*, **48**, 1077–1083 (1970).

[27] A.I. Eller, Fractional-harmonic frequency pairs in nonlinear systems, *J. Acoust. Soc. Am.*, **53**, 758–765 (1973).

[28] B.A. Korshak, I. Yu. Solodov and E.M. Ballad, DC-effects, sub-harmonics, stochasticity and "memory" for contact acoustic non-linearity, *Ultrasonics*, **40**, 707–713 (2002).

[29] S. Hirsekorn, A. Koka, A. Wegner, and W. Arnold, Quality assessment of bond interfaces by nonlinear ultrasonic transmission, in: *Rev. Progress in QNDE.*, v. **19**, Eds. D.O. Thompson, D.E. Chimenti (Plenum Press, New York) 1367–1374 (2000).

[30] A.K. Jain, Advances in mathematical models for image processing, *Proc. IEEE*, **69**, 502–508 (1981).

Generation Threshold of Subharmonic Modes in Piezoelectric Resonator

A. Alippi, A. Bettucci, M. Germano, and D. Passeri

Department of Energetics, University of Rome "La Sapienza"- Via A. Scarpa, 14-00161 Rome, Italy adriano.alippi@uniroma1.it.

1. Introduction

Intrinsic material elastic nonlinearity is usually very small even for high values of the acoustic strains: nonlinear effects are, indeed, quite negligible except for those taking place due to energy accumulation along the acoustic wave propagation distance. Recently, from the macroscopic point of view, highly nonlinear elastic response has been observed in the interaction between elastic waves and localized nonbonded contacts (grains, delaminations, cracks) [1]: the asymmetry in the contact stiffness for compressional and tensile stresses across the contact boundaries has been identified as an important source of nonlinearity that gives rise, besides harmonic and subharmonic generation, to nonclassical nonlinear phenomena such as instability [1], hysteresis [2], self-modulation, and parametric resonance of the contact oscillations [3] (see also Chapter 19, Part 2 of this book). These effects are being extensively used in nonlinear nondestructive evaluation applications [4,5] (see also Part 3 of this book).

Highly nonlinear oscillation effects have also been observed at a nanoscale level [6] (see Chapter 5, Section 3), mainly due to the variation of a characteristic distance between the tip of an atomic force microscope cantilever and the sample surface [7,8].

Although lattice anharmonicity in solids is generally small even at high values of the elastic strains, nonlinear coupling among modes in a multimode resonant structure can result in a rich variety of nonlinear phenomena. For example, it has been found that, in weakly nonlinear resonant finite structures subjected to a simple harmonic excitation at one of the normal modes (where there exists a special relationship between the driving frequency and the frequency of two or more linear modes), the stationary response may contain contributions in many modes of oscillation [9, 10]. Obviously, in such a case, the nonlinearity is distributed all over the structure, but experimenting on piezoelectric resonators some localized nonlinear mechanism has been found to be present as well [11]. Moreover, experimental evidence has been given of extremely low thresholds in subharmonic generation of ultrasonic waves in one-dimensional artificial piezoelectric plates with Cantor-like structure, as compared to the corresponding homogeneous and periodic plates [12].

In this chapter, a simple phenomenological theoretical model for subharmonic threshold generation in a monodimensional finite structure is presented and some experimental results of subharmonic threshold generation in a piezoelectric resonator are reported.

2. Theoretical Model for Subharmonic Generation

The phenomenon of subharmonic generation in a nonlinear oscillator is a threshold phenomenon, which takes place at an unpredictable moment any time the driving force exceeds a definite value, which is the threshold value. Successive evolution of subharmonic oscillations is a complex problem that may lead the system into chaotic dynamics, then to the unpredictability of its final state, whereas the threshold value can be easily evaluated and a brief account is given here, in order to model the behavior of more complex finite structures.

Let us write the general nonlinear dynamical equation of a mass-spring type oscillator as

$$m\ddot{x} + b\dot{x} + kx = F\cos\omega t + \beta\psi(x, \dot{x}), \tag{30.1}$$

where m is the mass, b the viscoelastic coefficient, k the elastic force per unit displacement, F the amplitude of the driving force at the angular frequency ω and ψ a generic nonlinear function of the displacement x from the equilibrium position and of its time derivative $\dot{x}$, and β the strength of the nonlinear interaction. For sufficiently small values of the driving force F, one may correctly suppose that the oscillation amplitude A and the nonlinear parameter β are small enough for the second term on the right side of Eq. (30.1) to be much smaller than the first one, so that one may firstly linearize the equation by neglecting the nonlinear term and obtain

$$x_{lin} = A\cos(\omega t + \varphi) \tag{30.2}$$

with A and φ the proper values of amplitude and phase of the linear solution. By means of an iterative procedure, then, one considers a correcting term $\xi(t) = (x - x_{lin}) \ll x_{lin}$ to be added to the linear solution (30.2), that solves the equation:

$$m\ddot{\xi} + b\dot{\xi} + k\xi = \beta\psi(x_{lin}, \dot{x}_{lin}). \tag{30.3}$$

The driving term ψ may then be developed in its Fourier spectral components, with frequencies multiple of the driving frequency $f = \omega/2\pi$, which will generate harmonic components in the system oscillations. In order to generate subharmonic oscillations at frequencies f/n, with n integer, the system should start oscillating by itself at one such frequency by a casual fluctuation $\epsilon(t)$, that will then be temporarily included in the linear solution x_{lin}, temporarily giving

$$x_{lin} + \epsilon(t) = x_{lin} + \epsilon_0\cos(\omega t/n). \tag{30.4}$$

In such a case, the driving term in Eq. (30.3) takes the form

$$\beta \psi (x_{lin} + \epsilon, \dot{x}_{lin} + \dot{\epsilon})$$

$$= \beta \left[\psi_0 + \epsilon_0 \cos(\omega t/n)\psi_1 + \epsilon_0^2 \cos^2(\omega t/n)\psi_2 + \text{higher} - \text{order terms} \right] \quad (30.5)$$

whose spectral components contain terms of frequency $f_{m,n} = \omega/2\pi (l + m/n)$ with l, m, n positive or negative integers. In particular, for subharmonic generation of frequency $f/2$, a component must be present in $\epsilon(t)$ at this frequency, such as to drive the system at this very same frequency if $n = 2$ and $l = 1, m = -1$, or $l = 2, m = -3$, and so on. Higher $|m|$-order terms, however, produce lower amplitudes of the driving term, because $|m| + l$ is the number of interactions needed to produce that term, and ϵ_0 is a small quantity. Therefore, when performing the harmonic balance of all the terms in Eq. (30.3) at the circular frequency $\omega/2$, one may correctly write

$$m\ddot{\epsilon} + b\dot{\epsilon} + k\epsilon = \beta \epsilon \psi_{11}(x_{lin}, \dot{x}_{lin}), \quad (30.6)$$

where ψ_{11} is the $\omega/2\pi$ frequency term in the harmonic expansion of term ψ_1. The threshold condition for the amplitude value that would set the system into oscillation at the first subharmonic frequency $\omega/4\pi$ will, then, be deduced from the equation:

$$\psi_{11} \geq \frac{\sqrt{m^2(\omega_0^2 - \omega^2/4)^2 + b^2\omega^2/4}}{\beta} \quad (30.7)$$

with $\omega_0^2 = k/m$ and will depend on the nonlinearity coefficient β as well as on the specific form of the nonlinear function $\psi(x_{lin}, \dot{x}_{lin})$. Equation (30.7) has been obtained by solving Eq. (30.6) as if ϵ were a known perturbation term in the driving term $\beta \epsilon \psi_{11}$ and then by setting the condition that the solution be greater than the perturbing term, thus producing a feedback ratio greater than one. In particular, if $\psi(x, \dot{x}) = x^2$, we have $\psi_{11} = 2x_{lin}$, and the threshold condition becomes:

$$F \geq \frac{\sqrt{m^2(\omega_0^2 - \omega^2)^2 + b^2\omega^2} \sqrt{m^2(\omega_0^2 - \omega^2/4)^2 + b^2\omega^2/4}}{2\beta}. \quad (30.8)$$

We turn now to the nonlinear generation of subharmonic waves in a finite system that is driven by an external force evenly distributed in the volume occupied by the system; in particular, the case is analyzed of a monodimensional homogeneous structure, defined in the space $0 \leq x \leq L$. The nonlinear wave equation in the general displacement variable u versus space x and time t can be written as

$$cu_{xx} = \rho u_{tt} + bu_t + \beta \psi(u, u_x) + Q(x, t), \quad (30.9)$$

where c is a proper elastic constant, ρ is the mass density per unit length, $Q(x, t)$ is the driving term, $\psi(u, u_x)$ the general nonlinear term depending both from the displacement u and its space derivative u_x that replaces the previously defined function $\psi(x, \dot{x})$, and finally, subscripts x and t stand for derivatives. As in the previous case of the pointlike oscillator, we suppose that $u_{lin}(x, t)$ that solves the linear equation Eq. (30.9) ($\beta = 0$), can be expanded in space harmonics as

$$u_{lin}(x, t) = \sum_n U_n(t) \sin n\frac{\pi}{L}x, \quad (30.10)$$

thus satisfying the boundary conditions $u_{lin}(0, t) = u_{lin}(L, t) = 0$. By introducing solution (30.10) into the wave propagation Eq. (30.9), the time equation for each space harmonic component $U_n(t)$ can be deduced and the solution subsequently obtained in the same form as the pointlike oscillator.

In the case of a piezoelectric plate, driven via the electric field uniformly produced within the entire structure by a sinusoidal voltage signal of amplitude Q_0 and circular frequency ω applied at its surfaces, and considering with no lack of generality just one direction of propagation x, one may write for each space Fourier component of the expansion (30.10):

$$U_n(t) = \frac{p \, Q_0(L/2n\pi^2)}{\sqrt{\rho^2(\omega^2 - n^2 c/\rho)^2 + b^2\omega^2}} \cos(\omega t + \varphi) \qquad \text{for odd } n$$

$$U_n(t) = 0 \qquad \qquad \qquad \text{for even } n, \qquad (30.11)$$

p being a proper piezoelectric coefficient that couples the driving voltage to the stress field in the plate, and φ the relative phase angle between the driving field and the strain response of the structure at frequency ω.

We limit the discussion of the nonlinear generation in a finite structure to the case of the first subharmonic mode of frequency $\omega/4\pi$, assuming for this purpose that a small perturbation is introduced into the system in the form:

$$\epsilon(x, t) = E_0 f(x) \cos(\omega t/2) \qquad (30.12)$$

with

$$f(x) = \sum_n \epsilon_n \sin n\frac{\pi}{L}x. \qquad (30.13)$$

The system will then be temporarily subject to the oscillation

$$u_{lin}(x, t) + E_0 f(x) \cos(\omega t/2) \qquad (30.14)$$

and the driving term in Eq. (30.9) be given by

$$\beta\psi(u + \epsilon, u_x + \epsilon_x)$$
$$= \beta\left[\psi_0 + E_0 \cos(\omega t/2)\psi_1 + E_0^2 \cos^2(\omega t/2)\psi_2 + \text{higher-order terms}\right]. \qquad (30.15)$$

By balancing each spatial Fourier component, one will then obtain:

$$- cn^2\epsilon_n = \rho\epsilon_{n,tt} + b\epsilon_{n,t} + \beta E_0\psi_{11}, \qquad (30.16)$$

where ψ_{11} is the $\omega/2\pi$ frequency term in the harmonic expansion of term ψ_1. The threshold condition for the amplitude value that would set the system into oscillation at the first subharmonic frequency $\omega/4\pi$ will, then, be deduced from the equation:

$$\psi_{11} \geq \frac{\sqrt{(cn^2 - \rho\omega^2/4) + b^2\omega^2/4}}{\beta}. \qquad (30.17)$$

The specific value of the ψ_{11} function and of the threshold value, therefore, obviously depend upon the media that are experimented on and validation of the model relies on the specific structure where subharmonic modes are generated.

3. Experimental Measurements

Some experimental features of the nonlinear effects due to subharmonic generation that may take place in a simple finite resonant structure such as a piezoelectric resonator are presented.

The phenomenological model previously described states that a threshold condition must exist for the vibration amplitude, and therefore for the amplitude of the driving force, at frequency $\omega/2\pi$ for setting the system into oscillation at the first subharmonic frequency $\omega/4\pi$. Figure 30.1 reports the amplitude of the first subharmonic at frequency $\omega/4\pi$ as a function of the applied voltage for a PZT piezoelectric hollow tube with wall Al electrodes, electrically excited at the thickness (radial) resonance ($\omega/2\pi = 634$ kHz). When increasing the driving voltage V_0 from low values, a sudden generation of the subharmonic (threshold) takes place at $V_0 = 33$ V: from this level onward, the subharmonic amplitude also increases with the applied voltage. The subharmonic threshold exhibits a hysteretic behavior: when decreasing the driving voltage, the amplitude of the subharmonic undergoes a drastic drop at $V_0 = 22.5$ V. Hysteretic phenomena are typical of systems exhibiting parametric resonance [13]. This confirmed by observing the displacement frequency spectra, reported in Figure 30.2, measured in the same sample for increasing values of the applied voltage: the displacement here has been optically measured (the experimental set-up has been described in [14]). The generation of the subharmonic oscillation at a frequency $\omega/4\pi$ is quite abrupt when the excitation exceeds the threshold: see Figures 30.2a,b below and above the threshold, respectively. More interestingly, as can be seen from Figures 30.2c and d, when increasing the applied voltage, the subharmonic is accompanied by frequency pairs: 282–352 kHz (Figure 30.2c), 228–406 kHz and 178–456 kHz (Figure 30.2d). The frequencies of the pair are a function of the applied voltage and their sum is equal

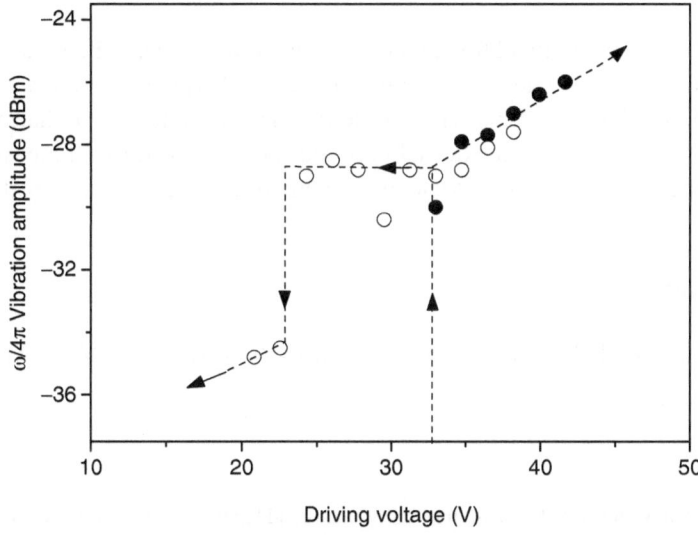

Fig. 30.1. Amplitude of the $\omega/4\pi$ frequency subharmonic oscillation versus the driving voltage: closed and open circles are for increasing and decreasing values of the applied voltage, respectively.

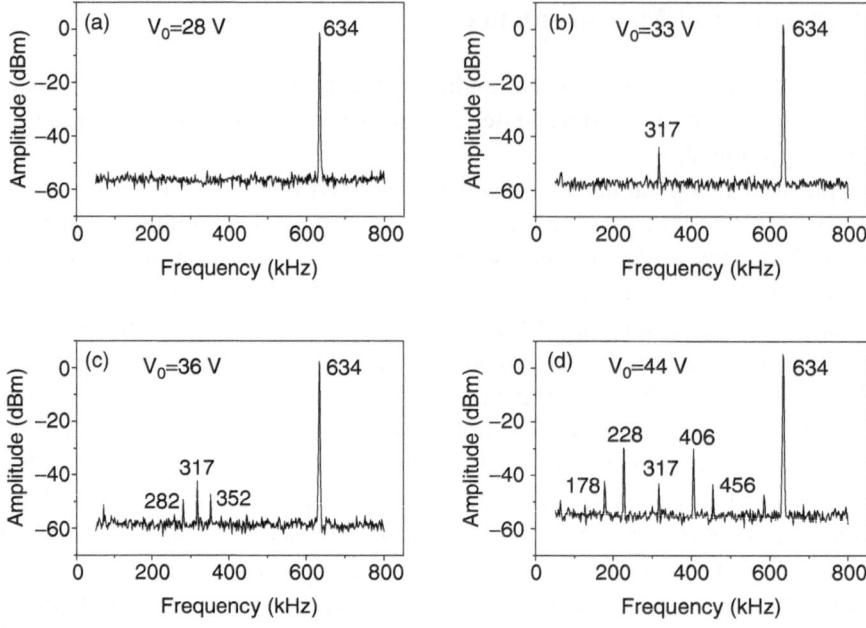

Fig. 30.2. Displacement frequency spectra for a PZT piezoelectric hollow tube electrically forced at the thickness resonance equal to 634 kHz; figures are for increasing values of the magnitude of the applied voltage V_0.

to the excitation frequency. This is a clear indication that a self-modulation oscillation is taking place in such a multimode structure.

4. Conclusions

A simplified theoretical model for subharmonic generation in a finite structure has been described: starting from a general nonlinear dynamical equation of a mass-spring type oscillator, the model then deals with the nonlinear generation of subharmonic waves in a finite system driven by an external periodic force. Experimental measurements of a subharmonic threshold in a piezoelectric resonator have also been reported.

Acknowledgment

The authors acknowledge the support of this work by ESF-PESC NATEMIS program.

References

[1] I. Yu. Solodov and B.A. Korshak, Instability, chaos and "memory" in acoustic-wave–crack interaction, *Phys. Rev. Lett.* **88**, 014303–1–014303–3 (2002).

[2] A. Moussatov, V. Gusev, and B. Castagnéde, Self induced hysteresis for nonlinear acoustic waves in cracked material, *Phys. Rev. Lett.* **90**, 124301–1–124301–4 (2003).

[3] I. Solodov, J. Wackerl, K. Pfleiderer, and G. Busse, Nonlinear self-modulation and subharmonic acoustic spettroscopy for damage detection and location, *Appl. Phys. Lett.* **84**, 5386–5388 (2004).

[4] I. Yu. Solodov, Ultrasonics of nonlinear interfaces in solids: new physical aspects and NDE applications, *Proc. of World Congress on Ultrasonics*, Paris, 2003, Vol. I, pp. 555–564.

[5] N. Krohn, R. Stoessel, and G. Busse, Acoustic non linearity for defect selective imaging, *Ultrasonics*, **40**, 633–637 (2002).

[6] N.A. Burnham, A.J. Kulik, G. Gremaud, and G.A.D. Briggs, Nanosubharmonics: The dynamics of small nonlinear contacts, *Phys. Rev. Lett.* **74**, 5092–5095 (1995).

[7] K. Iganaki, O. Matsuda, and O.B. Wright, Hysteresis of the cantilever shift in ultrasonic force microscopy, *Appl. Phys. Lett.* **80**, 2386–2388 (2002).

[8] V. Gusev, B. Castagnde, and A. Moussatov, Hysteresys in response of nonlinear bistable interface to countinously varying acoustic loading, *Ultrasonics*, **41**, 643–654 (2003).

[9] A.H. Nayfeh and B. Balachandran, Modal interaction in dynamical and structural systems, *ASME Appl. Mech. Rev.*, **42**, 175–202 (1989).

[10] A.H. Nayfeh, S.A. Nayfeh, T.A. Anderson and B. Balachandran, Transfer of energy from high frequency to low-frequency modes, in *Nonlinearity and Chaos in Engineering Dynamics*, Edited by J. M. Thompson and S. R. Bishop, Wiley, New York, pp. 39–58 (1994).

[11] A. Alippi, M. Albino, M. Angelici, A. Bettucci, and M Germano, Space distribution of harmonic mode vibration amplitudes in nonlinear finite piezoelectric transducer, *Ultrasonics*, **43**, 1–3 (2004).

[12] A. Alippi, G. Shkerdin, A. Bettucci, F. Craciun, E. Molinari, and A. Petri, Low-threshold subharmonic generation in composites structures with Cantor–like code, *Phys. Rev. Lett.*, **69**, 3318–3321 (1992).

[13] C. Hayashi, *Nonlinear Oscillations in Physical Systems*, Princeton University Press, Princeton, NJ, pp. 86–97 (1985).

[14] A. Alippi, M. Angelici, A. Bettucci, and M. Germano, Surface mapping of nonresonant harmonic and subharmonic generated modes in finite piezoelectric structures, *Proc. of 16th International Symphosium on Nonlinear Acoustics*, Moscow, 2002, Vol. II, pp. 759–762.

31

An Overview of Failure Modes and Linear Elastic Wave Propagation Damage Detection Methods in Aircraft Structures

Michele Meo[1] and Christophe Mattei

Material Research Center, Dep. of Mechanical Engineering, Bath University, Bath, BAZ 7AY, UK. Phone: 01225386708. Email: m.meo@bath.ac.uk. CSM Materialteknik AB. Box 1340, SE-581 13 Linköping, Sweden. Phone: (46)-13-16-91-71, Fax: (46)-13-16-90-00.
[1]To whom correspondence should be addressed.

Abstract

This chapter presents an overview of the typical failure scenarios suffered by aircraft structures. In particular, failure modes of composite and metal structures are presented. A description of general nondestructive techniques (NDT) used in the aircraft industry is also reported. Particular attention is devoted to wave propagation methods based on linear elastic waves. The limitations of currently used NDT methods, based on linear elastic wave propagation methods are described, showing the need to develop more robust damage detection methods for the implementation in a reliable aircraft structural health monitoring system.

Keywords: Corrosion, delamination, fatigue, lamb waves

1. Introduction

The history of aircraft has suffered several accidents due to structural and material failures, which have strongly influenced authority certification and policies. For aircraft structures, where safety is the main issue, to prevent catastrophes it is vital to have rapid estimation of the health of the load-bearing structures. The potential increase of structural failure scenarios in existing and future aircraft structures has added a greater degree of urgency to the ongoing need for reliable and efficient nondestructive evaluation (NDE) methods for detection and characterization of damage, flaws, and so on in aircraft structures. Among the past and recent incidents that have strongly affected the design and the maintenance procedures, two examples are reported.

   In the late 1950s, fierce competition was taking place among Lockheed, Douglas, and Boeing for the new "jet" aviation market. The Lockeed-188 was introduced into service in September, 1958. The aircraft was faster on shorter routes than its competitors the B-707 and DC-8. On September 29, 1959, near Buffalo, Texas, a L-188 while cruising at 15,000 feet, lost its left wing causing 27 fatalities. On March 17, 1960, near Cannelton, Indiana, another Lockeed-188 lost its left and right wings, with the loss of

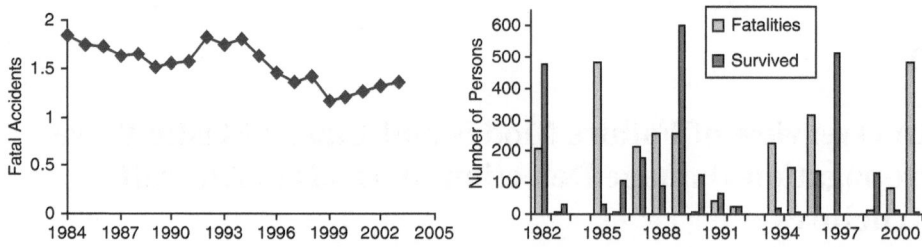

Fig. 31.1. (a) Number of fatal accidents per 100,000 flight hours; (b) Number of fatalities/survived persons.

63 persons. Both accidents involved wing separations during cruise. Initial attention of investigators was focused on basic flutter characteristics. The subsequent tests showed that the airplane satisfied the airworthiness requirements. After a detailed survey, it was found that both aircraft had been involved in the "hard landing" events sometime prior to accidents, and as a consequence damage was experienced by the engine mounting system. The undamaged mount had a 6 Hz "whirl" mode natural frequency, whereas after the damage, the natural frequencies were reduced to 3 cps. The wing bending and torsional natural frequencies were respectively 3.5 and 2 Hz, therefore vibratory coupling resulted and the aircraft engine/nacelle/wing developed oscillatory divergent deflections when subjected to moderate turbulence during cruise. Substantial wing modifications were required. After the accidents, changes to the airworthiness requirements were made to include fail-safe structural requirements in order to avoid aeroelastic instabilities.

Another striking incident occurred to a Boeing 737 of Aloha Airlines on April 28, 1988. The aircraft, while cruising at 8000 m, suffered extensive damage to the forward fuselage losing a large part of the structure (Figure 31.1b). Only one human life was lost, a stewardess sucked from the airplane. The plane subsequently made a safe emergency landing. The cause of the accident was attributed to the presence of significant corrosion, low lap joint strength and fatigue damage. The technique used to bond the overlapping fuselage skins together was inadequate, and in conjunction with exfoliation corrosion led to early debonding. As a result, many adjacent fastener holes started to crack. This form of cracking, known as multiple site damage, caused widespread fatigue damage (WFD). In these conditions, the aircraft fuselage structure was no longer able to carry the required residual strength loads. The cause of the accident was attributed to the failure of Aloha Airlines' maintenance program to detect the presence of structural damage. In the wake of the Aloha flight incident, the new requirements set by the Federal Aviation Administration (FAA) state that airlines must include an inspection requirement to detect exfoliation corrosion, low bond durability, and crack damage in older aircraft.

These two examples of aircraft structure failures highlight the growing concern about the need to increase the reliability, and therefore safety, of existing and future aircraft structures.

Another growing concern is the fact that military and civil aircraft are being used in service significantly longer than their original design life. This cost-driven trend is subjecting the structures of these aircraft to conditions that are increasing the probability of failure, particularly as a result of aging. Some aircraft models, which have already endured a long service life, are being considered for more years of service.

The effort of the aviation industry to reduce the number of fatal accidents is achieving substantial results, as shown in Figure 31.1a, however, the number of fatalities suffered is still a big unsolved problem for today's commercial aviation (Figure 31.1b).

Moreover, the growing need to use new and more efficient materials on future aircraft structures increases the need to develop methodologies to monitor and prevent structural and material failure modes. It is particular important to emphasize that although some well-established monitoring methodologies are available to monitor known damage presence, location, magnitude, and type, there is the need to develop methodologies capable of assessing unpredictable failure scenarios.

In light of these issues, this chapter presents an overview of the possible failure scenarios suffered by aircraft structures, NDT methodologies, and linear wave propagation-based NDT methodologies. In particular, limitations and issues related to current linear elastic wave propagation NDT techniques are presented highlighting the need to develop more robust and reliable NDT based on nonlinear elastic wave propagation NDT techniques.

1.1 Failure Scenarios

This section discusses some of the common failure modes suffered by aircraft structures. The majority of existing aircraft are mainly made of metal alloys; however, since the early 1990s the use of fiber-reinforced polymer (FRP) composites for aircraft structural components has increased significantly, especially for secondary structures. A review of common failure modes of metallic and composite structures is presented.

1.2 Metal Structures

A detailed summary of the frequency of failure modes [24] is presented in Table 31.1. The table reveals that for aircraft metal structures, fatigue is the predominant failure mode in service. The second more dangerous failure mode is corrosion. Although the number of fatigue failures is higher than corrosion failures, the detection and repair of corrosion damage on in-service aircraft requires more effort than the rectification of fatigue cracking. The high occurrence of fatigue failure observed probably reflects the destructive nature of this failure mode, whereas corrosive attack is generally slower than fatigue, and usually more easily spotted and rectified during routine maintenance.

An overview of the two most common failure modes is discussed below.

Table 31.1. Frequency of failure modes in aircraft metallic structures [24]

	Percentage of Failures
Fatigue	55
Corrosion	16
Overload	14
High temperature corrosion	2
SCC/Corrosion Fatigue/He	7
Wear/abrasion/erosion	6

1.2.1 Fatigue of Metal Structures

Aircraft structural components can fail at stresses below the tensile strength of the material if subjected to repeated or cyclic stresses. This phenomenon is called *metals fatigue*. A detailed survey of the number of failure and the common fatigue crack initiation sites observed in aircraft and helicopters that have led to failures [27, 28] is reported in Table 31.2. It shows that fatigue failures are strongly affected by the presence of discontinuities, flaws in aircraft structures, stress concentrations, and residual stresses (welds). Material surface discontinuities, defects, or surface cracking can increase the local stress, producing a concentration at these points that could initiate fatigue much more quickly than would be expected. Stress concentrations caused by surface defects such as scratches and wear tend to be more common as these may not be present at build, but can be introduced during service. Another common cause of stress concentration is corrosion, which can lead to fatigue crack initiation. In order to reduce fatigue failure, aircraft structures are designed with a safe or inspection-free life. Even if fatigue behavior of most metals and alloys is well understood, fatigue failures still occur, indicating the complex nature of this phenomenon and the difficulty in controlling the number of variables that affect fatigue. There are many factors that influence fatigue, some of which are the mean stress, peak stress, frequency of loading, temperature, environment, material microstructure, surface finish, and residual stresses. Many of these factors can be successfully taken into account when determining the safe life of a component and, therefore, the majority of fatigue failures in aircraft causing catastrophic failure tend to be those that initiate as the result of unforeseen circumstances.

The fatigue failure of ductile materials is preceded by characteristic changes in the material microstructure and occurs in a quasibrittle manner, that is, by crack propa-

Table 31.2. Summary of fatigue initiation location and number of failures in air vehicles [27, 28]

	Number of Failures	
Initiation Location	Fixed Wing	Rotary Wing
Bolt, stud or screw	108	32
Fastener hole or other hole	72	12
Fillet, radius or sharp notch	57	22
Weld	53	3
Corrosion	43	19
Thread (other than bolt or stud)	32	4
Manufacturing defect or tool mark	27	9
Scratch, nick or dent	26	2
Fretting	13	10
Surface or subsurface flaw	6	3
Improper heat treatment	4	2
Maintenance-induced crack	4	
Work-hardened area	2	
Wear	2	7

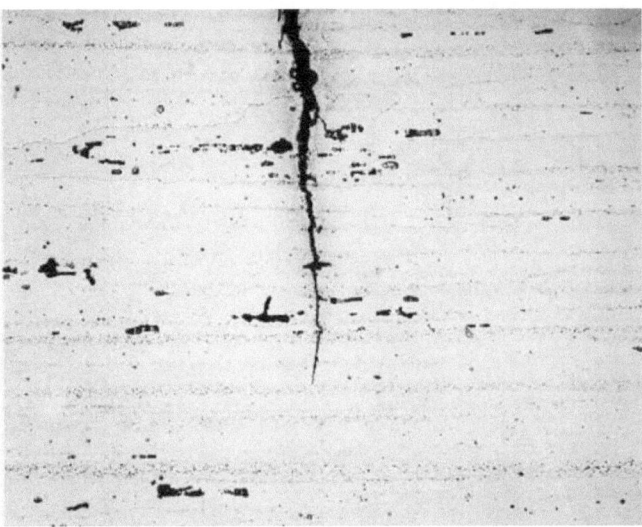

Fig. 31.2. Fatigue crack in Aluminum AA 7075.

gation. Usually, the structures that fail by fatigue experience three different stages of crack growth [24]:

- Initiation of a fatigue crack. This is strongly affected by the presence of residual stresses (i.e., welds etc.), and material flaws, and stress concentrations due to external loads.
- Crack growth. There is progressive crack propagation under repeated loadings.
- Final sudden rupture. The component cannot withstand the applied loads when the propagating crack reaches a critical size leading to sudden rupture, as shown in Figure 31.2.

1.2.2 Corrosion

Corrosion is the chemical degradation of metals as a result of a reaction with the environment. This type of damage onsets mostly at discontinuities, such as a rivet hole, where the protective surface treatment is particularly vulnerable. Failure occurs when the decrease of the thickness is such that the structures cannot withstand in-service loads or the corrosion makes the component susceptible to failure by some other mode (e.g., fatigue).

The corrosion depends on many factors such as the environment, protective treatments, and the inherent capacity of the materials themselves to resist corrosion. In particular, the most common external causes of corrosion are due to water intrusion into dry cavity areas or structural joints. This is generally caused by poor sealing and/or failure of the interface layer that protects the mating surface, lack of adequate drainage/ventilation, wearing of the protective coatings, contaminated fuels, and dissimilar metal components. The most susceptible areas to corrosion on an aircraft are fin skins/panels and leading edge areas, center fuselage tank rooms, air intake duct, bottom skin, and wingbox.

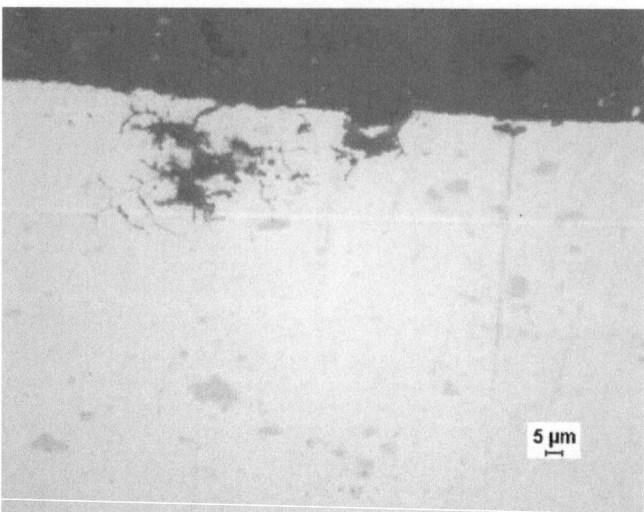

Fig. 31.3. Corrosion pitting and intergranualar cracks in Al 2024.

Detection of corrosion initiation is difficult and its effect increases in a nonlinear manner with the age of the aircraft [6]. There are various forms of corrosion that exist, each of which poses different problems to aircraft structures and are briefly discussed below.

- Uniform corrosion is characterized by a uniform decrease of the thickness of the structural component, usually without a localized attack.

- Pitting corrosion is a form of localized attack, causing localized perforation of the material. It can cause failure by perforation with very little weight loss to the material. The pitting corrosion damage area is usually very small and, therefore, difficult to detect during routine inspection. Moreover, it can cause failure by perforation with very little weight loss to the material. Under the presence of external mechanical loads, the pitting points facilitate the propagation of intergranular cracks, as shown in Figure 31.3. Extensive corrosion damage can also lead to exfoliation or the loss of flakes of materials that separates from the ground material.

- Crevice corrosion is a form of localized corrosion, generated by localized changes in the corrosive environment. Corrosion attack starts more easily in a narrow crevice that contains a stagnant environment resulting in a difference in concentration of the cathode reactant between the crevice region and the external surface of the material. Crevices, such as those found at flange joints, threaded, or welded connections are thus often the most critical sites for corrosion. A special form of crevice corrosion is filiform corrosion, characterized by the fact that chemical degradation of metals occurs under a protective film that has been breached. It usually occurs under painted or plated surfaces when moisture permeates the coating.

- Galvanic corrosion is due to the electric current flowing between two or more dissimilar metals, immersed in a corrosive solution. This results in corrosion of the less noble metal and protection of the more noble metal. Galvanic corrosion can

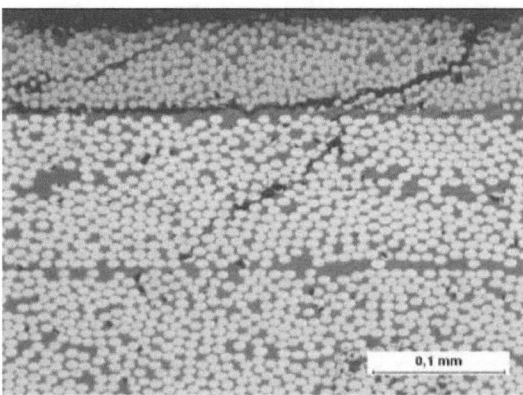

0,1 mm

Fig. 31.4. Matrix cracks and microdelamination in a CFRP composite.

be avoided by proper material selection and improved design of the connection between dissimilar materials.

- Stress-corrosion cracking occurs when applied stresses, such as in-service or residual stresses, and corrosion combine to initiate and propagate a fracture. One of the most common types is transgranular stress-corrosion cracking, SCC, that may develop in concentrated chloride-containing environments.

- Intergranular corrosion is a localized form of corrosion that attacks along the grain boundaries, keeping the bulk of the grains largely unaffected. This form of corrosion can proceed undetected through the material and may deteriorate the mechanical properties of the metal and cause fracture without any visible exterior signs of corrosion. A particular form of intergranular corrosion, called exfoliation corrosion, is typical of high-strength aluminum alloys that have been extruded or otherwise worked heavily. The damage often initiates at end grains encountered in machined edges, holes, or grooves and can subsequently progress through an entire section.

1.3 Composite Structures

Extensive efforts to identify the various modes of damage in composite materials have been undertaken in recent years. The primary finding of most of these investigations is that macroscopic fracture is usually preceded by an accumulation of the different types of microscopic damage and occurred by the coalescence of this small-scale damage into macroscopic cracks.

Macroscopic composite damage modes are usually divided in intraply and interply (interlaminar) failure modes.

The intraply damage modes are associated with the reduction of stiffness and/or strength of the single plies. The most common damage modes are matrix cracking or crazing, fiber breaking and buckling, and the failure of the fiber–matrix interface (fiber debonding). See Figure 31.4.

Damage progression in a ply will usually be initiated by matrix cracking due to tensile stress transverse to the fiber orientation. Damage growth may also involve matrix

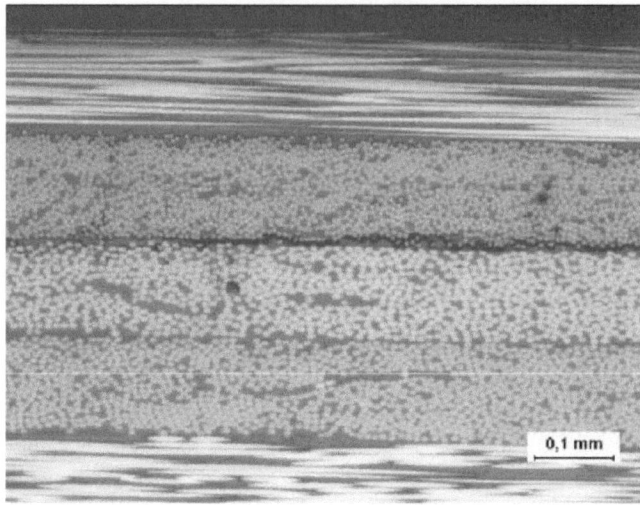

Fig. 31.5. Interlaminar delamination in a CFRP laminate.

Fig. 31.6. Failure of a CFRP laminate after bending fatigue test.

crushing due to transverse compression or fiber compressive failures due to longitudinal compression. Fiber compressive failures can be initiated by the loss of matrix stiffness. Fiber tensile failures may also occur but most composite structures are able to withstand some amount of fiber tensile failures prior to structural fracture [29].

It is important to underline that a local material failure (intraply) in a composite component will not usually mean immediate structural failure as is expected in metals with low fracture toughness. This is due to their higher damage-tolerant characteristics than homogeneous structures due to the inherent fail-safe characteristics (layered material).

The local or global separation of plies of material (interply failure) in layered structures, commonly referred to as delamination (Figure 31.5), is one of the most common failure modes suffered by aircraft structures. Laminated composites are especially susceptible to delamination owing to their weak transverse tensile and interlaminar shear strengths as compared to their in-plane properties and may arise from a wide variety of causes. Subsurface delaminations may be caused by accidental impacts by foreign objects during maintenance, repair, and flight missions. For thin laminates, damage usually occurs subsurface and is therefore not readily detected visually (this is known as barely visible impact damage, BVID).

Fig. 31.7. Adhesive failure of an aluminum bond.

Failure of composite structures is quite complex and usually it involves the concurrent action of several failure modes. An example of failure of a CFRP laminate under a fatigue failure test is shown in Figure 31.6. Some of the intraply and interlaminar damages modes are clearly visible such as fiber breakage on the bottom layer and local and widespread delamination.

In the case of sandwich structures, failure characteristics can be significantly different from conventional laminated structures, for example, skin–core debonding. The integrity of an adhesive bond is related to the cohesive part of the bond (presence and quality of an adhesive layer) and the adhesive part of the bond (quality of the interface between the adhesive and the composite or metallic substrate). Failure of bonded parts is usually related to the poor quality of the surface treatment at the production phase or to oxidation processes coupled to the ingress of water at the adhesive/adherent interface in service as shown in Figure 31.7.

Skin–core debonding also occurs under low-velocity impacts, where a permanent indentation in the impacted facesheet accompanied with localized core crushing beneath and around the impact site is produced. The facing skin will typically rebound to some degree after the impact event, therefore, the profile of the residual facesheet indentation does not necessarily correspond to that of the underlying crushed core. A comprehensive review of recent investigations of the failure in sandwich structures is given in Reference [30].

Due to the complex damage modes experienced by composite structures it is, therefore, important to have the capability to quantify damage initiation and damage propagation, and to estimate the remaining reliable life. This can be achieved only by developing robust, reliable, and user-friendly NDT to be implemented in a structural health monitoring system.

2. Nondestructive Testing (NDT) Techniques in Use in the Aerospace Industry

NDT of aerospace structures is commonly used as a quality control tool after manufacturing and throughout the service life of the structure to check for structural integrity. The number and frequency of inspections is a function of the part structural role and of the sensitivity of the NDT technique itself. If visual inspection may be the most-used inspection technique, it is limited to relatively large damage affecting the surface aspect of a part and often needs to be completed with techniques that give information on the integrity of the bulk of the material. The main techniques used in the aerospace industry are described below.

- *X-ray inspection*: The technique is based on the use of a source generating X-rays and a detector (film or digital detector). The technique is mainly used on metallic and composite parts but can be applied on composite materials with the use of a special contrast agent. Cracks and corrosion in metallic parts are the primary types of defects inspected with X-rays.

- *Ultrasonics and resonance-based techniques*: Ultrasonic-based techniques are routinely used as quality control for manufacturing of composite parts and bonded joints. Immersion or contact techniques are equally used in pulse–echo and transmission configuration. The typical frequency range is 1 MHz to 10 MHz. Quality control of composites with ultrasonics aims at the detection of potential delamination, debonding, porosity, and foreign objects such as plastic foils. Resonance techniques are mainly used to investigate bonded structures such as sandwich material (metal/honeycomb or composite honeycomb). The technique monitors variations in the first thickness resonances of the material and detects the local thickness and stiffness changes associated with disbond. Typical frequencies used with, for example, the Fokker bond tester (resonance-based equipment developed by Fokker) are from 50 kHz up to 300 kHz.

- *Eddy current*: EC is maybe the most commonly used technique for in-service inspection as its main field of application is the detection of fatigue cracks and corrosion in metallic structures. The technique is based on the monitoring of the electromagnetic properties of metals through the generation and detection of induced current at the near surface of the material. If EC aim at the same type of defect as X-ray, the technique focuses on portability and accessibility rather than rationality for large-area inspection. Another common application is the characterization of heat damage in aluminum by measurement of the material's conductivity.

- *Liquid-penetrant and magnetic particle inspection*: These two techniques are enhanced visual inspection based on the detection of surface breaking cracks and corrosion pits. They are used as quality control after manufacturing and as inspection techniques under maintenance tear-downs.

- *Shearography and thermography*: These full-field techniques are currently progressing in the aerospace industry because of the potential of large-area inspection. Shearography is an optical technique detecting local variations of stiffness by comparing the state of the surface before and after a slight loading of the structure applied with heat or vacuum. The comparison is performed using optical interferometry. Thermography is based on a visualization of the heat transfer mechanism in a structure that is exposed to a fast heat impulse. The techniques show good sensitivity to delamination in composite and disbonds in metal structures.

The aerospace industry has at its disposition a large toolbox of techniques that are used from the design phase, through manufacturing, to the in-service life of the aircraft. As this description shows, these techniques are sensitive to well-defined discontinuities in the material, that is, when the damage is already placed. The performance of inspection techniques, in term of sensitivity and threshold of detectability, play an indirect but nevertheless important role for the development of optimized design strategies. Traditional NDE techniques are generally not sufficiently sensitive to the presence and development of domains of incipient and progressive damage. The development of highly sensitive NDE techniques would improve not only the operational safety but also allow the design of more optimized structures.

2.1 Dynamic-Based Damage Detection Methods

Damage detection methods can be classified in two different branches: *vibration-based methods* and *wave propagation methods*.

The vibration-based approaches are based on the assumption that damage affects global structural stiffness, damping, and mass properties and, therefore, structural changes can be located by analyzing the changes of the dynamic properties such as natural frequencies, mode shapes, and damping. A detailed overview of vibration-based damage detection can be found in [3]. The main drawback of the vibration-based methods is that they detect structural changes that mainly affect the global modes of the structures. Moreover, they are based on the assumption that linear elastic structures remain linear elastic after damage and their sensitivity to detect defects such as cracks is very low.

The wave propagation-based methods detect structural damage by sending stress waves in the structures and measuring the changes in the received signal relative to the signal of the pristine structure [16, 17, 31, 32, 34]. A number of studies are under investigation by mainly using Lamb waves to detect structural changes in aircraft structures. The methods have been demonstrated to be effective in detecting corrosion [1], cracks in metallic structures [21], joints adhesion [2], debonding [15], or delamination in composites [18].

An overview of the Lamb wave-based methods, highlighting advantages and limitation of the damage detection methods, is presented below.

3. Lamb Waves

Conventional aircraft NDT are usually ground-based methods and used during periodic maintenance checks, whereas structural health monitoring continuously monitors aircraft structures in situ during flight. Of the currently used structural flaws detection techniques that are contenders for implementation into an integrated structural health-monitoring system, Lamb waves are one of the most promising techniques and are currently subjects of strong interest in academic and commercial laboratories. The major advantages for using Lamb waves are their capability of propagating a relatively long distance in thin plates, ability to follow curvature and penetrate into hidden and/or buried parts allowing detection of subsurface flaws, and wave structure dependence on frequency and phase velocity [1, 2]. Therefore, they are particularly suitable to inspect large aircraft areas and parts where direct access is not possible. Large areas, enclosed within a network of actuator sensors, can be inspected by analyzing the transmitted and/or reflected wave after interacting with the structure parts at boundaries or discontinuities. The presence and location of damage, flaws, and the like are usually identified by comparing the response of the undamaged configuration with the response signal of subsequent tests.

3.1 Lamb Waves for Damage Detection

Lamb waves are two-dimensional stress waves that can be generated in thin-walled structures with free boundaries, where the thickness is only a few wavelengths of the ultrasound wave. Lamb waves propagate through the entire thickness of a material in a symmetrical or antisymmetrical number of modes depending on their displacement pattern, as shown in Figure 31.8. The velocity of the wave propagation depends on the density, material properties, thickness of the material, and the wave frequency. There exist a finite number of modes that travel independently and satisfy the wave equation and the boundary conditions. Particle displacements and stresses in the Lamb waves occur throughout the thickness of the plate. Because Lamb waves produce stresses

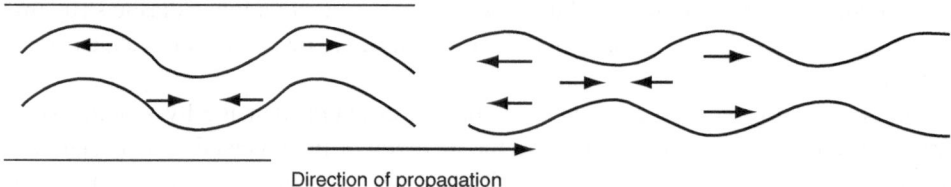

Direction of propagation

Fig. 31.8. (a) Symmetrical; (b) anti-symmetrical mode.

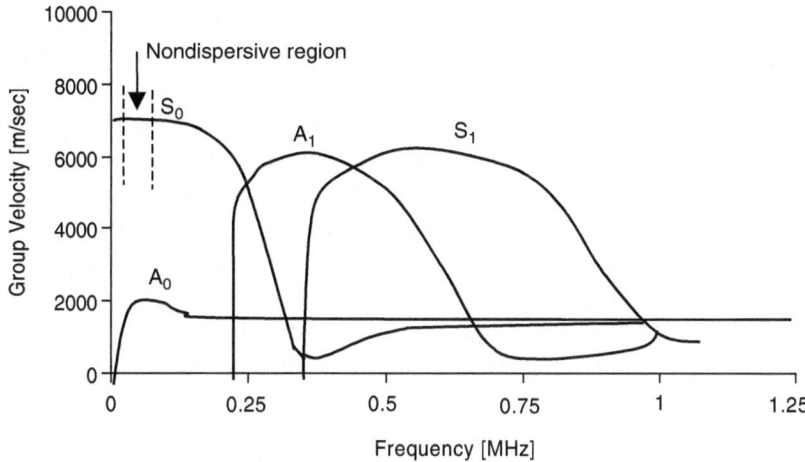

Fig. 31.9. Phase and group velocity.

throughout the plate thickness the entire thickness of the plate is interrogated, therefore they can be used to detect surface defects, and also internal defects.

Lamb waves can be generated by using various methods such as angled Perspex wedge, piezoelectric patches, air-coupled ultrasonic transducers, interdigital transducers, laser generation methods, and others.

For the application of active health monitoring, it is of primary importance to understand the Lamb waves that could be generated in the structures under investigation. An example of a typical dispersion curve of the first six Lamb modes propagating along a composite plate length with constant thickness is shown in Figure 31.9. Each curve represents a transient resonant mode with specific characteristics such as wave structure and energy distribution. By controlling both frequency and phase velocity different Lamb waves can be generated. However, for the creation of a robust damage detection methodology it is fundamental to know the type and size of damage experienced by the aircraft structure to be monitored, and it is important to identify and select Lamb wave modes that are sensitive to the presence of the specific structural anomaly under investigation. For example, some Lamb wave modes are sensitive to internal cracks, debonding, and delamination, whereas others are sensitive to corrosion, and so on.

The knowledge of the sensitive mode to a specific type of damage can be acquired by numerical technique [10, 11] and/or by comparing the Lamb wave propagation phenomena in the healthy and damaged structure [15]. Kundu et al. [11] have shown that a detailed analysis of the stress distribution inside the plate is important to select the Lamb mode to be used to detect a defect at a specific depth of the plate. By analyzing the influence of the defect on a particular component of the stress component, the Lamb wave mode with the highest stress component at the defect location should be selected.

The propagation of Lamb waves, however, is complicated due to unique characteristics such as dispersion and multimode and mode conversion [25].

The dispersive nature of waves causes the different frequency components of the Lamb waves to travel at different speeds, and the shape of the wavepacket to spread spatially and temporally, as it propagates through solid media. This broadening distribution of wave energy causes a drop in signal amplitude. It is, therefore, important to find a nondispersive point in such a way that dispersion between the transmitted input and the received signal is minimal so that the interpretation of response signals becomes easier because the difference between the input signal and the received signal can be wrongly attributed to presence of damage. An example of nondispersive mode points, are shown in Figure 31.9.

Moreover, although the input frequency should be high enough to make the wavelength of the Lamb wave comparable to the scale of local damage, the driving frequency also needs to be low so that higher modes are not excited (multimode excitation). To achieve this either the phase velocity (or wavenumber) and frequency bands of excitation must be chosen carefully to excite a single mode. A further reason for selecting modes that are well isolated in the dispersion curves is that mode conversion favors modes with similar phase velocity to that of the incident mode, and it is an important issue in the wave propagation analysis of composite material structures [36].

Another possible cause of the distortion of the excitation signal is the mode conversion that causes energy redistribution among multimodes. Mode conversion to other modes within the bandwidth of the incident signal may occur by any interference or discontinuity along the wave propagation and because the number of possible modes increases with frequency, lower bands may be preferred. A detailed study of the importance of single-mode operation possibility and the degree to which a mode can be isolated during excitation can be found in Reference [35].

For practical damage detection applications, a less dispersive frequency region should be selected so that the interpretation of response is clear and the frequency-thickness values should be kept below the cut-off frequency of the A_1 mode, where only the first two fundamental modes, namely, the first symmetrical (S_0) and antisymmetrical (A_0) propagation and modes, exist. Beyond this point, single-mode excitation becomes more difficult due to the numerous modes that the structure supports at higher frequencies.

Moreover, the A_0 mode is particularly attractive in ultrasonic NDE applications because of the associated predominance of out-of-plane displacements over in-plane displacements, because it is generally easier to generate and detect out-of-plane displacements with conventional ultrasonic transduction methods.

3.2 Failure Mode Detection Using Lamb Waves

The technology based on Lamb waves has been used to detect various types of damage in aircraft structures. Below is a summary of most common failure modes identified by Lamb wave technique are reported, which are discussed in detail in the next paragraphs.

Table 31.3. Most common failure modes identified by Lamb wave technique

Failure Type	Mode Selected
Delamination [17–19]	Antisymmetric mode A_0
Fatigue crack [21–31]	Symmetrical mode S_0
Sandwich debonding [15]	Antisymmetric mode A_0
Hidden corrosion [33]	Symmetrical mode S_0
Joints, adhesive strengths [23]	Antisymmetric mode A_0

3.2.1 Delamination

To detect this type of damage in composite structures, the antisymmetric mode is widely used due to its high sensitivity to delamination damage [16, 17]. Moreover, compared to other Lamb wave modes, the A_0 Lamb mode, for a given frequency, has slower speed and smaller wavelength, therefore it possesses better resolution.

Preliminary investigation of this type of damage was presented by Guo and Cawley [18]. Wang et al. [19] proposed an active diagnostic system to detect delamination by analyzing the diffracted wave energy caused by the presence of the delamination.

3.2.2 Hidden Corrosion Detection

Detection of corrosion damage in aircraft structures is an ongoing NDT challenge. Because Lamb mode velocities are a function of the frequency–thickness product, any structural or material changes such as corrosion/exfoliation or lack of adhesion between two layers will affect the propagating mode amplitude, velocity, and frequency spectrum. By sending a stress wave through a corroded area, a relatively low transmitted signal amplitude (Figure 31.10) will be recorded, whereas undamaged areas are associated with high received signal amplitude.

This methodology can also be applied to inspect lap splice joints by analyzing the low-amplitude signal when a nonperfect adhesion exists between the two bonded parts. Otherwise, if there is no damage, the excited mode will leak into the second joint producing relatively high-amplitude RF signal (Figure 31.10b).

3.2.3 Fatigue Crack Growth

The fundamental symmetric mode is currently used to detect surface crack growth in metallic structures due to its high sensitivity to cracks in structures. As an aircraft

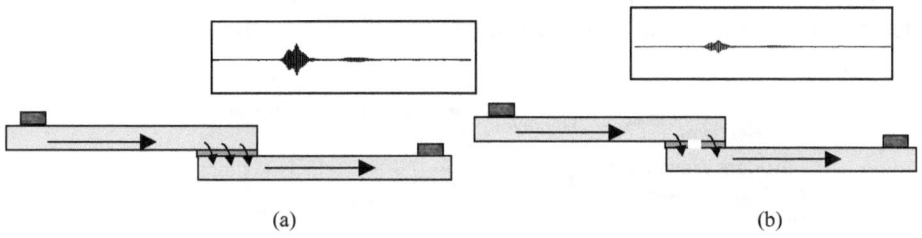

(a)   (b)

Fig. 31.10. (a) Undamaged joint; b) damaged joint.

structure undergoes increasing amounts of fatigue, an increasing number of cracks develop. These cracks degrade the Young modulus. Therefore, the local material exhibits a decrease in the velocity as the modulus degrades due to increasing fatigue. By monitoring the changes of velocity in various directions of wave propagations it is possible to understand damage location and damage size.

Current studies show [21] the efficiency of this mode in detecting closed and open cracks on an aluminum plate.

3.2.4 Joints and Adhesive Strength

Similar methodologies to the detection of corrosion can be applied to monitor the strength of bonds between two or more structural components. The use of guided wave inspection can give the possibility of maximizing the sensitivity to various types of bond conditions by exciting the appropriate cross-sectional mode shapes through selection of the corresponding frequency–velocity combinations. In particular, modes with large shear stresses at the bond layer, occurring generally for the antisymmetric modes, offer the highest sensitivity to bonded areas with reduced shear stiffness.

3.3 Currently Used Lamb Wave Damage Detection Methodologies

When an elastic wave travels through a region where there is a change of geometry and/or material properties, scattering occurs in all directions. The scattering is a typical phenomenon of a wave propagating in a material medium, in which the direction and frequency of the wave is changed when the wave encounters discontinuities in the medium. The scattering results in a disordered or random change in the incident energy distribution. In particular, the presence of damage or discontinuities produces a mode conversion process (Figure 31.11), where each produced mode has a characteristic propagation velocity and a characteristic frequency bandwidth.

This phenomenon can be used to locate the presence of structural anomalies in aircraft structures. In particular, by analyzing the scattered energy of the transmitted signals it is possible to detect the presence of damage. The decrease of energy in the direction of propagation should appear in other directions. An increase in the received energy can be detected in certain directions, showing that the damage can be considered as a scattering center [22].

Ihn and Chang [21] developed a damage index able to detect cracks in composite and metal aircraft panels. They calculated the scatter wave in the time domain by subtracting the reference data recorded for the structure with the initial damage size from the sensor data for the structure with the extended damage size. The damage index was based on the ratio between the scatter energy of the mode selected to the

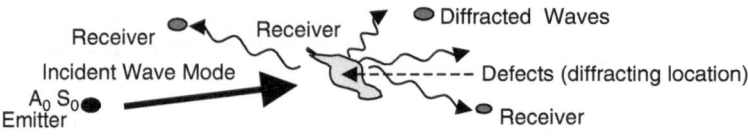

Fig. 31.11. Mode conversion process.

baseline energy. In particular, the damage index was calculated by dividing the time integration of the power scatter spectral density of the selected mode at a specified frequency, with the baseline signal

$$\text{Damage Index (DI)} = \left(\frac{\int_{t_i}^{t_f} |S_{scatter}(\omega_0, t)|^2 dt}{\int_{t_i}^{t_f} |S_{baseline}(\omega_0, t)|^2 dt} \right)^{\alpha}, \qquad (31.1)$$

where $S_{scatter}$, S_b are the spectral amplitude of the scatter, baseline signal ω_0 denotes the selected driving frequency, t_f and t_i are the upper and lower bounds of the fundamental symmetric mode in the time domain, respectively, and α is the gain factor. Clearly, the higher the values of the damage index are the higher the magnitude of the damage. Because the energy is directly related to the area under the time-varying power spectral amplitude, by monitoring the change of energy due to the structural changes' presence it possible to detect and localize damage.

3.4 Limitation of Lamb Waves

The linear acoustic–ultrasonic technique has many potential applications in the NDE of composite and metal aircraft structures. However, problems of poor reproducibility and of the sensitivity of the results to precise instrument settings have restricted its application in industry.

Multiple Lamb wave modes are created as the excitation frequency increases. In practice, it is difficult to generate a single pure mode, particularly above the cut-off frequency-thickness of the A_1 mode. Therefore, the received signal generally contains more than one mode, and the proportions of the different modes present can be modified by mode conversion which occurs at any interference or discontinuity along the wave propagation. The modes are also generally dispersive at high frequency, which means that the shape of a propagating wave changes with distance along the propagation path. This makes interpretation of the signals difficult and also leads to signal-to-noise problems because the peak amplitude in the signal envelope decreases rapidly with distance if the dispersion is high.

If amplitudes of the individual modes present in a multimode dispersive signal can be isolated, the relative amplitudes of the different modes generated by mode conversion at a defect could be measured, leading to possibilities not only of defect detection, but also of defect sizing.

Another important issue is the fact that the calculation of phase and group velocity dispersion curves can be obtained only for simple structures, but it becomes very complex and lengthy for structures with complex boundary conditions and interfaces.

4. Conclusions

The above-described difficulties have tended to reduce the attractiveness of Lamb wave testing and it is generally agreed that it is too complicated to be usable by technicians.

In light of these of technical and interpretation issues to be solved, it is difficult to envisage the implementation of a linear acoustic–ultrasonic technique in a structural health-monitoring system in the near future. This forces the scientific community to focus on finding alternative NDT methods that can be more reliable and robust, with ease of implementation and interpretation.

The new class of NDT based on the monitoring of nonlinear elastic wave propagation offers a promising solution to the many unsolved issues of NDT methods based on linear elastic wave propagation monitoring.

References

[1] Rose, J.L. and Barshinger, J.M. (1998). Using ultrasonic guided wave mode cutoff for corrosion detection, *IEEE International Ultrasonics Symposium*, 987–992.

[2] Lowe, M.J.S. Challis, R.E.C., and Chan, W. (2000). The transmission of Lamb waves across adhesively bonded lap joints, *J. Acoust. Soc. Am.*, 107(3): 1333–1345.

[3] Verrilli, M.J., Kantzos, P.T., and Telesman, J., Characterization of damage accumulation in a C/SiC composite subjected to mechanical loadings at elevated temperature, in the *Proceedings of Environmental, Mechanical, and Thermal Properties and Performance of Continuous Fiber Ceramic Composite (CFCC) Materials and Components*, ASTM STP 1392, M. G. Jenkins, Ed., American Society for Testing and Materials, West Conshohocken, PA, 2000.

[4] Cawley, P., Lowe, M.J.S., Alleyne, D.N., Pavlakovic, B., and Wilcox, P. Practical long range guided wave testing: Application to pipes and rails, *Mater. Eval.*, 61(1), 66–74 (2003).

[5] Rose, J.L., Standing on the shoulders of giants—An example of guided wave inspection, *Mater. Eval.*, 60(1): 53–59 (2002).

[6] Kautz, H.E., Acousto-Ultrasonics to Assess Material and Structural Properties, NASA/CR—2002-211881 (2002).

[7] Rose, J.L. and Soley, L.E., Ultrasonic guided waves for anomaly detection in aircraft components, *Mater. Eval.*, 58(9): 1080–1086 (2000).

[8] Gyekenyesi, A.L., Kautz H.E., and Cao, W., Damage Assessment of Creep Tested and Thermally Aged Udimet 520 Using Acousto-Ultrasonics, NASA/TM—2001-210988, (2001).

[9] Roth, D.J., Martin, R.E, Harmon, L.M., Gyekenyesi, A.L., and Kautz, H.E., Development of a High Performance Acousto-Ultrasonic Scan System, NASA/TM—2002-211913, (2002).

[10] Kundu, T., Potel, C., and de Belleval, J.F. (2001). Importance of the near Lamb mode imaging of multilayered composite plates, *Ultrasonics*, 39: 283–290.

[11] Maslov, K. and Kundu, T. (1997). Selection of Lamb modes for detecting internal defects in composite laminates, *Ultrasonics*, 35: 141–150.

[12] Cho, Y. (2000). Estimation of ultrasonic guided wave mode conversion in a plate with thickness variation, *IEEE Trans. Ultrasonics, Ferroelectr. Frequency Control*, 47(3): 591–603.

[13] Wei, Y. and Kundu, T. (1998). Guided waves in multilayered plates for internal defect detection, *J. Eng. Mech.*, 124(3): 141–150.

[14] Rose, J.L., Zhu, W. and Cho, Y. (1998). Boundary element modeling for guided wave reflection and transmission factor analyses in defect classification, In: *IEEE Ultrasonics Symposium Proceedings*, Vol. 1, pp. 885–888.

[15] Hay, T., Wei, L., and Rose, J. L. (2003). Rapid inspection of composite skin–honeycomb core structures with ultrasonic guided waves, *J. Composite Mater.*, 37(10).

[16] Chang, Z. and Mal, A. (1999). Scattering of Lamb waves from a rivet hole with edge cracks, *Mech. Mater.* 21: 197–204.

[17] Kessler, S.S. Spearing, S.M., and Soutis, C. (2002). Structural health monitoring in composite materials using Lamb wave methods, *Smart Mater. Struct.* 11: 269–78.

[18] Guo, N. and Cawley, P. (1993). The interaction of Lamb waves with delaminations in composite laminates *J. Acoust. Soc. Am.* 94 2240–6.

[19] Wang, C.S., Wu, F., and Chang, F.K. (2001). Structural health monitoring from fiber-reinforced composites to steel-reinforced concrete, *Smart Mater. Struct.* 10: 548–52.

[20] Wilcox, P.D., Lowe, M.J.S., and Cawley, P. (2001). Mode and transducer selection for long range Lamb wave inspection *J. Intell. Mater. Syst. Struct.* 12: 553–65.

[21] Ihn, J.-B. and Chang, F.-K. (2004). 'Detection and monitoring of hidden fatigue crack growth using a built-in piezoelectric sensor/actuator network: I Diagnostics.', *Smart Mater. Struct.* 13: 621–630.

[22] Kaczmarek, H. (2003). Lamb wave interaction with impact-induced damage in aircraft composite: Use of the A0 mode excited by air-coupled transducer, *J. Composite Mater.* 37: 217–232.

[23] Lanza di Scalea, F., Rizzo, P., and Marzani, A., (2004). Propagation of ultrasonic guided waves in lap-shear adhesive joints: Case of incident A0 Lamb mode, *J. Acoust. Soc. Am.*, 115(1): 146–156.

[24] Findlay, S.J. and Harrison, N.D. (2002). Why aircraft fail, *Mater. Today*, November.

[25] Viktorov, I.A. (1967). *Rayleigh and Lamb Waves*, Plenum Press: New York.

[26] Findlay, S.J. and Harrison, N.D. (2002). Why aircraft fail, *Mater. Today*, November.

[27] Campbell, G.S. and Lahey, R.T.C. (1983). *A Survey of Serious Aircraft Accidents Involving Fatigue Fracture, Vol. 1, Fixed-Wing Aircraft*, National Aeronautical Establishment, Canada.

[28] Campbell, G.S. and Lahey, R.T.C. (1983) *A Survey of Serious Aircraft Accidents Involving Fatigue Fracture, Vol. 2, Rotary-Wing Aircraft*, National Aeronautical Establishment, Canada.

[29] Minnetyan, L., Murthy, P.L.N., and Chamis, C.C. (1992). Progressive fracture in composites subjected to hygrothermal environment, *Int. J. Damage Mech.*, 1(1): 69–70.

[30] Abrate, S. (1997). Localized impact on sandwich structures with laminated facings. *Appl. Mech. Rev.*, 50(2): 69–81.

[31] Alleyne, D.N. and Cawley, P. (1992). The interaction of Lamb waves with defects, *IEEE Trans. Ultrason. Ferroelectr. Freq. Control*, 39: 381–97.

[32] Roh, Y.S. and Chang, F.-K. (1999). Built in diagnostics for identifying an anomaly in plates using wave scattering. PhD Dissertation. Department of Aeronautics and Astronautics, Stanford University, Stanford, CA.

[33] Zhu, W., Rose, J.L., Barshinger, J.N., and Agarwala, V.S. (1998). Ultrasonic guided wave NDT for hidden corrosion detection, *Res. Nondest. Eval.*, 10(4): 205–225.

[34] Wang, C.S. and Chang, F.-K. (1999). Built-in diagnostics for impact damage identification of composite structures, *Proc. 3rd Int. Workshop on Structural Health Monitoring* (Stanford, CA, 1999) pp. 612–21.

[35] Alleyne, D.N. and Cawley, P. (1992). Optimization of Lamb wave inspection techniques, *NDT. E Intl.* 25: 11–22.

[36] Dalton, R.P., Cawley, P., and Lowe, M.J.S. (2001). The potential of guided waves for monitoring large areas of metallic aircraft fuselage structure, *J. NDE*, 20: 29–46.

[37] Nayfeh, A.H. and Chimenti, D.E. (1989). Free waves propagation in plates of general anisotropic media, *J. Appl. Mech.* 56: 881–6.

[38] Nayfeh, A.H. and Chimenti, D.E. (1989). The propagation of horizontally polarized shear waves in multilayered anisotropic media *J. Acoust. Soc. Am.* 86: 2007–12.

[39] Nayfeh, A.H. and Chimenti, D.E. (1991). The general problem of elastic waves propagation in multilayered anisotropic media *J. Acoust. Soc. Am.* 89: 1521–31.

32

Inverse Problem Solution in Acoustic Emission Source Analysis: Classical and Artificial Neural Network Approaches

Zdenek Prevorovsky,[1] Milan Chlada, Josef Vodicka

Institute of Thermomechanics AS CR, Dolejskova 5, CZ - 18200 Prague 8, Czech Republic.
[1] To whom correspondence should be addressed: Zdenek Prevorovsky, e-mail: zp@it.cas.cz.

Abstract

Solution of nonlinear Inverse Problems (IPs) is a frequent task in nondestructive testing of materials and structures when structural defects or imperfections must be recognized. One among the most promising ultrasonic NDT techniques is the Acoustic Emission (AE) method, which can reveal a dangerous defect (e.g., cracks) growth in realtime. Two practical IP examples of AE source analysis are presented in this chapter: AE source location and identification. As a comparison, a classical approach to the identification IP is shown, whereas the source location and AE signal parameter correction IPs are treated by the use of the soft computing method based on Artificial Neural Networks (ANNs). A short introduction to the ANN approach is presented for that purpose.

Keywords: Acoustic emission, artificial neural networks, inverse problems

1. Introduction

The Forward Problem (FP) in physics is generally treated as the computation of data values given an appropriate model. On the other hand, the aim of Inverse Problems (IPs) is to reconstruct the model from a set of measurements. In the ideal case, an exact theory exists that prescribes how the data should be transformed in order to reproduce the model. For some selected examples, such a theory exists assuming that the required infinite and noise-free data sets are available. In NDT/NDE of materials and structures we often meet requirements on nonlinear IP solution. These IPs are solved either by classical IP treatment or by using soft computing methods as Artificial Neural Networks (ANN), which represent an alternative to the genetic algorithms (see Chapter 22, part 2, for more details on the GA approach to IPs). Many IP examples from various NDT topics are documented.[1]

Acoustic Emission (AE) source location and identification is an example of classical IP solutions in NDT. The AE method is a very effective NDT tool for diagnostics of materials and structures. In contrast to other, passive, ultrasonic defect detection methods, the AE is based on the detection of elastic waves emitted by the material itself during the defect growth (only active defects are detected). In addition to the realtime detection of the appearance of structural defects, AE monitoring should realize two other important tasks connected with IP solution:

(a) AE source (defect) location
(b) AE source identification and defect severity classification.

The task on AE source location is generally solved by the evaluation of i time arrival delays Δt_{ip} between p AE sensors properly dislocated on a tested structure, and then by solving the set of quadratic equations for the unknown source position.[2] Nevertheless, this approach requires good knowledge of the elastic wave propagation in the structure (in particular, a knowledge of wave propagation velocity), and often is not applicable to complicated anisotropic bodies or when dispersion effects and/or high background noise are present. In such complicated cases, current soft computing methods such as artificial neural networks offer a better solution.

2. Artificial Neural Networks Approach

Artificial neural networks originated as an attempt to imitate the processing capabilities of the human brain. In 1943, Pitts and McCulloch[3] introduced the Perceptron model, which represents a simplified model of the biological neuron, the basic unit of a human neural system. The artificial neuron model has since been mathematically generalized, but its fundamental properties remained: each neuron has a number of inputs, which are transformed into a single output as schematically illustrated in Figure 32.1a.

ANN consist of a number of interconnected neurons. Depending on the definition of the neurons and on their interconnection, many ANN architectures can be distinguished (feedforward, Hopfield, Kohonen, ART, etc.). This chapter deals almost exclusively with feedforward ANNs only, also called multilayer perceptrons or back-propagation networks (BP-networks). In BP-networks, the neurons are organized in a

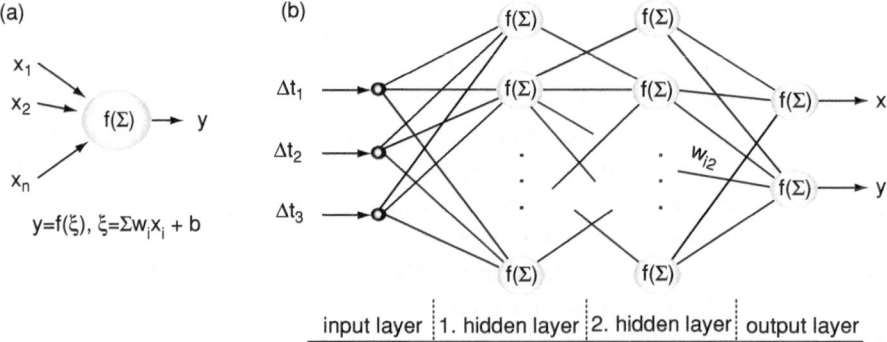

Fig. 32.1. a) A schematic description of a neuron with n inputs $x_1, x_2, \ldots, x_n$, and one output y. The output is determined by the function f acting on a linear combination of the inputs $t_1, \ldots, t_n$. Numbers are weights, b is bias. Selection of weights and bias determines the neuron behavior. b) Example of three-layer feed-forward ANN with three inputs Δt_1, Δt_2, Δt_3 and two outputs x, y.

number of layers, interconnected in such a manner that the output from each neuron of the first layer is connected to each neuron in the second one, each neuron output in the second layer is connected to each neuron in third one, and so on, as shown in Figure 32.1b. Interconnections, with different weights w_i are called *synapses*.

The inputs of BP-network coincide with the inputs to neurons in the input layer. Information then propagates forward through the network; weighted sums of the outputs from a previous layer are passed to neurons in the next layer. Finally, we get the network outputs in the last output layer. The layers between the input and output layer are called *hidden layers*.

Mathematically, feedforward ANN represent mappings from $\mathbf{R}^n$ to $\mathbf{R}^m$ (n, m are integers), where n is the number of neurons in the input layer, and m is the number of neurons in the output layer.

This mapping is determined by the network architecture and by the chosen weights. The process of weight adjustment is called ANN learning or training. The ability to learn from examples is the most significant feature of ANNs. Popularity of BP-networks stems from the existence of an effective learning algorithm called backpropagation.

The list of ANN applications during the last two decades covers a variety of different fields ranging from industry, transport, finances, business, telecommunications, and so on, up to medicine, speech, and entertainment. ANN can be applied to pattern recognition, identification, classification, nonlinear control systems, data analysis, optimization, scheduling, and so on. They also provide a very effective tool for forward and inverse problem solution. Nowadays, the design of an ANN application is substantially simplified by using very effective software development tools such as the Neural Network Toolbox in MATLAB,[4] used in the IP solution examples presented in next sections.

During training, parameters of BP-networks should be adjusted in order to achieve an appropriate structure for the particular task domain. Their desired behavior is usually formulated in the form of an objective function. For the standard BP-training algorithm, the objective (or error) function E represents the total error between the desired and actual output of all output neurons in the BP-network, taken for all training patterns from a fixed, finite training set : $T = \left\{ \left[\vec{x}_1, \vec{d}_1 \right] \ldots, \left[\vec{x}_P, \vec{d}_P \right] \right\}$

$$E = \frac{1}{2} \sum_{p=1}^{P} \sum_{i=1}^{N} \left(y_{ip} - d_{ip} \right)^2, \tag{32.1}$$

where p is an index over all P training patterns, i is an index over all N output neurons, y_{ip} denotes the actual and d_{ip} denotes the desired output value of the ith neuron for the pth training pattern. To minimize E by the gradient descent procedure, it is necessary to compute the partial derivatives of E with respect to each weight of the network (for further details, see, e.g., Rojas[5]). In this way, the particular weights of the network have to be adjusted by:

$$w_{ij}(t + 1) = w_{ij}(t) + \alpha \delta_j y_i, \tag{32.2}$$

where

$$\delta_j = \begin{cases} f'(\xi_j)(d_j - y_j) & \text{for an output neuron} \\[2mm] f'(\xi_j) \sum_k \delta_k w_{jk} & \text{for a hidden neuron.} \end{cases} \qquad (32.3)$$

k indexes neurons in the layer subsequent to the neuron j; d_j is the desired (in case of output layer only) and y_j the actual output value of the neuron j; $f'(\xi_j)$ is a derivative of transfer function at the point ξ_j, the potential of neuron j; $t + 1$ and t index of next and present weights, respectively; w_{ij} is the weight of the synapse directed from neuron i to neuron j; and a is a constant representing the learning rate.

The network is trained by initially choosing small random weights and thresholds and then presenting repeatedly all training patterns. Thresholds may be simulated by a fictive neuron with a constant output value equal to one. An essential component of the algorithm is an iterative method that propagates the error terms δ_j required for adjusting the weights, starting from the neurons in the output layer back to the neurons in the previous layers (backpropagation). These error terms correspond to the negative partial derivative of E with respect to the currently adjusted weight $-(\partial E / \partial w_{ij})$.

Faster variations of the backpropagation algorithm are based on other optimization techniques, belonging to two main categories. The first category uses heuristic techniques, as momentum learning, variable learning rate backpropagation, and resilient backpropagation, which are derived from performance analysis of the standard steepest descent algorithm. The second category of fast algorithms uses numerical optimization techniques such as conjugate gradient, quasi-Newton, and Levenberg–Marquardt.

Properly trained backpropagation networks should give reasonable answers when presented with unfamiliar input data. Typically, a new input leads to an output similar to the output obtained for training vectors similar to the new input presented. This generalization property makes it possible to train ANNs on a representative set of input/target pairs, and get good results without training the network on all possible input/output pairs. Therefore, the selection of an appropriate training data set has crucial importance for the final behavior of the multidimensional function, which represents the global output of a trained neural network. The set of training patterns should properly cover the whole domain of possible inputs.

The efficiency of neural network training can be increased by certain data preprocessing steps performed on the network inputs and targets, in order to ensure that they fall within a specified similar range. The most often used approach is to normalize the mean and the standard deviation of the training set. This procedure normalizes the inputs and targets to have zero mean and unity standard deviation. One of the problems that occurs during neural network training is called overfitting. The error on the training set is driven to a very small value, but when new data is presented to the network, the error is large. The network has memorized the training examples, but it has not learned how to generalize to new situations. Figure 32.2 shows a typical response of two neural networks that have been trained to approximate dependence between training data marked by the "+" symbols. A solid line shows the approximation obtained by a correctly trained network while the response of an overfitted network is drawn by

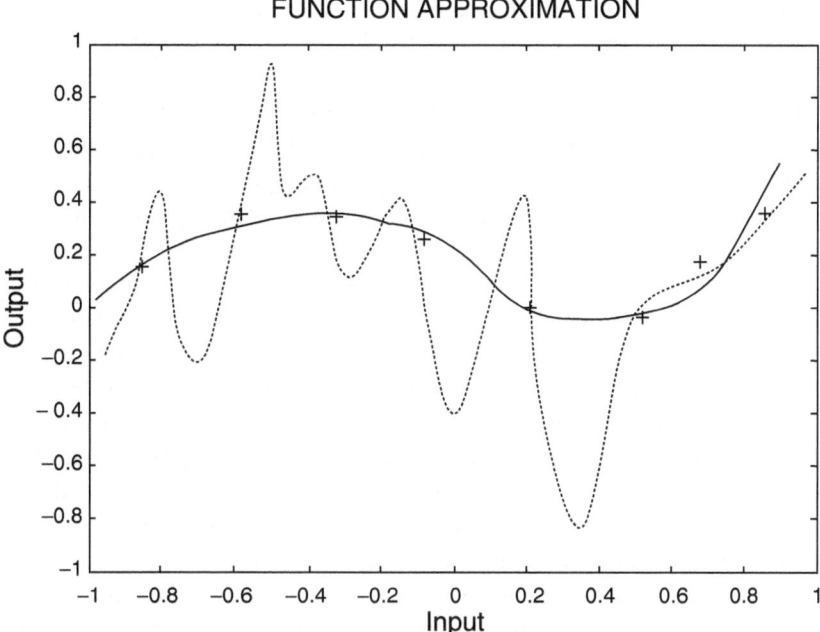

Fig. 32.2. Behavior of "well approximating" and "over fitted" network.

dotted line. It is evident that overfitting often implies fitting noise and the overfitted network will not generalize well.

The most powerful method for solving the overfitting problem is regularization. This method involves the modification of the performance function (which is normally sum of squares of network errors for training patterns) by adding a special term msw:

$$msw = \frac{1}{N_W} \sum_{j=1}^{N_W} w_j^2, \tag{32.4}$$

where N_W is the total number of network weights. It is clear that large weight values lead to high sensitivity of neurons to a variation of their inputs. Using the performance function E_{REG} leads to smaller weights and biases, which results in smoother network response and lower probability of over-fitting.

$$E_{REG} = \nu E + (1 - \nu)msw. \tag{32.5}$$

The difficulty of regularization lies in the optimal determination of performance ratio ν.

The ANN learning process is considerably affected by the selection of initial weights and biases. If the starting potentials of neurons lie within the interval of the highest sigmoidal function slope, the gradient of the error function acquires high values, and neurons still remain sensitive enough. If we consider neuron potentials as random variables, we can formulate the above hypothesis as a zero-mean requirement on the value of the initial weights. At the same time, the neuron weights should not be too small.

The absolute values of gradient components would dramatically fall during the back-propagation algorithm, due to their multiplication by small numbers, and the learning process would become unacceptably slow. Therefore, it is also useful to require unitary standard deviation of neuron potentials. When the selection of initial weights ensures zero mean and unitary variance of neuron potentials, the network learning will not start in a flat region of the error function, which results in higher learning speed.

Proper selection of input data (pattern features, e.g., some signal or image parameters) is of great importance when the ANN is used for pattern recognition or classification. It is important to recognize significant pattern features within an initially large redundant feature set, and to discard other features insignificant for the correct ANN decision. The feature selection methods examining the independence of attributes, such as correlation analysis or information gain, are simple and fast, but can fail in some applications. A common approach to feature selection for systems with one output is based on computing the marginal importance measure for each pattern feature. However, when we use some real data, input features may be interdependent and the system output then depends on relationships between inputs rather than on the input values themselves. When the number of input features is small, a quasiexhaustive search may be used to select the best feature subset. However, the number of possible combinations grows quickly with the number of attributes. An alternative, sensitivity analysis approach to feature selection (called FSS) is proposed in [6]. It comprises the following steps.

- Train the BP-network using all possible candidate features.
- For all training patterns p, for corresponding network outputs y_{jp} and inputs x_{ip}, compute the sensitivity coefficients s_{ji}, defined as:

$$s_{ji} = \frac{1}{P} \sum_{p=1}^{P} \left| \frac{\partial y_{jp}}{\partial x_{ip}} \right|. \tag{32.6}$$

- Eliminate "dummy" features with small values of coefficients s_{ji}. For trained BP-networks, high values of sensitivity coefficients indicate "important" features.

In any case, the differences of s_{ji} values between significant and dummy features become smaller when noise increases. If dummy features are eliminated, and training is repeated using only the remaining features, the performance error becomes lower. This confirms the advantage of eliminating dummy variables that just introduce noise. The backpropagation technique can be also applied to the calculation of the sensitivity coefficients of trained BP-networks. These coefficients express the sensitivity of ANN to the considered set of input patterns, which is formulated by means of the first derivatives of the BP-network outputs y_j with respect to its inputs x_i.

The sensitivity analysis of pretrained BP-networks allows the reduction of the network architecture. The achieved results show better results than many classical approaches to feature selection. A further advantage of sensitivity analysis consists in the possibility of comparing the efficiency of different BP-networks trained to solve a given problem with similar accuracy.

3. Acoustic Emission Source Location Using Artificial Neural Networks

Accurate AE source location is a necessary condition for precise damage area isolation in tested structures. Nevertheless, the problem of planar AE source location on, for example, a highly anisotropic body, cannot be solved in an easy way. Significant contribution to localization procedures represents an introduction of ANN algorithms.[7] Trained ANN with signal arrival time-difference inputs (see Figure 32.1b) can correctly solve that problem in all relatively complicated cases, for example, in both anisotropic and dispersive structures such as wound composite tubes under complex loading. Moreover, sometimes it is not necessary to perform detailed ANN training (calibration) on a real structure, and an easier training on calculated (numerically simulated) data is sufficient, as demonstrated in the following example.[8] ANN-based source location procedures give sufficiently accurate results in realtime and don't require any knowledge of wave propagation velocity or other material properties. One of the aims of the reported study[8] was to correlate areas on composite tubes, exhibiting high AE activity during multiaxial step loading, with the structural changes observed by subsequent ultrasonic C-scans. The used experimental setup is schematically drawn in Figure 32.3. Signals from six AE transducers were preamplified and recorded by a multichannel AE analyzer.

To solve the problem of AE source localization, an ANN approach with time-difference inputs was applied. Several ANN architectures were trained and tested on both numerically simulated and experimentally modeled AE sources. The resulting four-layer ANN architecture (one input layer, two hidden layers, and one output layer) was optimized on a minimal localization error. The inputs of the first layer are differences Δt_i of AE signal arrival times, and the outputs from the last layer are two AE source coordinates x, y (ref. to Figure 32.1b). The backpropagation algorithm was used for ANN training. The feedforward algorithm of the three-sensor ANN with 15 neurons in two hidden layers may be described by the following equations. Outputs a_{j1} of the first hidden layer are given by the equation

$$a_{j1} = \tanh\left(\sum_{i=1}^{3} w_{ij1} \Delta t_i + b_{j1}\right), \tag{32.7}$$

where $j = 1, \ldots, 15$ is the number of neurons in the first layer, w_{ijl} are weights and b_{jl} biases of neurons in that layer. The second hidden layer output is described by

$$a_{j2} = \tanh\left(\sum_{i=1}^{15} w_{ij2} a_{i1} + b_{j2}\right), \tag{32.8}$$

where $j = 1, \ldots, 15$ is now the number of neurons in the second layer and a_{i1} are the neuron outputs of the first hidden layer. The output of the network (two source coordinates) is expressed by the equation (linear transfer)

$$a_{j3} = \sum_{i=1}^{15} w_{ij3} a_{i2} + b_{j3}, \tag{32.9}$$

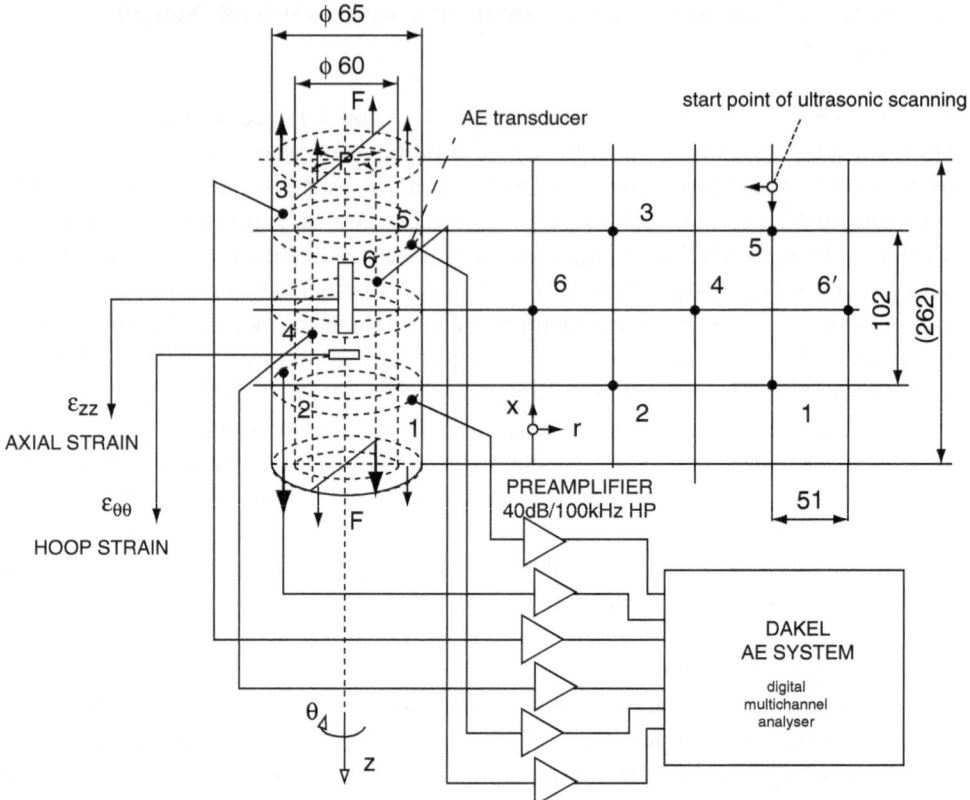

Fig. 32.3. Schematic view on experimental setup used for multiaxially loading of composite tubes.

where $j = 1, 2$ is the number of neurons in the last layer and a_{j2} are the outputs of the second hidden layer.

Six and three AE sensor array configurations were considered for high- and low-level signals, respectively. Relatively large wave attenuation in tested tubes (about 70 dB/m) and background noise allow only localization of higher amplitude AE events. As a consequence, not all six sensors will detect the low amplitude AE events, and only a three-transducer array is used for the source location. However, a lower localization error is reached by a six-transducer array due to redundancy. For three-sensor arrays, nine different networks must be trained. Eight similar networks are taught to localize AE source inside the triangle arrays, and one simpler network is trained to select the adequate sensor array.

Two different procedures were used for the ANN training : (1) the input time differences are determined using the Pen-Test experiments (breaking of pencil lead on a tube surface) in uniformly spaced training points, and (2) the directional diagram of group velocities is determined by means of Pen-Tests and then the training set of arrival time differences is calculated numerically. In both procedures, signal arrival times were precisely evaluated using adaptive signal filtering. Both of the above-mentioned ANN

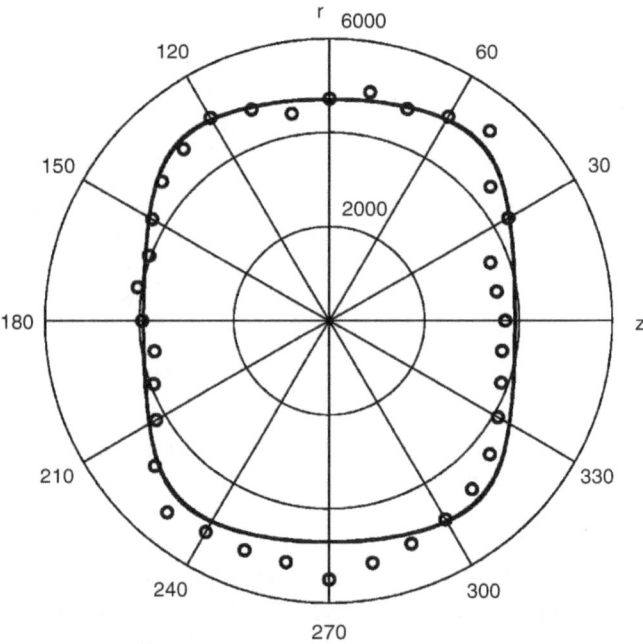

Fig. 32.4. Wave velocity directional diagram of GFRP composite tube. Circles are experimental values determined by Pen-Test. Regression curve is almost identical with theoretically computed diagram.

training procedures resulted in similar location errors lower than 8 mm. Figure 32.4 shows measured wave velocity diagram of the tubes.

The best fit of the experimental data in Figure 32.4 is obtained by using a regression function in the form of the theoretically predicted (computed) curve of the anisotropic group wave velocity. The theoretical curve follows from the knowledge of elastic properties of composite constituents. The regression and theoretical curves may not be identical due to the presence of guided waves in a relatively thin structure. However, in our case, the least square fitting curve differs from the theoretical one by only 2%. The maximal velocity $c_{90} = 4025$ m/s is in the hoop direction, and minimal velocity along the tube axis is $c_0 = 3120$ m/s so the anisotropy ratio is $c_{90}/c_0 = 1.29$. The standard deviation of measured data was 280 m/s. The determined velocity diagram may be used for numerical ANN training on another structure with any transducer configuration (similar anisotropy and wall thickness is required).

AE localization results are documented in Figure 32.5, which shows the spatial distribution of high-energy AE events(left part) during a tensile loading cycle of the tube up to $\sigma_{zz} = 47$ MPa, as compared to the differential ultrasonic C-scan of the tube (right part). The x- and r-coordinates represent, respectively, the longitudinal and radial coordinates. The AE sensor positions are marked by numbered squares 1 to 6 (see also Figure 32.3), and localized AE sources are represented by small circles. 4280 AE events were successfully localized from 18,647 registered events in the considered loading cycle. The top and bottom regions without AE detection are tube parts near the clamping where AE sources were not localized.

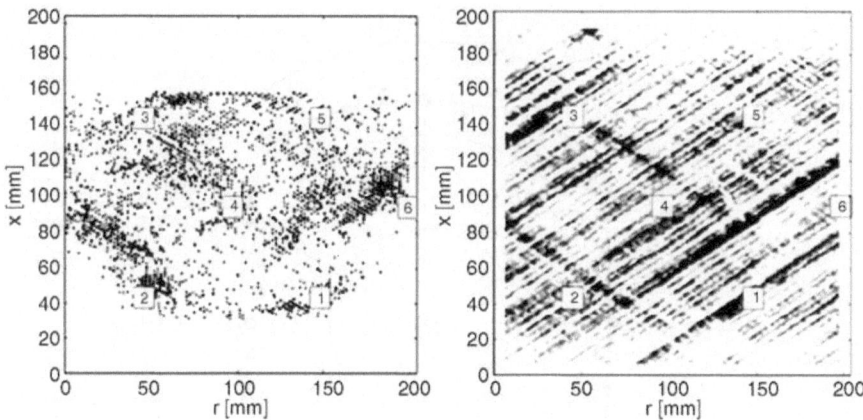

Fig. 32.5. AE source distribution map (left) and corresponding Differential C-scan (right). Numbered squares indicate AE transducer locations.

A concentration of localized AE sources in some particular regions and along fiber directions is evident in the left part of Figure 32.5. To understand the nature of such active regions, the source location map was correlated with other structural observations performed by ultrasonic C-scanning before and after the loading cycle. Both ultrasonic images were precisely shifted with respect to the maximum of their two-dimensional cross-correlation function and the difference was computed. The resulted differential C-scan is shown in the right part of Figure 32.5. The dark patterns in some regions are more pronounced, which indicates growing damage. The positions of six AE transducers were also marked on differential C-scan for comparison with the AE-source distribution. Significant black bands indicate damage concentration and delamination along some overstressed fiber bundles. Overlapping dark regions are remarkable on both parts of Figure 32.5.

From the comparison of the defective regions revealed by the two independent methods, it can be concluded that the AE localization algorithm, based on the ANN trained by the wave velocity data, is an appropriate and precise procedure for complicated anisotropic structures. Furthermore, this proves that an ANN trained on one body can satisfactorily localize AE sources on other bodies of similar geometry, made from the same material.

4. Acoustic Emission Source Identification

The second AE source identification problem represents one of the most difficult inverse problems in AE analysis. AE sources in stressed materials are mostly rapid dynamic processes at a microstructure level with duration from fractions of nanoseconds up to microseconds. Real AE sources, whose true physical description is almost always not available, are replaced by simple mechanical model wave-based approaches. From the wavelength point of view, the sources may be considered as nearly punctual, and the problem can be reduced to the study of a body response to point forces or quasipoint moments (force dipoles). The generalized theory of AE and AE source representation has been compactly formulated both by Ohtsu and One[9] in the field of AE, and by Aki and Richards in the field of quantitative seismology.[10]

AE sources may be modeled as an internal discontinuity displacement (rupture) caused by a dislocation or crack (seismic fault) or by volume changes (explosion). An AE signal then represents the kinematical response of the body (surface displacement or velocity field) to the dynamic forces and moments acting on a volume V. The dynamic displacement field $\vec{u}(\vec{x}, t)$, caused by the elastic wave propagating from the AE source represented by volume (point) forces $\vec{f}(\vec{x}, t)$, must generally fulfill the wave equation[10] (index summation rule is applied)

$$\frac{\partial}{\partial x_j} \left(c_{ijkl}(\vec{x}) \frac{\partial}{\partial x_l} u_k(\vec{x}, t) \right) + f_i(\vec{x}, t) = \rho(\vec{x}) \frac{\partial^2 u_i(\vec{x}, t)}{\partial^2 t}, \tag{32.10}$$

where $\vec{x} = (x_1, x_2, x_3)$ is the coordinate vector, t is time, c_{ijkl} the tensor of elastic constants, and ρ is the material density (assumed to be constant). In the abbreviated Cartesian tensor notation, Eq. (32.10) has the form

$$\left[c_{ijkl} \cdot u_{k,lj} \right] + f_i = \rho \cdot \ddot{u}_i. \tag{32.11}$$

Necessary initial and boundary conditions must be added to complete the problem definition. Standard conditions are expressed as a traction-free body surface $S(V)$ (i.e., homogeneous, zero-stress boundary conditions) and zero initial displacements and their derivatives in the whole volume V:

$$\begin{aligned} \tau_{ij}(\vec{x}, t) \, v_j^s &= 0, \quad \vec{x} \in S(V) \\ \vec{u}(\vec{x}, 0) &= 0, \quad \vec{x} \in V \\ \frac{\partial \vec{u}(\vec{x}, 0)}{\partial t} &= 0, \quad \vec{x} \in V, \end{aligned} \tag{32.12}$$

where $\tau_{ij}(\vec{x}, t)$ is the stress tensor and $\vec{v}^s$ denotes the normal unity vector (outward normal) to the boundary surface S of the volume V. The inverse problem consists in the determination of acting forces from the measured displacement or velocity signals. The problem is simplified if we know the Green's tensor function G_{ij} satisfying Eq. (32.10) with the standard conditions (32.12):

$$G_{ij} \left(\equiv \frac{u_i(\vec{x}, t)}{A} \right) = G_{ij}(\vec{x}, t; \vec{x}_0, 0). \tag{32.13}$$

The Green function represents the complex transfer function between the source and signal detector places, that is, displacement response in the i-direction at a point $\vec{x}$ to the unity pulse of the concentrated force $f_i(\vec{x}, t)$ acting in the j-direction at time t in the location $\vec{x}_0$

$$f_i(\vec{x}, t) = A\delta(\vec{x} - \vec{x}_0)\delta(t)\delta_{ij}, \tag{32.14}$$

where A is a unit constant of the force dimension, $\delta(\ )$ denotes the Dirac delta-function of space and time coordinates, and δ_{in} is the Kronecker delta. Finding the Green function of the tested body is the main task of elastodynamics. Its analytical form is known only in particular situations such as for partially unbounded bodies of simple geometry. In most cases, it must be determined numerically (e.g., by FEM, FDM, LISA, EFIT, etc.) or experimentally. Experimental determination of the Green function requires many measurements using artificial AE sources well approximating the delta function (e.g., laser pulse or the above-mentioned pencil-lead rupture test called PEN-Test). The numerical solution is also complicated, as huge computations must be performed, especially in 3-D cases. A promising method of computation and data reduction is the use of ANN for the Green function interpolation.[11] Knowledge of the Green function simplifies the solution of wave equation (32.10) to the convolution of the source function $F_j(t)$ with the Green function, assuming far-field displacement, that is, for $|\vec{x} - \vec{x}_0| > n\lambda$, where λ is the wavelength, and n an integer:

$$u_i(\vec{x}, t) = G_{ij}(\vec{x}, t; \vec{x}_0, 0) * F_j(t). \tag{32.15}$$

Single point forces, acting on the body surface, suit well as a mechanical substitution of the PEN-Test. Generally, AE source mechanisms are represented by the seismic moment tensor $M_{jk}(t)$ involving also coupled forces–dipoles and their moments acting inside the body. Equation (32.15) then becomes

$$u_i = G_{ij} * F_j + G_{ij,k} * M_{jk}. \tag{32.16}$$

The AE source is identified (i.e., substituted by its simple kinematic model) by the quantitative determination of all force and moment components. AE source representation by its mechanical model means that forces and moments, which induce equivalent displacement field as a real source, are determined. The physical source itself is not directly identified by this procedure; instead, its external effects are quantified. Assigning the detected quantities to some micromechanisms is a question for material sciences or other approaches.

Inverse solution (deconvolution) of Eq. (32.16) can be further simplified assuming a very simple form for the AE source function (kinematical representation),[12]

$$F_j(t) = P \cdot \varphi_j \cdot s(t) \tag{32.17}$$
$$M_{jk}(t) = Q \cdot \psi_j \cdot s(t).$$

Here P, Q are amplitude constants; φ_j, ψ_j express orientation vectors (source radiation diagrams); and $s(t)$ is a time function of the source activity (the same function for all components is assumed for simplicity). Under such simplified conditions, the IP solution may be treated as a multiple deconvolution procedure with a kernel expressed by a linear combination of the considered Green function components

$$u_i = (P \cdot G_{ij} \cdot \varphi_j + Q \cdot G_{ij,k} \cdot \psi_j) * s(t). \tag{32.18}$$

For digitally sampled AE signal (n signal samples) of one measured displacement component $u(n)$, Eq. (32.18) can be formally rewritten as a set of equations

$$u(n) = \sum_{i=1}^{R} c_i \sum_{k=1}^{n} g_i(n-k)s(k). \tag{32.19}$$

A multiple deconvolution procedure then consists in the solution of Eq. (32.19) for the unknown weight coefficients c_i (directional features of the AE source) and the time function $s(k)$. The set of nonlinear equations (32.19) often represents an ill-posed problem, and methods such as regularization or singular value decomposition should be used for its solution.

The above deconvolution procedure of the AE source identification has been successfully verified on numerically simulated data. Many other researchers have also reported good results of similar approaches to the AE source identification problems; see, for example, the work by Hamstad et al.[13] However, such procedures often partially fail when real AE sources must be quantified. This is mainly due to the many problems encountered in real situations: (a) the distance sensor-source is not much greater than a typical source dimension (point source approximation is unsatisfactory); (b) source location and its radiation pattern must not change during the AE signal detection; (c) the Green function must be determined with a very high precision which may be problematic; (d) the first signal arrival must be determined exactly and the source should be localized with approximately 1 mm accuracy (inverse procedures are strongly phase sensitive); (e) placement of transducers should respect the expected source radiation pattern; (f) signal filtering and transfer functions of sensors and devices must be taken into account (broadband, precisely calibrated transducers should be used); (g) presence of strong attenuation or wave dispersion can lead to partial or complete loss of information on the above-mentioned source characteristics. Inversion procedures may be improved by the averaging of signals detected by more transducers (information redundancy is desirable). From the direct comparison of numerically synthesized and measured signals, one can deduce which signal features may be due to, for example, dispersion effects. The time-frequency or wavelet domain seems to be the best for AE problem representation in such situations.[14]

The lack of exact knowledge about the influence of geometrical dispersion effects is one of the most important constraining factors in AE source classification. In many practical applications, the complete inverse solution is not necessary for diagnostic decision, and simplified AE source identification procedure is sufficient. Such a procedure is followed, for example, in the source description through significant signal parameters in statistical pattern recognition. Nevertheless, dispersion and attenuation effects also influence these parameters. For simple diagnostic purpose, the ANNs or other related methods of optimizing (genetic algorithms, cellular automata, fuzzy- and ANN-based classification and pattern recognition systems) could be used for the correction of extracted signal features. Application of advanced soft computing methods for AE data treatment simplifies and improves assessment of AE sources and their recognition, classification, and criticality assessment, which also facilitates better diagnostic conclusions. One possible way of performing parameter corrections using ANN is suggested in one of the reported work.[15] The proposed correction procedure was tested on numerically simulated AE experiments, schematically drawn in Figure 32.6.

Model AE signals were numerically generated at four sensor positions by the convolution of a model pulse, acting in selected source positions, with the Green function. The wave transfer Green function was calculated for thousands of source–sensor location pairs using numerical simulations performed by 2-D LISA code (Local Interaction

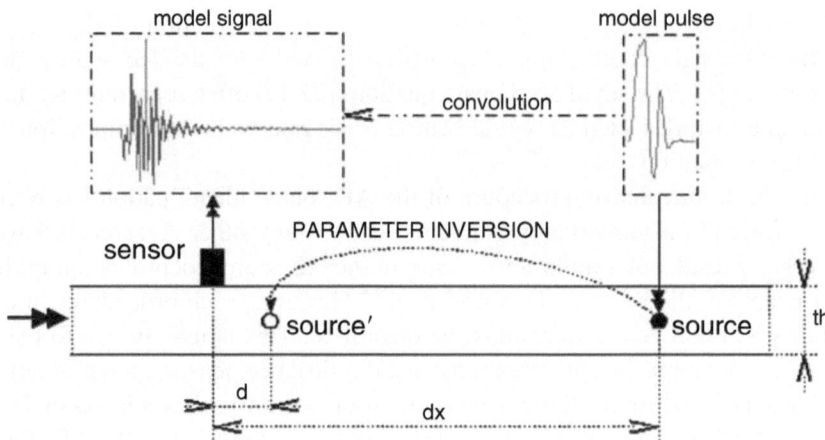

Fig. 32.6. Scheme of numerically simulated AE experiments for testing of signal parameter inversion procedure.

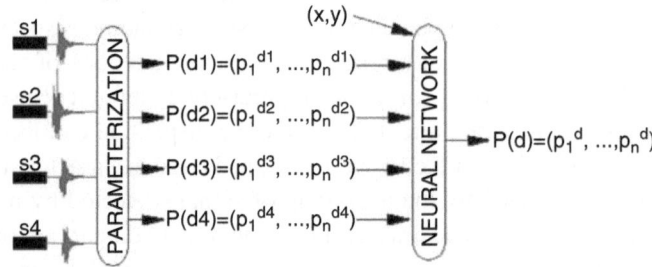

Fig. 32.7. Scheme of AE signal parameters correction procedure using ANN.

Simulation Approach[16]) on steel plates of various thickness th. Backpropagation ANN, representing a complex correction function of 17 extracted signal features was then properly trained at regularly spaced mesh nodes, which enabled determination of their values in "reference" positions placed in the neighborhood of the nearest AE sensor. The distance corrections were performed by ANN transfer of actual parameter (signal feature) values, detected at sensors, to the reference point, designated as *source'* in Figure 32.6. The ANN correction scheme is shown in Figure 32.7. Vectors of signal parameters $P(d1)$ to $P(d4)$ evaluated at sensors $s1$ to $s4$ are converted by the ANN, and AE sources are then classified using the already improved parameter values $P(d)$. Although numerical models have been increasingly successful in simulating wide ranges of real AE problems, advances in computer modeling must still be validated and verified against experiments. On the other hand, experimental results are required as input to computational models.

Acknowledgments

This work has been supported by the ESF program NATEMIS, and by the Czech Grant Agency GACR projects no. 205/03/0071 and 201/04/2102.

References

[1]   *Topics on NDE, vol. 3*, ASNT 1998.

[2]   *Nondestructive Testing Handbook, Vol. 5: Acoustic Emission*, ASNT 1987.

[3]   McCulloch, W.S., Pitts, W., A logical calculus of the ideas immanent in nervous activity, *Bulletin of Mathematical BioPhysics* 5, 115–133, 1943.

[4]   Demuth H, Beale M, *Neural Network TOOLBOX for Use with MATLAB*, The Mathworks, Inc., Natick, 1994.

[5]   Rojas R, *Neural Networks - A Systematic Introduction*, Springer Verlag, Berlin, 1996.

[6]   Fidalgo JN, Feature subset selection based on ANN sensitivity analysis—A practical study, in: N. Mastorakis, ed.: *Advances in Neural Networks and Applications*, WSES Press, pp. 206–211, 2001.

[7]   Grabec I, Sachse W, Solving AE problems by a neural network, *Journal of Acoustic Emission*, 6(1), 19–28, 1987.

[8]   Prevorovsky Z, Landa M, Blahacek M, Varchon D, Rousseau J, Ferry L, Perreux D, Ultrasonic Scanning and AE of Composite Tubes Subjected to Multiaxial Loading, *Ultrasonics*, 36 (1–5), 531–537, 1998.

[9]   Ohtsu M, Ono K, The generalized theory and source representation of AE, *Journal of Acoustic Emission*, 4(2), S50–S53, 1985, 5(4), 124–133, 1986.

[10] Aki K, Richards PG, *Quantitative Seismology: Theory and Methods*, W.H. Freeman and Comp., San Francisco, 1984.

[11] Grabec I, Sachse W, *Synergetics of Measurements, Prediction and Control*, Springer Verlag, Berlin, 1995.

[12] Landa M, A contribution of elastic wave propagation theory to AE signal analysis, Ph.D. Thesis, Institute of Thermomechanics AS CR, Prague, 1996.

[13] Hamstad MA, Kishi T, Ono K, eds., Progress in AE IX, *Proc of Internat. AE Conf.*, Big Island, Hawaii, Aug. 9–14, 1998.

[14] Blahacek M, Prevorovsky Z, Landa M, Processing of AE signals in dispersive media, *26th European Conf. on AE Testing "EWGAE 2004"*, Berlin, Sept. 15–17, 2004, *DGZfP-Proceedings BB 90-C, Lecture 66*, pp. 645–654.

[15] Chlada M, Prevorovsky Z, Vodicka J, Correction of AE signal parameters by neural networks, *34th Internat. Conf. on NDT "DEFEKTOSKOPIE 2004"*, Spindlerův Mlyn, 3–5 November 2004, Proc. ed. by P. Mazal, CNDT 2004, pp. 339–346.

[16] Delsanto PP, Connection machine simulation of ultrasonic wave propagation in materials, *Wave Motion* 20, 295–314, 1994.

Index